Lecture Notes in Computer Science 7097

Commenced Publication in 1973
Founding and Former Series Editors:
Gerhard Goos, Juris Hartmanis, and Jan van Leeuwen

W0227574

Mohamed Vall Mohamed Salem
Khaled Shaalan   Farhad Oroumchian
Azadeh Shakery   Halim Khelalfa (Eds.)

Information Retrieval Technology

7th Asia Information Retrieval
Societies Conference, AIRS 2011
Dubai, United Arab Emirates, December 18-20, 2011
Proceedings

 Springer

Volume Editors

Mohamed Vall Mohamed Salem
University of Wollongong in Dubai, Faculty of Computer Science and Engineering
Dubai Knowledge Village, P.O. Box 20183, Dubai, United Arab Emirates
E-mail: mohamedsalem@uowdubai.ac.ae

Khaled Shaalan
British University in Dubai, Faculty of Engineering and IT
Dubai International Academic City, Block 11, 1st and 2nd Floor
P.O. Box 345015, Dubai, United Arab Emirates
E-mail: khaled.shaalan@buid.ac.ae

Farhad Oroumchian
University of Wollongong in Dubai, Faculty of Computer Science and Engineering
Dubai Knowledge Village, P.O. Box 20183, Dubai, United Arab Emirates
E-mail: farhadoroumchian@uowdubai.ac.ae

Azadeh Shakery
University of Tehran, Faculty of Engineering
Department of Electrical and Computer Engineering
North Kargar Street, P.O. Box 14395-515, Tehran, Iran
E-mail: shakery@ut.ac.ir

Halim Khelalfa
University of Wollongong in Dubai, Faculty of Computer Science and Engineering
Dubai Knowledge Village, P.O. Box 20183, Dubai, United Arab Emirates
E-mail: halimkehlalfa@uowdubai.ac.ae

ISSN 0302-9743   e-ISSN 1611-3349
ISBN 978-3-642-25630-1   ISBN 978-3-642-25631-8 (eBook)
DOI 10.1007/978-3-642-25631-8
Springer Heidelberg Dordrecht London New York

Library of Congress Control Number: 2011941661

CR Subject Classification (1998): H.3, H.4, F.2.2, I.4-5, E.1, H.2.8

LNCS Sublibrary: SL 3 – Information Systems and Application, incl. Internet/Web
and HCI

Typesetting: Camera-ready by author, data conversion by Scientific Publishing Services, Chennai, India

Printed on acid-free paper

Springer is part of Springer Science+Business Media (www.springer.com)

Preface

The Asian Information Retrieval Societies Conference (AIRS) is one of the most established and competitive information retrieval conferences; the seventh edition of this conference (AIRS 2011) aimed to bring together international researches and developers to exchange new ideas and the latest results in information retrieval (IR). The scope of the conference encompassed theory and practice of all aspects of IR in text, audio, image, video and multimedia data. The call for papers invited submissions to the following areas of research:

- Arabic Script Text Processing and Retrieval
- IR Models and Theories
- Multimedia IR
- User Study, IR Evaluation, and Interactive IR
- Web IR, Scalability and Adversarial IR
- IR Applications
- Machine Learning for IR
- Natural Language Processing for IR

AIRS 2011 was the first edition to be organized in the western part of the Asian continent with a growing interest to foster IR research and communalities in natural language processing. A new track on Arabic Script Text Processing and Retrieval was added for the first time to the main areas of research in the conference.

Historically, AIRS 2011 is a continuation of the series of conferences that grew from the Information Retrieval with Asian Languages (IRAL) workshop series back in 1996. It has become a mature venue of IR work, finding support from the ACM Special Interest Group and Information Retrieval (SIGIR) and many other associations.

The Organizing Committee was very pleased with the quality and level of interest received to our call for contributions from the research community in the IR field. We received a total of 132 papers representing work by academics and practitioners from all over the world and we would like to thank all of them. The Program Committee used a double-blind reviewing process and as result 31 articles (23.5%) were accepted as full papers and 25 (19%) were accepted as short (poster) papers.

The success of this conference was only possible with the support of the extremely active Program Committee members without whom the present proceedings would not have been possible. We would like to acknowledge the contributions of Ali Farghaly (Oracle, USA), Minjie Zhang (University of Wollongong, Australia), Joemon M. Jose (University of Glasgow, UK), Tetsuya Sakai (Microsoft Research Asia), Min Zhang Tsinghua (University, China), Wang Bin (Chinese Academy of Sciences, China), Tie-Yan Liu (Microsoft Research Asia) and Chia-Hui Chang (National Central University, Taiwan).

For a conference to run smoothly, much behind-the-scene work is necessary, most of which is largely unseen by the authors and delegates. We would like to thank our Publication Chairs (Azadeh Shakery and Halim Khelalfa) who painstakingly worked with each individual author to ensure formatting, spelling, dictation and grammar were completely error-free.

October 2011

Khaled Shaalan
Farhad Oroumchian
Mohamed Vall Mohamed Salem

Organization

AIRS 2011 was organized by the Faculty of Computer Science and Engineering, University of Wollongong in Dubai in cooperation with ACM/SIGIR.

Executive Committee

Conference Co-chairs

Mohamed Val Salem	University of Wollongong in Dubai, UAE
Farhad Oroumchian	University of Wollongong in Dubai, UAE

Program Co-chairs

Khaled Shaalan	British University in Dubai, UAE
Farhad Oroumchian	University of Wollongong in Dubai, UAE

Publication Co-chairs

Azadeh Shakery	University of Tehran, Iran
Halim Khelalfa	University of Wollongong in Dubai

Publicity Chair

Asma Damankesh	University of Wollongong in Dubai
Abolfazl AleAhmad	University of Tehran, Iran

Program Committee

Program Co-chairs

Khaled Shaalan	British University in Dubai, UAE
Farhad Oroumchian	University of Wollongong in Dubai, UAE

Area Chairs

User Study, IR Evaluation, and Interactive IR	Tetsuya Sakai, Microsoft Research Asia
IR Models and Theories	Minjie Zhang, University of Wollongong, Australia
Multimedia IR	Joemon M Jose, University of Glasgow, UK

IR Applications	Kazunari Sugiyama, National University of Singapore, Singapore
Web IR, Scalability and Adversarial IR	Min Zhang, Tsinghua University, China Wang Bin, Chinese Academy of Sciences (CAS), China
Machine Learning for IR	Tie-Yan Liu, Microsoft Research Asia
Natural Language Processing for IR	Chia-Hui Chang, National Central University, Taiwan
Arabic Script Text Processing and Retrieval	Ali Farghaly, Oracle, USA

Referees

I. Sengor Altingovde	M. Diab	K. Jarvelin
I. Abu El-Khair	W. Ding	Wei Jin
M. Adriani	K. Eguchi	Hideo Joho
A. Aizawa	A. Elnagar	G. Jones
Z. Al Aghbari	M. Evens	Hanmin Jung
H. Aliane	H. Faili	Noriko Kando
H. Amiri	C. Gurrin	In-Su Kang
M. Asadpour	L. Gao	E. Kanoulas
E. Ashoori	R. Girju	Jaana Kekalainen
C. Man Au Yeung	Bo Gong	S. Khadivi
S. Bandyopadhyay	G. Grefenstette	Jungi Kim
L.H. Belguith	T. Gungor	F. Kimura
P. Bhattacharyya	Jiafeng Guo	K. Kishida
W. Bin	N. Habash	P. Kolari
M. Blondel	A. Hamdi-Cherif	K. Kuriyama
R. Cai	Y. Harada	M. Lalmas
T.W. Cai	F. Harrag	A. Lampert
F. Can	K. Hatano	F. Lazarinis
S. Carberry	Ben He	Seungwoo Lee
V. Cavalli-Sforz	Sam Yin He	Yeong Su Lee
Chia-Hui Chang	Yulan He	Hyowon Lee
Chien Chin Chen	Xuanjing Huang	Guoliang Li
Hsin-Hsi Chen	S. Hussain	Peng Li
Sung-Kwon Choi	A. Jatowt	Sujian Li
G. Dupret	A.M. Jaoua	Xiaojun Lin

Yiqun Liu
Zhiyuan Liu
Wen-Hsiang Lu
Yuanhua Lv
Qiang Ma
A. Maeda
M. Melucci
M.R. Mustaffa
C. Mahlow
M. Maragoudakis
K. Megerdoomian
D. Metzler
R. Mitkov
M.F. Moens
A. Moffat
T. Mori
Seung-Hoon Na
P. Nakov
H. Nanba
Kenta Oku
M. Okumura
F. Oroumchian
I. Ounis

W. Radford
F. Radlinski
A. Rafaa
R. Ren
K. Riaz
Hae-Chang Rim
I. Ruthven
T. Sakai
M. Sanderson
J. Savoy
F. Scholer
K. Seki
S. Sekine
A. Shakery
M. Shamsfard
H. Shima
M. Shokouhi
T. Shumizu
Sa-Kwang Song
Young-In Song
A. Soudi
K. Sugiyama
A. Sun

L. Sun
Y. Suzuki
Sim-Hui Tee
U.S. Tiwary
M. Vilares-Ferro
J. Vilares-Ferro
Jingdong Wang
Mingwen Wang
Taifeng Wang
W. Webber
R. White
Kam-Fai Wong
Yunqing Xia
Zheng Ye
E. Yilmaz
M. Yoshioka
N. C. Zamorani
A. ZarehBidoki
Lanbo Zhang
Yin Zhang
Jun Zhao
I. Zitouni

Sponsoring Institutions

Dubai Knowledge Village
Micorsoft Middle East

Table of Contents

Information Retrieval Models and Theories

Information Retrieval Applications and Multimedia Information Retrieval

User Study, Information Retrieval Evaluation and Interactive Information Retrieval

Web Information Retrieval, Scalability and Adversarial Information Retrieval

Machine Learning for Information Retrieval

Natural Language Processing for Information Retrieval

Arabic Script Text Processing and Retrieval

Query-Dependent Rank Aggregation with Local Models

Hsuan-Yu Lin, Chi-Hsin Yu, and Hsin-Hsi Chen

Department of Computer Science and Information Engineering, National Taiwan University
#1, Sec.4, Roosevelt Road, Taipei, 10617 Taiwan, ROC
Tony@widelab.org, jsyu@nlg.csie.ntu.edu.tw, hhchen@ntu.edu.tw

Abstract. The technologies of learning to rank have been successfully used in information retrieval. General ranking approaches use all training queries to build a single ranking model and apply this model to all different kinds of queries. Such a "global" ranking approach does not deal with the specific properties of queries. In this paper, we propose three query-dependent ranking approaches which combine the results of local models. We construct local models by using clustering algorithms, represent queries by using various ways such as Kullback-Leibler divergence, and apply a ranking function to merge the results of different local models. Experimental results show that our approaches are better than all rank-based aggregation approaches and some global models in LETOR4. Especially, we found that our approaches have better performance in dealing with difficult queries.

Keywords: Local model, query dependent ranking, distributed IR, LETOR.

1 Introduction

Many popular machine learning approaches such as RankSVM [1], AdaRank [2], ListNet [3], and so on have been proposed for information retrieval. These models result in a single linear ranking function. We call such a model a **global model** because it is used for all different queries. A single global ranking model is simple and stable but it is not easy to perform well for all kinds of queries. On the other hand, different queries have their ranking models in query dependent ranking [4]. We call such a query-dependent model a **local model** because it uses only local information to build a ranking function. The local model is good for some sort of queries but unstable for the other queries. To select a good ranking model is crucial and challenging, and the selection depends on the semantics of each query and documents.

For example, a searcher wants to search information about Apple's products such as iPad and Macbook. If the searcher uses keyword "apple" to search, it is safer to use a global model. Using "tech-Apple" local model or "fruit-apple" local model is risky because we have no idea about the user intention. But if the searcher uses the keyword "apple ipod" to search, using the "tech-Apple" local model to rank documents is better to fit the user intention. This example illustrates the difficulty and benefit of selecting good local models to rank documents.

There are two possible ways to use local models. First, we can use the information of a query to find **one** best local model to rank documents for this query. Second, we

M.V.M. Salem et al. (Eds.): AIRS 2011, LNCS 7097, pp. 1–12, 2011.

can aggregate the results of many local models to balance the risk of using a bad local model and use the information of this query to determine the weights of aggregation. In these two alternatives, how to extract the information of a query is fundamental and crucial. We call this a "**query representation**" problem in this paper. Finding a good query representation is very challenging because we don't know what kind of information is interesting to a user for an unseen query. However, if the framework could represent queries correctly, it would improve the performance much.

In this paper, we will explore three approaches of using local models and investigate two different query representation schemes. We also analyze the upper-bound of selecting the best local model, estimate the challenging of these approaches, and discuss the results from the viewpoint of query difficulty. The experimental results show that we propose the best aggregation approaches in LETOR4 datasets, and our approach has a better performance to deal with hard queries.

This paper is organized as follows. Section 2 presents our approaches. Section 3 shows the performance of each approach and gives the detail analyses. Section 4 summarizes the results.

2   Query Dependent Ranking

A query dependent ranking framework consisting of training and testing steps describes how to derive and use local models. Three query dependent ranking approaches, named as *Naïve*, *SelectRanker* and *Transform*, are proposed under the framework. In the following, we first introduce the framework and then describe each approach in details.

2.1   Framework

A query dependent ranking framework is shown in Figure 1. The corresponding training and testing steps are described as follows.

Training Step. The first step of our framework is to build local models. For this purpose, we use K-means clustering algorithm to separate a training set into small clusters, and use RankSVM [1] to train local models from the individual clusters. In other words, each cluster has its own local model. The details of the clustering algorithm depending on which approach, i.e., the *Naïve*, *SelectRanker* or *Transform* approaches is used, will be discussed later.

Testing Step. For the aggregation, we measure the weight of each local model by using the similarity between test query q' and cluster i (i=1,...,m). The definition of similarity, which depends on the *Naïve*, *SelectRanker* and *Transform* approaches, will be discussed later. The rank aggregation function is defined by Equation (1).

$$fr\big(\varphi(q',d)\big) = \frac{1}{m}\sum_{i=1}^{m} w_i \times \langle\varphi(q',d),r_i\rangle \tag{1}$$

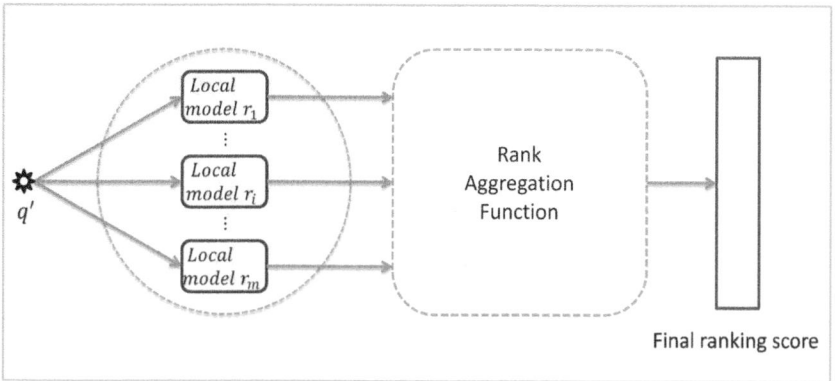

Fig. 1. The query dependent ranking framework

where m is the number of clusters, $\varphi(q', d) \in R^n$ is a feature vector of document d related to test query q', $r_i \in R^n$ is a local model produced by RankSVM using cluster i, w_i is the weight of local model r_i, and $\langle \varphi(q', d), r_i \rangle$, the inner product of $\varphi(q', d)$ and r_i, represents the predicted score of d using vector r_i. The final ranking score of a document d is calculated by averaging the weighted local predicted scores.

2.2   Three Approaches

The first approach named *Naïve* uses heuristics to represent a query and then selects a best local model to rank the documents of the query. Comparatively, the other two approaches, i.e., *SelectRanker* and *Transform*, aggregate scores from local models. The *Naïve* approach, which selects the best model, can be viewed as a special case of Equation (1) where the best local model i has weight $w_i = 1$ and the other local models have weight $w_j = 0$ $(j \neq i)$. Therefore the framework in Figure 1 and Equation (1) can be used to describe different query-dependent ranking approaches. We specify these approaches as follows respectively.

Naïve. Algorithm 1 selects a best local model for document ranking. Then we use it to determine the ranks of documents related to this query in the similar way as we do in global model.

The m input clusters in Algorithm 1 are generated in the training step as follows. First, the relevant query-document (q-d) pairs of the training set are collected. Second, these relevant q-d pairs are clustered by using K-means clustering algorithm with Euclidean distance as a similarity metric. Third, if any of q_j's relevant q-d pairs are in cluster i, add all relevant and non-relevant q-d pairs of query q_j to cluster i. Last, the cluster is represented by a vector $C_i \in R^n$ where C_i is the average of relevant q-d pairs in this cluster, and the local model r_i of cluster i is generated by using RankSVM with cluster i as input.

Algorithm 1. *Naïve* approach

Input:	(1)	Test query q'
	(2)	A set of clusters i (i=1,...,m) with representation $C_i \in R^n$
Steps:	(1)	For test query q', use the global RankSVM model to find query's top-t ranked documents, and use the mean of t vectors of q-d pairs to represent query q'. The query representation is denoted as $V_{q'}$.
	(2)	Calculate the similarities between $V_{q'}$ and each cluster representation C_i using Euclidean distance.
	(3)	Find the most similar cluster.
Output:		r: the best local model, i.e., the local model of the most similar cluster.

SelectRanker. This approach aggregates scores from local models. The weight of cluster i is determined by the top-k most similar queries in this cluster. Algorithm 2 describes the procedure to calculate weight w_i of cluster i.

Algorithm 2. *SelectRanker* approach

Input:	(1)	Test query q'
	(2)	A set of clusters i (i=1,...,m) with representation $C_i \in R^n$
	(3)	The base ranking model r_b and local ranking model r_i for each cluster
	(4)	$KL(r_b||r_i, q_j)$ for each query q_j in cluster i (using Equation (2))
Steps:	(1)	For test query q', calculate $KL(r_b||r_i, q')$ for each cluster i.
	(2)	Calculate $distance(q', q_j) = KL(r_b||r_i, q') - KL(r_b||r_i, q_j)$ for each query q_j in cluster i.
	(3)	In each cluster i, find the top-k queries with the smallest $distance(q', q_j)$.
	(4)	Calculate $w_i = \sum_{j=1}^{k} AP(q_j, r_i)$, where q_j is a query in cluster i, and $AP(q_j, r_i)$ is the average precision of query q_j using local model r_i.
Output:		w_i: i=1,...,m

The Kullback-Leibler divergence shown in Equation (2) is used to measure the distance between local model and base model for a given query. Peng et al. [5] showed that KL divergence is useful in information retrieval. For the n retrieved documents of a given query q, $r_b(d)$ and $r_i(d)$ are the normalized relevance scores [5] of document d in r_b and r_i, respectively.

$$KL(r_b||r_i, q) = \sum_{d=1}^{n} r_b(d) \cdot \log_2 \frac{r_b(d)}{r_i(d)} \tag{2}$$

Step 3 in Algorithm 2 filters out non-relevant queries. Step 4 computes the average precision as the confidence of a query, and sums the confidences as the weight of this cluster. We choose BM25 as the base ranking model r_b in the experiments.

After determining the weights of local models, we apply Equation (1) to merge the ranking scores of each local model to determine the final scores of documents.

Transform. This approach aggregates scores from local models too, but it uses different scheme to represent queries and feeds the transformed queries into a clustering algorithm. Because it is hard to find a good representation on the form of $\varphi(q', d)$, we adapt two global ranking models to transform query's representation. This is also motivated by Peng et al. [5]. Algorithm 3 describes how to determine the weight of each cluster in the testing step. In the training step, we first transform each query q_i in training set to a $\left(KL(r_a||r_b, q_i), JS(r_a||r_b, q_i)\right) = x^{q_i} \in \mathcal{R}^2$ vector, where r_a and r_b are two base ranking models, and Equation (3) shows the Jensen-Shannon divergence. In the experiments, we let r_a and r_b be BM25 and global RankSVM model.

$$JS(r_a||r_b, q) = \frac{1}{2} \cdot KL(r_a||r_b, q) + \frac{1}{2} \cdot KL(r_b||r_a, q) \tag{3}$$

Then we cluster those 2-dimension query vectors using K-means clustering with Euclidean distance. We use $C_i \in R^2$, which is generated by averaging 2-dimension query vectors in a cluster i, to represent cluster i. After we have the input C_i , we run Algorithm 3 to get the weights of the clusters for a test query q', and the documents of query q' are ranked using Equation (1). In this approach, the representation C_i of cluster i acts as a local model, and the weight w_i is calculated by measuring the distance between C_i and test query $x^{q'}$.

Algorithm 3. *Transform* approach

Input:	(1) Test query q'
	(2) A set of cluster representations : $- C_i \in R^2$: i=1,…,m "
	(3) Two base ranking models r_a and r_b
Steps:	(1) For test query q', calculate $x^{q'} = \left(KL(r_a||r_b, q'), JS(r_a||r_b, q')\right)$.
	(2) Calculate $w_i = EuclideanDistance\left(x^{q'}, C_i\right)$.
Output:	w_i: i=1,…,m

2.3 Time Complexity

The time complexity of *SelectRanker* is the largest one of the three approaches. Comparing to the ordinal global ranking, this approach has three more steps. The first extra step is step 2 of Algorithm 2 which takes $O(n \times$ #q-d $\times m)$, where n is the dimensions of q-d pairs, #q-d is the maximum number of q-d pairs of queries in training set, and m is the number of local models. The other two extra steps are steps 3 and 4 of Algorithm 2, which takes $O(m^2 \times \ln(m))$. In general settings, m is a predefined constant. Thus it does not add too much cost in response time.

3   Experimental Results

3.1   Dataset and Parameter Selection

We adopt LETOR4 dataset [6] for our experiments. It contains two collections: MQ2007 and MQ2008. MQ2007 and MQ2008 have 1,692 and 784 queries with total

69,622 and 15,211 documents, respectively. Each q-d pair has 46 features. In our approaches, some parameters, including the number m of clusters in the three algorithms, the t in step 1 of Algorithm 1, and the k in step 3 of Algorithm 2, are determined by the validation set to have the best mean average precision (MAP), and are applied to the testing set.

3.2   Performance Comparison with LETOR4 Baselines

We compare our aggregation approaches to LETOR4 rank aggregation baselines, and the results are shown in Table 1. The performance is in terms of the average of 5 folds. The score highlighted in **bold** means the best performance of all approaches. We can see that the proposed three approaches are better than all baselines. The two aggregation approaches, i.e., *SelectRanker* and *Transform*, have better performance than *Naïve* approach. We will analyze the results later in details.

Table 1. Comparison between our approaches and LETOR4 rank aggregation baselines

Dataset	MQ2007/MQ2007-agg		MQ2008/MQ2008-agg	
Metric	MAP	MeanNDCG	MAP	MeanNDCG
BordaCount	0.3252	0.3216	0.3945	0.3895
CPS-KendallTau	0.3891	0.4088	0.4219	0.4209
CPS-SpearmanFootrule	0.3898	0.4094	0.4027	0.3977
CPS-SpearmanRankCorrelation	0.4069	0.4330	0.4102	0.4126
Naïve	0.4175	0.4461	0.4349	0.4391
SelectRanker	0.4654	**0.4991**	**0.4751**	**0.4874**
Transform	**0.4655**	0.4967	0.4747	0.4831

In Table 1, the four LETOR4 rank aggregation baselines adopted the approaches proposed by Qin et al. [7]. They used MQ2007-agg and MQ2008-agg while we use MQ2007 and MQ2008 in our experiments. MQ2007/MQ2008 and MQ2007-agg/MQ2008-agg have the same query set respectively, but their features are different. MQ2007-agg and MQ2008-agg used order as feature, where the order is from individual rankers. Qin et al. used order-based rank aggregation approaches [7] in MQ2007-agg and MQ2008-agg. Comparatively, we use score-based rank aggregation approaches in MQ2007 and MQ2008. The performance of our approaches is better.

Although the performances of our approaches are far beyond the LETOR4 rank aggregation baselines, we want to know further the comparison to LETOR4 global models. We show the results in Table 2, where "+" after a score means our proposed approach outperforms the RankSVM baseline, and N.@k, MeanN., and AdaRank denote abbreviations of NDCG@k, MeanNDCG, and AdaRank-MAP algorithm, respectively. Although the results with + are better than the RankSVM baseline, they did not pass the significant test.

Table 2. Comparison between our local models and LETOR 4 global models

Dataset	MQ2007				MQ2008			
Metric	MAP	P@1	P@2	P@5	MAP	P@1	P@2	P@5
RankSVM	0.4645	0.4746	0.4496	0.4135	0.4696	0.4273	0.4069	**0.3474**
AdaRank	0.4577	0.4392	0.4301	0.4054	0.4764	0.4426	0.4165	0.3419
ListNet	0.4652	0.4640	0.4471	0.4126	**0.4775**	**0.4451**	0.4120	0.3426
Naïve	0.4175	0.4421	0.4105	0.3695	0.4349	0.4184	0.3756	0.3087
SelectRanker	0.4654+	**0.4782+**	**0.4540+**	**0.4136+**	0.4751+	0.4413+	**0.4171+**	0.3454
Transform	**0.4655+**	0.4764+	0.4466	0.4099	0.4747+	0.4438+	0.4056	0.3428
Metric	MeanN.	N.@1	N.@2	N.@5	MeanN.	N.@1	N.@2	N.@5
RankSVM	0.4966	0.4096	0.4073	0.4142	0.4832	0.3626	0.3984	0.4695
AdaRank	0.4891	0.3821	0.3900	0.4070	**0.4915**	**0.3754**	**0.4141**	**0.4794**
ListNet	0.4988	0.4002	0.4063	0.4170	0.4914	**0.3754**	0.4112	0.4747
Naïve	0.4461	0.3826	0.3658	0.3683	0.4391	0.3503	0.3635	0.4172
SelectRanker	**0.4990+**	**0.4139+**	**0.4103+**	**0.4174+**	0.4873+	0.3715+	0.4072+	0.4744+
Transform	0.4967+	0.4078	0.4034	0.4124	0.4831	0.3639+	0.3960	0.4696+

We can see that the *Transform* approach has the best MAP and the *SelectRanker* approach has the best P@1, P@2 and P@5 in MQ2007. If the metric is NDCG, the *SelectRanker* approach has the best performance in MQ2007, and its performance in MQ2008 is better than RankSVM baseline, which is used to build our local models.

In MQ2008, only the *SelectRanker* approach has the best P@2. If we compare our results to RankSVM, most performance of the *SelectRanker* and *Transform* approaches are better than that of RankSVM. On the other hand, the *Naïve* approach does not perform well when we compare it with the global models. Selecting the best local model did not perform well because of our poor query representation scheme. We show our analysis in next section.

3.3   Analysis of Selecting the Best Local Model

We represent a query by averaging top-t documents in the *Naïve* approach. Is this a good query representation scheme? If we have a better query representation scheme, is it possible for the *Naïve* approach to achieve better performance? To answer the above two questions, we replace the query representation scheme in Algorithm 1 by averaging the relevant documents only. Of course, that is not feasible in testing, but the analysis will give us some interesting insights. The result is shown in Fig. 2.

In Fig. 2, the x-axis is the number of clusters, and the line of RelevantOnly is the query representation scheme that uses only relevant documents. We can see that if we have a better query representation scheme, the *Naïve* approach can beat RankSVM baseline. We can also find that the performance increases along with the number of clusters. That means we can find a more specific local model to fit users need. For example, if the training set is clustered into 85 clusters, we can find a very specific cluster to rank the documents of a test query precisely. From this analysis, we can

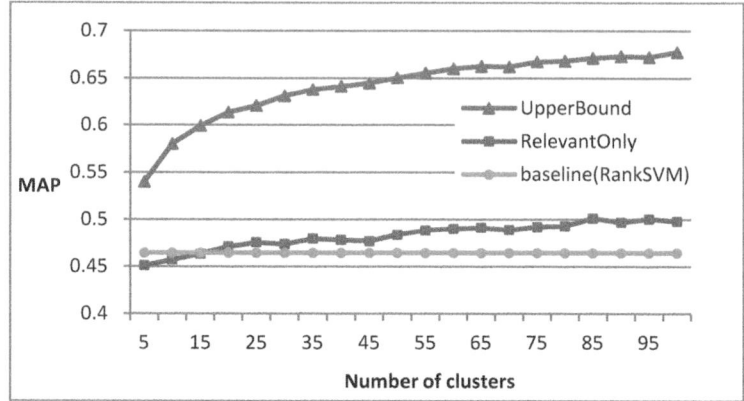

Fig. 2. Performance analysis of selecting best model

conclude that our query representation scheme that uses the top-*t* documents may introduce a lot of noise, and query representation scheme is the main reason why the *Naïve* approach did not perform well.

We are also interested in the upper bound of selecting the best local model approach. In the *Naïve* approach, if we can always select the best cluster that results in the best average precision (AP) of a query, what is the performance? The line of UpperBound in Fig. 2 shows the situation. We can see that there is a lot of room for improvement. But is it easy to achieve improvement? We further answer this question in Fig. 3.

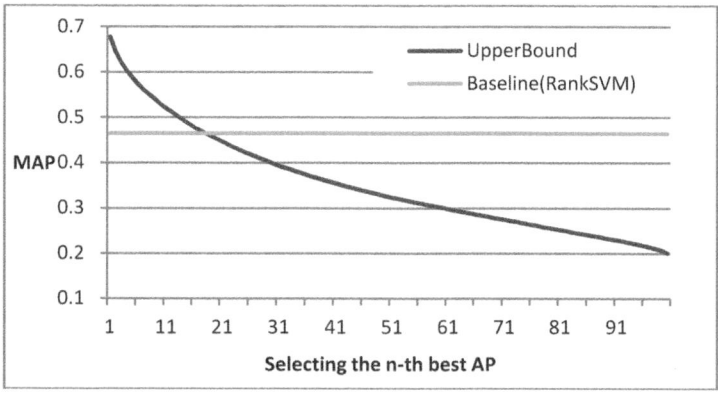

Fig. 3. The performance of selecting the *n*-th best AP in 100 clusters

In Fig. 3, we fix the number of clusters to be 100. We calculate AP of a query by using the local model of a cluster, and order the 100 clusters by their AP scores. In this way, we know the best AP, the 2nd best AP, the 3rd best AP, and so on, of a query among the 100 local models. If we always select the best AP, we get the upper bound. If we always select the *n*-th best AP, we know the challenging of this approach. In Fig. 3, the x-axis is the *n* when the *n*-th best AP is selected. The performance of RankSVM baseline is around in selecting the 19-th best AP out of 100 clusters. In other words, if we can always select one of the top-19 clusters out of 100 clusters, we can beat the baseline. This is neither an easy task nor a very hard task. We believe that it is achievable, but we leave it as future work.

3.4   Analysis in Terms of Query Difficulty

TREC held a hard track[1] in 2003 and 2005. The goal of the hard track is to improve the retrieval technology by focusing on the poorly performing topics. The topics of the testing queries are selected from the topics that had low average effectiveness across the runs in previous TRECs.

In this paper, the proposed algorithms heavily rely on the relevant documents of a query, so we define the difficulty of queries in terms of the number of relevant documents. A hard query is defined to be a query with the number of relevant documents less than 4; an easy query is defined to be a query with the number of relevant documents more than 20; and a query with the number of relevant documents between 4 and 20 is regarded as a normal query. We compare the performance of RankSVM and our approaches at the three difficult levels. The query distribution with different number of relevant documents of MQ2007 is shown in Fig. 4, and the results in terms of various difficulty levels are shown in Table 3.

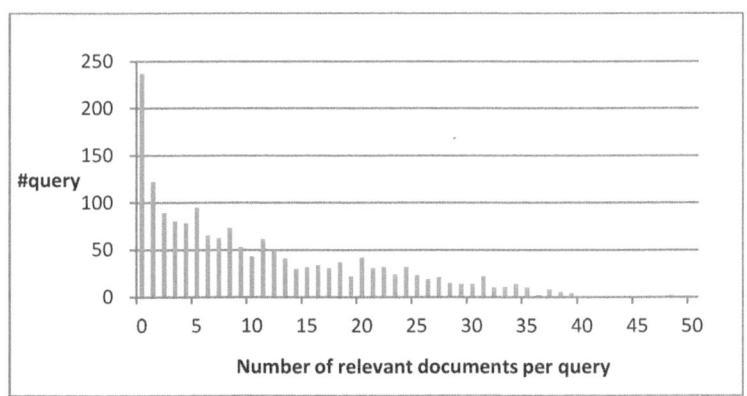

Fig. 4. Query distribution with different number of relevant documents in MQ2007

In Fig. 4, we can find there are lots of queries with zero relevant documents, so we could ignore these queries in Table 3. According to our statistics, the ratios of queries at the hard, normal, and easy levels in MQ2007 are 17%, 47% and 21%, respectively, and their distributions in MQ2008 are 38%, 31%, and 2%, respectively.

---

[1]   TREC hard track: http://trec.nist.gov/data/hard.html

Table 3. Comparison of performance (MAP) at different levels of query difficulty in LETOR 4

	MQ2007		
	Hard (17%)	Normal (47%)	Easy (21%)
RankSVM	0.3287	0.5085	0.8204
SelectRanker	**0.3296**	**0.5106**	0.8192
Transform	**0.3303**	**0.5107**	0.8187

	MQ2008		
	Hard (38%)	Normal (31%)	Easy (2%)
RankSVM	0.5697	0.7425	0.7935
SelectRanker	**0.5790**	**0.7490**	0.7918
Transform	**0.5871**	0.7394	0.7820

In Table 3, we use **bold** to emphasize the cases in which our approaches are better than the baseline. We can find that our approaches perform worse than RankSVM in easy queries. But the performances of our approaches are better than RankSVM in hard and normal queries. This is an important property because doing well in hard and normal queries will have higher user satisfaction. If we can find relevant documents when the relevant documents are paucity, user may feel happier. We believe that this advantage comes from the nature of aggregation approaches. Aggregation approaches can utilize many local models to prevent very bad ranking results especially when relevant document is scarce. We discuss this point by presenting the results of some queries shown in Table 4.

Table 4. The average precision of some queries and their query terms

#	Query terms	#relevant document	RankSVM	*SelectRanker*	*Transform*
1	i wish i wa in the land of cotton	1	0.14	**0.20**	**0.25**
2	venom mammal	1	0.25	**0.33**	**0.50**
3	north dakota deer hunt	1	1.00	0.50	0.50
4	polyest carpet	2	0.19	**0.32**	**0.36**
5	jame patrick connelli	5	0.21	**0.33**	**0.32**
6	how to stop smoke	11	0.29	**0.34**	**0.45**
7	thing to do Georgia	24	0.79	0.58	0.58

In Table 4, the AP in **bold** means our approach outperforms RankSVM, and underline of word in query terms means the query term is misspelled. We can find that the performance of our approaches is better than RankSVM when query term lose alphabets. For example, the word "*was*" is misspelled as "*wa*", "*polyester*" as "*polyest*", and "*connelly*" as "*connelli*". Those examples demonstrate an aggregation approach can prevent very poor performance. On the other hand, an aggregation approach is not as good as the baseline when the baseline has very high AP. This is also the nature of aggregation approaches because some non-relevant local models are introduced. How to filter out non-relevant local models is the key to improve the aggregation approaches.

4   Related Work

Many learning to rank techniques have been proposed such as RankSVM [1], Ada-Rank [2], Topical RankSVM [8], ListNet [3] and RankBoost [9]. These algorithms can be classified in terms of different criteria. The first four algorithms are classified as linear models because they result in a linear ranking function. On the other hand, RankBoost is classified as non-linear model because it combines base learners to rank documents. Researchers also classify the learning to rank algorithms into point-wise, pair-wise, and list-wise approaches [10] based on the way the learners use the training data. RankSVM uses the information of pair of q-d pairs to train learner, so it is a pair-wise approach. ListNet use the ranked list information to train learner, so it is a list-wise approach. Aggregation algorithm, which builds on the top of base learners and integrates base learners' results to produce final ranking, is a special type of algorithms among these classifications.

Dwork et al. [11], Peng et al. [5] and Farah and Vanderpooten [12] proposed rank aggregation approaches to enhance web systems. The meta-search approaches [11] can be viewed as aggregation approaches, in which they aggregate results from different search engines. The difference between these aggregation approaches and our approaches is the use of base learners. They all use global models as base models. In general, if base learners have good results for all different kinds of queries, aggregation approach can produce better results. Our approaches use local models as base models. Although we do not have a rigid experimental comparison, we believe that the performance of aggregating local models has a higher upper bound than that of aggregating global models. We leave this issue as future work.

Bian et al. [8, 13] and Fan et al. [14] also adopt aggregation approaches. Their works used learning approaches to find best weighting scheme in clusters [8, 14] and pre-defined query categories [13].

Geng et al. [4] proposed a good local model approach named as query-dependent approach. Their approach of generating query representation is much like the algorithm we used in Algorithm 1, but they used K-nearest-neighbor to select similar queries and dynamically generated local models from those similar queries. We use fixed number of local models instead.

5   Conclusion

In this paper, we propose three query dependent ranking approaches to improve retrieval performance, and the experiments show our approaches are better than all LETOR rank aggregation approaches and are comparable to global machine learning techniques. We conduct many detail analyses on the proposed approaches, and find that our approaches perform well in hard and normal queries. We also found that selecting the ideal local model has a very high upper bound, and there is a lot of room to improve this kind of approaches. Finally, our analyses showed that query representation and filtering out non-relevant local models are two key issues to improve performance of using local models.

Several issues need to be further studied in the aggregation framework. The alternative ways to represent a query, the other learning approaches to select local models, and some further methods to filter out non-relevant local models will be explored in the future work.

Acknowledgment. Research of this paper was partially supported by National Science Council (Taiwan) under the contract NSC 98-2221-E-002-175-MY3.

References

1. Joachims, T.: Optimizing search engines using clickthrough data. In: Proceedings of the Eighth International Conference on Knowledge Discovery and Data Mining (ACM SIGKDD), pp. 133–142 (2002)
2. Xu, J., Li, H.: AdaRank: a boosting algorithm for information retrieval. In: SIGIR 2007, pp. 391–398 (2007)
3. Cao, Z., Qin, T., Liu, T.-Y., Tsai, M.-F., Li, H.: Learning to rank: from pair-wise approach to list-wise approach. In: ICML 2007, pp. 129–136 (2007)
4. Geng, X., et al.: Query dependent ranking using K-nearest neighbor. In: SIGIR 2008, pp. 115–122 (2008)
5. Peng, J., Macdonald, C., Ounis, I.: Learning to Select a Ranking Function. In: Gurrin, C., He, Y., Kazai, G., Kruschwitz, U., Little, S., Roelleke, T., Rüger, S., van Rijsbergen, K. (eds.) ECIR 2010. LNCS, vol. 5993, pp. 114–126. Springer, Heidelberg (2010)
6. Liu, T.-Y., Xu, J., Qin, T., Xiong, W., Li, H.: LETOR: Benchmark Dataset for Research on Learning to Rank for Information Retrieval. In: LR4IR 2007, in Conjunction with SIGIR 2007 (2007)
7. Qin, T., Geng, X., Liu, T.-Y.: A New Probabilistic Model for Rank Aggregation. In: NIPS 2010 (2010)
8. Bian, J., Li, X., Li, F., Zheng, Z., Zha, H.: Ranking specialization for web search: a divide-and-conquer approach by using topical RankSVM. In: WWW 2010, pp. 131–140 (2010)
9. Freund, Y., Iyer, R., Schapire, R.E., Singer, Y.: An efficient boosting algorithm for combining preferences. J. Machine Learning Research 4, 933–969 (2003)
10. Liu, T.-Y.: Learning to Rank for Information Retrieval. Foundations and Trends in Information Retrieval 3, 225–331 (2009)
11. Dwork, C., Kumar, R., Naor, M., Sivakumar, D.: Rank aggregation methods for the Web. In: WWW 2001, pp. 613–622 (2001)
12. Farah, M., Vanderpooten, D.: An outranking approach for rank aggregation in information retrieval. In: SIGIR 2007, pp. 591–598 (2007)
13. Bian, J., Liu, T.-Y., Qin, T., Zha, H.: Ranking with query-dependent loss for web search. In: WSDM 2010, pp. 141–150 (2010)
14. Li, F., Li, X., Bian, J., Zheng, Z.: Optimizing Unified Loss for Web Ranking Specialization. In: CIKM 2010, pp. 1593–1596 (2010)

On Modeling Rank-Independent Risk
in Estimating Probability of Relevance

Peng Zhang[1], Dawei Song[1], Jun Wang[2],
Xiaozhao Zhao[3], and Yuexian Hou[3]

[1] School of Computing, Robert Gordon University, UK
{p.zhang1,d.song}@rgu.ac.uk
[2] Department of Computer Science, University College London, UK
[3] School of Computer Sci & Tec, Tianjin University, China

Abstract. Estimating the probability of relevance for a document is fundamental in information retrieval. From a theoretical point of view, risk exists in the estimation process, in the sense that the estimated probabilities may not be the actual ones precisely. The estimation risk is often considered to be dependent on the rank. For example, the probability ranking principle assumes that ranking documents in the order of decreasing probability of relevance can optimize the rank effectiveness. This implies that a precise estimation can yield an optimal rank. However, an optimal (or even ideal) rank does not always guarantee that the estimated probabilities are precise. This means that part of the estimation risk is rank-independent. It imposes practical risks in the applications, such as pseudo relevance feedback, where different estimated probabilities of relevance in the first-round retrieval will make a difference even when two ranks are identical. In this paper, we will explore the effect and the modeling of such rank-independent risk. A risk management method is proposed to adaptively adjust the rank-independent risk. Experimental results on several TREC collections demonstrate the effectiveness of the proposed models for both pseudo-relevance feedback and relevance feedback.

Keywords: Probability of Relevance, Estimation, Risk Management, Ranking-Independent Risk, Language Modeling.

1   Introduction

A main aim of IR is to determine the relevance of each document in a collection with respect to the user's information need (represented as a query). Relevance has been regarded as a concept in a probabilistic view for decades [8]. The probability ranking principle (PRP) justified that ranking documents in the order of decreasing probability of relevance can optimize the rank effectiveness [11,12]. This implies that a precise estimation for the probability of relevance can yield an optimal rank and the rank effectiveness can empirically indicate the quality of the estimation. Many retrieval models that explicitly or implicitly estimate the probability of relevance are mainly for the ranking purpose. For

M.V.M. Salem et al. (Eds.): AIRS 2011, LNCS 7097, pp. 13–24, 2011.

instance, the classical probabilistic models [12] usually estimate the odds-ratio of the probability of relevance. The final relevance scores are often obtained via some rank-equivalent and approximate calculations. The language modeling (LM) approaches [10,15] can also be considered as estimating the probability of relevance under the generative relevance framework [4].

However, a fundamental research problem arises: can an optimal (or even ideal) rank guarantee that the estimated probabilities are precise and without a risk? For instance, suppose that based on the actual relevance judgements of a group of users, the probabilities of relevance for two documents d_1 and d_2 are $p_1 = 0.74$ and $p_2 = 0.26$, respectively. Therefore, the correct rank is d_1 at first and then d_2. Assume that we have two sets of estimated probabilities by two models. One model gives $p_1 = 0.71$ and $p_2 = 0.29$, while the other gives $p_1 = 0.92$ and $p_2 = 0.08$. Both models give a correct rank. However, the second model overestimates d_1 and underestimates d_2. Theoretically, this example indicates that part of the estimation risk could be independent of the rank.

The rank-independent risk is not only of theoretical importance for risk modeling, but also for a wide range of retrieval tasks, where the initial estimation for document relevance is not the final decision. For example, in pseudo-relevance feedback (PRF), the estimated relevance probabilities from the first-round retrieval largely determine the document weights used for query expansion and thus play an important role in the PRF models [5]. In meta-search [7], as another example, the relevance scores or probabilities obtained from different search engines should be fused before the final estimation for the document relevance. Therefore, it is necessary to control the estimation risk at the very early stage before it spreads and gets more complicated in the later stages.

It is important to clarify that the *rank-dependent risk* refers to the relevance probability estimation risk that can influence the rank, while the *rank-independent risk* does not. Since in practice the ideal rank is usually unavailable, both types of risks may exist in the estimated relevance probabilities. In this paper, we focus on the latter, which, to our best knowledge, has not yet been paid much attention in the literature. Therefore, we aim to single out the effect of the rank-independent risk associated to the different estimated relevance probabilities when two resultant ranks are identical.

We propose an easy-to-implement risk management method to adjust the rank-independent risk adaptively for an estimated probability distribution. For a given retrieval model, the proposed method can be regarded as the micro-level adjustment, as opposed to the re-ranking approaches (tackling the rank-dependent risk). The latter can be regarded as the macro-level adjustment and is out of the scope of this paper. Our proposed method is applied and evaluated in the pseudo-relevance feedback and relevance feedback. The hypothesis is that the management of the rank-independent risk associated to the estimated probabilities in the first-round retrieval can improve the performance of the second-round retrieval. Experimental results on several large-scale TREC collections have shown the effectiveness of our method.

2   Literature Review

Researchers have paid much attention to the probabilistic characteristics of relevance [8] over decades. The probabilistic ranking principle (PRP) [11] suggests that the document ranking in the order of decreasing *probability of relevance* of documents can give the optimal rank effectiveness (e.g., in terms of the expected precision [11]) and minimize the overall risk [13]. The risk here refers to the retrieval risk [11,13], which is based on the loss function associated with a decision on whether or not to retrieve a document. Therefore, the retrieval risk is closely related to the rank effectiveness. The risk minimization framework [3,16] suggests that the optimal ranking strategies can be obtained through considering suitable loss functions in different IR tasks, and the retrieval risks are formulated not only in terms of relevance, but also other factors such as novelty and redundancy. In this paper, we focus on relevance only and leave the extension to other factors as future work.

Recently, several approaches have been proposed for modeling the risk in estimating relevance probabilities or scores. In [17], it is argued that the formulation in most estimates of document relevance only provide the point estimation, i.e., the mean, but ignoring the second-moment estimation, i.e., the variance. The variance in computing the relevance score of each individual document can, however, reflect the uncertainty of the corresponding estimation [17]. Wang and Zhu [14] integrated the similar relevance estimation's uncertainty and the inter-document dependency into a Portfolio Theory (PT) based framework. In the above two models, a parameter is involved to adjust the level of uncertainty of the relevance estimation. Different parameter settings can yield different document rankings, thus presumably satisfying different kinds of user preference, or different performance metrics for different IR tasks [17,14]. It turns out that in the literature, little attention has been paid to the modeling of the rank-independent risk, which is the aim of this paper. Our main contributions are:

- We propose to study the rank-independent risk in estimating the probability of relevance.
- A risk management method is proposed to control such risk.
- The above method has been effectively applied to pseudo-relevance feedback and relevance feedback.

3   Rank-Independent Risk Modeling

The probability of relevance of each document corresponds to one basic retrieval question [4]: what is the probability of this document d being relevant to a query q? Accordingly, it can be formulated as $p(r|d,q)$ [12]. Let $\hat{p}(r|d,q)$ denote the estimated probability of $p(r|d,q)$. Once we obtain $\hat{p}(r|d,q)$, assuming a uniform prior $p(d)$, we can normalize it as

$$S_q(d) = \frac{\hat{p}(r|d,q)}{\sum_{d' \in D} \hat{p}(|r|d',q)} \tag{1}$$

where D is the document set. This normalization is for the further analysis on the estimated relevance probabilities for all the documents in D. $S_q(d)$ denotes the estimated relevance probability (after normalization) of the document d with respect to the query q. Let S_q denote the estimated relevance distribution for all the documents in D.

Our proposed rank-independent risk modeling is expected to be applicable to most retrieval models that can estimate the probability of relevance. In this paper, our focus is on the language modeling (LM) approaches [10,15]. Lafferty and Zhai [3,4] linked the LM approaches to the probability of relevance $p(r|d, q)$. As explained in the introduction, we are going to explore the rank-independent risk associated with any two rank-equivalent relevance distributions. Therefore, we first show two rank-equivalent LM approaches as follows.

3.1   Rank-Equivalent LM Approaches

The query-likelihood (QL) approach [10,15] is a standard language modeling (LM) approach for the first-round retrieval. It is formulated as:

$$p(q|\theta_d) = \prod_{i=1}^{m_q} p(q_i|\theta_d) \tag{2}$$

where $p(q|\theta_d)$ is the query-likelihood, $q = q_1 q_2 \cdots q_{m_q}$ is the given query, m_q is q's length, and θ_d is a smoothed language model for a document d.

The Negative KL-Divergence (ND) [3] between the query language model θ_q and document language model θ_d is formulated as

$$-D(\theta_q|\theta_d) = -H(\theta_q, \theta_d) + H(\theta_q) \tag{3}$$

where $H(\theta_q, \theta_d)$ is the cross entropy between θ_q and θ_d, and $H(\theta_q)$ is the entropy of the θ_q. According to the deviation in [3,9], if a maximum-likelihood estimator is used to estimate the query language model θ_q, then

$$-H(\theta_q, \theta_d) = \frac{1}{m_q} \log p(q|\theta_d). \tag{4}$$

The above equation shows that $-H(\theta_q, \theta_d)$ is logarithmically proportional to the query-likelihood $p(q|\theta_d)$. This means that $-H(\theta_q, \theta_d)$ and $p(q|\theta_d)$ are equivalent in terms of ranking documents. Since in Eq. 3, the $H(\theta_q)$ is independent of document ranking, it turns out that negative KL-divergence is rank-equivalent to the query-likelihood approach.

3.2   Difference between the Two Rank-Equivalent Estimations

We now present the difference between the two document relevance distributions estimated by the QL model and ND model. For a given q, the document relevance distribution estimated by the QL model is denoted as:

$$S_q^{QL}(d) = \frac{p(q|\theta_d)}{\sum_{d' \in D} p(q|\theta_{d'})} \tag{5}$$

where D is a set consisting of all concerned documents.

The document relevance distribution estimated by the ND model can be defined as the normalized exponential of the negative KL-divergence:

$$S_q^{ND}(d) = \frac{\exp\{-D(\theta_q|\theta_d)\}}{\sum_{d' \in D} \exp\{-D(\theta_q|\theta_{d'})\}} \tag{6}$$

The exponential transformation (i.e $\exp\{\}$) is to transform the divergence value to a probability value. Since the $H(\theta_q)$ in Eq. 3 is a constant for every $d \in D$, it can be eliminated in the normalization process of Eq. 6. We then get

$$S_q^{ND}(d) = \frac{\exp\{-H(\theta_q, \theta_d)\}}{\sum_{d' \in D} \exp\{-H(\theta_q, \theta_{d'})\}} = \frac{[p(q|\theta_d)]^{\frac{1}{m_q}}}{\sum_{d' \in D} [p(q|\theta_{d'})]^{\frac{1}{m_q}}} \tag{7}$$

After normalizing $p(q|\theta_d)$ by Z_{QL} (i.e. $\sum_{d' \in D} p(q|\theta_{d'})$), we have

$$S_q^{ND}(d) = \frac{[p(q|\theta_d)/Z_{QL}]^{\frac{1}{m_q}}}{\sum_{d' \in D} [p(q|\theta_{d'})/Z_{QL}]^{\frac{1}{m_q}}} = \frac{[S_q^{QL}(d)]^{\frac{1}{m_q}}}{\sum_{d' \in D} [S_q^{QL}(d')]^{\frac{1}{m_q}}} \tag{8}$$

It shows that in the estimated ND distribution S_q^{ND}, the relevance probabilities are raised to the powers of $\frac{1}{m_q}$ of $S_q^{QL}(d)$, turning to $[S_q^{QL}(d)]^{\frac{1}{m_q}}$ before normalization. As a result, compared with the QL relevance distribution in Eq. 5, the ND relevance distribution in Eq. 6 is often more smooth, in a sense that there are less very large or very small probabilities. We will show in the next section that the rank-independent risk of a relevance distribution is related to its smoothness. The powers-based idea in Eq. 8 then motivates the distribution remodeling process of our proposed risk management method (detailed in Section 3.4).

3.3 Entropy-Based Risk Measurement

In our work, the concerned probability distribution is the estimated document relevance distribution S_q (see Eq. 1) generated by a retrieval model. For instances, S_q can be S_q^{QL} (see Eq. 5) or S_q^{ND} (see Eq. 6). The entropy of an estimated S_q is defined as:

$$H(S_q) = -\sum_{d \in D} S_q(d) \log S_q(d) \tag{9}$$

where H is the Shannon entropy of the distribution S_q, and D is the document set, which can be the whole document collection or the top n ranked documents. The entropy $H(S_q)$ generally indicates the smoothness of the distribution S_q. In general, the larger entropy of S_q implies a higher degree of smoothness, i.e., there are less probabilities which are relatively too large or too small in S_q.

Our assumption here is that the larger the entropy (i.e., the higher degree of smoothness) is, the less rank-independent risk would be with the corresponding distribution, if true relevance judgements are not available to support the rationality of some too large or too small relevance probabilities. In this case, a higher risk would be posed, if a document with relatively too small/large probability is actually relevant/irrelevant.

Let us further illustrate the intuition of this assumption through an example. Given two documents d_1 and d_2, suppose two models give different estimations (p_1=0.71 p_2=0.29) and (p'_1=0.92 p'_2=0.08), respectively, for (d_1, d_2). The first distribution (0.71, 0.29) is more smooth than the second one (0.92, 0.08). If we do not have any relevance judgements of document d_1 and d_2, the possible (binary) relevance judgements for (d_1, d_2) can be (1, 1), (1,0), (0,1) and (0,0), where 1 denotes relevance and 0 denotes irrelevance. We can see that only in the second case, i.e., (1,0), it is sure that d_1 is more relevant than d_2 and it would be reasonable that d_1's probability is much bigger than that of d_2. However, in all other possible cases, the first distribution (which is more smooth) is better than the second one. Specifically, in the cases (1,1) and (0,0), there is no distinction between two relevance judgements, suggesting that the smoother distribution is safer. In the case (0,1), it turns out that d_1 should not have too large probability. Thus, the first distribution has a less risk since it is better in most (3 out 4) cases.

3.4   Powers-Based Risk Management (PRM) Method

We will present a novel risk management method and provide a theoretical analysis to show that the method can make every pair of probabilities in an estimated distribution become more smooth so as to reduce (overall) rank-independent risk (without changing the original document rank). This method can remodel an estimated distribution and the remodeling method is motivated by the powers-based idea described in Eq. 8 Specifically, given a retrieval model and its estimated document relevance distribution S_q, the remodeling method will raise every probability in S_q to the powers ($f(q)$) and then normalize the revised probabilities. It can be formulated as:

$$\widetilde{S_q}(d) = \frac{[S_q(d)]^{\frac{1}{f(q)}}}{\sum_{d' \in D}[S_q(d')]^{\frac{1}{f(q)}}} \tag{10}$$

where $\widetilde{S_q}$ denotes the remodeled distribution, and the powers $f(q)(> 0)$ is a function for the query q. $f(q)$ can not only be m_q in Eq. 8, but also can be other functions (detailed later). Here, we first explain the relations between this remodeling method and the rank-independent risk measurement. This remodeling algorithm preserves the original document rank and has a property described in Proposition 1 (See Appendix A for the formal Proof). This proposition proves that in Eq. 10, the bigger $f(q)$ value (i.e. b in the Proposition), the smaller the relative difference between any two probabilities in the distribution $\widetilde{S_q}$ and thus the higher degree of overall smoothness of the distribution.

Proposition 1. *Given a distribution S_q, suppose $S_q(d_i)$ and $S_q(d_j)$ are the estimated relevance probabilities of any two document d_i and d_j, respectively.*

If $0 < a < b$, then the relative difference between $[S_q(d_i)]^{\frac{1}{b}}$ and $[S_q(d_j)]^{\frac{1}{b}}$ should be smaller than that between $[S_q(d_i)]^{\frac{1}{a}}$ and $[S_q(d_j)]^{\frac{1}{a}}$.

Intuitively, a bigger $f(q)$ value will exclude too large or too small probabilities in the distribution $\widetilde{S_q}$, making the distribution become smoother. Thus, we can draw the observation that in Eq. 10, the bigger the $f(q)$ value is, the larger the entropy $H(\widetilde{S_q})$ of the distribution $\widetilde{S_q}$ will be. This has been verified based on several probability distributions (e.g., exponential distribution) and estimated document relevance distributions from the TREC data.

In this paper, we adopt two options for $f(q)$, each corresponding to an instantiated algorithm of our method. The first option is m_q as used in the Eq. 8, where m_q is the length of query q. We denote this option as

$$f_{ND}(q) = m_q \tag{11}$$

Since m_q is often greater than 1, it turns out that the estimated distribution (i.e. $S_q^{ND}(d)$ in Eq. 8) by the ND model is often more smooth than the one (i.e. $S_q^{QL}(d)$ in Eq. 8) by the QL model.

The second option of $f(q)$ can be an adjustable parameter λ as follows:

$$f_\lambda(q) = \lambda \ (\lambda > 0) \tag{12}$$

This option allows us have different remodeled distributions and a bigger λ generally leads to a smoother remodeled distribution.

4   Application

The proposed Powers-based Risk Management (PRM) method (in Eq. 10) can be viewed to some extent as a micro adjustment (or called fine adjustment) for the estimated document relevance distribution. Generally, the applications of the proposed method are those tasks where the initial estimation of the document relevance is not the final decision. In this paper, the tasks we focus on are the pseudo-relevance feedback (PRF) and the relevance feedback (RF), where the relevance estimation in the first-round retrieval can indicate feedback documents' weights used in the second-round retrieval.

Relevance Model (RM) [5] is a typical language modeling approach for the second-round retrieval. For each query q, based on the given document set D ($|D| = n$), the RM[1] is formulated as:

$$p(w|\theta_R) = \sum_{d \in D} p(w|\theta_d) \frac{p(q|\theta_d)p(\theta_d)}{\sum_{d' \in D} p(q|\theta_{d'})p(\theta_{d'})} \tag{13}$$

where $p(w|\theta_R)$ is the estimated relevance model, $p(\theta_d)$ is d's prior probability. A number of terms with top probabilities in $p(w|\theta_R)$ will be selected to estimate

---

[1] This formulation is equivalent to RM1 in [5].

the expanded query model, which is then used for the second-round retrieval. In the RM [5], the document prior $p(\theta_d)$ is often assumed to be uniform. It turns out that the estimated document relevance distribution (i.e. $S_q^{QL}(d)$ in Eq. 5) by the QL model plays an important role in the RM, since theoretically it distinguishes RM from a mixture of document language models (say $\sum_{d \in D} p(w|\theta_d)$).

The Relevance Model was initially derived for the PRF task, where the document set D is set as the top n retrieval documents in the first-round retrieval. It can also be used in the RF task [1], by selecting all the truly relevant documents (based on the relevance data available for the standard benchmarking collections) in top n documents as the document set D in Eq. 13.

For both PRF and RF tasks, the risk management method will remodel the distribution S_q^{QL} (see Eq. 5) obtained from the first round retrieval by the QL model. Our hypothesis is that the management of the rank-independent risk in the distribution S_q^{QL} could improve the retrieval performance of both tasks.

We would like to mention that other factors, e.g., the query-drifting after query expansion [2] or the combination coefficient for the feedback model [6], etc., also have a direct influence on the rank performance of the PRF and RF tasks. However, in this paper, we mainly focus on the usefulness of managing the rank-independent risk in estimating relevance probabilities of documents.

5   Empirical Evaluation

5.1   Evaluation Configuration

Evaluation Data. The evaluation involves three standard TREC collections, including WSJ (87-92, 173,252 docs), AP (88-89, 164,597 docs) in TREC Disk 1 & 2, and ROBUST 2004 (528,155 docs) in TREC Disk 4 & 5. Both WSJ and AP data sets are tested on queries 151-200, while the ROBUST 2004 is tested on queries 601-700. The *title* field of the queries are used. Lemur[9] 4.7 is used for indexing and retrieval. All collections are stemmed using the Porter stemmer and stop words are removed in the indexing process.

Evaluation Set-up. The first-round retrieval is carried out by a baseline language modeling (LM) approach, i.e., the query-likelihood (QL) model [15,10] in Eq. 2. The smoothing method for the document language model is the Dirichlet prior [15] with $\mu = 1000$, which is a default setting in Lemur toolkit, and also a typical setting for query-likelihood model.

After the first-round retrieval, the top n ranked documents are selected as the pseudo-relevance feedback (PRF) documents for the PRF task. The truly relevant documents in the PRF documents are selected as the relevance feedback (RF) documents for the RF task. We report the results with respect to $n = 30$. Nevertheless, we have similar observations on other n (e.g., 50, 70, 90). The Relevance Model (RM) in Eq. 13, is selected as the second baseline method, where the document prior is set as uniform. The number of expanded terms is fixed as 100. 1000 retrieved documents by the KL-divergence model are used for performance evaluation in both the first-round retrieval and second-round retrieval.

Table 1. Overall Pseudo-relevance Feedback Performance

MAP (chg)	WSJ8792	AP8889	ROBUST04
LM	0.3127	0.3058	0.2880
RM	0.3720 (+0.1896$^{\alpha}$)	0.3864 (+0.2636$^{\alpha}$)	0.3324 (+0.1542$^{\alpha}$)
PRM_ND	0.3765 (+0.2040$^{\alpha}$)	0.4034 (+0.3192$^{\alpha\beta}$)	**0.3443 (+0.1955$^{\alpha\beta}$)**
PRM_λ	**0.3837 (+0.2271$^{\alpha\beta}$)**	**0.4064 (+0.3290$^{\alpha\beta}$)**	0.3439 (+0.1941$^{\alpha\beta}$)

Improvements at significance level 0.05 over LM and RM are marked with α and β, respectively.

Table 2. Overall Relevance Feedback Performance

MAP (chg)	WSJ8792	AP8889	ROBUST04
RM	0.4420	0.4493	0.4228
PRM_ND	0.4844 (+0.0959$^{\beta}$)	0.4913 (+0.0935$^{\beta}$)	0.4736 (+0.1202$^{\beta\gamma}$)
PRM_λ	**0.4929 (+0.1152$^{\beta\gamma}$)**	**0.4972 (+0.1066$^{\beta\gamma}$)**	**0.4818 (+0.1395$^{\beta\gamma}$)**

Improvements at significance level 0.05 and 0.01 over RM are marked with β and γ, respectively.

The Mean Average Precision (MAP), which reflects the overall rank performance, is adopted as the primary evaluation metric. The Wilcoxon signed rank test is the measure of the statistical significance of the improvements over baseline methods.

Evaluation Procedure. We aim to test the performance of different Powers-based Risk Management (PRM) algorithms (see Section 3.4). We denote these algorithms (corresponding to $f_{ND}(q)$ and $f_{\lambda}(q)$) as PRM_ND and PRM_λ, respectively. For both PRF and RF tasks, the risk management method will remodel the estimated document relevance distribution, i.e., the S_q^{QL} in Eq. 8. Then, the remodeled document relevance distribution will be input to the RM to construct the expanded query model for the second-round retrieval. Note that the PRM_ND is to remodel the S_q^{QL} in the first-round retrieval.

5.2   Evaluation on Risk Management Method for PRF Task

The experimental results for different PRM algorithms are summarized in Table 1. We can easily observe that RM significantly outperforms LM on every collection, which demonstrates its effectiveness for the second-round retrieval.

For PRM_ND, we can observe that PRM_ND can improve the RM on every collection, and the improvements are statistically significant on AP8889 and ROBUST2004 collections. This indicates that if we transform the negative KL-divergences to probabilities, these transformed probabilities can be used as the document weights in the RM, for which the document weights are usually from the query-likelihood model.

For PRM_λ, the results show that the PRM_λ with its best λ can significantly improve RM. It is also necessary to test PRM_λ's performance on different λ. Recall that the bigger the λ is, the smoother (i.e., with larger entropy) the remodeled probabilities are. The pseudo-relevance feedback performance of PRM_λ are shown in Fig. 1, from which we can generally conclude that when $\lambda > 1$ (i.e. to

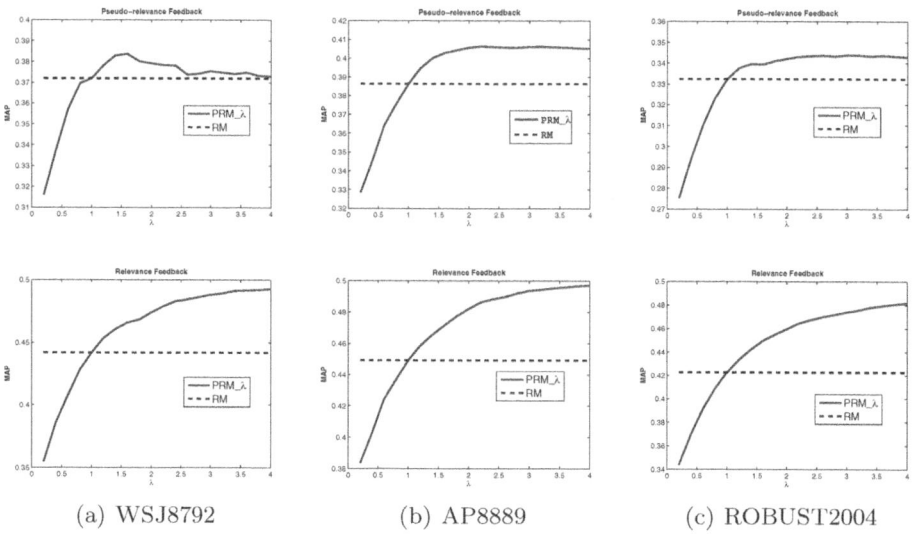

(a) WSJ8792   (b) AP8889   (c) ROBUST2004

Fig. 1. The performance $(y-axis)$ of PRM_λ in the PRF task (the first row) and the RF task (the second row), where the value of λ $(x-axis)$ is in $[0.2, 4]$ with step 0.2

smooth the original probabilities in $S_q^{QL}(d)$ in Eq. 5), PRM_λ can significantly improve RM. On the other hand, when $\lambda < 1$ (i.e., to force the original probabilities to be less smooth), the performance is always below that of the RM. Since the results usually reach a peak and then drop down, we can not say that the smoother estimated relevance probability distribution entails the better PRF performance.

5.3   Evaluation on Risk Management Method for RF Task

This experiment is to test the performance of our PRM algorithms in the RF context. Note that the involved documents in the RF task are all relevant (i.e. with the same relevance status 1). Therefore, a more smooth relevance distribution over documents would have less risk.

The results are summarized in Table 2. We can observe that PRM_ND can significantly improve RM on all collections. In some cases, the significance level is 0.01. For PRM_λ, the results show that the PRM_λ with $\lambda = 4$ can significantly (with significance level 0.01) improve RM on all collections. Concerning the influence of different λ on PRM_λ's performance, Fig. 1 shows that the larger the λ is, the less the rank-independent risk would be and hence the better performance can be achieved.

This further indicates the importance of managing the rank-independent risk. In above experiments, we can observe that with the same ranking different estimated probabilities can have quite different impact on the next-round retrieval performance. It also shows that the remodeling algorithm is effective in reducing the rank-independent risk.

6   Conclusions and Future Work

In this paper, we propose to look at the rank-independent risk in estimating the probability of relevance. This paper aims to answer how to compare the estimation risks between two rank-equivalent retrieval models, and how properly modeling of such risk can improve the retrieval performance.

Specifically, we first show that even though two language modeling approaches (i.e., QL and ND models) are rank-equivalent, their estimated relevance distributions are different and the distribution of the ND model is more smooth than the one of the QL model. In addition, a risk management method, which is based on the powers-based remodeling idea motivated from the distribution difference (see Eq. 8) of QL and ND models, is proposed to generally manage the rank-independent risk for a given retrieval model. We apply the proposed risk management method to the pseudo-relevance feedback and relevance feedback. Experimental results on several TREC collections demonstrate the effectiveness of the proposed method.

In the future, it would be very interesting to derive an optimization method to automatically obtain the optimal λ of PRM_λ for different query. To this end, machine learning algorithms could be helpful . In addition, we would like to incorporate relevance judgements from users/assessors into the risk modeling and management process. For example, we can add some constraints designed by using relevance judgements or implicit feedback features that can indicate the document relevance. Moreover, in the pseudo-relevance feedback task, we would consider how to implement the rank-independent risk management after the first-round retrieved documents are re-ranked. Furthermore, we are also interested to incorporate the proposed risk modeling into the score distribution calibration for the classical probabilistic models.

Appendix A: Proof for Proposition 1

Proof. For simplicity, in this proof, let S_i and S_j denote $S_q(d_i)$ and $S_q(d_j)$, respectively. Without loss of generality, we assume that $S_i > S_j > 0$. Then, we have

$$\frac{(S_i^{\frac{1}{b}} - S_j^{\frac{1}{b}})/S_j^{\frac{1}{b}}}{(S_i^{\frac{1}{a}} - S_j^{\frac{1}{a}})/S_j^{\frac{1}{a}}} = \frac{(S_i/S_j)^{\frac{1}{b}} - 1}{(S_i/S_j)^{\frac{1}{a}} - 1} \tag{14}$$

Since $(S_i/S_j) > 1$ and $0 < \frac{1}{b} < \frac{1}{a}$, we get $1 = (S_i/S_j)^0 < (S_i/S_j)^{\frac{1}{b}} < (S_i/S_j)^{\frac{1}{a}}$. This means that the right hand side of Eq. 14 is less than 1. Therefore, we have

$$\frac{(S_i^{\frac{1}{b}} - S_j^{\frac{1}{b}})/S_j^{\frac{1}{b}}}{(S_i^{\frac{1}{a}} - S_j^{\frac{1}{b}})/S_j^{\frac{1}{b}}} < 1 \tag{15}$$

The proposition then follows.

Acknowledgments. This research is funded in part by the UK's EPSRC (EP/F014708/2), the China's NSFC (61070044) and the EU's Marie Curie Actions-IRSES (247590).

References

1. Croft, W.B., Cronen-Townsend, S., Lavrenko, V.: Relevance feedback and personalization: A language modeling perspective. In: DELOS Workshop: Personalisation and Recommender Systems in Digital Libraries (2001)
2. Dillon, J.V., Collins-Thompson, K.: A unified optimization framework for robust pseudo-relevance feedback algorithms. In: CIKM, pp. 1069–1078 (2010)
3. Lafferty, J.D., Zhai, C.: Document language models, query models, and risk minimization for information retrieval. In: SIGIR, pp. 111–119 (2001)
4. Lafferty, J.D., Zhai, C.: Probabilistic relevance models based on document and query generation. In: Language Modeling and Information Retrieval, pp. 1–10 (2003)
5. Lavrenko, V., Croft, W.B.: Relevance-based language models. In: SIGIR, pp. 120–127 (2001)
6. Lv, Y., Zhai, C.: Adaptive relevance feedback in information retrieval. In: CIKM, pp. 255–264 (2009)
7. Manmatha, R., Rath, T., Feng, F.: Modeling score distributions for combining the outputs of search engines. In: SIGIR, pp. 267–275 (2001)
8. Maron, M.E., Kuhns, J.L.: On relevance, probabilistic indexing and information retrieval. J. ACM 7, 216–244 (1960)
9. Ogilvie, P., Callan, J.: Experiments using the lemur toolkit. In: TREC 2002, pp. 103–108 (2002)
10. Ponte, J.M., Croft, W.B.: A language modeling approach to information retrieval. In: SIGIR, pp. 275–281 (1998)
11. Robertson, S.E.: The probability ranking principle in IR. Journal of Documentation, 294–304 (1977)
12. Robertson, S.E., Zaragoza, H.: The probabilistic relevance framework: Bm25 and beyond. Foundations and Trends in Information Retrieval 3(4), 333–389 (2009)
13. van Rijsbergen, C.J.: Information Retrieval. Butterworths (1979)
14. Wang, J., Zhu, J.: Portfolio theory of information retrieval. In: SIGIR, pp. 115–122 (2009)
15. Zhai, C., Lafferty, J.D.: A study of smoothing methods for language models applied to ad hoc information retrieval. In: SIGIR, pp. 334–342 (2001)
16. Zhai, C., Lafferty, J.D.: A risk minimization framework for information retrieval. Inf. Process. Manage. 42(1), 31–55 (2006)
17. Zhu, J., Wang, J., Cox, I.J., Taylor, M.J.: Risky business: modeling and exploiting uncertainty in information retrieval. In: SIGIR, pp. 99–106 (2009)

Measuring the Ability of Score Distributions to Model Relevance

Ronan Cummins

Department of Information Technology
National University of Ireland, Galway
ronan.cummins@nuigalway.ie

Abstract. Modelling the score distribution of documents returned from any information retrieval (IR) system is of both theoretical and practical importance. The goal of which is to be able to infer relevant and non-relevant documents based on their score to some degree of confidence.

In this paper, we show how the performance of mixtures of score distributions can be compared using inference of query performance as a measure of *utility*. We (1) outline methods which can directly calculate average precision from the parameters of a mixture distribution. We (2) empirically evaluate a number of mixtures for the task of inferring query performance, and show that the log-normal mixture can model more relevance information compared to other possible mixtures. Finally, (3) we perform an empirical analysis of the mixtures using the recall-fallout convexity hypothesis.

1 Introduction

Analysing the document scores returned from information retrieval (IR) systems is a very useful, yet challenging problem. Work in this area can be dated back to the early days of IR [16]. Modelling the document scores returned for different queries (and from different systems) is an important task because it has been noticed that the distribution of relevant document scores is different than that of non-relevant document scores. For example, if given the entire score distribution (SD) returned from a system, the distribution of relevant documents could be accurately determined, it would be particularly useful for automatic query performance prediction and/or meta-search (fusion) tasks [8,4]. Regardless, the problem of correctly modelling the distribution of relevant and non-relevant documents remains an open, and theoretically important, area in IR.

Over the last decade, the predominant distributions [1] for modelling relevant and non-relevant document scores have been a normal and an exponential respectively. There has been relatively little justification as to why relevant and non-relevant document scores should be drawn from two different families of distributions. Nevertheless, these distributions have best fit the data for many years now. More recently, it has been suggested that the normal-exponential mixture is not theoretically valid under certain assumptions [14], and in fact, a

M.V.M. Salem et al. (Eds.): AIRS 2011, LNCS 7097, pp. 25–36, 2011.

more theoretically valid approach might be to model the scores using two gamma distributions [9].

This paper deals with determining the *best* distribution for use in a mixture model by using the inference of performance (i.e. average precision) as a measure of *utility*, when relevance information is available. While the task of inferring average precision might be viewed as only one measure of *utility*, it is one of the most important tasks in IR. This measure of *utility* is very important as it is linked to a model's ability to accurately model the relevance contained within, and is not only of practical concern but is of theoretical importance. We show that the log-normal distribution is the best distribution to use in a mixture model of score distributions for both *goodness-of-fit* and *utilty*.

The remainder of the paper is organised as follows: Section 2 reviews related work on modelling document score distributions. Section 3 outlines four mixture distributions used in this paper, before the formulae for calculating the average precision from a mixture model are introduced. Section 4 presents empirical results comparing the four mixture models for a number of metrics when relevance information is known. Section 5 presents an empirical analysis of the four mixtures based on Robertson's recall-fallout convexity hypothesis. Finally, section 6 outlines our conclusions and future work.

2   Related Research

In this section, we review related work in document score distributions.

2.1   Related Work

Early work into SD modeling has shown that the distribution of relevant documents somewhat follows a normal curve [16]. Approaches over the years have tried various curves and *'fits'* to try and uncover the underlying distributions. More recent work has shown that modelling relevant and non-relevant document scores using a normal and exponential distribution respectively, fits for the scores at the head of the ranked list (i.e. top 1000 documents) [1]. Indeed, this has been the predominant trend over recent years [14].

Others have addressed more theoretical aspects of the underlying distributions, and have developed hypotheses under which certain distributions can be theoretically rejected [14]. The aforementioned work develops a recall-fallout hypothesis which states that the recall-fallout curve for *good* systems should be upper convex and has shown that if the probability ranking principle [13] holds, then certain distributions should be rejected on theoretical grounds. Further work [2] has hypothesised that a theoretically valid distribution must be able to approach the Dirac delta function (i.e. it must be able to approach an impulse under which the entire mass of documents can reside).

Some of the theoretical problems associated with the infinite support that some distributions allow were addressed recently [1] using truncated forms of distributions. Some novel approaches [10] to modelling the score distribution have used multiple normal distributions for the relevant documents and a gamma distribution of the non-relevant ones. This approach uses these distributions because they are a good *'fit'* given the data. Important work in analysing the generation process (i.e. ranking functions) of document scores and their resultant distributions has also been conducted [9]. On a practical note, research has been conducted to use the score distributions for data fusion [12] and score threshold optimisation [1].

2.2   Contributions

This work has a number of contributions. Firstly we show how average precision can be inferred from a mixture distribution. Secondly, we conduct an extensive evaluation of several mixture models for a number of metrics (one of which is the task of inferring average precision accurately), and advocate the use of the log-normal model in particular. Interestingly, we show that the best method of estimating parameters for the task of inferring average precision, is the method of moments (MME), rather than maximum likelihood estimates (MLE). Finally, we show that the despite its superior performance the log-normal mixture does not adhere to Robertson's recall-fallout convexity hypothesis as well as the gamma mixture.

3   Models

In this section, we present four mixture distributions used in this paper to model the scores of both relevant and non-relevant documents.

3.1   Assumptions and Restrictions

Consider an IR system that retrieves a returned set of N documents, and thus N scores given a query (Q). Firstly, we assume that an IR system ranks documents independently of each other, in accordance with the probability ranking principle (PRP) [13]. While this may not be true for certain systems (e.g. for those that wish to promote diversity), it is a widely held principle in IR. Secondly, we assume a binary view of relevance. While score distributions can be modelled as mixtures of a multiple of differently graded relevance distributions, this work only models a binary view of relevance.

   We used the following two criteria to select the distributions that are presented in section 3.2. Firstly, under on the strong SD hypothesis [2], the distribution of both relevant and non-relevant documents should be able to approach Dirac's delta function (these distributions are valid under that hypothesis). And secondly, there is no theoretically valid reason why relevant and non-relevant documents should be drawn from two different families of distributions, given that the document score of relevant and non-relevant documents is generated using the same process (ranking function) within an IR system.

3.2 Mixture Distributions

The distributions that we consider are the normal distribution (N), the exponential distribution (E), the log-normal distribution (L), and the gamma distribution (G) [11]. For most of the mixtures outlined in this section both relevant and non-relevant documents are modelled using the same distribution. We only include the normal-exponential ($N_1 E_0$) mixture as it has been used in many studies to model score distributions for various tasks. Therefore, the next step is to outline the mixture model that can be used in conjunction with any distribution. For most mixtures, we model both sets of documents using the same distribution, where $P(s|1)$ is the pdf (probability density function) for the scores (s) of relevant documents, and $P(s|0)$ is the pdf for the scores of non-relevant documents. Therefore, similar to previous approaches, the document score distribution can be thought of as a mixture of relevant and non-relevant documents as follows:

$$P(s) = (\lambda) \cdot P(s|1) + (1 - \lambda) \cdot P(s|0) \tag{1}$$

where $\lambda = \frac{R}{N}$ is the proportion of relevant documents R in the entire returned set N. In practice, no form of document score normalisation is necessarily needed for the upper limit for any of the distributions. Although, negative values are not supported for the log-normal or gamma distributions, for the information retrieval models used in this work, the issue of supporting negative scores is not a problem in practice[1].

In this paper, we study four mixtures. Table 1 outlines the mixtures and the parameters that need to be estimated for each model. For the parameters of each model, we use the subscript 1 to imply that the parameter is used with the distribution of relevant document scores, whereas we use the subscript 0 to imply that the parameter is used with the distribution of non-relevant document scores. For three of the mixtures, there are a total of five parameters (i.e. the mixture parameter, two parameters to model the relevant scores and two parameters to model the non-relevant scores), while the normal-exponential model ($N_1 E_0$) has only four parameters. This is important for comparison purposes, as models (and distributions) with more parameters have more flexibility in modelling the observed data. Therefore, some models may have less flexible in terms of their ability to model scores from different systems. Although we have included the normal-exponential ($N_1 E_0$) model in this study, it is in the authors opinion that document scores of relevant and non-relevant documents should not be drawn from two different families of distribution. For the $N_1 E_0$ and $N_1 N_0$ mixtures, the MME (method of moments estimates) and MLE (maximum likelihood) estimates are equivalent. However, for the $L_1 L_0$ and $G_1 G_0$ mixtures, the MME and MLE estimates will lead to different parameter settings.

---

[1] The occurrence of negative score can easily be overcome in practice by simply shifting all scores by some constant factor. In theory, as scores are generated from term-frequency evidence (bounded by zero), there are some arguments as to why negative scores should not occur in an IR model.

Table 1. Composition of Mixtures

Label	Relevant	Non-Relevant	# of parameters	parameters	$MME = MLE$
N_1E_0	Normal	Exponential	4	$\mu_1,\sigma_1,\beta_0,\lambda$	yes
N_1N_0	Normal	Normal	5	$\mu_1,\sigma_1,\mu_0,\sigma_0,\lambda$	yes
L_1L_0	Log-Normal	Log-Normal	5	$\mu_1,\sigma_1,\mu_0,\sigma_0,\lambda$	no
G_1G_0	Gamma	Gamma	5	$k_1,\theta_1,k_0,\theta_0,\lambda$	no

3.3   Inferring Average Precision

In this section, we will show how average precision (a standard metric for the effectiveness of a query) can be calculated directly from the mixture of continuous distributions. Firstly, it is worth noting that average precision is an informative measure. As average precision can be viewed geometrically as the area under the precision-recall curve [3][2], we know that it summarises the performance over a large portion of the ranked list, and therefore, conveys a broad view of the effectiveness of a query. Secondly, it is a stable measure [5], and is probably the most prevalent metric of both query and system performance used in IR literature. The interested reader is referred to research which strongly outlines the theoretical importance of average precision [15].

As recall is the proportion of relevant returned documents compared to the entire number of relevant documents, the recall at score s can be defined as follows:

$$recall(s) = \int_s^\infty \frac{\lambda \cdot P(s|1) \cdot ds}{\lambda} = \int_s^\infty P(s|1) \cdot ds \qquad (2)$$

which is the cumulative density function (cdf) of the distribution of relevant documents (viewed from ∞). Under the distributions outlined earlier for our model, we know that $recall(s)$ will vary between 0 and 1, (i.e. when $s = 0$, $recall(s) = 1$ as ensured by the cdf). Similarly, the precision at s (the proportion of relevant returned documents over the number of returned documents) can be defined as follows:

$$precision(s) = \frac{\int_s^\infty \lambda \cdot P(s|1)}{\int_s^\infty (\lambda) \cdot P(s|1) + (1 - \lambda) \cdot P(s|0)} \qquad (3)$$

Now that we can calculate the precision and recall at any score s in the range $[0 : \infty]$, we can create a precision-recall curve. Furthermore, as average precision can be estimated geometrically by the area under the precision-recall curve [3], the average precision ($avg.prec$) of a query can be calculated as follows:

$$avg.prec() = \int_0^1 precision(s) \cdot dr(s) \qquad (4)$$

---

[2] Preliminary experiments have shown that the linear correlation between the actual average precision and the area under the interpolated precision-recall curve is greater than 0.95.

where $r(s) = recall(s)$ which is in the range $[0:1]$. This formulation is an elegant and intuitive way of calculating average precision using the score distributions. As these expressions are not closed-form, they can be calculated using relatively simple geometric numerical integration methods.

4   Mixture Performance

In this section we perform a comparative analysis of the four mixture models across a number of different IR models (i.e. vector space, classic probabilistic, language model, learned model, and axiomatic model). First, however, we will motivate our choice of comparison metrics.

4.1   Goodness-of-Fit, Correlation, and RMSE

Usually, the performance of a mixture model is determined by how well the model '*fits*' actual data. For different fields of study and for different problems, different metrics may be applicable. Usually, *goodness-of-fit* tests (e.g. Kolmogorov-Smirnov test) are used to either accept or reject certain models as a '*good fit*'. However, in IR, it is well-known that documents, and therefore document scores, at the head of a ranked list are more important than those further down the list [3]. These *goodness-of-fit* tests do not make a distinction between observations (i.e. scores) at various locations and they do not measure the amount of relevant information that can be correctly maintained in the model.

We propose that better mixture models are better able to model the information regarding relevance. An intuitive way of measuring this is by trying to infer the average precision of a query using the model (and its known parameters). Average precision is a natural candidate for capturing the performance (as discussed earlier). Therefore, over a set of topics, the correlation between the inferred average precision from the mixture model and the actual average precision of the query from the IR system, gives us a measure of how much relevance information is contained in each model. From an information theoretic point of view, it also gives us an indication of how much relevance information is lost when modelling each ranking with a particular mixture model.

Table 2. Test Collection Details

Collection		# docs	# topics	range
Test	AP	242,918	149	051-200
	FT	210,158	188	251-450
	WT2G	221,066	50	401-450
	WT10G	1,692,096	100	451-550

---

[3] Looking at only a part of the ranked list (e.g. documents up to rank 1000) does not effectively solve this problem.

4.2   Comparative Analysis

We now compare the four mixture models introduced earlier (i.e. Table 1) over a range of IR systems and settings. Different distributions may better be able to model different IR systems and so for a broader comparison, we compared the four mixtures across five IR models. We chose the vector space model using pivoted document normalisation (PIV) [7], the probablistic model (BM25) [7], a language modelling (LM) approach (Jelinek-Mercer smoothing) [17], a learned approach (ES) [6], and the axiomatic approach (F2EXP) [7], as these represent a broad range of classical and more modern ranking functions. Table 2 shows the test collections used in this research.

Goodness-of-fit. Table 3 shows the Kolmgorov-Smifnoff D-statistic[4] (a measure of *goodness-of-fit*) on each of the collections averaged over the five systems. The D-statistic measures the maximum distance between the cumulative density function of the theoretical distribution (i.e. one of the mixtures) and the empirical distribution (i.e. the actual scores). Firstly, we can see that Table 3 shows that the log-normal model has a significantly[5] better fit compared to the gamma model on two collections for the entire returned set of document scores. The results also show that the log-normal models fits non-web collections very well, but the gamma model has a better fit for some IR systems on web collections. The normal-exponential model is the third best model in terms of fit, while the normal-normal model is particularly poor. We can also see from Table 3 that the MLE parameter estimation technique provides better fits, in general, than MME.

Table 3. Average Kolmgorov-Smifnoff D-statistic for queries across all systems for title queries using entire returned set of document scores

Collection	N_1E_0	N_1N_0	MME		MLE	
			L_1L_0	G_1G_0	L_1L_0	G_1G_0
AP	0.4580	0.7062	0.1676 $\dagger_5$	0.2096	0.1549 $\dagger_5$	0.1901
FT	0.3690	0.6946	0.1316 $\dagger_5$	0.1554	0.1181 $\dagger_5$	0.1405
WT2G	0.3058	0.7464	0.1197 $\dagger_2$	0.1172 $\dagger_3$	0.1225 $\dagger_2$	0.1126 $\dagger_3$
WT10G	0.3113	0.7517	0.1315 $\dagger_1$	0.1253 $\dagger_4$	0.1349 $\dagger_1$	0.1241 $\dagger_4$

Correlations and RMSE. Now we analyse the amount of relevance information that can be correctly contained within each mixture model across the five IR systems using correlation measures. Using the MME and MLE approaches

---

[4] As the parameters of the model are estimated from the observed samples, the critical values of the Kolmgorov-Smifnoff test are invalid. However, we use the D-statistic as a relative measure to compare the mixtures, and not as a statistical test to accept or reject the validity of the distribution.

[5] $\dagger_x$ denotes that the statistic is significantly lower than the next best model using the same parameter estimation technique for x of the five systems.

Table 4. Average Spearman (and Pearson in parentheses) correlation between mixture model's inferred average precision and actual average precision across five IR systems for title queries using entire returned set of documents

Mixture	$N_1 E_0$	$N_1 N_0$	MME		MLE	
			$L_1 L_0$	$G_1 G_0$	$L_1 L_0$	$G_1 G_0$
AP	0.47 (0.26)	0.56 (0.32)	**0.89** (0.84)	0.84 (0.76)	0.80 (0.71)	0.77 (0.66)
FT	0.33 (0.24)	0.55 (0.30)	**0.89** (0.81)	0.86 (0.75)	0.83 (0.75)	0.80 (0.67)
WT2G	0.45 (0.32)	0.49 (0.35)	**0.83** (0.83)	0.81 (0.81)	0.72 (0.67)	0.73 (0.70)
WT10G	0.39 (0.33)	0.40 (0.07)	**0.74** (0.61)	0.66 (0.55)	0.62 (0.46)	0.58 (0.44)

we can estimate the five parameters in each mixture model assuming relevance information is known (i.e. labelled data). We then compare the correlation of the inferred average precision (calculated from equation 4) for the mixture model with the actual average precision from the IR system in question.

Table 4 shows the average Spearman and Pearson correlation of the four mixture models averaged across the five systems[6]. Firstly, it is worth noting that the correlation coefficients for some of the mixtures are quite high, indicating that much of the information regarding average precision (relevance) are correctly modeled by some of the mixtures. We can also see that the mixture model comprised of a normal and exponential (i.e. the predominant model over the last decade) is the lowest performing model of the four that we have studied. The normal-normal model outperforms the exponential-normal model in terms of utility despite having a worse fit (see Table 3). In general, we can also see that the log-normal mixture model tends to outperform the gamma model across a variety of settings and parameter estimation techniques (i.e. for both MME and MLE estimates). In general, the results show that the log-normal model is the more general and consistent model for preserving relevance information across a variety of IR systems.

Table 5 shows the *root mean squared error (RMSE)*[7] of the inferred average precision compared to the actual average precision for a set of queries for both the BM25 and LM systems (the other systems tested showed comparable results). We can see that the actual average precision predicted by the log-normal model is closer to the true average precision. While the RMSE is not of major importance in terms of the predictive quality of a model, it does inform us that the raw output of the log-normal mixture model is closer to the actual average precision of a query. The RMSE results of all other IR systems are comparatively similar to those in Table 5. One reason for this error is that the formulae given for inferring average precision from score distributions (Section 3.3) will actually over-estimate the actual average precision value on TREC data due to the fact

---

[6] The bold font indicates that the average correlation is higher than the next highest across all five systems. Statistical tests do not show any difference between the top two performing mixture models. Statistical tests do show a higher correlation for the gamma and log-normal models compared to the other mixtures.

[7] The † denotes that the reduced error is significant compared to the gamma mixture. A Wilcoxon ranked sign test at the 0.01 level was used.

Table 5. RMSE of Inferred Average Precision (using MME) compared to two System's (BM25 and LM) Actual Average Precison for title queries using entire returned set of document scores

Mixture	N_1E_0	N_1N_0	L_1L_0	G_1G_0	N_1E_0	N_1N_0	L_1L_0	G_1G_0
	BM25				LM			
AP	0.227	0.176	0.115 †	0.232	0.207	0.159 †	0.170	0.275
FT	0.315	0.270	0.143 †	0.179	0.303	0.361	0.182 †	0.260
WT2G	0.310	0.263	0.159	0.122 †	0.309	0.235	0.091 †	0.141
WT10G	0.234	0.209	0.164 †	0.220	0.214	0.167	0.113 †	0.239

that recall is calculated as the number of relevant documents in the returned set, rather than the total number of relevant documents in the collection.

MME vs MLE. Another interesting point is that the MME approach to parameter estimation consistently outperforms the MLE approach in terms of *utility* (i.e. for the task of inferring performance as measured by the correlations in Table 4). However, when all sample observations are treated equally (as for *goodness-of-fit* tests), the D-statistic in Table 3 shows that models derived from MLE are closer to the observed samples. This provides further proof that the correlation coefficients and *goodness of fit* tests measure different aspects. As we are dealing with IR systems, and models of relevance, we argue that a standard measure of *utility* is more apt.

5   Recall-Fallout Convexity Analysis

Of the mixtures studied in this paper, we have empirically determined that the mixture of two log-normals is one of the better mixtures for modelling document scores for a number evaluation metrics. Furthermore, our results suggest that it is very robust and can accurately model rankings returned from many systems. However, it is unclear if this mixture adheres to useful theoretical properties. In this section, we analyse all of the mixtures using the recall-fallout convexity hypothesis [14]. Interestingly, we show that the gamma mixture violates the recall-fallout hypothesis less often than the log-normal mixture near the head of the ranked list (i.e. where it is more important).

5.1   Locating Points of Non-convexity

The recall-fallout hypothesis states that as we traverse a ranked-list, the recall should always be greater than fallout. This seems theoretically justifiable, as IR systems should at least provide a better than random ranking. Therefore, when modelling document rankings as continuous distributions, the recall-fallout hypothesis can be more formally stated as $\int_s^\infty P(s|1) \cdot ds > \int_s^\infty P(s|0) \cdot ds$ for all s. A detailed analysis of the recall-fallout convexity hypothesis for all of the mixtures studied in this paper (except the two log-normal mixture) can be found in the original work [14]. Using notation similar to the original work, the convexity condition that must be satisfied to ensure that recall is greater than fallout for all scores, can be written as follows:

$$g_1(s) > g_0(s) \tag{5}$$

where

$$g(s) = \frac{1}{f(s)} \frac{df(s)}{ds} \tag{6}$$

where $f(s)$ is the probability density function of a particular distribution. Now assuming this hypothesis to be valid, it would be interesting to see how closely the better mixture models adhere it.

Gamma Mixture. Therefore, as $g(s) = ((k-1)/s) - 1/\theta$ for the gamma mixture [14], the score at which the condition is violated is found by simplifying the following:

$$\frac{k_1 - 1}{s} - \frac{1}{\theta_1} = \frac{k_0 - 1}{s} - \frac{1}{\theta_0} \tag{7}$$

which simplifies to

$$s = \frac{\theta_1 \theta_0 k_0 - \theta_1 \theta_0 k_1}{\theta_1 - \theta_0} \tag{8}$$

We can see that if $\theta_1 = \theta_0$, there are no roots for s, and so no violations occur. Furthermore, if $k_1 = k_0$, $s = 0$ and so, the violation occurs at the point at which both recall and fallout are 1 (which is acceptable). For the two-gamma mixture, if $s > 0$, the violation occurs at a score that can be encountered by the mixture, otherwise the violation does not occur.

Log-Normal Mixture. For the log-normal mixture $g(s) = (\mu - log(s) - \sigma^2)/(s \cdot \sigma^2)$, and therefore, the score at which the convexity condition is violated is found at:

$$s = e^{(\mu_1 \sigma_0^2 - \mu_0 \sigma_1^2)/(\sigma_0^2 - \sigma_1^2)} \tag{9}$$

by following a similar simplification process as the gamma mixture. We can see that if $\sigma_1 = \sigma_0$, the function has no roots, and therefore, no violations (similar to the normal distribution [14]). If the variances are not exactly equal, a violation of the convexity condition, will occur at a score above zero. The score at which a violation occurs can be translated to a point of recall using equation (2).

5.2   Empirical Results and Discussion

We analysed the four mixtures models by calculating the points of recall at which the convexity condition was violated for each query on the test collections. It is reasonable to assume that a violation at the head (i.e. low point of recall) of the ranked list is more serious than if it occurs at high recall. However, if the convexity condition is violated at a score that is rarely, or never, encountered

Table 6. Average recall at which convexity violations occur for different models

Mixture	$N_1 E_0$	$N_1 N_0$	$L_1 L_0$	$G_1 G_0$
AP	0.001	0.401	0.178	0.540
FT	0.003	0.309	0.199	0.556
WT2G	0.000	0.236	0.159	0.695
WT10G	0.000	0.309	0.147	0.594

by an IR metric (at high recall), it is deemed less serious. Table 6 reports the average point of recall at which a violation of the convexity condition occurs for a set of queries averaged across the five IR systems. The results in Table 6 are from the four models when using MME as the parameter estimation technique.

In general, we can see that the two-gamma model is the more theoretically sound as violations occur, on average, at a higher point of recall (e.g. at a lower score s). Surprisingly, violations occur at a low point of recall for the log-normal model (even lower than the two-normal model), which suggests that it is theoretically less sound that either the two-gamma model or the two-normal model. The exponential-normal mixture has violations at both ends of the relevant distribution (i.e. both high and low recall) 100% of the time, and therefore, we can see from Table 6 that the violations occur very early on in the ranking (i.e. low point of recall). The results from Table 6 confirm previous analysis [14] with regard to many of these models.

The average results across the five IR systems in Table 6 are highly representative of each individual system. More work is needed to understand the reasons for the apparent shortcoming in the theoretical behaviour of the two log-normal model (especially as it outperforms other mixtures in terms of *goodness-of-fit* and *utility*).

6   Conclusion

In this work, we have performed a comparative analysis of different distributions that comprise mixtures for document score distributions in IR systems. We have determined that the log-normal distribution is the best performing model in terms of both its accuracy in inferring average precision, and its *goodness-of-fit*. The log-normal model has been used in relatively few practical works. Interestingly, we have shown despite its good performance the log-normal model is theoretically less sound than the two-gamma model towards the head of a ranking. Interesting future work would be to create mixture models that unconditionally adhere to the recall-fallout convexity hypothesis (e.g. by ensuring $\sigma_1 = \sigma_0$ for the two log-normal model) and then compare the *utility* of those valid models.

References

1. Arampatzis, A., Kamps, J., Robertson, S.: Where to stop reading a ranked list?: threshold optimization using truncated score distributions. In: SIGIR, pp. 524–531 (2009)

2. Arampatzis, A., Robertson, S.: Modeling score distributions in information retrieval. Inf. Retr. 14(1), 26–46 (2011)
3. Aslam, J.A., Yilmaz, E.: A geometric interpretation and analysis of r-precision. In: CIKM, pp. 664–671 (2005)
4. Baumgarten, C.: A probabilistic solution to the selection and fusion problem in distributed information retrieval. In: ACM SIGIR Conference on Research and Development in Information Retrieval, SIGIR 1999, pp. 246–253. ACM, New York (1999)
5. Buckley, C., Voorhees, E.M.: Evaluating evaluation measure stability. In: SIGIR, pp. 33–40 (2000)
6. Cummins, R., O'Riordan, C.: Learning in a pairwise term-term proximity framework for information retrieval. In: SIGIR, pp. 251–258 (2009)
7. Fang, H., Zhai, C.: An exploration of axiomatic approaches to information retrieval. In: SIGIR, pp. 480–487 (2005)
8. He, B., Ounis, I.: Query performance prediction. Inf. Syst. 31(7), 585–594 (2006)
9. Kanoulas, E., Dai, K., Pavlu, V., Aslam, J.A.: Score distribution models: assumptions, intuition, and robustness to score manipulation. In: SIGIR, pp. 242–249 (2010)
10. Kanoulas, E., Pavlu, V., Dai, K., Aslam, J.A.: Modeling the Score Distributions of Relevant and Non-Relevant Documents. In: Azzopardi, L., Kazai, G., Robertson, S., Rüger, S., Shokouhi, M., Song, D., Yilmaz, E. (eds.) ICTIR 2009. LNCS, vol. 5766, pp. 152–163. Springer, Heidelberg (2009)
11. Hastings, N., Evans, M., Peacock, B.: Statistical distributions, third edition. Measurement Science and Technology 12(1), 117 (2001)
12. Manmatha, R., Rath, T., Feng, F.: Modeling score distributions for combining the outputs of search engines. In: Proceedings of the 24th Annual International ACM SIGIR Conference on Research and Development in Information Retrieval, SIGIR 2001, pp. 267–275. ACM, New York (2001)
13. Van Rijsbergen, C.J.: Information Retrieval, 2nd edn. Butterworth-Heinemann, Newton (1979)
14. Robertson, S.: On Score Distributions and Relevance. In: Amati, G., Carpineto, C., Romano, G. (eds.) ECiR 2007. LNCS, vol. 4425, pp. 40–51. Springer, Heidelberg (2007)
15. Robertson, S.E., Kanoulas, E., Yilmaz, E.: Extending average precision to graded relevance judgments. In: Proceeding of the 33rd International ACM SIGIR Conference on Research and Development in Information Retrieval, SIGIR 2010, pp. 603–610. ACM, New York (2010)
16. Swets, J.A.: Information retrieval systems. Science 141(3577), 245–250 (1963)
17. Zhai, C., Lafferty, J.: A study of smoothing methods for language models applied to information retrieval. ACM Trans. Inf. Syst. 22, 179–214 (2004)

Cross-Language Information Retrieval with Latent Topic Models Trained on a Comparable Corpus

Ivan Vulić, Wim De Smet, and Marie-Francine Moens

Department of Computer Science, K.U. Leuven, Belgium
{ivan.vulic,wim.desmet,marie-francine.moens}@cs.kuleuven.be

Abstract. In this paper we study cross-language information retrieval using a bilingual topic model trained on comparable corpora such as Wikipedia articles. The bilingual Latent Dirichlet Allocation model (BiLDA) creates an interlingual representation, which can be used as a translation resource in many different multilingual settings as comparable corpora are available for many language pairs. The probabilistic interlingual representation is incorporated in a statistical language model for information retrieval. Experiments performed on the English and Dutch test datasets of the CLEF 2001-2003 CLIR campaigns show the competitive performance of our approach compared to cross-language retrieval methods that rely on pre-existing translation dictionaries that are hand-built or constructed based on parallel corpora.

Keywords: Cross-language retrieval, topic models, comparable corpora, document models, multilingual retrieval, Wikipedia.

1 Introduction

With the ongoing growth of the World Wide Web and the expanding use of different languages, the need for cross-language models that retrieve relevant documents becomes more pressing than ever. Cross-language information retrieval deals with the retrieval of documents written in a language different from the language of the user's query. At the time of retrieval the query in the source language is typically translated into the target language of the documents with the help of a machine-readable dictionary or machine translation system. Translation dictionaries do not exist for every language pair, and they are usually trained on large parallel corpora, where each document has an exact translation in the other language, or are hand-built. Parallel corpora are not available for each language pair. In contrast, comparable corpora in which documents in the source and the target language contain similar content, are usually available in abundance. In this paper we address the question whether suitable cross-language retrieval models can be built based on the interlingual topic representations learned from comparable corpora. We accomplish this goal by means of a cross-language generative model, i.e., bilingual Latent Dirichlet Allocation (BiLDA), trained on a comparable corpus such as one composed of Wikipedia articles. The resulting probabilistic translation model is incorporated in a statistical language model for information retrieval. The language models for retrieval have a sound statistical foundation and can easily incorporate probabilistic evidence in order to optimize the cross-language retrieval process.

M.V.M. Salem et al. (Eds.): AIRS 2011, LNCS 7097, pp. 37–48, 2011.

The contributions of the paper are as follows. Firstly, we show the validity and the potential of training a bilingual LDA model on bilingual comparable corpora. Secondly, we successfully integrate the topic distributions resulting from training the bilingual LDA model in several variant retrieval models and perform a full-fledged evaluation of the retrieval models on the standard CLEF test collections. We show that the results obtained by our retrieval models, which do not exploit any linguistic knowledge from a translation dictionary, are competitive with dictionary-based models. Our work makes cross-language information retrieval portable to many different language pairs.

2   Related Work

Probabilistic topic models such as probabilistic Latent Semantic Indexing [9] and Latent Dirichlet Allocation [1] are both popular means to represent the content of a document. Although designed as generative models for the monolingual setting, their extension to multilingual domains follows naturally. Cimiano et al. [6] use standard LDA trained on concatenated parallel and comparable documents in a document comparison task. Roth and Klakow [23] try to use the standard LDA model trained on concatenated Wikipedia articles for cross-language information retrieval, but they do not obtain decent results without the additional usage of a machine translation system.

Recently, the bilingual or multilingual LDA model was independently proposed by different authors ([17,14,7,2]) who identify interlingual topics of different languages. These authors train the bilingual LDA model on a parallel corpus. Jagarlamudi and Daumé III [10] extract interlingual topics from comparable corpora, but use additional translation dictionary information. None of these works apply the bilingual LDA model in a cross-lingual information retrieval setting.

Cross-language information retrieval is a well-studied research topic (e.g., [8,19,24,18]). As mentioned, existing methods rely on a translation dictionary to bridge documents of different languages. In some cases interlingual information is learned based on parallel corpora and correlations found in the paired documents [13], or are based on Latent Semantic Analysis (LSA) applied on a parallel corpus. In the latter case, a singular value decomposition is applied on the term-by-document matrix, where a document is composed of the concatenated text in the two languages, and after rank reduction, document and query are projected in a lower dimensional space ([3,15,5,29]). Our work follows this line of thinking, but uses generative probabilistic approaches. In addition, the models are trained on the individual documents in the different languages, but paired by their joint interlingual topics. Cross-language relevance models [12] have also been applied for the task, but they still require either a parallel corpus or a translation dictionary. LDA-based monolingual retrieval has been described by Wei and Croft [28].

Transfer learning techniques, where knowledge is transfered from one source to another, are also used in the frame of cross-language text classification and clustering. Transfer learning bridged by probabilistic topics obtained via pLSA was proposed by Xue et al. [29] for the task of cross-domain text categorization. Recently, knowledge transfer for cross-domain learning to rank the answer list of a retrieval task was described by Chen et al. [4]. Takasu [26] proposes cross-language keyword recommendation using latent topics. Except for Wang et al. [27], where the evaluation is vague and unsatisfactory (the same dataset is used for training and testing), and relies solely on 30 documents and 7 queries, none of the above works use LDA-based interlingual topics in cross-language retrieval.

3 Bilingual LDA

The topic model we use is a bilingual extension of a standard LDA model, called *bilingual LDA* (BiLDA) ([17,14,7,2]).

As the name suggests, it is an extension of the basic LDA model, taking into account bilingualism and initially designed for parallel document pairs. We test its performance on a collection of comparable texts where related documents are paired, and therefore share their topics to some extent. BiLDA takes advantage of the document alignment by using a single variable that contains the topic distribution θ. This variable is language-independent, because it is shared by each of the paired bilingual comparable documents. Algorithm 3.1 summarizes the generative story, while Figure 1 shows the plate model.

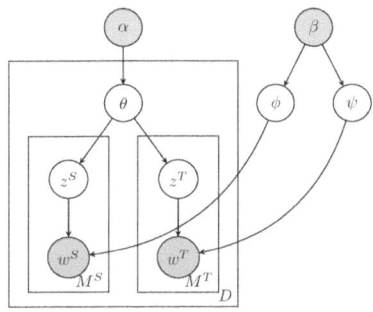

Algorithm 3.1. GENERATIVE STORY FOR BILDA()

for each document pair d_j

do $\begin{cases} \textbf{for each} \text{ word position } i \in d_{jS} \\ \quad \textbf{do} \begin{cases} \text{sample } z_{ji}^S \sim Mult(\theta) \\ \text{sample } w_{ji}^S \sim Mult(\phi, z_{ji}^S) \end{cases} \\ \textbf{for each} \text{ word position } i \in d_{jT} \\ \quad \textbf{do} \begin{cases} \text{sample } z_{ji}^T \sim Mult(\theta) \\ \text{sample } w_{ji}^T \sim Mult(\psi, z_{ji}^T) \end{cases} \end{cases}$

Fig. 1. Generative description and plate model of the bilingual BiLDA model

Having one common θ for both of the related documents implies parallelism between the texts, which might not always be the case. Still, we later show that the BiLDA model can provide satisfactory results when trained on a comparable corpus such as Wikipedia.

The described BiLDA model serves as a framework for modeling our retrieval models. After the training using Gibbs sampling ([25]), two sets of probability distributions are obtained for each of the languages. One set consists of per-topic word probability distributions, calculated as $P(w_i|z_k) = \phi_{k,i}^S = \frac{n_k^{(w_i)}+\beta}{\sum_{j=1}^{|W^S|} n_k^{(w_j)}+W^S\beta}$, where $n_k^{(w_i)}$ denotes the total number of times that the topic z_k is assigned to the word w_i from the vocabulary W^S. The formula for a set of per-topic word probability distributions ψ for the target side of a corpus is computed in an analogical manner.

The second set consists of per-document topic probability distributions, calculated as $P(z_k|D_J) = \theta_{J,k} = \frac{n_J^{(k)}+\alpha}{\sum_{j=1}^{K} n_J^{(j)}+K\alpha}$, where for a document D_J and a topic z_k, $n_J^{(k)}$ denotes the number of times a word in the document D_J is assigned to the topic z_k.

4   LDA-Based CLIR

This section provides a theoretical insight to cross-language retrieval models relying on per-topic word distributions and per-document word distributions.

4.1   LDA-only CLIR Model

Given the set $\{D_1, D_2, \ldots, D_L\}$ of documents in a target language T, and a query Q in a source language S, the task is to rank the documents according to their relevance to the query. We follow the basic approach for using language models in monolingual information retrieval [28]. The probability $P(Q|D_J)$ that the query Q is generated from the document model D_J, is calculated based on the unigram language model:

$$P(Q|D_J) = P(q_1, \ldots, q_m|D_J) = \prod_{i=1}^{m} P(q_i|D_J). \tag{1}$$

The main difference between monolingual IR and CLIR is that documents are not in the same language as the query. Thus, one needs to find a way to efficiently bridge the gap between languages. The common approach is to apply translation dictionaries, translate the query and perform monolingual retrieval on the translated query. If a translation resource is absent, one needs to find another solution. We propose to use sets of per-topic word distributions and per-document topic distributions, assuming the shared space of latent topics. We calculate the right-hand side of equation (1) as

$$P(q_i|D_J) = \delta_1 \sum_{k=1}^{K} \overbrace{P(q_i|z_k^S)}^{Source\ z_k} \underbrace{P(z_k^T|D_J)}_{Target\ z_k} + (1 - \delta_1) P(q_i|Ref)$$

$$= \delta_1 \sum_{k=1}^{K} \phi_{k,i}^S \theta_{J,k}^T + (1 - \delta_1) P(q_i|Ref), \tag{2}$$

by using the two BiLDA-related probability distributions $\phi_{k,i}^S$ and $\theta_{J,k}^T$. The parameter δ_1 is an interpolation parameter, while $P(q_i|Ref)$ is the maximum likelihood estimate of the query word q_i in a monolingual source language reference collection Ref. It gives a non-zero probability for words unobserved during the training of the topic model in case it occurs in the query. Here, we use the observation that latent topics constitute a language-independent space shared between the languages.

The per-topic word distributions for the source language are used to predict the probability that the word q_i from the query Q will be sampled from the topic z_k^S, and the per-document topic distributions for the target language to predict the probability that the same topic z_k^T (but now in the other language[1]) is assigned to a token in the target

---

[1] z_k^S and z_k^T basically refer to the same cross-language topic z_k, but z_k^S is interpreted as a cross-language topic used by source language words, and z_k^T by the target language words.

document D_J. As LDA is a generative model, we may infer the source or target language part of a pre-trained bilingual model on any monolingual collection in the source or the target language, using the same formulas for $\phi_{k,i}^S$ or $\psi_{k,i}^T$ and θ_J, k as in Section 3.

We can now merge all the steps into one coherent process to calculate the probability $P(Q = q_1, q_2, \ldots, q_m | D_J)$, where Q denotes a query in the source language, and D_J denotes a document in the target language. We name this model the **LDA-only model**:

1. Infer the trained model on a test corpus in the target language to learn $P(z_k^T | D_J)$
2. For each word $q_1 \ldots q_m$ in the query, do:
    (a) Compute $P(q_i | z_k^S)$ for all source language topics, $k = 1, \ldots, K$
    (b) Sum the products of per-topic word and per-document topic probabilities:

$$P'(q_i | D_J) = \sum_{k=1}^{K} P(q_i | z_k^S) P(z_k^T | D_J)$$

3. Compute the whole probability score for the given query and the current document D_J:

$$P(Q|D_J) = \prod_{i=1}^{m} \left(\delta_1 \sum_{k=1}^{K} \phi_{k,i}^S \, \theta_{J,k}^T + (1 - \delta_1) P(q_i | Ref) \right) \qquad (3)$$

This gives the score for one target language document D_J. Finally, documents are ranked based on their scores. If we train a bilingual (or a multilingual) model and wish to reverse the language of queries and the language of documents, the retrieval is performed in an analogical manner after the model is inferred on a desired corpus.

4.2   LDA-Unigram CLIR Model

The LDA-only CLIR model from Subsection 4.1 can be efficiently combined with other models for estimating $P(w|D)$. If we assume that a certain amount of words from the query does not change across languages (e.g. some personal names) and thus could be used as an evidence for cross-language retrieval, the probability $P(q_i | D_J)$ from (1) may be specified by a document model with the Dirichlet smoothing. We adopt smoothing techniques according to evaluations and findings from [30]. The Dirichlet smoothing acts as a length normalization parameter and penalizes long documents. The model is then:

$$P_{lex}(q_i | D_J) = \delta_2 \left(\frac{N_d}{N_d + \mu} P_{mle}(q_i | D_J) + (1 - \frac{N_d}{N_d + \mu}) P_{mle}(q_i | Coll) \right) \\ + (1 - \delta_2) P(q_i | Ref), \qquad (4)$$

where $P_{mle}(q_i | D_J)$ denotes the maximum likelihood estimate of the word q_i in the document D_J, $P_{mle}(q_i | Coll)$ the maximum likelihood estimate in the entire collection $Coll$, μ is the Dirichlet prior, and N_d the number of words in the document D_J. δ_2 is another interpolation parameter, and $P(q_i | Ref)$ is the background probability of q_i, calculated over the large corpus Ref. It gives a non-zero probability for words that have zero occurrences in test collections. We name this model the **simple unigram model**.

We can now combine this document model with the *LDA-only* model using linear interpolation and the Jelinek-Mercer smoothing:

$$P(q_i|D_J) = \lambda P_{lex}(q_i|D_J) + (1 - \lambda)P_{lda}(q_i|D_J) \tag{5}$$

$$= \lambda \Big(\delta_2 \big(\frac{N_d}{N_d + \mu} P_{mle}(q_i|D_J) + (1 - \delta_2)P(q_i|Ref) \big) \Big)$$
$$+ (1 - \lambda)P_{lda}(q_i|D_J) \tag{6}$$

where P_{lda} is the *LDA-only* model given by (2), P_{lex} the simple unigram model given by (4), and λ is the interpolation parameter. We call this model the **LDA-unigram model**.

The combined model presented here is straightforward, since it directly uses words shared across a language pair. One might also use cognates (orthographically similar words) identified, for instance, with the *edit distance* ([16]) instead of the shared words only. However, both approaches improve retrieval results only for closely related language pairs, where enough shared words and cognates are observed. We believe that a more advanced "non-LDA" part[2] of the document model may result in even higher scores, since knowledge from other translation resources may be used to model the probability $P_{lex}(q_i|D_J)$.

5   Experimental Setup

5.1   Training Collections

The data used for training of the models is collected from various sources and varies strongly in theme, style and its "comparableness". The only constraint on the training data is the need for document alignment, and it is the only assumption our BiLDA model utilizes during training.

The first subset of our training data is the Europarl corpus [11], extracted from proceedings of the European Parliament and consisting of 6,206 parallel documents in English and Dutch. We use only the evidence of document alignment during the training and do not benefit from the "parallelness" of the sentences in the corpus.

Another training subset is collected from Wikipedia *dumps*[3] and consists of paired documents in English and Dutch. Since the articles are written independently and by different authors, rather than being direct translations of each other, there is a considerable amount of divergence between aligned documents. Our Wikipedia training sub-corpus consists of 7,612 documents which vary in length, theme and style[4].

As a preprocessing step we remove stop words, and our final vocabularies consist of 76,555 words in English, and 71,168 words in Dutch.

---

[2] By the "LDA-part" of the retrieval model, we assume the part of the model in equation (2).
[3] http://dumps.wikimedia.org/
[4] We will make the corpus publicly available at
http://www.cs.kuleuven.be/groups/liir/software.php

5.2 Test Collections

Our experiments have been conducted on three data sets taken from the CLEF 2001-2003 CLIR campaigns: the LA Times 1994 (**LAT**), the LA Times 1994 and Glasgow Herald 1995 (**LAT+GH**) in English, and the NRC Handelsblad 94-95 and the Algemeen Dagblad 94-95 (**NC+AD**) in Dutch. Statistics of the collections are given in Table 1.

Table 1. Statistics of the experimental setup

(a) Statistics of test collections

Collection	Contents	# of Docs
LAT	LA Times 94 (EN)	110,861
LAT+GH	LA Times 94 (EN) Glasgow Her.95 (EN)	166,753
NC+AD	NRC Hand. 94-95 (NL) Alg. Dagblad 94-95 (NL)	190,604

(b) Statistics of used queries

CLEF Topics (Year: Topic Nr.)	# Queries	Used for
NL '01: 41-90	47	LAT
NL '02: 91-140	42	LAT
NL '03: 141-200	53	LAT+GH
EN '01: 41-90	50	NC+AD
EN '02: 91-140	50	NC+AD
EN '03: 141-200	56	NC+AD

Queries are extracted from the *title* and *description* fields of CLEF topics for each year. Stop words have been removed from queries and documents. Table 1(b) shows the queries used for the test collections.

Parameters α and β for the BiLDA training are set to values $50/K$ and 0.01 respectively, where K denotes the number of topics following [25]. The Dirichlet parameter μ in the LDA-unigram retrieval model is set to 1000. The parameters δ_1 and δ_2 are set to negligible values[5], while we set $\lambda = 0.3$, which gives more weight to the topic model.

6 Results and Discussion

This section reports our experimental results for both English-Dutch CLIR and Dutch-English CLIR. The cross-language topic model is trained just once on a large bilingual training corpus. After training, it can be used for both retrieval directions, after we infer it on the appropriate test collection. We have carried out the following experiments: (1) we compare our LDA-only model to several baselines that have also tried to exploit latent concept spaces for cross-language information retrieval, such as cross-language Latent Semantic Indexing (cLSI) and standard LDA trained on concatenated paired documents. We want to prove the soundness and the usefulness of the basic LDA-only model and, consequently, other models that might later build upon the foundation established by the LDA-only model. (2) We provide an extensive evaluation over all

---

[5] These parameters contribute to the theoretical soundness of the retrieval models, but, due to the computational complexity, we did not use counts over a large monolingual reference collection. We used a fixed small-value constant in all our models instead, since we detected that it does not have any significant impact on the results.

CLEF test collections with all our retrieval models, and provide a comparison of the best scoring LDA-unigram model with some of the best CLIR systems from the CLEF 2001-2003 campaigns. We have trained our BiLDA model with a different number of topics (400, 1000 and 2200) on the combined **EP+Wiki** corpus. The main evaluation measure we use for all experiments is the *mean average precision* (MAP). For several experiments, we additionally provide precision-recall curves.

6.1   Comparison with Baseline Systems

The LDA-only model serves as the backbone of other, more advanced BiLDA-based document models. Since we want to make sure that the LDA-only model constructs a firm and sound language-independent foundation for building more complex retrieval models, we compare it to state-of-the-art systems which try to build a CLIR system based around the idea of latent concept spaces: (i) the cross-language Latent Semantic Indexing (cLSI) as described by [3], which constructs a reduced (latent) vector space trained on concatenated paired documents in two languages, and (ii) the standard LDA model trained on the merged document pairs [23].

We have trained the cLSI model and the standard LDA model on the combined *EP+Wiki* corpus with 400 and 1000 dimensions (topics) and compared the retrieval scores with our LDA-only model which uses the BiLDA model with the same number of topics. The LDA-only model outscores the other two models by a huge margin. The MAP scores for cLSI and standard LDA are similar and very low, and vary between the MAP of 0.01 and 0.03 for all experiments, which is significantly worse than the results of the LDA-only model. The MAP scores of the LDA-only model for NL 2001, NL 2002, and NL 2003 for K=1000 are 0.1969, 0.1396, and 0.1227, respectively, while the MAP scores for EN 2001, EN 2002, and EN 2003 for K=1000 are 0.1453, 0.1374, and 0.1713, respectively.

One reason for such a huge difference in scores might be the ability to infer the BiLDA model on a new test collection (due to its fully generative semantics) more accurately. Cross-language LSI for CLIR reported in the literature always uses the same corpus (or subsets of the same corpus) for training and testing, while this setting asks for inferring on a test corpus which is not by any means content-related to a training corpus. BiLDA has a better statistical foundation by defining the common per-document topic distribution θ, which allows inference on new documents based on the previously trained model and also avoids the problem of overfitting inherent to the pLSI model and, consequently, the cLSI model. Another problem with the baseline methods might be the concatenation of document pairs, since one language might dominate the merged document. On the other hand, BiLDA keeps the structure of the original document space intact.

6.2   Comparison of Our CLIR Models

Using a Fixed Number of Topics (K=1000)   In this subsection, the LDA-only model, the simple unigram model and the combined LDA-unigram model have been evaluated on all test collections, with the number of topics initially fixed to 1000. Table 2 contains MAP scores for the LDA-unigram model, Figure 2(a) shows the precision-recall

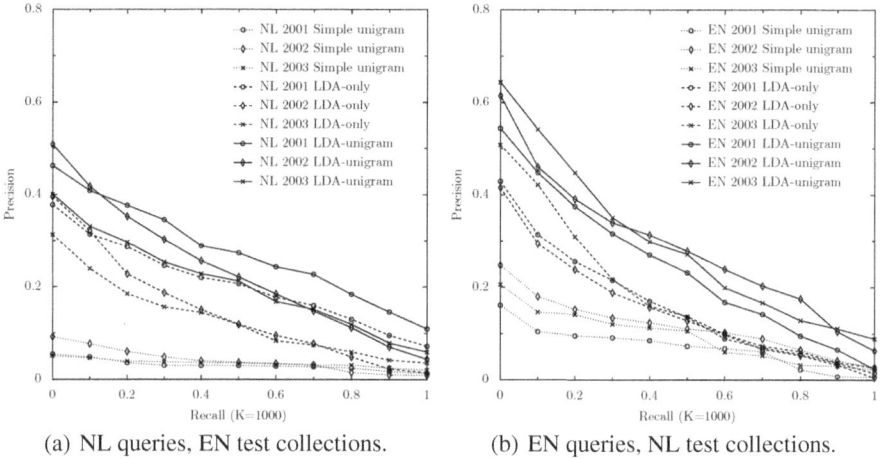

(a) NL queries, EN test collections. (b) EN queries, NL test collections.

Fig. 2. Precision-recall for all models. K=1000, training corpus is EP+Wiki.

Table 2. MAP scores of the LDA-unigram model for all test collections and different number of topics K. Training corpus is EP+Wiki.

Queries	K=400	K=1000	K=2200
NL 2001	0.2330	0.2673	0.2813
NL 2002	0.2093	0.2253	0.2206
NL 2003	0.1608	0.1990	0.1658
EN 2001	0.2204	0.2275	0.2398
EN 2002	0.2455	0.2683	0.2665
EN 2003	0.2393	0.2783	0.2450

values obtained by applying all three models to the English test collections and the Dutch queries, while Figure 2(b) shows the precision-recall values for the Dutch test collections and the English queries.

Varying the Number of Topics. The main goal of the next set of experiments was to test the performance of our models if we vary the number of topics set for BiLDA training. We have carried out experiments with the CLIR models relying on BiLDA trained with different numbers of topics (400, 1000 and 2200). Figure 3 shows the precision-recall values of the LDA-only and the LDA-unigram model, while the associated MAP scores of the best scoring LDA- unigram model are presented in Table 2.

Discussion. As the corresponding figures show, the LDA-only model seems to be too coarse to be used as the only component of an IR model (e.g., due to its limited number of topics, words in queries unobserved during training). However, the combination of the LDA-only and the simple unigram model, which allows retrieving relevant documents based on shared words across the languages (e.g. personal names), leads to much better scores which are competitive even with models which utilize cross-lingual dictionaries or machine translation systems. For instance, our LDA-unigram model would

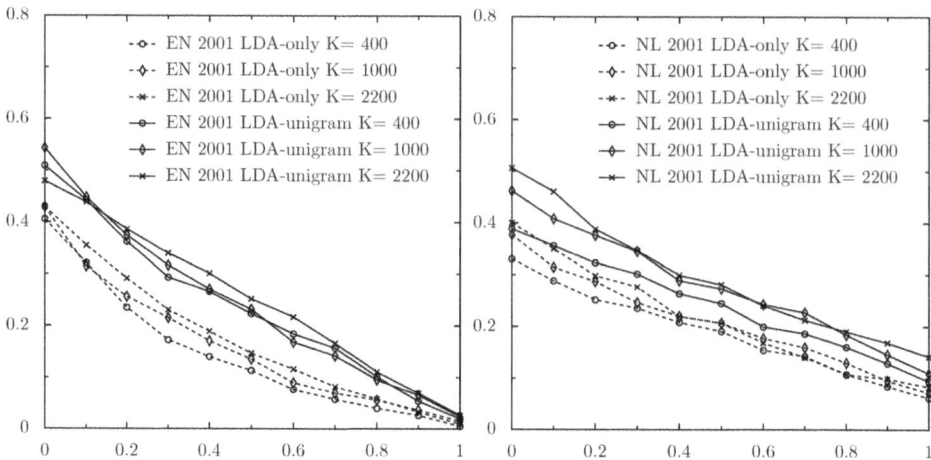

Fig. 3. Precision-recall for the LDA-only and the LDA-unigram model for the 2001 test collections. Training corpus is EP+Wiki.

have been placed among the top 5 retrieval systems for the CLEF 2002 Bilingual to Dutch task, would have been placed among the top 3 retrieval systems for the CLEF 2001 Bilingual to Dutch task, and outperforms the only participating system in the CLEF 2002 Dutch to English task (MAP: 0.1495) [20,21]. All these state-of-the-art CLEF systems operated in a similar settings as ours and constructed queries from *title* and *description* or *title*, *description* and *narrative* fields from the CLEF topics. They, however, rely on translation resources which were hand-built or trained on parallel corpora. We obtain competitive results by using the BiLDA model trained on comparable corpora. We believe that our results could still improve by training the BiLDA model on a corpus which is topically related with the corpus on which we perform the retrieval.

7   Conclusions and Future Work

We have proposed a novel language-independent and dictionary-free framework for cross-language information retrieval that does not use any type of a cross-lingual dictionary or translation system. The framework is built upon the idea of cross-language topic models obtained by applying a bilingual Latent Dirichlet Allocation model (BiLDA), where the only prerequisite is the availability of abundant training data consisting of comparable document-aligned documents.

We have thoroughly evaluated this cross-language retrieval model using standard test collections from the CLEF 2001-2003 CLIR campaigns and have shown that our combined model, which fuses evidence from the BiLDA model and the unigram model, is competitive with the current top CLIR systems that use translation resources that are hand-built or are trained on parallel corpora.

In future work, we will accumulate more comparable document-aligned data, exploiting Wikipedia and other sources. We also plan to construct other models that will combine topical knowledge with other evidences (for instance, using cognates instead of exactly the same words shared across languages). Additionally, we plan to expand the standard BiLDA to fit more divergent comparable training datasets. In addition, the cross-language knowledge transfer based on the proposed generative topic models that are trained on comparable corpora might be useful in many other multilingual information management tasks including categorization and summarization.

References

1. Blei, D.M., Ng, A.Y., Jordan, M.I.: Latent Dirichlet Allocation. Journal of Machine Learning Research (3), 993–1022 (2003)
2. Boyd-Graber, J., Blei, D.M.: Multilingual topic models for unaligned text. In: Proceedings of the 25th Conference on Uncertainty in Artificial Intelligence, pp. 75–82 (2009)
3. Carbonell, J.G., Yang, J.G., Frederking, R.E., Brown, R.D., Geng, Y., Lee, D.: Translingual information retrieval: A comparative evaluation. In: Proceedings of the 15th International Joint Conference on Artificial Intelligence. pp. 708–714 (1997)
4. Chen, D., Xiong, Y., Yan, J., Xue, G.R., Wang, G., Chen, Z.: Knowledge transfer for cross domain learning to rank. Information Retrieval (13), 236–253 (2010)
5. Chew, P.A., Bader, B.W., Kolda, T.G., Abdelali, A.: Cross-language information retrieval using PARAFAC2. In: Proceedings of the 13th ACM SIGKDD International Conference on Knowledge Discovery and Data Mining, pp. 143–152 (2007)
6. Cimiano, P., Schultz, A., Sizov, S., Sorg, P., Staab, S.: Explicit versus latent concept models for cross-language information retrieval. In: Proceedings of the 21st International Joint Conference on Artifical Intelligence, pp. 1513–1518 (2009)
7. De Smet, W., Moens, M.F.: Cross-language linking of news stories on the Web using interlingual topic modeling. In: Proceedings of the CIKM 2009 Workshop on Social Web Search and Mining, pp. 57–64 (2009)
8. Grefenstette, G.: Cross-Language Information Retrieval, Norwell, MA, USA (1998)
9. Hofmann, T.: Probabilistic Latent Semantic Indexing. In: Proceedings of the 22nd Annual International ACM SIGIR Conference on Research and Development in Information Retrieval, pp. 50–57 (1999)
10. Jagarlamudi, J., Daumé III, H.: Extracting Multilingual Topics from Unaligned Comparable Corpora. In: Gurrin, C., He, Y., Kazai, G., Kruschwitz, U., Little, S., Roelleke, T., Rüger, S., van Rijsbergen, K. (eds.) ECIR 2010. LNCS, vol. 5993, pp. 444–456. Springer, Heidelberg (2010)
11. Koehn, P.: Europarl: A parallel corpus for statistical machine translation. In: Proceedings of the MT Summit 2005, pp. 79–86 (2005)
12. Lavrenko, V., Choquette, M., Croft, W.B.: Cross-lingual relevance models. In: Proceedings of the 25th Annual International ACM SIGIR Conference on Research and Development in Information Retrieval, pp. 175–182 (2002)
13. Mathieu, B., Besançon, R., Fluhr, C.: Multilingual document clusters discovery. In: Proceedings of the 7th Triennial Conference on Recherche d'Information Assistée Par Ordinateur (RIAO), pp. 116–125 (2004)
14. Mimno, D., Wallach, H.M., Naradowsky, J., Smith, D.A., McCallum, A.: Polylingual topic models. In: Proceedings of the 2009 Conference on Empirical Methods in Natural Language Processing, pp. 880–889 (2009)
15. Muramatsu, T., Mori, T.: Integration of pLSA into probabilistic CLIR model. In: Proceedings of NTCIR 2004 (2004)

16. Navarro, G.: A guided tour to approximate string matching. ACM Computing Surveys 33(1), 31–88 (2001)
17. Ni, X., Sun, J.T., Hu, J., Chen, Z.: Mining multilingual topics from Wikipedia. In: 18th International World Wide Web Conference, pp. 1155–1156 (2009)
18. Nie, J.Y.: Cross-Language Information Retrieval. Synthesis Lectures on Human Language Technologies (2010)
19. Nie, J.Y., Simard, M., Isabelle, P., Durand, R.: Cross-language information retrieval based on parallel texts and automatic mining of parallel texts from the Web. In: Proceedings of the 22nd Annual International ACM SIGIR Conference on Research and Development in Information Retrieval, pp. 74–81 (1999)
20. Peters, C., Braschler, M., Gonzalo, J., Kluck, M. (eds.): CLEF 2001. LNCS, vol. 2406. Springer, Heidelberg (2002)
21. Peters, C., Braschler, M., Gonzalo, J., Kluck, M. (eds.): CLEF 2002. LNCS, vol. 2785. Springer, Heidelberg (2003)
22. Platt, J.C., Toutanova, K., Yih, W.T.: Translingual document representations from discriminative projections. In: Proceedings of the Conference on Empirical Methods in Natural Language Processing, pp. 251–261 (2010)
23. Roth, B., Klakow, D.: Combining Wikipedia-Based Concept Models for Cross-Language Retrieval. In: Cunningham, H., Hanbury, A., Rüger, S. (eds.) IRFC 2010. LNCS, vol. 6107, pp. 47–59. Springer, Heidelberg (2010)
24. Savoy, J.: Combining multiple strategies for effective monolingual and cross-language retrieval. Information Retrieval 7(1-2), 121–148 (2004)
25. Steyvers, M., Griffiths, T.: Probabilistic topic models. Handbook of Latent Semantic Analysis 427(7), 424–440 (2007)
26. Takasu, A.: Cross-lingual keyword recommendation using latent topics. In: Proceedings of the 1st International Workshop on Information Heterogeneity and Fusion in Recommender Systems, pp. 52–56 (2010)
27. Wang, A., Li, Y., Wang, W.: Cross-language information retrieval based on LDA, pp. 485–490 (November 2009)
28. Wei, X., Croft, W.B.: LDA-based document models for ad-hoc retrieval. In: Proceedings of the 29th Annual International ACM SIGIR Conference on Research and Development in Information Retrieval, pp. 178–185 (2006)
29. Xue, G.R., Dai, W., Yang, Q., Yu, Y.: Topic-bridged pLSA for cross-domain text classification. In: Proceedings of the 31st Annual International ACM SIGIR Conference on Research and Development in Information Retrieval, pp. 627–634 (2008)
30. Zhai, C., Lafferty, J.: A study of smoothing methods for language models applied to information retrieval. ACM Transactions on Information Systems 22, 179–214 (2004)

Construct Weak Ranking Functions for Learning Linear Ranking Function*

Guichun Hua, Min Zhang, Yiqun Liu, Shaoping Ma, and Hang Yin

State Key Laboratory of Intelligent Technology and Systems,
Tsinghua National Laboratory for Information Science and Technology,
Department of Computer Science and Technology, Tsinghua University, Beijing 100084, China
huaguichun@gmail.com, {z-m,yiqunliu,msp,yhang10}@tsinghua.edu.cn
http://www.thuir.cn

Abstract. Many Learning to Rank models, which apply machine learning techniques to fuse weak ranking functions and enhance ranking performances, have been proposed for web search. However, most of the existing approaches only apply the $Min - Max$ normalization method to construct the weak ranking functions without considering the differences among the ranking features. Ranking features, such as the content-based feature $BM25$ and link-based feature $PageRank$, are different from each other in many aspects. And it is unappropriate to apply an uniform method to construct weak ranking functions from ranking features. In this paper, comparing the three frequently used normalization methods: $Min - Max$, Log, $Arctan$ normalization, we analyze the differences among three normalization methods when constructing the weak ranking functions, and propose two normalization selection methods to decide which normalization should be used for a specific ranking feature. The experimental results show that the final ranking functions based on normalization selection methods significantly outperform the original one.

Keywords: Learning to Rank; Weak Ranking Function; Final Ranking Function; Normalization.

1 Introduction

We often refer to information retrieval techniques to find information from a vast dataset or Internet. Ranking is an important part of information retrieval. Nowadays, more and more features used to construct ranking functions have been proposed e.g. content-based features such as $TFIDF$, $BM25$; link-based features such as $PageRank$, $HITS$; user behavior features based on clickthrough data. There are hundreds of parameters to tune when constructing a ranking function and it is unpractical to tune these parameters manually. So 'Learning to Rank' (shown as LTR for simplicity in the rest of the paper), an interdisciplinary field of information retrieval and machine learning, has gained increasing attention. The LTR method optimizes loss function to tune the parameters for each weak ranking functions and fuses them into a final ranking function, such as the linear ranking function. The final linear ranking function is what this paper focuses on.

* Supported by Natural Science Foundation (60736044, 60903107, 61073071) and Research Fund for the Doctoral Program of Higher Education of China (20090002120005).

M.V.M. Salem et al. (Eds.): AIRS 2011, LNCS 7097, pp. 49–60, 2011.

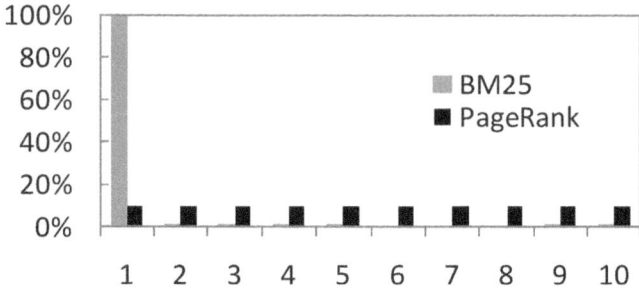

Fig. 1. Distribution Comparison Between $BM25$ and $PageRank$ on the Proportion of the Unique-Feature-Value Number in each interval

Almost all LTR methods, which output the linear ranking functions, need constructing the weak ranking functions by normalizing the ranking features before the training process. And the $Min - Max$ Normalization is almost the only method considered by the field of LTR for the best of our knowledge. The differences among the ranking features are neglected when constructing the weak ranking functions. $Min - Max$ Normalization performs a linear transformation on the original data values and preserves the relationships among the original data values.

However, ranking features are different from each other in many aspects, such as data value distribution. For example, we conduct the statistics on the dataset used in Section 4.1. The ranges of $BM25$ and $PageRank$ are divided into 10 equal intervals respectively, and the proportion of the unique-feature-value number in each interval to the unique-feature-value number in the range is calculated, shown in the Figure 1. The results show: (1) the proportion is 99% and 10% respectively for $BM25$ and $PageRank$ in the first interval; (2) there is only 1% of the unique values for $BM25$ in the last 9 intervals; (3) the differences among the number of unique values for $PageRank$ in each interval could be almost ignored. When used to construct the weak ranking functions from $BM25$ and $PageRank$, the $Min - Max$ normalization method does not change the distribution showed in the Figure 1. And 99% of the unique $BM25$ values are located in the interval of $[0, 0.1)$ when the range is $[0, 1]$ after the normalization.

Is it appropriate to just apply the $Min - Max$ normalization method when constructing the weak ranking functions from the ranking features for the final linear ranking functions? This question is the motivation of this paper, and the main contributions of this work are: (1) three typical normalization methods are analyzed and compared to construct the weak ranking functions: $Min - Max$, Log and $Arctan$ normalization methods. The experiments show that for some ranking features, the final ranking functions could achieve significant improvements by using Log and $Arctan$ normalization instead of $Min - Max$ normalization. (2) two intuitive normalization selection methods are proposed to handel the question above: Feature Distribution Selection Method and One-Switch Selection Method. The experimental results show that final ranking functions based on normalization selection methods significantly outperform the original one.

The rest of the paper is organized as follows. In section 2, related work is presented. Section 3 introduces the normalization methods and normalization selection methods. In section 4, experimental results are reported and discussed. In the last section, the conclusion and future work are briefly discussed.

2 Related Work

The field of LTR is related to our work. In the LTR task, the data is composed of queries, the retrieved documents for every query, and the relevance levels labeled by human for the document and query pairs. The LTR methods proposed now could be classified into two categories according to the formation of final ranking functions: LTR models using linear ranking functions and LTR models using nonlinear ranking functions. For LTR models using nonlinear ranking functions, the formation of the weak ranking function is nonlinear, such as $RankBoost$ [2], the formation is bipartite. For LTR models using linear ranking functions, the formation of weak ranking functions is linear, such as $Ranking\ SVM$ [3,9], $SVMMap$ [14,12] and $listMLE$ [13]. $Ranking\ SVM$ [3,9](shown as $RankSVM$ for simplicity in the rest of the paper) is a very effective algorithm proved by many previous studies [15,10]. The optimization formulation of $RankSVM$ is as follows:

$$\min_{\omega,\xi>0} \tfrac{1}{2}\overrightarrow{\omega}\cdot\overrightarrow{\omega} + C\sum\varepsilon_{i,j,k}$$
$$\forall\left(d_k^i, d_k^j\right)\in \boldsymbol{d_i}\times \boldsymbol{d_i}: \overrightarrow{\omega}\Phi\left(q_k, d_k^i\right) > \overrightarrow{\omega}\Phi\left(q_k, d_k^j\right) + 1 - \varepsilon_{i,j,k}$$

This paper focuses on the LTR model using linear ranking functions and $RankSVM$ is used to verify the effectiveness of different methods to construct the weak ranking functions from specific ranking features.

3 Construct the Weak Ranking Functions from the Ranking Features

3.1 Normalization Methods

There are three typical normalization methods used to normalize the data into same range: $Min - Max$ Normalization, Log Normalization and $Arctan$ Normalization Method.

$Min - Max$ Normalization performs a linear transformation on the original data values and preserves the relationships among the original data values, showing as Equation 1 where x_i is the value of ith ranking feature; the range of ith ranking feature is $[X_{i,MIN}, X_{i,MAX}]$, and $X_{i,MIN}$ and $X_{i,MAX}$ are the maximum and minimum of ith ranking feature respectively; $f_{i,MM}$ is regarded as the value of the ith weak ranking function constructed from the ith ranking feature.

$$f_{i,MM} = \Psi_m(x_i, X_{i,MIN}, x_{i,MAX}) = \frac{x_i - X_{i,MIN}}{X_{i,MAX} - X_{i,MIN}} \tag{1}$$

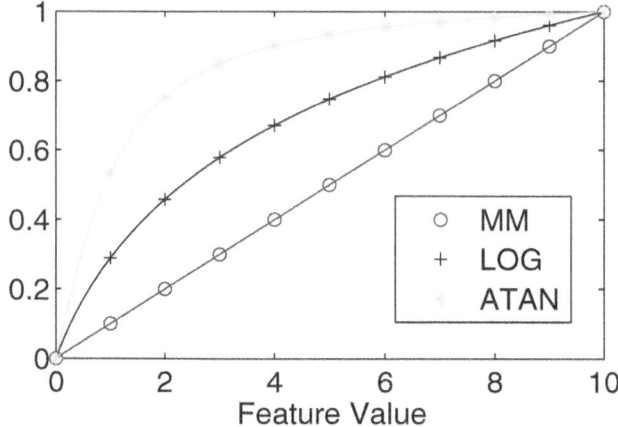

Fig. 2. The Comparison of Three Normalization Methods from the Same Ranking Feature whose range is $[0, 10]$

Log Normalization and *Arctan* Normalization perform a nonlinear transformation on the original data values, shown as Equation 2 Equation 3 respectively where $f_{i,LOG}$ and $f_{i_{ATAN}}$ depict the value of ith weak ranking function after the *Log* Normalization and *Arctan* Normalization respectively.

$$f_{i,LOG} = \Psi_l(x_i, X_{i,MIN}, x_{i,MAX}) = \frac{\log(x_i - X_{i,MIN} + 1)}{\log(X_{i,MAX} - X_{i,MIN} + 1)} \qquad (2)$$

$$f_{i,ATAN} = \Psi_a(x_i, X_{i,MIN}, x_{i,MAX}) = \frac{arctan(x_i - X_{i,MIN})}{arctan(X_{i,MAX} - X_{i,MIN})} \qquad (3)$$

Three Normalization Methods map the original data into the same range and the comparison is shown in Figure 2. The comparison result shows: (1) the *Log* and *Arctan* Normalization methods could scatter the small values of the ranking feature and cluster the large values; (2) the *Log* and *Arctan* normalization methods are useful to uniformly distribute the original data when values are clustered around small values with few large values.

3.2 Normalization Selection Method

There are three normalization methods: $Min - Max$, *Log* and *Arctan* normalization methods. In this section, two intuitive normalization selection methods are proposed based on the feature value distribution: Feature Distribution Selection Method(FDM) and One-Switch Selection Method(OSM) summarized in Figure 3 and Figure 4 respectively.

FDM selects the features of which data values are clustered around small/large value with few large/small values, while OSM selects the features with better performances through evaluation measure MAP. After the feature selection with FDM or

---

Procedure. `Feature Distribution Selection Method`

Input 5-fold-cross-validation datasets without normalization: Set, and the number of ranking features is N;

Initialize feature set S_f with all features in Set;

for $i = 1$ *to* N **do**

Statistics the training set in Set to get following information: the number of unique feature values N_i; the maximum and minimum feature values MAX_i and MIN_i; Divide range $[MIN_i, MAX_i]$ equally into ten intervals, and the number of unique feature values in each interval is $\alpha_{i,1}, \alpha_{i,2}, ..., \alpha_{i,10}$;

for $j = 1$ *to* 10 **do**

if $\frac{\alpha_{i,j}}{N_i} > 0.2$ **then**

Delete i^{th} feature from S_f;

Break loop;

end

end

end

Output feature set: S_f;

Fig. 3. Feature Distribution Selection Method

Procedure. `One-Switch Selection Method`

Input 5-fold-cross-validation datasets without normalization: Set, and the number of ranking features is N;

Initialize feature set $S_f = \Phi$;

Normalize all ranking features in Set with $Min - Max$ normalization to get normalized set: Set_0 ;

Train and test the $RankSVM$ model on Set_0 and gain MAP(showing in Section 4.1) value: map_0;

for $i = 1$ *to* N **do**

Normalize all ranking features in Set with $Min - Max$ normalization except for i^{th} ranking feature with Log or $Arctan$ normalization, then get normalized set: Set_i;

Train and test the $RankSVM$ model on Set_i and obtain MAP value map_i;

if $map_i > map_0$ **then**

Add i_{th} feature into S_f

end

end

Output feature set: S_f;

Fig. 4. One-Switch Selection Method

OSM, the weak ranking functions are constructed through LOG or $ATAN$ normalization from selected ranking features, and through $Min - Max$ normalization from other ranking features.

4   Experiment

4.1   Experimental Setting

Data Information. The experiment dataset is Microsoft Learning to Rank Datasets [4, 6, 11] which are latest and one of the largest benchmark datasets in the field of LTR. There are two large scale datasets: $MSLR-WEB30k$ with more than $30,000$ queries and $MSLR-WEB10K$ with $10,000$ queries, and the queries in latter dataset are randomly sampled from the ones in the former dataset. The relevance judgments, depicting the relevance degree between a query and a document, are obtained from a labeling set of Microsoft Bing search engine, which take 5 values from 0 (irrelevant) to 4 (perfectly relevant). And the larger the judgement value is, the more relevant the document is with respect to a specific query. 136 ranking features, widely used in the research community, are extracted such as $BM25$, $PageRank$, $HITS$ and $Language\ Model$ [5]. $MSLR-WEB30k$ is applied in the experiment.

Five-fold cross validation is adopted to train and test the ranking models with three sets for training, one set for validation and one set for test. The evaluation measure value means the average among five test sets in the rest of this paper.

Evaluation Measure. We use $P@N$, $NDCG@N$, MAP and ERR as our evaluation measures.

$P@N$ is the precision at top N returned results, which is defined as:

$$P@N = \frac{1}{N} \sum_{i=1}^{N} rel(d_i), \text{ where } rel(d_i) = \begin{cases} 1 & \text{if } d_i \text{ is relevant to the query} \\ 0 & \text{otherwise} \end{cases}$$

$NDCG@N$(Normalized Discounted Cumulative Gain) [7,8], considering the position (rank) of the document with different relevance degrees in the returned result list, is quite an useful and popular measure for evaluating web search and related tasks. The $NDCG$ score at top n returned results is defined as:

$$NDCG@N = \frac{1}{Z_N} \sum_{i=1}^{N} \begin{cases} 2^{R(i)} - 1 & i = 1 \\ \frac{2^{rel(i)} - 1}{\log(i)} & i > 1 \end{cases}$$

where $R(i)$ is the relevance degree which equals to the relevance judgement in this paper; Z_N is the normalization constant that makes the perfect list get a $NDCG$ score of 1.In this paper, $N = 1, 2, \ldots, 10$.

MAP is the mean average precision for the queries. It is a comprehensive measure which takes both precision and recall into consideration. The definition can be shown as:

$$MAP = \frac{1}{m} \sum_{j=1}^{m} AP_j = \frac{1}{m} \sum_{j=1}^{m} \frac{1}{R_j} (\sum_{i=1}^{k} rel_j(d_i) \cdot (P@i)_j)$$

where m is the total number of queries, R_j is the total number of relevant documents for the jth query, k is the total number of returned documents for the query, $rel_j(d_i)$ and $(P@i)_j$ are the $rel(d_i)$ and $P@i$ scores for the jth query respectively.

ERR(Expected Reciprocal Rank) [1], calculates the gain and discount for a document in a position considering the documents shown above it. And the definition could be shown as:

$$ERR = \sum_{i=1}^{N} \frac{1}{i} \prod_{j=1}^{i-1} (1 - \frac{2^{R(j)} - 1}{2^{R_{MAX}}})$$

where R_{MAX} is the largest value of the relevance degree, and $R_{MAX} = 4$ in this paper.

4.2   Dataset Statistics Information

The statistics is conducted on the dataset, and we obtain following information for each ranking feature(e.g. i^{th} ranking feature):

1. the maximum of minimum values for i^{th} ranking feature: MAX_i and MIN_i;
2. the number of unique values for i^{th} ranking feature: N_i;
3. the range of ith ranking feature is equally divided into 10 intervals, and the number of unique values for each interval: $\alpha_{i,1}$, $\alpha_{i,2}$, , ..., $\alpha_{i,10}$.
4. the proportion of ith ranking feature for j^{th} interval is: $\frac{\alpha_{i,2}}{N_i}$; then we get proportion distribution for i^{th} ranking feature.

The ranking features could be classified into four categories based on the differences of the proportion distribution: (1) the first category, there does not exist an interval of which the proportion is larger than 20%. the ranking features belonging to this category: $6 \sim 10$, $130 \sim 133$. (2) the second category, the sum of first five intervals is larger than 80%. the ranking features belonging to this category: $1 \sim 5$, $11 \sim 95$, $106 \sim 110$, 126, 127, 129, $134 \sim 136$. (3) the third category, the sum of last five intervals is larger than 80%. the ranking features belonging to this category: $101 \sim 105$, $111 \sim 125$, 128. (4) the fourth category, the number of unique values is smaller than 3. the ranking features belonging to this category: $96 \sim 100$.

The intuitive idea is that: for the first category, the $Min - Max$ normalization is more useful; for second and third categories, the Log and $Arctan$ normalization are more useful, and the feature value should be reversed first before normalization for the third category; for the last category, all normalization methods perform the same as each other. Is this intuitive idea correct? The experiments as follows will give the answer to this question.

4.3   Experiments Results

Two experiments are conducted in this section: normalization methods comparison without selection, and the experiment with normalization selection. The former experiment is to verify that other normalization methods, such as Log and $Arctan$ normalization methods, could enhance the performance of the final ranking function. And it is unappropriate to just use $Min - Max$ normalization method to construct the weak ranking functions. The latter experiment is to verify the effectiveness of the normalization selection method.

Table 1. Notations of four Ranking Functions

Name	Number of Weak Ranking Functions	Normalization Methods
MM	136	$Min - Max$ Normalization
LOG	136	Log Normalization
$ATAN$	136	$Arctan$ Normalization
$MERGE$	$136 * 3 = 408$	$Min - Max$, Log and $Arctan$ Normalization

Table 2. T-Test Results among four Ranking Functions with NDCG@1

	LOG	$ATAN$	$MERGE$
MM	$5.84E - 61$	$2.72E - 21$	$5.19E - 56$
LOG	N/A	$1.67E - 11$	0.49
$ATAN$	N/A	N/A	$7.65E - 13$

Normalization Methods Comparison Without Selection. Three normalization methods are applied to construct the weak ranking function, and we just apply the normalization methods to construct the weak ranking function without considering which one is more appropriate for a specific ranking feature. Then four ranking functions are trained in the new datasets: MM, the ranking function trained on the weak ranking functions normalized from the ranking features through the $Min - Max$ normalization method; LOG, the ranking function through the Log normalization method; $ATAN$, the ranking function through the $Arctan$ normalization method; $MERGE$, the ranking function trained on the datasets which are composed of three datasets used by MM, LOG and $ATAN$. The comparison of these four ranking functions shows in Table 1 and the experimental results show in Figure 5. T-Test is conducted among four ranking functions with $NDCG@1$, and the result is shown in Table 2($p - value < 0.05$ means the difference is significant);

From the results, we could see that: (1) four ranking functions conduct significantly different performances from each other except for $MERGE$ and LOG with $p - value$ 0.49. (2) $MERGE$ and LOG significantly outperform the other two ranking functions; (3) MM performs the poorest in the four ranking functions; (4) $ATAN$ performs better than MM and worse than LOG, so $Arctan$ normalization is ignored in the normalization selection method in the next section.

$Min - Max$ normalization is the poorest method to construct the weak ranking functions from the ranking features, and it is unappropriate to just apply the $Min - Max$ normalization when the ranking features are quite different from each other in many aspects such as the number distribution of the unique ranking feature values. Log normalization method is the best method, which means that Log normalization method could fit the distribution of most ranking features and normalize them to construct the weak ranking functions better than $Min - Max$ and $Arctan$ normalization methods.

The Experiment with Normalization Selection. Which normalization method should be used for a specific ranking feature? In this section, two normalization selection methods are applied to choose an appropriate normalization method for a ranking feature: FDM and OSM. After normalization selection with either selection method, the Log

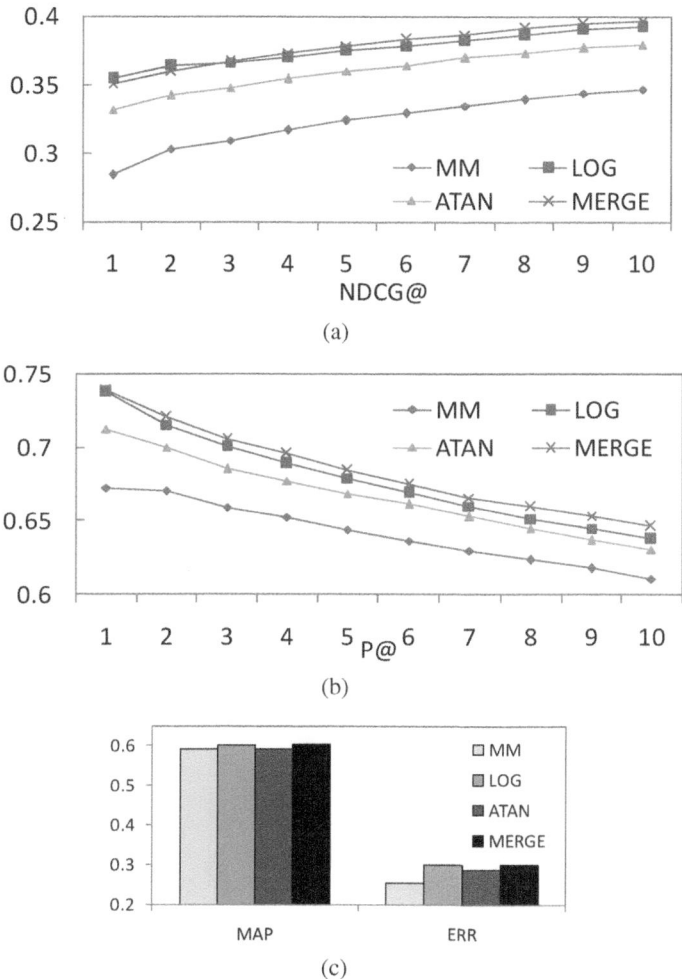

Fig. 5. The Comparison Result among Four Ranking Functions with $NDCG@1 \sim 10$ in (a), $P@1 \sim 10$ in (b), MAP and ERR in (c)

normalization method is used to construct the weak ranking functions from the selected ranking features and the $Min - Max$ normalization method from the other ranking features. Then four final ranking functions based on different weak ranking functions are trained and tested: MM, LOG, RF_FDM and RF_OSM shown in Table 3. MM and LOG are the same ranking functions used in Section 4.3.

$Arctan$ and Log normalization methods are useful when the data values are clustered around small values with few large values, and Log normalization is better than $Arctan$ normalization according to the results in Section 4.3. So only $Min - Max$ and Log normalization methods are used to verify the effectiveness of the normalization selection methods.

Table 3. Notations of four Ranking Functions

Name	Normalization Selection Method	Normalization Methods
MM	N/A	$Min - Max$ Normalization
LOG	N/A	Log Normalization
RF_FDM	FDM	Log Normalization for selected ranking features and $Min - Max$ normalization for others
RF_OSM	OSM	Log Normalization for selected ranking features and $Min - Max$ normalization for others

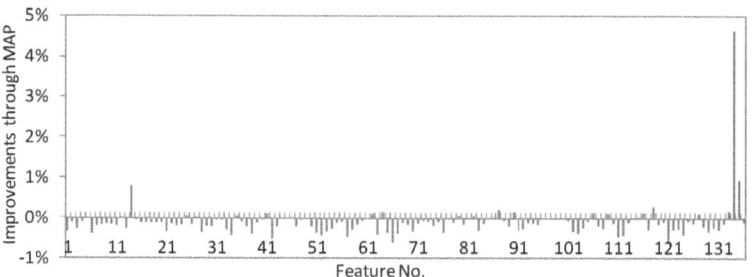

Fig. 6. The Improvements of One-Switch Normalization Selection

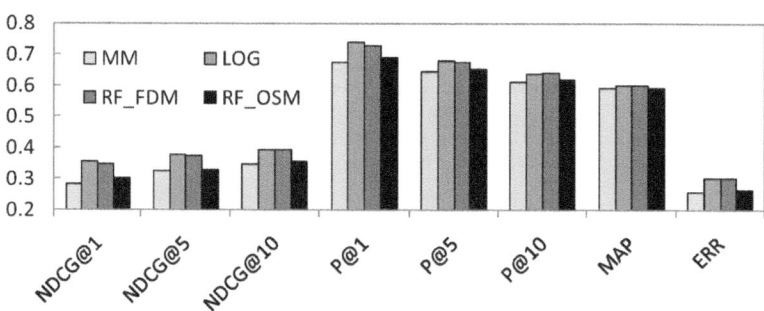

Fig. 7. The Comparison Results of Four ranking functions with $NDCG@1, 5, 10$, $P@1, 5, 10$, MAP and ERR

After the normalization selection, the ranking feature set, not chosen by FDM, contains all the ranking features which are the second and third categories in Section 4.2; the ranking feature set, chosen by OSM, is: { 5, 12, 14, 25, 27, 28, 35, 41, 46, 61, 62, 64, 79, 82, 86, 87, 90, 106, 109, 115, 116, 118, 127, 133, 134, 135 }, in which 133 belongs to the first category and all others belong to the second or third category. OSM applies the MAP values to measure whether to choose a ranking feature, and the improvement percent of map_i over map_0 in Figure 4 for i^{th} ranking shows in Figure 6. T-Test is conducted among four ranking functions with $NDCG@1$, and the result shows in Table 4.

Table 4. T-Test Results among four Ranking Functions with NDCG@1

	LOG	RF_FDM	RF_OSM
MM	$5.84E-61$	$3.26E-47$	$8.62E-3$
LOG	N/A	0.15	$7.92E-52$
RF_FDM	N/A	N/A	$1.72E-39$

The comparison results among four final ranking functions show in Figure 7. From the results, we could see that: (1) four ranking functions perform differently from each other significantly, except for LOG and RF_FDM with $p-value$ 0.15. (2) LOG and RF_FDM significantly outperform the other two ranking functions; (3) MM performs the poorest in the four ranking functions; (4) RF_OSM performs significantly better than MM and worse than LOG and RF_FDM.

After normalization selection with FDM, RF_FDM performs similarly to LOG and better than MM, which means that: (1) the ranking features, selected by FDM, are normalized with Log normalization in RF_FDM and with $Min-Max$ normalization in MM, and Log normalization is more useful for them than $Min-Max$ normalization. So, FDM is effective and could distinguish the ranking features whether Log or $Min-Max$ normalization is more useful. (2) the ranking features, not chosen by FDM, are normalized with Log normalization in LOG and with $Min-Max$ normalization in RF_FDM, and Log normalization performs almost the same as $Min-Max$ normalization for these ranking features.

After normalization selection with OSM, RF_OSM performs better than MM, but significantly worse than LOG and RF_FDM. The reasons we analyze are that: ranking features are not independent from each other, so the weak ranking functions affect the performances of the final ranking functions as a sub feature set instead of an individual ranking feature. That makes the OSM not so useful as FDM.

According to the experimental results in this section, we could answer the question in Section 4.2: the intuitive idea is almost correct except that: $Min-Max$ and Log normalization are both useful for the first category. Based on the proportion distribution, we could choose an appropriate normalization method to construct the weak ranking function for a specific ranking feature.

5   Conclusion and Future Work

The weak ranking functions are mostly constructed through $Min-Max$ normalization method for the final linear ranking functions in the field of LTR without considering the differences among the ranking features. There do exist differences among the ranking features from many aspects. In this paper, we analyze the ranking features and apply three normalization methods to construct the weak ranking functions: $Min-Max$, Log and $Arctan$ normalization methods, and find that Log normalization could significant improve the performances of the final ranking functions. Then two intuitive normalization selection methods are proposed to try to handle the problem which normalization is appropriate to construct the weak ranking function for a specific ranking feature.

The experimental results show that the ranking functions based on the normalization selection methods significantly outperform the original one.

Future study could follow these aspects: (1) other normalization methods could be attempted to constructed the weak ranking functions based on the specific distribution of the ranking features; (2) a ranking feature depicts a special aspect of the relevance between a query and a document, and how to design a normalization selection method considering this to choose the best normalization method.

References

1. Chapelle, O., Metlzer, D., Zhang, Y., Grinspan, P.: Expected reciprocal rank for graded relevance. In: Proceeding of the 18th ACM Conference on Information and Knowledge Management, CIKM 2009, pp. 621–630. ACM, New York (2009)
2. Freund, Y., Iyer, R., Schapire, R.E., Singer, Y.: An efficient boosting algorithm for combining preferences. J. Mach. Learn. Res. 4, 933–969 (2003)
3. Herbrich, R., et al.: Large margin rank boundaries for ordinal regression. In: Advances in Large Margin Classifiers, pp. 115–132 (2000)
4. http://research.microsoft.com/enus/projects/mslr/
5. http://research.microsoft.com/enus/projects/mslr/feature.aspx
6. http://research.microsoft.com/enus/um/people/letor/
7. Järvelin, K., Kekäläinen, J.: Ir evaluation methods for retrieving highly relevant documents. In: SIGIR 2000: Proceedings of the 23rd Annual International ACM SIGIR Conference on Research and Development in Information Retrieval, pp. 41–48. ACM, New York (2000)
8. Järvelin, K., Kekäläinen, J.: Cumulated gain-based evaluation of ir techniques, vol. 20, pp. 422–446. ACM, New York (2002)
9. Joachims, T.: Optimizing search engines using clickthrough data. In: KDD 2002: Proceedings of the Eighth ACM SIGKDD Internatiounal Conference on Knowledge Discovery and Data Mining, pp. 133–142. ACM, New York (2002)
10. Liu, T.-Y.: Learning to rank for information retrieval. In: Foundation and Trends on Information Retrieval, pp. 641–647 (2009)
11. Qin, T., Liu, T.-Y., Xu, J., Li, H.: Letor: A benchmark collection for research on learning to rank for information retrieval. Information Retrieval Journal 13, 346–374 (2010)
12. Tsochantaridis, I., Joachims, T., Hofmann, T., Altun, Y.: Large margin methods for structured and interdependent output variables. J. Mach. Learn. Res. 6, 1453–1484 (2005)
13. Xia, F., Liu, T.-Y., Wang, J., Zhang, W., Li, H.: Listwise approach to learning to rank: theory and algorithm. In: ICML 2008: Proceedings of the 25th International Conference on Machine Learning, pp. 1192–1199. ACM, New York (2008)
14. Yue, Y., Finley, T., Radlinski, F., Joachims, T.: A support vector method for optimizing average precision. In: SIGIR 2007: Proceedings of the 30th Annual International ACM SIGIR Conference on Research and Development in Information Retrieval, pp. 271–278. ACM, New York (2007)
15. Zhang, M., et al.: Is learning to rank effective for web search. In: SIGIR 2009 Workshop: Learning to Rank for Information Retrieval, pp. 641–647 (2009)

Is Simhash Achilles?

Qixia Jiang, Yan Zhang, Liner Yang, and Maosong Sun

State Key Laboratory on Intelligent Technology and Systems
Tsinghua National Laboratory for Information Science and Technology
Department of Computer Science and Technology, Tsinghua University, Beijing 100084, China
{qixia.jiang,lineryang}@gmail.com,
zhang-y-05@mails.tsinghua.edu.cn, sms@mail.tsinghua.edu.cn

Abstract. Simhash generates compact binary codes for the input data thus improves the search efficiency. Most recent works on Simhash are designed to speed-up the search, generate high-quality descriptors, etc. However, few works discuss in what situations Simhash can be directly applied. This paper proposes a novel method to quantitatively analyze this question. Our method is based on Support Vector Data Description (SVDD), which tries to find a tighten sphere to cover most points. Using the geometry relation between the unit sphere and the SVDD sphere, we give a quantitative analysis on in what situations Simhash is feasible. We also extend the basic Simhash to handle those unfeasible cases. To reduce the complexity, an approximation algorithm is proposed, which is easy for implementation. We evaluate our method on synthetic data and a real-world image dataset. Most results show that our method outperforms the basic Simhash significantly.

Keywords: Information Retrieval, Simhash, Support Vector Data Description.

1 Introduction

Simhash [1], just like Achilles the Greek hero of the Trojan War, is the hero to solve the large-scale similarity search problem such as text retrieval [2], duplicate detection [3], etc. Simhash hashes similar objects to similar hash values such that searching over the obtained hash values is much more efficient [2,3]. A natural question is whether Simhash can be applied to any similarity search problems? Obviously, the answer is NO. In this paper, we will give a quantitative analysis on this problem. We will also give one feasible method for the case that the basic Simhash is infeasible.

The basic Simhash [1] randomly generates some hyperplanes and objects' hash values are determined by the sides of these planes that they locate on. Just like the heel is the deadly weakness of Achilles in spite of his overall strength, Simhash loses efficiency when the data is ill-distributed. This is mainly because when data are ill-distributed, a lot of points tend to lie on the same side of hyperplanes thus are hashed to the same hash value. However, we can handle this problem by incorporating both similarity between two objects and the location of the whole dataset.

Motivated by above analysis, we propose a novel analysis method of data distribution based on Support Vector Data Description (SVDD)[4]. It involves two steps: (1) using SVDD to find a tighten sphere (called SVDD sphere) to cover most of the data

M.V.M. Salem et al. (Eds.): AIRS 2011, LNCS 7097, pp. 61–72, 2011.

points, and (2) calculating the intersection circle between the unit sphere and the obtained SVDD sphere. Then, we can quantitatively measure whether the basic Simhash is suitable or not. For those unfeasible cases, this paper proposes two methods, model- and feature- based Simhash. Specifically, model-based Simhash tries to generate some random hyperplanes crossing through the data area, which makes the obtained hyperplanes be more discriminative. An approximation of this method is designed for large-scale data, which is very simple for implementation. Differently, feature-based Simhash alters the data representation so that the data points can be well distributed thus the basic Simhash can be directly used.

We evaluate our method on a synthetic dataset and a real-world image dataset. We also design some evaluation criterions to quantitatively compare our method with the basic Simhash. Most results show that our method outperforms the basic Simhash.

This paper is organized as follows: Section 2 presents our spacial analysis method. In Section 3, we present our spacial analysis based Simhash algorithm in details. Experimental results and some analysis are shown in Section 4. Section 5 reviews some related works. In Section 6, we make a conclusion.

2   Spatial Data Analysis

To begin, we fix some notation. We assume each data point is described by a column vector $\mathbf{x} \in \mathbb{R}^K$. Since Simhash measures the cosine similarity between two points, we normalize each data point such that all the points are located on a unit sphere, i.e. $\|\mathbf{x}\| = 1$. We have a data set $\mathcal{X} = \{\mathbf{x}_i\}$, $i = 1, \ldots, N$.

2.1   Support Vector Data Description

Support Vector Data Description (SVDD) [4] concerns the characterization of a data set but includes almost no superfluous space. Inspired by the Support Vector Machine (SVM), SVDD describes a data set in terms of a sphere which tries to enclose all the data. The sphere is characterized by the center $\mathbf{a}$ and the radius $R > 0$. SVDD obtains the optimal description via solving the following optimization problem:

$$\arg\min_{R,\mathbf{a}} R^2 + C \sum_{i=1}^{N} \xi_i$$
$$s.t. \quad \|\mathbf{x}_i - \mathbf{a}\|_2^2 \leq R^2 + \xi_i, \quad \forall i \qquad (1)$$
$$\xi_i \geq 0, \quad \forall i$$

where ξ_i is the slack variable and control parameter C is set to be a constant value in pervious which controls the trade-of between the volume and the errors.

The intuitive interpretation of above problem is that (1) we hope that all the data points can be involved in a tightened sphere, and (2) by introducing the slack variables, SVDD can tolerate the outliers in the data set.

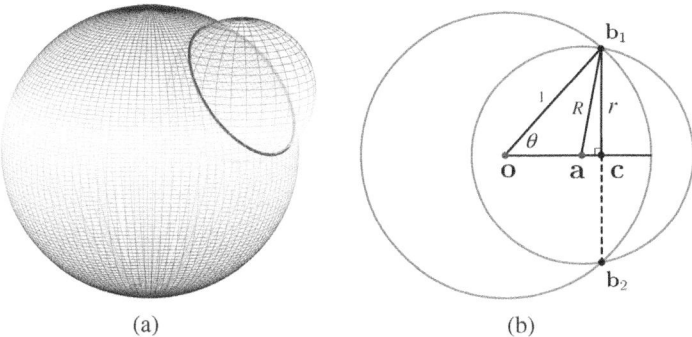

(a)   (b)

Fig. 1. Spatial data analysis. (a) An illustration of the relationship between the unit sphere (the larger) and the SVDD sphere (the smaller) in 3-dimensional space. The red circle is the intersection of such two spheres. (b) An illustration of the unit sphere and the SVDD sphere on a 2-dimensional plane. The point $\mathbf{o} = (0,0)$ and $\mathbf{a}$ are the centers of two circles respectively. The redias of two circles are 1 and R respectively. Two circles intersection at point $\mathbf{b}_1$ and $\mathbf{b}_2$.

2.2   Accurate Spatial Data Analysis

Given a data set, we can obtain a SVDD sphere as described in Section 2.1. The relation between the unit sphere and the SVDD sphere is illustrated in Figure 1(a). Since all the points are located on the unit sphere, the volume of the SVDD sphere is no larger than the unit sphere. Obviously, such two spheres must intersect and the intersected region is the location of the given data. For description simplicity, as shown in Figure 1(b), we assume our data is 2 dimensional and the center of the unit sphere $\mathbf{o} = (0,0)$. Given the center $\mathbf{a}$ and the radius R of the SVDD sphere [1], it is convenient to obtain the radius r of the intersection circle by solving the following equation:

$$\|\mathbf{a}\|_2 = \sqrt{1 - r^2} - \sqrt{R^2 - r^2}, \tag{2}$$

and the center of the intersection circle is

$$\mathbf{c} = \frac{\|\mathbf{a}\|_2 + \sqrt{R^2 - r^2}}{\|\mathbf{a}\|_2}\mathbf{a}. \tag{3}$$

Therefore, the intersection circle is:

$$\begin{cases} \mathbf{c}^T \cdot \underline{\mathbf{x}} = \|\mathbf{c}\|_2 \\ \|\mathbf{x} - \mathbf{c}\|_2 = r \end{cases} \tag{4}$$

and the cone vertex angle is $2\theta = 2\arcsin(r)$.

In short, we can summarize one guideline: the larger the distance from the center of the intersection circle to the center of the unit sphere (the distance equals to $\|\mathbf{c}\|_2$), the worse the data distribution is.

---

[1] In fact, the geometric relationship between the unit sphere and the SVDD sphere can be divided into two classes. One is that the center of the SVDD sphere is on the "left" hand of the point $\mathbf{c}$, and the other is that the center of SVDD sphere is on the "right" hand. We only illustrate the first case which is described in Figure 1(b). The result of the second case is similar.

3   Spatial Data Analysis Based Simhash (SDA-Simhash)

In this section, we firstly briefly introduce Simhash and figure out the Achilles hill of it. Then, based on the analysis in Section 2, we will describe our approach.

3.1   Simhash

Charikar's Simhash [1] is an effective method for similarity search. A notable implementation is *random hyperplane based Simhash* [1,3]. Specifically, it randomly generates some hyperplanes and hash objects according to the sides of these hyperplanes that they lie on. Given an object $\mathbf{x}$ and a random hyperplane $\mathbf{r} = \{r_1, r_2, \ldots, r_n\}$, the hash function can be specified as:

$$h(\mathbf{x}) = \text{sign}(\mathbf{r}^T \mathbf{x}) = \begin{cases} +1, & if\ \mathbf{r}^T \mathbf{x} \geq 0 \\ -1, & otherwise \end{cases} \tag{5}$$

The Achilles's Heel of Simhash. The simplicity and effectiveness of simhash leads to its great success in many applications in web search and data mining area. However, just like the heel is the deadly weakness of Achilles in spite of his overall strength, a shortcoming of Simhash makes it inefficient and even infeasible in some cases especially when the data is seriously distributed. For example, as described in Figure 1(a), if most of the points are located in a small region of the unit sphere, Simhash is very inefficient. In fact, Simhash would generate very similar fingerprints for most of points such that further processing is quite time consuming. In short, Simhash prefers the well-distributed data.

3.2   SDA-Simhash

In this section, we will present our improved Simhash algorithm which is based on the spacial analysis. Inspired by the guideline described in Section 2.2, we hope to:

– Make the random hyperplanes cross through the intersection circle;
– Minimize the distance from the center of the intersection circle to the center of the unit sphere, i.e. $\|\mathbf{c}\|_2$.

In brief, we have two ways to extend the basic Simhash. One (called *model-based Simhash*) is to generate the random hyperplanes with the constraint that they should cross through the intersection circle. Another (called *feature-based simhash*) is to alter the original representation such that the distance from the center of the intersection circle to the center of the unit sphere, i.e., $\|\mathbf{c}\|_2$, is minimum.

Model-Based Simhash. Model-based Simhash makes the hyperplanes cross through the intersection circle. This is carried out by adding a constraint to the generation process. Specifically, after the spacial analysis, we can obtain a SVDD sphere $(\mathbf{a}, R)$, an intersection circle $(\mathbf{c}, r)$ and the cone vertex angle 2θ. We hope to choose a random hyperplane that crosses through the intersection circle. Some feasible methods include coordinate-by-coordinate strategy [5] and a simple rejection method. Details are omitted here for the space limitation.

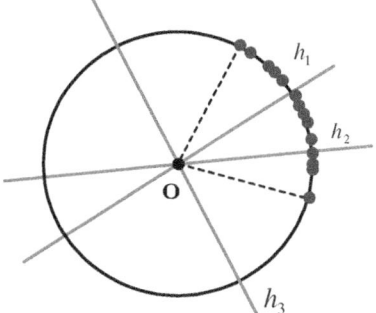

Fig. 2. An illustration of the approximation. The circle represents the unit sphere, and red points represent the data. There are three hyperplanes in this figure, h_1, h_2 and h_3. Two hyperplanes h_1 and h_2 cross through the shell while the hyperplane h_3 does not.

Approximated Model-based Simhash. Although the model-based Simhash described above can follow the guideline, the rejection method is inefficient for high-dimensional data. This part presents an approximation version of the model-based Simhash for high-dimensional data.

For description simplicity, we assume there do not exist outliers, i.e. the parameter C in Equation 1 is set to a very large value. As shown in Figure 2, some hyperplanes cross through the shell while others do not. Therefore, we divide all the random hyperplanes into two groups:

$$\mathcal{U}_+ = \{\mathbf{r} \mid \exists \mathbf{x}_i, \mathbf{x}_j \in \mathcal{X}, \operatorname{sign}(\mathbf{r}^T \mathbf{x}_i) \neq \operatorname{sign}(\mathbf{r}^T \mathbf{x}_j)\} \tag{6}$$

$$\mathcal{U}_- = \{\mathbf{r} \mid \forall \mathbf{x}_i, \mathbf{x}_j \in \mathcal{X}, \operatorname{sign}(\mathbf{r}^T \mathbf{x}_i) = \operatorname{sign}(\mathbf{r}^T \mathbf{x}_j)\} \tag{7}$$

We measures the percentage of the points lying on the two different sides of a hyperplane $\mathbf{r}$ as follows:

$$\Gamma(\mathbf{r}, \mathcal{X}) = \left| \frac{\sum_{\mathbf{x} \in \mathcal{X}} \mathbb{I}\{\operatorname{sign}(\mathbf{r}^T \mathbf{x}) = 1\}}{|\mathcal{X}|} - 0.5 \right| \tag{8}$$

where $\mathbb{I}\{\cdot\}$ equals to 1 when the condition is true otherwise 0, and $|\mathcal{X}|$ the size of the data set. Intuitively, if all the data lie on the same side of a hyperplane $\mathbf{r}$, $\Gamma(\mathbf{r}, \mathcal{X}) = 0.5$. On the other hand, if the numbers of points lying on the two sides of a hyperplane are equal, $\Gamma(\mathbf{r}, \mathcal{X}) = 0$. Obviously, we can find that $\Gamma(\mathbf{r}, \mathcal{X}) > 0, \forall \mathbf{r} \in \mathcal{U}_+$, while $\Gamma(\mathbf{r}, \mathcal{X}) = 0, \forall \mathbf{r} \in \mathcal{U}_-$.

Furthermore, we assume all the points in the data set $\mathcal{X}$ are normally distributed on the shell (determined by a SVDD sphere). Then, we observe that if the $\Gamma(\cdot)$ value of a hyperplane equals to 0, such a hyperplane splits the shell into two parts of equal size. The smaller the $\Gamma(\cdot)$ value is, the closer the hyperplane is to the boundary of the shell. This offers a good inspiration for speed up the model-based Simhash.

If we choose a candidate set of hyperplanes uniformly distributed on the unit sphere with size N, sort this set of hyperplanes in terms of the $\Gamma(\cdot)$ values and select $\frac{2\theta N}{\pi}$

hyperplanes with smallest $\Gamma(\cdot)$ values, we can ensure to a certain extent that the selected hyperplanes should cross through the shell. The size of the candidate set can be determined according to Theorem 1.

Theorem 1. *Given the cone vertex angle 2θ, we need to generate N candidate random hyperplanes such that, with the confidence interval $(\frac{2\theta}{\pi} - \varepsilon, \frac{2\theta}{\pi} + \varepsilon), \varepsilon > 0$ and the confidence level α, we can ensure that there are $\frac{2\theta N}{\pi}$ hyperplanes cross across the shell. The N can be determined by the following inequality:*

$$\exp\{-D(\frac{2\theta}{\pi} - \varepsilon||\frac{2\theta}{\pi}) \cdot N\} + \exp\{-D(\frac{2\theta}{\pi} + \varepsilon||\frac{2\theta}{\pi}) \cdot N\} \geq \alpha \qquad (9)$$

where $D(x||y) = x \log \frac{x}{y} + (1-x) \log \frac{1-x}{1-y}$ is the Kullback-Leibler divergence between Bernoulli distributed random variables with parameters x and y respectively.

Proof. For description simplicity, a random hyperplane $\mathbf{r}$ crossing through the shell is considered as an event A. The probability that A occurs ($A = 1$) is $p = \Pr[h(\mathbf{x}) \geq h(\mathbf{y})] = 2 \Pr[\text{sign}(\mathbf{r}^T\mathbf{x}) > 0, \text{sign}(\mathbf{r}^T\mathbf{y}) < 0] = 2\theta/\pi$. Assume random variables $A_1, A_2, \ldots, A_N \in \{0, 1\}$ are i.i.d. According to Chernoff-Hoeffding theorem, we have

$$\Pr\left[\frac{1}{N}\sum A_i \geq p + \varepsilon\right] \leq \exp\{-D(p + \varepsilon||p) \cdot N\} \qquad (10)$$

$$\Pr\left[\frac{1}{N}\sum A_i \leq p - \varepsilon\right] \leq \exp\{-D(p - \varepsilon||p) \cdot N\} \qquad (11)$$

Furthermore,

$$\Pr\left[p - \varepsilon \leq \frac{1}{N}\sum A_i \leq p + \varepsilon\right]$$

$$= \Pr\left[\frac{1}{N}\sum A_i \geq p - \varepsilon\right] - \Pr\left[\frac{1}{N}\sum A_i \geq p + \varepsilon\right] \qquad (12)$$

$$\geq 1 - \exp\{-D(p - \varepsilon||p) \cdot N\} - \exp\{-D(p + \varepsilon||p) \cdot N\}. \qquad (13)$$

On the other hand, according to the definitions of confidence level and interval, we have

$$\Pr\left[p - \varepsilon \leq \frac{1}{N}\sum A_i \leq p + \varepsilon\right] = 1 - \alpha. \qquad (14)$$

Put it all together and finally we can get

$$\exp\{-D(p - \varepsilon||p) \cdot N\} + \exp\{-D(p + \varepsilon||p) \cdot N\} \geq \alpha \qquad (15)$$

$\square$

In short, the complete algorithm is illustrated in Algorithm 1. The structure of our method is simple, so it is quite easy to implement. We should emphasize that, as the method shown in [5], the generation of hyperplanes normally distributed on the unit sphere is very efficient.

Algorithm 1. Approximated Model-Based Simhash

INPUT: M, $\mathcal{X}$;
calculate the SVDD sphere;
calculate the cone vertex angle 2θ;
calculate N as Inequality 9;
generate N random hyperplanes normally distributed on the unit sphere;
for *each* **r do**
| calculate $\Gamma(\mathbf{r}, \mathcal{X})$ as Equation 8;
end
sort N hyperplanes according to the $\Gamma(\cdot)$ value;
select top M hyperplanes with the smallest $\Gamma(\cdot)$ values;
for *each* $\mathbf{x} \in \mathcal{X}$ **do**
| generate M-bit fingerprint according to the selected M hyperplanes;
end

Feature-Based Simhash. In this part, we present an alternative way to follow the guideline. Different from the model-based Simhash, we directly convert the data representation. As we know, if a K-dimensional variable $\mathbf{x}$ follows the standard normal distribution $\mathcal{N}(0, \mathbf{I})$, $\mathbf{x}/\|\mathbf{x}\|_2$ is normally distributed on the unit sphere. This implies that if we can convert a given distribution to the standard normal, we can re-locate the data uniformly on the unit sphere. For example, if a data set $\mathcal{X}$ is normally distributed and the features are independent, we can standardizing it via scaling transform and translation transform. Specifically, we firstly calculate the mean E_j and the variance V_j of the j-th component. Secondly, we standardize each component of $\mathbf{x}$ as follows:

$$x'_j = \frac{x_j - E_j}{\sqrt{V_j}}. \tag{16}$$

Then, we normalize each object $\mathbf{x}'$ such that its L2-norm equals to 1, i.e. $\|\mathbf{x}'\|_2 = 1$. Finally, the basic Simhash is directly applied to process the normalized data.

4 Experiments

To reflect different properties of our SDA-Simhash, we evaluate our method on a toy case and a real-world data set. To begin, we will firstly describe the evaluation criterions and then show the experimental results.

4.1 Evaluation Criterions

To demonstrate the effectiveness of our method, we compare SDA-Simhash with the basic Simhash.

Simhash involves two periods, pre-processing and query. In the pre-processing period, Simhash generates fingerprints for data points by concatenating the outputs of Equation 5 with M randomly generated hyperplanes $\{\mathbf{r}_1, \ldots, \mathbf{r}_M\}$. Then the algorithm permutes the fingerprints and sort them lexicographically to form T sorted orders. In the

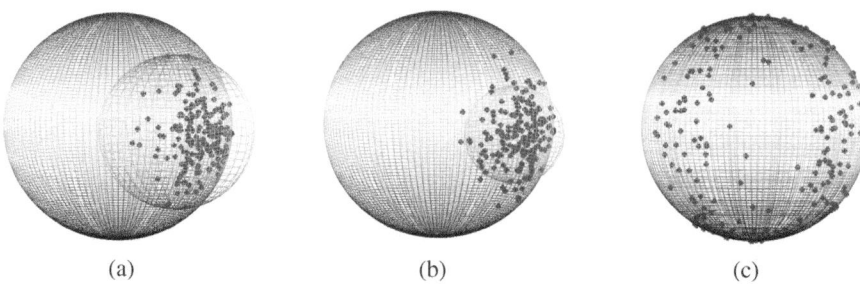

           (a)   (b)   (c)

Fig. 3. A toy case. The big sphere represents the unit sphere and the red points represent the data points. (a),(b) The small spheres in (a) and (b) are the SVDD spheres obtained by solving the Equation 1 when the control parameter C equals to 0.5 and 0.01 respectively. It shows that the radius of the obtained SVDD decreases with the decrease of the parameter C. (c) Data distribution after standardization. Obviously, standardization makes the data well-distributed on the unit sphere.

query period, we firstly perform an approximation search in M-dimensional Hamming space to generate a candidate set. The candidate set is defined as follows:

$$\mathcal{V} = \{\mathbf{x} \mid \|\mathbf{x} \text{ XOR } \mathbf{q}\|_1 < k, \mathbf{x} \in \mathcal{X}\}. \tag{17}$$

where $\mathbf{q}$ is a query fingerprint. Then, we need to compute the actual similarity between the query object and objects in $\mathcal{V}$. Finally, the items with the largest similarity are returned as results. Obviously, if the candidate set is very large, the further processing is quite inefficient. Therefore, the size of candidate set $|\mathcal{V}|$ is a good measurement of algorithms' efficiency.

Alternatively, we can alter the data representation to relocate the data uniformly on the unit sphere such that the basic Simhash can be directly applied. As the analysis in Section 2, we calculate the distance from the intersection circle to the center of the unit sphere $\|\mathbf{c}\|_2$.

4.2   A Toy Case

To better understand the properties of our method, we utilize a toy case to prove the correctness of our method. We will generate some synthetic points in $\mathbb{R}^3$. The visualization of the results can help us intuitively understand our method.

We generate the synthetic data in two steps: (1) choosing an area (shell) on the 3-dimensional sphere, and (2) randomly generating 200 points uniformly distributed on the shell. The synthetic data are represented as the red points in Figure 3(a). We can see that the data is ill-balanced distributed.

Figure 3(a)(b) present the influence of the control parameter C to our spatial data analysis method. As described in Section 2.1, we compute the SVDD sphere with different values of C. When C is set to 0.5 and 0.01, the corresponding SVDD spheres are displayed in Figure 3(a)(b) respectively. It shows that the radius of the SVDD sphere decreases with the decrease C, which consists with the analysis in Section 2.1. The data

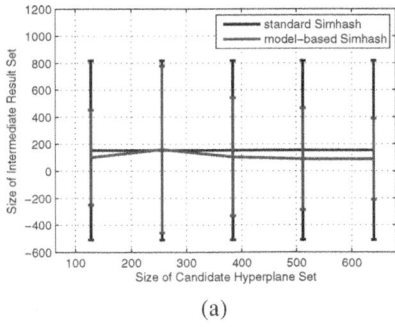
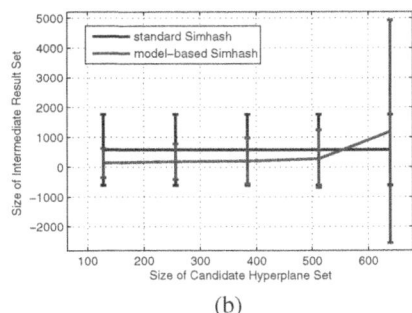

(a)   (b)

Fig. 4. The mean and the variance of $|\mathcal{V}|$. (a) Gray-Scale Histogram. (b) Local Binary Pattern (LBP). It shows that our model-based Simhash can almost consistently decrease the mean and the variance of $|\mathcal{V}|$.

after standardization is displayed in Figure 3(c). Obviously, the data is well distributed on the unit sphere thus the basic Simhash can be directly applied.

4.3 Real-World Data

In this section, we evaluate our method on a real-world image data set. This image set is collected as follows: (1) crawling a large number of images from the web, (2) resizing the images such that all the images are in the same size, and (3) removing all the duplicated images. In short, we get an image set with 253,083 distinct images with the same size.

Data Representation. We use two common features to represent each data, gray-scale histogram and Local Binary Pattern (LBP) [6]. Gray-scale histogram is a representation of the gray-scale distribution in the images. Given an image, we convert it to 256-level gray scale, statistic its gray-scale histogram and finally normalize the histogram such that the sum of all the components equals to 1. Local Binary Pattern feature has been widely used in various applications, such as image retrieval, texture classification, etc [6]. The most important property of LBP is its robustness to monotonic illumination changes and rotation invariance. In our experiments, we use 8-neighborhood. Therefore each image is represented as the histogram of 59 labels, which is a 59-dimensional vector. Please see [6] for more details.

Experimental Results. Considering the dimension of images is relatively high, we utilize the approximated model-based Simhash. We compare our method with the basic Simhash [1] in terms of the size of candidate set $|\mathcal{V}|$. We generate a 64-bit fingerprint (M=64) for each image. To obtain the set $\mathcal{V}$, we set the threshold k=1 when using LBP feature and k=3 when using gray-scale histogram feature. The reason why we use different thresholds is that we hope the intermediate set $\mathcal{V}$ to be large enough for comparison. In our experiments, each time, we randomly select 1,000 images from the data set as queries. Then we calculate the mean and the variance of $|\mathcal{V}|$. To reduce the influence of the random errors, this process is repeated 10 times.

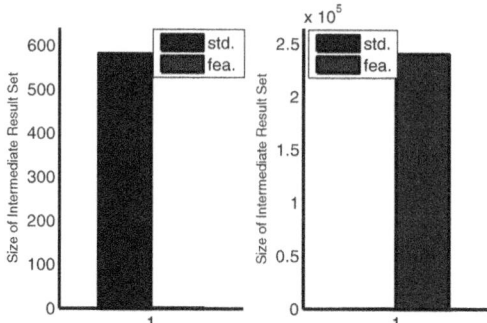

Fig. 5. The size of the candidate set $\mathcal{V}$ obtained by the feature-based Simhash. The left side of the figure is LBP and the right side corresponds to gray-scale histogram. It shows the feature-based Simhash works well when using LBP feature but totally fails when using gray-scale histogram.

Table 1. The mean and the deviation of $|\mathcal{V}|$ of different algorithms when using LBP: basic Simhash, model-based Simhash, feature-based Simhash and the combination of model- and feature- based Simhashs

	$E[|\mathcal{V}|]$	$SD[|\mathcal{V}|]$
basic Simhash	580.0	1184.230
mod. Simhash	187.457	595.943
fea. Simhash	1.790	7.902
fea.+mod. Simhash	1.430	1.874

The results of approximated model-based Simhash are presented in Figure 4. It shows that for both gray-scale histogram and LBP, our algorithm can effectively decrease $E[|\mathcal{V}|]$ and $D[|\mathcal{V}|]$. This illustrates that in the further processing, our method is more efficient thus is more discriminative.

From Figure 4(a), we find when the size of candidate hyperplane set is large (640 here), the basic Simhash outperforms our method. This is mainly because of the hyperplane selection process. Specifically, we select the random hyperplanes with the smallest $\Gamma(\cdot)$. According to the definition of the $\Gamma(\cdot)$ value, random hyperplanes with smaller $\Gamma(\cdot)$ values tend to cross through the center of the shell. So when the N is very large, our method prefers to choosing those hyperplanes that just cross through the center of the shell, which makes the selected hyperplanes tend to be similar.

Figure 5 reports the performance of the feature-based Simhash. The left side of Figure 5 shows our feature-based Simhash can effectively decrease $E[|\mathcal{V}|]$. But unfortunately, as presented in the right side of the Figure 5, the feature-based Simhash totally fails when using gray-scale histogram feature. We examine the data set carefully and find that most images have a large area of white background and the original value distribution is not normal or near normal. In this case, some other techniques should be considered to convert the given distribution to a standard normal.

A natural question is whether the combination of the model-based and the feature-based Simhash can improve the final search performance. Here, we combine such two

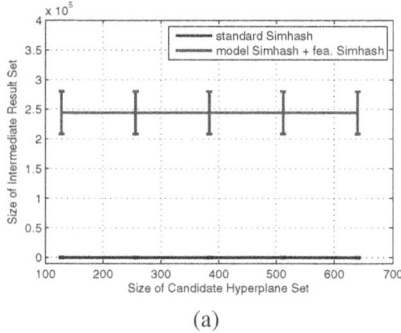 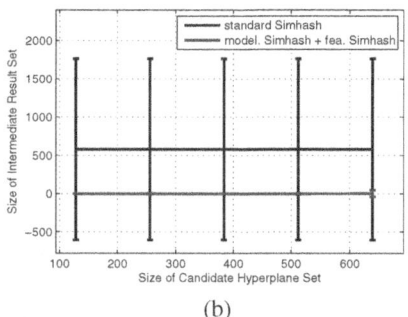

(a)   (b)

Fig. 6. The mean and the variance of $|\mathcal{V}|$. (a) Gray-Scale Histogram. (b) Local Binary Pattern (LBP). It shows that our model-based Simhash can almost consistently decrease the mean and the variance of $|\mathcal{V}|$.

models and test the combination on the image set. Specifically, we firstly process the original data by the standardization transformation. Then, our approximated model-based Simhash is carried out.

The results are shown in Figure 6. Figure 6(b) shows that when using LBP feature, the combination significantly outperforms the basic Simhash. However, affected by the feature-based simhash, the combination of model- and feature- based Simhashs completely fail (as presented in Figure 6(a)). As analysis above, Normal standardization in feature-based Simhash needs to be replaced by some other standardization techniques.

Considering the result in Figure 6, we make a quantitative analysis of different algorithms when using LBP feature. The result is listed in Table 1. It shows that model- and feature- based Simhash are complementary and the combination can improve the performance.

Most results illustrate that our model-based Simhash can significantly decrease $E[|\mathcal{V}|]$ and $D[|\mathcal{V}|]$, which demonstrates that our method is more effective, robust and stable than the basic Simhash.

5   Related Work

Many works have proposed on fast similarity search. For example, KD-tree [7] returns accurate results. But when the dimension of the feature space is high (> 10), it becomes quite inefficient and even slower than brute-force approach [8]. For high-dimensional data, a notable method is Locality Sensitive Hashing (LSH) [9]. Just like Simhash, LSH also performs a random linear projection to map similar objects to similar hash codes. However, in practice, LSH suffers from the efficiency problem, since LSH may tend to generate long codes [10]. Another notable work is Spectral Hashing [11], which has been proven significantly improvement on LSH. However, the assumption made by spectral hashing of input vectors with a known probability distribution restricts its application. Some other works involve reducing the storage of LSH/Simhash [12,13], kernelized Simhash [14] and so on.

6 Conclusions

In this paper, we have presented a novel method for quantitatively measuring whether the basic Simhash can be directly applied to perform similarity search in a given data set. Our method is mainly based on the analysis on the location of the whole data set via Support Vector Data Description (SVDD). Based on the analysis, we propose two methods to improve the search performance of the basic Simhash algorithm. Most results show that our method outperforms the basic Simhash significantly.

Acknowledgments. We would like to thank the anonymous reviewers for their insightful comments. This work is supported by the National Natural Science Foundation of China under Grant No. 60873174.

References

1. Charikar, M.S.: Similarity estimation techniques from rounding algorithms. In: Proceedings of the Thiry-Fourth Annual ACM Symposium on Theory of Computing, p. 388 (2002)
2. Stein, B.: Principles of hash-based text retrieval. In: Proceedings of the 30th Annual International ACM SIGIR Conference on Research and Development in Information Retrieval, p. 534 (2007)
3. Manku, G.S., Jain, A., Das Sarma, A.: Detecting near-duplicates for web crawling. In: Proceedings of the 16th International Conference on World Wide Web, p. 150 (2007)
4. Tax, D.M.J., Duin, R.P.W.: Support vector data description. Machine Learning 54, 45–66 (2004)
5. Poland, J.: Three different algorithms for generating uniformly distributed random points on the n-sphere (2000)
6. Ojala, T., Pietikäinen, M., Mäenpää, T.: Gray Scale and Rotation Invariant Texture Classification with Local Binary Patterns. In: Vernon, D. (ed.) ECCV 2000. LNCS, vol. 1842, pp. 404–420. Springer, Heidelberg (2000)
7. Bentley, J.L.: K-d trees for semi-dynamic point sets. In: Proc. of the 6th ACM Symposium on Computational Geometry (SCG), pp. 187–197 (1990)
8. Weber, R., Schek, H.J., Blott, S.: A quantitative analysis and performance study for similarity-search methods in high-dimensional spaces. In: Proceedings of the International Conference on Very Large Data Bases, pp. 194–205 (1998)
9. Indyk, P., Motwani, R.: Approximate nearest neighbors: towards removing the curse of dimensionality. In: Proceedings of the Thirtieth Annual ACM Symposium on Theory of Computing, pp. 604–613 (1998)
10. Salakhutdinov, R., Hinton, G.: Semantic hashing. International Journal of Approximate Reasoning 50, 969–978 (2009)
11. Weiss, Y., Torralba, A., Fergus, R.: Spectral hashing. In: Advances in Neural Information Processing Systems, vol. 21, pp. 1753–1760 (2009)
12. Lv, Q., Josephson, W., Wang, Z., Charikar, M., Li, K.: Multi-probe lsh: Efficient indexing for high-dimensional similarity search. In: Proceedings of the 33rd International Conference on Very Large Data Bases, pp. 950–961 (2007)
13. Bawa, M., Condie, T., Ganesan, P.: Lsh forest: self-tuning indexes for similarity search. In: Proceedings of the 14th International Conference on World Wide Web, p. 660 (2005)
14. Kulis, B., Grauman, K.: Kernelized locality-sensitive hashing for scalable image search. In: Proc. 12th International Conference on Computer Vision, ICCV (2009)

XML Information Retrieval through Tree Edit Distance and Structural Summaries

Cyril Laitang, Mohand Boughanem, and Karen Pinel-Sauvagnat

IRIT-SIG,
118 route de Narbonne,
31062 Toulouse Cedex 9,
France

Abstract. Semi-structured Information Retrieval (SIR) allows the user to narrow his search down to the element level. As queries and XML documents can be seen as hierarchically nested elements, we consider that their structural proximity can be evaluated through their trees similarity. Our approach combines both content and structure scores, the latter being based on tree edit distance (minimal cost of operations to turn one tree to another). We use the tree structure to propagate and combine both measures. Moreover, to overcome time and space complexity, we summarize the document tree structure. We experimented various tree summary techniques as well as our original model using the SSCAS task of the INEX 2005 campaign. Results showed that our approach outperforms state of the art ones.

1 Introduction

XML documents are organized through semantically meaningful elements. As content is distributed over different levels of the document structure, using this information should improve the overall search process as well as it enables the returned information to be more focused on the expressed needs. In this context, Semi-Structured information retrieval (SIR) models aim at combining content and structure search processes.

XML documents are structured through nested tags. This hierarchical organization is naturally expressed through a specific graph representation, i.e. trees[1], in which nodes are elements and edges hierarchical dependencies. In an XML tree, the text is located in the *leaves* which are the bottom nodes of the hierarchy. In structured retrieval, queries can be expressed using either *Content Only* constraints (CO) or both *Content And Structure* constraints (CAS). As for XML documents, the structural constraints expressed in CAS queries can be visualized through a tree representation, and might contain two parts: the *target element* indicates the tag element we want to retrieve and the rest of the structural constraints is called *support* or *environment*.

Figure 1 shows a conversion example of an XML document and a query[2]. The

---

[1] Trees are a particular type of graphs which do not contain any cycles. Cycles are paths starting and ending with the same node.

[2] This query is expressed in the NEXI [20] (Narrowed Extended XPath) language used in the context of the INEX evaluation campaign.

M.V.M. Salem et al. (Eds.): AIRS 2011, LNCS 7097, pp. 73–83, 2011.

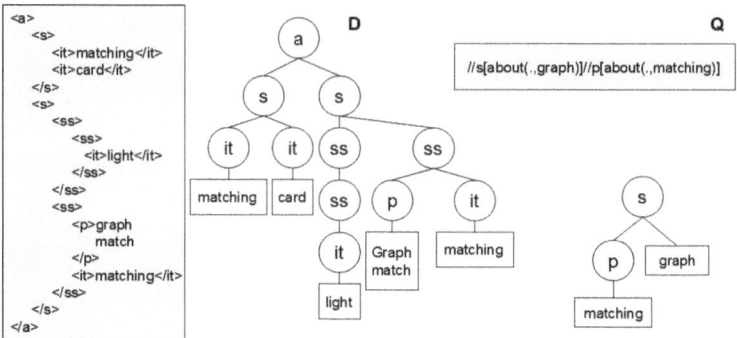

Fig. 1. Tree representation of an XML document and a query in which *we want a "p"*
element about "matching"' contained in an "s" element about "graph"

target element is "p". For clarity reasons we shorten the tags. In real case the
"p" is equivalent to the semantically richer "paragraph".

Based on these representations we propose a SIR model based on both graph
theory properties and content scoring. To our knowledge, only few SIR models
explicitly use tree matching algorithms between documents and queries. Most of
them reduce the structure of documents to strings which impoverish the amount
of information available to the matching process. Some authors such as Alilaouar
et al. [1], Ben Aouicha et al. [2] and Popovici et al. [16], lose a high level of
details to get back to their domain known approaches. Our approach uses the
well-known tree edit distance combined with a traditional IR model for content.
This allows to use the whole document structure to select relevant elements.
Moreover to improve space and time complexity we propose three methods to
summarize documents.

The rest of this paper is organized as follows: Section 2 presents existing
tree matching algorithms and XML retrieval approaches using trees; Section 3
presents our model and finally Section 4 discusses the experiments and results
obtained by our approach using the SSCAS task of the INEX 2005 campaign.

2   Related Works

In this section we will first overview some tree matching algorithms and then
give a brief survey on some SIR approaches based on tree matching.

2.1   Structural Similarities between Trees

Two graphs are called isomorphic if they share the same nodes and edges. Evaluating how isomorphic are two graphs is called graph matching. We make the
distinction between approximate matching and exact matching. The first one attempts to find a degree of similarity between two structures while exact matching
tries to validate the similarity. Because of the context of our work, we will focus
here on approximate tree matching. There are three main families of approximate tree matching: *alignment, inclusion,* and *edit distance.* As the later offer

the most general application we will focus on this one. Tree edit distance algorithms [18] generalizes Levenshtein *edit distance*[14] to trees. The similarity is the minimal set of operations (adding, removing and relabeling) to turn one tree to another. Given two forests (set of trees) F and G, Γ_F and Γ_G their rightmost nodes, $T(\Gamma_F)$ the tree rooted in Γ_F and the cost functions $c_{del}()$ and $c_{match}()$ for removing (or adding) and relabeling, the distance $d(F, G)$ is evaluated according to the following recursive lemma:

$$d(F, \emptyset) = d(F - \Gamma_F, \emptyset) + c_{del}(\Gamma_F)$$
$$d(\emptyset, G) = d(\emptyset, G - \Gamma_G) + c_{del}(\Gamma_G)$$
$$d(F, G) = min \begin{cases} (a) \ d(F - \Gamma_F, G) + c_{del}(\Gamma_F) \\ (b) \ d(F, G - \Gamma_G) + c_{del}(\Gamma_G) \\ (c) \ \begin{array}{l} d(T(\Gamma_F) - \Gamma_F, T(\Gamma_G) - \Gamma_G) \\ + d(F - T(\Gamma_F), G - T(\Gamma_G)) + c_{match}(\Gamma_F, \Gamma_G) \end{array} \end{cases} \tag{1}$$

Operations *(a)* and *(b)* are respectively the cost $c_{del}()$ of removing Γ_F or Γ_G while *(c)* is the cost $c_{match}()$ of relabeling the Γ_F by Γ_G. Later, Klein et al. [13] reduced the overall complexity in time and space by splitting the tree structure based on the heavy path (defined in Section 3.3). Demaine et al. [6] further improved this algorithm by storing substrees scores in order to reduce calculation time. Finally Touzet et al. [8] used a decomposition strategy to dynamically select always the best nodes to recurse on between rightmost and leftmost which reduce the number of subtrees in memory. The best tree edit distance algorithms uses $\Theta(nm)$ on space complexity and between $\Theta(n \times log(n).m \times log(m))$ [8] and $\Theta(n.m(1 + log(\frac{n}{m})))$ [6] on time for n and m the respective sizes in nodes of two trees T_1 and T_2. Tree edit distance algorithms are efficient but slow on large trees. One way to overcome this issue in SIR is to reduce the document tree size by pruning or summaries. Used mainly on indexing and clustering, the latter is based on the intuition that the document underlying structure can be captured efficiently with a smaller size representation. As we will see in Section 2.2, the summarizing algorithm proposed in [4] provides satisfying results in improving the runtime. In this paper, we will apply this algorithm to XML retrieval.

2.2   Semi-structured Information Retrieval

XML documents as well as CAS queries can be represented as trees. It then seems natural to apply tree matching algorithms for the document-query matching process. However, following INEX proceedings overviews ([7] and [9]), tree matching is uncommon in SIR. One can however find two main categories of retrieval approaches in the literature using XML trees matching.

The first one uses *relaxation*, which means reducing the links or constraints. Relaxing is a research space expansion process. The general idea is to translate the tree structure into a set of binary weighted edges. This simplified representation allows to use traditional IR models. For example, Alilaouar et al. [1] combined relaxation and minimal covering trees. Similarly Ben Aouicha et al. [2] relaxed the whole hierarchical constraints by adding virtual edges. These edges are weighted based on the initial hierarchical distance in the document structure.

The resulting arcs are then projected on a vector space and the document-query matching is done using a traditional *Vector Space Model*. Some other approaches as for example [5] use *fuzzy closing* in which a set of virtual edges representing all tree's transitive relationships is created based on the closing property. One can also found approaches that directly use *edit distance algorithms*. Popovici et al. [16] translated documents tree structure in a set of paths and evaluated the document-query similarity using the Levenshtein [14] string edit distance. Other works using tree matching algorithms can be found in IR related domains, like error detecting and clustering. In their papers, Boobna [3] and Rougemont [17] check XML documents conformity to their DTD thought tree edit distance. Similarly, Dalamagas et al. [4] summarize documents before applying Tai [18] algorithm to cluster them.

The fact that there are relatively few SIR models using explicitly tree matching algorithms could be due to the complexity of such approaches. All of the previously presented models always applies trees conversion to restrict the problem to a smaller search space. This removes a lot of details. In this paper we attempt to overcome the complexity drawback by summarizing the structure without loosing too much of the structural information.

3   Tree-Edit Distance for Structural Document-Query Matching

We assume that a query is composed of content (keywords) and structure conditions, as shown in Figure 1. The document-query similarity is evaluated by considering content and structure separately, and we then combine these scores to rank relevant elements. In this section, we first describe the content evaluation. We then present our subtree extraction and summary algorithms. Finally we detail our structure matching algorithm based on tree edit distance.

3.1   Content Relevance Score Evaluation

First, we used a $tf \times idf$ (Term Frequency × Inverse Document Frequency [11]) formula to score the document leaf nodes according to query terms contained in content conditions. To score inner nodes, our intuition is that a node score must depend on three elements. First, it should take into account its leaves scores, that form what we call intermediate score. Second we should score higher a node located near a relevant element than a node located near an irrelevant one. Finally, there must be a way to balance the hierarchical effect on the node score. Based on these constraints we define the content score $c(n)$ of an element n as the *intermediate content score of the element itself plus its father's intermediate score plus all its father's descendants score*. Recursively, and starting from the document root:

$$
c(n) = \begin{cases}
\underbrace{\dfrac{p(n)}{\mid leaves(n) \mid}}_{(i)} + \underbrace{\dfrac{p(a_1) - p(n)}{\mid leaves(a_1) \mid}}_{(ii)} + \underbrace{\dfrac{c(a_1) - \frac{p(a_1)}{\mid leaves(a_1) \mid}}{\mid children(a_1) \mid}}_{(iii)} & \text{if } n \neq root \\[3em]
\dfrac{p(n)}{\mid leaves(n) \mid} & \text{otherwise}
\end{cases}
\tag{2}
$$

(i) is the *intermediate content score* part with $\mid leaves(n) \mid$ the number of leaf nodes descendants of n and $p(n)$ the intermediate score of the node based on the sum of the scores of all its leaf nodes: $p(n) = \sum_{x \in leaves(n)} p(x)$, with $p(x)$ evaluated using a $tf \times idf$ formula.

(ii) is the *neighborhood score* part which allows us to convey a part of the relevance of a sibling node through its father a_1. $p(a_1)$ is the intermediate score of a_1 and $\mid leaves(a_1) \mid$ the number of leaves of a_1.

(iii) is the *ancestor scores*, with $c(a_1)$ the final content score of the father a_1 minus its intermediate score.

3.2 Extracting and Summarizing Subtrees

For each relevant leaf node (i.e. with a score $p(x) > 0$), we extract all the subtrees rooted by all its ancestors starting from the first one whose tag is similar to a tag in the query. In Figure 2, five subtrees are extracted.

Even the best edit distance algorithm runs with the minimal time complexity $O(n^3)$ [8], it thus remains costly on large trees. In order to reduce the matching space, we choose Dalamagas [4] summary rules which seem to be the best compromise between preserving the overall structure and speed. As illustrated in Figure 3 these rules are "remove nesting" which is equivalent to move the subtree of a node having the same label than one of its ancestor; and "remove duplicates from father-child relationship" which removes the siblings having the same labels. Based on these rules we create four different summary versions. The first one, illustrated on top, is the strict adaptation of these rules to all nodes while the second version, illustrated at the bottom, apply the summary rules only for the nodes which labels are not in the query. Thanks to this version

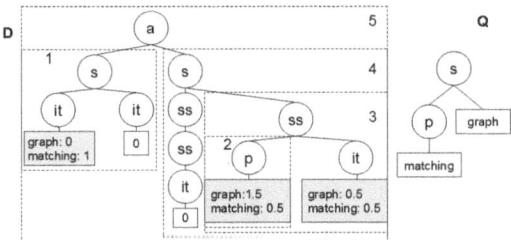

Fig. 2. Subtrees extraction based on relevant leaf nodes

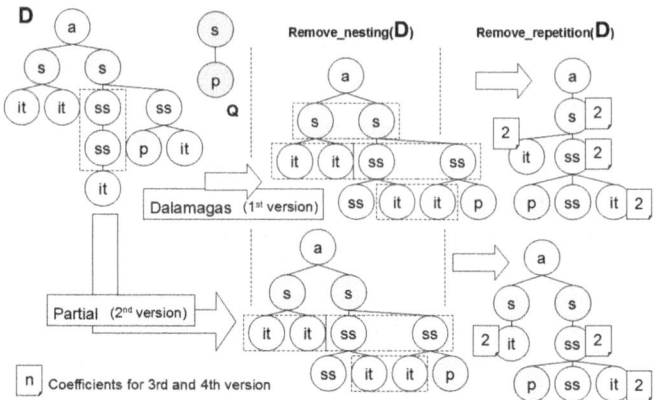

Fig. 3. The original document tree D and it's tree summaries. On top the rules applied for all node, at the bottom the rules applied only for tags that are not in the query.

we keep as many relevant nodes as possible while reducing the document size. Finally, in the third and fourth versions, we adapt the first two ones by keeping a record of the number of removed nodes. This will be used to multiply the edit costs $c_{del}()$ or $c_{match}()$ during the recursion (eg. $c_{del}(it)$ will be multiplied by 2).

3.3   Structure Relevance Score Evaluation

As seen in Section 2.1, the tree edit distance is a way of measuring similarity based on the minimal cost of operations to transform one tree to another. The number of subtrees stored in memory during this recursive algorithm depends on the direction we choose when applying the operations. Our algorithm is an extension of the optimal cover strategy from Touzet et al. [8]. The difference is that the optimal path is computed with the help of the *heavy path* introduced by Klein et al. [13]. The heavy path is the path from root to leaf which pass through the rooted subtrees with the maximal cardinality. This means that selecting always the most distant node from this path allows us create the minimal set of subtrees in memory during the recursion creating the *optimal cover strategy*. Formally a heavy path is defined as a set of nodes $[n_1, ..., n_i, ..., n_z]$ satisfying:

$$\forall (n_i, n_{i+1}) \in heavy \begin{cases} n_{i+1} \in children(n_i) \\ \forall x \in children(n_i), x \notin n_{i+1}, \mid T(n_{i+1}) \mid \geq \mid T(x) \mid \end{cases}$$

This strategy is used on the document and the query as input to our tree edit distance algorithm (Algorithm 1). F, G are two forests (i.e. the document and the query as first input), and p_F and p_G are positions in O_F and O_G the *optimal paths* (i.e. paths of the *optimal cover strategy*). Function $O.get(p)$ returns the node in path O corresponding to position p. .

To evaluate the final structure score of a node n, we average the tree edit distances between its subtrees and all of its ancestor ones and the query. These distance are averaged by their cardinality in order to reduce the gap size between the subtrees. With $Anc(n)$ the set of n ancestors; $a \in Anc(n)$; $T(a)$ the subtree rooted in a; $d(T(a), Q)$ the edit distance between the tree rooted in a $T(a)$ and Q, the structure score $s(n)$ is formally:

$$s(n) = \frac{\sum_{a \in \{n, Anc(n)\}} (1 - \frac{d(T(a), Q)}{|T(a)|})}{| Anc(n) |} \tag{3}$$

Algorithm 1. Edit distance using optimal paths

$\mathsf{d}(F, G, p_F, p_G)$ **begin**
    if $F = \oslash$ **then**
        if $G = \oslash$ **then**
            return 0;
        else
            return $\mathsf{d}(F - O_F.\mathsf{get}(p_F)), \oslash, p_F{+}{+}, p_G) + c_{del} (O_F.\mathsf{get}(p_F);$
        end
    end
    if $G = \oslash$ **then**
        return $\mathsf{d}(\oslash, G - O_G.\mathsf{get}(p_G)), p_F, p_G{+}{+}) + c_{del} (O_G.\mathsf{get}(p_G);$
    end
    $\mathsf{a} = \mathsf{d}(F - O_F.\mathsf{get}(p_F), G, p_F{+}{+}, p_G) + c_{del} (O_F.\mathsf{get}(p_F);$
    $\mathsf{b} = \mathsf{d}(F, G - O_F.\mathsf{get}(p_F), p_F, p_G{+}{+}) + c_{del} (O_G.\mathsf{get}(p_G));$
    $\mathsf{c} = \mathsf{d}(F - O_F.\mathsf{get}(p_F), G - O_G.\mathsf{get}(p_G), p_F{+}{+}, p_G{+}{+}) + \mathsf{d}(F - T(O_F.\mathsf{get}(p_F)), G - T(O_G.\mathsf{get}(p_G))) + c_{match} (O_F.\mathsf{get}(p_F), O_G.\mathsf{get}(p_G));$
    return $\min(\mathsf{a, b, c});$
end

3.4 Final Combination

The final score $score(n)$ for each candidate node n extracted as explained in Section 3.2 is evaluated through the combination of the previously normalized scores $\in [0, 1]$. Then the elements corresponding to the target nodes are filtered and ranked. Formally, with $\lambda \in [0, 1]$:

$$score(n) = \lambda \times c(n) + (1 - \lambda) \times s(n). \tag{4}$$

4 Experiments and Evaluation

We evaluated our approach on both summarized and unsummarized subtrees on the INEX 2005 collection and compared our results with the official participants.

4.1 INEX Collection

INEX (*Initiative for the Evaluation of XML Retrieval*) is the reference evaluation campaign for XML retrieval. To evaluate our approach we used the 2005 collection which is composed of 16000 XML documents from the IEEE Computer Society scientific papers. These documents have an average of 1500 elements for a hierarchical depth of 6.9.

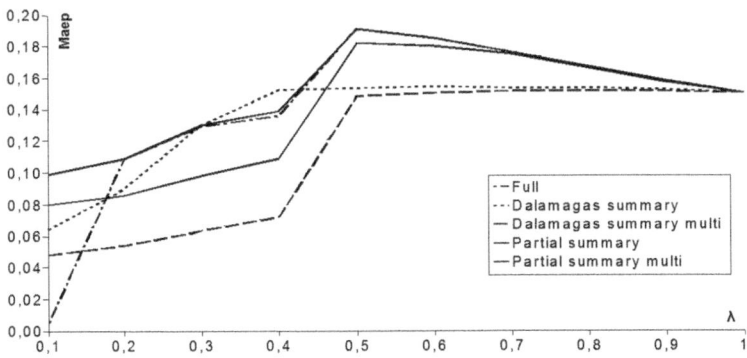

Fig. 4. MAeP score evolution over λ parameter variation for content and structure

Two main types of queries are available, namely *Content Only* (CO) and *Content And Structure* (CAS). Tasks using CAS queries are centered on structural constraints and were not reconducted in later campaigns[3]. Four subtasks were proposed. To evaluate queries in which structural constraints are semantically relevant, we use in our experiments the SSCAS subtask, in which constraints are strict on the target element and its environment.

There are two measures for the CAS subtasks [12]: *Non-interpolated mean average effort-precision* (MAeP) which is used to average the effort-precision measure at each rank and *Normalized cumulated gain* (nxCG). This last one corresponds to the cumulative gain at a threshold and is based on the ideal ranking over the sum of all score to that threshold.

Finally we will use the "strict" quantization to aggregate specificity and exhaustivity relevance judgments as it only takes into account fully relevant elements.

4.2 Experiments

We run our algorithm with four summary versions of document and query trees. Our baseline is unsummarized version (*full*). Regarding the summary versions, *dalamagas summary* corresponds to the run for the exact summary rules while in *partial summary* nodes containing tags from the query are not summarized. The *multi* extension for the both previous versions are for the cases in which we keep a record of the number of removed nodes as a coefficient in the edit distance. We set the removing cost $c_{del}()$ to 0.5 for a label in the query and 1 otherwise and the relabeling cost $c_{match}()$ to 0 for similar tags (eg: *section* and *subsection* are considered as semantically similar) and 1 otherwise.

Figure 4 shows the results for the *MAeP* metric depending on the λ parameter of equation (4). The best tuning for our system is around 0.5 which means an

---

[3] Since 2010 a new task called *Datacentric* appeared. Centered on rich structure it is based on the IMDB database which contains around 1 590 000 documents about movies and several millions about actors, producers, etc. Our next experimentations will of course use this collection.

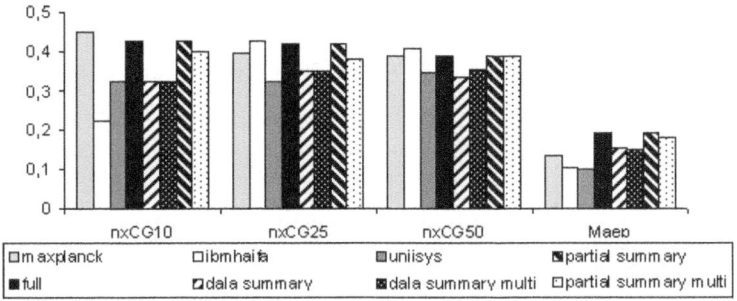

Fig. 5. Results for $\lambda = 0.5$ compared to INEX 2005 participants

equal part of content and structure. However, if all of our solutions return better results around the same value of λ it is clear that two of them are well over the others (more than 80% for the *MaeP*): our full tree algorithm and the *partial summary* algorithm in which nodes that have tags equivalent to the query ones are not summarized. We further notice that the results are equivalent for these two cases which means that we can gain on runtime while keeping the score of our intial solution. Moreover, keeping track of the removed nodes does not improve our results. Even if it seems natural for the *dalamagas summary* in which the tree structure is strongly altered through the removal of duplicate node descendants, it seems more surprising for our *partial summary*. It can however be explained by the fact that it artificially increases the number of nodes for a substitution score equal to 0. It means that it gives the same score for a summarized tree with several noisy children nodes than for a subtree with only the correct number of relevant nodes. Considering now efficiency, *Dalamagas summary* reduces the number of nodes to 48%, while our *partial summary* version reduces their number to 45%. Regarding the running time, it is improved 20 times on the average.

Figure 5 shows the results for $\lambda = 0.5$ over the various INEX 2005 metrics compared to the best participants, namely the Max Planck institute with TopX [19] a database-centered approach; IBM Haifa Research Lab [15] with a vector space model and the University of Klagenfurt [10] which also uses a vector space model. While our *full tree* and *partial summary* runs score slightly the same than the first participants for the nxCG (an average of 5% under the first while 13% over the second), our *MAeP* results are overall better (+45 % compared to the first one and +90% for the second) for all our runs.

4.3   Conclusions and Future Work

In this paper we presented an XML retrieval approach whose main originality is to use tree edit distance. This solution allowed us to outperform other approaches on the *MAeP* measure by 45% while proving the usefulness of structural information in SIR process. However its main drawback, i.e. complexity, was time consuming on the runs. We overcame this issue with a new set of summary rules which allowed us to keep the effectiveness of our solution while improving efficiency though search space reduction. In future work we will improve our content score as well as the overall edit distance cost system in order to better use the tag semantics. Finally we are currently working on the INEX 2010 *Datacentric* collection which will allow us to confirm our results on a bigger and semantically richer collection.

References

1. Alilaouar, A., Sedes, F.: Fuzzy querying of XML documents. In: International Conference on Web Intelligence and Intelligent Agent Technology, Compigne, France, pp. 11–14. IEEE/WIC/ACM (September 2005)
2. Ben Aouicha, M., Tmar, M., Boughanem, M.: Flexible document-query matching based on a probabilistic content and structure score combination. In: Symposium on Applied Computing (SAC), Sierre, Switzerland. ACM (March 2010)
3. Boobna, U., de Rougemont, M.: Correctors for XML Data. In: Bellahsène, Z., Milo, T., Rys, M., Suciu, D., Unland, R. (eds.) XSym 2004. LNCS, vol. 3186, pp. 97–111. Springer, Heidelberg (2004)
4. Dalamagas, T., Cheng, T., Winkel, K.-J., Sellis, T.: A methodology for clustering XML documents by structure. Information Systems 31, 187–228 (2006)
5. Damiani, E., Tanca, L., Arcelli, F.: Fuzzy XML queries via context-based choice of aggregation. Kybernetika 36, 635–655 (2000)
6. Demaine, E.D., Mozes, S., Rossman, B., Weimann, O.: An optimal decomposition algorithm for tree edit distance. ACM Trans. Algorithms 6, 2:1–2:19 (2009)
7. Demartin, G., Denoyer, L., et al.: Report on INEX 2008. SIGIR Forum 43, 17–36 (2009)
8. Dulucq, S., Touzet, H.: Analysis of tree edit distance algorithms. In: Proceedings of the 14th Annual Symposium of Combinatorial Pattern Matching, pp. 83–95 (2003)
9. Geva, S., Kamps, J., Lethonen, M., Schenkel, R., Thom, J.A., Trotman, A.: Overview of the INEX 2009 Ad Hoc Track. In: Geva, S., Kamps, J., Trotman, A. (eds.) INEX 2009. LNCS, vol. 6203, pp. 4–25. Springer, Heidelberg (2010)
10. Hassler, M., Bouchachia, A.: Searching XML Documents – Preliminary Work. In: Fuhr, N., Lalmas, M., Malik, S., Kazai, G. (eds.) INEX 2005. LNCS, vol. 3977, pp. 119–133. Springer, Heidelberg (2006)
11. Sparck Jones, K.: Index term weighting. Information Storage and Retrieval 9(11), 619–633 (1973)
12. Kazai, G., Lalmas, M.: INEX 2005 Evaluation Measures. In: Fuhr, N., Lalmas, M., Malik, S., Kazai, G. (eds.) INEX 2005. LNCS, vol. 3977, pp. 16–29. Springer, Heidelberg (2006)
13. Klein, P.N.: Computing the Edit-Distance Between Unrooted Ordered Trees. In: Bilardi, G., Pietracaprina, A., Italiano, G.F., Pucci, G. (eds.) ESA 1998. LNCS, vol. 1461, pp. 91–102. Springer, Heidelberg (1998)
14. Levenshtein, V.I.: Binary Codes Capable of Correcting Deletions, Insertions and Reversals. Soviet Physics Doklady 10, 707 (1966)
15. Mass, Y., Mandelbrod, M.: Component Ranking and Automatic query Refinement for XML Retrieval. In: Fuhr, N., Lalmas, M., Malik, S., Szlávik, Z. (eds.) INEX 2004. LNCS, vol. 3493, pp. 73–84. Springer, Heidelberg (2005)

16. Popovici, E., Ménier, G., Marteau, P.-F.: SIRIUS: A Lightweight XML Indexing and Approximate Search System at INEX 2005. In: Fuhr, N., Lalmas, M., Malik, S., Kazai, G. (eds.) INEX 2005. LNCS, vol. 3977, pp. 321–335. Springer, Heidelberg (2006)
17. Rougemont, M., Vieilleribière, A.: Approximate schemas, source-consistency and query answering. J. Intell. Inf. Syst. 31, 127–146 (2008)
18. Tai, K.-C.: The tree-to-tree correction problem. J. ACM 26, 422–433 (1979)
19. Theobald, M., Schenkel, R., Weikum, G.: TopX and XXL at INEX 2005. In: Fuhr, N., Lalmas, M., Malik, S., Kazai, G. (eds.) INEX 2005. LNCS, vol. 3977, pp. 282–295. Springer, Heidelberg (2006)
20. Trotman, A., Sigurbjörnsson, B.: Narrowed Extended XPath I (NEXI). In: Fuhr, N., Lalmas, M., Malik, S., Szlávik, Z. (eds.) INEX 2004. LNCS, vol. 3493, pp. 16–40. Springer, Heidelberg (2005)

An Empirical Study of SLDA for Information Retrieval

Dashun Ma[1], Lan Rao[2], and Ting Wang[1]

[1] College of Computer, National University of Defense Technology
410073 Changsha, Hunan, P.R. China
{madashun,tingwang}@nudt.edu.cn
[2] College of Humanities and Social Sciences, National University of Defense Technology
410073 Changsha, Hunan, P.R. China
raolan21@21cn.com

Abstract. A common limitation of many language modeling approaches is that retrieval scores are mainly based on exact matching of terms in the queries and documents, ignoring the semantic relations among terms. Latent Dirichlet Allocation (LDA) is an approach trying to capture the semantic dependencies among words. However, using as document representation, LDA has no successful applications in information retrieval (IR). In this paper, we propose a single-document-based LDA (SLDA) document model for IR. The proposed work has been evaluated on four TREC collections, which shows that SLDA document modeling method is comparable to the state-of-the-art language modeling approaches, and it's a novel way to use LDA model to improve retrieval performance.

Keywords: Information Retrieval (IR), Language Model, Document Model, Pseudo-Feedback, Latent Dirichlet Allocation (LDA).

1 Introduction

The language modeling approach has been successfully applied to many IR tasks [16]. However, the state-of-the-art language model is a unigram language model because of the computational complexity. Although various heuristics (e.g. proximity [13]) and resources (e.g. WordNet [3]) have been used to improve it, the unigram language model can hardly capture semantic information in an article. For example, considering trying to match the following query in a set of articles -- *pianist*, the unigram language modeling approach intends to find documents that include words "pianist", "piano", or "musician". A sentence such as "Her hands mercilessly pounded the keys, notes cascading into the surrounding stairway." would be likely assigned a poor score, but obviously, this sentence is closely related to *pianist*.

Using topic models for document representation is an interesting and exciting research in IR. The Latent Semantic Indexing (LSI) model [5] and the probabilistic Latent Semantic Indexing (pLSI) model [9], especially the recent Latent Dirichlet Allocation (LDA) model [2], all focus on reducing high-dimensional data vectors to lower-dimensional representations. Compared with the unigram language model, LDA model has several advantages: (1) It creates a topical level between words and

M.V.M. Salem et al. (Eds.): AIRS 2011, LNCS 7097, pp. 84–92, 2011.

documents [2], which gives a better generalization performance.[1] (2) Unlike the unigram language model often using interpolated score, LDA integrates syntax [8], specific information [4] and word burstiness [12] into the document generative process naturally. (3) It offers a method for using semantic information in IR; it is likely to highly rank documents that are related to the topic (even if they don't necessarily contain the exact query terms or their synonyms [4]). However, it is not optimistic about directly using the LDA modeling approach as the document representation in the IR literatures. Wang et al. [14] tested the TNG (topical n-gram model, a variant of LDA) and LDA for IR on the SJMN (San Jose Mercury News) collection, and pointed out that when the two models are directly applied to do ad-hoc retrieval, the performance is very poor (their average precisions are 0.0709 and 0.0438, which are much lower than the state-of-the-art language modeling approaches). We believe that there are two reasons which restrict the application of LDA model in IR. First, good document model does not always bring good retrieval performance [1], other factors (e.g. retrieval method, smoothing strategy, etc.) are also important. Second, training LDA model for a corpus is too inefficient and the corpus-level topics are not fit for each document in the set.

This paper proposes a generative probabilistic model for a document, which tries to deal with the constraints of applying LDA model in IR mentioned above. The model, which we call the single-document-based LDA (SLDA) model, is an extension to the LDA model. In this paper, we further investigate the parameter setting and retrieval method of SLDA, and compare it with the state-of-the-art language modeling approach (the KL-divergence retrieval model [10]) on four typical TREC collections. The experiment results show that (1) the appropriate topic number of SLDA model is less than five; (2) the query likelihood retrieval method is suitable for SLDA model; (3) compared with using the LDA model directly as the document representation, using the SLDA model can obtain better retrieval performance, which competes with the current state-of-the-art approaches.

This paper is organized as follows. The related work is reviewed in Section 2. The SLDA model is defined in Section 3. In Section 4, the experiments are presented. Finally we conclude the work in Section 5.

2   Related Work

Wei and Croft [15] believed that the LDA itself may be too coarse to be used as the representation for IR, so they proposed three ways to integrate the LDA model into the language modeling framework. Their method made the LDA-based document model consistently outperform the cluster-based approach [11] and is close to the Relevance Model.

Chemudugunta et al. [4] proposed a mixture model named SWB, modeling the special words into generative model. Based on the modified AP and FR collections, SWB improves the retrieval performance, and beats the TF-IDF retrieval method.

Wang et al. [14] presented a topical n-gram (TNG) model that automatically determines unigram words and phrases according to the context and assigns mixture

---

[1] The document model evaluated by LDA has lower *perplexity* score than by the unigram language model.

of topics to both individual words and n-gram phrases. Although directly employing TNG gets poor retrieval performance, significant improvements still can be achieved through a combination with the basic query likelihood model.

All the previous approaches used LDA (or modified LDA) as an assistant to language model. Directly using LDA as document representation hurts retrieval performance badly. In our work, we employ SLDA document model alone, and get the comparable results to the state-of-the-art. Our goal is not to argue that SLDA model can take the place of language modeling approach in IR, but to prove that the LDA modeling approach has been underexploited, and show a novel way to use LDA model to improve retrieval performance.

3   SLDA Modeling Framework

In information retrieval, most of existing works on LDA model are set for the corpus and assume that all the documents are consistent with the same probability distribution. Figure 1(a) shows the graphical model representation of the standard LDA model. There are C documents and K is the number of topics. θ represents the document-topic multinomial and ϕ represents the topic-word multinomial. α and β are parameters of Dirichlet priors for θ and ϕ. For each document D, the N_D words are generated by drawing a topic t from the document-topic distribution θ and then drawing a word w from the topic-word distribution ϕ.

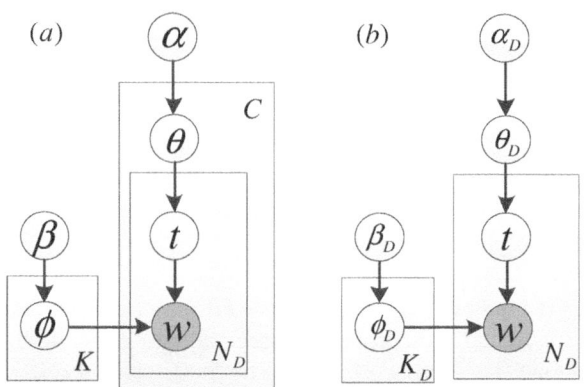

Fig. 1. Graphical models for LDA (a) and SLDA (b)

3.1   SLDA Model

For each document in the collection, corpus-level topics are too coarse. Intuitively, the number of topics in a corpus is much larger than that in single document. It is clear that using a large number of topics to represent the semantics of a document is not unavailable, but the semantic representation of the document would be too generalized, which will make no obvious semantic difference between two documents. What's more, the LDA model of whole corpus implies relationships among words of different documents, but these relationships would be useless or even harmful to the retrieval score. Therefore, we decide to establish LDA model on single document. Figure 1(b) shows the graphical model for SLDA model. It should be noted that SLDA has a similar structure to the LDA model while the difference is that all the parameters are for single document, not corpus.

3.2   Relevant Metrics

Query Likelihood Method. The basic approach for language modeling for IR is the query likelihood method, which takes the maximum likelihood of the document model generating the query terms under the "bag-of-words" assumption as the relevance between query and document. Given a query Q, the retrieval score of a document D is:

$$Score_{QL}(D,Q) = p(Q \mid \varphi_D) \quad . \tag{1}$$

where φ_D is a document model of D (i.e. $p(w \mid D)$ for each word w of D). So, we can use SLDA to create φ_D for each document D, and then compute the relevant score via query likelihood method.

Like language model, SLDA need to take some smoothing strategy to handle the sparseness problem of assignment zero probability to unseen words. In this work, we take the Jelinek-Mercer (fixed coefficient interpolation) smoothing method for SLDA.

$$p(w \mid D) = (1 - \lambda) p_{slda}(w \mid D) + \lambda p_{slda}(w \mid Ref) \quad . \tag{2}$$

where $p_{slda}(w \mid Ref)$ is the reference model, i.e. the maximum likelihood estimate of word w in the background collection. And $p_{slda}(w \mid D)$ represents the SLDA model for a document, its construction process is that: at first, we estimate the parameters θ and ϕ using Gibbs Sampling [6, 7], after get their posterior estimates $\hat{\theta}$ and $\hat{\phi}$, then we calculate the probability of a word in a document as follows,

$$p_{slda}(w \mid D) = p(w \mid D, \hat{\theta}, \hat{\phi}) = \sum_{t=1}^{K} p(w \mid t, \hat{\phi}) p(t \mid \hat{\theta}, D) \quad . \tag{3}$$

Negative KL-divergence and JS-divergence. LDA model can infer a new model on a different set of data using existing model on old dataset, so we get another idea to compute the relevance of document and query. After get each document SLDA model, we use it to infer the query SLDA model, so we get document-topic and query-topic multinomial distributions on the same topic collections, then we can use the Kullback-Leibler (KL) divergence (a non-symmetric measure of the difference between two probability distributions) of these two models to measure how close they are to each other and use their negative distance as a score to rank documents as follows:

$$Score_{nKL}(D,Q) = -D(\theta_Q \parallel \theta_D) = -\sum_{t=1}^{K} p(t \mid Q) \log \frac{p(t \mid Q)}{p(t \mid D)} \quad . \tag{4}$$

where θ_Q represents query-topic distribution and θ_D represents document-topic distribution.

In probability theory and statistics, the Jensen-Shannon (JS) divergence is a popular method of measuring the similarity between two probability distributions. It is a symmetrized and smoothed version of the KL-divergence☐

$$Score_{JS}(D,Q) = (D(\theta_Q \| \theta_M) + D(\theta_D \| \theta_M))/2 \ . \tag{5}$$

where $\theta_M = (\theta_Q + \theta_D)/2$, we also use it as the metric.

4   Experiments

4.1   Dataset and Experiment Setup

The proposed work has been evaluated on four collections from TREC: AP (Associated Press News 1988-1989), FR (Federal Register), SJMN (San Jose Mercury News) and TREC8 (the ad hoc data used in TREC8) with three different TREC topic sets, TOPIC 51-100, TOPIC 101-150 and TOPIC 401-450. Queries are taken from the *title* field of topics. Table 1 shows some basic statistics on these data sets.

Table 1. Statistics of data sets

Collection	query	avgdl	#docs	#qrels
AP	51-100	462	164,597	6101
FR	51-100	1495	45,820	502
SJMN	51-150	408	90,257	4881
TREC8	401-450	480	528,155	4728

All the experiments make use of Lemur toolkit[2] and Gibbs Sampling LDA toolkit[3] for implement. Both the queries and documents are stemmed with the Porter stemmer. Besides stemming, a total of 418 stop words from the Lemur stoplist are removed.

4.2   Comparison of Relevant Metrics

First of all, we compare the relevant metrics on AP dataset. Besides the query likelihood method (QL), negative KL-divergence method (nKL) and JS-divergence method (JS) mentioned above, we add two additional strategies. The motivation is that when we take document-topic and query-topic as two K-dimensional vectors, we can use angle and Euclidean distance of these two vectors (named VA and VD) to measure their relevance. We use Dirichlet priors in the SLDA estimation with

---

[2] http://www.lemurproject.org/
[3] http://gibbslda.sourceforge.net/

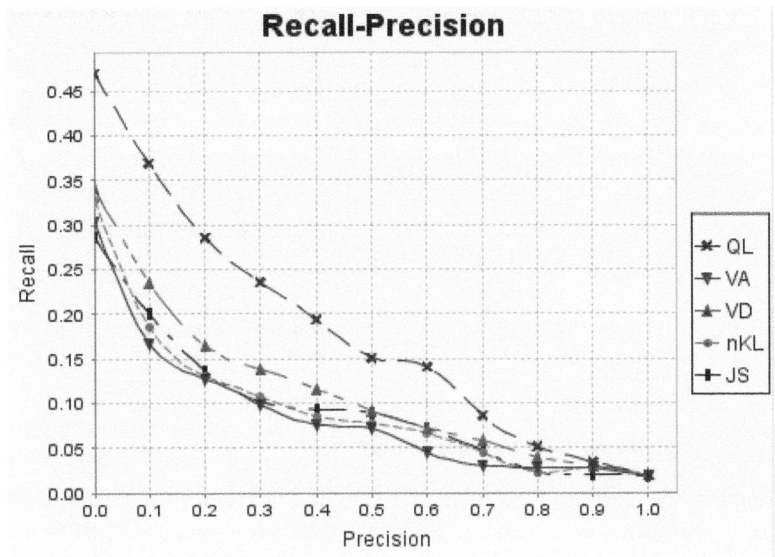

Fig. 2. The comparison of five metrics

$\alpha = 50 / K$ and $\beta = 0.1$, which are the common and default settings in current research [7]. We will try different values of K and Gibbs sampling iterations in following experiments, so that we just fix the number of topics with $K=100$ and set the number of iterations with 2000. The interpolated Recall-Precision curve represents the results in Figure 2, which shows that the QL method significantly outperforms the others, so we use query likelihood as the retrieval strategy in the rest experiments.

4.3   Parameter Settings

There are several parameters that need to be determined in our experiments. For the SLDA model, the number of topics and the number of iterations are very important in topic modeling. At the current stage of our work, we select these parameters through exhaustive search manually. We have tried different iteration numbers with different numbers of topics to see the MAP (Mean Average Precision) values of retrieval results on the AP set. [4]

Table 2 shows the retrieval results on AP with different number of topics (K) and iterations. We find that the performance impact of different numbers of iterations is not very obvious; and the best selection of K is less than five, performance is significantly lower when there are more than 10 topics. Therefore, 50 iterations and $K=3$ are a good tradeoff between accuracy and efficiency.

In order to choose a suitable value of λ on (2), we take a similar experiment process as above on the AP collection and find 0.5 to be the best value for performance.

---

[4] The results on others corpus show the same trends, so we only list results on AP.

Table 2. Results (MAP) on AP with different K and iterations

K	iterations		
	50	200	1000
2	0.2460	0.2461	0.2458
3	0.2467	0.2465	0.2467
5	0.2442	0.2445	0.2421
10	0.2398	0.2372	0.2371
20	0.2327	0.2315	0.2303
30	0.2307	0.2268	0.2266
50	0.2261	0.2257	0.2249

4.4   Comparison with Language Model

We compare the performance of the SLDA model (with query likelihood retrieval method) with the KL divergence language model [10] (noted as LM) with Dirichlet prior smoothing (we set the smoothing parameter to 2000, which are the common settings in current research [17]) on the TREC collections. There are two comparison experiments named Rank and Re-rank. For the "Rank" experiment, we use the two models to retrieve top-ranked 1000 documents on the whole collections for each query and compare their retrieval performance. In the other "Re-rank" experiment, we first use the baseline model (i.e. LM) to retrieve 2000 documents for each query, and organize these initial retrieved documents as a subset of the corpus, then re-rank the subset and use top 1000 documents for all runs to compare performance.

Table 3. The comparison of LM and SLDA retrieval results

Metrics	Rank							
	AP		FR		SJMN		TREC8	
	LM	SLDA	LM	SLDA	LM	SLDA	LM	SLDA
MAP	0.2568	0.2464	0.1127	0.0982	0.1921	0.1755	0.2312	0.2209
P@10	0.3980	0.3780	0.0680	0.0640	0.2760	0.2530	0.4360	0.4160
#rel_ret	3454	3553	232	272	2987	2943	2764	2752
Metrics	Re-rank							
	AP		FR		SJMN		TREC8	
	LM	SLDA	LM	SLDA	LM	SLDA	LM	SLDA
MAP	0.1810	0.1686	0.0708	0.0575	0.1592	0.1182	0.1823	0.1641
P@10	0.3280	0.2700	0.0560	0.0440	0.2630	0.1960	0.2820	0.3100
#rel_ret	2928	2940	153	176	2590	2349	2398	2269

The re-rank step could be considered as using the pseudo-feedback technology in retrieval task. Generally, the language modeling approach using pseudo-feedback documents to re-estimate the query model. However, in our experiment, we employ the query likelihood retrieval model which cannot accommodate the feedback information naturally [16]. Therefore, unlike the traditional method, we use the subset to train the background model (reference model) and re-estimate the document model. In order to facilitate a fair comparison, we also use the pseudo-feedback documents to re-estimate document model in language modeling approach. Table 3 shows the comparison of the two models.

In Table 3, we can observe that on the AP and TREC8, SLDA is comparable to the LM; on the FR and SJMN, SLDA falls a bit behind. Fortunately, on all the collections, the recall of SLDA method is good, even higher than LM approach. Therefore, there are much room for improvement. What's more, on SJMN, SLDA (MAP 0.1755, recall 2943) is more superior to TNG (MAP 0.0709, recall 2450) [14] and LDA (MAP 0.0438, recall 2257), which probably means that, SLDA has an advantage over traditional LDA-like models on document representation in IR.

5   Conclusions and Future Work

Using LDA as an aid can improve the retrieval performance; however, directly using LDA as representation of document hurts the retrieval performance [14, 15]. In this paper, we propose the SLDA model which employs LDA model directly on single document representation. Our experiment results show that SLDA document model is close to the current state-of-the-art language modeling approaches, which is better than traditional LDA models to improve information retrieval performance. We think that the LDA modeling approach has been underexploited, our goal is not to argue that SLDA model can take the place of language modeling approach in IR, but to show a novel way to use LDA model to improve the retrieval performance.

We further study the parameter settings and retrieval model of SLDA. Experiment results on four TREC test collections show that the appropriate topic number of SLDA model is less than five and the query likelihood retrieval method is suitable.

Our work can be extended in several directions: First, although we have found empirically that document-level topics are better than corpus-level topics for document representation, how to determine the number of topics is still a very important problem. Second, for different documents, setting different number of topics instead of a fixed number of topics for all the docs is an interesting direction. Finally, it is challenging to develop a method to define the Gibbs Sampling iterations.

Acknowledgments. The research is supported by National Natural Science Foundation of China (60873097, 60933005). We thank the anonymous reviewers for their useful comments.

References

1. Azzopardi, L., Girolami, M., van Risjbergen, K.: Investigating the Relationship between Language Model Perplexity and IR Precision-Recall Measures. In: Proc. of 26th SIGIR, pp. 367–370 (2003)

2. Blei, M., Ng, A., Jordan, M.: Latent Dirichlet Allocation. Journal of Machine Learning Research 3, 993–1022 (2003)
3. Cao, G.H., Nie, J.Y., Bai, J.: Integrating Word Relationships into Language Models. In: Proc. of 28th SIGIR, pp. 298–305 (2005)
4. Chemudugunta, C., Smyth, P., Steyvers, M.: Modeling General and Specific Aspects of Documents with a Probabilistic Topic Model. In: Proc. of 19th NIPS, pp. 241–248 (2006)
5. Deerwester, S., Dumais, S.T., Furnas, G.W., Landauer, T.K., Harshman, R.: Indexing by latent semantic analysis. Journal of the American Society for Information Science 41(6), 391–407 (1990)
6. Geman, S., Geman, D.: Stochastic Relaxation, Gibbs Distributions, and the Bayesian Restoration of Images. IEEE Transactions Pattern Analysis and Machine Intelligence 6, 721–741 (1984)
7. Griffiths, T.L., Steyvers, M.: Finding scientific topics. Proc. of the National Academy of Sciences 101(suppl. 1), 5228–5235 (2004)
8. Griffiths, T., Steyvers, M., Blei, D., Tenenbaum, J.: Integrating topics and syntax. In: Proc. of 17th NIPS, pp. 537–544 (2005)
9. Hofmann, T.: Probabilistic latent semantic indexing. In: Proc. of 22nd SIGIR, pp. 35–44 (1999)
10. Lafferty, J.D., Zhai, C.X.: Document language models, query models, and risk minimization for information retrieval. In: Proc. 24th of SIGIR, pp. 111–119 (2001)
11. Liu, X., Croft, W.B.: Cluster-based retrieval using language models. In: Proc. of 27th SIGIR, pp. 186–193 (2004)
12. Madsen, R.E., Kauchak, D., Elkan, C.: Modeling Word Burstiness Using the Distribution. In: Proc. of 22nd ICML, pp. 298–305 (2005)
13. Tao, T., Zhai, C.X.: An Exploration of Proximity Measures in Information Retrieval. In: Proc. of 30th SIGIR, pp. 295–302 (2007)
14. Wang, X.R., McCallum, A., Wei, X.: Topical N-grams: Phrase and Topic Discovery, with an Application to Information Retrieval. In: Proc. of IEEE 7th ICDM, pp. 697–702 (2007)
15. Wei, X., Croft, W.B.: LDA-Based Document Models for Ad-hoc Retrieval. In: Proc. of 29th SIGIR, pp. 178–185 (2006)
16. Zhai, C.X.: Statistical Language Models for Information Retrieval: A Critical Review. Foundations and Trends in Information Retrieval 2(3), 137–213 (2008)
17. Zhai, C.X., Lafferty, J.: A Study of Smoothing Methods for Language Models Applied to Ad Hoc Information Retrieval. In: Proc. of 24th SIGIR, pp. 334–342 (2001)

Learning to Rank by Optimizing Expected Reciprocal Rank

Ping Zhang[*], Hongfei Lin, Yuan Lin, and Jiajin Wu

Information Retrieval Laboratory, Dalian University of Technology,
Dalian 116024
{pingzhang,yuanlin,wujiajin}@mail.dlut.edu.cn, hflin@dlut.edu.cn

Abstract. Learning to rank is one of the most hot research areas in information retrieval, among which listwise approach is an important research direction and the methods that directly optimizing evaluation metrics in listwise approach have been used for optimizing some important ranking evaluation metrics, such as MAP, NDCG and etc. In this paper, the structural SVMs method is employed to optimize the Expected Reciprocal Rank(ERR) criterion which is named SVMERR for short. It is compared with state-of-the-art algorithms. Experimental results show that SVMERR outperforms other methods on OHSUMED dataset and TD2003 dataset, which also indicate that optimizing ERR criterion could improve the ranking performance.

Keywords: Information Retrieval, Learning to rank, Listwise, ERR.

1 Introduction

Learning to rank is one of the popular research fields in Information Retrieval. At present, learning to rank has been divided into three categories in general, they are the pointwise approach that is based on single document with respect to a given query, the pairwise approach that is based on document preference pair according to a given query and the listwise approach that is based on document list w.r.t a given query[1].

The pointwise approach takes the feature vector of each single document as input instance, and takes the similarity between document and query as output. Here, ranking model could be regarded as regression model or classification model. The representative algorithms are McRank[2] and Pranking[3]. The disadvantage of pointwise approach is that the relative order between documents cannot be considered in learning process. Pairwise approach takes the document pair as input instance, the representative algorithms are RankNet[4] and Ranking SVM[5]. The input of listwise approach contains a group of documents that are relevant to the given query and the

---

[*] This work is supported by grant from the Natural Science Foundation of China (No.60673039 and 60973068), the National High Tech Research and Development Plan of China (No.2006AA01Z151), National Social Science Foundation of China (No.08BTQ025), the Project Sponsored by the Scientific Research Foundation for the Returned Overseas Chinese Scholars, State Education Ministry and The Research Fund for the Doctoral Program of Higher Education (No.20090041110002).

M.V.M. Salem et al. (Eds.): AIRS 2011, LNCS 7097, pp. 93–102, 2011.

output is a ranked list. Listwise approach can be divided into two categories, one of which measures the difference between predict labels and ground truth labels using loss function, such as ListNet[6] and ListMLE[7], the other one regards loss function as the upper bound of evaluation metric, such as SVMMAP[8].

The experimental results on Letor3.0 indicate that listwise approach generally performs better than pointwise approach and pairwise approach[1]. Hence listwise approach could improve ranking performance. DCG metric is an important measure for evaluating web retrieval results, which assumes that whether user chooses some document in a ranked list only depends on position information of the document. Nevertheless, the assumption ignores the fact that whether user chooses i-th document in ranked list also depends on other factors. Therefore a new evaluation metric named Expected Reciprocal Rank(ERR)[9] is proposed, which is employed for results evaluation in learning to rank challenge organized by Yahoo! at 2010. Structural SVMs[10] could find the global optimal solution, hence this paper proposes SVMERR method for the above reasons, which employs structural SVMs to optimize ERR metric. We expect this method may improve the retrieval performance.

This paper is organized as follows. It starts with the introduction of the new evaluation metric ERR in Section 2. Section 3 describes our approach named SVMERR. We make the experiments to test the performance of our method in Section 4. Section 5 gives the conclusion and future work.

2   Ranking Evaluation Criteria

2.1   Expected Reciprocal Rank (ERR)

ERR is an improvement of DCG metric, DCG metric assumes that whether user chooses some document in ranked list only depends on position information of the document. However, in real scenario of a search engine, the browsing behavior of a user is determined by diverse factors. Whether user choses i-th document in ranked list also depends on the satisfaction that user thinks of the documents less than i, this model is called cascade model, which can model the user's real browsing behaviour, and ERR metric is a ranking measure that based on cascade model.

For the given query q, we assumes that document i satisfies the user with probablity R_i, then for a given set of R_i, we assumes that user browses ranked list from top to bottom, the likelihood for which the user is satisfied and stops at position r is calculated as follows:

$$\prod_{i=1}^{r-1} (1 - R_i)R_r \tag{1}$$

which denotes the probability that the user is not satisfied with the first r-1 documents and is satisfied with the r-th document.

The satisfied probability R_i could be estimated by maximum likelihood on the click logs, and also could be set as a function of documents' relevance degree. Let g_i be the relevance degree of the i-th document, then:

$$R_i = R(g_i) \tag{2}$$

where R is a mapping from relevance degree to relevance probability. Similar to the gain function of DCG, R could be defined as follows:

$$R(g) = \frac{2^g - 1}{2^{g_{max}}}, \qquad g \in \{0, ..., g_{max}\} \qquad (3)$$

where g is the relevance degree, g_{max} denotes that the document is extremely relevant. For example, if a 5 point scale is used, then g equals to 0 denotes that the document is irrelevant, and the most relevant document could be marked as g equals to 4. Equation (3) indicates that the probability that the user will be satisfied and stop browsing is big when the relevance degree of document is high.

Give a utility function φ, which is similar with the discount function in DCG, should meet the condition that $\varphi(1)$ equals to 1, and $\varphi(r)$ tends to 0 when position r tends to infinity. Given function φ, a cascade based metric is the expectation of $\varphi(r)$, the variable r is the ranking position where the user finds the document that he want.

Expected Reciprocal Rank (ERR) is defined as a cascade based metric that using function $\varphi(r) = 1/r$:

$$ERR = \sum_{r=1}^{n} \frac{1}{r} P_r \qquad (4)$$

where P_r stands for the probability that the user stops at position r, and n denotes the number of documents in the ranked list. P_r is defined in equation (1). ERR criterion is generated by taking equation (1) into equation (4):

$$ERR = \sum_{r=1}^{n} \frac{1}{r} \prod_{i=1}^{r-1} (1 - R_i) R_r \qquad (5)$$

Average the $ERR(q)$ over all queries, we could get the final ERR value.

3   Optimize ERR Metric Using Structural SVMs

3.1   Structural SVMs

Ranking is composed of three parts, the first part is ranking model w, and the second part is a feature map $\varphi(x_k, y)$ that maps ranked list x_k and relevance labels y to a d-dimensional space vector, in which d is the number of feature vectors. A score for rank y is noted as $w^T \varphi(x_k, y)$, through which we could evaluate the merits of rank y. Given the document list for a query, we assume that the ideal ranking list is y_k (i.e. it places all relevant documents at top ranks and irrelevant documents after that). y is the ranked list that generated by model w, and the difference between y_k and y is the third part in ranking, which is called loss function and noted as $\Delta(y_k, y)$. We use the structural SVMs to learn a ranking model w, the optimization problem is described as follows:

$$\min_{\omega, \xi \geq 0} \frac{1}{2} \| \omega \|^2 + \frac{C}{m} \sum_{k=1}^{m} \xi_k$$

$$s.t. \quad \forall k, \forall y \in Y \setminus y_k : \qquad (6)$$

$$\omega^T \varphi(x_k, y_k) \geq \omega^T \varphi(x_k, y) + \Delta(y_k, y) - \xi_k$$

where parameter C is a penalty factor. Parameter m is the number of queries in training dataset, w denotes ranking model. If predicted ranked list y is bad, then $\Delta(y_k, y)$ will be very large, to meet the constraint in equation (6), the slack variable ξ_k also needs to be increased.

3.2   Optimize ERR Metric Using Structural SVMs

For any query k, the number of possible predicted ranked lists y is $n!$, where n is the number of documents corresponding to k, which will increase the complexity of finding solution. For this reason, cutting plane algorithm is employed to optimize ERR metric, which is described in Table 1.

Table 1. Cutting plane algorithm

1: Input: $(x_k,y_k),k \in [1,n]$, C and tolerance ε
2: Set W_i to zero vector, for all $i = 1, \dots, n$
3: **Repeat**
   3.1: **for** $i = 1, \dots, n$ do
      3.1.1: compute $H(y, w)=\Delta(y_k, y)+ w^T\varphi(x_k, y)-w^T\varphi(x_k, y_i)$
      3.1.2: find $y^*=\text{argmax}_y \in Y \{H(y,w)\}$
      3.1.3: find $\xi_i =\max\{0,\max_{y' \in w_i} H(y',w)\}$
      3.1.4: if $H(y^*,w) >\xi_i+\varepsilon$ then
          $W_i = W_i \cup (y^*)$
       optimize equation (12) on $W=\cup_i W_i$
Until no W_i has changed during iteration

There are four problems to solve when structural SVMs is employed to optimize ERR criteria. The first one is the selection of feature map $\varphi(x_k, y)$; the second problem is how to find the most violated constraint y^* that maximums the $H(y,w)$; the third one is how to define the loss function of ERR during training process; and the final problem to solve is the definition of variation quantity generated by exchanging documents. These will be explained in the following subsections.

In this paper, we choose the partial order feature map which is defined as follows:

$$\varphi(x_k, y) = \frac{1}{|C^+||C^-|}\sum_{i=1}^{|C^+|}\sum_{j=1}^{|C^-|} y_{ij}(x_{ki} - x_{kj}) \tag{7}$$

where $|C^+|$ is the number of relevant documents in document list x_k, and $|C^-|$ is the number of irrelevant documents; for any $y \in Y$, $y_{ij}=1$ if relevant document d_i is ranked ahead of irrelevant document d_j, $y_{ij}=-1$ otherwise. x_{ki} denotes the feature vector of document d_i in the document list corresponding to query k.

For each iteration of cutting plane algorithm, there is a working set. For the k-th query, the ranked list y that maximizes $H(x_k, y; w)$ need to be find, where $y \neq y_k$; then the ranked list y will be added to the working set, and equation (6) will be optimized on the new working set. Its mathematical expression is listed below:

$$\arg\max_{y \ne y_k^*} H(x_k, y; \omega) = \arg\max_{y \ne y_k^*}[\Delta(y_k, y) + \omega^T \varphi(x_k, y) - \omega^T \varphi(x_k, y_k)] \qquad (8)$$

Since for the given query k, the score $w^T\varphi(x_k, y_k)$ of ideal ranked list y_k is constant with respect to y, equation (8) or equivalently,

$$\arg\max_{y \ne y_k^*} H(x_k, y; \omega) = \arg\max_{y \ne y_k^*}[\Delta(y_k^*, y) + \omega^T \varphi(x_k, y)] \qquad (9)$$

where $w^T\varphi(x_k, y)$ is the score of ranked list y, according to the feature map in 3.2.1, the $w^T\varphi(x_k, y)$ in equation (9) is:

$$\omega^T \varphi(x_k, y) = \frac{1}{|C^+||C^-|}\sum_{i=1}^{|C^+|}\sum_{j=1}^{|C^-|} y_{ij}(s_i - s_j) \qquad (10)$$

where s_i equals to $w^T x_{ki}$, which denotes the score of document i in the ranked list corresponding to query k. The loss function of ERR criterion $\Delta(y_k, y)$ in equation (9) is defined as follows:

$$\Delta(y_k, y)_{err} = 1 - ERR(y) = 1 - \sum_{r=1}^{n}\frac{1}{r}\prod_{i=1}^{r-1}(1 - R_i)R_r \qquad (11)$$

For given query k, the value of $H(x_k, y; w)$ will be changed when we exchange two documents. Fist, the relevant documents and irrelevant documents will be both sorted in descending order by $w^T x_{ki}$ and the ideal ranked list is generated (i.e. all relevant documents are ahead of irrelevant documents).Then we insert each irrelevant document into relevant documents list and find the positions that maximize $H(x_k, y; w)$, so we get the most violated constraint y. In the following, we will consider the situation that the i-th relevant document and j-th irrelevant document are swapped with each other in query k. The $\delta_{exchange}$ denotes the variation quantity generated by exchanging the i-th relevant document and the j-th irrelevant document, whose definition is:

$$\delta_{exchange} = \delta[\omega^T \varphi(x_k, y)] + \delta[\Delta(y_k, y)] \qquad (12)$$

$\delta_{exchange}$ is made up of two parts, one of which is generated by the change of feature map score $w^T\varphi(x_k, y)$, the other is generated by the change of loss function $\Delta(y_k, y)_{err}$. Their definitions are listed in equation (13) and equation (14).

$$\delta[\omega^T \varphi(x_k, y)] = -\frac{2}{|C^+||C^-|}\omega^T(x_{ki} - x_{kj}) \qquad (13)$$

$$\delta[\Delta(y_k, y)] = \frac{1}{2i}\prod_{t=1}^{j-1}(1 - R_t) - \frac{1}{2j}\prod_{t=1}^{j-1}(1 - R_t) - \frac{1}{2}\sum_{r=i+1}^{j-1}\frac{1}{r}\prod_{t=1}^{r-1}(1 - R_t)R_r \quad (t \ne i) \qquad (14)$$

Equation (14) denotes the variation quantity generated by the loss function of ERR $\Delta(y_k, y)_{err}$. Comparing with the change generated by DCG, equation (14) could reflect the influence generated by exchanging documents well, because ERR metric considers that the documents ranked in front of document i will influence user's choice behavior. In equation (14), variable i and j are the ranking positions of document i and document j in the ranked list. R_t is the probability that user satisfied with the t-th document, and R_r is the probability that user satisfied with the r-th document (reference to equation (1)).

After we defined the loss function of ERR and the variation quantity generated by exchanging documents, the most violated constraint y is generated by the method of exchanging documents. Then ranking model w is computed by using cutting plane algorithm. Further, the documents for query q will be scored by model w and be ranked by scores in descending order.

4   Experiments

We use OHSUMED and TD2003 dataset in LETOR3.0 released by MSRA[12] as our experimental data collection. The exact set of features is the set of standard features proposed by LETOR3.0. OHSUMED dataset contains 106 queries, 11303 irrelevant documents and 4837 relevant documents and TD2003 dataset contains 50 queries, 516 relevant documents and 48655 irrelevant documents. The two dataset are both evenly divided into five groups for five times cross validation. We perform the experiments on these dataset, and the results are compared with Regression (pointwise approach), Ranking SVM (pairwise approach), SVMMAP, SVMNDCG and ListNet (listwise approach).

4.1   Experiment on OHSUMED Data

Figure 1 shows the precision curve for each algorithm on OHSUMED dataset, from which we could observe that ListNet algorithm gets the best result at P@1and P@3; SVMNDCG algorithm get the best result at P@2, but SVMERR algorithm does not perform well and only outperforms others at the P@5 to P@10. From Figure 2, we could see that at NDCG metric level, SVMERR performs normally compared with other algorithms.

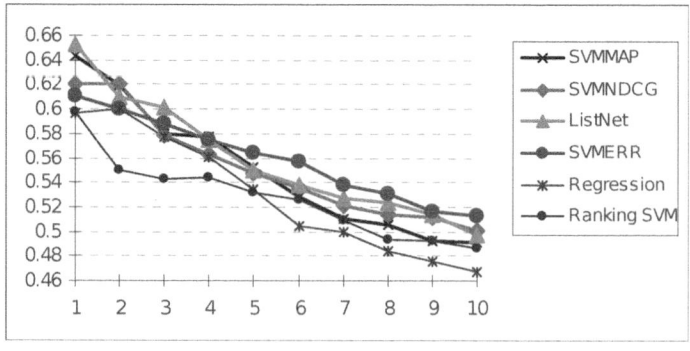

Fig. 1. Average P@k Curve on OHSUMED dataset

Table 2 shows the MAP value for each algorithm, and table 3 presents the compared result on ERR value between SVMNDCG and SVMERR.

From table 2, we could see that SVMERR does not make much improvement on MAP score level compared with other methods, and table 3 indicates that SVMNDCG method outperforms SVMERR on ERR level.

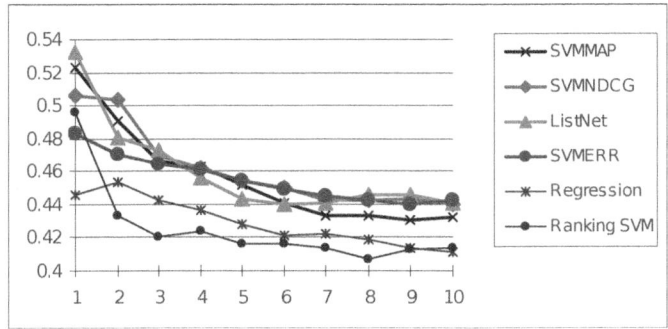

Fig. 2. Average NDCG@k Curve on OHSUMED dataset

Table 2. Comparison with other methods by MAP

Algorithms	Fold1	Fold2	Fold3	Fold4	Fold5	avgMAP
SVMMAP	0.3423	0.4540	0.4620	**0.5179**	0.4500	0.4453
SVMNDCG	**0.3492**	**0.4730**	0.4610	0.5112	0.4487	0.4485
ListNet	0.3464	0.4500	0.4610	0.5106	0.4611	0.4457
Regression	0.2979	0.4302	0.4398	0.4978	0.4442	0.4220
Ranking SVM	0.3038	0.4468	**0.4648**	0.4990	0.4528	0.4334
SVMERR	0.3483	0.4500	0.4620	0.5121	**0.4710**	**0.4487**

Table 3. Comparison with NDCG method by ERR

Algorithms	Fold1	Fold2	Fold3	Fold4	Fold5	avgERR
SVMNDCG	0.4239	0.5190	0.5300	0.5052	0.5205	0.4995
SVMERR	**0.4251**	0.4650	**0.5350**	**0.5157**	**0.5209**	0.4923

4.2   Experiment on TD2003 Data

Figure 3 presents the precision curve for each algorithm, where x-axis denotes the position k and Y-axis denotes the precision for each position k, and Figure 4 is the NDCG@k curve for each algorithm, where x-axis also denotes the position k and Y-axis denotes the NDCG@k value.

From Figure 3 and Figure 4, we could observe that the performance of SVMERR is better than others at P@1, P@2 as well as NDCG@1, NDCG@3 and so on, because the metric ERR, which is based on cascade model, takes user's browsing behavior into consideration, and exactly corresponds with the real retrieval situation.

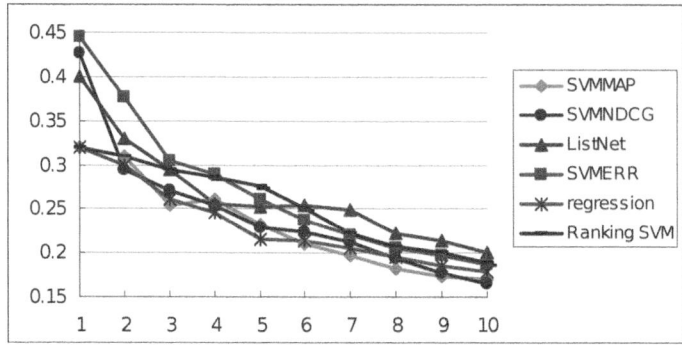

Fig. 3. Average P@k curve on TD2003 dataset

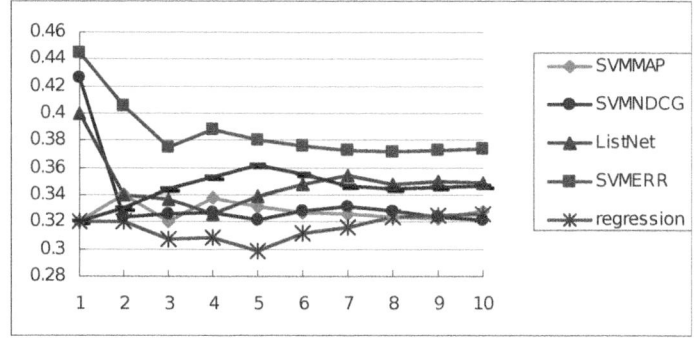

Fig. 4. Average NDCG@k Curve on TD2003 dataset

Table 4 shows the MAP values of each algorithm at each fold in TD2003 dataset, and table 5 presents the comparison of SVMERR method and SVMNDCG method on ERR metric.

Table 4 presents that at MAP metric' level, SVMERR obtains the best result on Fold1 and Fold5 compared with other methods and make a improvement at average MAP value. Table 5 shows that compared with SVMNDCG, SVMERR gets the best value on Fold2, Fold3 and Fold5, and the average ERR value outperforms that of SVMNDCG.

Table 4. Comparison with other methods by MAP

Algorithms	Fold1	Fold2	Fold3	Fold4	Fold5	avgMAP
SVMMAP	0.1719	0.2369	0.3421	0.2760	0.1958	0.2445
SVMNDCG	0.1822	0.2206	**0.4463**	**0.2979**	0.1685	0.2631
ListNet	0.1925	**0.3249**	0.3813	0.2755	0.2023	0.2753
Regression	0.1262	0.3009	0.2665	0.2658	0.2450	0.2409
Ranking SVM	0.1637	0.2576	0.4079	0.2356	0.2490	0.2628
SVMERR	**0.2020**	0.2888	0.3704	0.2753	**0.2873**	**0.2847**

Table 5. Comparison with SVMNDCG method by ERR

Algorithms	Fold1	Fold2	Fold3	Fold4	Fold5	avgERR
SVMNDCG	**0.2740**	0.3351	0.3606	**0.2911**	0.3427	0.3207
SVMERR	0.2489	**0.3741**	**0.3699**	0.2735	**0.4595**	**0.3451**

In general, SVMERR method performs normally on OHSUMED dataset but performs better on TD2003 dataset. We analyse the reason and think that the most likely reason for this phenomenon is the scale of OHSUMED dataset is small, by contrast, TD2003 dataset's scale is large. Therefore experimental result on TD2003 is more obvious. Anyway, SVMERR method definitely improves the performance in information retrieval.

4.3   Experimental Analysis

We analysed the experimental results and concluded the reasons as follows.

First of all, the input of listwise approach contains a group of document that are relevant with given query and the output is a ranked list, however, pointwise approach only considers single document with respect to a given query, which is more like a traditional classification or regression problem, and pairwise approach concerns with document preference pair according to a given query. Both of which couldn't consider the whole document list, so in general, listwise approach outperforms the other two approaches. Therefore the experimental results of the algorithms that belongs to listwise approach are good, such as SVMMAP, SVMNDCG, ListNet as well as SVMERR proposed in this paper.

Secondly, DCG metric, which is an evaluation measure of multi-level relevance degree, could model the user's browsing behaviour better. Hence the SVMNDCG method performs well.

Finally, the DCG metric, which is based on position model, does not consider the relevance of documents above the document of interest. However, ERR metric considers this factor. In a way, ERR metric is an improved version of DCG, and SVMERR is the method that optimizes ERR metric, therefore SVMERR performs better than other methods.

Optimizing ERR metric brings the improvement of ranking performance, i.e. making the ranked list better and increasing the user's satisfaction. Nevertheless, SVMERR improves the performance at the expense of time. More calculation will be produced when calculating ERR loss function $\Delta(y_k, y)_{err}$. So the real systems should balance the relationship between effectiveness and efficiency.

5   Conclusion

In this paper, we propose an algorithm named SVMERR, which adopts structural SVMs to optimize ERR metric. We perform the experiments on LETOR3.0 dataset. Experimental results show that optimizing ERR metric could help improve ranking performance. Our contribution contains two parts as follows. Firstly, structural SVMs is employed to optimize ERR metric; secondly, we define the loss function of ERR metric and the variation quantity generated by exchanging documents. Our future work will focus on optimizing other evaluation measures.

References

[1] Liu, T.Y.: Learning to Rank for Information Retrieval. Foundations and Trends in Information Retrieval 3(3), 225–331 (2009)

[2] Li, P., Burges, C., Wu, Q.: McRank: Learning to rank using multiple classification and gradient boosting. In: Proceedings of the 21st Annual Conference on Neural Information Processing Systems (NIPS 2007), Vancouver, British Columbia, Canada, pp. 845–852 (2007)

[3] Crammer, K., Singer, Y.: Pranking with ranking. In: Proceedings of the 14th Annual Conference on Neural Information Processing Systems (NIPS 2002), Vancouver, British Columbia, Canada, pp. 641–647 (2001)

[4] Burges, C., Shaked, T., Renshaw, E.: Learning to Rank using Gradient Descent. In: Proceedings of the 22nd International Conference on Machine Learning (ICML 2005), Bonn, Germany, pp. 89–96 (2005)

[5] Joachims, T.: Optimizing Search Engines using Clickthrough Data. In: Proceedings of the 8th ACM SIGKDD International Conference on Knowledge Discovery and Data Mining (KDD 2002), Edmonton, Alberta, Canada, pp. 133–142 (2002)

[6] Cao, Z., Qin, T., Liu, T.Y.: Learning to rank: from pairwise approach to listwise approach. In: Proceedings of the 24th International Conference (ICML 2007), Corvalis, Oregon, USA, pp. 129–136 (2007)

[7] Xia, F., Liu, T.Y., Wang, J.: Listwise approach to learning to rank: theory and algorithm. In: Proceedings of the 25th International Conference (ICML 2008), Helsinki, Finland, pp. 1192–1199 (2008)

[8] Yue, Y.S., Finley, T., Radlinski, F.: A support vector method for optimizing average precision. In: Proceedings of the 30th Annual International ACM SIGIR Conference on Research and Development in Information Retrieval (SIGIR 2007), Amsterdam, The Netherlands, pp. 271–278 (2007)

[9] Chapelle, O., Metzler, D., Zhang, Y.: Expected Reciprocal Rank for Graded Relevance. In: Proceedings of the 18th ACM Conference on Information and Knowledge Management (CIKM 2009), Hong Kong, China, pp. 621–630 (2009)

[10] Tsochantaridis, I., Hofmann, T., Joachims, T.: Large margin methods for structured and interdependent output variables. Journal of Machine Learning Research (JMLR) 6(9), 1453–1484 (2005)

[11] Joachims, T.: A support vector method for multivariate performance measures. In: Proceedings of the 22nd International Conference (ICML 2005), Bonn, Germany, pp. 377–384 (2005)

[12] Qin, T., Liu, T.Y., Xu, J.: LETOR: Benchmark Collection for Research on Learning to Rank for Information Retrieval. In: SIGIR 2007 Workshop on Learning to Rank for Information Retrieval (LR4IR 2007), Amsterdam, The Netherlands, pp. 3–10 (2007)

[13] Chakrabarti, S., Khanna, R., Sawant, U.: Structured learning for non-smooth ranking losses. In: Proceedings of the 14th ACM SIGKDD International Conference on Knowledge Discovery and Data Mining (KDD 2008), Las Vegas, Nevada, USA, pp. 88–96 (2008)

Information Retrieval Strategies for Digitized Handwritten Medieval Documents

Nada Naji and Jacques Savoy

University of Neuchatel, Computer Science Department, Rue Emile-Argand 11,
2000 Neuchatel, Switzerland
{Nada.Naji,Jacques.Savoy}@unine.ch

Abstract. This paper describes and evaluates different IR models and search strategies for digitized manuscripts. Written during the thirteenth century, these manuscripts were digitized using an imperfect recognition system with a word error rate of around 6%. Having access to the internal representation during the recognition stage, we were able to produce four automatic transcriptions, each introducing some form of spelling correction as an attempt to improve the retrieval effectiveness. We evaluated the retrieval effectiveness for each of these versions using three text representations combined with five IR models, three stemming strategies and two query formulations. We employed a manually-transcribed error-free version to define the ground-truth. Based on our experiments, we conclude that taking account of the single best recognition word or all possible top-k recognition alternatives does not provide the best performance. Selecting all possible words each having a log-likelihood close to the best alternative yields the best text surrogate. Within this representation, different retrieval strategies tend to produce similar performance levels.

Keywords: Medieval manuscripts, IR with noisy text, OCR, handwritten text IR, Middle High German, text recognition, digital libraries.

1   Introduction

During the last decade, there has been a growing interest in building large digital libraries with the largest projects receiving national (e.g., Gallica) or international support (The European Library, or Europeana). The main motivation behind such projects is the preservation of our cultural heritage and allowing a worldwide user-friendly access to this valuable material. From a technical perspective, handling old historical documents and in particular medieval manuscripts represents a difficult task. During the image processing and text recognition phases, we faced with the artifacts surrounding the handwritten text, ink bleeding, holes and stitches on parchments, etc. Our main objective is however to perform effective searches on the transcriptions generated from these phases. Unfortunately, it is almost impossible to obtain a perfect digital transcription of the original documents, meaning that we must accept the fact that recognition errors will always reside. The level of the error rate depends on various factors such as the accuracy of the recognition system, the quality

M.V.M. Salem et al. (Eds.): AIRS 2011, LNCS 7097, pp. 103–114, 2011.

of the contrast between the background and the ink, the regularity of the handwriting, etc., keeping in mind that a high error rate may result in low retrieval effectiveness.

The medieval manuscripts were written in a non-standardized spelling, this aspect introduced an additional challenge. By inspecting some passages we can easily find different spellings referring to the same entity. This issue will reduce the retrieval effectiveness. Moreover, the grammar used in medieval languages was clearly different once compared to that of our modern days, thus allowing more flexibility to the writer, varying from one region to another, or even from one writer to another residing in the same region. In this context, our research group is participating in the HisDoc[1] project wherein a large set of medieval handwritten manuscripts has been carefully digitized and automatically transcribed (work done at the University of Fribourg and the University of Bern respectively).

The rest of this paper is structured as follows. Section 2 provides an overview of related work while Section 3 describes the corpora used in our experiments. Section 4 outlines the indexing strategies and describes the selected IR models. Section 5 evaluates and analyzes the results obtained from applying these IR strategies.

2   Related Work

Performing effective retrieval on historical manuscripts is still an unsolved issue despite the increasing need of museums, libraries, and even the general public for easy access to historical manuscripts [1], [2]. Many studies and experiments, in both commercial and academic frameworks, have tackled the task of retrieving noisy text.

The first challenge is having to deal with the effectiveness loss caused by imperfect character recognition [3]. In this context, TREC-5 (confusion track) constitutes a useful starting point [4]. During this evaluation campaign, three different versions of a corpus written in English were made available. The first and clean version of the corpus forms the baseline, the second and third versions are the output of scanning the printed corpus having character error rates of around 5% and 20% respectively. To measure the retrieval effectiveness with the presence of noisy text, the TREC evaluation campaign chose the MRR (mean reciprocal rank) metric, which is based on the inverse of the rank of the first relevant item retrieved [5]. Such a measure reflects the concern of users wishing to find one or a few good responses to any given request.

Using this measure, the best system in TREC-5 [6] had an MRR of 0.7353 for the clean corpus, and an MRR of 0.5737 (relative difference of -22%) when facing with an error rate of 5%. For the corpus having a 20% character error rate, the MRR value was as low as 0.4978 (-32%). Similar degradation levels were obtained by other participants [4]. However, these results should be moderated, given that Tagva *et al.* [7] had shown that when using high-definition images and a high quality OCR system, the error rate could be limited to around 2%.

Dealing with medieval handwritten manuscripts instead of typed text as well as colored paper or writing media with stains, holes and stitches instead of a

---

[1] http://hisdoc.unine.ch

high-contrast black-and-white would certainly generate an error rate higher than the estimated 2%. Previous studies [1], [2], [3], [4] were also limited to the English language, with, basically, documents dating to the last decades of the eighteenth century (e.g. George Washington manuscripts [1], [2]). Working with older languages means that we need to face with both spelling and grammar variations. In such cases, various approaches have been suggested to deal with spelling variability as for example the plays and poems in Early Modern English (1580 – 1640) [8], [9]. Shakespeare for instance had his name spelled as "Shakper", "Shakspe", "Shaksper", "Shakspere" or "Shakspeare", but never as the current spelling. The German language is known for its compound construction (e.g., worldwide, handgun). For instance, the word *Kühlschrank* (refrigerator) is made up of two words, namely *kühl* (cold) and *Schrank* (cupboard). According to CLEF evaluation campaigns [10], splitting compound words has shown to be effective for IR purposes as the same concept can be expressed using different forms (e.g., *Computersicherheit* vs. *Sicherheit mit Computern*). It is worth mentioning that compounding was not used as frequently in Middle High German as it is in modern German. The percentage of compound nouns to nouns in the first half of the thirteenth century was around 6.8%, this ratio increased over the centuries reaching 25.2% in the modern German language [11].

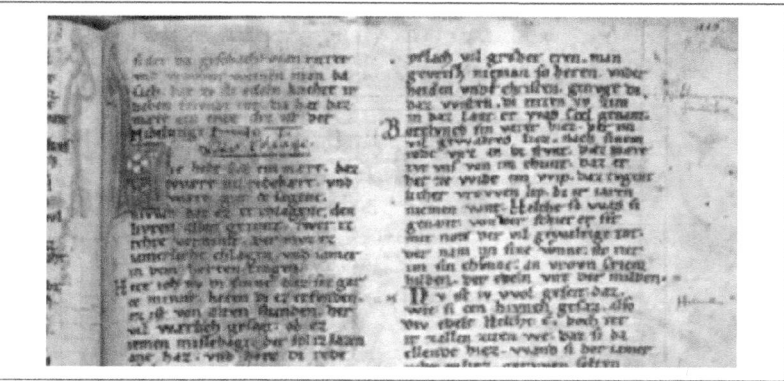

Fig. 1. A small excerpt of the Parzival manuscript

3   Evaluation Corpora and Methodology

The corpus used in our experiments is based on a well-known medieval epic poem called Parzival and is attributed to Wolfram von Eschenbach. The first version dates to the first quarter of the thirteenth century and was written in the Middle High German language. Currently, we can find several versions (with variations) but the St. Gall collegiate library *cod.* 857 is the one used for experimental evaluation [12]. An excerpt is shown in Figure 1. An error-free transcription of the poem was created manually and made available by experts. This version forms our ground-truth (GT) text and was used to assess the performance levels in our experiments.

3.1   Handwritten Recognition

In HisDoc, the manuscripts have been transcribed to the digital format using a Hidden Markov Model whose basic features are described in [12]. This recognition system is based on the closed vocabulary assumption implying that each word in the test set is known or has already appeared in the training set. This recognition system has evolved in terms of performance and achieved a word-accuracy close to 94%. Thus, the produced transcription has a word-error rate of around 6% and represents the noisy version of Parzival employed in our IR experiments. Each image corresponding to a whole page has been automatically subdivided into smaller images, each representing a single line (verse) of the poem. The line images are then processed for recognition during which the system determines, for each word, the best set of possible recognitions. Instead of being limited to a single candidate, the recognition system provides a set of seven possible words. Within each set, the seven alternatives are graded and sorted in terms of their likelihood to be correct. Thus, for each word, the seven resulting recognition#likelihood pairs are stored as $[w_1\#L_1, w_2\#L_2, \ldots, w_7\#L_7]$ where w_1 is the most likely word with L_1 having the highest log-likelihood value. As a concrete example, we can inspect the word *"man"* in the verse *"dem man dirre aventivre giht"* for which the recognition pairs are: [*"man"*#36006.7, *"min"*#35656.8, *"mat"*#35452.5, *"nam"*#35424.7, *"arm"*#35296.2, *"nimt"*#35278.2, *"gan"*#35265.7]. In this case, the system succeeded to recognize this occurrence of the word *"man"* correctly as it appeared in the first position of the recognition set.

3.2   The Generation of Various Evaluation Corpora

Following medievalists' tradition, each verse (line) of the poem represents a document. The current version used in our IR experiments contains 1,328 documents corresponding to only a subset of the complete Parzival transcription. The remaining 4,477 verses form the training set used during the recognition phase and thus are not included in the search evaluation. The number of tokens per verse ranges between 2 and 9 with a mean length equal to 5.3 words.

Having access to the internal representation of the recognition system, we can investigate the quality of different output formats from an IR perspective. When facing with a word error rate of around 6%, the simplest solution is to consider only the most likely recognition for each word (e.g., using only *"man"* in our example). We will refer to this version as BW1, which represents the classical output of a recognition system. Considering that each document (verse) is quite short, the existence of a recognition error will make the retrieval of the corresponding verse an unattainable task without some sort of spelling correction or soft matching between the search keywords and the text representation. Therefore, we generated three additional corpora denoted BW3, BW7 and BWδ. The BW3 version is similar to BW1 except that the best three possible alternatives for each word were automatically included. For the same example above, the first token of the verse will be represented by the three words *"man"*, *"min"* and *"mat"*. Following the same vein, we generated

BW7 by incorporating all of the seven alternatives for each word. This version corresponds to the highest intensity of spelling (and recognition) variants.

Finally, we produced BWδ with a wiser strategy for incorporating term substitutes. Alternatives are included as long as the difference between the candidate's log-likelihood value and that of the most likely term is less than or equal to δ (where δ = 1.5% in the current study). Using our previous example, only the alternative "*min*" is present in addition to the term "*man*" in the BWδ version.

The benefit sought from incorporating more than one recognition alternative is to overcome the word recognition errors as well as the non-normalized spelling. For instance, the term "*Parzival*" appeared in the original manuscript as "*Parcifal*", "*Parcival*" and "*Parzifal*". All of these variants are possible and must be considered as correct spellings. Another example of spelling variants would be "*vogel*" vs. "*fogel*" (bird) and "*fisch*" vs. "*visch*" (fish) - unlike modern German, capitalizing nouns' initials was not used at that era. During the recognition phase, some of these variants may appear in the recognitions' list and using the BW3, BW7 or even BWδ corpora, some or all of them can be retained in the text representation. At this lexical level, one should also consider the inflectional morphology where various suffixes were possibly added to nouns, adjectives and names to indicate their grammatical case (e.g. as in Latin and Russian). With this additional aspect, the proper name "*Parcival*" may appear as "*Parcivale*", "*Parcivals*" or "*Parcivalen*", increasing the number of potential correct spelling variants.

3.3   Known-Item Query Generation

The manual construction of user queries together with their relevance judgments is a very costly task. A cheaper alternative is to generate them automatically. This issue had been the subject of many studies in order to obtain comparable quality between the automatically generated queries and those built by real users [13], [14], [15]. In the context of simulated query building and known-item search, we have adopted the approach suggested by [13]. This approach is based on a probabilistic framework (see Table 1) simulating the behavior of a user who wants to retrieve a known document, trying to aggregate terms that s/he recollects from the target item.

Table 1. The basic known-item query generation algorithm according to [13]

Initialize an empty query Q = { }
Select the document d_k to be the known-item with probability Prob[d_k]
Select the query length s with probability Prob[s]
Repeat s times {
Select a term t_i from the document model of d_k with probability Prob[$t_i|\theta_d$].
Add t_i to the query Q. }
Record d_k and Q to define the (known-item / query) pair.

In adopting this algorithm, we excluded all short words (whose length is less than four characters) from being potential search terms. Moreover, words belonging to the list of the 150 most frequent terms in the corpus were eliminated automatically. Finally, we generated three sets of 60 queries each, these are denoted QT1, QT2 and QT3 and contain single-, 2- and 3-term queries respectively.

To define the probability of selecting a given document (Prob[d_k]), we used a uniform distribution. For choosing the query terms (Prob[t_i/θ_d]), each word has a chance proportional to its length (in characters), the higher the length, the higher the probability of being selected. This random process was used to generate the QT1 set.

For longer queries, we decided to augment the source of the search keywords. The underlying language model would consider the verse defining the known-item itself, and possibly the preceding and the following lines. In this random process, the verses were not assigned the same probability since the words occurring in the central verse were given twice the chance to be selected. As a final modification, when generating the QT3 set, the third query term had the possibility to originate from the short words list or the 150 frequent terms. This makes it possible to obtain nominal and prepositional phrases by having two informative terms combined with a preposition.

4   Indexing Strategies and Retrieval Models

In this paper we present a broad view of the performance achieved by various combinations of document representations and retrieval models [16]. As a first representation of verses and queries we adopted the word-based model without any stemming normalization. We used the same representation again but removed the four most common forms, namely: *der, daz, ir* and *er*. Another possible variation would be to remove all common words (e.g., the 150 most frequent forms).

As a second approach, we applied a light stemmer especially adapted for the Middle High German language. Following the principles used in an English light stemmer [17], this solution will remove a limited number of suffixes (namely "-e", "-en", and "-er") under the constraint that the resulting stem is longer than three characters. We also implemented a more aggressive stemmer to remove a larger set of both inflectional and derivational suffixes. We expect a higher effectiveness level using this approach since the best performing run in TREC-5 confusion track [6] implemented an aggressive stemmer (Porter).

Alternatively, text can be represented by overlapping sequences of *n* letters [18]. When setting *n* = 4 for example, the word "computing" generates the indexing terms "comp," "ompu," ... and "ting." When adopting this representation strategy, the stemming procedure can thus be ignored, letting the weighting scheme assign low weights to the most frequent sequences (e.g., "ting," or "ably"). As an alternative we suggest considering only the first *n* letters of each word (trunc-*n*). When specifying *n* = 4, the word "computing" would generate the single indexing term "comp".

The terms extracted from a document can then be weighted using the classical *tf idf* formula [16]. In this case, we account for the term frequency tf_{ij} of a term t_j in a document d_i and its document frequency df_j. More precisely, we define the *idf* component as $idf_j = \log(n/df_j)$ with *n* indicating the number of documents in the corpus. Cosine normalization can then be applied to obtain better performance levels.

Several variants of this vector-space model have been suggested, given that the occurrence of a new term must be regarded as a rare event. In this case the first occurrence of certain term should be given more importance than its following occurences, thus the *tf* component can be evaluated as log(*tf*)+1.

Moreover, to account for document length differences, Buckley *et al.* [19] suggest using the *Lnu-ltu* weighting method where *Lnu* and *ltu* correspond to the term weighting of the document and the query respectively (Eq. 1). This scheme was used in the best performing TREC-5 system [6].

$$w_{ij} = \frac{\left(\dfrac{\left(\ln(tf_{ij})+1\right)}{(\ln(\text{mean } tf)+1)}\right)}{(1-slope)\cdot pivot + slope\cdot nt_i} \qquad w_{qj} = \frac{\left(\ln(tf_{qj})+1\right)\cdot idf_j}{(1-slope)\cdot pivot + slope\cdot nt_i} \qquad (1)$$

where nt_i indicates the number of terms in a document d_i, *pivot* and *slope* are two constants used to normalize the weights as a function of the mean document length.

As a complement to these two vector-space models, we considered two IR probabilistic schemes. First, we used the Okapi approach [20] based on the following formulation for the term t_j in document d_i.

$$w_{ij} = [(k_1+1)\cdot tf_{ij}] / (K + tf_{ij}) \quad \text{where } K = k_1\cdot [(1-b) + ((b\cdot l_i) / \text{mean } dl)] \qquad (2)$$

where l_i is the length of the d_i document, b, k_1 and *mean dl* are constants whose values are set to 0.55, 1.2 and 5.3 respectively. The second probabilistic scheme is the $I(n_e)B2$ model, a member of the *Divergence from Randomness* (DFR) family [21]. In this case, the weight w_{ij} reflecting the weight of term t_j in document d_i was a combination of two information measures as follows:

$$w_{ij} = \text{Inf}^1_{ij}\cdot \text{Inf}^2_{ij} = \text{Inf}^1_{ij}\cdot (1-\text{Prob}^2_{ij}) \quad \text{and}$$
$$\text{Prob}^2_{ij} = 1 - [(tc_j +1) / (df_j\cdot (tfn_{ij}+1))]$$
$$\text{Inf}^1_{ij} = tfn_{ij}\cdot \log_2[(n+1) / (n_e+0,5)] \quad \text{with } n_e = n\cdot [1 - [(n-1)/n]^{tc_j}] \qquad (3)$$

where tc_i represents the collection frequency of the term t_j.

Finally, we used the language model (LM) [22] in which the probability estimates were based directly on the occurrence frequencies in a document d_i, or in the corpus C. In this paper, we chose to implement Hiemstra's model [22] (Eq. 4), combining an estimate based on a document (Prob[$t_j|d_i$]) and a corpus (Prob[$t_j|C$]).

$$\text{Prob}[d_i\,|\,q] = \text{Prob}[d_i]\cdot \prod_{t_j\in Q} [\lambda_j\cdot \text{Prob}[t_j\,|\,d_i] + (1-\lambda_j)\cdot \text{Prob}[t_j\,|\,C]] \qquad (4)$$

$$\text{Prob}[t_j\,|\,d_i] = tf_{ij} / nt_i \quad \text{and} \quad \text{Prob}[t_j\,|\,C] = df_j / lc \quad \text{with } lc = \textstyle\sum_k df_k \qquad (5)$$

In this formula, λ_j is a smoothing factor (set to a constant value equal to 0.35 for all terms t_j), and lc is an estimation of the size of the corpus C.

5   Evaluation

As a retrieval effectiveness measure, the TREC-5 evaluation campaign selected the MRR (mean reciprocal rank) approach which is based on the inverse of the rank of the first relevant item retrieved [4], [5]. Such a measure reflects the user concern wishing to find one or a few good responses to a given request. In our case, we can apply the same evaluation methodology. As the relevant items are verses, returning the verse immediately before or after the correct one should not be regarded as fully-impertinent, particularly when knowing that the user usually reads a passage (a few verses) instead of a single line. Considering the immediate neighbor verses to be fully-relevant is inadequate either. To allow some degree of relevance for adjacent lines, we need to adapt the strict MRR computation to support graded relevance.

In this vein, Eguchi *et al.* [23] proposed a metric for the task of finding "one highly relevant document" called *Weighted Reciprocal Rank* (WRR), a solution improved in [24]. The word "highly" obviously implies some sort of graded relevance, but this metric imposes that a "*partially* relevant document at rank 1 is more important than ranking a *highly* relevant document at rank 2."

In this context, there exist three possible cases (where the first two are identical to the classical MRR scheme). When the search is unsuccessful; the query is evaluated as 0, while if the rank of the fully relevant document is better than these of its neighbors, the query is then evaluated as 1/R (the reciprocal of the rank of the relevant item). Third, when one of the two neighbors appears in a better rank than that of the target item (ranks denoted as R_n and R_t respectively), the query is evaluated as $Max[1/R_t; \gamma 1/R_n]$. With $\gamma=0.5$, suppose having $R_n = 2$ and $R_t = 3$, thus $Max[1/3; 0.5\cdot 1/2] = 1/3$. In this example, the rank of the relevant item is not really far from its neighbor, the evaluation is based on the former. Having $R_n = 2$ and $R_t = 5$, the query evaluation is then $Max[1/5; 0.5\cdot 1/2] = 0.25$. In this case, the evaluation depends on the rank of R_n as R_t is too late to be favored to R_n. We will refer to this evaluation scheme as G(M)RR for *Graded (Mean) Reciprocal Rank*.

5.1   Evaluation of the Recognition Corpora

Since our evaluation measure focuses on high precision, we deemed a word-based representation without any morphological normalization (performance shown under the label "No stem" in Table 2) would be adequate. Alternatively, a light stemmer should also provide comparable or even better performance levels since it simply removes a very limited number of plural suffixes (labelled "Light"). Finally, the evaluation when eliminating some derivational suffixes, as it is the norm in many IR empirical studies, is shown under the label "Aggressive" in Table 2. As an alternative to the word-based model, we selected the *n*-gram model with a value of $n = 4$ which usually provides good performance levels. Another option was to apply the trunc-*n* scheme with $n = 4$, a strategy found effective for different corpora [18].

Concerning the IR models, we selected two vector-space schemes (*Lnu-ltu* and *tf idf*), two probabilistic models (Okapi, DFR-I(n_e)B2), and a language model (LM). Based on other experiments, we selected the BWδ as the source for building the text representation since this corpus has generally demonstrated the best performance. We also implemented an approach to automatically decompose compound terms which resulted in a slight improvement. The performance values shown in Table 2 & Table 3 and Table 4 were obtained using the 3-term and single-term queries respectively. Table 2 reports results from experiments based on the BWδ corpus. As can be seen, the Okapi model yields the best results regardless of the stemmer (columns "No stem", "Light" or "Aggressive") or representation used. As indicated in the last line of

Table 2, the mean performance differences between the various text representations are rather small, varying from -1.1% (4-gram) to -2.34% (trunc-4) when compared to an approach ignoring stemming normalization. In this table, statistically significant differences in performance based on the t-test (significance level $\alpha = 5\%$) compared to the best approaches (depicted in bold) are marked with an asterisk (*). As shown, the performance differences between the Okapi and the LM models are rather small and non-significant. However, with the *tf idf* model, the performance differences are usually statistically significant. It can be seen from Table 3 that BW1 is the best performing corpus followed by BWδ where the baseline is the error-free ground-truth corpus (GT) with differences in performance equal to -1.10% and -4.19% respectively. The performance differences between BW1 and GT are always non-significant. On the other hand, BWδ provides better retrieval effectiveness than BW1 for the QT1 set. This leads us to the conclusion that BWδ and BW1 represent the best text surrogates and that BWδ outperforms BW1 when considering single-term queries. From the MRR values shown in Table 4 for the single-term queries (QT1) and the various corpora, we can also conclude that BW3 and BW7 do not constitute pertinent alternatives. In Tables 3 and 4, we have, once again, applied the t-test (significance level $\alpha = 5\%$) using the performance achieved by the GT as a baseline. Significant performance differences compared to the GT levels are denoted with an asterisk (*). As shown in these tables, the performance differences between the GT and the BW3 and BW7 corpora are usually statistically significant.

Table 2. GMRR for the BWδ corpus with five IR models, using three stemming strategies, 4-gram, and trunc-4 representations, with the QT3 (60 queries)

Representation	word-based			4-gram	trunc-4
IR Model	No stem	Light	Aggressive		
Okapi	**0.6706**	**0.6586**	**0.6528**	0.6471	**0.6503**
DFR-I(n_e)B2	0.6161*	0.6145	0.6183	0.6473	0.6336
LM (λ=0,35)	0.6647	0.6485	0.6466	**0.6504**	0.6479
Lnu-ltu	0.6544	0.6315	0.6379	0.6393	0.6240
tf idf	0.6313	0.6146*	0.6193	0.6176*	0.6057*
Difference %		-2.15%	-1.92%	-1.10%	-2.34%

Table 3. GMRR for different recognition corpora together with the ground-truth (GT) using five IR models, aggressive stemmer (QT3, 60 queries)

IR Model	GT	BWδ	BW1	BW3	BW7
Okapi	0.6594	**0.6528**	0.6555	0.6406	0.5866*
DFR-I(n_e)B2	0.6572	0.6183	0.6528	0.6358	0.5808*
LM (λ=0,35)	0.6642	0.6466*	**0.6596**	**0.6491**	**0.5875***
Lnu-ltu	0.6649	0.6379	**0.6596**	0.6130	0.5866*
tf idf	**0.6679**	0.6193*	0.6495	0.5956*	0.5405*
Difference %		-4.19%	-1.10%	-5.42%	-13.03%

Table 4. MRR for different recognition corpora together with the ground-truth (GT) using five IR models, aggressive stemmer (QT1, 60 queries)

IR Model	GT	BWδ	BW1	BW3	BW7
Okapi	0.6123	0.6030	0.5839	0.4001*	0.3184*
DFR-I(n_e)B2	0.6123	0.5950	0.5839	0.4001*	0.3184*
LM (λ=0,35)	0.6123	0.5951	0.5839	0.4078*	0.3085*
Lnu-ltu	0.6121	0.6210	0.5837	**0.4100***	0.3218*
tf·idf	**0.6271**	**0.6365**	**0.5890**	0.3956*	**0.3271***
Difference %		-0.83%	-4.94%	-34.54%	-48.18%

5.2   Selected Query-by-Query Analyses

Single-term Queries #4, #11, #25, and #43 are composed of a rare term each, having, according to the GT corpus, *df* values equal to 1 (term appearing only in the known-item). Using the BW1 corpus and Okapi model, none of these four queries got any pertinent items retrieved. Query #4 *"machete"*, for instance, has the verse *"er machete ê daz er gein ir sp(ra)ch"* as its known-item. Using the BW1 corpus, the most likely recognition of the term *"machete"* is *"machen"* which is an incorrect recognition. Since this is the only occurrence of the term in the collection, the system failed to retrieve any documents in that case. Using the BWδ corpus, the known-item was retrieved at Rank 1 since the term *"machete"* was the second likely recognition with a log-likelihood difference of less than 1.5%. This second possible recognition was therefore included in the text representation. Using the BW3 or BW7 corpora, the known-item was also retrieved but in lower ranks, in Rank 3 for BW3 and 18 for BW7. This poor performance is due to the fact that BW3 and BW7 include more alternatives which merely acted as noise in this case. Non-relevant documents containing the search term alternatives are now competitors as the final ranking depends on the *tf* values. A non-relevant item yet with a higher *tf* will thus appear before the target verse in the ranked list of retrieved items. For the same examples using the classical *tf idf* vector-space model, the known-items were retrieved in all cases (BW1, BW3, BW7, and BWδ) in acceptably high rankings: 3, 2, 2 and 7 or in positions 7, 7, 2 and 15 using DFR-I(n_e)B2. Applying an automatic decompounding strategy may provide some successful improvement, particularly for longer queries (three terms). The use of either a light or a more aggressive stemming approach was also usually more beneficial for longer queries.

6   Conclusion

In this study we investigated the underlying issues when facing with a noisy corpus having a word error rate of around 6%, originating from medieval manuscripts handwritten in Middle High German and digitized via a text recognition device. In addition to this issue, difficult matching between queries and documents can be caused by the non-standardized spelling used in medieval languages and the presence of less strict grammatical rules. Having access to the internal representation of the recognition phase, we generated the BW1 corpus by considering only the best recognition for each input word. We also created the BW3 and BW7 corpora retaining, respectively, the best three and seven alternatives for each word. The fourth version, BWδ, includes all possible word recognition(s) having a log-likelihood less than or equal to 1.5% compared to the highest value. The error-free ground-truth transcription had been manually transcribed by the experts and served as the evaluation baseline. Based on three text representation formats (word-based, n-gram, trunc-n), five IR models (*tf idf*, *Lnu-ltu*, Okapi, DFR-I(n_e)B2, LM), three stemmers (none, light, and aggressive), and two query formulations (single-term, 3-term), we found that the best retrieval effectiveness was usually produced by the Okapi and LM models. We cannot clearly determine the best text representation as the mean differences among them are rather small. Regarding the stemming procedure, we suggest applying either an aggressive stemming that tends to produce slightly better retrieval performance when facing with longer queries. On the other hand, ignoring the stemming normalization with short queries offers usually the best performance. We have assessed the various recognition corpora against the ground-truth version. Compared to this error-free version, the BWδ shows a mean degradation in retrieval performance ranging from -4.19% (3-term queries) to -6.05% (single-term queries). When using the classical output of the recognition process (recognition output limited to a single term, or BW1), the degradation in mean performance ranges from 1.1% (3-term queries) to 10.24% (single-term queries). Considering systematically three (BW3) or seven (BW7) alternatives per input word usually tends to cause the retrieval effectiveness to decrease significantly (from -5.42% for 3-term queries to 64.34% with single-term queries).

With very few media written in older languages (newspapers did not exist in the thirteenth century), the corpus size is limited to a subset of the manuscript pages with their corresponding error-free transcriptions serving as the ground-truth text during the evaluation. Another issue that arises is that manual transcription of medieval manuscripts is a very costly task as it is quite time-consuming and can solely be performed by the experts. The best practices found and the conclusions drawn from our experiments can be applied to further manuscripts, hence making them digitally accessible and effectively searchable with the least cost possible by partially or totally eliminating the need to having them manually transcribed which will result in saving a lot of resources (time, human effort, money, etc.). With these documents being totally searchable via digital means, real user needs (queries) will thus be obtained via search requests from experts as well as public users, which in turn, will help improve the performance a great deal.

Acknowledgment. This research is supported by the Swiss NSF under Grant CRSI22_125220.

References

1. Toni, M., Manmatha, R., Lavrenko, V.: A Search Engine for Historical Manuscript Images. In: Proceedings of the ACM-SIGIR, pp. 369–376. The ACM Press, New York (2004)
2. Nicholas, R., Toni, M., Manmatha, R.: Boosted Decision Trees for Word Recognition in Handwritten Document Retrieval. In: Proceedings of the ACM-SIGIR, pp. 377–383. The ACM Press, New York (2005)
3. Callan, J., Kantor, P., Grossman, D.: Information Retrieval and OCR: From Converting Content to Grasping Meaning. SIGIR Forum 36(2), 58–61 (2002)

4. Voorhees, E.M., Garofolo, J.S.: Retrieving Noisy Text. In: Voorhees, E.M., Harman, D.K. (eds.) TREC, Experiment and Evaluation in Information Retrieval, pp. 183–197. The MIT Press, Cambridge (2005)
5. Buckley, C., Voorhees, E.: Retrieval System Evaluation. In: Voorhees, E.M., Harman, D.K. (eds.) TREC, Experiment and Evaluation in Information Retrieval, pp. 53–75. The MIT Press, Cambridge (2005)
6. Ballerini, J.P., Büchel, M., Domering, R., Knaus, D., Mateev, B., Mittendorf, E., Schäuble, P., Sheridan, P., Wechsler, M.: SPIDER Retrieval System at TREC-5. In: Proceedings of TREC-5, pp. 217–228. NIST Publication #500-238 (1997)
7. Tagva, K., Borsack, J., Condit, A.: Results of Applying Probabilistic IR to OCR Text. In: Proceedings of the ACM-SIGIR, pp. 202–211. The ACM Press, New York (1994)
8. Craig, H., Whipp, R.: Old Spellings, New Methods: Automated Procedures for Indeterminate Linguistic Data. Literary & Linguistic Computing 25(1), 37–52 (2010)
9. Pilz, T., Luther, W., Fuhr, N., Ammon, U.: Rule-Based Search in Text Databases with Nonstandard Orthography. Literacy & Linguistic Computing 21(2), 179–186 (2006)
10. Peters, C., Gonzalo, J., Braschler, M., Kluck, M. (eds.): CLEF 2003. LNCS, vol. 3237. Springer, Heidelberg (2004)
11. Gardt, A., Hauss-Zumkehr, U., Roelcke, T.: Sprachgeschichte als Kulturgeschichte. Walter de Gruyter, Berlin (1999)
12. Fischer, A., Wüthrich, M., Liwicki, M., Frinken, V., Bunke, H., Viehhauser, G., Stolz, M.: Automatic Transcription of Handwritten Medieval Documents. In: 15th International Conference on Virtual Systems and Multimedia (2007)
13. Azzopardi, L., de Rijke, M.: Automatic Construction of Known-Item Finding Test Beds. In: Proceeding ACM SIGIR, pp. 603–604. The ACM Press, New York (2006)
14. Callan, J., Connell, M.: Query-Based Sampling of Text Databases. Information Systems 19(2), 97–130 (2001)
15. Jordan, C., Watters, C., Gao, Q.: Using Controlled Query Generation to Evaluate Blind Relevance Feedback Algorithms. In: Proceedings of the Sixth ACM/IEEE-CS Joint Conference on Digital Libraries, pp. 286–295. The ACM Press, New York (2006)
16. Manning, C.D., Raghavan, P., Schütze, H.: Introduction to Information Retrieval. Cambridge University Press, Cambridge (2008)
17. Harman, D.: How Effective is Suffixing. Journal of the American Society for Information Science 42(1), 7–15 (1991)
18. McNamee, P., Mayfield, J.: Character n-gram Tokenization for European Language Text Retrieval. IR Journal 7(1-2), 73–97 (2004)
19. Buckley, C., Singhal, A., Mitra, M., Salton, G.: New Retrieval Approaches using SMART. In: Proceedings of TREC-4, pp. 25–48. NIST Publication #500-236 (1996)
20. Robertson, S.E., Walker, S., Beaulieu, M.: Experimentation as a Way of Life: Okapi at TREC. Information Processing & Management 36(1), 95–108 (2000)
21. Amati, G., van Rijsbergen, C.J.: Probabilistic Models of Information Retrieval Based on Measuring the Divergence from Randomness. ACM Transactions on Information Systems 20(4), 357–389 (2002)
22. Hiemstra, D.: Using Language Models for Information Retrieval. CTIT Ph.D. Thesis (2000)
23. Eguchi, K., Oyama, K., Ishida, E., Kando, N., Kuriyama, K.: Overview of the Web Retrieval Task at the Third NTCIR Workshop. NII Publication (2003)
24. Sakai, T.: Bootstrap-Based Comparisons of IR Metrics for Finding One Relevant Document. In: Ng, H.T., Leong, M.-K., Kan, M.-Y., Ji, D. (eds.) AIRS 2006. LNCS, vol. 4182, pp. 374–389. Springer, Heidelberg (2006)

Query Phrase Expansion Using Wikipedia
in Patent Class Search

Bashar Al-Shboul and Sung-Hyon Myaeng

Korea Advanced Institute of Science and Technology
373-1 Guseong-dong, Yuseong-gu, Daejeon 305-701, South Korea
{bashar,myaeng}@kaist.ac.kr

Abstract. Relevance Feedback methods generally suffer from topic drift caused by words ambiguity and synonymous uses of words. As a way to alleviate the inherent problem, we propose a novel query phrase expansion approach utilizing semantic annotations in Wikipedia pages, trying to enrich queries with context disambiguating phrases. Focusing on the patent domain, especially on patent search where patents are classified into a hierarchy of categories, we attempt to understand the roles of phrases and words in query expansion in determining the relevance of documents and examine their contributions to alleviating the query drift problem. Our approach is compared against Relevance Model, a state-of-the-art, to show its superiority in terms of MAP on all levels of the classification hierarchy.

Keywords: Pseudo-Relevance Feedback, Patent Information Retrieval, Wikipedia Categories, Query Expansion, Phrase-based Query Expansion.

1 Introduction

Query Expansion (QE) is one of the Information Retrieval (IR) techniques to enhance effectiveness of document retrieval. It is generally used to disambiguate the context of a user query. One of the heavily researched approaches is Pseudo-Relevance Feedback (PRF): an automatic query expansion method based on the assumption that top ranked documents retrieved for a query are the most relevant ones. While the terms in top-ranked documents are considered the best resource for selecting expansion terms, past research shows that PRF-like models suffer from several drawbacks such as query-topic drift [6], [7] and inefficiency [5]. There have been attempts to use lexical resources, most notably WordNet, for QE. As WordNet has been exploited in various IR tasks including QE [10], [13], [14], a common conclusion resulting from its limited coverage of words and relations, and the lack of contextual information for each word, is low effectiveness [11]. While words' ambiguity is a main reason behind using QE to disambiguate the query context, phrases can play the same role because the surrounding word(s) in a phrase provide additional contextual information. When phrases were used alongside with words for IR, however, the results have been disappointing with only a slight improvement or even a decrease in effectiveness [16]. A reason is that phrases have different distributions over documents when compared to words [15]. Motivated by the topic

M.V.M. Salem et al. (Eds.): AIRS 2011, LNCS 7097, pp. 115–126, 2011.

drift problem in PRF, inability to improve retrieval effectiveness with automatically identified phrases, and the limitations of using WordNet for word-based QE [10],[17], this paper proposes a novel QE approach utilizing Wikipedia semantic annotations (i.e. categories) for query phrase expansion. Our approach concentrates on reducing query topic drift using both WordNet and Wikipedia, to handle word synonyms and concurrently compensate the limitation of WordNet by enriching queries with phrases to disambiguate query context.

Wikipedia is an online collaborative contribution of the Web community to building an encyclopedia. Generally, it was utilized for IR-related tasks including Word-Sense Disambiguation (WSD) [9], Named-Entity Retrieval [24], Document Clustering/Classification [20], Question Answering [21], and QE [24], [28]. In Wikipedia, a page describing concept expressed as a word or phrase belongs to one or more categories, each of which contains a category title (i.e. Information Retrieval), titles of the pages in this category (i.e. Discount Cumulative Gain, Generalized Vector Space Model, etc.), and other related categories (i.e. Information Science). Our work differs from all other works in its utilization of individual categories, page titles under each category, and links to other related categories.

We test our method on patent search, which has arisen as one of the important information retrieval fields, especially for legal IR [1], [8]. As current patent search systems use a keyword-based approach, effectiveness of retrieval relies on the quality of search keywords [2]. Patents are particularly suitable for testing our method because they usually contain a large number of phrases because they often deal with technical vocabulary. Aside from several unique characteristics such as vocabulary, usage, and structure, patent search is unique in that each patent is manually assigned to one or more classes from the IPC (International Patent Classifications). A group of patents belonging to a class are said to be relevant to each other. IPC has a hierarchy of three levels, Sub Class (SC), Main Group (MG), and Sub Group (SG), with SC being most general. For example, when a patent is assigned to the IPC "G06F 17/30", its SC, MG, and SG are "G06F", "G06F 17", and "G06F 17/30", respectively. The IPC hierarchy allows us to study generalization/specialization capabilities of phrases as our retrieval task is to classify patents with or without query expansion. Our proposed query expansion approach has been tested on US Patent & Trademark Office (USPTO) patents provided by NTCIR.

In our experiment, comparisons are made against Relevance Model (RM) [4] because it has often been used as a comparison benchmark [3], [4]. Among many PRF-based efforts [4], [5], [6], [11], RM has drawn much attention with its strong probabilistic ground. Details of RM are omitted for brevity.

The contributions of this work are: (1) a thorough study of the effect of adding phrases to the baseline and RM in ranking documents. (2) an evaluation of the effect of RM and WordNet for word-based query expansion, their roles in generalizing/specifying the query topic, and their relevance to the query topic. (3) an exploration of the effect of utilizing Wikipedia for query phrase expansion and its role in generalizing/specifying the query topic, and (4) an analysis of the interaction between WordNet-based expanded words and Wikipedia-based expanded phrases, in addition to studying their effects on ranking documents and then finding an optimal weight ratio between them for achieving the best retrieval effectiveness.

This paper is organized as follows. In section 2, related work is presented. Our proposed framework is described in section 3. Experiment goals, environment and results are discussed in section 4. Finally, we conclude in section 5.

2   Related Work

Among the several automatic QE approaches, PRF has shown its effect in improving retrieval effectiveness [19]. While several attempts to enhance PRF were reported [12], [18], [4], others tried different approaches (i.e. expansion based on query characteristics [22], mining user logs [25], using external resources [26]). QE based on local collection statistics may fail due to the lack of relevant documents for a query, necessitating external collections were used for query enrichment [26]. In this work, we follow the strategy of using external resources as a way of enriching and disambiguating queries with relevant words and phrases.

Relevance feedback models have been proposed as query expansion methods. However, it is well known that it suffers from *topic drift* [6], [7], especially in short queries. Major causes are the ambiguity of query terms and the method adopted by PRF in weighting and selecting query candidate expansion terms (i.e. IDF). We attempt to alleviate these problems by expanding query phrases using external resources rather than collection-dependent phrases. RM was adopted in Lemur IR Toolkit, which we used to implement and test our model. Generally, in Lemur toolkit an RM query takes the form of *#weight(w_1 Baseline_Query w_2 Expansion_Terms)*, where w_1 and w_2 are normalized weights, set to 0.5 by default. This way of scoring seems unfair considering that expansion terms are not guaranteed to be relevant to the query topic in the first place. In this paper, we try to find the optimal weight balance between the baseline query and the expansion terms for producing the best effectiveness and then compare our proposed model with the best possible results obtained from RM.

WordNet has been utilized for QE in different ways. For example, WordNet semantic relations have been used to solve the problem of vocabulary mismatch [10]. In [17], the query words were expanded separately by intersecting each word's synsets sharing a lexical relation in different resources. An interesting result shows that using WordNet synsets in addition to *"glosses"* as expansion words contributed significantly towards effectiveness. We compare our model with best reported results in [17].

Wikipedia has been utilized for QE in different ways. A link-based expansion approach is [28], where PRF is modified to rank links based on their target documents scores, and then use link texts as expansion terms. RM is then used for another round of expansion. This approach has shown a significant improvement over RM on blog documents. Our work is different from the state of the art works in that we utilize different parts from Wikipedia (i.e. categories, category links to relevant categories, and page titles under category pages), trying to expand different type of query terms (i.e. Phrases).

Several types of phrases (i.e. noun phrases, head/modifier, bigrams, and others) have been used for different IR applications [23], [3], [27], but very rarely for query expansion. In [3], for example, semantic information extracted from DBpedia is utilized. A phrase is run against DBpedia index using cosine similarity. SKOS (Simple Knowledge Organization System) of top ranked documents are used as thesauri to expand phrases. Additionally, all synonyms extracted from WordNet are added as expansion terms to the query used for research article classification over the IPC hierarchy. Our work shares the same idea of using an external resource for phrase-based QE but proposes a new method with a different resource, i.e. Wikipedia, and examines various options with a greater depth.

3   Proposed Method

Our method concentrates on providing contextual information to the query and avoiding vocabulary mismatches. Query words are expanded considering the most likely synset returned from searching WordNet, where synsets' ranking is based on word frequencies in the British National Corpus. In addition, phrases are also expanded as they contain contextual information for individual query words. Phrases are used to search the Wikipedia index to extract a set of Wikipedia categories (primary categories), their related categories (secondary categories), and the Wikipedia pages belonging to the primary categories, which are later processed to generate a set of candidate expansion phrases. Merging the original query with the candidate expansion words and phrases generates our expanded query.

3.1   Page Similarity

In Wikipedia, each titled page belongs to one or more categories. In computing page similarities, we consider pages belonging to the same category are similar to each other, instead of using the text in them. In addition, we use the related (secondary) categories linked to a category page of the primary category. Two categories with different titles can be seen similar to each other when they share related categories. For example, two pages whose categories are "Green Automobiles" and "Green Vehicles" are considered relevant to each other when the corresponding category pages share some secondary category names. However, their similarity is not as strong as the pages belonging to the same category. Further, page titles can be used for similarity calculation when the pages do not match in terms of their primary and secondary category names. We consider two pages are similar to some extent when there is an overlap between the titles of the pages belonging to the primary and secondary categories of the original pages being compared. For example, while "Precision & Recall" page and "Accuracy & Precision" page do not share any primary and secondary categories, they share several page titles belonging to their categories. Similarity between two Wikipedia pages is computed as follows:

$$Sim(P_1, P_2) = 0.5 \times PPC_{P_1, P_2} + 0.3 \times PSC_{P_1, P_2} + 0.2 \times PT_{P_1, P_2} \tag{1}$$

Where PPC is percentage of primary category phrases match to all distinct primary categories, PSC is percentage of secondary category phrases match to all distinct secondary categories, and PT is percentage of page title phrases match to all distinct page titles. Phrases considered matching iff an exact string match exists. Weights assigned arbitrarily upon our assumption that PPC is the most important among all, followed by PSC then PT. Weight optimization is left as future work. For this computation, each Wikipedia page is indexed with primary and secondary categories and page titles of the categories. Indexing in this way will allow us to estimate Maximum Likelihood Estimate (MLE) of each query phrase under language models of each field (i.e. primary category field, secondary categories field, category page titles field), as will be explained later.

3.2   Query Term Extraction and Expansion

In patent search, terms are extracted from a query patent to form a search query. Key terms are extracted from different parts of query patents to see their roles in experiments as will be explained in section 4. We first apply a stop word filter that uses a customized list collected from different sources (i.e. KEA, InQuery, Lemur). We then apply Stanford POS Tagger to the resulting text and Regular Expressions (RegEx) to extract keywords and keyphrases. This process is particularly useful for short fields like titles and abstract as the brevity might not be enough for statistics-based methods (e.g. TFIDF, CHI, IG) of feature selection. Nouns, verbs and adjectives are extracted as keywords, and then phrases using RegEx. Further, all unigrams that are covered by a phrase are removed, as they are already disambiguated by additional contextual information in the phrase. The RegEx used to extract keyphrases is given by: (VBG|VBN|JJ|JJR|JJS)(NN|NNS|NNP|NNPS)+ where VBG and VBN are verbs, JJ, JJR and JJS represent adjectives, and NN, NNS, NNP and NNPS represent nouns and proper nouns in singular and plural forms. This RegEx was built based on our observations over the tagged query patents. After generating a query by extracting key terms from a query patent document, it is split into a bag of words (BOW) and a bag of phrases (BOP). Each entry in BOP and in BOW is expanded using Wikipedia and WordNet, respectively. A query phrase is expanded in order to alleviate any vocabulary mismatch by, for example, finding an alternative name for a technology (e.g. "speaker identification" and "speaker verification"). Often times, the same technology can be expressed in different phrases. A phrase is expanded using the Wikipedia index. To retrieve Wikipedia pages, we employ the following model:

$$P\left(ph_j|D\right) = \sum_{i \in G}\lambda_i \; p(ph_j \,|\, \theta_{i,j}) \tag{2}$$

where G = {Primary Categories, Secondary Categories, Titles of the pages under Primary and Secondary Categories} represents three background language models, and $\sum_{i \in G}\lambda_i = 1$ are the mixture weights which were empirically set to 0.5, 0.3, and 0.2, respectively. Additionally, language models are estimated as:

$$p(ph_j \,|\, \theta_{i,j}) = \frac{c(ph_j, X_i)}{c(X_i)} \tag{3}$$

where $c(ph_j, X_i)$ is the count of phrase ph_j in the X_i field of the index (with cosine similarity higher than 0.7), and $c(X_i)$ is the count of distinct categories and page titles in the LMs. Further, top ranked 5 documents categories and titles are intersected to find common phrases. Phrases exist in 3 or more documents are added to the query. Finally, another cycle of word filtering is done to remove words covered by phrases trying to avoid duplicates, considering that those words are already disambiguated through a phrase or more.

In our method, every query word is expanded, where the first synset found in WordNet is considered as an expansion candidate term. That is because different feature selection approaches (i.e. Chi, IG, DF,..etc) have shown different weaknesses in weighting terms (i.e. preference of terms with specific characteristics). For example, a statistical model such as IDF weights the term based on its discriminating power assuming that terms with low document frequency are more discriminating than others. To alleviate such an issue, query words are expanded separately aiming to reduce the effect of terms with higher weights over others with low weights, granting the chance for all query words to be expand regardless of their distribution over the corpus. This method of word expansion was followed hoping that WordNet expansion words will help in the case of vocabulary mismatch between the query and the collection. Furthermore, in patent domain IPC skewness is a very serious problem for patent search tasks (i.e. State-of-Art). In our indexed USPTO collection, less than 50 IPCs embrace more than 13% of the documents, showing a long-tailed distribution where term sparseness for the majority of IPCs is the major problem.

3.3   Patent Retrieval

Using the expanded queries generated by our model, they were run against patent index using Okapi BM25 with default variables set in Lemur where, k_1 and b are BM25 variables set to *1.2* and *0.75* respectively, while *IDF* is described in [29].

After a ranked list of patents R_Q is returned from a query, we re-rank them using the IPC information associated with them. The main idea is that if a retrieved patent belongs to the IPC to which the query patent belongs, it should be considered relevant to the query. A re-ranking algorithm should consider the fact that a search result is likely to contain multiple documents belonging to the same IPC and that each document may have multiple IPCs itself. Our re-ranking process is divided into three successive stages: *IPC Mapping*, *IPC Expansion*, and *IPC Scoring*. The process of generating a new list of documents' IPC(s) with their scores is called *IPC Mapping*. Next, patents with multiple IPCs are expanded into several entries holding the same score, and rank, as the whole patent. This stage is called *IPC Expansion*. The expanded list is denoted as R_{IPC}. Final stage is re-scoring IPCs as an IPC will probably occur several times in R_{IPC}. For that purpose list $L_Q = \{$*distinct IPCs in* R_{IPC} *for query Q*$\}$. IPCs are re-scored as follows:

$$Score\,(ipc) = \frac{\sum_{ipc \in L_Q} rank\,(ipc, R_{IPC})}{count(ipc, R_{IPC})} \qquad (4)$$

In this case, IPCs with higher frequency on higher scores in R_{IPC} are favored as they are assumed to be highly relevant compared to other IPCs. IPCs are further re-ranked based on their new scores.

4   Experiments

Our experiments aim at answering the following series of research questions:

- What is the effect of using phrases in the baseline query in addition to words?
- What are the best weights assigned to baseline and expansion parts in RM query?
- What is the effect of query word expansion using WordNet?
- What is the effect of using Wikipedia for query phrase expansion?
- What is the effect of combining both Wikipedia and WordNet expansions?
- What is the best weight balance between words and phrases in our expansion model?

For our experiments, we used Wikipedia Dump of August 2010, in addition to Indri, which was used for indexing Wikipedia articles and NTCIR-6 USPTO patents of 1993-2002 (~1.3M documents). Queries were patents selected randomly from those contained in the IPCs having at least two patents. The query patents were then removed from the Index. A total of 1,780 patents were selected as the queries for experiments. Relevance judgments of the queries were made based on IPCv9 crawled from the USPTO website. Query patents were mapped to their corresponding IPCv9 to generate the relevance judgment. In all our experiments, we used Okapi BM25 as the retrieval model for the combined query of words, phrases, and their expansions in our model. To determine which parts of a patent will generate the best query terms *preliminary experiments* with various combinations of patent parts were undertaken to find that a combination of Titles and Abstracts was the best. Thus all the results reported in this paper are for queries extracted from those two parts. During experiments, Precision and Recall were used to evaluate our work; however, depending on IPCs instead of documents. They are given as follows:

$$Precision(Q) = \frac{Relevant\_Retrieved\_IPCs(Q)}{All\_Retrieved\_IPCs(Q)} \qquad (5)$$

$$Recall = \frac{Distinct\_Relevant\_Retrieved\_IPCs}{All\_Relevant\_IPCs} \qquad (6)$$

where Q is Query, Relevant_Retrieved_IPCs represents the number of documents with relevant IPCs in R_Q, All_Retrieved_IPCs is set to number of top ranked documents (i.e. 1000 in our case), Distinct_Relevant_Ret-rieved_IPCs is the number of distinct IPCs correctly retrieved, while All_Relevant_IPCs represents number of IPCs assigned to the query patent. Same precision was used in evaluating NTCIR Patent Mining tasks; however, Recall was modified to evaluate based on IPCs. The number of relevant IPCs for each classification level is given beside the classification level notation (i.e. "SC (3155)" indicates that at Sub-Class level, there are 3,155 relevant IPCs for the whole set of 1,780 query patents used for experiments)

4.1   Baseline Queries

In our experiments two different baseline queries were used: unigram baseline queries and the unigram baseline queries plus phrases after deleting the unigrams involved in the phrases (denoted as "word/phrase" in Table 1). Table 1 shows a significant decrement in MAP (Mean Average Precision) with the word/phrase queries. This can be attributed to the generality of added phrases, while unigrams succeeded to rank relevant documents higher than word/phrase did. However, the recall drop was less severe, meaning that phrases disambiguated the context almost similarly to unigrams.

Table 1. A comparison between the baseline, RM, and our method (OM)

Query	SC (3155)		MG (3801)		SG (5391)	
	MAP	Recall	MAP	Recall	MAP	Recall
Unigram	39.19	84.40	23.99	80.03	15.64	72.71
Word/Phrases	36.75	84.12	22.15	78.19	13.32	70.69
OM Wikipedia	33.97	84.21	20.9	78.37	12.23	68.72
Wiki-Links	50.57	77.65	37.94	73.30	22.06	78.52
OM WordNet	40.19	84.43	24.66	80.53	15.84	75.71
WN-Gloss	22.39	82.12	12.95	75.24	9.25	61.07
RM	52.99	83.99	36.07	81.76	22.86	87.01
Our Model	54.34	84.97	38.37	82.92	25.82	86.95

4.2   Relevance Model

In our work, as in Lemur, RM expanded queries generally follow a weighting structure that provides 50% of the weight to the original query terms, and the other 50% to the expansion terms. To tune weights for RM queries, a *preliminary set of experiments* for variants of weights have been undertaken, and 0.6/0.4 was selected for Baseline/Relevance parts of the queries as they generated the best result (Experiment results were not listed in the paper due to page limit). As can be seen in Table 1, our model gave significantly better MAP values compared to RM but only slightly better results in recall. With further analysis it was found that RM performance worsens as participation from RM expansion part in scoring documents increases (i.e. 78% or more of query weight determined by RM expansion part). However, performance enhanced after weight participation decreased to less than 77%.

4.3   WordNet

In our method (an experiment of expanding baseline query using WordNet, and skipping Wikipedia expansion part), 76% of words added by WordNet (representing more than 23% of all query words) had very high IDF; however, many of them were irrelevant to the patent topic. Further analysis revealed that WordNet expansion terms exist twice as many in the irrelevant retrieved documents as in the relevant ones. Those terms had the highest weights in the query and affected the search negatively, causing the query topic to drift. Table 1 also shows that queries consisting of words and their WordNet expansions work slightly better than those phrase queries at the general level (SC) but much better at a specific retrieval level (SG Level). An interesting finding is that in each WordNet expanded query there are in average 8 words overlapped with phrases in the title-abstract queries, representing 26% of average query length, and having a good chance covering general query topic. Furthermore, our method of WordNet expansion was compared to the WordNet expansion using glosses described in [17], denoted as WN-Gloss in Table 1. The result shows that QE using WordNet words from the first synset retrieves a better result than that of using WordNet glosses.

4.4   Wikipedia

In this section, two different sets of experiments were performed. First, to understand the effect of our proposed idea of Wikipedia expansion on query performance, this expansion has been performed on the baseline queries (annotated as OM Wikipedia). Even though this experiment's MAP result was the second worst amongst all, it is important to note its recall result, which is only slightly lower than those of most other cases at the general level of IPC, and worsened at more specific classification levels. Examining Wikipedia expansion phrases, it was found that these results can be attributed to one (or more) of the following reasons:

- Wikipedia titling policy which requires that category titles are topic descriptors specific enough to be distinguished from each other but general enough to cover more specific concepts in each page.
- The number of relevant/irrelevant expansion phrases. As in retrieval process all phrases has exactly the same weight. This weighting policy was basically followed aiming not to judge a phrase based on its existence in Wikipedia index because it might be a new or uncommon phrase. On the other hand, the equal weighting policy might seem have helped irrelevant phrases to drift the query topic; however, this seems to be a low possibility considering that relevant phrases appear, in average, more frequently in relevant documents (i.e. In average, 2.7 expansion phrases exist in relevant documents compared to 2.1 phrases in irrelevant ones)
- Wikipedia categories can be unreliable sometimes because it is possible to arrive at complete different sets of categories from similar pages (i.e. Wikipedia pages "precision and recall" and "accuracy and precision").

Furthermore, our model was compared to another state-of-the-art method of utilizing Wikipedia for query expansion [14]. As can be seen, the state-of-the-art model (annotated as Wiki-Links in Table 1) performed better in terms of MAP than our proposed method of Wikipedia expansion (without WordNet part) at all the classification levels and also better in recall at the SG classification level. However, it performed worse in recall at the more general classification levels (i.e. SC, MG). This indicates that expansion phrases added by our method are more general than terms added from hyperlinked texts, because they performed better on recall over higher classification levels; however, they are not enough for retrieving relevant documents higher in rank as can be concluded from MAP results. Note that the Wiki-Links based query expansion performed comparably to RM. This indicates that selected hyperlinked texts contain good expansion terms comparable to the ones selected using RM.

4.5   Proposed Method

Our proposed method depends on the interaction between Wikipedia and WordNet expansions. All words and phrases were considered for expansion under the hypothesis that phrases provide good contextual information and hence valuable assistance to context disambiguation. Words and phrases were merged together in order to generate a more coherent, and a less ambiguous query than a RM expanded query. Our analysis reveals that phrases play an important role of promoting relevant documents in the ranked list since IDF values for phrases are usually higher than those of the words. It is worth mentioning that baseline queries consisted on average of 31 terms, and after applying our expansion method, the number of terms was increased to become 51. As shown in Table 1, our method performed better than RM at all levels of classification. As can be seen, the combination of WordNet-expanded words, and Wikipedia-based expanded phrases exploited the best of both to achieve the best MAP and recall. As in Table 2, our method shows significantly better precision@N, using a t-test at $p = 0.05$, especially at the specific classification levels where precision is significantly higher.

Table 2. Precision@N, Our Method (OM) vs. Relevance Model (RM)

	SC		MG		SG	
	RM	OM	RM	OM	RM	OM
P@5	18.81	19.84	15.21	16.88	14.29	16.21
P@10	11.23	11.74	9.74	10.64	9.81	11.16

4.6   Phrase Weight Balance

As our proposed method performed the best amongst all other cases, it seems important to find the best weight balance between words and phrases in the expansion method. A *preliminary experiment with a subset of queries* showed that fixing phrase weights twice as big as that of words gave the best results amongst all other combinations. However, giving higher weights to phrases over a certain limit (i.e. twice as much in our experiment) will generalize the query more, and decrease effectiveness as well (Experiment detailed results were not listed in the paper due to page limit). The same experiment was performed on RM by modifying phrase weights in the baseline part of the expanded query; however, the results shows that the higher weight assigned to phrases in the baseline part worsens the results in terms of both MAP and Recall. The reason is that RM expansion part still dominates the weighting of the query, with minor participation from the baseline part. Furthermore, increasing the weights of phrases in the baseline part decreases their weighting effect severely.

5   Conclusions and Future Works

We have proposed a new method of using Wikipedia categories and WordNet at the same time for query word and phrase expansion in the task of search for patents. In a series of experiments using IPC categories and USPTO patents, our proposed method has shown the usefulness of expanding query phrases with Wikipedia categories of two kinds and titles, when they are used together with expanded words. Our analysis of the experimental results reveals that added phrases control topic drift caused by PRF. Our future work includes incorporating the Wiki-Links idea to our proposed expansion method, an IPC re-ranking model is being researched, as well as devising a more accurate method for keywords expansion by using a more accurate synset recommender method, so that we can pinpoint where the expanded phrases play an essential role.

Acknowledgement. This research was partially supported by the MKE (The Ministry of Knowledge Economy), Korea, under the ITRC (Information Technology Research Center) support program supervised by the NIPA (National IT Industry Promotion Agency) (NIPA-2011-C1090-1111-0008).

References

1. Azzopardi, L., Vanderbauwhede, W., Joho, H.: Search system requirements of patent analysts. In: Proc. of SIGIR 2010 (2010)
2. Xue, X., Croft, W.B.: Transforming patents into prior-art queries. In: Proc. of SIGIR 2009 (2009)
3. Al-Shboul, B., Myaeng, S.H.: IRNLP@KAIST in the subtask of Research Papers Classification in NTCIR-8. In: Proc. of NTCIR-8 (2010)
4. Lavrenko, V., Croft, W.B.: Relevance-based language models. In: Proc. of SIGIR 2001 (2001)
5. Yin, Z., Shokouhi, M., Craswell, N.: Query Expansion Using External Evidence. In: Boughanem, M., Berrut, C., Mothe, J., Soule-Dupuy, C. (eds.) ECIR 2009. LNCS, vol. 5478, pp. 362–374. Springer, Heidelberg (2009)
6. Lang, H., Metzler, D., Wang, B., Li, J.T.: Improved latent concept expansion using hierarchical markov random fields. In: Proc. of CIKM 2010 (2010)
7. Lv, Y., Zhai, C.: Adaptive relevance feedback in information retrieval. In: Proc. of CIKM 2009 (2009)
8. Maxwell, K., Schafer, B.: Concept and Context in Legal Information Retrieval. In: Proc. of JURIX 2008 (2008)
9. Navigli, R.: Word sense disambiguation: A survey. ACM Comput. Surv. 41(2), Article 10 (2009)
10. Voorhees, E.: Query expansion using lexical-semantic relations. In: Proc. of SIGIR 1994 (1994)
11. Bai, J., Nie, J.Y.: Adapting information retrieval to query contexts. Inf. Process. Manage. 44(6), 1901–1922 (2008)
12. Lee, K., Croft, B., Allan, J.: A cluster-based resampling method for pseudo-relevance feedback. In: Proc. of SIGIR 2008 (2008)

13. Vechtomova, O., Karamuftuoglu, M., Robertson, S.: On document relevance and lexical cohesion between query terms. Information Processing & Management 42(5), 1230–1247 (2006)
14. Vechtomova, O., Karamuftuoglu, M.: Query expansion with terms selected using lexical cohesion analysis of documents. Information Processing & Management 43(4), 849–865 (2007)
15. Lewis, D., Croft, B.: Term clustering of syntactic phrases. In: Proc. of SIGIR 1990 (1989, 1990)
16. Koster, C., Beney, J.: Phrase-based document categorization revisited. In: Proc. of PaIR 2009 (2009)
17. Navigli, R., Velardi, P.: An analysis of ontology-based query expansion strategies, workshop on adaptive text extraction and mining (ATEM 2003). In: 14th European Conference on Machine Learning (ECML 2003) (2003)
18. Cao, G., Nie, J.Y., Gao, J., Robertson, S.: Selecting good expansion terms for pseudo-relevance feedback. In: Proc. of SIGIR 2008 (2008)
19. Xu, J., Croft, B.: Query expansion using local and global document analysis. In: Proc. of SIGIR 1996 (1996)
20. Banerjee, S., Ramanathan, K., Gupta, A.: Clustering short texts using Wikipedia. In: Proc. of SIGIR 2007 (2007)
21. Ganesh, S., Varma, V.: Exploiting structure and content of Wikipedia for Query Expansion in the context of Question Answering. In: Recent Advances in Natural Language Processing (RANLP 2009), Bulgaria (2009)
22. Xu, Y., Jones, G., Wang, B.: Query dependent pseudo-relevance feedback based on Wikipedia. In: Proc. of SIGIR 2009 (2009)
23. Kapalavayi, N., Murthy, S., Hu, G.: Document classification efficiency of phrase-based techniques. In: IEEE/ACS International Conference on Computer Systems and Applications (2009)
24. Li, Y., Luk, W., Ho, K., Chung, F.: Improving weak ad-hoc queries using Wikipedia as external corpus. In: Proc. of SIGIR 2007 (2007)
25. Cui, H., Wen, J., Nie, J., Ma, W.: Query Expansion by Mining User Logs. IEEE Transactions on Knowledge and Data Engineering 15(4), 829–839 (2003)
26. Kwok, K., Chan, M.: Improving two-stage ad-hoc retrieval for short queries. In: Proc. of SIGIR 1998 (1998)
27. Arampatzis, A., Tsoris, T., Koster, C., Van Der Weide, T.: Phrase-based information retrieval. Information Processing & Management 34(6), 693–707 (1998)
28. Arguello, J., Elsas, J.L., Callan, J., Carbonell. J.G.: Document Representation and Query Expansion Models for Blog Recommendation. In: Proc. of ICWSM 2008 (2008)
29. Robertson, S., Jones, K.: Relevance weighting of search terms. Journal of the American Society for Information Science 27, 129–146 (1976)

Increasing Broadband Subscriptions for Telecom Carriers through Mobile Advertising

Chia-Hui Chang and Kaun-Hua Huo

Dept. of Computer Science and Information Engineering,
National Central University, Taoyuan, Taiwan

Abstract. Mobile devices are popular. However, mobile broadband subscriptions are 20 percent of the mobile subscriptions, due to the high payments requirement for broadband subscriptions. On the other hand, US mobile ad spend will exceed US$1 billion in 2011 according to emarketer.com. Therefore, to increase broadband subscribers by providing free or discounted fees through the deployment of mobile advertising framework by the telecommunication system is important. Telecommunication runs ads agent platform to attract investments from advertisers. Subscribers read promotional advertisements to get discounted payment. While the advertisers pay a reasonable price for advertising, the possible commercial activities will bring revenues. As a result, this framework is a triple-win for telecommunication, advertisers and subscribers. We describe a framework for delivering appropriate ads of the ideal time at the ideal place to the ideal subscriber is the three key issues on how/when to show the ads and what potential ads clicked by subscribers.

1 Introduction

In recent years, we have seen the trends in the growing popularity of mobile device (e.g. mobile phone, PDAs, vehicle phone, and e-books). According to the report of International Telecommunication Union (ITU), mobile cellular subscriptions have reached an estimated 5.3 billion (over 70 percent of the world population) at the end of 2010. Intuitionally, mobile device is now playing an increasingly important role in human life. Moreover, it is becoming a special market potential. According to IAB[1] (Interactive Advertising Bureau), mobile advertising is accepted by a lot of consumers, comparing to online advertising on World Wide Web due to its highly interactive with users, which is hard to achieve for other media. Mobile advertising is an alternative way of web monetization strategies, especially for telecommunication corporations to expand revenue. Furthermore, it is predicted that the average annual growth will rose by 72% in the next five year.

There are many ways of mobile communication, such as WAP, GPRS, 3G modem, and PHS. However, due to high charge of internet access, not all users have mobile internet access. Mobile broadband subscriptions are 1 billion (about 20 percent of the mobile subscriptions)[2]. In Taiwan, for example, the ratio of

---

[1] http://www.iab.net/media/file/StateofMobileMarketing.pdf
[2] http://www.mobiletechnews.com/info/2011/02/02/131112.html

M.V.M. Salem et al. (Eds.): AIRS 2011, LNCS 7097, pp. 127–136, 2011.

the mobile web users to the mobile users is less than 50%. Meanwhile, VAS (Value-Added Services) is deeply influenced by prices. There are about 75% customers, who will consider prices rather than interest when choose VAS [9]. Therefore, the lower VAS prices and internet accessing charges the customers get, the more customers we find. Thus, a way to reduce the internet accessing fee mobile devices is important.

In this paper, we propose a framework for mobile advertising, which involves the telecommunication, the advertiser, and the subscriber. It is dubbed "triple-win" that all of them get benefits from each other. The telecommunication provides the Ads Agent Platform to attract campaign for advertisers. The advertiser registers commercial ads with the telecommunication, and pays a reasonable price. The users subscribe for promotional ads, which are sent to their mobile devices by the Ads Agent Platform in order to get discounted internet accessing from the telecommunication. Within this framework, the advertiser gets repay when the ads bring potential customers; the telecommunication gets benefit by increasing the number of internet subscribers; and the subscriber also gets coupons and discounted internet accessing. As a result, this framework is triple-win for the telecommunication, the advertiser and the subscriber.

However, delivering appropriate ads of the ideal time at the ideal place to the ideal subscriber are key issues. We consider HWW (How, When and What), the three key issues that we really care about: (1) How do we show the ads in subscribers' mobile phone? (2) When should we show the ads to ensure subscribers will read ads without redundancy? (3) What potential ads will be clicked by the subscribers?

In this paper, we describe our design for mobile advertising platform based on three aspects: context, content and user preference. The system makes utilization of velocity-detection and content-match to allocate personalized ads. In order to decrease redundant ads and to increase business value, mobile ad matching system should consider subscriber's current status and activities before delivering personalized mobile ads.

2   Related work

2.1   Consumer Behavior and Personalized Advertising

Marketing researchers have studied consumer behaviors for decades from analyzing the factors of consumer behaviors and to model building. Turban et al. [7] analyzed the main factors on consumer's decision, including consumer's individual characteristics, the environment and the merchant's marketing strategy (such as price and promotion). Varshney and Vetter [10] also proposed the use of demographics, location information, user preference, and store sales and specials for mobile advertising and shopping application. Rao and Minakakis [5] also suggest that customer profiles, history, and needs are important for marketing. Xu et al. [9] proposed a user model based on experimental circumstances studies of the restaurants, and used Baysian Network to analyze which variables would affect the attitude of consumers toward mobile advertising.

In summary, the most effective factors can be divided into three parts: Content, Context and User Preference. Content factor includes brand, price, marketing strategy, etc. Context factor includes user location, weather, time, etc. At last, User Preference includes brand, interest, user activity, etc.. Online (Web) advertising is shown to be more effective than traditional media since it predicts user's intent by the pages that the user visits. However, for mobile advertising, more contextual information and user preference can be obtained based on the location, speed of users' mobile device.

2.2 Web Contextual Advertising

In contextual advertising, an ad generally features a title, a text-based abstract, and a hyperlink. The title is usually depicted in bold or a colorful font. The latter of the abstract is generally clear and concise due to space limitations. The hyperlink links to an ad web page, known as the landing page.

Several studies pertaining to advertising research have stressed the importance of relevant associations for consumers and how irrelevant ads can turn off users and relevant ads are more likely to be clicked [1], [2], [3]. Ribeiro-Neto et al [6]. proposed a number of strategies for matching pages to ads based on extracted keywords. To identify the important parts of the ad, the authors explored the use of different ad sections (e.g., bid phrase, title and body) as a basis for the ad vector. The winning strategy required the bid phrase to appear on the page, and then ranked all such ads using the cosine of the union of all the ad sections and the page vectors. While both pages and ads are mapped to the same space, there exists a discrepancy (called "impedance mismatch") between the vocabulary used in the ads and on the pages. Hence, the authors improved matching precision by expanding the page vocabulary with terms from similar pages.

Besides, in contextual advertising, Fan et al. proposed the utilization of sentiment detection for blogger-centric contextual advertising. The results clearly indicated that their proposed method can effectively identify those ads that are positively correlated with the given blogp ages [2], [3].

2.3 Mobile Advertising

Mobile advertising is predicted to will become an important telecommunication's revenue and monetization strategies. Advertisers can associate each user with fully personalized ads to increase large value of mobile ads. The existing mobile advertising methods can be divided into 3 categories including SMS, Applets, and Browser. SMS typically contain one or more commercial offers or ads that invite users to subscribe or purchase products and services. Mobile ads embedded in applets are contextual ads that are set pop-up when users are using the applets. The browser is a particular applet for retrieving, presenting, and traversing information resource on the World Wide Web. Mobile ads showed in browser are more similar contextual advertising on the web. Note, both applets and browsers require data transmission through the internet. This leads to additional payment of internet charges to users, while receiving SMS is free for users.

SMS is the most common mobile advertising for enterprises. However, due to infrastructure limitations, it does not support customized mobile ads per individual. Thus, many users treat these SMS as spam. As indicated by Giuffrida et al. in [4], even though the mobile advertising makes a substantial improvement in overall business performance by targeting users with most relevant offers based on user purchase histories, the hard limit imposed by the carrier forces them to target clusters of consumers and send to all users in a cluster.

As mentioned earlier, mobile broadband adoption using 3G are about 20 percents of all mobile subscriptions worldwide due to high price. Thus, it seems promising to explore the mobile advertising market for telecom industry. In fact, the telecom industry has used SMS for mobile advertising since a long time ago. However, such advertising mechanism does not work well since users do not gain any benefit. On the other hand, web-based advertising provided by Google and Apple is more acceptable by users for the reason of free applications. By providing discounted broadband subscriptions, users are more willing to read advertisements. After all, the idea of paying broadband subscription fee to receive ads will not make users feel good, even for free applications. For example, VIBO Telecom Inc. in Taiwan, provides MobiBon service for users to obtain bonus by reading online ads. However, the data transmission of MobiBon is paid by users. We believe that the proposed framework in this paper could solve the dilemma and increase the revenue for telecom industry.

3   Design Methodology

3.1   Addresses the 3 Key Issues for Mobile Advertising

SMS or Broadband Ads? There are 2 mechanisms for existing mobile advertising: SMS and broadband advertisement. SMS is a common way of mobile advertising. It often contains textual ads to promote products and services. However, the infrastructure limitation of SMS makes full customization infeasible [4]. On the other hand, broadband ads are not restricted by this limitation and can be supported by cloud computing technique.

For broadband advertising, the access prices/fees of mobile broadband services will also affect the advertising mechanism. If the access fees are charged by packages, the style of the ads and the number of ads to be delivered can be limited. If, however, the subscribers have unlimited internet access, various style of ads can also be considered. Here, we consider a design of ad allocator which can be applied for both situations via mobile broadband service. While delivering ads to subscribers, the potential ad allocator also records the time that user spent on ads, the number of mobile ads that are clicked/closed by the user in exchange for discounted mobile broadband access. This kind of design can work for either kinds of charging methods.

In addition, due to the limited size of smartphone screen and data transmission charge, we use the textual ads as the mobile ads to save the screen space. We show the mobile ads on the top of screen to get subscribers attention. If the subscribers are interested in the ads, he/she can read the detail by clicking on the ads. The subscribers can also submit query to the mobile ad allocator for promotions and select products by themselves. In the meantime, the mobile ad allocator provides maps and the location information of the products. He/she can visit the nearby stores. We consider such activities to be more effective, since the subscriber is more likely to buy their favorite products.

When to Show the Ads? Two timing often used in mobile advertising are trigger-based and fixed-schedule. The trigger-based timing refers to activities which happen after certain user events. For example, the mobile ads are sent to the user when the user sends or receives SMS in [4]. In our framework, the mobile ad allocator is designed with both fixed schedule as well as trigger-based advertising. To avoid redundant ads and to reduce the system loads, the mobile ad allocator does not match or show ads when the phone is on standby mode. When a smart phone is running on standard power mode, mobile ads are shown in an interval of 30 minutes. However, ads will also be triggered when the smart phones wake up from standby mode for five minutes or by submitted queries.

Ads Allocation Factors. As mentioned earlier, mobile advertising has a wide variety of factors to be considered, including: context, content and user preference. In our framework, we consider one context factor (locality) and three dimensions of content factor including product category and promotion activity as well as short-term interest represented by user query. These four factors and the mapping between the user and the content are used for personalized ads allocation.

- *Locality* We consider the reachibility of ads based on user location and velocity. That is, the system calculates the available range of a user and then selects mobile ads from the available range for recommendation. The motivation is to attract customers who are near the store.

- *Product Category* The system categorizes ads can be classified into 5 categories: delicacies, clothing, residence, transportation and life service. The ad category is paired with the long-term interest of the user.

- *Promotion* Price and discount are the most significant factor for customers' decision. For users who make decisions based on price and discount, such information is a key factor.

- *User Query* It reflects the factor for short-term requirement that is trigger-based advertising. Such information can be used to locate ads that are of short-term interest to users.

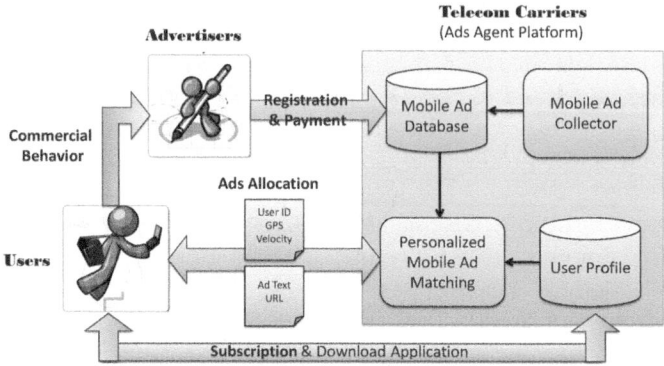

Fig. 1. The Ad Agent Platform consists of three components: the mobile ad database, the user profile and the personalized ad matching

3.2   System Architecture

As shown in figure 2, the Ad Agent Platform consists of three components: the mobile ad database, the user profile and the personalized ad matching. When the advertiser registers commercial mobile ads, we obtain the data for the mobile ad database. Similarly, we can attain subscriber's information as user profile when they subscribe for the mobile advertising service and download the mobile ad allocator which is developed by carrier to get discounted broadband access. The ad allocator matches mobile ads with user profiles based on subscriber's location information and current velocity.

3.3   Personalized Ad Matching

Personalized ad matching can be regarded as an information retrieval problem. In other words, we can calculate similarity between the user information and the ad information with combination of various IR models. Our ad matching algorithm is given in Algorithm 1, which contains 3 steps: ad filtering, content similarity computation and adjustment. In our opinion, we try to find the ad which is satisfying the user needs from context information. That is, we suppose probably the user needs, and then we select the most related ad though the scoring function of the candidate ads.

First, we calculate the radius, the available distance, $Radius$, which a user u can arrive in m minutes. Assuming the velocity of the user u is v, then we get $Radius = v * m$. The mobile ads in this range is denoted by $R(u) = \{a | d(a, u) < Radius\}$ where $d(a, u)$ represents the distance between the user u and the store of the ad a.

Next, for each candidate ad, we calculate the content similarity score between ad a and user u. The long-term factors include the promotion activity and five product category, which are binary attributes, so we calculate them by Jaccard similarity $Jaccard(u, a)$. Furthermore, the short-term factor, represented by user query, $q(u)$ is compared with ad titles $t(a)$ and landing pages $p(a)$ to get the cosine similarity, denoted by $Cos(v_1, v_2) = v_1 \bullet v_2$. Note that Jaccard similarity scores can sometimes dominate the content similarity score. So, the Jaccard similarity score is multiplied by a weighted value α, $\alpha \in [0.1, 0.2]$.

Finally, after the calculation of content similarity, if the highest score of the filtered ads in R(u) is less than α, then the top k ads with discounted content similarity score are added based on distance average. For ads with distance less than average distance, the discounted score is calculated by $Sim_{discounted}(u, a) = 0.5 * Sim_{score}(u, a)$; while, for ads with distance larger than average distance, the discounted score is calculated by $Sim_{discounted}(u, a) = 0.3 * Sim_{score}(u, a)$. On completion of ordering the scores, we can get the ads that are most appropriate for the user u.

Algorithm 1. The personalized ad matching algorithm contain ad filtering, content similarity computation and adjustment.

---

1. Ads filtering based on GPS and velocity v. $Radius = v * m$
2. Content similarity scoring based on long-term and short-term:

$$Sim_{score}(\overrightarrow{u}, \overrightarrow{a}) = \alpha * Jaccard(u, a) + Cos(q(u), t(a)) + Cos(q(u), p(a)); \quad (1)$$

   Jarccard similarity is used for binary product categories and promotion
   Cosine similarity is used for user query and ad titles / landing pages
3. If the highest score of filtered ads in step 1 is less than α, then add top ranked 50 ads by discounted score.

---

4   Mobile Ad Collector

In general, the information of mobile ads is acquired from the advertiser when the advertiser registers the advertisement with the telecommunication. However, we didn't have a massive amount of mobile ads in live data because this framework does not run in a real business environment. Thus, we build a mobile ad collector, collecting ads from online advertisements.

4.1   Ad-crawler Platform

We proposed a mobile ad collector, which collect online advertisements automatically from Google AdSense[3] as shown in [4], [5]. First, we choose some general topic words as the query term to request web pages from search engines such as Google and Yahoo! About 200,000 web pages are retrieved as our web page set. Next, we place these web pages on the ad-clawer platform to obtain the

---

[3] http://www.labnol.org/google-adsense-sandbox/

corresponding online ads assigned by Google AdSense. The information of these online advertisements consisted of hyperlink, title and abstract. We collect 54,709 online ads. In order to calculate the similarity between the user and the ad, we extract features including postal address, promotion activity, product category and user query (or the brand name) as the mobile ads factors in our system. We extracted these features for each online ad from its landing page as follows.

4.2   Ad Feature Extraction

Postal Address Extraction. The landing page of each online advertisement may contain the location information for consumers. A landing page could even contain more than one postal address. For example, the location search page for KFC[4] lists ten postal addresses in one web page. Thus, an ad could be associated with more than one postal address. Unfortunately, only 4,003 online advertisements contain postal addresses (a total of 9,327 postal addresses are extracted). Hence, we randomly assign a geographic coordinates around a user to each online advertisement. Using Google Map API, we convert each geographic coordinates into a postal address.

Text Preprocessing. Before introducing the promotion activity identification and the product category classification, the landing page of each online advertisement is processed for term representation. The preprocessing steps include HTML tag removal, tokenization, stemming, etc. Finally, we count term frequency in each landing page. To be brief, the text preprocessing is a process to translate the raw landing pages into term features and term frequency.

Promotion Activity Identification. The promotion activity identification is regarded as a classification task. The promotion classifier is implemented with the tool WEKA[5] using bag of words representation. We train a model to classify whether an ad contains promotional information with supervised learning. We manually label 550 online ads with postal addresses in Illinois for 10-fold cross validation. The (weighted) average precision, recall and F-Measure are about 88.9%.

Category Classification. Five classes including delicacies, clothing, residence, transportation and life service are used for product categories. To prepare training data, we define some query keywords for each category (except for the last category: life service), and use an IR system to retrieval top relevant ads for manual labeling. Relevant ads are used ad positive examples, while negative examples are chosen randomly. For each categorization task, around 300 training examples including equal number of positive and negative examples are collected for training data. Then, a binary classifier is trained by state-of-the-art learning algorithm using 10-fold cross-validations. Note that if an ad is not classified to any category, we attribute it with the life service category. In brief, we make use

---

[4] http://www.kfcclub.com.tw/Story/Store/
[5] http://www.cs.waikato.ac.nz/ml/weka/

Table 1. The training result for category classification with ten-fold cross-validations shows the performance is acceptable

Class	Examples	Precision	Recall	F-measure
Delicacies	313	0.911	0.911	0.911
Clothing	302	0.984	0.983	0.983
Residence	302	0.815	0.815	0.814
Transportation	302	0.931	0.930	0.930

of an IR system to prepare the training data for each product. Table 1 shows the performance (weighted average) of the categorization tasks are acceptable (above 90%) except for the category "residence" (81.4%).

5 Experiment Result

We built a simulated platform to evaluate the effectiveness of our approach. The location of the experiment is set at Illinois, USA. Two situations are copied in the simulated platform: surrounding situation is when the user goes around without particular destination, while route situation is when the user travels from a start location to some end location. The simulated platform is a web site written in HTML, Java Script and PHP.

Table 2 shows preliminary result of the proposed approaches with 30 subjects, each with 20-25 runs of tests. At each run, the user is presented with four ads selected by four approaches shown randomly. The evaluation is measured by precision, recall and F-measure per user base and then averaged over all subjects. The first row shows the performance of the proposed approach, which includes all four factors, while the second row shows the performance of user information involving long-term interests (promotion activity and product category) and short-term interests (user query) but excluding locality factor. We also show the performance of the locality factor and random selection as a comparison. The proposed approach has the best precision (49.4%), recall (36.9%) and F-measure (40.1%).

Table 2. Performance with various approaches

Approach	Precision	Recall	F-measure
Proposed Approach	0.494	0.369	0.401
User Information	0.428	0.311	0.313
Locality	0.310	0.199	0.232
Random	0.200	0.118	0.141

6 Conclusion and Future Work

Mobile devices are more and more popular and vital for people's life. So, mobile advertising will be an important market for web monetization. Google and Apple have launched their mobile advertising strategies through AdMob and iAd, which will bring revenue for the OS providers by sharing profits with programmers. While telecommunication still have a hard time increasing the number of broadband subscriptions since many student users still consider the cost to be the main issue. In this paper, we propose the framework that is a triple-win for the telecommunication, the mobile advertisers and the subscribers. We address the three key issues: (1) how to show the mobile ads, (2) when to show the mobile ads and (3) what potential mobile ads will be click by the subscribers. Our system recommends mobile ads to the users based on the factors of velocity-detection and content-match. In order to conduct experiments, we also design a mobile ad collector, which crawls online advertisements automatically. The preliminary result shows that ad allocation based on all factors (locality, product category, promotion and user query) can achieve the best performance.

References

1. Fan, T.-K., Chang, C.-H.: Blogger-Centric Contextual Advertising. Journal of Expert Systems With Applications 38(3), 1777–1788 (2011)
2. Feng, J., Bhargava, H.K., Pennock, D.: Comparison of Allocation Rules for Paid Placement Advertising in Search Engine. In: Proceedings of the 5th International conference on Electronic Commerce, ACM EC 2003, pp. 294–299. ACM, Pittsburgh (2003)
3. Giuffrida, G., Sismeiro, C., Tribulato, G.: Automatic content targeting on mobile phones. In: EDBT 2008 (2008)
4. Penev, A., Wong, R.K.: Framework for Timely and Accurate Ads on Mobile Devices. In: CIKM, Hong Kong, November 2-6 (2009)
5. Rao, B., Minakakis, L.: Evolution of Mobile Location-based Services. Communications Of The ACM (2003)
6. Ribeiro-Neto, B., Cristo, M., Golgher, P.B., de Moura, E.S.: Impedance coupling in content-targeted advertising. In: ACM SIGIR 2005, pp. 496–503 (2005)
7. Turban, E., Lee, N., King, D., Chung, H.M.: Electronic Commerce Managerial Perspective. Prentice Hall (2000)
8. Varshney, U., Vetter, R.: Mobile Commerce: Framework, Applications and Networking Support. Mobile Networks and Applications (2002)
9. Xu, D.J., Liao, S.S., Li, Q.: Combining empirical experimentation and modeling techniques: A design research approach for personalized mobile advertising applications. Decision Support Systems 44 (2008)

Query Recommendation by Modelling
the Query-Flow Graph

Lu Bai, Jiafeng Guo, and Xueqi Cheng

Institute of Computing Technology, Chinese Academy of Sciences, Beijing, China
{bailu,guojiafeng}@software.ict.ac.cn,
cxq@ict.ac.cn

Abstract. Query recommendation has been widely applied in modern search engines to help users in their information seeking activities. Recently, the query-
flow graph has shown its utility in query recommendation. However, there are
two major problems in directly using query-flow graph for recommendation. On
one hand, due to the sparsity of the graph, one may not well handle the recommendation for many dangling queries in the graph. On the other hand, without
addressing the ambiguous intents in such an aggregated graph, one may generate
recommendations either with multiple intents mixed together or dominated by
certain intent. In this paper, we propose a novel mixture model that describes the
generation of the query-flow graph. With this model, we can identify the hidden
intents of queries from the graph. We then apply an intent-biased random walk
over the graph for query recommendation. Empirical experiments are conducted
based on real world query logs, and both the qualitative and quantitative results
demonstrate the effectiveness of our approach.

1 Introduction

Nowadays, query recommendation has been recognized as an important tool that helps
users seek their information needs. Many approaches have been proposed to generate
query recommendations by leveraging query logs. Different types of information in the
query logs have been taken into account, including search results (11), clickthrough
(12) and search sessions (5).

Recently, the query-flow graph (2) has been introduced as a novel representation of
session information in query logs. It integrates queries from different search sessions
into a directed and homogeneous graph. Nodes of the graph represent unique queries,
and two queries are connected by a directed edge if they occur consecutively in a search
session. The Query-flow graph has shown its utility in query recommendation (2–4).

However, there are several problems in directly using the query-flow graph for recommendation as in existing approaches. Firstly, due to the information sparsity, lots
of dangling queries which have no out-links exist in the query-flow graph[1]. Therefore,
recommendation approaches based on random walks (2, 3) over the directed graph may
not well handle such dangling queries. Moreover, queries are often ambiguous in their

---

[1] In our experiment, we observe that the dangling queries account for nearly 9% of the total
queries, which is not negligible in real application.

M.V.M. Salem et al. (Eds.): AIRS 2011, LNCS 7097, pp. 137–146, 2011.
© Springer-Verlag Berlin Heidelberg 2011

search intent and thus the aggregated query-flow graph in fact is a mixture of multiple search intents. Most existing approaches (2–4) do not take into account the ambiguous intents in the query-flow graph when generating recommendations. Therefore, for ambiguous queries, one may either produce recommendations with multiple intents mixed together which are difficult for users to consume, or provide recommendations dominated by certain intent which cannot satisfy different user needs.

In this paper, we propose to model the query-flow graph for better query recommendation. Specifically, we introduce a novel mixture model for the query-flow graph. The model employs a probabilistic approach to interpret the generation of the graph, i.e., how the queries and the transitions between queries are generated under the hidden search intents. We then apply an intent-biased random walk over the graph for query recommendation. In this way, we can well resolve the recommendation problems for dangling queries and ambiguous queries in using query-flow graph.

We conducted empirical experiments based on a collection of query logs from a commercial search engine. Both the qualitative and quantitative results demonstrate the effectiveness of our approach as compared with existing baseline methods.

2   Related Work

2.1   Query Recommendation

Query recommendation is a widely accepted tool employed by search engines to help users express and explore their information needs. Beeferman et al. (1) applied agglomerative clustering algorithm over the clickthrough bipartite graph to identify related queries for recommendation. Ma et al. (7) developed a two-level query recommendation method based on both the user-query graph and the query-URL graph. Zhu et al. (13) generated diverse query recommendations based on the query manifold structure.

Recently, query-flow graph was introduced by Boldi et al. (2), and they applied personalized random walk (2, 3) over the query-flow graph to recommend queries. Unlike previous work on query-flow graph, our approach explores the query-flow graph with a mixture model for query recommendation, so that we can well resolve the recommendation problems for dangling queries and ambiguous queries in using query-flow graph.

2.2   Mixture Models

Recently, there have been different mixture models applied on graphs for community discovery. For example, Newman et al. (8) proposed a probabilistic mixture model to discover the overlapped communities in graph. Ramasco et al. (9) introduced a more general mixture model on graph for the same purpose. Ren et al. (10) described a mixture model for undirected graph, where each edge in the graph is assumed to be from the same community. Inspired by the above work, we propose a novel mixture model to interpret the generation of the query-flow graph under multiple hidden intents.

3   Our Approach

In this section, we first briefly introduce the query-flow graph. We then describe the proposed mixture model in detail, which learns the hidden intents of queries by modelling the generation of the query-flow graph. Finally, we show how to leverage the learned intents for better query recommendation with an intent-biased random walk.

3.1 Query-Flow Graph

The query-flow graph integrates queries from different search sessions into a directed and homogeneous graph. Formally, we denote a query-flow graph as $G = (V, E, w)$, where $V = Q \cup \{s, t\}$ is the set of unique queries Q in query logs plus two special nodes s and t, representing a starting state and a terminal state of any user search session. $E \subseteq V \times V$ denotes the set of directed edges, where two queries q_i and q_j are connected by an edge if there is at least one session of the query log in which q_j follows q_i. w is a weighting function that assigns to every pair of queries $(q_i, q_j) \in E$ a weight $w_{\overrightarrow{ij}}$. The definition of the weight w may depend on the specific applications. In our work, we simply consider the weight to be the frequency of the transition in the query log. Fig. 1 shows a query-flow graph that is constructed from the a set of search sessions.

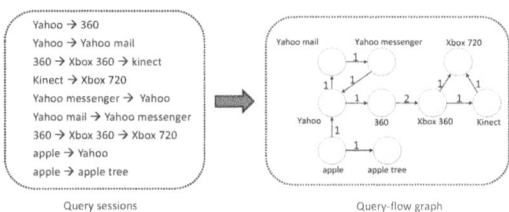

Query sessions   Query-flow graph

Fig. 1. A simple query-flow graph constructed from query sessions

3.2 Mixture Model on Query-Flow graph

We propose a novel mixture model to interpret the generation of the query-flow graph. In essentials, our model is based on the following assumption: Queries are generated from some hidden search intents, and two queries occurred consecutively in one session if they are from the same search intent. The above assumption is quite natural and straightforward. Typically, users submit a query to search according to their potential information needs (i.e., search intent). Users may consecutively reformulate their queries in a search session until their original needs are fulfilled (or exit with a failure). Therefore, without loss of generality, queries occurred consecutively in a search session can be viewed as under the same search intent.

Specifically, given a query-flow graph G which consists of N nodes and M directed edges, we assume the graph G is generated under K potential search intents, where each intent is characterized by a distribution over queries. Let $e_{\overrightarrow{ij}} \in E$ denote a directed edge from query q_i to query q_j. We then assume the following generative process for each edge $e_{\overrightarrow{ij}}$ in the query-flow graph:

1. Draw an intent indicator $g_{\overrightarrow{ij}} = r$ from the multinomial distribution π.
2. Draw query nodes q_i, q_j from the same multinomial intent distribution β_r, respectively.

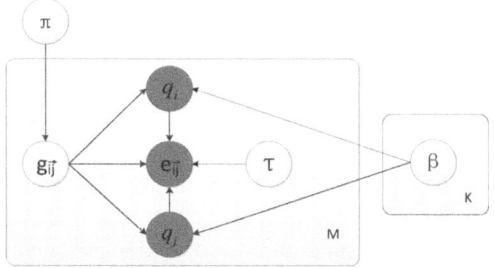

Fig. 2. The graphic model of generation of query-flow graph

3. Draw the directed edge $e_{\overrightarrow{ij}}$ from a binomial distribution $\tau_{\overrightarrow{ij},r}$ under a hidden intent $g_{ij} = r$.

Here, the K-dimensional multinomial distribution π reflects the proportion of different search intents over the whole query-flow graph, the multinomial distribution β over queries describes the hidden search intents, and the binomial distribution τ captures the probability of the edge direction between two queries under a given search intent. The Fig. 2 shows the graphic model of generating query flow graph.

Based on the above process, the probability of an observed directed edge $e_{\overrightarrow{ij}}$ belonging to the r-th search intent can be obtained by

$$\Pr(e_{\overrightarrow{ij}}, g_{\overrightarrow{ij}} = r | \pi, \beta, \tau) = \pi_r \beta_{r,i} \beta_{r,j} \tau_{\overrightarrow{ij},r}$$

In this way, the likelihood of the whole query-flow graph G is

$$\Pr(G|\pi, \beta, \tau) = \prod_{i=1}^{N} \prod_{j:j \in C(i)} \left(\sum_{r=1}^{K} \pi_r \beta_{r,i} \beta_{r,j} \tau_{\overrightarrow{ij},r} \right)^{w_{\overrightarrow{ij}}} \tag{1}$$

where $w_{\overrightarrow{ij}}$ denotes the weight of edge $e_{\overrightarrow{ij}}$, and $C(i)$ denotes the set of nodes pointed by query q_i.

The parameters to be estimated in our model are π, β, and τ. We maximize the likelihood shown in Equation (1) to estimate these parameters. The sum in the bracket makes the direct estimation difficult, but with the help of Expectation Maximization (EM) algorithm the problem can be solved easily.

As we can see, the hidden variables in our mixture model are intent indicators $g_{\overrightarrow{ij}}$. In E-step, the posterior probabilities of hidden variables are calculated as

$$q_{\overrightarrow{ij},r} = \Pr(g_{\overrightarrow{ij}} = r | e_{\overrightarrow{ij}}) = \frac{\Pr(e_{\overrightarrow{ij}}, g_{\overrightarrow{ij}} = r)}{\Pr(e_{\overrightarrow{ij}})} = \frac{\pi_r \beta_{r,i} \beta_{r,j} \tau_{\overrightarrow{ij},r}}{\sum_{r=1}^{K} \pi_r \beta_{r,i} \beta_{r,j} \tau_{\overrightarrow{ij},r}}$$

In M-step, we update the parameters by the following formulas:

$$\pi_r = \frac{\sum_{i=1}^{N} \sum_{j:j \in C(i)} w_{\overrightarrow{ij}} q_{\overrightarrow{ij},r}}{\sum_{r=1}^{K} \sum_{i=1}^{N} \sum_{j:j \in C(i)} w_{\overrightarrow{ij}} q_{\overrightarrow{ij},r}}, \quad \tau_{\overrightarrow{ij},r} = \frac{w_{\overrightarrow{ij}} q_{\overrightarrow{ij},r}}{w_{\overrightarrow{ij}} q_{\overrightarrow{ij},r} + w_{\overrightarrow{ji}} q_{\overrightarrow{ji},r}}$$

$$\beta_{r,i} = \frac{\sum_{j:j\in C(i)} w_{\overrightarrow{ij}} q_{\overrightarrow{ij},r} + \sum_{k:i\in C(k)} w_{\overrightarrow{ki}} q_{\overrightarrow{ki},r}}{\sum_{i=1}^{N} \left(\sum_{j:j\in C(i)} w_{\overrightarrow{ij}} q_{\overrightarrow{ij},r} + \sum_{k:i\in C(k)} w_{\overrightarrow{ki}} q_{\overrightarrow{ki},r} \right)}$$

The E-step and M-step are repeated alternatively until the log-likelihood does not increase significantly. Note that the EM algorithm will not necessarily find the global optimal. We resolve this by trying several different starting points to get an good solution in practice.

3.3 Intent-Biased Random Walk

As aforementioned, directly applying the traditional personalized random walk on query-flow graph into recommendation may not well handle the dangling queries and ambiguous queries. Here we further introduce our intent-biased random walk to recommend queries based on the learned intents above. The basic idea of our model is to integrate the learned intents of queries into the prior preference of the personalized random walk, and apply the random walk under different search intent respectively.

Formally, let W denote the weight matrix of the query-flow graph G with row normalized. An intent-biased random walk over the query-flow graph G under the r-th search intent given the original query q_i is then determined by the following transition probability matrix $A_{i,r}$, which is defined as following:

$$A_{i,r} = (1 - \lambda)W + \lambda \mathbf{1} P_{i,r}$$

where λ denotes the teleportation probability, and $P_{i,r}$ denotes the preference vector of intent-bias random walk under the r-th intent defined as

$$P_{i,r} = \rho \cdot \mathbf{e}_i^T + (1 - \rho) \cdot \beta_r$$

where $\mathbf{e}_i^T$ is the vector whose entries are all zeroes, except for the i-th whose value is 1, β_r is our learned r-th intent distribution over queries, and $\rho \in [0, 1]$ is the weight balancing the original query and its intent.

The intent-biased random walk has a unique stationary distribution $R_{i,r}$ such that $R_{i,r} = A_{i,r}^T R_{i,r}$ (called the personalized PageRank score relative to q_i under the r-th intent). Such a personalized PageRank can be computed using the power iteration method. We can then employ the personalized PageRank score to rank queries with respect to q_i for recommendation.

We apply our intent-biased random walk under each intent of query q_i, and obtain the corresponding recommendations. Finally, the recommendations are grouped by intent and represented to users in a structured way, where the intent groups are ranked according to the intent proportion of the given query q_i calculated by

$$\Pr(r|i) \propto \Pr(r) \Pr(i|r) = \pi_r \beta_{r,i}$$

As we can see, if we set the parameter ρ to 1, our intent-biased random walk will degenerate into the traditional personalized random walk as applied in previous work (2, 4). Obviously, under such a personalized random walk, we may not obtain any recommendations for dangling queries. To avoid non-recommendation, one may add a small uniform vector to the preference vector (i.e., teleportation to every node). However, in that case, the recommendations for the dangling query will be mostly popular queries which may not related to original query at all. While in our model, we set ρ less than 1 so that we can smooth the preference vector with the learned intents, which can provide rich background information of the original query in a back-off way. If the original query is a dangling query, the preference vector will be reduced to the corresponding intent distribution so that we can still obtain related recommendations for the original query.

Moreover, previous approaches usually applied the personalized random walk on the graph for recommendation without addressing the hidden intents. In this way, for ambiguous queries, they may either produce recommendations with multiple intents mixed together which are difficult for users to consume, or provide recommendations dominated by certain intent which cannot satisfy different user needs. In our model, we can naturally generated recommendations for the original query with respect to its different intents. The structured recommendation results would be easy to understand and diverse search intents can be covered.

4   Experiments

4.1   Data Set

The experiments are conducted on a 3-month query log from a commercial search engine. We split the query stream into query sessions using 30 minutes timeout, and construct the query-flow graph as section 3.1 described. To decrease the noise in search sessions, we get rid of those edges with frequencies lower than 3. We then draw the biggest connected component of the graph for experiment. After these steps, the obtained graph consists of $16,980$ distinct queries and $51,214$ distinct edges.

4.2   Evaluation of Intents

Fig. 3 shows how the log likelihood varies over iterations under different number of hidden intents. According to Fig. 3, the increase of log likelihood turns slow when the

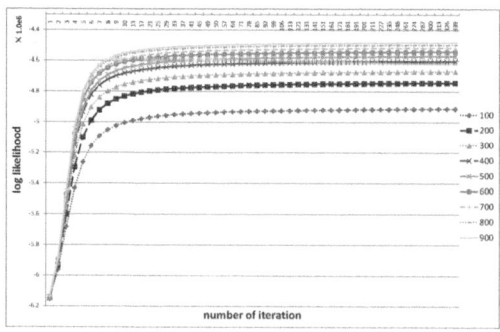

Fig. 3. Log-likelihood over iterations under different mixture Numbers

Table 1. Top 10 queries for 3 randomly sampled learned intents

lyrics	cars	poems
lyrics	bmw	poems
song lyrics	lexus	love poems
lyrics com	audi	poetry
a z lyrics	toyota	friendship poems
music lyrics	acura	famous love poems
azlyrics	nissan	love quotes
lyric	infiniti	sad poems
az lyrics	mercedes benz	quotes
rap lyrics	volvo	mother s day poems
country lyrics	mercedes	mothers day poems

Table 2. Recommendations for dangling queries

Query = "yamaha motor"		Query = "radio disney"	
baseline	ours	baseline	ours
mapquest	yamaha	mapquest	disney
american idol	honda	american idol	disney channel
yahoo mail	suzuki	yahoo mail	disney com
home depot	kawasaki	home depot	disneychannel com
bank of america	yamaha motorcycles	bank of america	disney channel com
target	yamaha motorcycle	target	disneychannel

intent number is larger than 600. It indicates that the mixture model with 600 hidden dimensions is basically sufficient to capture the potential search intents over this graph. Larger number of intents are very probably to be redundant and may cause the problem of over-fitting. Therefore, we set the intent number to 600 in our experiments.

We randomly sample 3 learned intents to demonstrate the effectiveness of our mixture model, as shown in Table 1. For each intent, we list the top 10 ranked queries according to their probabilities under the corresponding intent. The first, second, and third columns are about lyrics, cars, and poems, respectively. The labels of each intent are created by human judge for illustration.

4.3 Evaluation of Query Recommendation

In this part, we evaluate the recommendation performance of our approach by comparing with traditional personalized random walk (2). For our intent-biased random walk, the parameter λ is set to 0.8, and ρ is set to 0.3.

Qualitative Comparison. We take the randomly selected dangling queries "yamaha motor" and "radio disney" as examples to demonstrate the effectiveness of our approach. The recommendation results from our approach and baseline method are demonstrated in the Table 2 . We can see the recommendations from our methods are much more related to the initial queries . On the contrary, the recommendations from baseline method are mostly queries that are popular in the whole date set but unrelated to the original queries. This is because for the dangling queries, the traditional random walk based approaches can only find recommendations with the help of the uniform teleport.

Table 3. Recommendations for Ambiguous Queries

Query = "hilton"		Query = "we"	
baseline	ours	baseline	ours
marriott		wwe	
expedia	marriott	wells fargo	wells fargo
holiday inn	holiday inn	weather	bank of america
hyatt	sheraton	wellsfargo com	wellsfargo
hotel	hampton inn	we channel	wamu
mapquest	embassy suites	tna	
hampton inn	hotels com	bank of america	weather
sheraton		yahoo mail	weather channel
hilton com	paris hilton	weather channel	accuweather
hotels com	michelle wie	wellsfargo	noaa
embassy suites	nicole richie	espn	
residence inn	jessica simpson	usbank com	wwe
choice hotels	pamela anderson	wwe com	tna
marriot	daniel dipiero	www wellsfargo com	wrestleview
hilton honors	richard hatch	bankofamerica com	ecw

We also compared our approach with the baseline method on ambiguous queries. We randomly selected two queries with multiple hidden search intents based on our learned model as shown in the Table 3. We can see that structured query recommendations can be provided by our approach for ambiguous queries. Take the query "we" as an example, the top three categories of recommendations provided by our approach correspond to "financial", "weather" and "wrestling", respectively. The labels here are also human annotated for illustration. However, the baseline method only produces one recommendation list which is a mixture of several intents. Query "hilton" is another interesting example with multiple intents. In this case, the recommendations generated by the baseline method are dominated by queries related to the hotel. In contrast, our approach can obtain two categories of recommendations, one about the hotel and the other about the celebrity. Therefore, our approach may better satisfy users' needs by covering diverse intents of the query.

Quantitative Comparison. Furthermore, we conducted quantitative experiments for evaluating the performance of our recommendation approach. As aforementioned, our proposed approach naturally provides *structured* query recommendation to users . While the baseline method using personalized random walk provides the traditional *list-based* query recommendations to users. We thus follow the way proposed in (6) to compare the performances of different recommendation methods by users' click behaviour.

We randomly sampled 100 queries as our test set. For each query, top 15 recommendations are used for performance comparison. For each recommendation, human judges are required to label how likely he/she would like to click it with a 6-point scale (0, 0.2, 0.4, 0.6, 0.8, 1) as the willingness measure. 3 human judges were asked to participate the labelling process .

We also adopted the *Clicked Recommendation Number (CRN)*, *Clicked Recommendation Score(CRS)*, and *Total Recommendation Score(TRS)* as our evaluation measures. For each query q, let $R = \{r_1, \ldots, r_k\}$ denote the k recommendations generated by a certain method, and $L = \{l_1, \ldots, l_k\}$ denote the corresponding label scores on these

Table 4. Comparisons between Our Approach and Baseline Approach

	|Baseline|	Ours
Average CRN	4.09	4.21(+2.9%)
Average CRS	0.598	0.652(+9.0%)
Average TRS	0.181	0.194(+7.1%)

recommendations, where k is the size of recommendations. The three measures for query q are then defined as follows

$$CRN_q =| \{r_i|l_i > 0, i \in [1,k]\} |, \quad CRS_q = \frac{\sum_{i=1}^{k} l_i}{CRN_q}, \quad TRS_q = \frac{\sum_{i=1}^{k} l_i}{k}$$

where $| * |$ denote the set size. As we can see, CRN reflects the adoption frequency of query recommendations, CRS shows the preference on adopted recommendations, and TRS indicates the effectiveness of overall query recommendations.

Table 4 shows the quantitative evaluation results of the two approaches. The numbers in the parentheses are the relative improvements of our approach over the baseline method. The results show that by providing structured query recommendations based on our intent-biased random walk, we can largely improve both the click number and click willingness on recommendations.

5 Conclusions

In this paper, we propose to explore the query-flow graph for better query recommendation. Unlike previous methods, our novel mixture model identifies the hidden search intents from the query-flow graph. An intent-biased random walk is then introduced to integrate the learned intents for recommendation. Experimental results show the effectiveness of our approach. For the future work, it would be interesting to try to combine the query words, search session and clickthrough information in a unified model to help generate better query recommendations.

Acknowledgments. This research work was funded by the National High-tech R&D Program of China under grant No. 2010AA012500, and the National Natural Science Foundation of China under Grant No. 61003166 and Grant No. 60933005.

References

1. Beeferman, D., Berger, A.: Agglomerative clustering of a search engine query log. In: Proceedings of the Sixth ACM SIGKDD International Conference on Knowledge Discovery and Data Mining, KDD 2000, pp. 407–416. ACM, New York (2000),
   http://doi.acm.org/10.1145/347090.347176
2. Boldi, P., Bonchi, F., Castillo, C., Donato, D., Gionis, A., Vigna, S.: The query-flow graph: model and applications. In: Proceeding of the 17th ACM Conference on Information and Knowledge Management, CIKM 2008, pp. 609–618. ACM, New York (2008),
   http://doi.acm.org/10.1145/1458082.1458163

3. Boldi, P., Bonchi, F., Castillo, C., Donato, D., Vigna, S.: Query suggestions using query-flow graphs. In: Proceedings of the 2009 Workshop on Web Search Click Data, WSCD 2009, pp. 56–63. ACM, New York (2009),
   http://doi.acm.org/10.1145/1507509.1507518
4. Bordino, I., Castillo, C., Donato, D., Gionis, A.: Query similarity by projecting the query-flow graph. In: Proceeding of the 33rd International ACM SIGIR Conference on Research and Development in Information Retrieval, SIGIR 2010, pp. 515–522. ACM, New York (2010),
   http://doi.acm.org/10.1145/1835449.1835536
5. Cucerzan, S., White, R.W.: Query suggestion based on user landing pages. In: Proceedings of the 30th Annual International ACM SIGIR Conference on Research and Development in Information Retrieval, SIGIR 2007, pp. 875–876. ACM, New York (2007),
   http://doi.acm.org/10.1145/1277741.1277953
6. Guo, J., Cheng, X., Xu, G., Shen, H.: A structured approach to query recommendation with social annotation data. In: Proceedings of the 19th ACM International Conference on Information and Knowledge Management, CIKM 2010, pp. 619–628. ACM, New York (2010),
   http://doi.acm.org/10.1145/1871437.1871518
7. Ma, H., Yang, H., King, I., Lyu, M.R.: Learning latent semantic relations from clickthrough data for query suggestion. In: Proceeding of the 17th ACM Conference on Information and Knowledge Management, CIKM 2008, pp. 709–718. ACM, New York (2008),
   http://doi.acm.org/10.1145/1458082.1458177
8. Newman, M.E.J., Leicht, E.A.: Mixture models and exploratory analysis in networks. Proc. Natl. Acad. Sci. USA 104, 9564 (2007), doi:10.1073/pnas.0610537104
9. Ramasco, J.J., Mungan, M.: Inversion method for content-based networks. Phys. Rev. E 77(3), 036122 (2008)
10. Ren, W., Yan, G., Liao, X., Xiao, L.: Simple probabilistic algorithm for detecting community structure. Phys. Rev. E 79, 036111 (2009),
    http://link.aps.org/doi/10.1103/PhysRevE.79.036111
11. Sahami, M., Heilman, T.D.: A web-based kernel function for measuring the similarity of short text snippets. In: Proceedings of the 15th International Conference on World Wide Web, WWW 2006, pp. 377–386. ACM, New York (2006),
    http://doi.acm.org/10.1145/1135777.1135834
12. Yi, J., Maghoul, F.: Query clustering using click-through graph. In: Proceedings of the 18th International Conference on World Wide Web, WWW 2009, pp. 1055–1056. ACM, New York (2009),
    http://doi.acm.org/10.1145/1526709.1526853
13. Zhu, X., Guo, J., Cheng, X., Du, P., Shen, H.W.: A unified framework for recommending diverse and relevant queries. In: Proceedings of the 20th International Conference on World Wide Web, WWW 2011, pp. 37–46. ACM, New York (2011),
    http://doi.acm.org/10.1145/1963405.1963415

Ranking Content-Based Social Images Search Results with Social Tags

Jiyi Li, Qiang Ma, Yasuhito Asano, and Masatoshi Yoshikawa

Department of Social Informatics, Graduate School of Informatics, Kyoto University,
Yoshida-Honmachi, Sakyo-ku, Kyoto 606-8501, Japan
jyli@db.soc.i.kyoto-u.ac.jp, {qiang,asano,yoshikawa}@i.kyoto-u.ac.jp

Abstract. With the recent rapid growth of social image hosting websites, such as Flickr, it is easier to construct a large database with tagged images. Social tags have been proven to be effective for providing keyword-based image retrieval and widely used on these websites, but whether they are beneficial for improving content-based image retrieval has not been well investigated in previous work. In this paper, we investigate whether and how social tags can be used for improving content-based image search results. We propose an unsupervised approach for automatic ranking without user interactions. It propagates visual and textual information on an image-tag relationship graph with a mutual reinforcement process. We conduct experiments showing that our approach can successfully use social tags for ranking and improving content-based social image search results, and performs better than other approaches.

1 Introduction

Social image hosting websites, e.g., Flickr, have recently become very popular. On these websites users can upload and tag their images for sharing them to others. This social tagging is similar to keyword annotation in traditional image retrieval systems. One difference is that keyword annotation requires several experts for annotating images. This requires too much time and labor if the image database is large. Social tagging does not have this problem because a large number of users can participate in tagging task. It is easier to construct a large database with tagged images. Another difference is that social tags are user-generated and folksonomy tags [1]. Compared with taxonomy keywords in keyword annotation which uses a number of specific fixed words, social tags have an open vocabulary in which the words are free and are neither exclusive nor hierarchical. This results in social tags having lots of noises.

Social tags have been proven to be effective for providing keyword-based image retrieval and widely used on social image hosting websites. It is regarded that textual information can naturally improve the results of keyword-based image retrieval, but whether social tags are beneficial for improving content-based image retrieval (CBIR) has not been well investigated in previous work. CBIR has a long history and a large amount of research has gone into it, but its performance still needs to be improved for practical application. In CBIR, for a query image sample, systems search for content-based similar images from a specific

M.V.M. Salem et al. (Eds.): AIRS 2011, LNCS 7097, pp. 147–156, 2011.

148   J. Li et al.

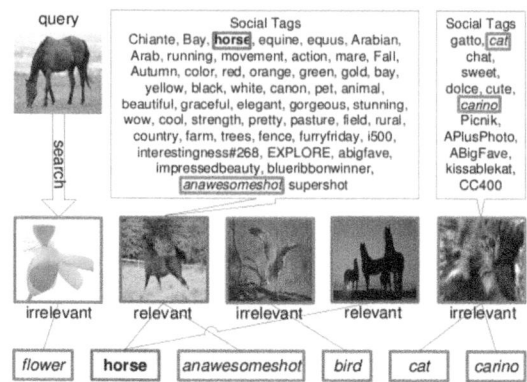

Fig. 1. Query by Example, Content-Based Similar Image Results and Social Tags

multimedia database by image visual information. Since the query image does
not include any textual information, the relationships between the query image
and the textual information of other images in the database are hard to be evaluated because of the well-known semantic gap problem. For example, for the
query "horse" image in Fig. 1, it's hard to know the relationship between it and
the "cat" tag of a "cat" image in the database. The effectiveness of textual information, especially social tags, for improving content-based similar image results
is unknown.

We observe that in content-based similar image results of a given query image
and database, relevant images are relatively few while irrelevant images are many.
There is a characteristic followed by image semantics that the image semantics
of relevant images are alike while the image semantics of irrelevant images are
diverse. For example, for a query "horse" image, its relevant images have alike
"horse" concept while its irrelevant image have diverse concepts such as "cat",
"bird", and so on. However in most cases content-based similar image results do
not follow this characteristic. Fig. 1 shows a query image and its content-based
similar images by SIFT feature [13]. These diverse similar images are regarded
as "relevant" images in content-based similar image results by CBIR. This is
one of the reasons that why the performance of CBIR is unsatisfactory. On
the other hand, social tags sometimes follow this characteristic. In Fig. 1, the
relevant images have alike tag sets including a "horse" tag, while the irrelevant
images have diverse tag sets. It shows that social tags may be able to be used
for improving content-based image search results.

We propose an unsupervised approach which automatically ranks the images
in content-based similar image results. We construct an image-tag relationship
graph model with both images and their tags as vertices, and using image similarity, tag co-occurrence and image-tag annotation relationships as edges. The
approach propagates visual and textual information on the graph with a mutual
reinforcement process. Fig. 1 gives a brief overview of the graph. It shows some
of the content-based similar images and social tags on the graph. In the mutual reinforcement ranking process, the good tags (in red and bold) of relevant

images contribute more scores on the graph; the bad tags of relevant images, and the good and bad tags (in blue and italic) of irrelevant images contribute less scores on the graph; the irrelevant images contribute less to their tags, while the relevant images contribute more. In other words, a high-ranked image is one to which many high-ranked tags point; a high-ranked tag is a tag that points to many high-ranked images. After several iterations, the relevant images can obtain higher rank scores.

The contributions of this paper are as follows.

- We investigate whether social tags can be used for improving content-based search results. We successfully propose an approach which can use social tags to improve the results effectively. To the best of our knowledge, it has not been well investigated in previous work. We conduct experiments showing that our proposed approach performs better than other approaches.
- The mutual reinforcement process is not so novel and some approaches based on it have been proposed in other areas. We also conduct experiments showing that a naive mutual reinforcement approach does not perform well for our topic. Our proposed improved mutual reinforcement approach performs better than a naive one.

The remainder of this paper is organized as follows. In Section 2 we give a brief review of related work. In Sections 3, we propose our social image ranking approach. In Section 4, we report and discuss the experimental results, and present a summary and discuss future work in Section 5.

2 Related Work

There have been studies related to image ranking for keyword-based image retrieval in unsupervised scenarios. Lin et al. [2] proposed an approach only based on text information. Several approaches [3,4,5] only based on visual information have also been proposed. The well-known visualrank approach proposed by Jing and Baluja [5] applies a random walk method for ranking images. We have different goals than the above-mentioned studies based on keyword-based image retrieval. We concentrate on image ranking with social tags for content-based image retrieval. It has not been well investigated in previous work. The graph-based mutual reinforcement approach we propose efficiently uses both visual and textual information in the refining process, and performs better.

Relevance feedback (RF) has been widely used in image ranking in supervised scenarios[6]. In early work some approaches [7,8] adjust the weights of different components of the queries or change the query representation to better suit the user's information need. On the other hand many approaches use RF instances as training sets and include a offline learning process for learning a query-independent ranking model to classify image search results into relevant and irrelevant images, e.g. the approaches [9,10] using support vector machines (SVM). We have different goals than the above-mentioned studies based on relevance feedback. We concentrate on automatically image ranking in unsupervised scenarios without user interactions.

3   Automatic Social Image Ranking

Our ranking task could be formulated as follows. For a given query image q, the content-based image retrieval system computes the content-based similar image results $\mathcal{A} = \{a_1, ..., a_n\}$ from a specific multimedia database $\mathcal{D}$. Let s_{iq} as the similarity between q and a_i. We regard $\mathcal{A}$ as the candidate image set and the social tags of images in $\mathcal{A}$ as the candidate tag set $\mathcal{T} = \{t_1, ..., t_m\}$. We define $\mathcal{T}_{a_i}$ as the tag set of each image $a_i \in \mathcal{A}$. Our task is to rank the image set $\mathcal{A}$ with the tag set $\mathcal{T}$.

There is no user interaction information available for the ranking task in our unsupervised scenario. The approach should automatically gather visual and textual information to rank the content-based similar image results. We analyze the relationships among the images and social tags to construct our image-tag relationship model. We propose an approach with a mutual reinforcement process base on the characteristics of this graph model.

3.1   Image-Tag Relationship Model

To leverage social image visual information as well as social tag textual information for ranking, we construct a graph model in Fig. 2 with candidate set $\mathcal{A}$ and $\mathcal{T}$ for analyzing the image-tag relationships. The vertices of the graph model denote social images which represent visual information and their tags which represent textual information. Note that query image q has no textual information.

The edges of the graph model denote the relationships among images and tags. There are three kinds of image-tag relationships: image-to-image relationship based on image similarity, tag-to-tag relationship based on tag co-occurrence to images, and image-to-tag annotation relationship. The first two kinds of relationships reflect the intra relationships among images or tags. The third one reflects the inter relationship between images and tags.

3.2   Visual and Textual Descriptor

To make use of visual and textual information in our approach, we convert them into visual and textual descriptors. The visual descriptors are based on image similarity. To compute image similarity, we use the following six types of low level features [13]: 64-D color histogram, 144-D color correlogram, 73-D edge direction histogram, 128-D wavelet texture, 225-D block-wise color moments and 500-D bag of words based on SIFT. The distance between image a_i and a_j on low level feature k is computed using a correlation distance $d(\mathcal{H}_{ik}, \mathcal{H}_{jk})$ defined as

$$d(\mathcal{H}_{ik}, \mathcal{H}_{jk}) = \frac{\sum_x (H'_{ik}(x) * H'_{jk}(x))}{\sqrt{(\sum_y H'^2_{ik}(y)) * (\sum_y H'^2_{jk}(y))}},$$

$$\mathcal{H}'_{ik}(x) = \mathcal{H}_{ik}(x) - \sum_y \mathcal{H}_{ik}(y)/\mathcal{N}_k,$$

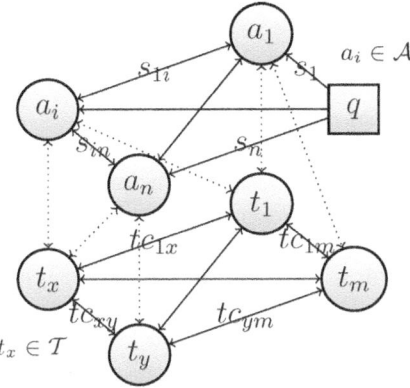

Fig. 2. Image-Tag Relationship Model

where $\mathcal{H}_{ik}$ and $\mathcal{H}_{jk}$ are feature vectors. $\mathcal{N}_k$ is the size of the feature vector k. The image similarity s_{ij} between two images with multi-feature is computed using a weighted sum.

$$s_{ij} = s(a_i, a_j) = \frac{\sum_k w_k d(\mathcal{H}_{ik}, \mathcal{H}_{jk})}{\sum_k w_k}, \quad vd_i = s_{iq}.$$

We use $w_k = 1$ for any k in our work. It means that all low level features have same weights. This strategy has usually been used in existing work such as [11]. For each image a_i in candidate image set $\mathcal{A}$, we propose visual descriptor vd defined as vd_i.

To leverage social tags information, different from some existing work in other topics such as [12] which use some paired tag co-occurrence measures tc_{xy} for each pair of tags t_x and t_y, we propose a single tag co-occurrence measure tc_x for each tag t_x. This measure considers the local tag frequency in candidate tag set $\mathcal{T}$ as well as the global tag frequency in database $\mathcal{D}$. It can evaluate how important tag t_x is to current candidate tag set $\mathcal{T}$ in database $\mathcal{D}$. We propose textual descriptor td defined as

$$tc_x = \frac{|t_x|_{\mathcal{T}}}{|t_x|_{\mathcal{D}}}, \quad td_x = \begin{cases} tc_x, & \text{if } |t_x|_{\mathcal{T}} > \delta, \\ 0, & \text{if } |t_x|_{\mathcal{T}} \leq \delta. \end{cases}$$

Here, $|t_x|_{\mathcal{T}}$ means the number of images in candidate tag set $\mathcal{T}$ that contain t_x, $|t_x|_{\mathcal{D}}$ means the number of images in database $\mathcal{D}$ that contain t_x. δ is a local frequency threshold for ignoring the noisy tags which have low frequency in $\mathcal{T}$ as well as in $\mathcal{D}$ and therefore have high value on tc_x.

3.3   Social Image Ranking

Following the image-tag annotation relationships in the graph model, we propagate the rank scores of images in $\mathcal{A}$ and tags in $\mathcal{T}$ along the links between images and tags. We observe a phenomenon that for an image a_i, when propagating the rank scores from images to tags, if a_i has a high rank score, its related tags will obtain higher rank scores. When propagating the rank scores is from tags to images, if the related tags of a_i have high rank scores, a_i will obtain a higher rank score. On the other hand, for a tag t_x, it also has similar phenomenon. Therefore, we naturally come to the following mutual reinforcement assumption: a high-ranked image for q is one to which many high-ranked tags point; a high-ranked tag for q is a tag that points to many high-ranked images. The iterative formulas for computing the rank scores are defined as follows:

$$Initialization:\ \mathcal{Q}_0'(a_i) = \Phi(vd_i),\ \mathcal{Q}_0'(t_x) = \Phi(td_x);\ 0 \le \alpha, \beta \le 1$$

$$Iteration: \begin{cases} \mathcal{Q}_{k+1}(t_x) = \alpha\Phi(td_x) + (1-\alpha)\sum_{\vee a_i : t_x \in T_{a_i}} \Phi(vd_i)\mathcal{Q}_k'(a_i) \\ \mathcal{Q}_{k+1}(a_i) = \beta\Phi(vd_i) + (1-\beta)\sum_{\vee t_x : t_x \in T_{a_i}} \Phi(td_x)\mathcal{Q}_k'(t_x) \\ \mathcal{Q}_{k+1}'(t_x) = \Phi(\mathcal{Q}_{k+1}(t_x)),\ \mathcal{Q}_{k+1}'(a_i) = \Phi(\mathcal{Q}_{k+1}(a_i)) \end{cases}$$

$$\Phi(\mathcal{Q}_k(t_x)) = \frac{Q_k(t_x) - min\{Q_k(t_y)\}}{max\{Q_k(t_y)\} - min\{Q_k(t_y)\}}$$

The iteration parameters α and β are damping factors. k is the number of iteration steps. We initiate $\mathcal{Q}_0'(t)$ of tags with textual descriptors and $\mathcal{Q}_0'(a)$ of images with visual descriptors. $\mathcal{Q}_k'(\cdot)$ is the normalized rank score of $\mathcal{Q}_k(\cdot)$.

Content-based image similarity to the query image is an inherent property of a candidate image a_i. The images which have high similarity can be regarded as more important on the graph. For a candidate tag t_x, it is also similar. We therefore use visual descriptors and textual descriptors as the weights in the iterations. These weights represent the importance of these images and tags on the graph.

4   Experiment

4.1   Experimental Settings

The dataset we use for experiment is NUS-WIDE [13]. It is created by downloading images and their social tags from social image hosting website Flickr. It has 269,648 images and about 425,000 unique original tags. For images, it provides six types of low-level features extracted from the images, which we have introduced in section 3.2. For tags, the authors of this dataset set several rules to filter the original tag set. They delete the tags with too low frequency. The low frequency threshold is set to 100. They also remove the tags that does not exist in WordNet. At the end, they provide 5,018 unique tags. We keep this filtering in our experiment for the following reasons. It reduces the noises in the tag set. It also reduces the size of candidate tag set $\mathcal{T}$ and the number of links between images and tags, which can reduce the time cost in the ranking computation.

NUS-WIDE also provides image annotation ground-truth of 81 concepts for the entire dataset, but it doesn't appoint query sample set and provide ground-truth for content-based image retrieval. We need to construct them by ourselves

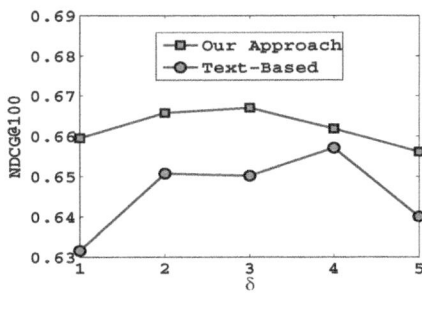

Fig. 3. δ of td_x

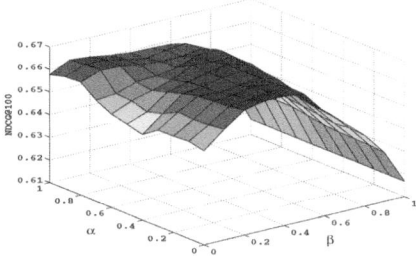

Fig. 4. (α, β) for Iterations

for our experiment. In our experiment, we randomly choose 100 images as a query image set from the entire dataset for our evaluation. Note that there is no textual information available for these queries. For each query, we rank the images in top-n content-based similar image results, $n = 100$. The images in content-based similar image results are labeled with five levels by human beings. The range of levels is from 0 (irrelevant) to 4 (relevant). The evaluation metric used in our experiment is Normalized Discounted Cumulative Gain (NDCG) [14]. NDCG is an effective metric often used in information retrieval for evaluating the rank results with relevance levels. It is defined as follows.

$$NDCG@k = Z_k \sum_{j=1}^{k}((2^{r(j)} - 1)/log(1+j))$$

$r(j)$ is the relevance level of the image at rank j. Z_k is a normalization constant and equal to the maximum DCG value that the top-k ranked images can reach, so that NDCG score is equal to 1 for the optimal results. We evaluate the performance with the average NDCG value of all query images.

4.2 Parameter Selections

The proper local frequency threshold δ of td is different for different approaches. According to Fig 3 which shows NDCG@100 value of the approaches with different δ, we set $\delta = 3$ for our approach. Note that the NDCG@100 of the content-based similar images results is 0.6168.

To select proper iteration parameters α and β in our mutual reinforcement approach, we choose their candidate values by an interval of 0.1 in the range of $[0, 1]$ and obtain 121 pairs of candidate values. We run our approach with these pairs on the query image set and observe the performance on the NDCG@100 metric. According to Fig.4, we choose (α, β) as $(0.0, 0.4)$ in our experiments.

When $\beta = 1$, it means the iteration formula of an image rank score has degenerated into only depending on the visual descriptor. Because of using image similarity as the visual descriptor, the ranking result is equal to the content-based similar image results. The figure also shows that for any (α, β) in the range, our approach performs not worse than content-based similar image results.

4.3   Experimental Results

We compare our approach with three other approaches as well as with the content-based similar image results. The parameters selection and tuning is carried out for all these approaches. We compare the best results these approaches can generate.

1. Visual-based Approach (VisualRank): This approach does not use any social tag information. It is based on VisualRank but with modifications to make it appropriate for our scenario. VisualRank uses a random walk method on the image complete graph in which vertices are the candidate images, and uses content-based image similarity for computing the transition matrix. The iteration formula is as follows.

$$\mathcal{Q}_{k+1}(a_i) = (1 - \gamma) * \frac{1}{n} + \gamma * \sum_j (\mathcal{Q}_k(a_j) * \frac{s_{ij}}{\sum_x s_{xj}}), \quad \mathcal{Q}_0(a_i) = \Phi(vd_i)$$

We follow the settings in VisualRank and set damping factor γ to 0.85. n is the size of the candidate image set $\mathcal{A}$. We want to confirm that social tags are beneficial for ranking content-based similar image results and show that our approach performs better than VisualRank in the content-based social image ranking scenario.

2. Text-based Approach: To show that our mutual reinforcement process can use social tag information more effectively, we design a text-based approach that uses social tag information but without a mutual reinforcement process. We compute the rank score of candidate image a_i by using the following formula. Note that pure text-only-based approach in our scenario because the candidate image and tag set is generated by visual information. According to Fig 3, we set $\delta = 4$.

$$\mathcal{Q}(a_i) = \sum_{\vee a_i : t_x \in T_{a_i}} \Phi(td_x)$$

3. Naive Mutual Reinforcement Approach: The mutual reinforcement process is not so novel and some approaches based on it have been proposed in other areas. We design this naive mutual reinforcement approach to show that our proposed mutual reinforcement approach performs better than a naive one. There are two important differences between this naive one and our approach. One is that the iterative formulas of this naive one do not consider the importance of images and tags on the graph. The other is that we observe the performance among different parameters and select them cautiously in our approach.

Initialization: $\mathcal{Q}'_0(a_i) = \Phi'(vd'_i), \; \mathcal{Q}'_0(t_x) = \Phi'(td'_x); \; 0 \leq \alpha, \beta \leq 1;$

Iteration: $\begin{cases} \mathcal{Q}_{k+1}(t_x) = \alpha \Phi'(td'_x) + (1 - \alpha) \sum_{\vee a_i : t_x \in T_{a_i}} \mathcal{Q}'_k(a_i) \\ \mathcal{Q}_{k+1}(a_i) = \beta \Phi'(vd'_i) + (1 - \beta) \sum_{\vee t_x : t_x \in T_{a_i}} \mathcal{Q}'_k(t_x) \\ \mathcal{Q}'_{k+1}(t_x) = \Phi'(\mathcal{Q}_{k+1}(t_x)), \; \mathcal{Q}'_{k+1}(a_i) = \Phi'(\mathcal{Q}_{k+1}(a_i)) \end{cases}$

$$\Phi'(\mathcal{Q}_k(t_x)) = \frac{\mathcal{Q}_k(t_x)}{\sqrt{\sum_y \mathcal{Q}_k(t_y)^2}}, vd'_i = vd_i, td'_x = \sum_{t_y \in \mathcal{T}} \frac{|t_x \cap t_y|_{\mathcal{T}}}{|t_x \cup t_y|_{\mathcal{T}}}.$$

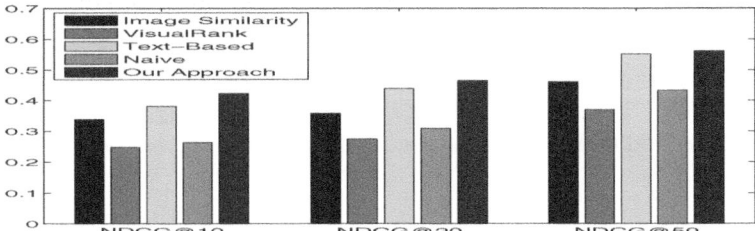

Fig. 5. Approaches Comparison

$|t_x \cap t_y|_{\mathcal{T}}$ means the number of the images that contain both of t_x and t_y, and $|t_x \cup t_y|_{\mathcal{T}}$ means the number of the images that contain t_x or t_y. (α, β) is also chosen as $(0.0, 0.4)$. Note that it is not too naive because it still uses good visual descriptor and damping factors. The rule of parameters chosen here is to choose some intuitive parameters.

Fig. 5 illustrates the evaluation of NDCG@10, NDCG@20 metrics on ranking the top-100 images in content-based similar image results. Compared with content-based similar image results, all metrics are improved with our approach. Our approach performs best among all of the approaches here. Table 1 illustrates a social image ranking samples of our approach.

Our approach performs better than the visual-based approach in our content-based social image ranking scenario. Although VisualRank performs well in [5], the initial image results in that work are from keyword-based image retrieval, and the ranking is based on image similarity. In other words, the approach uses both visual and textual information to generate the final results. But in our scenario, since the initial image results are content-based, the visual-based approach can only use visual information. It illustrates that using social tag information for ranking the content-based image search results is beneficial. Furthermore our approach also has better time complexity than VisualRank because our approach computes less links on the graph in real time iterations. In our experiments, on the average running time for all queries, VisualRank costs 0.076 seconds while ours costs 0.031 seconds.

Compared with the text-based approach which does not use our mutual reinforcement process, our approach which is also based on textual information performs better. The text-based approach also performs better than VisualRank. It shows that social tags are beneficial for ranking content-based similar image results and our approach can use them more effectively.

Furthermore, our approach performs better than the naive mutual reinforcement approach which outperforms than VisualRank. It shows that a mutual reinforcement process is useful in our ranking scenario, but a naive mutual reinforcement approach without an optimized design still can not generate better rank results than the content-based similar image results from a statistical point. Our proposed mutual reinforcement approach can improve the content-based similar image results effectively.

Table 1. Social Image Ranking Example

5 Conclusion

In this paper, we confirm that we can successfully use social tags to improve content-based social image search results. We propose an approach with a mutual reinforcement process using both visual and textual information on a image-tag relationship graph model. The experiments illustrate that our approach can reach the goals and performs better than other approaches. For future work, we will extend our work to a keyword-based image retrieval scenario.

Acknowledgments. This work was partially supported by Grant-in-Aid for Scientific Research(B) (20300036) from Japan Society for the Promotion of Science (JSPS).

References

1. Bischoff, K., Firan, C.S., Nejdl, W., Paiu, R.: Can all tags be used for search? In: CIKM, pp. 193–202 (2008)
2. Lin, W.H., Jin, R., Hauptmann, A.: Web Image Retrieval Re-Ranking with Relevance Model. In: IEEE/WIC, pp. 242–248 (2003)
3. Zitouni, H., Sevil, S., Ozkan, D., Duygulu, P.: Re-ranking of web image search results using a graph algorithm. In: ICPR, pp. 1–4 (2008)
4. Zhou, W.G., Tian, Q., Yang, L.J., Li, H.Q.: Latent visual context analysis for image re-ranking. In: CIVR, pp. 205–212 (2010)
5. Jing, Y., Baluja, S.: VisualRank: Applying PageRank to Large-Scale Image Search. TPAMI 30(11), 1877–1890 (2008)
6. Yong, R., Huang, T.S., Ortega, M., Mehrotra, S.: Relevance feedback: a power tool for interactive content-based image retrieval. IEEE Trans. CSVT 8(5), 644–655 (1998)
7. Porkaew, K., Mehrotra, S., Ortega, M.: Query reformulation for content based multimedia retrieval in MARS. In: ICMCS, pp. 747–751 (1999)
8. Porkaew, K., Chakrabarti, K., Mehrotra, S.: Query Refinement for Multimedia Similarity Retrieval in MARS. ACM Multimedia, 235–238 (1999)
9. Zhang, L., Lin, F.Z., Zhang, B.: Support vector machine learning for image retrieval. In: ICIP, pp. 721–724 (2001)
10. Chen, Y.Q., Zhou, X.S., Huang, T.S.: One-class SVM for learning in image retrieval. In: ICIP, pp. 34–37 (2001)
11. van de Sande, K.E.A., Gevers, T., Snoek, C.G.M.: Evaluating Color Descriptors for Object and Scene Recognition. TPAMI 32(9), 1582–1596 (2010)
12. Sigurbjörnsson, B., van Zwol, R.: Flickr tag recommendation based on collective knowledge. In: WWW, pp. 327–336 (2008)
13. Chua, T.S., Tang, J.H., Hong, R.C., Li, H.J., Luo, Z.P., Zheng, Y.T.: NUS-WIDE: A Real-World Web Image Database from National University of Singapore. In: CIVR (2009)
14. Jarvelin, K., Kekalainen, J.: Cumulated gain-based evaluation of IR techniques. ACM Transactions on Information Systems (TOIS) 20(4) (2001)

Profiling a Non-medical Professional Searcher on a Medical Domain: What Do Search Patterns and Demographic Details Reveal?

Anushia Inthiran[1], Saadat M. Alhashmi[1], and Pervaiz K. Ahmed[2]

[1] School of Information Technology Monash University Sunway Campus,
[2] School of Business Monash University Sunway Campus
anushia.inthiran@monash.edu

Abstract. Previous research is able to distinguish search patterns of domain experts and non-domain experts. However, little is known about the finer details of a non-domain expert searcher. This is especially so when a non-domain expert performs a search on an expert type domain. Do non-domain experts search similarly? What can we learn and infer from their search patterns? More importantly can we identify the searcher behind the search? In this paper, we perform a study of non-domain experts search behavior on an expert domain. Our results indicate search patterns can be used to generally classify a non-domain expert searcher.

Keywords: interactive search, medical domain, user study.

1 Introduction

At some point of their life, most people have searched for medical information. An online medical search behaviour survey reports people between the ages of 18-34 go online to find healthcare information while people above the age of 50 seek for medical information online after consulting their physician [10]. As more health care questions and challenges arise, many turn to online resources to obtain information [15]. There is also a trend among patients today where they prefer to be more informed before, during and after consulting their physicians [3].

Previously, medical search engines were exclusive for medical professionals; other types of users were only passively exposed to medical information online [4]. Today, non-medical professionals are aware of many medical search engines and utilize them to search for medical based information. Medical and health queries have declined as a proportion of all web queries, as the use of specialised medical/health websites have increased [15]. In some cases, users solely rely on the Web for healthcare diagnosis and treatment [16,11].

Searching for medical information is certainly not as straightforward as other types of search task. A health information seeker must know within the realm of language, the near specific location of the knowledge they seek [12] otherwise it is likely that a search might become unsuccessful. Medical terms have multiple sense and can be

M.V.M. Salem et al. (Eds.): AIRS 2011, LNCS 7097, pp. 157–168, 2011.

discussed from different perspectives [3]. Performing a medical search on a huge and information intensive domain such as a medical search domain where there are massive amounts of information available at disposal [3] could intimidate a non-medical professional.

Little is known about the search behaviour of non-medical professionals on this domain. This information will enable medical search engines to categories users based on search behaviour and subsequently tailor relevant information retrieval strategies to enhance a user's search session.

In this paper, we present results of an exploratory survey of studying interactive search behaviour for non-medical professionals on *Medline Plus[1]*. To better understand the search behaviour of non-medical professionals we study query and browsing patterns using simulated work task scenarios. We attempt to report and classify non-homogenous search behaviour of non-medical professionals when searching on a medical domain. We believe these classifications will allow for better understanding of search behaviour of non-domain experts. The rest of the paper is organised as follows: in Section 2 we review related work, in Section 3 we describe our research methodology, in Section 4 we provide demographic details of our participants, in Section 5 we provide our results and analysis, in Section 6 we discuss our findings and in Section 7 we conclude the paper with future work.

2   Related Work

Domain experts search differently compared to non-domain experts. They prefer technical sites, complete tasks faster and use specialised vocabulary [18]. Research has also found when domain experts perform a search task outside their domain or topic expertise only general purpose strategies were utilised [1]. There were obvious differences in site selection and search goal sequencing. Software engineers demonstrate distinct and unique search behaviour when searching on work-related tasks in comparison to when a search is performed for a non-work related task [5]. They issue longer queries, used technical terminology and acronyms in 66% of their work related tasks. Domain knowledge has also been found to influence search behaviour [1,8]. Users with domain knowledge used elaborate reformulation strategies in comparison to users without domain knowledge. Users without domain knowledge used simple stemming and backtracking modification techniques in a search session. Specific topic knowledge has also been found to influence search behaviour. Searchers with high topic knowledge expressed queries more effectively and found more search results [7].

Self-rated domain experts also used different search strategies in comparison to non-self rated domain experts. Self-rated domain experts used significantly more unique query terms while non-self rated domain experts required supplemental features to augment their search experience [6]. Search experience also affects search behaviour. When an experienced searcher had little knowledge on a domain, more

---

[1] http://www.nlm.nig.gov/medlineplus

time is spent preparing queries and examining search results [8]. Search novices were more likely to reformulate queries and access websites through search engines rather than accessing websites directly [7].

Hu, Zeng and Niu [9] were able to determine the gender of a searcher simply by analysing words on a page that is clicked on and topic classifiers used. Similarly, Jones [14] was able to determine the age bracket of searchers based on queries used. Webber and Castello [17] conducted an in-depth study to relate specific search behaviours with demographic details. They found older and younger searchers use different terms to arrive at the same result page. Searchers from higher income per-capita brackets locate web site addresses using navigational queries. Demographic information based on age and ethnic background also helped to effectively determine the second query term for the purpose of query expression.

In summary, previous work focused on the differences amongst domain experts and non-domain experts. However, little is known about the finer details of a search session of a non-domain expert on an expert domain (i.e. medical domain). Nor is there research to relate medical searching to demographic details. This deeper understanding will allow medical search engines to categorize search behaviour of searchers and provide relevant assistance during a search session. An interactive study such as this provides initial understanding of search behaviour and eliminates issues with noisy and data sparcity from search logs.

3   Research Methodology

We performed an exploratory survey on a convenient sample of 30 participants in a university setting. In this paper, to qualify a participant as a non-medical professional we ensured participants did not have any formal medical education. We utilized the following data gathering techniques in our survey: pre-experiment interview, simulated work task scenario, observation and post-experiment interview. Participants were told that the entire search session had to be performed on *Medline Plus*. *Medline Plus* was chosen as the domain of search as it currently only enables basic text matching techniques and spelling correction. This domain enables us to obtain true user interaction when searching for medical information without being affected by additional information retrieval strategies.

The search was limited to *Medline Plus* as we did not want participants to perform medical searching on non-medical domains which could interfere or influence data gathered. There was no time limit for each search task and participants were told that they may stop searching once they have found satisfactory results or were not able to locate satisfactory results (user perceived satisfaction). Each participant was given an instruction letter to explain their involvement in the experiment. None of the participants had prior search experience on *Medline Plus*. Hence, we gave them some time to familiarize themselves with this medical domain.. Each participant had to perform 3 searches using simulated work task scenarios. These scenarios were rotated for each participant.

3.1   Pre-experiment Interview

The pre-experiment interview was used to obtain demographic details, information on general search experience, medical knowledge and medical search experience. We asked participants to self rate their medical knowledge using one of three options: expert, average or poor.

3.2   Simulated Work Task

We developed 3 (Task A,B,C) simulated work task scenarios according to [2]. We used simulated work task scenarios to study interactive search behavior as it not only invokes a common information need amongst participants but also allows participants to engage in the task. We developed clinical- based scenarios as this type of medical search is typically performed by non-medical professionals. We provide the scenario for Task A, B and C in Figures 1,2 and 3. Task A describes a common health complain amongst young and old adults in a surveyed health clinic. Task B describes a prevalent health condition surveyed at a public hospital and Task C describes a condition experienced by one of out of 25 children between the ages of 6 months and 5 years. Each work task focuses on different medical topics to provide variety. Each work task also invokes several search goals within a search session.

Simulated work task scenario: Today morning after getting out of bed you noticed that you could not move your neck. You can't move it left or right. There is swelling on the left side of your neck. The swelling seems to be near lymphatic nodes. You want to find out what is wrong.

Indicative Request: Find for an instance, information to inform you of your condition, what caused you to experience pain and what can be done to relief you of this condition.

Fig. 1. Simulated Work Task A

Simulated work task scenario: Your colleague had just undergone a health test and found out that his kidney is enlarged and there is a stricture. A procedure called URS&RPG was performed. After this procedure, he then experienced urine retention. He is in pain and has been told that surgery is required. You are concerned for your colleague and would like to use Medline Plus to provide him with some information.

Indicative Request: Find for an instance, information about his condition and determine if alternatives are available to treat his condition or is surgery the only solution.

Fig. 2. Simulated Work Task B

Simulated work task scenario: Yesterday at the mall you witnessed a child about the age of five fitting/having seizure. The child's parents who were nearby were in a state of shock. On arriving home, you were curious about the situation at the mall and would like to find out more about the child's condition. Use Medline Plus to help you find out what could have been done to assist the child.

Indicative Request: Find for an instance, information to provide you with some background about fits/seizures, why does it take place and what first aid measures you could have taken to assist the child.

Fig. 3. Simulated Work Task C

3.3   Observation

We observed the following query activities: the number of queries issued, the type of query issued (medical/non-medical), usage of query operators, ineffective queries (queries which returned no search results), length of the query (number of terms) and query re-issues (exact). We also observed the following search results evaluation behavior: number of search results clicked, sub-links clicked and going beyond the first page of results. We also kept track of unsuccessful search session (users ended the search session without arriving at satisfactory results) and task completion time. These observations were performed for each task.

3.4   Post- experiment Interview

The post experiment interview took place after participants completed each individual search task. At this stage, we obtained feedback from the search session.

4   Demographic Details

There were 30 participants in our experiment. There were 16 female participants and the rest were males. The average age of was 33.7 years (SD=9.6). The youngest participant was 19 and the oldest was 57 years. Participants ranged from undergraduate students from the business and information technology discipline (13.3%) and post-graduate students from the information technology, business and health sciences discipline (40%). Other participants include industry researchers, university lecturers, managers, consultants and retirees with undergraduate or/and post-graduate degrees in the following disciplines: Physics, Business, Health Science, Information Technology, Bio-medical and Engineering. 30% of participants who were working professionals had postgraduate education. Participants were of different nationalities: 60% of participants were Malaysians, 6% were from the Middle East, 6% were Sri Lankans, and the remaining 28% were made up of Indonesians, Burmese, Africans, Japanese, Indians and Bangladeshi.

The average general search experience was 10.3 years (SD=3.5). The most experienced searcher had been searching on the Internet for 16 years and the least experienced searcher had been searching on the Internet for 2 years. 67% of participants self rated their medical knowledge as average, 27% self-rated their medical knowledge as poor and 6% self-rated their medical knowledge as expert. Only 2 participants had never searched for medical information. The rest have searched for medical information using general search engines or medical search engines. 9 participants had searched for medical information using medical search engines. The average search experience on a medical search engine was 6.2 years (SD=3.5). The minimum and maximum time participants had searched on a medical search engine was 2 and 12 years respectively. Other participants who had been searching for medical information on general search engines had been doing so for an average of 9.9 years (SD=3.5). The least experienced medical searcher on a general search engine had been doing so for one year and the most experienced searcher had been doing so for 10 years.

Participants from the health sciences and biomedical discipline performed medical searches on an average of 3 hours per day. Typically they searched for research articles and treatment options for clients. Other participants performed medical searching as and when there was a need to do so. Typically, they searched for information on diagnosis, medication options and general healthcare.

5   Results and Analysis

We provide an overview of results from 90 search sessions conducted by 30 participants using 3 simulated work task scenarios in Table 1. We analyze search behavior based on querying behavior, search results evaluation behavior and querying versus result browsing ratio. These search features were investigated as they represent typical interactions during a search session.

5.1   Querying Behavior

In this section we analyze querying details in Table 1.

5.1.1   Medical Query

We qualify a query as a medical query if the query is found in the *Medical Subject Heading Database (MeSH)*[2] database and not used to describe our work task. We provide samples of medical queries issued by participants: *epilepsy, hydronephrosis, retrograde pyelogram, ureteroscopy, tuberous sclerosis, polycystic kidney.* Our observation data show medical queries are only issued in the middle of a search session. Participants who issued medical queries discovered the medical query as a result of thorough browsing and evaluation of search results. This explains why a

[2] http://www.nlm.nih.gov/mesh/MBrowser

Table 1. Overall Search Session Details Based on 90 Search Sessions

Search Feature		Value
Total Number of Queries		395
Query Details	Natural Language	8.5%
	Medical Query	10%
	Keyword	47%
	Structured	42.7%
	Operators	0.8%
	Re-Issue	2.5%
	Query Without Clicks	13.6%
	Domain Information	14.5%
	Ineffective	7.6%
	Mean Number of Queries Per Search Session [SD]	4.0 [1.8]
	Mean Query Length Per Search Session [SD]	8.6 [4.9]
Total Search Time (Mean) [SD]		218 minutes (7.9) [4.8]
Total Search Results Clicked (Mean) [SD]		240 (4) [2.3]
Search Results Details	Sub-links Clicked	7.5%
	2nd Page of Results Evaluated	0.9%
Unsuccessful Search Sessions		6.5%

Note: A query may fall into more than one query detail category

medical query is only issued mid-way in a search session. Participants continued to use the medical query as part of their query iteration process for subsequent queries.

Participants who issued medical queries all had experience searching for medical information on medical domains (mean=6.2 SD=3.5) regardless of age and education background. Participants who issued medical queries did so at least once while searching on Task B and C. Other participants without medical search experience on medical domains did arrive at result pages that contained medical queries but did not 'discover' the medical query in the results page. Participant's prior search experience on medical domains allowed for an active learning process to take place during the search session. These participants had the ability to acquire and use 'new' knowledge discovered during the search process. This factor led to the ability of expressing medical queries within this group of participants. We note medical queries were not issued for Task A.

5.1.2   Natural Language

We define natural language queries as queries that form a complete English sentence. We provide samples of natural language queries: *what is a stricture, what are causes of enlarged kidneys, what has to be done for a 5 year old with fits, what first aid measure for children with seizure, what is the care for swelling of lymphatic nodes on the left side of the neck, how to help a child with seizure.* There were 10 natural language queries observed in total.

Whenever a natural language query was issued, participants did not click on returned search results. Instead, they continued to issue natural language queries or immediately ended the search session. In search sessions where natural language queries were issued, participants ended the search without arriving at satisfactory search results.

In the 10 observations of natural language usage, 20% of these search session only consisted of natural language queries. In the remaining 80% of search sessions keyword based queries were issued first before natural language queries. Participants who issued natural language queries were above the age of 50 years. There were 3 participants above the age of 50 and all three participants issued natural language queries in all three work task at least once. These three participants had an average general search experience of 9 years (SD=6.5). These participants have never searched on a medical domain but have performed medical searching on general search engines (mean= 2.6, SD= 2.0). The use of natural language queries is usually related to novice Internet searches. Our results indicate an experienced searcher searching on an unfamiliar domain could also demonstrate search behaviour akin to novice Internet searchers. Domain specific search experience plays a more important factor in influencing search behaviour.

The use of natural language queries in a search session had two implications. Firstly, a search session is coming to an end and secondly a search session was going to be unsuccessful.

5.1.3 Structured Queries

We define a structured query as a semi-complete English sentence. We provide some examples of structured queries: *precautionary measures for fitting, solutions to enlarged kidney, alternatives to kidney surgery, cure for urine retention, consequences for kidney enlarged, alternative for urine retention, pain kidney surgery treatment, treatment for swelling neck, relief for stiff neck, medication for lymph nodes, first aid for seizure for children.* Participants who issued structured queries were between the ages of 27 and 32 regardless of education background. Only three participants out of 15 participants who fell within this age group did not issue structured queries. Participants within this age range had an average general search experience of 11 years   (SD= 1.8) and medical search experience of 6.2 years (SD=3.7).

Participants outside this age range did not issue structured queries but used keyword queries instead. Participants within this age group did not only use structured queries in their search session, but also issued keyword based queries (10%). Structured queries appeared mid-way of a search session. Once a structured query is issued participants did not revert to using keyword based queries. Structured queries were issued by participants within this age group for all three tasks.

5.1.4 Branching Away from Query Usage

Some participants branched out from using queries to use *MedlinePlus* built-in search features such as *Health Topics, Drugs and Supplements*. After briefly using these features, participants continued the search process by using the *Search MedlinePlus* feature and did not continue to use these built-in features. Participants who used these features were mapped into two categories. The first category was participants with experience using medical search domains (mean=6.5 SD=0.7) and had health sciences and bioinformatics background. The second category was participants who have never searched for medical information. The post-experiment interview reveals other participants were aware of this search feature but opted not to use them. Participants with medical domain search experience and background in health sciences and biomedical used the same search utilities as an in-experienced medical searchers.

5.2   Search Results Evaluation Behaviour

In this section we analyze search results evaluation details in Table 1.

5.2.1 Preference for Non-text Based Search Results

Majority of participants favoured search results that contained only text information, while 26.7% participants preferred search results with images, embedded video files and interactive tutorials. In the post-experiment interview participants who had selected search results with these multimedia elements made comments to say they preferred such search results. Participants who favoured these results were between the ages of 27 – 32 had post-graduate education or were pursuing post-graduate education. Other participants outside this age group and education level did arrive at the same page but favoured text-based results instead. Multimedia based pages were not available for Task C hence, for this task participants within this age group and education level favoured text results instead.

5.2.2 Interaction on Search Results Page

Half of participants used the *Find on this page* function (*Ctrl-F*) which is a default function available on web-browsers. In all cases the searched item within the page was domain information. We provide samples of items searched for : *URS, kidney stricture, fits, seizure, lymphatic nodes*. This search behaviour was demonstrated for 70% of result pages that were clicked. Participants who demonstrated this behaviour either had or were pursuing post-graduate degrees in Information Technology, Biomedical or Health Sciences.

5.2.3 Medline Plus Features Used in Refining the Query

Among *MedlinePlus* features used to refine search results were: *Refine by keyword, Refine by topic and Refine by type*. These features were made available after search results were displayed. Participants used these features to further refine returned search results. This behaviour occurred in 30% of search sessions. Participants who used these *MedlinePlus* features were either experienced medical searchers on medical search domains (mean= 6.2 years SD=3.5) or were students or post-graduate degree holders from the Information Technology field. The post-experiment interview reveal other participants were aware of these features but opted not to use them.

5.3   Querying versus Results Browsing Behaviour

There were three patterns of querying to browsing ratios: high and equal querying to browsing ratio, high querying and low browsing ratio and low and equal querying and browsing ratio. Participants who had search experience using medical domains (Mean=7.5 SD= 3.1) demonstrated high and similar ratios for querying and browsing behaviour (87:97) Participants with more than 10 years of general search experience (mean= 13.3 SD=1.8) demonstrated uneven and high querying ratios compared to browsing ratios (16:5). Other participants who did not fit into these two categories had relatively low and almost equal querying and browsing behaviour (4:5).

6   Discussion

Our study dwells into the finer details of search behavior of non-medical professionals on a medical domain. We were able to identify non-homogenous search patterns amongst non-medical professionals. Our results are not without limitation. The use of operators, query re-issues, clicking on sub-links, going beyond the first search result page were too few to analyze. Queries without clicks and ineffective queries were related to resources available on *Medline Plus* thus not covered as part of our study. We were not able to relate the number of queries issued, query length and task completion time to a particular category of users. Neither were we able to determine which category of user is prone to experience unsuccessful search sessions. There were two patterns as to how search results are viewed. Half of participants viewed results serially while the other half viewed results in parallel. In serial browsing results are not re-visited and the search session stops as soon as a satisfactory result is found or when participants decide to terminate the search. In parallel browsing, pages are often re-visited and compared. We were not able to classify this behavior to any particular category.

We considered all possible methods of classification based on: age, education level, self-rated medical knowledge, general search experience, medical search experience, nationality, gender and native language and only report classifications where participants matched 100%. Hence, gender, nationality, native language, self-rated medical knowledge did not affect a non-medical professional's search behavior.

There were many interesting outcomes of this research study. We were able to generally classify users into demographic groups based on search behavior. Medical search engines could exploit these findings to enhance a user's search session. We suggest several personalization approaches based on our findings. The usage of a medical query suggests users with above average medical knowledge and search experience. Similarly this group of users could also be identified by the high querying to browsing ratio. Search engines could return pages that require higher level understanding to these users. Providing these users with medical query suggestions could minimize search time and effort required to locate and issue medical queries. Participants between the ages of 27 – 32 issue structured queries and favor multimedia based search results. Hence, once a structured query is identified search engines should aim to provide multimedia based pages to these searchers. These pages should appear at the top of a page with some information on the content of the page.

The use of natural language queries should immediately signal the retrieval strategy that the user is an inexperienced medical search and is having difficulty in the search session. It is likely that the user is already dissatisfied and unhappy with the session. To enhance the users experience we suggest that retrieval strategies exploit the search strategies of participants from the Information Technology background or had prior experience on medical search engines (branching away from query, highlighting keywords in the results page, using query refining features). These participants were able to utilize search features to achieve higher search efficacy in comparison to other participants.

7   Future Work and Conclusion

We conducted an exploratory survey on medical information search behavior of non-medical professionals.. Our study is not without limitation, our results is limited to the population of study and does not necessarily represent medical search behavior of non-medical professionals as a whole. In addition, results obtained are specific to the domain where the search is performed.

For future work we intend to run the same study on a larger group to further substantiate our findings. Similarly we would also like to study the search behavior of medical professionals and medical students.

Acknowledgement. We thank participants of this survey.

Reference

1. Bhavnani, S.K.: Important Cognitive Components of Domain Specific Search Knowledge. In: Proceedings TREC, pp. 571–578 (2002)
2. Borlund, P.: The IIIR evaluation model of interactive information retrieval systems. Journal of Information Research 8(3) (April 2003)
3. Can, A.B., Baykal, N.: MedicoPort: A medical search engine for all. Computer Methods and Programs in Biomedicine 86, 73–86 (2007)
4. Coiera, E.: Four rules for the reinvention of health care. British Medical Journal 328, 1197–1199 (2004)
5. Freund, L., Toms, E.G.: Enterprise search behavior of software engineers. In: Poster Session of the 29th Annual International ACM SIGIR Conference, Seattle, Washington (2006)
6. Hembrooke, H.A., Gay, G.K., Granka, L.A.: The effects of expertise and feedback on search term selection and subsequent learning. JASIST 56(8), 861–871 (2005)

7. Hoelscher, C., Strube, G.: Web search behavior of Internet experts and newbies. Computer Networks 33(1-6), 337–346 (2000)
8. Hsieh-Yee, I.: Effects of Search Experience and Subject Knowledge on the Search Tactics of Novice and Experienced Searchers. Journal of The American Society for Information Science 44(3), 161–174 (1993)
9. Hu, J., Zeng, H.-J., Li, H., Niu, C., Chen, Z.: Demographic prediction based on user's browsing behavior. In: WWW 2007, pp. 151–160 (2007)
10. Kantar Media: E-patients' Online Search Behavior Influenced by Gender, Condition Type, `http://community.pathoftheblueeye.com/print/658` (accessed on January 11, 2011)
11. Khoo, K., Bolt, P., Babi, F.E.: Health information seeking by parents in the internet age. Journal of Pediatric Child Health 44, 419–423 (2004)
12. Nadkarni, P.M.: Information retrieval in medicine. Overview and applications. Journal of Postgraduate Medicine 46, 116–122 (2000)
13. Jones, R., Kumar, R., Pang, B., Tomkins, A.: I know what you did last summer: query logs and user privacy. In: CIKM, pp. 909–914 (2007)
14. Spink, A., Yang, Y., Jansen, J., Nykanen, P., Lorence, D.P., Ozmutlu, S., Ozmutlu, H.C.: A study of medical and health queries to the web search engines. Health Information and Libraries Journal 21, 44–51 (2004)
15. Wainstien, B.K., Sterling-Levis, K., Baker, S.A.: Use of the Intern by parents of pediatric patients. Journal of Pediatrics Child Health 42, 528–532 (2006)
16. Webber, I., Castello, C.: The Demographics of Web Search. In: SIGIR 2010, pp. 78–82 (2010)
17. White, R.W., Dumais, S.T., Teevan, J.: How Medical Expertise Influences Web Search Interaction. In: Proceedings of the 26th Annual SIGCHI Conference on Human Factors in Computing, pp. 179–181 (2008)

Prioritized Aggregation of Multiple Context Dimensions in Mobile IR

Ourdia Bouidghaghen[1], Lynda Tamine-Lechani[1], Gabriella Pasi[2],
Guillaume Cabanac[1], Mohand Boughanem[1], and Célia da Costa Pereira[3]

[1] IRIT- University Paul Sabatier,
118 Route de Narbonne, 31062, Toulouse, France
{bouidgha,tamine,cabanac,bougha}@irit.fr
[2] Università degli Studi di Milano Bicocca, DISCO
Via Bicocca degli Arcimboldi 8, 20126 Milano (MI), Italy
pasi@disco.unimib.it
[3] Università degli Studi di Milano, DTI
celia.pereira@unimi.it

Abstract. An interesting aspect emerging in mobile information retrieval is related to the several contextual features that can be considered as new dimensions in the relevance assessment process. In this paper, we propose a multidimensional ranking model based on the three dimensions of topic, interest, and location. The peculiarity of our multidimensional ranking lies in a "prioritized combination" of the considered criteria, using the "prioritized scoring" and "prioritized and" operators, which allow flexible personalization of search results according to users' preferences. In order to evaluate the effectiveness of our model, we propose a simulation based evaluation framework that investigates the integration of the contextual dimensions into the evaluation process. Extensive experimental results obtained by using our simulation framework show the effectiveness of our multidimensional personalized ranking model.

Keywords: mobile IR, multidimensional personalization, contextual criteria, prioritized aggregation, simulation-based evaluation.

1   Introduction

Every user has a distinct context and a specific background when searching for information. The goal of contextual information retrieval (CIR) is to tailor search results to a particular user according to his/her specific context and preferences [1]. This implies the need to go beyond the topical relevance assessment to a multi-dimensional relevance assessment, where the considered relevance dimensions encompass besides the topical relevance, some contextual relevance [2].

We are specifically interested in the emerging mobile IR field, where mobile users' queries are known to be sensitive to several interdependent contextual criteria (e.g., users' interests, location, time) and where users can be differently interested in each context criterion depending on their information needs [3]. For

M.V.M. Salem et al. (Eds.): AIRS 2011, LNCS 7097, pp. 169–180, 2011.
© Springer-Verlag Berlin Heidelberg 2011

example, let us consider the query *"cultural event."* Document relevancy of this query may depend on a user's implicit interest *"jazz events"* and location *"in Paris."*

Traditionally, IR algorithms have been evaluated primarily at a system level with little reference to the user. This discrepancy has led to criticism of the IR community for relying on relevance criteria that are solely objective, considering only the relationship between a retrieved document and the query with respect to a topical perspective, rather than considering subjective dimensions of relevance related to the person whose individual information needs led to the query being conducted [2]. Several researchers [9,10] have thus argued for the multidimensional nature of the concept of relevance. Mizzaro [10], for example, proposed a relevance model in which relevance is represented as a four dimensional relationship between an information resource and a representation of the user's problem. A further judgment is made according to the topic, task, or context, at a particular point in time. These dimensions pointed out by Mizzaro were extended by Coppola et al. [11] in an attempt to define the concept of relevance in mobile IR settings. The authors argued for the necessity to move the notion of relevance into the "real/physical world" so that it will be closer to what users want and need.

Multidimensional personalization approaches that include aspects of the mobile user's contextual environment have then been proposed. For instance, authors in [12,5] integrated users' interests as a second criterion besides the topical relevance, to personalize search results. Given the importance of location for mobile users, other works have integrated the user's location as a criterion to select or to rank the retrieved search results according to their spatial distance from the user's location [13,6,14]. Some works attempt to go besides the location context, and also handle time context [15,16] or social context [4,17]. The main limitations of the aforementioned works is that these new personalization dimensions are considered as filters or are combined in a linear model independently of users' preferences over the relevance dimensions. None of these works has attempted to formulate a functional relationship between the combined criteria and the user's perception of relevance in a multi-criteria setting. An interesting research direction related to personalized search is to make the user an actor in determining such an aggregation model.

The contribution of this paper is twofold. First, we propose a multidimensional ranking model based on the three dimensions of topic, interest, and location. The multidimensional rank has the peculiarity of exploiting some "prioritized aggregation operator" [7,8] allowing a flexible personalization of search results according to users' preferences. In this way, for a same query and a same user, different document rankings can be computed based on the user's preference over the different relevance criteria. The proposed prioritization is modeled by making the weights associated with a criterion dependent upon the satisfaction of the higher-priority criteria. Hence, it is possible to take into account the fact that the weight of a less important criterion should be proportional to the satisfaction degree of more important criteria. This combination has the merit of being user-dependent, by allowing the user to express his/her preference order on the considered criteria. To illustrate this, let us come back to our introductory example query "*cultural event*," depending on user's preferences, we can imagine a scenario in which the user, who does not want to move, will favor document about any cultural event in "*Paris*," independently of being a "*jazz event*," and dismiss documents about "*jazz events*" that are outside "*Paris.*" In another scenario, the user may favor documents about his/her interest in "*jazz events*" although the event location is different from "*Paris.*"

Second, in the absence of a standard evaluation collection suited to evaluate an IR system which is user-dependent, we propose to build a simulated user-centered evaluation framework in order to test the effects of the prioritized aggregation operators on the final system performance in comparison with a standard linear combination approach.

The paper is organized as follows. Section 2 presents the prioritized multi-criteria aggregation background. Section 3 presents our multidimensional relevance ranking model. In section 4, we present our simulation-based framework to evaluate our approach and discuss the obtained results. In the last section, we conclude and outline future work.

2   Background: Prioritized Multi-criteria Aggregation

The problem of prioritized aggregation is typical in situations when one wants to model a relationship between multiple criteria. In such a case, the lack of satisfaction of a higher priority criterion cannot be compensated with the satisfaction of a lower priority criterion. In this section we apply the approach proposed in [7,8] for a priority-based aggregation of distinct relevance assessments. In sect. 2.1 the problem representation is introduced as a multi-criteria decision making problem where the possible alternatives are the documents in the considered document collection. In sect. 2.2 the priority-based aggregation operators "prioritized scoring" and "prioritized and" are described.

2.1   Problem Representation

Let us consider a decision making setting in which we have the following components:

- The set C of the considered criteria: $C = \{C_1, \ldots, C_n\}$. In order to simplify the notation, we denote by C_i also the function evaluating the ith criterion. After the user preference reordering of the n considered criteria we denote by C_1 the most preferred criterion, by C_n the least preferred criterion (i.e., the last in the user preference list), and we assume that C_i is preferred to C_j if and only if $i < j$.
- The collection of documents D.
- The $C_j(d)$ satisfaction score (of document d with respect to relevance criterion j).

- An aggregation function F to calculate for each document $d \in D$ an overall score $RSV(d) = F(C_1(d), \dots, C_n(d))$ on the basis of the evaluation scores of the considered criteria.

For each criterion $C_i \in C$, an importance weight is computed in a way that is both document and user-dependent. In fact, the weight computation depends both on the preference order expressed by the user over the criteria, and also on both the weight computed for criterion C_{i-1} (of greater priority with respect to C_i), and the satisfaction degree of the document with respect to C_{i-1}. In other words, for a considered document d, for each criterion C_i an importance weight $\lambda_i \in [0,1]$ is computed, which varies in accordance with the considered documents. The weights associated with the ordered criteria (criteria are ordered by users on the basis of their preferences), are computed as follows:

- For each document d, the weight of the most important criterion C_1 is set to 1 (i.e., by definition we have: $\forall d \; \lambda_1 = 1$).
- The weights of the other criteria C_i for $i \in [2, n]$ are calculated as follows:

$$\lambda_i = \lambda_{i-1} \cdot C_{i-1}(d) \tag{1}$$

where $C_{i-1}(d)$ is the degree of satisfaction of criterion C_{i-1} by document d, and λ_{i-1} is the importance weight of criterion C_{i-1}.

What is changing in this aggregation model is the way in which function F is defined, as is explained in the next section.

2.2 Prioritized Aggregation Operators

In this section we present two alternative formalizations of the proposed prioritized aggregation operator F: "prioritized scoring" and "prioritized and."

- **Prioritized "Scoring"** (F_s). This operator allows to calculate the overall score value from several criteria evaluations, where the weight of each criterion depends both on the weights and on the satisfaction degrees of the most important criteria. The higher the satisfaction degree of a more important criterion, the more the satisfaction degree of a less important criterion influences the overall score. It is defined by: $F_s : [0,1]^n \to [0,n]$ which is such that, for a given document d,

$$F_s(C_1(d), \dots, C_n(d)) = \sum_{i=1}^{n} \lambda_i \cdot C_i(d) \tag{2}$$

where all the C_i represent the considered relevance dimensions.

- **Prioritized "And"** (F_a). The peculiarity of such an operator, which also distinguishes it from the traditional "min" operator, is that the extent to which the least satisfied criterion is considered depends on its importance for the user. If it is not important at all, its satisfaction degree should not be considered, while if it is the most important criterion for the user, only its satisfaction degree is considered. This way, if we consider a document d for which the least satisfied criterion C_k is also the least important one, the overall satisfaction degree will be greater than $C_k(d)$; it will not be C_k as it would be the case with the traditional "and" operator, since the less important is the criterion, the lesser its chances to represent the overall satisfaction degree. The aggregation operator F_a is defined by: $F_a : [0,1]^n \rightarrow [0,1]$ which is such that, for all document d,

$$F_a(C_1(d), \ldots, C_n(d)) = \min_{i \in [1,n]} (\{C_i(d)\}^{\lambda_i}) \qquad (3)$$

3   Multidimensional Personalization of Mobile IR

In this paper, we apply the proposed prioritized relevance model as a ranking model within personalized IR in a mobile environment. Our personalization approach is multidimensional, considering within this context the set of three main relevance criteria: $C = \{topic, interest, location\}$. The final ranking of a document d, given a query Q, a user's interest I and a location L will be represented by his overall score $RSV(d)$ defined by:

$$RSV(d) = F(topic(d, Q), interest(d, I), location(d, L))$$

where $topic(d, Q)$ (respectively $interest(d, I)$, $location(d, L)$) is the function evaluating the $topic$ (respectively $interest$, $location$) criterion, and F is the prioritized aggregation operator. In this section we present a description and a formal definition for each of these criteria and their associated relevance functions.

3.1   Topic

The "Topic" criterion refers to the standard topical relevance computed by IR systems. The topical relevance is generally measured with an IR model. One of the prominent models is the probabilistic model [18] with the BM25 weighting scheme as a ranking function. For this reason, we adopt this model although topical relevance could be also computed based on alternative models. BM25 is a bag-of-words retrieval function that ranks a set of documents based on the query terms occurring in each document. More precisely, given a query Q containing keywords $t_1, \ldots, t_n$, the $topic$ relevance score of a document d is:

$$topic(d, Q) = \sum_{i=1}^{n} IDF(t_i) \cdot \frac{f(t_i, d) \cdot (k_1 + 1)}{f(t_i, d) + k_1 \cdot \left(1 - b + b \cdot \frac{|d|}{avgdl}\right)} \qquad (4)$$

where $f(t_i, d)$ is the frequency of term t_i in the document d, $|d|$ is the number of words occurring in document d, and $avgdl$ is the average document length in the text collection from which documents are retrieved. k_1 and b are free parameters usually chosen such that $k_1 = 2.0$ and $b = 0.75$. $IDF(t_i)$ is the inverse document frequency of the query term t_i, usually computed as:

$$IDF(t_i) = \log \frac{N - n(t_i) + 0.5}{n(t_i) + 0.5} \tag{5}$$

where N is the total number of documents in the collection, and $n(t_i)$ is the number of documents containing t_i.

3.2 Interest

The "Interest" criterion measures how strongly a retrieved document is similar to the user's interest. Users' interests are known to be the most important contextual factor that can be used to personalize web search in an ad hoc retrieval task [19]. The Interest criterion is measurable when the system makes use of a user profile. In this paper, we use the semantic user profile model proposed in our previous work [20], where user's interests are represented as a list of weighted concepts from the ODP[1]. Each concept c_j in the ODP is represented by a term vector $\vec{c_j}$ extracted from the web pages classified under that concept, as well as of its sub-concepts. Each term's weight w_i in a concept $\vec{c_j}$ is computed using $tf \times idf$ weighting scheme. In order to compute the interest score of a document d according to the user profile I, document d is represented by a vector of weighted terms. The *interest* relevance function of the document d is computed according to a term-based similarity measure, namely the cosine similarity measure between the document d and the top k ranked concepts of the user profile I as follows:

$$interest\,(d, I) = \sum_{c_j \in I \wedge j \in [1,k]} sw\,(c_j) \times \cos\left(\vec{d}, \vec{c_j}\right) \tag{6}$$

where $sw\,(c_j)$ is the similarity weight of the concept c_j in the user profile I.

3.3 Location

In this paper, we recognize the importance of location information in mobile search (more than 31% according to a recent study [21]), and propose to incorporate the user's location in addition to user's interest in the personalized search. Dealing with geographical information needs and localizing search results is a known problem within the field of geographical IR [22]. Of the various geographical ranking functions defined in the literature, we adapted the geographical weighting function presented in [23], as a geographical relevance score of a document. Given a geographic hierarchy GH, a geographical place L of a user query, and a document d, the *location* relevance function is given by:

$$location\,(d, L) = f(L) + \sum_{L_i \in offspring(L)} f(L_i) \tag{7}$$

where $f(L)$ refers to the number of occurrences of location L in d and the *offspring* locations L_i of the given location L are identified from GH.

---

[1] Open Directory Project (ODP): http://www.dmoz.org/

4   Experimental Evaluation

In the absence of a standard evaluation collection suited to evaluate an IR system which is user-dependent, we propose an evaluation protocol that integrates the user context in the evaluation by means of simulation. Our experimental setup is in line with simulation-based evaluation [24]. The objective of our experimental evaluation is twofold: 1) to study the effectiveness of the combination of the multi-relevance criteria in comparison with the traditional topical relevance assessment standard, and 2) to compare the effect of the prioritized aggregation operators in the retrieval performance in comparison with the linear combination schemes. In the following, we first present our experimental settings then we discuss the obtained results.

4.1   Experimental Settings

For experimental purposes, we use a branching part of the ODP Ontology consisting of the set of web pages classified under the US region. The ODP is the most widely distributed data base of web content classified by humans. These web pages are in fact classified under concepts but also under geographical places allowing us to study all our criteria topicality, interest and location. We crawled 171,541 web pages that we divided into two sets: a test set (T) representing 2/3 of the crawled web pages, and a profile set (P) composed of the remaining web pages. The documents were randomly assigned to one of the two subsets.

Document collection. The (T) set is used as the document collection for search, it is indexed using the Terrier[2] search engine [25], and is used as the search collection. For all these documents, we kept track of which concepts and locations these documents were originally classified under. This information is exploited as an evidence source in the relevance judgment as described below.

Users' interests. We simulated 30 users profiles. For simplicity, we assigned one interest to each user ($k = 1$ and $sw(c_1) = 1$ in formula 5). To simulate users' interests we randomly selected 30 concepts from the ODP. More specifically, each interest is represented as a concept from the ODP using the set of documents from (P) classified under this concept as described in sect. 3.2.

Users' locations. Users' locations are chosen from the US cities. We suppose a different location for each user. They are simply added to the users' queries.

Users' Queries. To simulate search, we designed a set of 6 queries for each user profile. They are constructed using different strategies. As a result, a query may be formulated in each one of the following ways:

1. A set of terms describing a particular information need about the concept.
2. The most frequent term in the concept.
3. The two most frequent terms in the concept.

---

[2] http://terrier.org/

Table 1. An example of queries constructed according to each strategy

Strategy	Query terms
1	" book room spa wifi San Francisco"
2	"hotel San Francisco"
3	"hotel service San Francisco"
4	"hotel service chain San Francisco"
5	"location reserve resort San Francisco"
6	"park activity San Francisco"

4. The three most frequent terms in the concept.
5. Two or more overlapping terms within highest weighting 10 terms among different concepts.
6. An information need expressed using terms from another concept (different from the current considered one).

Our goal behind these strategies was twofold: (1) since the queries of mobile users tend to be short and ambiguous [26], we wanted to cover such queries, and (2) to cover at the same time situations in which the user formulates queries in line with his/her interest, but also queries formulated on new interests (not yet present in his/her profile). We finally obtained a set of 180 queries (30 × 6). Table 1 gives an example of queries constructed according to each strategy. Queries 1 to 5 are constructed on the concept "Hotels and Motels" and query 6 is on an another concept "Parks."

Evaluation scenarios and relevance judgment assignment. The importance order ($\succ$) of a user on the three relevance criteria (topic, interest, and location) allows us to define six possible evaluation scenarios:

- TIL: Topical$\succ$Interest$\succ$Location.
- TLI: Topical$\succ$Location$\succ$Interest.
- ITL: Interest$\succ$Topical$\succ$Location.
- ILT: Interest$\succ$Location$\succ$Topical.
- LTI: Location$\succ$Topical$\succ$Interest.
- LIT: Location$\succ$ Interest$\succ$Topical.

Each pair (query, document) returned in the result list was judged according to each one of these evaluation scenarios. Relevance judgments were made automatically by exploiting the locations and concepts from which the documents were originally classified in the ontology. Relevance assignment of each individual criterion was done like this:

- If a document was classified under the query concept, it was judged topically relevant to the query.
- If a document was classified under a concept which corresponds to the current user's interest, it was judged relevant to the user's profile.
- If a document was classified under the user's location, it was judged relevant to the user's location.

Algorithm 1. Assigning relevance judgment depending on the ordered criteria satisfaction degree

---

```
if the document does not satisfy the first order criterion then
    - the document is not relevant.
else
    if the document does not satisfy the second order criterion then
        - the document is partially relevant.
    else
        if the document does not satisfy the third order criterion then
            - the document is relevant.
        else
            - the document is very relevant.
        end if
    end if
end if
```

The final relevance judgment of a document is made by combining its individual relevances using a four-level relevance scale. It is done according to Algorithm 1 depending on the ordered criteria satisfaction degree.

Evaluation metrics. To estimate the quality of the produced ranks, three measures are used. The MAP, the nDCG at n and the Precision at n, which are usually used to represent the system performance. We computed them with standard *trec_eval*[3] program. The computed values were averaged over all the evaluated queries results and/or over the different evaluation scenarios.

4.2   Results and Discussions

Effectiveness of the Combination of the Multi-Relevance Criteria. In this first series of experiments we performed for each query a standard *baseline* search using a topical based relevance system Terrier [25]; we then computed an interest score and a location score, for all the documents in the collection like described in sect. 3.2 respectively in sect. 3.3. In order to combine these scores, we computed a normalization of these individual scores. We then performed a combination relevance scheme using standard operators *min*, *average* and *weighted average* (denoted *w-average*) where we assigned weights for the criteria in accordance with their importance order in the evaluated scenario. We have carried out experiments with the aforementioned evaluation scenarios. Figure 1(a) and 1(b) show the results performance, measured using Precision respectively nDCG at different cut-off points, averaged over all the evaluation scenarios. Results show that in general, the combination of relevance criteria outperforms the topical relevance assessment standard, with the *weighted average* combination achieving best performance in terms of nDCG and Precision at different cut-off points. This confirms the effectiveness of the multi-relevance based model.

Effectiveness of the Prioritized Aggregation Operators. In this second series of experiments we combined the multi-relevance criteria using the prioritized aggregation operators. We then compared the obtained result ranks with

---

[3] http://trec.nist.gov/trec_eval/

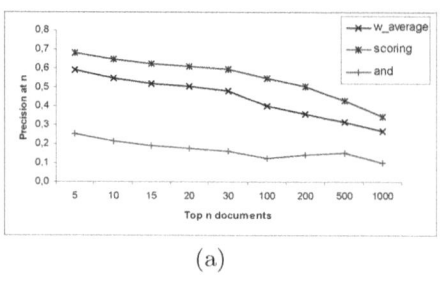 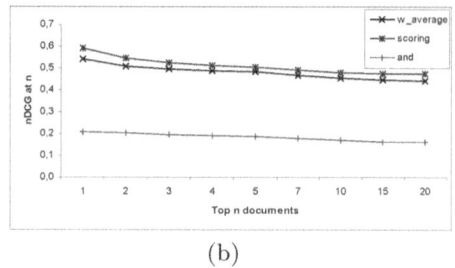

(a)   (b)

Fig. 1. Average Precision at n (a) and nDCG at n (b) comparison between the combination search using standard operators and the baseline search using a topical relevance ranking scheme, over all scenarios

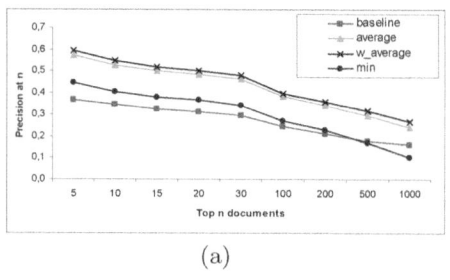 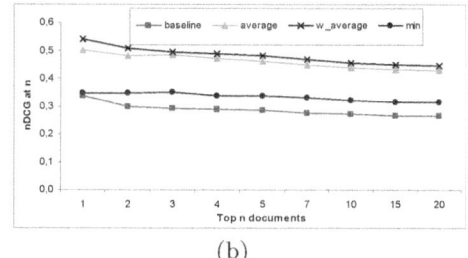

(a)   (b)

Fig. 2. Average Precision at n (a) and nDCG at n (b) comparison between the combination search using prioritized operators and the combination search using standard weighted average operator, over all scenarios

Table 2. MAP comparison between the scoring and the standard weighted average ranks for each evaluation scenario

Table 3. MAP comparison between the scoring and the standard weighted ranks for each query strategy

Scenario	W-average	Scoring	Improvement
TIL	0.0419	0.0478	14.08%(*)
TLI	0.0316	0.0413	30.70%(*)
ITL	0.0378	0.1150	204.23%(*)
ILT	0.0265	0.1086	309.81%(*)
LTI	0.8255	0.8312	0.69%(*)
LIT	0.8315	0.8315	0.00%(-)

Strategy	W-average	Scoring	Improvement
1	0.2915	0.3207	10.02%(*)
2	0.3033	0.3305	8.96%(*)
3	0.3113	0.3387	8.83%(*)
4	0.3152	0.3437	9.04%(*)
5	0.2822	0.3141	11.31%(*)
6	0.2912	0.3277	12.52%(*)

the rank of the standard *weighted average* as the *baseline*. Figure 2(a) and 2(b) show results performance in terms of Precision respectively nDCG at different cut-off points, averaged over all the evaluation scenarios. Results show that the *"prioritized scoring"* operator outperforms its counterpart standard *weighted average*. However, the *"prioritized and"* operator degrades the results.

In order to evaluate significance of the *"scoring"* operator improvement, we conducted a paired two-tailed t-test. Table 2 shows the performance results in terms of MAP computed for each evaluation scenario. Results show that the improvement of the *"scoring"* operator comparatively to the *baseline* was found to be statistically significant (noted * in the table) with p-values < 0.01 for the majority of the scenarios. However improvements are different between the tested scenarios. We notice that best improvements are obtained on scenarios where user's interest is the first order criterion. Little and no improvement was noticed on the MAP of the two scenarios LTI respectively LIT where the location criterion is the first order criterion. We notice also that the MAP of the two scenarios LTI and LIT are somewhat better than the other scenarios. This is likely due to the fact that the score of the location criterion is computed with higher precision than the interest and the topical based scores.

Further, we analyzed the results performance for the different types of queries issued from the different construction strategies. Table 3 shows comparison on MAP results averaged over the six evaluation scenarios for each query construction strategy, obtained by the *"scoring"* and the *weighted average* ranks. Results show that the *"scoring operator"* shows its superiority in all query construction strategies, with a statistically significant improvement (noted * in the table) with p-values < 0.01 over the standard *weighted average* operator.

5 Conclusion and Outlook

In this paper, we have proposed a multi-criteria relevance model for personalizing mobile IR. The main contribution of this work concerns the adoption of a "prioritized operators" for aggregating the considered relevance criteria. Thanks to this aggregation scheme, it is possible to take the user's preference order over the criteria in the aggregated score of a document. Experimental results show that: (1) the "prioritized scoring" aggregation scheme allows to improve the ranking of the documents for the majority of the considered preference orders and all the query strategies, and (2) the "and operator" is not suited for the aggregation of criteria within our retrieval mobile IR settings. In future work, we plan to enhance our multidimensional model to include other contextual criteria and to conduct experiments with real mobile users and queries.

References

1. Ingwersen, P., Järvelin, K.: The Turn: Integration of Information Seeking and Retrieval in Context. Springer, Dordrecht (2005)
2. Borlund, P.: The concept of relevance in IR. JASIST 54(10), 913–925 (2003)
3. Sohn, T., Li, K.A., Griswold, W.G., Hollan, J.D.: A diary study of mobile information needs. In: 26th Annual SIGCHI Conference on Human Factors in Computing Systems, pp. 433–442. ACM (2008)
4. Rhodes, B.: Using Physical Context for Just-in-Time Information Retrieval. IEEE Transactions on Computers 52(8), 1011–1014 (2003)
5. Goenka, K., Arpinar, I.B., Nural, M.: Mobile web search personalization using ontological user profile. In: Proc. of the 48th ACM SE 2010, pp. 13–21 (2010)
6. Yokoji, S., Takahashi, K., Miura, N.: Kokono: a location-based search engine. In: Proc. of the 10th WWW Conference (2001)

7. da Costa Pereira, C., Dragoni, M., Pasi, G.: Multidimensional Relevance: A New Aggregation Criterion. In: Boughanem, M., Berrut, C., Mothe, J., Soule-Dupuy, C. (eds.) ECIR 2009. LNCS, vol. 5478, pp. 264–275. Springer, Heidelberg (2009)
8. da Costa Pereira, C., Dragoni, M., Pasi, G.: A Prioritized "And" Aggregation Operator for Multidimensional Relevance Assessment. In: Serra, R., Cucchiara, R. (eds.) AI*IA 2009. LNCS, vol. 5883, pp. 72–81. Springer, Heidelberg (2009)
9. Cooper, W.S.: On selecting a measure of retrieval effectiveness. JASIST 24(2), 87–100 (1973)
10. Mizzaro, S.: Relevance: the whole history. JASIST 48(9), 810–832 (1997)
11. Coppola, P., Della Mea, V., Di Gaspero, L., Mizzaro, S.: The Concept of Relevance in Mobile and Ubiquitous Information Access. In: Crestani, F., Dunlop, M.D., Mizzaro, S. (eds.) Mobile HCI International Workshop 2003. LNCS, vol. 2954, pp. 1–10. Springer, Heidelberg (2004)
12. Varma, V., Sriharsha, N., Pingali, P.: Personalized web search engine for mobile devices. In: Int'l. Workshop on Intelligent Information Access, IIIA 2006 (2006)
13. Mountain, D., MacFarlane, A.: Geographic information retrieval in a mobile environment: evaluating the needs of mobile individual. JIS 33(5), 515–530 (2007)
14. Göker, A., Myrhaug, H.: Evaluation of a mobile information system in context. IPM 44(1), 39–65 (2008)
15. Tezuka, T., Tanaka, K.: Temporal and Spatial Attribute Extraction from Web Documents and Time-Specific Regional Web Search System. In: Kwon, Y.J., Bouju, A., Claramunt, C. (eds.) W2GIS 2004. LNCS, vol. 3428, pp. 14–25. Springer, Heidelberg (2005)
16. Church, K., Smyth, B.: Who, What, Where & When: A New Approach to Mobile Search. In: Proc. of IUI 2008, pp. 309–312. ACM (2008)
17. Church, K., Neumann, J., Cherubini, M., Oliver, N.: Socialsearchbrowser: A novel mobile search and information discover tool. In: Proc. of the 14th Int'l Conference on Intelligent User Interfaces, pp. 101–110. ACM (2010)
18. Jones, S., Walker, K.S., Robertson, S.E.: A probabilistic model of information retrieval: Development and comparative experiments. IPM 36(6), 779–808 (2000)
19. Park, T.: Toward a theory of user-based relevance: A call for a new paradigm of inquiry. JASIST 45(3), 135–141 (1994)
20. Daoud, M., Tamine, L., Boughanem, M.: Learning user interests for session-based personalized search. ACM Information Interaction in Context, 57–64 (2008)
21. Church, K., Smyth, B.: Understanding the intent behind mobile information needs. In: 13th Int'l. Conference on Intelligent User Interfaces, pp. 247–256. ACM (2009)
22. Jones, C.B., Purves, R.S.: Geographical information retrieval. IJGIS 22(3), 219–228 (2008)
23. Wang, L., Wang, C., Xie, X., Forman, J., Lu, Y., Ma, W.Y., Li, Y.: Detecting Dominant Locations from Search Queries. In: Proc. of SIGIR 2005, pp. 424–431 (2005)
24. Sanderson, M.: The Reuters collection. In: Proc. of the BCS IRSG Colloquium (1994)
25. Ounis, I., Amati, G., Plachouras, V., He, B., Macdonald, C., Johnson, D.: Terrier Information Retrieval Platform. In: Losada, D.E., Fernández-Luna, J.M. (eds.) ECIR 2005. LNCS, vol. 3408, pp. 517–519. Springer, Heidelberg (2005)
26. Kamvar, M., Baluja, S.: Deciphering trends in mobile search. Computer 40(8), 58–62 (2007)

Searching for Islamic and Qur'anic Information on the Web: A Mixed-Methods Approach

Rita Wan-Chik [*], Paul Clough, and Nigel Ford

The Information School, University of Sheffield, United Kingdom
{lip08rzw,p.d.clough,n.ford}@sheffield.ac.uk,
ritazaharah@miit.unikl.edu.my

Abstract. This paper seeks to understand and describe web searching patterns for Islamic and Qur'anic information, an area receiving little attention in past research. A mixed-methods approach has been taken to data collection utilizing both quantitative and qualitative techniques. Query logs collected in 2006 from the Microsoft Live search engine were analysed for Islamic-related terms. Characteristics such as query frequency, term frequency, query length, and session length were derived from the data. To complement these quantitative data, interview data were collected from 25 users who had experienced searching for Islamic and/or Qur'anic materials on the web. The interviews gave a deeper understanding of aspects of information seeking including search processes, challenges and opinions on locating Islamic and Qur'anic information on the web.

Keywords: user studies, interactive IR, information seeking, query log analysis, Islamic and Qur'anic information.

1 Introduction

Increasingly the Internet is being used as a common means of transmitting information of a religious nature. Højsgaard and Warburg (2005) reported that by the year 2004, there were approximately 51 million religious websites on the Internet disseminating information and communicating with their followers. A number of further studies have looked at how religions use Internet technologies in disseminating their beliefs (Helland, 2000; Dawson & Cowan, 2004; Karaflogka, 2006). Members of religious communities use the Internet to undertake activities, such as listening to sermons, asking for advice, networking and even shopping for religious merchandise (Helland, 2002; Foltz & Foltz, 2003; Campbell, 2005a; Cheong et al., 2009).

An area of increasing interest in the field of information seeking is the study of the relationship between religion and Internet use (Campbell, 2005b; Krueger, 2005). Studies have found that an important component of online religious activities is

---

[*] Rita Wan-Chik is a lecturer at The Malaysian Institute of Information Technology, Universiti Kuala Lumpur, Malaysia.

M.V.M. Salem et al. (Eds.): AIRS 2011, LNCS 7097, pp. 181–192, 2011.
© Springer-Verlag Berlin Heidelberg 2011

searching for religious information (Casey, 2001; Larsen & Rainie, 2001; Ho et al., 2008). A study in 2002 by the *Pew & Internet American Life Project* reported that 25% of Americans used search engines to search for religious information (Fox, 2002). In a more recent study, Hoover et al. (2004) found that 64% of 128 million American online users had used the Internet for spiritual and religious purposes where 28% of the online users had searched for their own religion and 26% for religion of others.

Dawson (2000) suggests that, in addition to analyses of online religious content, it is important to investigate people's search purposes and processes. This is supported by Jansen et al. (2009) who claim that few studies have investigated how people search specifically for religious-related information. Understanding the users of search tools is important for the design of systems that help people with different cultural and linguistic backgrounds to search for religious and culturally-related information (Neelameghan & Raghavan, 2005). However, the studies mentioned above do not focus on searching processes and issues specifically relating to searching for religious information.

Therefore, given the growth of Islamic and Qur'anic content on the web, and the paucity to date of user-oriented studies in this field, the study reported here aimed to complement existing literature on religious searching by investigating search purposes and processes specifically for relating to Islamic and Qur'anic information, and also to better understand the nature of religious queries on the web.

Two different approaches have been used to collect data and inform the findings: (1) a quantitative approach based on real user-system interactions from a large web search engine; (2) a qualitative approach based on interviews with a purposive sample of users selected to reflect the diversity of Islamic and Qur'anic information seekers on the web. The findings provide a picture of what users are actually searching for on the web and of their current experiences with web search.

The paper is structured as follows: Section 2 discusses previous work on analysing patterns of web searching; Section 3 presents the methodology used to gather data including both the datasets used in the log analysis study, and the interview data; Section 4 discusses the results; and Section 5 concludes with a summary of the paper and directions for future study in this area.

2   Related Work

2.1   Patterns of Web Searching

Studies of user searching patterns have a long history going way back to the late '70s where one of the first researchers to investigate how people search was Bates (1979). Her study looked at how people perform searches, and she proposed ways to characterize the overall search process. After web search engines became available and popular, many studies were conducted to understand how people perform searches on the web. These include studies by Jansen et al. (1998), Silverstein et al.

(1999), Spink et al. (2001) and Baeza-Yates et al. (2005), which explored how online users conduct their searches, and proposed various search behavior models.

2.2   Islam and Web Searching

There have been a number of previous studies on information seeking and Islam. For example, Bunt (2003) and Lawless (2004) found that there are various kinds of information on Islam on the web, e.g. Qur'an translations, recitations, opinions, sermons, fatwas and discussions. These are intended for different kinds of readers i.e. Muslim or non-Muslim. There have also been studies investigating the presence of Muslim and Islamic documents on the web and Muslim's online activities in general (Bunt, 2003, 2004; Adamu, 2002; Brouwer, 2004).

The number of Islamic and Qur'anic websites appearing online has shown to be steadily increasing (Bunt, 2003). However, as Shoaib et al. (2009) mention in their study, in the case of keyword searching for verses in the Qur'an, there are three basic problems: (1) in most cases all the relevant verses are not retrieved; (2) the sequence of retrieved verses does not appeal to the reader; and (3) some irrelevant verses are also retrieved. Baqai et al. (2009) believe that it *"remains a challenge to reach out to the learner and research community of the Qur'an, using the emerging technologies, to help create better, user-centric means to facilitate the learning, exploration, management and retrieval of Qur'anic knowledge resources"*.

Overall, there have been very few studies undertaken from the perspective of end user of their real experiences of retrieving information related to Islam and the Qur'an. This is an area which the current study aimed to fill.

3   Methodology

To gather data for analysing patterns of web searching behaviour for Islam-related information and Qur'anic materials on the web, a mixed methods approach was adopted (Tashakkori & Teddlie, 2010). This entails the complementary use of both quantitative and qualitative methods. The quantitative analysis (Section 3.1) was based on queries from a large web search engine, and explored search terms. The qualitative analysis (Section 3.2) was based on interviews with people who commonly search for Islamic/Qur'anic information, and aimed to discover their preferences and levels of success when searching for information online.

3.1   Query Logs

We used the Live Search query log released by Microsoft in 2006 which contains 12,251,068 queries originating from users located in the United States. The logs contain data about each query: a unique identifier, the query string itself, timestamp, the URLs of clicked results and the click position of items selected by the user. From all queries we extracted only those with Islamic-related terms as a part of the query string. We used a list of terms gathered from a combination of the top 10 religious queries (Islamic-related) as suggested by Jansen et al. (2009) and terms compiled from directories in the Islamic sections of Google, Yahoo! and DMOZ. We extracted queries with any or a combination of the terms listed in Table 1. This also included queries with possible spelling variations.

Table 1. Islamic-related terms used to identify queries from the query logs

Allah	Fiqh	Haram	Madinah	Muslim	Religion	Sufi
Aqsa	God	Holy	Makkah	Prayer	Salat	Sunni
Daawah	Hadith	Islam	Masjid	Prophet	School	Tafsir
Eid	Hajj	Jihad	Mosque	Qur'an	Shari'ah	Umrah
Fatwa	Halal	Ka'aba	Muhammad	Ramadan	Shiite	Zakat

2,089 queries were found to contain at least one of the terms listed in Table 1. As suggested by Jansen and Pooch (2001), query logs can be analysed at three levels: the *session*, the *query*, and the *term*. At the session level we derived statistics on the number of unique sessions and the average session length (the number of queries in each session). At the query level we computed the number of queries submitted with any combination of the terms listed in Table 1 above (query frequency). At the term level we computed the total number of terms, unique terms, and the number of terms per query (query length).

3.2   Interviews

In-depth, face-to-face interviews were conducted with 25 users who were willing to share on their experience of searching for Qur'anic materials on the web. In accordance with established qualitative research principles (Silverman, 2009) a purposive sampling approach was used. Email invitations were sent to selected groups including Islamic student groups, new (converted from other beliefs) Muslim societies, and cultural societies from around the UK. Invitations were also broadcast to similar groups and societies via Facebook, the online social network. A range of respondents were selected for the interview based on a purposive sampling approach (Mason, 2002) in order to reflect the diverse range of web-based Qur'anic information seekers.

We used open-ended questions to emphasize issues pertaining to the online users' search processes and purposes, their perceived challenges, levels of relevance, satisfaction and desired improvement. Transcriptions of the interviews were analysed inductively to identify emergent themes (Boyatzis, 1998).

4   Results

4.1   Query Log Analysis

Of the 12.2 million queries from the Microsoft Live Search, 2089 (0.02%) contained at least one of the Islamic-related terms listed in Table 1. There were 1220 unique sessions.

Next, we examined the number of queries in each unique session (session length) where more than 65% of searchers for Islamic-related terms submitted only one query per session. This is within range with the Jansen et al. (2009) study on the religious and religious-related searches (between 61 to 76 percent) for one to two queries per session. We recorded 12 queries as the largest number of queries submitted in a session. The average number of queries per session was 1.72, which is lower than the 2.84 reported in the Jansen et al. (2000) study on general web search.

Next, we analysed the number of terms used in each query (query length) where two-term queries were most frequent (26.47%). The maximum number of terms in a query was 17. The average (mean) number of terms per query was 3.14 terms. This is higher than that of the general web population which is 2.16 terms per query (Spink et al., 2001).

Table 2. Searching trends for Islamic queries

All queries	12,251,068 (100%)
Queries with Islamic terms (query frequency)	2,089 (0.02%)
Unique sessions	1,220
Terms (total)	5,398 (100%)
Unique Terms	942 (17.5%)
Number of queries per session (session length)	
1 query	795 (65.2%)
2 queries	211 (17.3%)
3 queries	111 (9.1%)
4+ queries	103 (8.4%)
mean (average)	1.72
median	1.00
mode	1.00
Number of terms per query (query length)	
1 term	357 (17.1%)
2 terms	553 (26.5%)
3 terms	454 (21.7%)
4+ terms	725 (34.7%)
mean (average)	3.14
median	3.00
mode	2.00

Table 3. Top 10 most frequent queries

Query	Frequency	%
islam	71	3.40
koran	36	1.72
muslim names	27	1.29
quran	20	0.96
99 names of allah	18	0.86
islamic prayers for repentence	16	0.77
islamic calligraphy	16	0.77
nation of islam	14	0.67
qur'an	13	0.62
islamic art	13	0.62

The queries issued by users was analysed and result are shown in Table 3 which shows the top 10 most frequent queries. There are three spelling variations of Qur'an (*koran, quran* and *qur'an*) which are listed in the top 10 queries frequently used by the users. The combination of all three queries makes up 3.3% of all the Islamic queries issued.

Query terms were also analysed to establish the most frequently occurring terms across all queries. The top 10 most frequently occurring terms is shown in Table 4.

Table 4. Top 10 most frequently occuring terms (across all queries)

Term	Frequency	%
islam	608	11.26
islamic	380	7.04
muslim	351	6.50
mosque	93	1.72
muslims	82	1.52
koran	75	1.39
quran	72	1.33
allah	71	1.32
names	67	1.24
halal	54	1.00

From the list of terms in Table 4, combining the terms with similar concepts or spelling variations into groups, the most common terms are muslim/muslims (8%) and koran/quran (2.7%). This follows Islam/Islamic (18.3%) which is the most popular query topic. The distribution of query terms generally follows a Zipfian curve: relatively small set of terms are used frequently, whilst large set of terms are used infrequently (Jansen et al., 2000). The 100 most frequently utilized terms were found to account for 61.8% of the total number of terms, which is relatively higher than the 18-22% for general web search reported in the study by Jansen et al. (2000).

4.2   Interview data analysis

Table 5 summarises characteristics (age, gender, occupation, country of origin, ethnicity and religion) about the 25 interview participants.

Table 5. Profile of the interview sample

Age range	21 – 61
Gender	13 Male, 12 Female
Occupation	Student (Undergraduates, Postgraduates) (11), Researchers (5), University Professors/Lecturers (2), University Faith Advisors (2), Computer-related Executives (2), Housewives (3)
Country of origin	Algeria (1), France (1), Germany (1), Indonesia (1), Jordan (1), Kashmir (1), Kuwait (1), Malaysia (3), Morocco (1), Netherlands (1), Pakistan (3), Philippines (1), Saudi Arabia (1), Sudan (1), Syria (1), United Kingdom (6)
Ethnicity	Caucasian (7), Asian (11), Arab (5), African (2)
Religion	Islam [born (18), convert (3)], Christianity (1), Atheist (1), Other (2)

Emergent themes

Almost all of the participants interviewed used the English version of Google in their searches for Islamic and/or Qur'anic information, with the exception of one who used the French version of Google. Most searched using English keywords, but there were a number that searched in Arabic using Arabic characters, or Arabic keywords spelled using Roman alphabets. Interviewees commented that searching in other languages, such as French, Malay, German, Urdu, and Persian would retrieve fewer results compared to English and Arabic. More than half of the participants believed that there are not enough reliable websites in the language of their choice, and that this situation needs to be improved. Many believe that a lot of good links and content are in Arabic and therefore queries using non-Arabic languages would not be able to retrieve them unless they had been translated. A couple of participants believed that with some countries imposing filters on search engines, many good links from the Arabic and Muslim countries are being filtered out, thus hiding 'innocent' good links.

Whilst Qur'anic verses, translations, explanations and recitations were the most searched for items, almost all the Muslim participants reported that they would also search for Hadith (sayings and teachings by the Prophet Muhammad) when searching for Qur'anic information, since the Qur'an and Hadith are very closely related to each other. Most of the explanations for verses in the Qur'an came from Hadith. However, the construction of Hadith is different from that of the Qur'an. The Qur'an only has one version with a standard structure (i.e. chapters and verses) whereas Hadith is a collection of Prophet Muhammad's sayings and teachings, and has been compiled by a number of people. The structure varies among the compilers. Therefore, searching for Hadith is more difficult than searching for Qur'anic information, especially for those with little knowledge of Hadith. Searching for Hadith in non-Arabic language is even more problematic. This was also seen as one of the most important aspects of Qur'anic searching that needs to be improved. As one of the Muslim participants said, *"As the Qur'an verses have Hadith connected to them, not getting the related Hadith means not getting the whole knowledge. Results should retrieve the related Hadith and also the Asbab (occasions/circumstances of Qur'an revelation)."* Other topics being searched for are listed in Table 6.

Table 6. List of the most searched for topics by participants

Qur'an	translation, Tafsir (explanation), Ruqyah (verses to cure illness), motivational verses, miracles of Qur'an, Asbab-ul-Qur'an (occasions/circumstances of Qur'an revelation), history, learning the Qur'an, tajwid/tarteel (proper pronounciation when reciting Qur'an) and how to read the Qur'an,
Hadith	Hadith & Sunnah (practice of Prophet Muhammad), translation, sanad (chain of narration), matn (text of the Hadith), explanation
Other Islamic topics	Fatwa (religious opinion issued by scholars concerning Islamic law) and rulings on current issues, Prophets, Seerah (historical biography of Prophets), Sahabah (Prophet Muhammad's Companions), Fiqh (jurisprudence, an expansion under the Sharia law), talks, Islamic Finance, digitized Islamic books, masa'el (issues and problems), women/feminism in Islam, about Islam, Shariah law, Naat (poetry to praise Prophet Muhammad)

Among the reasons cited by the participants for why they used the web to search for the topics listed in Table 6 (especially learned Muslims and those that already have a large collection of physical copies of the books) included: speed and ease of access to many versions of Tafsirs and Hadiths; access to more collections; and being able to retrieve more and deeper explanations for Qur'anic verses. They also use retrieved information to prepare for studies, assignments, exams or papers; for teaching, talks and sermons; to gather scholars' opinions as solutions to problems; for work; out of self-interest or curiosity to learn more or to find the truth concerning misconceptions; to help in reciting or memorizing the verses by listening to it online, to copy the sounds of recitation for non Arabic speakers; to see what incorrect information about Islam has been put up on the Web and to correct the misinformation; to acquire the full version of half-remembered knowledge; to discuss/debate issues; and to copy verses into their own writings.

However, the main issue for Muslim participants, and also for a few non-Muslims, was the reliability and credibility of Islamic and Qur'anic websites. This affected their selection processes when viewing search results, where the brief descriptions given with each result (snippet) did not help them verify the credibility of the link. The only way was to click on each hit and to browse through the websites to see if it was an authentic Islamic or Qur'anic website. This entailed a long, and sometimes frustrating, process. Some would only look at websites that were recommended by scholars or friends. Most would regard links from universities, established institutions or foundations, and URLs ending with *.org* as reliable, especially those that are associated with famous scholars names in the particular field. Most of the Muslims agreed that familiar information which corroborates their existing knowledge tended to be considered relevant and correct. However, in some cases it was recognized that in order to establish credibility one would need to read a number of different parts of a website, or visit a website frequently in order to be able to assess its balance and reliability.

In terms of the credibility of information, most of the Muslim participants would cross-check retrieved Qur'anic verses or Hadith with the printed materials in their collections, or verify them with a learned friend, an accessible scholar, or other trusted Qur'anic/Hadith websites. However, some non-Muslims did admit that they would not be able to differentiate between authentic Islamic or Qur'anic websites and non-authentic ones. One of them even believed that there would not be any fake Qur'anic information on the web as it would be such a waste of time for anyone to put them there. A couple of Muslims said that they would be suspicious of contents without the Arabic verses. Other issues and challenges mentioned by the participants are listed in Table 7.

Table 7. A list of issues and challenges emerging from the interviews

Information overload	Unavailable information
Misleading information	Incorrect information
Disorganized information	Too much sectarian information
Low volume of information	Dissatisfaction with retrieved information

Among the many types of website mentioned and discussed by the participants, Islamic forums are one that received mixed reviews. Although more than half thought that the information from forums is not reliable since they tend to be more opinion-based, a couple of Muslim participants thought that forums actually give better information because it is common to find that answers in forums tend to be supported with Qur'an or Hadith references, and sometimes by recommended links to further information. In a way, forums can be a one-stop resource where one obtains answers or solutions to problems, as well as recommendations to other links specific to set topics.

Most participants who thought that this type of searching is different from general searching, or searching for other topics, considered prior knowledge (of Islam or Qur'an or Hadiths) and the ability to specify keywords accurately as helping to retrieve more relevant results. However, query recommendation given by Google during the query formulation process does not necessarily retrieve better or more relevant results. Almost all of the Muslim participants expressed concern about those with little or no prior knowledge retrieving non-authentic websites without realizing it, and not knowing how to verify their contents. As one Muslim participant said, *"One also needs to have some prior knowledge. Those with little knowledge should ask around as to which ones are good sites, as they wouldn't know how to differentiate authentic from non-authentic sites."*

Participants made a number of suggestions for improvement. These include the need for us to find ways where online users can collectively rate or recommend links that are believed to be authentic with reliable contents or to find ways on how to push good links up in the search result lists. Another suggestion was that there should be a governing body to verify the authenticity of those Islamic and Qur'anic links.

We end this section with a quote from one of the participants which we think summarizes the main finding of this study perfectly, *"The issue here is evaluating your results hits. How do you decide which one is reliable? ... It's a problem for everybody who is not Islamic to be able to find something that they are able to trust. I cannot see what I can trust."*

5 Conclusion

This study is the first to combine analysis of search engine query logs and user interviews in order to understand how people search for Islamic and Qur'anic information on the web. The study has revealed the frequency of Islamic and Qur'anic related searches amongst 12.2 million searches by people using a large general web search engine, and has identified the most frequently used queries and most frequently occurring terms relating to these queries, as well as the spelling variations of Qur'an being used in the queries. The study complements the quantitative study with qualitative information gathered from interviews in order to illuminate the purposes and thoughts of people searching for Islamic and Qur'anic related information on the web.

From a diverse purposive sample, a number of key issues emerged as of concern to users. These included: information relevance and credibility; retrieval of Hadith in relation to the Qur'an; and the language of search queries in relation to the language of the web content itself. It was also found through the interviews that all of the Muslim participants only use the terms *qur'an* or *quran* in their queries for Qur'anic information whereas other participants would also use the terms *koran, kuran* and *coran*. For the Muslim participants, cross-checks between items retrieved from the web and the printed materials (Qur'an, Hadith, Tafsir, etc.), or with learned persons, are almost always needed to be done to avoid retrieving wrong information. However, for the non-Muslim participants they put more trust in the results from web search engines and therefore would not commonly cross-check the results.

In the future we plan to study particular groups of people interested in searching for Islamic and Qur'anic related information and searchers from other parts of the world, in order to discover if there are, for example, age-related or cultural differences in searching or if there are any other differences in the searching behaviour and activities amongst the groups. Also, further research is needed to discover whether religious searching of this type differs significantly in any way compared to more general searching. Such results might be able to inform both search engine designers and the providers of religious materials on the web.

References

Adamu, A.U.: Islam and the Internet. Weekly Trust (2002), http://www.kanoonline.com/publications/islam_and_the_internet.htm (retrieved July 10, 2009)

Baeza-Yates, R., Hurtado, C., Mendoza, M., Dupret, G.: Modeling user search behavior. In: Proceedings of the Third Latin American Web Congress, October 31-November 02, p. 242 (2005)

Baqai, S., Basharat, A., Khalid, H., Hassan, A., Zafar, S.: Leveraging semantic Web technologies for standardized knowledge modeling and retrieval from the Holy Qur'an and religious texts. In: ACM Special Interest Group on Artificial Intelligence, Proceedings of the 7th International Conference on Frontiers of Information Technology. ACM, New York (2009)

Bates, M.J.: Information search tactics. Journal of the American Society for Information Science, 205–214 (July 30, 1979)

Boyatzis, R.E.: Transforming qualitative information: thematic analysis and code development. Sage, London (1998)

Brouwer, L.: Dutch-Muslims on the Internet: A new discussion platform. Journal of Muslim Affairs 24(1), 47–55 (2004)

Bunt, G.R.: Islam in the digital age: E-Jihad, online fatwas and cyber Islamic environment. Pluto Press, London (2003)

Bunt, G.R.: Rip. Burn. Pray.: Islamic expression online. In: Dawson, L.L., Cowan, D.E. (eds.) Religion Online Finding Faith on the Internet, pp. 123–134. Routledge, New York (2004)

Campbell, H.: Exploring religious community online: We are one in the network (Digital Formations). Peter Lang Publishing, New York (2005a)

Campbell, H.: Making space for religion in Internet studies. The Information Society 21(4), 309–315 (2005b)

Casey, C.A.: Online religion and finding faith on the Web: An examination on Beliefnet. Org. Media Ecology Association 2, 32–40 (2001)

Cheong, P.H., Poon, J.P.H., Huang, S., Casas, I.: The Internet highway and religious communities: Mapping and contesting spaces in religion-online. The Information Society 25(5), 291–302 (2009)

Dawson, L.L.: Researching religion in cyberspace: Issues and strategies. In: Hadden, J.K., Cowan, D.E. (eds.) Religion and the Social Order. Elsevier Science Inc., London (2000)

Dawson, L.L., Cowan, D.E.: Religion online: Finding faith on the Internet. Routledge, London (2004)

Foltz, F., Foltz, F.: Religion on the Internet: Community and virtual existence. Bulletin of Science, Technology & Society 23(4), 321–330 (2003)

Fox, S.: Search Engines. Pew Internet project data memo. Pew Internet and American Life Project (2002), http://www.pewinternet.org/reports/toc.asp (retrieved September 10, 2009)

Helland, C.: Online-religion/religion-online and virtual communitas. In: Hadden, J., Cowan, D.E. (eds.) Religion on the Internet: Research Prospects and Promises. JAI Press/Elsevier Science, London (2000)

Helland, C.: Surfing for salvation. Elsevier Science 32, 293–302 (2002)

Ho, S.S., Lee, W.P., Hameed, S.S.: Muslim surfers on the Internet: Using the theory of planned behaviour to examine the factors influencing engagement in online religious activities. New Media & Society 10(1), 93–113 (2008)

Højsgaard, M.T., Warburg, M.: Religion and Cyberspace. Routledge, London (2005)

Hoover, S., Clark, L.S., Rainie, L.: Faith online: 64% of wired Americans have used the Internet for spiritual or religious information. Pew Internet and American Life Project (2004), http://www.pewinternet.org/pdfs/ PIP_Faith_Online_2004.pdf (retrieved February 17, 2009)

Jansen, B.J., Pooch, U.: A review of web searching studies and a framework for future research. Journal of the American Society for Information Science and Technology 52(3), 235–246 (2001)

Jansen, B.J., Spink, A., Bateman, J., Saracevic, T.: Real life information retrieval: A study of user queries on the Web. ACM SIGIR 32(1), 5–17 (1998)

Jansen, B.J., Spink, A., Saracevic, T.: Real life, real users, and real needs: a study and analysis of user queries on the web. Information Processing and Management 36, 207–227 (2000)

Jansen, B.J., Tapia, A., Spink, A.: Searching for salvation: An analysis of religious searching on the World Wide Web. Religion 40(1), 39–52 (2009)

Karaflogka, A.: E-religion a critical appraisal of religious discourse on the World Wide Web. Equinox, London (2006)

Krueger, O.: Methods and theory for studying religion on the internet: Introduction to the special issue on theory and methodology. Heidelberg Journal of Religions on the Internet 1(1), 1–7 (2005), http://www.ub.uni-heidelberg.de/archiv/5822 (retrieved March 9, 2009)

Larsen, E., Rainie, L.: Cyberfaith: How Americans pursue religion online, p. 21 (December 23, 2001)

Lawless, A.: Islam on-line: Adapting to the digital age. Three Monkeys Online Magazine (2004), http://www.threemonkeysonline.com/als/ Islam%20on%20the%20net.html (retrieved February 10, 2009)

Mason, J.: Qualitative researching, 2nd edn. Sage, London (2002)

Neelameghan, A., Raghavan, K.S.: An online multi-lingual, multi-faith thesaurus: A progress report on f-thes. Webology 2(4) (2005)

Shoaib, M., Nadeem Yasin, M., Hikmat, U.K., Saeed, M.I., Khiyal, M.S.H.: Relational WordNet model for semantic search in Holy Quran. Paper Presented at the International Conference on Emerging Technologies, ICET 2009 (2009)

Silverman, D.: Doing qualitative research, 3rd edn. Sage, London (2009)

Silverstein, C., Henzinger, M., Marais, H., Moricz, M.: Analysis of a very large Web search engine query log. ACM SIGIR 33(1), 6–12 (1999)

Spink, A., Wolfram, D., Jansen, B.J., Saracevic, T.: Searching the Web: The public and their queries. Journal of the American Society for Information Science and Technology 52(3), 226–234 (2001)

Tashakkori, A., Teddlie, C.: Sage handbook of mixed methods in social & behavioral research, 2nd edn. Sage, London (2010)

Enriching Query Flow Graphs
with Click Information

M-Dyaa Albakour[1], Udo Kruschwitz[1], Ibrahim Adeyanju[2], Dawei Song[2],
Maria Fasli[1], and Anne De Roeck[3]

[1] University of Essex, Colchester, UK
malbak@essex.ac.uk
[2] Robert Gordon University, Aberdeen, UK
[3] Open University, Milton Keynes, UK

Abstract. The increased availability of large amounts of data about
user search behaviour in search engines has triggered a lot of research in
recent years. This includes developing machine learning methods to build
knowledge structures that could be exploited for a number of tasks such
as query recommendation. Query flow graphs are a successful example of
these structures, they are generated from the sequence of queries typed
in by a user in a search session. In this paper we propose to modify the
query flow graph by incorporating clickthrough information from the
search logs. Click information provides evidence of the success or failure
of the search journey and therefore can be used to enrich the query flow
graph to make it more accurate and useful for query recommendation. We
propose a method of adjusting the weights on the edges of the query flow
graph by incorporating the number of clicked documents after submitting
a query.

We explore a number of weighting functions for the graph edges using
click information. Applying an automated evaluation framework to assess
query recommendations allows us to perform automatic and reproducible
evaluation experiments. We demonstrate how our modified query flow
graph outperforms the standard query flow graph. The experiments are
conducted on the search logs of an academic organisation's search engine
and validated in a second experiment on the log files of another Web site.

Keywords: Search Log Analysis, Query Suggestions, Automatic
Evaluation.

1 Introduction

User interfaces of modern search engines have evolved rapidly in recent years.
Modern web search engines do not only return a list of documents as a response
to a user's query but they also provide various interactive features that help
users in quickly finding what they are looking for or assist them in browsing the
information. Google, for example, provides a list of query suggestions while a
user is typing in her queries in the search box. Beyond Web search we also observe more interaction emerging as illustrated by the success of AquaBrowser[1] as

---

[1] http://serialssolutions.com/aquabrowser/

M.V.M. Salem et al. (Eds.): AIRS 2011, LNCS 7097, pp. 193–204, 2011.
© Springer-Verlag Berlin Heidelberg 2011

a navigation tool in digital libraries. Such interfaces rely on a wealth of knowledge that characterise the domain and specify relations between the different concepts and entities. A number of approaches have been developed to extract knowledge structures that could be exploited to enrich these interfaces. One promising approach is to perform search log analysis which captures the community knowledge about the domain. Query flow graphs extracted from query logs are an example of these approaches which have proven to be useful for providing query recommendations.

In this study, we extend the query flow graph model which relies on query flows as implicit source of feedback by incorporating the post-query user browsing behaviour in the form of clicks. We explore various settings of this model by running an automatic evaluation on actual search logs to understand the impact of various interpretations of click information on the quality of query recommendations.

The paper is structured as follows. We will give a short review of related work in Section 2. Section 3 will describe how we extend the query flow graph model by adding click information using query logs. The experimental setup is explained in Section 4. Results are presented and discussed in Section 5. We will draw conclusions in Section 6 and outline future work in Section 7.

2   Related Work

Query recommendations have become ubiquitous in modern search engines. This is true for Web search engines but also for more specialised search engines. The challenge is to identify the right suggestions for any given search request, and this may depend on a number of factors such as the actual user who is searching, the context, the time of the day etc. A promising route for deriving query recommendations appears to be the exploitation of past interactions with the search engines as recorded in the logs. Several approaches have been proposed in the literature to provide query modification suggestions. Studies have shown that users want to be assisted in this manner by proposing keywords [19], and despite the risk of offering wrong suggestions they would prefer having them rather than not [16].

With the increasing availability of search logs obtained from user interactions with search engines, new methods have been developed for mining search logs to capture "collective intelligence" for providing query suggestions as it has been recognised that there is great potential in mining information from query log files in order to improve a search engine [9,15].

Given the reluctance of users to provide explicit feedback on the usefulness of results returned for a search query, the automatic extraction of implicit feedback has become the centre of attention of much research. Clickthrough data is one form of the implicit feedback left by users which can be used to learn the retrieval ranking function [10], [11], [1]. Queries and clicks can be interpreted as "soft relevance judgements" [6] to find out what the user's actual intention is and what the user is really interested in. Query recommendations can then be derived, for example, by looking at the actual queries submitted and building query flow graphs [4], [5], query-click graphs [6], cover graphs [3] or association rules [8]. Jones *et al.* combined mining query logs with query similarity measures to derive query modifications [12].

Mining post-query click behaviour has also been studied and applied in information retrieval tasks. For example, Cucerzan *et. al.* [7] used landing page information to derive query suggestions. White *et. al.* [18] mined user search trails for search result ranking, where the presence of a page on a trail increases its query relevance. Click graphs were used by White and Chandrasekar to derive labels to shortcut search trails to help users reach target pages efficiently [17].

Given the successful application of both the query flow graph model as well as post-query click information we explore the potential of extending the query flow graph with click information for deriving query recommendation suggestions.

3   The Model

3.1   The Query Flow Graph

The query flow graph was introduced in Boldi *et al.* [4] and applied for query recommendations.

The query flow graph G_{qf} is a directed graph $G_{qf} = (V, E, w)$ where:

- V is a set of nodes containing all the distinct queries submitted to the search engine and two special nodes s and t representing a *start state* and a *terminate state*;
- $E \subseteq V \times V$ is the set of directed edges;
- $w : E \to (0..1]$ is a weighting function that assigns to every pair of queries $(q, q') \in E$ a weight $w(q, q')$.

The graph can be built from the search logs by creating an edge between two queries q, q' if there is one session in the logs in which q and q' are consecutive. A session is simply defined as a sequence of queries submitted by one particular user within a specific time limit.

The weighting function of the edges w depends on the application. Boldi *et al.* [4] developed a machine learning model that assigns to each edge on the graph a probability that the queries on both ends of the edge are part of the same chain. The chain is defined as a topically coherent sequence of queries of one user. This probability is then used to eliminate less probable edges by specifying some threshold. For the remaining edges the weight $w(q, q')$ is calculated as:

$$w(q, q') = \frac{freq(q, q')}{\Sigma_{r \in R_q} freq(q, r)} \qquad (1)$$

Where:

- $freq(q, q')$ is the number of the times the query q is followed by the query q'.
- R_q is the set of all reformulations of query q in the logs.

Note that the weights are normalised so that the total weights of the outgoing edges of any node is equal to 1.

3.2   Enriching the Query Flow Graph

In this section we explain how we extend the query flow graph model with click data. The intuition here is to use implicit feedback in the form of clickthrough data left by users when they modify their queries which has been shown to be powerful feedback, e.g. [6]. We consider the number of clicked documents by a user after submitting a query as an indication of how useful the results are. This is line with previous work on evaluating search engines with clickthrough data [14].

Let $\phi(q, q') = \{\varphi_0(q, q'), \varphi_1(q, q'), \varphi_2(q, q'), ..\}$ be an array of the frequencies of the reformulation (q, q'), where $\varphi_k(q, q')$ is the number of the times the query q is followed by the query q' and the user has clicked k (and only k) documents on the result list presented to the user after submitting query q'. We aggregate over all users here.

We modify the weighting function in equation 1 to incorporate the click information as follows

$$w(q, q') = \frac{\Sigma_i C_i.\varphi_i(q, q')}{\Sigma_{r \in R_q} \Sigma_i C_i.\varphi_i(q, r)} \tag{2}$$

Where C is an array of co-efficient factors for each band of click counts. Choosing different values for C_i allows us to differentiate between queries that resulted in more or fewer clicks. For example queries which result in a single click might be interpreted as more important than the ones which resulted in no clicks or more than one click as the single click may be an indication of quickly finding the document that the user is looking for.

In our experiments we investigate how different values of the co-efficient C_i affect the quality of the query recommendations. Note that the weighting function of the standard graph in Equation 1 is the special case where $C_0 = C_1 = C_2 = .. = 1$.

3.3   Query Recommendations

Query recommendation is the problem of finding for a given query q relevant query suggestions. If we want to recommend only a single query, then we try to identify the "most important" query q'. The query flow graph can be used for this purpose by ranking all the nodes in the graph according to some measure which indicates how reachable they are from the given node (query). Boldi et al. [4] proposed to use graph random walks for this purpose and reported the most promising results by using a measure which combines relative random walk scores and absolute scores. This measure is

$$\overline{s}_q(q') = \frac{s_q(q')}{\sqrt{r(q')}} \tag{3}$$

where:

- $s_q(q')$ is the random walk score relative to q i.e. the one computed with a preference vector for query q.

– $r(q')$ is the absolute random walk score of q' i.e. the one computed with a uniform preference vector.

In our experiments, we adopted this measure for query recommendation and used the random walk parameters reported by Boldi *et al.*

4   Experimental Setup

The aim of the experiments is to investigate whether the query flow graph can be enhanced and how the performance of query recommendations can be affected by different values of the coefficient factors of click counts presented in Equation 2.

The experiments conducted try to answer these questions:

1. Using search logs of a local search engine[2], can we achieve better query recommendations over the standard query flow graph by boosting certain co-efficient factors of click counts and eliminating others?
2. Does the same observation hold true when we use the search of another organisation?

In this section we first provide a description of the search logs used in these experiments. Then we introduce our experimental design and illustrate the different models being tested.

4.1   Search Logs

The main search log data in our experiments are obtained from the search engine of the Web sites of the University of Essex (UOE). In this search log we can obtain the query that has been entered, a time stamp of the transaction and the session identifier. In addition to that the clicked documents from the result lists by users following each query can also be obtained. We used a period of 10 weeks of logs between February and May 2011. During this period a total number of 142,231 queries were submitted to the search engine in 90,684 user sessions and 99,733 clicks on the results were logged. Figure 1 illustrates a histogram of the frequency of queries corresponding to the resulting number of clicks following each query as recorded in the logs of that search engine.

To validate the findings of our experiments on those search logs we conducted further experiments on search logs of another academic institution, the Open University (OU), where the same sort of data can be obtained. Figure 2 shows the corresponding histogram for the logs of the OU search engine using exactly the same 10-week period. It has a similar shape with much higher values of counts. In both histograms, for most cases the users either click on one result or do not click at any.

---

[2] Here we investigate a search engine of an academic organisation.

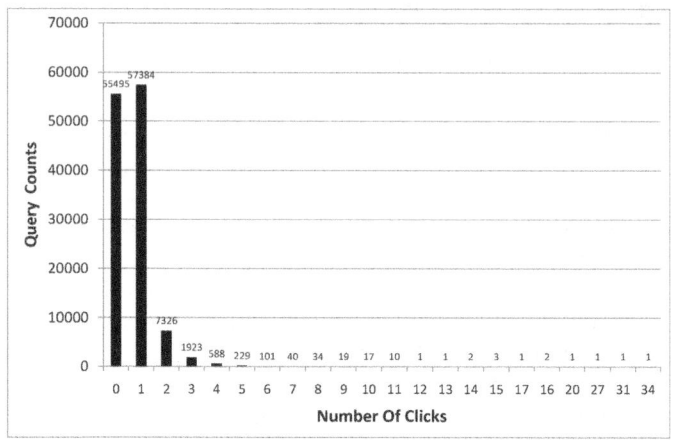

Fig. 1. Frequency of queries for each click counts band - UOE Search Engine

4.2   Query Flow Graphs

To assess the quality of query recommendations that can be achieved using our enriched query graph model we used an automatic evaluation approach based on the search logs to compare the quality of recommendations for various combinations of co-efficient factors of click counts.

Based on the fact that less than 2% of all queries result in more than 2 clicks, we simplified Equation 2 for the experiments as follows:

$$w(q,q') = \frac{C_0.\varphi_0(q,q') + C_1.\varphi_1(q,q') + C_k.\varphi_k(q,q')}{\Sigma_{r \in R_q} \Sigma_i C_i.\varphi_i(q,r)} \tag{4}$$

where C_k is the co-efficient factor of all click counts which are larger than 1. i.e. no matter whether a query has resulted in 2 or more clicks on resulting documents we treat all cases the same.

Table 1 lists all the combinations we considered in running the automatic evaluation framework.

We adopted the frequency weighting used by Boldi et al. [4] without incorporating the learning step as our goal is to show how we can enrich the query flow graph with click data. The learning step can always be added to the enriched version of the graph.

$QFG_{standard}$ is the standard query flow graph where no click information are incorporated. QFG_{no_zero} is an enriched query flow graph where reformulations which result with no clicks on the presented document list to the user are not considered. Both QFG_{boost_one} and $QFG_{boost\_one\_more}$ are enriched graphs that boost queries with a single click on the presented list. $QFG_{penalise\_many}$ penalises queries which attract 2 clicks or more.

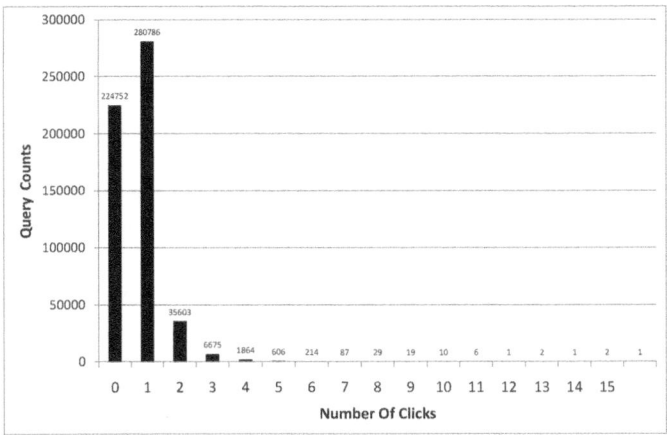

Fig. 2. Frequency of queries for each click counts band - OU Search Engine

Table 1. Experimental Graphs

	C_0	C_1	C_k
$QFG_{standard}$	1.0	1.0	1.0
QFG_{no_zero}	0.0	1.0	1.0
QFG_{boost_one}	1.0	2.0	1.0
$QFG_{boost\_one\_more}$	1.0	3.0	1.0
$QFG_{penalise\_many}$	1.0	2.0	0.5

4.3   The Evaluation Framework

The automatic evaluation framework assesses the performance of query recommender systems over time based on actual query logs by comparing suggestions derived from a query recommender to query modifications actually observed in the log files. The validity of the framework has been confirmed with a user study [2].

The evaluation is performed on arbitrary intervals, e.g. on a weekly basis. For all Q query modifications in a given week, we can calculate the system's Mean Reciprocal Rank (MRR) score as

$$MRR_w = (\sum_{i=1}^{Q} \frac{1}{r_i})/Q \qquad (5)$$

where r_i is the rank of the actual query modifications in the list of modifications recommended by the system. Note that in the special case where the actual query modification is not included in the list of recommended modifications then $1/r$ is set to zero. The above evaluation process results in a score for each logged week. So overall, the process produces a series of scores for each query recommendation system being evaluated. These scores allow the comparison between different system. One query recommender system can therefore be considered superior over another if a statistically significant improvement can be measured over the given period.

In our experiments we start with an empty query flow graph and we go through the search log data. At the end of each interval, we calculate the MRR score for that interval by producing a ranked list of query suggestions using the process described in Section 3.3 and then we use that interval data to update the graph adding necessary edges and adjusting the weights.

Producing query suggestions from the graph is computationally expensive as it requires performing a random walk on the nodes in the graph. Due to computing limitations, when calculating the MRR score we consider only a sample of the query modifications in the batch by taking every tenth query modification.

5   Results and Discussion

The automatic evaluation framework has been run on the various enriched query flow graphs listed in Table 1. We used the log files collected on the UOE search engine for our first experiments. We ran the evaluation on the entire 10-week period and used weekly batches to calculate the MRR scores for each graph.

Using the MRR scores, we can assess the graph performance over time in generating query recommendations and compare the performance of different graphs.

Table 2. Average Weekly MRR scores obtained for the query flow graphs in UOE search logs. The graphs are ordered by their scores.

Graph	Avg. Weekly Score
QFG_{boost_one}	0.0820
$QFG_{boost\_one\_more}$	0.0817
$QFG_{penalise\_many}$	0.0812
$QFG_{standard}$	0.0789
QFG_{no_zero}	0.0533

Table 2 presents the average weekly MRR scores obtained (ordered by average score). We observe that the enriched query flow graphs are outperforming the standard query flow graph. Apart from QFG_{no_zero} all enriched graphs are producing higher average MRR scores. To perform a statistical analysis on the differences between the enriched query flow graphs, in Table 3 we compare the query flow graphs using the average percent increase of MRR scores and the p value of a two-tailed t-test.

We observe that when boosting the co-efficient factor of single clicks, statistically significant improvements are obtained. Both QFG_{boost_one} and

Table 3. Comparison of the query flow graphs (UOE search engine)

	per. increase(%)	paired t-test
QFG_{boost_one} vs. $QFG_{standard}$	2.3%	< 0.05
$QFG_{boost\_one\_more}$ vs. $QFG_{standard}$	2.2%	< 0.05
$QFG_{boost\_one\_more}$ vs. QFG_{boost_one}	-0.1%	0.91
$QFG_{penalise\_many}$ vs. QFG_{boost_one}	-0.8%	0.16
QFG_{no_zero} vs. $QFG_{standard}$	-61.2%	< 0.01

$QFG_{boost\_one\_more}$ are significantly better than the standard query flow graph $QFG_{standard}$. However no further improvement can be observed when we further boost the co-efficient factor of single clicks. In fact $QFG_{boost\_one\_more}$ is slightly worse than QFG_{boost_one}.

Comparing $QFG_{penalise\_many}$ to QFG_{boost_one} would inform us about the impact of reducing the co-efficient factor of more than one click counts. The results show that this does not have a positive impact on the quality of recommendations. $QFG_{penalise\_many}$ is worse than QFG_{boost_one}.

Only enriched graph QFG_{no_zero} failed to improve the MRR scores, and in fact it was significantly worse than the standard graph with a high average percentage decrease. This appears to be counter-intuitive as we would assume that queries resulting in no clicks are not good candidates for query recommendation suggestions, and this finding warrants further analysis in future experiments.

In any case, this last finding suggests that completely eliminating reformulations with no user clicks affects the query recommendation quality negatively. Note that in QFG_{boost_one} and $QFG_{boost\_one\_more}$ we are considering these reformulations but we are also penalising them as they have a smaller co-efficient factor.

To validate the findings we obtained the log files of another academic search engine. To get a comparable number of interactions we decided to run this experiment in daily batches over 10 days of the April 2011 logs, i.e. we now use daily intervals to update the graph and calculate the MRR scores.

Table 4 presents the results obtained in this experiment. The corresponding t-test results can be found in Table 5.

Table 4. Average Daily MRR scores obtained for the query flow graphs in OU search logs. The graphs are ordered by their scores.

Graph	Avg. Daily Score
QFG_{boost_one}	0.0488
$QFG_{penalise\_many}$	0.0480
$QFG_{boost\_one\_more}$	0.0478
$QFG_{standard}$	0.0476
QFG_{no_zero}	0.0425

Table 5. Comparison of the query flow graphs (OU search engine).

	per. increase(%)	paired t-test
QFG_{boost_one} vs. $QFG_{standard}$	2.1%	0.15
$QFG_{boost\_one\_more}$ vs. $QFG_{standard}$	0.1%	0.88
$QFG_{boost\_one\_more}$ vs. QFG_{boost_one}	-2.0%	< 0.05
$QFG_{penalise\_many}$ vs. QFG_{boost_one}	-1.4%	< 0.05
QFG_{no_zero} vs. $QFG_{standard}$	-9.9%	< 0.01

Despite some minor differences we can see the same pattern. The ordering of the graphs according to their average MRR scores is similar. Only positions 2 and 3 ($QFG_{boost\_one\_more}$ and $QFG_{penalise\_many}$) are swapped. The enriched query flow graphs are outperforming the standard query flow graph but no statistical significant was observed this time.

Like before, reducing the co-efficient factor of many click counts did not have a positive impact on this dataset either. In fact $QFG_{penalise\_many}$ is now significantly worse than QFG_{boost_one}. Again, we find that eliminating queries that result in no clicks does not improve performance but instead results are significantly worse.

6   Conclusions

Query flow graphs built from query logs are a common and efficient technique to learn useful structures that can be utilised in query recommendation. We presented a new approach for incorporating user post-query browsing behaviour in the query flow graph. This is done by taking into account the number of documents that have been clicked by the user after submitting a query.

In this paper we explored variations of interpreting the number of clicked documents by conducting controlled, deterministic and fully reproducible experiments. which are based on an automatic evaluation framework that uses real world data to assess the performance of different models. Our experiments allowed us to quantitatively answer our research question and to draw very useful conclusions.

Boosting queries which result in a single document click has a positive impact on query recommendation. A single click can be interpreted as quickly reaching a landing page and rewarding these queries significantly improved the automatic evaluation scores. This is line with previous findings on using landing pages to generate query recommendations [7].

Eliminating queries which result in no clicks negatively impacted query recommendation. One possible explanation (but certainly only one single aspect) could be that some users found what they are looking for in the result snippets and as a result they would not continue clicking on the right document. Therefore, the graph will miss those useful suggestions. Penalising these reformulations without completely eliminating them though would have a positive effect as graphs QFG_{boost_one}, $QFG_{boost\_one\_more}$ have a smaller co-efficient factors for zero clicks.

We also show the observation made on one dataset was similar on a different dataset. The performance of the experimented graphs was similar on both datasets. However no statistical significance was observed for the enriched query flow graph over the standard query flow graph on the OU search engine. This may be due to the higher sparsity of the OU search engine logs.

7   Future Work

There is much room for future work. One area we will investigate is to automatically optimise the parameters. An extension of that work will then also allow us to look at building a machine learning model which can be trained on actual search log data taking as features the post query browsing behaviour including the click information to optimise the graph weighting function. Other browsing behaviour features can be further explored.

The appeal of an automated evaluation framework is that we can re-run experiments and explore a large search space without any user intervention. The shortcoming is that any automated evaluation makes some simplifying assumptions, and end users will ultimately need to be involved to assess the real impact of the query recommendation suggestions being employed. We see our evaluation as a first step in assessing what methods are promising and select those that promise the highest impact. We are about to incorporate a number of these models in a live Web site where we interleave recommendations coming from different models in the spirit of the active exploration approach presented by Radlinksi et al. [13]

Acknowledgments. This research is part of the AutoAdapt[3] research project. AutoAdapt is funded by EPSRC grants EP/F035357/1 and EP/F035705/1.

References

1. Agichtein, E., Brill, E., Dumais, S.: Improving web search ranking by incorporating user behavior information. In: Proceedings of SIGIR 2006, pp. 19–26. ACM, New York (2006)
2. Albakour, M.-D., Kruschwitz, U., Nanas, N., Kim, Y., Song, D., Fasli, M., De Roeck, A.: AutoEval: An Evaluation Methodology for Evaluating Query Suggestions Using Query Logs. In: Clough, P., Foley, C., Gurrin, C., Jones, G.J.F., Kraaij, W., Lee, H., Mudoch, V. (eds.) ECIR 2011. LNCS, vol. 6611, pp. 605–610. Springer, Heidelberg (2011)
3. Baeza-Yates, R., Tiberi, A.: Extracting semantic relations from query logs. In: Proceeding of KDD 2007, San Jose, California, pp. 76–85 (2007)
4. Boldi, P., Bonchi, F., Castillo, C., Donato, D., Gionis, A., Vigna, S.: The query-flow graph: model and applications. In: Proceeding of CIKM 2008, pp. 609–618. ACM, New York (2008)

---

[3] http://autoadaptproject.org

5. Bordino, I., Castillo, C., Donato, D., Gionis, A.: Query similarity by projecting the query-flow graph. In: Proceedings of SIGIR 2010, Geneva, pp. 515–522 (2010)
6. Craswell, N., Szummer, M.: Random Walks on the Click Graph. In: Proceedings of SIGIR 2007, Amsterdam, pp. 239–246 (2007)
7. Cucerzan, S., White, R.W.: Query suggestion based on user landing pages. In: Proceedings of SIGIR 2007, pp. 875–876. ACM, New York (2007)
8. Fonseca, B.M., Golgher, P.B., de Moura, E.S., Ziviani, N.: Using association rules to discover search engines related queries. In: Proceedings of the First Latin American Web Congress, Santiago, Chile, pp. 66–71 (2003)
9. Jansen, J., Spink, A., Taksa, I. (eds.): Handbook of Research on Web Log Analysis. IGI (2008)
10. Joachims, T., Granka, L., Pan, B., Hembrooke, H., Gay, G.: Accurately interpreting clickthrough data as implicit feedback. In: Proceedings of SIGIR 2005, Salvador, Brazil, pp. 154–161 (2005)
11. Joachims, T., Radlinski, F.: Search engines that learn from implicit feedback. IEEE Computer 40(8), 34–40 (2007)
12. Jones, R., Rey, B., Madani, O.: Generating query substitutions. In: Proceedings of WWW 2006, pp. 387–396 (2006)
13. Radlinski, F., Joachims, T.: Active exploration for learning rankings from click-through data. In: Proceedings of KDD 2007, pp. 570–579. ACM, New York (2007)
14. Radlinski, F., Kurup, M., Joachims, T.: How does clickthrough data reflect retrieval quality? In: CIKM, pp. 43–52 (2008)
15. Silvestri, F.: Mining query logs: Turning search usage data into knowledge. Foundations and Trends in Information Retrieval 4, 1–174 (2010)
16. White, R.W., Bilenko, M., Cucerzan, S.: Studying the Use of Popular Destinations to Enhance Web Search Interaction. In: Proceedings of SIGIR 2007, Amsterdam, pp. 159–166 (2007)
17. White, R.W., Chandrasekar, R.: Exploring the use of labels to shortcut search trails. In: Proceeding of SIGIR 2010, pp. 811–812. ACM, New York (2010)
18. White, R.W., Huang, J.: Assessing the scenic route: measuring the value of search trails in web logs. In: Proceeding of SIGIR 2010, pp. 587–594. ACM, New York (2010)
19. White, R.W., Ruthven, I.: A Study of Interface Support Mechanisms for Interactive Information Retrieval. JASIST 57(7), 933–948 (2006)

Effect of Explicit Roles on Collaborative Search in Travel Planning Task

Marika Imazu[1], Shin'ichi Nakayama[2], and Hideo Joho[2]

[1] College of Knowledge and Library Sciences, School of Informatics
University of Tsukuba
s0711565@u.tsukuba.ac.jp
[2] Graduate School of Library, Information and Media Studies
University of Tsukuba
{nakayama,hideo}@slis.tsukuba.ac.jp

Abstract. This paper presents a task-based user study carried out to investigate how explicit roles assigned to group members affected collaborative information seeking behaviour during a travel planning task. 24 pairs participated our study where half of them were given a specific instruction to divide the roles into a searcher and writer, while others were given no such instruction. The evaluation looked at travel plans generated, search interaction logs, task perceptions, and dialogues between members. The results suggest that explicit division of roles can have significant effect son a group's knowledge building during the collaborative search task. The paper also discusses experimental designs of task-based collaborative search studies.

1 Introduction

Collaborative search can be seen as an activity where a group of people carry out search for shared goals [3]. According to a survey [6], collaborative search is often conducted in professional context such as academic literature survey, but also in personal context such as travel planning. As people's information seeking tasks increases its complexity, they are more likely to pursue a collaboration to complete the problems [11]. However, collaborative search itself can be more complex than single-person search which has been a main assumption made in existing Information Retrieval (IR) research.

This paper presents a user study investigating effects of explicit roles on collaborative search. Taking a role is a typical and effective way to perform a complex task [5]. Roles have an effect of labour division in group activities, clarifying individual contributions to the goal, and increasing awareness during collaboration [2], and thus, it is an important research agenda in collaborative search. Algorithmic support for different roles has been suggested [10], but behavioural analysis of role-based collaborative information seeking has been limited. Therefore, our main research question was to examine effects of explicit human roles on collaborative search tasks.

M.V.M. Salem et al. (Eds.): AIRS 2011, LNCS 7097, pp. 205–214, 2011.
© Springer-Verlag Berlin Heidelberg 2011

We chose travel planning as our evaluation task for several reasons. First, it is one of the most popular motivations for people to perform collaborative search [6]. Second, finding travel information (e.g., transportations, accommodations, visiting places) and organising them into a single plan is a fairly complex task. Finally, travel planning is a fun and familiar task, and thus, we can expect a high level of user commitment to the task which is an important element in interactive IR studies. However, along the way of our investigation, it became clear that we needed to develop a fair part of evaluation methodology for a task-based evaluation of collaborative search. More specifically, we devised a set of dependent variables for travel planning tasks. Therefore, this paper should also be seen as a report of an exercise that considered various aspects of experiments when travel planing was used as an evaluation task in collaborative search.

The rest of this paper is structured as follows. Section 2 describes the experimental design we devised to examine effects of explicit roles. Section 3 presents the results of our experiment. Finally, Section 4 discusses the implications of our main findings to conclude the paper.

2   Experiment

This section presents the user study designed to examine effects of explicit roles on collaborative search task.

2.1   Travel Planning Task

As discussed earlier, travel planning task is a frequent collaborative search task with a high level of complexity and familiarity. A set of task conditions was devised to ensure its complexity and to facilitate user engagement in the experiment. One, a travel plan was for five to seven days where the main destinations were abroad. Two, a pair had to agree on decisions made during the task. Three, it was not allowed to take a package tour which covered most of the plan (but a short guided tours were allowed). Four, the plan should have place names (to visit, eat, or stay), transportation methods and their schedule and costs, and any activities. Fifth, the upper limit of the travel cost was set to 5,000 USD per person. We set a relatively high budget to encourage participant pairs to consider a wide range of opportunities during the task.

Participants were given 90 minutes to complete the task. However, since task completion time was not our main concern in this study, we gave extra time when participant pairs asked. Some pairs asked for few minutes extra time to wrap up the planning. No noticeable difference of average task completion time was found between two conditions, which will be described in the following section.

2.2   Independent Variable

Our research question looks at the effect of explicit roles assigned to a group member on their collaborative information seeking behaviour. Therefore, our independent variable is the presence (or absence) of explicit instruction to divide the roles during the travel planning task. The instruction of role division was given to participant pairs as a part of the description of travel planning task. In our task, all members involved in travel planning. However, the instruction explicitly asked a pair to take roles of a *searcher* and *writer*. The searcher role operates a PC to gather information needed to make a travel plan, while the writer role takes notes and write down travel information on a given sheet. This role assignment has been shown to be effective in solving a complex task [5]. Other types of role models were left for future work. Participant pairs were asked to decided the role in advance, but they were allowed to swap the role anytime during the task. Another difference between the two conditions was the number of PCs. Two laptops (one for each) were used by the pairs who were given no specific role instruction, while only one laptop (for searcher) was used by those who were given the explicit role instruction. This was to ensure that assigned roles were taken place in the latter. 12 participant pairs were given the task description with the explicit role division while the other 12 participant pairs were given the task description without the explicit role division.

2.3 Dependent Variables

The effect of explicit roles on collaborative information seeking behaviour was measured at different levels, which are described below.

Executability of travel plan. The first question raised in our experimental design was "what was a good travel plan?". This was not trivial to answer. Existing work [7,9] which used travel planning as an evaluation task did not either ask participants to make a travel plan or make careful consideration on a performance of travel plan as dependent variables. There are some performance measures of travel plans [8] in the context of workplace, and are not necessarily applicable to our experimental design. After a series of discussion among the authors, we decided to employ the concept of *executability* as a performance measure of travel plans. The assumption here was that a good travel plan should define necessary information to complete a trip. The necessary information includes the name of places to visit, accommodation to stay, restaurant to eat, transfer methods (e.g., air plane, train, bus, taxi, walk), time of transportation if used, and the cost. Although a real travel plan is likely to have uncertainty in these aspects for a good reason, we defined that more travel information was defined the better the plan was.

Search actions and subjective assessments The next two groups of dependent variables were common to interactive IR studies. Search actions include the number of queries submitted to search engines and other web sites, number of web pages visited, and domains. These objective variables specifically looked at search efforts made by participants to complete the task. Search behaviour which recorded by the QT-Honey system (`http://cres.jpn.org/?QT-Honey`). QT-Honey is an extension of the Lemur Query Log Toolbar (`http://www.lemurproject.org/querylogtoolbar/`)

Table 1. Dialogue development scheme

Levels	Travel	Search
1: Experience	I heard Carib irelands are great	
2: Clarification	Is Hawaii warm in the winter?	What's there in Germany?
3: Opinion	We aren't opera people, are we?	Should we find the way to get there?
4: Formulation	How can we get there, ..., it's far away.	I can't find with "london train"!
5: Suggestion	Shall we do some sightseeing around here before taking the bus?	Why don't we search for X bus Y because we don't know the schedule?

which works as a Firefox add-on to record action events detected on the web browser. Subjective assessments, on the other hand, measures how participants perceived various aspects of the travel planning tasks. This study asked many questions but will report the perceptions about task completion, satisfaction, and other general feedback on collaborative tasks.

Development of dialogues. The last category of dependent variables was a level of dialogue development between team members. Our objectives of analysing dialogue data were to gain insight into the verbal communication during the task. While it has been suggested that investigating relevant information or knowledge shared by team members is an important aspect of collaborative information seeking and retrieval [3], little study has been carried out to examine the verbal communication. In our study, participant pairs conversations were video-recorded and the dialogue data were coded by a variant of a scheme proposed by Chan, et al. [1]. A brief description of the dialogue development levels are as follows. Level 1 is an utterance regarding participant's experience or its generalisation. Level 2 is regarding clarification of term's definition or its explanation. Level 3 is regarding confirmation of a group member's opinion or perspective. Level 4 is regarding awareness and formulation of a problem to solve. Finally, Level 5 is regarding presentation of an idea or suggestion about the problem recognised during the task. A summary of the dialogue development model is shown in Table 1 along with sample conversations. The scheme allowed us to measure a level of dialogue development between team members. Due to lack of resources, we were only able to code the first 10 minutes, middle 10 minutes, and last 10 minutes of the task. The coding was performed by one of the authors, and when necessarily, the authors jointly coded ambiguous cases.

2.4 Participants

Participants were recruited as a pair to ensure comfortable communication and collaboration during the task. Recruitment was carried out by multiple channels such as mailing lists, notice board, and word-of-month in our university. As a result, 24 pairs (48 people) participated the experiment. The entry sheet established that all participants were either undergraduate or postgraduate students in our university. Their academic majors ranged from Pure Science, Informatics, to Social Science, Arts and Humanities. Of 24 pairs, 12 pairs (50%) had known each other for over 3 years, 8 pairs (33%) for over 1.5 years, and 4 pairs (17%) for less than 6 months. Participants had on average 7.5 years (SD: 2.3) of search experience on the Web and frequent users of search engines. 11 pairs (46%) had had a collaborative search task with the pair member at least once. In summary, our participants were frequent search engine users with various academic background although they were all university students. All pairs had a friendly relationship which helped to perform a collaborative task in the experiment.

2.5   Procedure

The whole session took approximately two hours per participated team. Participants were rewarded with 10 USD per an hour of work. The procedure of the user study was as follows. 1) Welcoming a pair and asked them to read Information Sheet which described the aim of the experiment and participating conditions; 2) Asked the pair to sign Agreement Form; 3) Asked the pair to fill Entry Questionnaire individually to collect demographic information and search experience; 4) Asked the pair to read the task instruction (with/without explicit role division) along with an example travel planning sheet; 5) Asked the pair to perform the travel planning task for 90 minutes, writing their plan on a blank sheet; 6) When the task completed, we asked the pair to fill in Task Questionnaire individually to capture their subjective assessments on the task; and finally, 7) Asked the pair to fill in Exit Questionnaire individually to capture an overall feedback on collaboration and experiment. Pilot tests were carried out with three pairs to verify and adjust the experimental conditions prior to the final experiments whose results will be presented in the next section.

3   Results

This section presents the results of the experiments. We use the notation of *ER* group to denote the pairs who were given an explicit role instruction while *NR* group to denote those who were given no role instruction.

3.1   Travel Plans

The first investigation looked into the travel plans generated by participant pairs as the outcome of the task. Since participants were asked to write a travel plan in a free format, the generated schedule was first converted to a simple graph model where a node represented a place and edge represented a path between the places. The following analyses were based on the converted models.

The first analysis on the travel plans counted the number of places identified in the sheet. The types of places counted were transportation stops (airports, train stations, bus stops), accommodations, and places to sightsee. In addition, participants sometimes formulated only a vague idea about the places to sightsee, and wrote a description like "somewhere to go" tentatively. We counted these

210   M. Imazu, S. Nakayama, and H. Joho

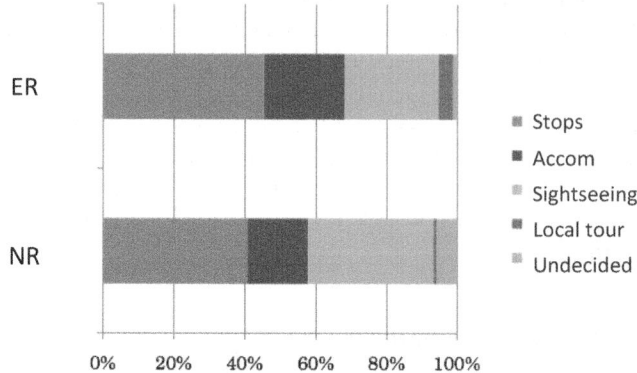

Fig. 1. Percentage of traveling places

cases as *undecided* in the analysis. Finally, some pairs chose to take a local tour
as a part of plan. This was also included in the analysis. The average number of
places identified in the travel plans was 17.3 (Standard Deviation: 5.1) in the ER
group and 18.3 (SD: 7.2) in the NR group. The Mann-Whitney U test showed
that the difference was not significant ($p \leq .862$). Figure 1 shows the distribution
of the places in the ER and NR groups. As can be seen, the EP group tended
to have a higher proportion of transportation stops and accommodation than
the NR group, while the NR group tended to have a higher proportion of places
to sightsee than the ER group. The NR group also tended to have a higher
proportion of undecided places than the ER group.

The next analysis looked at the travel methods participants planed to used for
moving between the places. We categorised the travel methods into six groups,
namely, walking, buses, trains, flights, taxies, and others. The average percentage
of the travel methods is shown in Figure 2. As can be seen, the most frequent
travel methods identified in the plans was flights in both groups. A relatively
large difference was observed in the use of buses, trains, and taxies. In particular,
the use of buses was very rare in the ER group, while the NR group had a similar
percentage of all travel methods except flights.

The last analysis on the travel plans investigated the amount of travel information written in the sheet. More specifically, we counted the number of
information about time (schedule) and costs of travel. In the ER group, 38.6%
(SD: 14.6) of travel information contained time data while it was 27.0% (SD:
14.3) in the NR group. 28.1% had cost data in the ER group while it was 20.7%
(SD: 15.5) in the NR group. Therefore, the ER group tended to have a higher
ratio of recording the time and cost information during the travel planning task.
The t test shows that the difference was not significant in both cases. The results
also suggest that both groups appeared to be more aware of recording the time
information than the cost information.

To summarise, this section looked at the travel plans generated by participant
pairs. Noticeable differences were observed in the types of places identified in the

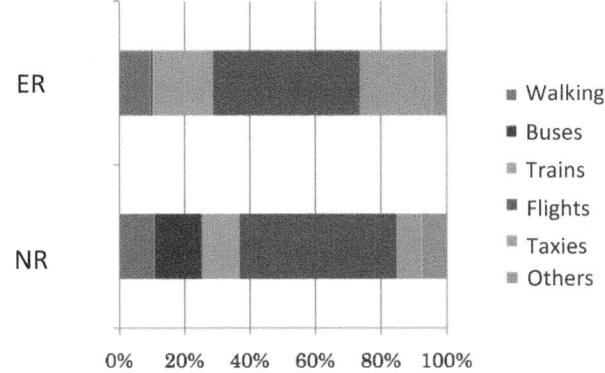

Fig. 2. Percentage of traveling methods

Table 2. Search Behaviour (N=12)

	ER Group	NR Group	p-value
Result pages	54.9 (30.3)	86.0 (24.3)	.011
Web pages	142.4 (39.1)	240.1 (100.7)	.005
Input forms	11.8 (7.0)	22.3 (11.7)	.014
URI domains	30.7 (8.7)	44.6 (9.1)	.001

plans, their choices of travel methods, and records of travel information. However, these differences were not statistically significant.

3.2 Search Behaviour

The second set of investigation looked at participant information searching behaviour during the travel planning task. To gain understanding of participant pairs search behaviours, we analysed the number of search result pages viewed, number of web pages browsed, number of domains (e.g., www.booking.com), and number of input form pages submitted from QT-Honey logs. The results are shown in Table 2. The numbers in the NR group are a union of both members who had a PC each. As can be seen, the NR group had a greater level of search activities than the ER group. The difference was found to be significant by the t test in all aspects. However, as we observed earlier, these difference did not seem to have a significant impact on travel plans.

3.3 Dialogue Development

As described in Section 2.3, our dialogue analysis was carried out at the first 10 minutes, middle 10 minutes, and last 10 minutes of the task.

First, we counted the number of utterances made by participants. The total number of utterances in the ER group was on average 276.2 (SD: 52.2) while

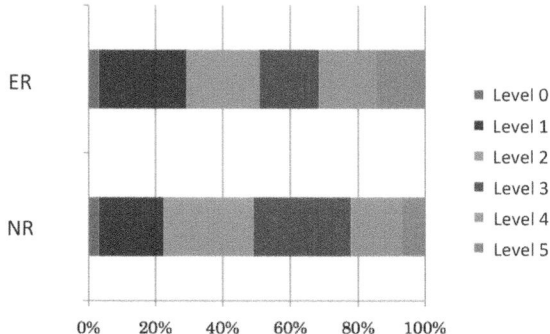

Fig. 3. Percentages of dialogue development levels

the NR group was 191.3 (SD: 64.0). The Mann-Whitney test showed that the difference was significant ($p \leq .004$). This suggests that having an explicit role can facilitate the communication in a group.

Next, we analysed the development of conversation using a variant of Chan's dialogue development scale [1]. We added Level 0 to Chen's scale to annotate the utterances which were judged to be clearly irrelevant to the task. The results are shown in Figure 3. We can observe that the percentage of Level 1 and 5 in the ER group was larger than the NR group while Level 2 and 3 of the NR group were larger than the ER group. The Mann-Whitney tests showed that the difference between the two groups were significant at Level 3 ($p \leq .037$) and Level 5 ($p \leq .015$). This suggests that the conversations in the ER group tended to develop further than the NR group. According to Chan's scale, the NR group tended to utter to confirm a partner's opinions more frequently than the ER group who, on the other hand, tended to utter to explain their opinions and make suggestions based on an issue commonly recognised by the group.

We also analysed objectives of utterances from two perspectives: travel and search. We observed, on average, 97.0% of utterances were about travel in the ER group and 98.4% in the NR group. Therefore, most utterances were made to discuss travel itself and only 1.6 to 3% were made to discuss specific search issues. This was common to both groups and the difference was found to be insignificant. This reinforces our intension of looking at collaborative search in the context of work task. A large proportion of group's communication concerned with the goal of work task, and thus, they were likely to affect search behaviour which played a significant role to complete the travel planning task.

To summarise, this section analysed the dialogue data to examine the effects of explicit roles on travel planning tasks. The results suggest that explicit roles have significant effect to facilitate the communication between members. It also suggests that the dialogues in the ER group tended to be more advanced than the NR group.

3.4   User Perceptions

The last part of our investigation looked at participants' perceptions on the task they performed and collaboration in general.

At the end of the task, we asked participants to indicate the degree of task progress and level of satisfaction about the outcome of the task. We used a 5-point Likert scale to capture participants' perceptions. As for the task progress, ER group had 3.6 (SD: 1.1) on average while NR had 3.3 (1.1). As for the satisfaction, ER had 2.6 (SD: 1.1) while NR had 2.4 (1.1). Therefore, the ER group appears to have a higher level of task completion and satisfaction when compared to the NR group. However, the t tests show that the difference was not significant.

In Exit Questionnaire, we asked a series of questions about the task and collaboration. One of the questions where we observed a noticeable difference was about potential ideas to improve the collaborative task. In the NR group, 11 participants a need for more facilitated communication, 8 participants stated a need for more efficient labour division, and 3 participants stated a need for a better planning strategy. In the ER group, on the other hand, 6 participants stated a need for a better planning strategy, 6 participants stated a need for a better use of search engines, others mentioned different aspects of planning strategy. This supports two aspects of our study. One was that the effect of independent variable. There was some observations in the search logs which suggests that there was some labour division strategy in the NR group. However, the feedback from the NR group indicates that the labour division was not effectively executed while no one mentioned about labour division in the ER group. Another was the effect of explicit roles on the communication. In the dialogue analysis, we observed that the ER group had more frequent and advanced communication than the NR group. The most frequently mentioned feedback in the NR group was about communication.

4   Conclusive Discussion

This paper presented a task-based user study to examine the effects of explicit human roles on collaborative information seeking and retrieval. There are two major contributions in this paper which are discussed below.

The first contribution of this paper was to gain insight into the effects of explicit roles given to group members on the performance of a collaborative search task. There are multiple findings. First, explicit roles can facilitate the communication between group members with advanced level of development. In our results, the NR group had more conversations for confirming a partner's opinions while less for suggesting ideas. The resulted advanced communication appears to enable the ER group to be more aware of the details of a given task. On the other hand, we observed an effect of explicit roles to limit a range of opportunities that can be found by the group. This might be an artifact of our experimental set and other combinations of roles might remedy the effect. However, it is also clear that there are many remaining questions that our study did not address. One such aspect is the relationship between the shared knowledge (via dialogues) and subsequent search strategies. We are currently performing further analysis on the dialogue data along with search logs to address the problem.

The second contribution of this paper was the research design for a task-based evaluation of collaborative search. Travel planning has been known as one of the most frequently performed collaborative search task [6]. In addition, as we discussed in Section 1, travel planning is an intellectually complex task, and therefore, it is worth a careful consideration on research methodology. In this paper, we proposed a set of dependent variables that can be used to measure effects of independent variables in the context of travel planning tasks. Those includes the properties of travel plans generated and dialogue analyses in addition to conventional searching behaviour and subjective task assessments. We do not claim that this is a definitive set of dependent variables, but we consider that it is significantly more holistic than existing work in this area. Collaborative search is a complex research problem, and a task-based investigation is crucial to understand underlying issues of the activity. We are interested in revising what we deviced in this study, as well as, developing dependent variables for other types of collaborative search tasks, as future work.

Acknowledgements. Funding was provided by KAKENHI Grant-in-Aid for Young Scientists (B) (Ref: 22700244). Any opinions, findings, and conclusions described here are the authors and do not necessarily reflect those of the sponsor.

References

1. Chan, C., Burtis, J., Bereiter, C.: Knowledge building as a mediator of conflict in conceptual change. Cognition and Instruction 15, 1–40 (1997)
2. Forsyth, D.R.: Group dynamics, 4th edn. Wadsworth (2006)
3. Foster, J.: Collaborative information seeking and retrieval. Annual Rev. Info. Sci. & Technol. 40, 329–356 (2006)
4. Joho, H., Hannah, D., Jose, J.M.: Comparing collaborative and independent search in a recall-oriented task. In: Information Interaction in Context, pp. 89–96 (2008)
5. Miyake, N.: Constructive interaction and the iterative process of understanding. Cognitive Science 10, 151–177 (1986)
6. Morris, M.R.: A survey of collaborative web search practices. In: Proceeding of the 26th SIGCHI Conference, pp. 1657–1660 (2008)
7. Morris, M.R., Horvitz, E.: Searchtogether: an interface for collaborative web search. In: Proceedings of the 20th UIST Symposium, pp. 3–12 (2007)
8. Rye, T.: Travel plans: do they work? Transport Policy 9(4), 287–298 (2002)
9. Saito, H., Takaku, M., Egusa, Y., Terai, H., Miwa, M., Kando, N.: Connecting Qualitative and Quantitative Analysis of Web Search Process: Analysis Using Search Units. In: Cheng, P.-J., Kan, M.-Y., Lam, W., Nakov, P. (eds.) AIRS 2010. LNCS, vol. 6458, pp. 173–182. Springer, Heidelberg (2010)
10. Shah, C., Pickens, J., Golovchinsky, G.: Role-based results redistribution for collaborative information retrieval. Inf. Process. Manage. 46(6), 773–781 (2009)
11. Sonnenwald, D.H., Pierce, L.G.: Information behavior in dynamic group work contexts: interwoven situational awareness, dense social networks and contested collaboration in command and control. Inf. Process. Manage. 36, 461–479 (2000)

A Web 2.0 Approach for Organizing Search Results Using Wikipedia

Mohammadreza Darvish Morshedi Hosseini[1], Azadeh Shakery[1],
and Behzad Moshiri[2]

[1] School of Electrical and Computer Engineering, College of Engineering
University of Tehran, Tehran, Iran
[2] Control and Intelligent Processing Center of Excellence, ECE, Faculty of Engineering,
University of Tehran, Tehran, Iran
`mmorshedi@gmail.com, {shakery,moshiri}@ut.ac.ir`

Abstract. Most current search engines return a ranked list of results in response
to the user's query. This simple approach may require the user to go through a
long list of results to find the documents related to his information need. A
common alternative is to cluster the search results and allow the user to browse
the clusters, but this also imposes two challenges: 'how to define the clusters'
and 'how to label the clusters in an informative way'. In this study, we propose
an approach which uses Wikipedia as the source of information to organize the
search results and addresses these two challenges. In response to a query, our
method extracts a hierarchy of categories from Wikipedia pages and trains
classifiers using web pages related to these categories. The search results are
organized in the extracted hierarchy using the learned classifiers. Experiment
results confirm the effectiveness of the proposed approach.

Keywords: Wikipedia, Classification, Search result Organization.

1 Introduction

With the current growth of the World Wide Web, it is now possible to access huge
amounts of data and information from different sources all over the world. Web
search engines aim to help users to find appropriate data for their information needs.
Most current search engines return a list of results in response to the user's query.
Although this simple method is sufficient in many cases, it may be inefficient in
others, requiring the user to sift through a long list of results to reach the relevant
documents. This issue is especially annoying when the keywords that the user has
chosen as the query have other more popular senses. For example, suppose that a user
is using the query "Java" to get some information about Java Island. Since the
"programming language" sense of Java is much more popular, the user will have
difficulty finding results related to his information need.

There has been limited research on organizing search results and efficiently
presenting them to the user compared to the vast amount of work on ranking search
results. Most work on organizing search results is focused on clustering the results
and presenting them to the user in groups. Hearst [24] provided a clustering algorithm
as an alternative to simple ranked list. Sanderson [12] introduced a method to
automatically create a hierarchical organization of concepts. Although this work is

M.V.M. Salem et al. (Eds.): AIRS 2011, LNCS 7097, pp. 215–225, 2011.
© Springer-Verlag Berlin Heidelberg 2011

not focused on organizing search results, it can be applied to such a problem as well. Also [3, 4, 5, 6, 11, 15, 18, 19, 20, 21 and 25] introduced new techniques to cluster search results. There are also real-world clustering engines like Clusty[1] and Carrot[2] that cluster search results before presenting them to the user. Although all these approaches show some benefit of clustering search results, all these approaches are faced with two challenges: 1) how to identify appropriate clusters and 2) how to label the clusters properly.

To address these two challenges, in this paper we propose to use Wikipedia to organize the search results. Specifically, our proposed method first identifies the appropriate categories using Wikipedia. It then extracts a set of training documents for each category and builds a soft classifier to classify the search results in the categories. We also create a hierarchical structure, so if a user needs categories to be more detailed, she can go through the hierarchical structure. Since the articles in Wikipedia are written and edited manually, the extracted categories are expected to better match the ideal partitioning of documents for the user.

Recently there has been a lot of research on using Wikipedia for different purposes. [7, 8, 9 and 15] have used different features of Wikipedia for clustering and/or cluster labeling, [2, 13, 10, 17, 22 and 23] have used Wikipedia for query processing purposes and [1 and 16] have used Wikipedia for query expansion.

Although a lot of research has been conducted on Wikipedia for different purposes, our perspective for using Wikipedia for hierarchical search result organization seems to be new.

In our study, experiment results show that using Wikipedia, we can build a hierarchical category structure that is more accurate than the structures constructed by the clustering methods in most cases. Also we discovered that by using Wikipedia, we can build appropriate training sets for the purpose of training the classifiers. Our proposed method is shown to be especially successful for simple queries and one word queries.

The rest of the paper is organized as follows: We describe some of Wikipedia's features and the details of our proposed method for organizing search results in section 2, discuss the experiment results in section 3 and finally conclude in section 4.

2   Search Results Organization

Wikipedia is a free collaborative online multilingual encyclopedia. The basic unit of Wikipedia is article. Each article is about a specific entity, a concept or an incident. Usually important words in each article are linked to their corresponding pages. Each article is divided into a number of sections, and a specific concept is discussed in each section. Usually, these concepts have their own pages in Wikipedia and there is a link to that article called *Main Article*. In some articles, there exists a link to a list of related topics. Usually this list is organized well and can be very helpful for extracting related categories. For example in the list of Iran related articles[3], a number of articles

---

[1] http://clusty.com
[2] http://search.carrot2.org/stable/search
[3] http://en.wikipedia.org/wiki/List_of_Iran-related_topics

about Iran are categorized into meaningful categories. Some of these categories are: "Economy", "Historical", "Politics" and "Geographic". Another type of Wikipedia pages that we are interested in is *Template* pages. In template pages, like the list of related topics, we can find articles related to a specific topic categorized into meaningful categories. Each Wikipedia article is placed in one or more categories. These categories have a hierarchical structure and each one contains a set of related topics.

Search queries vary in semantics and user's need. Similar to Xu [2] we define three types of queries:

1. Ambiguous queries (AQ)
2. Queries about a specific entity (EQ)
3. Broader queries (BQ).

By AQ we mean a type of query that contains exactly one term which has more than one potential sense, e.g. "Java". These queries may or may not be semantically ambiguous. Usually there exists a disambiguation page for ambiguous terms. By EQ, we mean a type of query that has specific meaning and covers a narrow topic, e.g. "Persian history". This kind of query should have an article in Wikipedia with the exact title. We should note that this definition does not restrict the entity queries to queries about entities; they could be queries about an entity, a concept, or an event. By BQ we refer to the rest of query types. These queries are neither ambiguous nor focused on a specific entity, and there exist no article in Wikipedia with the same title. The difference between our definitions of query types and the definition propose by Xu [2] is that we count the queries about specific subjects with no corresponding Wikipedia pages as BQ. For example a query such as "Persia History" with no article in Wikipedia is treated as BQ.

We organize the search results in five steps, as shown in Figure 1.

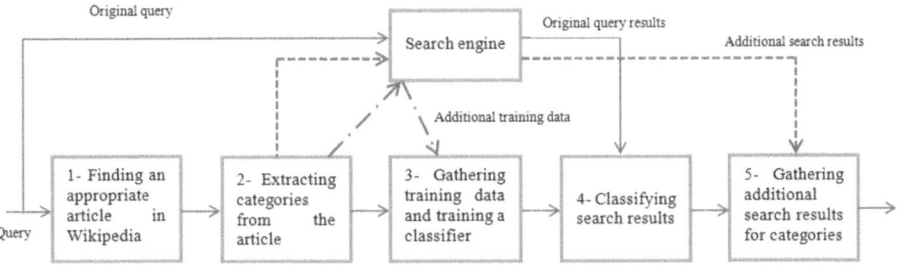

Fig. 1. Search result organization flow

In the first step we try to find a matching article in Wikipedia for the query. After finding the matching article we extract categories from that article. The extraction process is described in details in the following sections. In the third step, we gather training data for each category and train a classifier for the categories. We also expand the training set to obtain better results. The expansion process is described in

2.2. In the fourth step we give the original query to a search engine and categorize returned result into the extracted categories using the trained classifier. In order to avoid sparse categories, in the fifth step, we expand the original query for each category to provide additional search results. The result will be a hierarchy of organized search results, with at least a minimum number of pages in each category.

2.1   Finding Wikipedia Page and Extracting Categories

Although the overall algorithm is the same for all types of queries, it differs in details especially in finding a matching Wikipedia page and extracting categories. These details are described in the following sub sections.

- **Finding Wikipedia Page and Extracting Categories for Ambiguous Queries**

For this type of query, it would be desirable to categorize the results in different categories corresponding to different senses of the ambiguous term. This introduces two challenges: how to find all the meanings of the term, and how to categorize search results in the groups

As stated before, Wikipedia contains disambiguation pages that cover the possible senses of a specific term. Even some people believe that Wikipedia is more suitable than WordNet for disambiguation [14]. For example while WordNet contains three different senses for the word "Java", Wikipedia's disambiguation page for this term contains eight possible senses. Each section in a disambiguation page refers to a specific sense of that word. Also each section contains a number of links to some articles related to that sense. For Example in the disambiguation page for 'Java' the title of the first section is "Animals" which contains links to articles with titles "Java Pipistrelle", "Java Shark", "Java Sparrow" and "Java Chicken". The title of each section perfectly describes that sense and thus we propose to use the section's titles as the category names. For the query Java, the extracted groups from Wikipedia will be: 'Java-Animals', 'Java-Literature', 'Java-Computer Science', 'Java-Consumables', 'Java-Entertainment', 'Java-Geography', 'Java-Plants' and 'Java-Transportations'. In order to train a classifier corresponding to these categories we use existing links in each section as training data set with the intuition that the links in each section are pointing to related articles.

- **Finding Wikipedia Page and Extracting Categories for Entity Queries**

By EQ, we mean a type of query that has specific meaning and covers a narrow topic. Also there should exist an article about these queries in Wikipedia. When the query exactly matches the title of an article, the corresponding Wikipedia article can be easily located. Similar to the case of ambiguous queries, we use section's titles as group names. Usually each section contains some subsections. We use these subsections to create a hierarchical structure. To gather training data, we use links in each sections or subsections. For example for query "Iran", to gather training data about the group "Iran-history", we use links in history section of article with the title of "Iran". The links in each section are about a topic in that section and that topic somehow relates to that section's title, thus we expect to gather an appropriate dataset for training this way. Each section may contain a list of links, and the theme changes smoothly in each section. Thus we use uniform sampling for link selection in each section, both to avoid the large number of training articles, and to make sure that we have articles related to all parts of the section in our training data.

• **Finding Wikipedia Page and Extracting Categories for Broader Queries**

Since for some queries the corresponding page cannot be found easily, we try to use other features of Wikipedia for this purpose. Consider a query like "Iran energy". To find the appropriate categories, we consider each query term in turn and try to find the appropriate page from the list of related topics to the term. In each list of related topics, the topics are presented in a categorized manner. Each category is considered as a group name and a classifier is trained for the categories using the linked pages for training. We then classify the original query in the extracted categories. In the above example and among the categories extracted from the list of related topics to "Iran", the query will fall in the "Economy" category. Now we create a new classifier whose groups are the links in the selected category. We train this new classifier by content of linked article. Also we gather our training data for each group by extracting texts from a uniform sample of links in that article. In the above example, candidate groups are "Airlines of Iran", "Economy of Iran", "Energy in Iran", "Iranian cars" and etc. We use existing text in each article as training data and classify the word "energy" using that classifier. We select the top ranked article and consider that page as the target Wikipedia page. Extracting the categories from this page is similar to the method presented for EQ.

At the end of some articles there is a link to a template page related to that page. If no article matches our need in related pages we can use these template pages just like list of related topics to find the target page.

This algorithm only works if a corresponding page in Wikipedia exists whose content is about the query or is similar to it. This method will not work for queries with more than one ambiguous term. For example it will succeed for query "Iran energy" but it will not find a matching page for "north Iran" or "Java Brand Iran" and so it cannot organize results for these queries. Focusing on these kinds of queries is postponed to our future works.

2.2   Expanding the Training Set and Gathering Search Results

Since in some articles there may not exist enough links for training the classifier, we propose an expansion technique to expand our training set. To this end, we construct a query for each category, which is an expansion of the original query biased toward the category. We use popular information retrieval query expansion techniques for this purpose. To expand the query for each category, we use the links in the corresponding section. We gather linked documents as document set for query expansion, search the query within this document set, and then we use top ranked results to expand the query. We expand our query using lemur toolkit language modeling methods. We do expansion three times with different expansion strategies and then fuse the expansion results. To do so, we simply add the score of each term in all three result sets and select top five results as the biased query terms. We use these terms along the original query as our new query and submit it to a search engine. We add top 10 returned results' content to our training set. Expanding the training set had a great influence on the classification accuracy. In our experiments we noticed that even when there are lots of links available in the original document, the expansion step makes the classification more accurate.

Having a category structure and a training set makes it easy to categorize search results. But what if after categorizing all results we end up with some empty categories? For example when we searched "java" in Google, there were no result related to the island of java in the first 5 pages. To avoid this situation we try to change the query and collect a minimum number of results for our categories. To do so, we create a new query using our original query along with the category name and use the returned results to fill empty categories. Despite changing the query usually there exist some results that aren't related to our target category. So we use a binary classifier to determine if a returned result belongs to our target category or not. We use a binary language model classifier for this purpose and train it with the collected training data. After gathering all the results we simply re-rank the results so that user can find high quality results at top.

3   Experiments and Results

Since Wikipedia's articles are created by human, we expect that our results be more accurate and suitable than any clustering algorithm. To illustrate, Figure 2 shows the clustered result returned by carrot for the query "tiger" as of May 2011. As like many other queries, the query, tiger has multiple meanings, we expect the organizing engine, to cover all possible senses. But returned results by carrot do not do so.

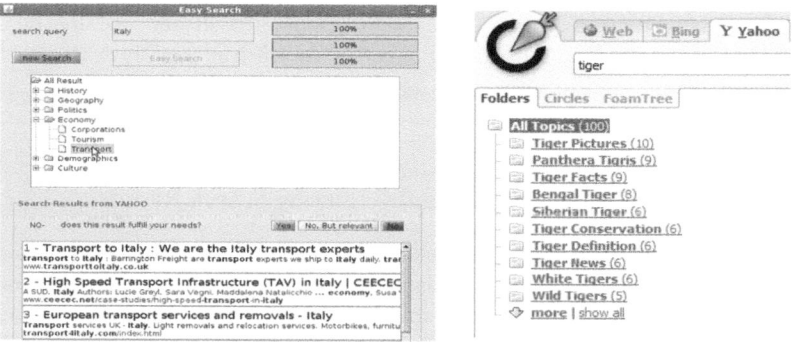

Fig. 2. Our evaluation system (left) and carrot clustering engine (right)

Consider that we want some information about a company with the named "tiger". By current result of this clustering engine, it is difficult to find some pages related to this information need.

It will be easy to find the needed information with our proposed method. Considering the query "tiger", the Categories returned by our method is: "Zoology", "People", "Places", "Vehicles", "Sports", "Media and fiction", "Game character", "Music", "Business", "Technology and mathematics", "Other".

Our method almost returns all possible meaning for that query. With these categories, it will be easier to find some information about a company named "tiger".

In order to do a more thorough evaluation of our system, we did a user study, and evaluated our method based on participants' judgments. A screenshot of the evaluation system is shown on the left side of Figure 2

3.1   Experiment Settings

In our experiments we downloaded our needed page from Wikipedia website using "htmlparser" library[4], and created a Naïve Bayes classifier using "lingpipe" framework. Also we used KL-divergence method in lemur toolkit[5] for expanding query. All the coding was done with Java language. To obtain some search results for training and testing purpose, we used "Yahoo BOSS"[6]. Yahoo boss is a free API that allows developers to access yahoo search.

We evaluate our method from two aspects. 1) Provided categories suitability. 2) Classification accuracy

We had 9 participants in our experiments from age 22 to 28. Four of them were female and five were male. Three of them had BS degrees, two of them had MS Degree and the remaining Four Person was MS student. Also one of them had BS in Horticulture, one of them had MS in Food Science and another participant was MS student of Tourism management. The remaining six participants had BS or were MS students in computer related fields.

Each participant has tested the system with at least 5 queries with one term, 3 ambiguous Queries and 3 queries with more than one term. In First steps designed software asked participants to provide a query term and a preferred category structure. After that System shows retrieved categories and asks users to compare retrieved categories with theirs and give our category a grade by selecting among these options:

1) Just like mine, or even Better
2) Good. But from different point of view
3) Not bad. But incomplete
4) Not relevant

Also we asked them if there exist a category that exactly matches their seeking category. After that system shows them search results in a categorized view.

For Evaluating the classification and re-ranking performance, System asks participants to judge top 10 results by answering a question about does this result fulfill their information needs by selecting among following options: 'Yes', 'No, But

---

[4] http://htmlparser.sourceforge.net/
[5] http://www.lemurproject.or
[6] http://developer.yahoo.com/search/boss/

Relevant', 'No'. By first option they declare that the item is interesting and fulfills their information needs. By selecting second option, user tell us that the result is relevant to subject but it wasn't exactly what they were looking, and by third option they tell the system that the search result is not related to the subject.

After evaluating top 10 results we ask them 2 additional questions. First we ask them to search the exact query in carrot (by selecting yahoo in carrot) and one system or both as their preferred categories. Second question was that does our category give them a hint to improve their query. They could answer this question by a simple yes/no.

- **Ambiguous queries**

As it is shown in Figure3 in 70.4% of questions user answered that our categories are similar or even better than their provided categories. 22.2% of answer was "good. But from different point of view" and 7.4% answered that the categories are incomplete. Some of these queries were: "jaguar", "Tourist"," mint" and etc.

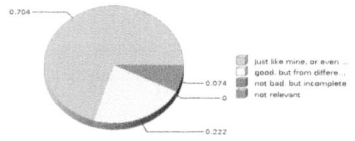

Fig. 3. AQ category suitability

Table 1. Ambiguous Query results

Choice	Result
Our is better	70.3%
Cannot make decision	22.2%
Carrot is better	7.4%

In table 1 we show Participant preference between our category and carrot categories. In answer to existence of users preferred category within the presented categories, in 56% of queries the answer was "yes". Also in question about does system gave them a hint to improve their query the answers for 59% of queries were "yes". We also asked participant to judge top 10 results in their selected categories. 84% of these judged results were exactly what they were looking for. 9% of judged results were categorized correctly but were not exactly what they were looking for. And the remaining 7% was not relevant to the selected category.

- **Entity Queries**

Participants evaluated out method by answering the questions introduced in 4.1. As it is shown in Figure3 in 75.6% of questions user answered that our categories is similar to or even better than their provided categories. 11.1% of answer was "good. But from different point of view" and 13.3% answered that the categories are incomplete. Some of these queries were: "Dennis Ritchie", "data mining", "flowers", "JavaScript", "tourism", "data warehouse", "Yoghurt" and etc.

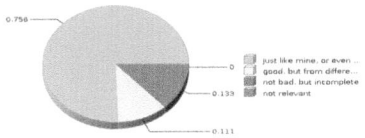

Fig. 4. EQ category suitability

Table 2. Entity Queries results

Choice	Result
Our is better	58%
Cannot make decision	20%
Carrot is better	22%

Table 2 shows Participant preference between our categories and carrot categories. In answer to existence of users preferred category within the presented categories, in 73% of queries the answer was "yes". Also in question about does system gave them a hint to improve their query the answers for 78% of queries were "yes". We also asked participant to judge top 10 results in their selected categories. 90.1% of these judged results were exactly what they were looking for. 6% of judged results were categorized correctly but were not exactly what they were looking for. And the remaining 0.9% was not relevant to the selected category.

- **Broader Queries**

In 44% of questions system wasn't able to provide category structure for user queries.

The result below is about those 56% of queries which our system could find category structure for them.

As Figure 4 shows in 73.3% of questions user answered that our categories is similar to or even better than their provided categories. 20% of answer was "good. But from different point of view" and in 6.7% they said that the results is not relevant to their query. Some of these queries were: "Maldives music", "fiber made of glass", "Revolution in Egypt", "Middle east war" and etc.

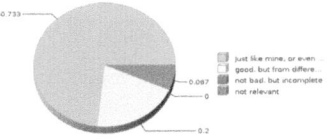

Fig. 5. BQ category suitability

Table 3. BQ results

Choice	Result
Our is better	47%
Cannot make decision	13%
Carrot is better	40%

In table 3 we show Participant preference between our category and carrot categories. In answer to existence of users preferred category within the presented categories, in 53% of queries the answer was "yes". Also in question about does system gave them a hint to improve their query the answers for 60% of queries were "yes". We also asked participant to judge top 10 results in their selected categories. 74% of these judged results were exactly what they were looking for. 18% of judged results were categorized correctly but were not exactly what they were looking for. And the remaining 8% was not relevant to the selected category.

4   Conclusions and Future Works

In this paper, we proposed a search result organization method with dynamically extracted categories for each query. Our algorithm uses Wikipedia for extracting categories and gathering training data. By comparing our system's result with carrot's, we noticed that when the query is simple, usually there is a well-formed and complete article in Wikipedia that leads to better human defined categories which covers the subject from different aspects. Carrot uses its own techniques to automatically generate labels that does not cover all aspects of query and may not be relevant to each other. But when the query becomes much more complex, the quality of Wikipedia's articles decreases and the article is no longer complete. This situation makes users to prefer carrot over our system. It shows that our proposed method still needs to be improved so that it could be able to handle more Complex Queries.

Our proposed algorithm is still a prototype and we are planning to do some functional improvement in it. For example Wikipedia has lots of features that can help us to improve our algorithm like Wikipedia redirections and Wikipedia's category structure. Another interesting feature of Wikipedia is "Main article". Using this feature will probably help constructing the hierarchy.

Acknowledgments. This research is partially supported by Iran Telecommunication Research Center (ITRC).

References

[1] Alemzadeh, M., Karray, F.: An Efficient Method for Tagging a Query with Category Labels Using Wikipedia towards Enhancing Search Engine Results. In: WI-IAT, pp. 192–195 (2010)

[2] Xu, Y., Jones, G.J., Wang, B.: Query dependent pseudo-relevance feedback based on wikipedia. In: SIGIR, pp. 59–66 (2009)

[3] Wang, X., Zhai, C.: Learn from web search logs to organize search results. In: SIGIR, pp. 87–94 (2007)

[4] Zeng, H., He, Q., Chen, Z., Ma, W., Ma, J.: Learning to cluster web search results. In: SIGIR, pp. 210–217 (2004)

[5] Chen, H., Dumais, S.: Bringing order to the Web: automatically categorizing search results. In: CIII, pp. 145–152 (2000)

[6] Xing, D., Xue, G., Yang, Q., Yu, Y.: Deep classifier: automatically categorizing search results into large-scale hierarchies. In: WSDM, pp. 139–148 (2008)

[7] Hu, X., Zhang, X., Lu, C., Park, E.K., Zhou, X.: Exploiting Wikipedia as external knowledge for document clustering. In: SIGKDD, pp. 389–396 (2009)

[8] Hu, J., Fang, L., Cao, Y., Zeng, H., Li, H., Yang, Q., Chen, Z.: Enhancing text clustering by leveraging Wikipedia semantics. In: SIGIR, pp. 179–186 (2008)

[9] Banerjee, S., Ramanathan, K., Gupta, A.: Clustering short texts using wikipedia. In: SIGIR, pp. 787–788 (2007)

[10] Meij, E., de Rijke, M.: Supervised query modeling using wikipedia. In: SIGIR, pp. 875–876 (2010)

[11] Chen, M., Hearst, M., Hong, J., Lin, J.: Cha-Cha: a system for organizing intranet search results. In: USENIX, pp. 5–5 (1999)

[12] Sanderson, M., Croft, B.: Deriving concept hierarchies from text. In: SIGIR (1999)

[13] Tan, B., Peng, F.: Unsupervised query segmentation using generative language models and wikipedia. In: WWW, pp. 347–356 (2008)

[14] Mihalcea, R.: Using Wikipedia for Automatic Word Sense Disambiguation. In: Proc. NAACL HLT, pp. 196–203 (2007)

[15] Carmel, D., Roitman, H., Zwerdling, N.: Enhancing cluster labeling using Wikipedia. In: SIGIR, pp. 139–146 (2009)

[16] Cao, H., Hu, D.H., Shen, D., Jiang, D., Sun, J., Chen, E., Yang, Q.: Context-aware query classification. In: SIGIR, pp. 3–10 (2009)

[17] Hu, J., Wang, G., Lochovsky, F., Sun, J., Chen, Z.: Understanding user's query intent with wikipedia. In: WWW, pp. 471–480 (2009)

[18] Toda, H., Kataoka, R.: A search result clustering method using informatively named entities. In: WIDM, pp. 81–86 (2005)

[19] Kules, B., Kustanowitz, J., Shneiderman, B.: Categorizing web search results into meaningful and stable categories using fast-feature techniques. In: JCDL, pp. 210–219 (2006)

[20] Käki, M.: Findex: search result categories help users when document ranking fails. In: CHI, pp. 131–140 (2005)

[21] Bernardini, A., Carpineto, C., D'Amico, M.: Full-Subtopic Retrieval with Keyphrase-Based Search Results Clustering. In: WI-IAT, pp. 206–213 (2009)

[22] Zamir, O., Etzioni, O.: Grouper: a dynamic clustering interface to Web search results. In: WWW, pp. 1361–1374 (1999)

[23] Wang, S., Hauskrecht, M.: Effective query expansion with the resistance distance based term similarity metric. In: SIGIR, pp. 715–716 (2010)

[24] Hearst, M.A., Pedersen, J.O.: Reexamining the cluster hypothesis: scatter/gather on retrieval results. In: SIGIR, pp. 76–84 (1996)

[25] Zamir, O., Etzioni, O.: Web document clustering: a feasibility demonstration. In: SIGIR, pp. 46–54 (1998)

Recommend at Opportune Moments

Chien-Chin Su and Pu-Jen Cheng

Department of Computer Science and Information Engineering
College of Electrical Engineering and Computer Science
National Taiwan University
{r98922131,pjcheng}@csie.ntu.edu.tw

Abstract. We propose an approach to adapt the existing item-based (movie-based) collaborative filtering algorithm based on the timestamp of ratings to recommend movies to users at opportune moments. Over the last few years, researchers focused recommendation problems on rating scores mostly. They analyzed users' previous rating scores and predicted those unknown rating scores. However, we found rating scores are not the only problem we have to concern about. When to recommend movies to users is also important for a recommender system since users' shopping habits vary from person to person. To recommend movies to users at opportune moments, we analyzed the rating distribution of each movie by the timestamps and found a user tending to watch similar movies at similar moments. Several experiments have been conducted on MovieLens Data Sets[1]. The system is evaluated by different recommendation lists during a specific period of time - $t_{specific}$, and the experimental results show the usefulness of our system.

Keywords: Recommender Systems, Collaborative Filtering, Advertising.

1   Introduction

In film market analysis, we find rating scores are not the only problem we have to solve. Fig. 1 is an example to illustrate our problem. One user is used to watching the latest action movie in the first few weeks of the release date and giving it a high rating score. Of course, a recommender system would give the latest action movie to the user at start because of his previous high rating scores on action movies. Nevertheless, after weeks, if the user still did not watch the movie, should we assume that the user just missed the movie information and keep recommending the movie to the user? Or should we assume that the user already knew the movie and did not like it and then stop recommending the movie to the user?

People watch a movie for several reasons. If we group people by their reasons and analyze their timestamps, we would find an interesting cascading process. First, film critics would watch a movie at the release date. Then stars' fans and people appealed

---

[1] http://www.cs.umn.edu/Research/GroupLens

M.V.M. Salem et al. (Eds.): AIRS 2011, LNCS 7097, pp. 226–237, 2011.

by the previews would watch it in the first few weeks. Next, people who accept friends' recommendation or positive feedback from the public would watch it. Afterward people who enjoy watching movies at home can rent a DVD or watch it on Cable TVs. In the previous example of the first paragraph, the user is one of the 2nd group users of action movies. If the user does not watch the movie in the first few weeks, we will assume that the user does not like the movie and then stop recommending the movie to the user. In other words, if he liked the movie, he would have already watched it. In our system, a user would be dropped from the recommendation list of the new movie if he belongs to the early group but does not watch in the early time.

Coincidentally, Diffusion of Innovations theory [3], one of the most frequently cited books in Management Science[2], demonstrated a similar idea. The theory divides users of a product into 5 categories by their adoption time: Innovators, Early Adopters, Early Majority, Late Majority and Laggards and the proportions of them are 2.5%, 13.5%, 34%, 34% and 16%. It inspired researchers [8][14] to discover the trust relationship between the innovators and followers wherein the innovators and early adopters influence the community while the early majority, late majority and laggards follow their lead.

However, what attracts our attention more is not only the trust relationship between the innovators and followers. First, the theory infers that similar products have similar innovators. Namely, similar products may have similar innovators and similar followers. Second, users of a product are categorizable. That is, if the categorization is good enough, partial users may represent the composition of whole users of a product. Fig. 2 illustrates the main idea of this paper. Similar products may have similar users in each corresponding category and vice versa.

In this paper, we propose a new approach rather than the trust network. We adapt the existing item-based (movie-based) CF algorithm to demonstrate that people are latent suggested to be users of the corresponding categories among similar movies.

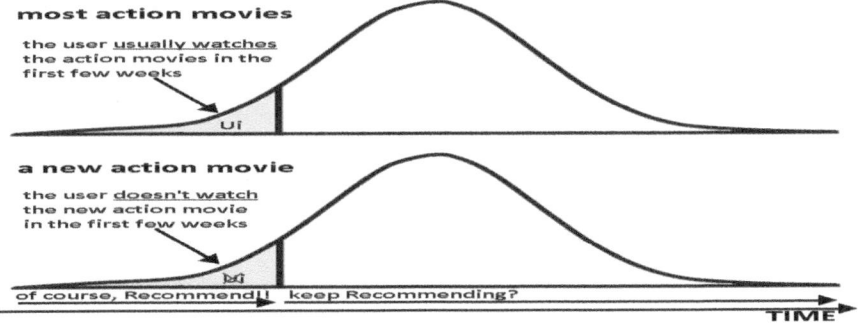

Fig. 1. Our problem

---

[2] http://mansci.journal.informs.org/

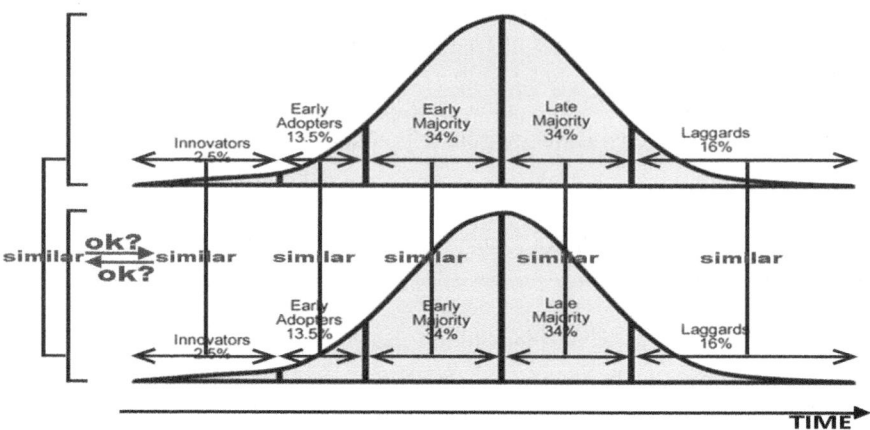

Fig. 2. The main idea

2   Related Work

Most existing works [1][4][12] on recommendation problems focus on the like / dislike problem. A recommender system is measured by MAE and RMSE of its predictions on the unknown rating scores so that it can recommend users' likes accurately. Other works focus on the purchase / not purchase problem by optimizing some evaluation metrics such as Precision and Recall. [5] used the item-to-item collaborative filtering algorithm, "customers who bought items in your shopping cart also bought", to improve Precision of a recommendation list. Our work focuses on the purchase / not purchase problem and uses time factors to improve a recommendation list. Due to the lack of real "purchase" logs, we consider each rating to be that the user really purchases the item.

There were two related papers applying the Diffusion of Innovations theory on solving recommendation problems recently. Both of them transfer the rating information into graphical network. S. K. Tyler [14] modeled user adoption behaviors by creating a total ordering. It showed that by using order sequences, a recommender system could more accurately predict the category of a new product as well as predict which users will follow. Noriaki Kawamae [8] proposed a recommendation algorithm that focuses on the search time that, in the absence of any recommendation, each user would need to find a desirable and novel item by himself. Following the hypothesis that the degree of user's surprise is proportional to the estimated search time, the algorithm considers both innovators' preferences and trends for identifying items with long estimated search times. In our work, other than finding relationship between users, we discover relationship between user groups. We take early users to be a leader group. A leader group influences a follower group; similar leader groups may have similar follower groups.

3   Problem Specification

3.1   Motivation

People have their own shopping behaviors, so we mine the rules of their behaviors and then apply the rules to a recommender system. Here we bring up 2 types of shopping behaviors: favorite & habit.

For favorite, people give each movie a rating with a score to show their feelings of it. A recommender system analyzes users' previous ratings and predicts that one user would have similar favorites to other users (User-Based CF Algorithm) or one user would like other similar movies (Item-Based CF Algorithm). For Habit, people have different attitudes toward new products: some people rush to the store whenever it has new products, and some people would like to try after many people saying it is good.

Obviously, "Rating scores" should not be the only feature to classify users and movies anymore because they do not explain the habits we defined here. A mature recommender system should consider both "Rating Scores" and "Rating Timestamps." Rating scores represent personal favorites, which most conventional recommender systems addressed. Rating timestamps represent personal shopping habits which we address in this work.

The following formula describes the function of our recommender system:

$f'(m_{id}, u_{id}, t_{specific}) \rightarrow$ Yes/No

1. Opportuneness during a specific period of time - $t_{specific} = [t_{lb}, t_{ub}]$
2. Recommend m_{id} to u_{id} or not during $t_{specific}$
3. Evaluated by Precision

Other than conventional recommender systems which predict personal favorites:

$f(m_{id}, u_{id}) \rightarrow$ rating scores

1. Rating prediction
2. Recommend users' likes
3. Evaluated by MAE & RMSE of rating scores

The Basic concept of our system is whether the user likes the movie or not, the "purchase" does happen according to logs. Beyond the reasons of the purchase, should we recommend movies to user all the time or recommend movies to user during a specific period of time - $t_{specific}$? The application of our system is to reduce the recommendation cost. Since every recommendation has an opportunity cost, so we have to recommend at opportune moments.

3.2   Group Dependency

Here we divide users of each movie averagely into 10 groups (10% users for each group), $G_{m_{id}} = \{g_1, g_2, ..., g_{10}\}$, and the boundaries of $G_{m_{id}}$ are $T_{m_{id}} = \{t_{0\%}, t_{10\%}, t_{20\%}, ..., t_{100\%}\}$, and use vectors composed of partial users of each movie to compute the similarity. Fig. 3 illustrates how we divide users of each movie.

In the following, we will choose a boundary $t_{x\%} \in \{t_{10\%}, t_{20\%}, ..., t_{90\%}\}$ and analyze the relation between the early part similarity - $CosSimilarity(m_{i(t_{0\%}, t_{x\%}]}, m_{j(t_{0\%}, t_{x\%}]})$ and the late part similarity - $CosSimilarity(m_{i(t_{x\%}, t_{100\%}]}, m_{j(t_{x\%}, t_{100\%}]})$ of any two distinct movies in the observation dataset. We use a parenthesis instead of a bracket because there is no user at $t_{0\%}$ so set a period of time which excludes the left boundary and includes the right boundary.

For every movie $m_i \neq$ movie m_j, we generate a <tuple> = <x,y> where x = the early part similarity and y = the late part similarity. Then we plot these tuples on a scatter graph to be Fig. 4 and found the group dependency between the early parts and late parts of movies through the trend line. Fig. 4 is the summary of this section. Scatter graphs with the boundaries from $t_{10\%}$ to $t_{90\%}$ show the robustness of our observation of the group dependency. That is, the early part is positive related to the late part.

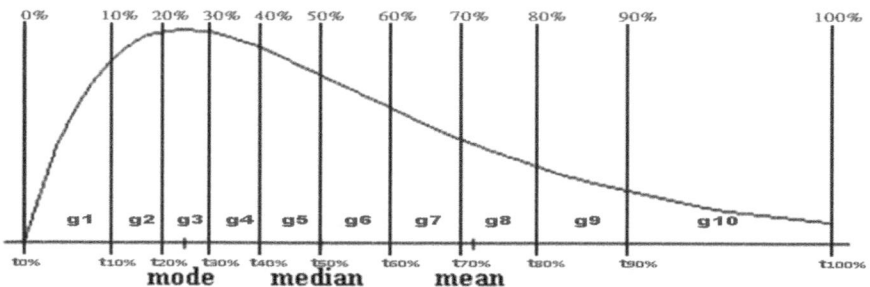

Fig. 3. Divide users into 10 groups

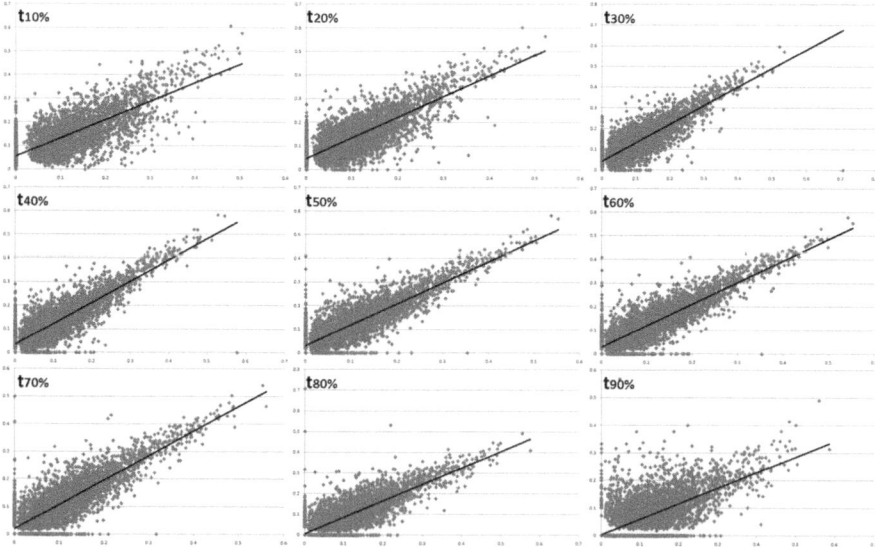

Fig. 4. Scatter graphs with the boundaries from $t_{10\%}$ to $t_{90\%}$

3.3   Main Idea

Due to the group dependency, If m_i (a new movie) and m_j (an old movie) are highly similar to each other, u_{id} who saw m_j in $(t_{x\%}, t_{100\%}]$ should have higher priority to be recommended to watch m_i during $(t_{x\%}, t_{100\%}]$. In other words, we prefer recommending movies to users in the corresponding time - $t_{specific}$.

Here we assume the late part of each movie is $t_{specific}$. In our experiments, each leave-one-out new movie would have 3 recommendation lists during $t_{specific}$:

1. $List_{OUR}$ is composed of users of the late parts of similar movies.
2. $List_{CV}$ is composed of all users of similar movies
3. $List_{CT}$ is composed of users of the early parts of similar movies.

Of course, users who watched the new movie in the early time would be deleted from the recommendation lists. We take $List_{CV}$ to be the baseline for our experiments since it is a conventional method and want to verify "Is the performance $List_{OUR} > List_{CV} > List_{CT}$?"

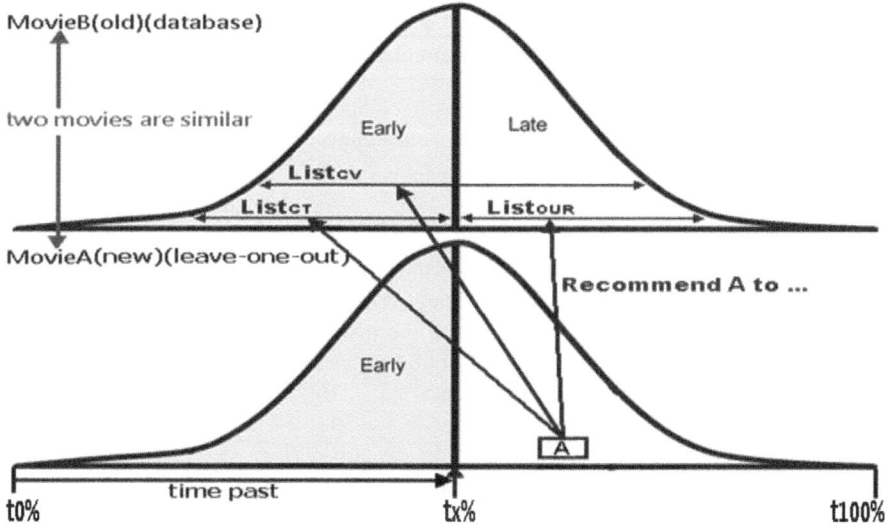

Fig. 5. Recommendation lists: $List_{OUR}$, $List_{CV}$ and $List_{CT}$

4   Methodology

4.1   The Proposed Approach

Given a movie and a period of time t_{period}, the system will generate an opportune recommendation list. The t_{period} decides the boundary $t_{x\%}$ and $t_{specific}$ where $t_{x\%}$ should be early than t_{period}. After knowing the boundary $t_{x\%}$, the system will compute the similarity between movies by the early parts and select Top-K (Given K=47) similar movies to generate the recommendation list by the late parts. The ground truth is composed of users of the late part of the leave-one-out new movie. Afterward, the recommendation list is evaluated by the evaluation metrics and reported the experimental results.

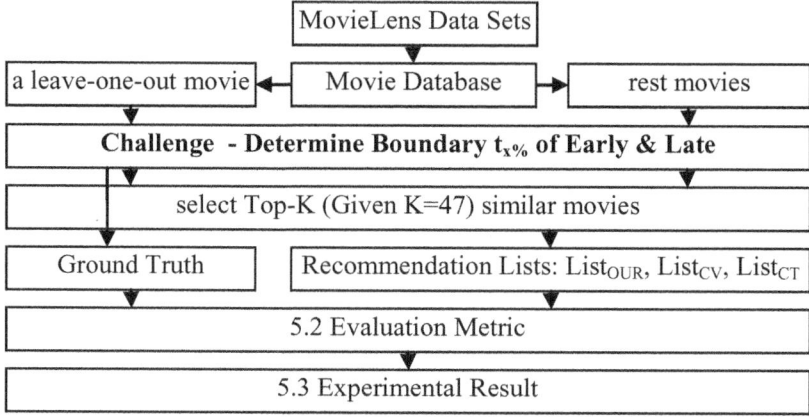

Fig. 6. Our system architecture

4.2   Theory-Based Boundary ($t_{2.5\%}$, $t_{16\%}$, $t_{50\%}$, $t_{84\%}$)

The inspiration of the theory-based boundary is from Diffusion of Innovations theory [3]. It divides users of a product into 5 categories: Innovators, Early Adopters, Early Majority, Late Majority and Laggards and the proportions of them are 2.5%, 13.5%, 34%, 34% and 16% and provides four tentative boundaries: $t_{2.5\%}$, $t_{16\%}$, $t_{50\%}$ and $t_{84\%}$ where the accumulated user amount reaches 2.5%, 16%, 50% and 84% respectively.

4.3   Global Optimal Boundary

The global optimal boundary is a heuristic boundary. We test all possible boundaries from $t_{2\%}$ to $t_{98\%}$ with a fixed interval – 2%. There are two problems have to be solved in this experiment. On the one hand, we want to get an optimal boundary for this movie database. So once the accumulated user amount of the new movie reaches x%, we can make an optimal recommendation list. Of course, $t_{x\%}$ is the earlier the better. On the other hand, we want to know whether the more information of the early part - $(t_{0\%}, t_{x\%}]$, the better Precision we can get on predicting the late part - $(t_{x\%}, t_{100\%}]$.

4.4   Local Optimal Boundary

This section, we use a dynamic boundary other than a fixed one. Fig. 7 illustrates the idea of this section. First, we use a fixed boundary – $t_{50\%}$ on the leave-one-out new movie i and then use a dynamic boundary – $t_{x\%}$ (adjusting from $t_{2\%}$ to $t_{98\%}$ with a fixed interval – 2%) on the rest old movies j in the database to maximize the $sim\left(m_{i(t_{0\%},t_{50\%}]}, m_{j(t_{0\%};t_{x\%}]}\right)$.

The idea of the local optimal boundary is that not all movies are ideal distributed. Here we divide users of each movie into two groups – the leader group and the follower group. The boundary of two groups is $t_{x\%}$, and the principle is that similar leader groups affects similar follower groups to purchase the item. Finally, we merge users of the follower groups to be $List_{OUR}$.

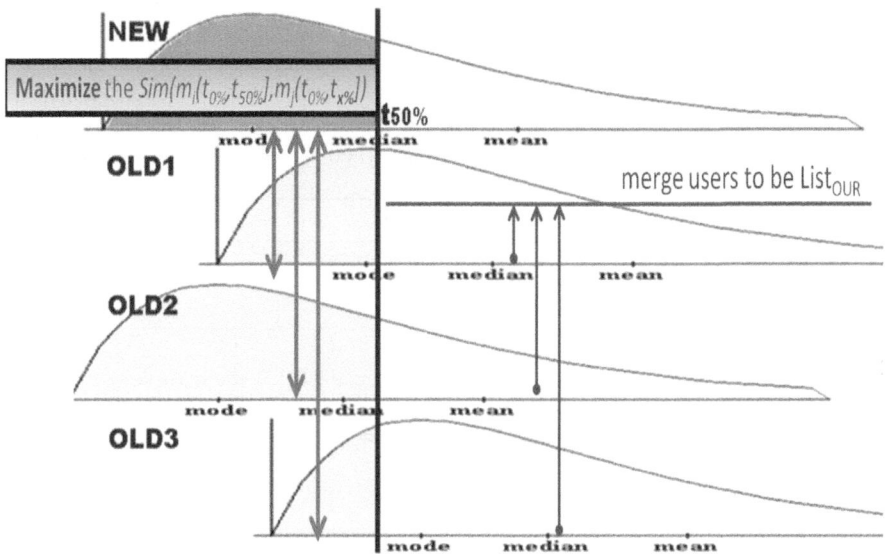

Fig. 7. Local Optimal Boundary

5   Experiment

5.1   Dataset

We use 1-million MovieLens Data Sets for our experiments. Since we try all the boundaries between $t_{2\%}$ to $t_{98\%}$ in 4.3 and 4.4, the minimal rating amount is necessarily required for each interval. We filter out all movies with ratings less than 100, so every movie in the filtered dataset has at least one user in its 1% interval. Table 1 is the information of the filtered dataset.

5.2   Evaluation Metric

In our experiments, we use Precision and Recall to evaluate the three recommendation lists. A perfect precision score of 1.0 means every user in the recommendation list can

Table 1. The filter 1-million MovieLens Data Sets

	1-million MovieLens Data Sets	Filtered 1-million MovieLens Data Sets
DEMOGRAPHIC	UserID::Gender::Age::Occupation::Zip-code	
MOVIES FILE DESCRIPTION	MovieID::Title(IMDB) (with year)::Genres	
RATING SCALE	5-star scale (whole-star ratings only)	
TIME	Unix Timestamp(s)	
# of ratings	1,000,209	942,225
Movies	3706	2019
Movie(MAX ratings)	3428	3428
Movie(min ratings)	1	100
Users	6040	6040
User(MAX ratings)	2314 (UserID:4169)	1568 (UserID:4169)
User(min ratings)	20	12

be found in the ground truth. A perfect recall score of 1.0 means every user in the ground truth can be found in the recommendation list. We will analyze Precision and Recall of the top 1, 5 and 20 users in the recommendation lists of each movie.

5.3   Experimental Result

In the following experiments, we assume that we are at the time $t_{50\%}$ of each leave-one-out new movie - m_{new}, so our recommender system only knows the rating information in $m_{new(t0\%,t50\%]}$.

Table 2. The theory-based boundary

Theory-based Boundary	$t_{2.50\%}$		$t_{16\%}$		$t_{50\%}$		$t_{84\%}$	
	Precision	Recall	Precision	Recall	Precision	Recall	Precision	Recall
TOP1	0.8694	0.0038	**0.9142**	0.0061	0.8893	0.0095	0.7469	0.0177
TOP5	0.7801	0.0154	**0.8567**	0.0216	0.8312	0.0345	0.6457	0.0685
TOP20	0.6896	0.0504	**0.7723**	0.0688	0.7385	0.1072	0.5064	0.1947

From the results, we can see the Precision is maximized with the boundary $t_{20\%}$, hence we would choose the boundary $t_{20\%}$ to be a heuristic boundary in this movie database. Of course, the boundary is necessary before the boundary $t_{50\%}$ (we are at the time $t_{50\%}$),

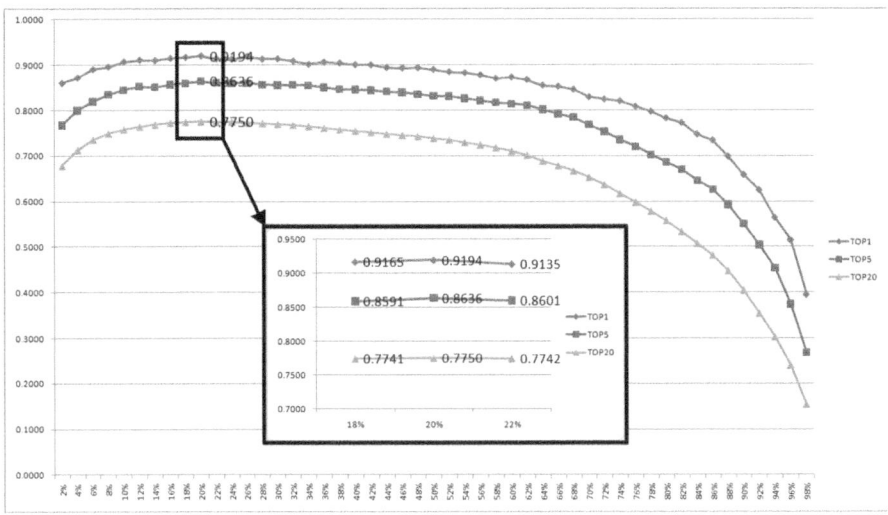

Fig. 8. The global optimal boundary

Table 3. The local optimal boundary

	Global: $t_{50\%}$		Local Optimal	
	Precision	Recall	Precision	Recall
TOP1	0.8893	0.0095	0.8921	0.0095
TOP5	0.8312	0.0345	0.8421	0.0351
TOP20	0.7385	0.1072	0.7486	0.1092

Table 4. The three recommendation lists under the global optimal boundary $t_{20\%}$

Global: $t_{20\%}$	List$_{OUR}$		List$_{CV}$		List$_{CT}$	
	Precision	Recall	Precision	Recall	Precision	Recall
TOP1	0.9194	0.0064	0.9169	0.0066	0.0998	0.0004
TOP5	0.8636	0.0231	0.8536	0.023	0.07	0.0014
TOP20	0.775	0.0729	0.7594	0.0712	0.0542	0.0044

so the boundary $t_{20\%}$ is proper. Finally, we check out the experiment results of the three recommendation lists in Table 4.

In Table 4, we can see $List_{OUR}$ has better results than $List_{CV}$, and the difference does pass the t-test with p-value<0.01. Moreover, $List_{OUR}$ obviously outperforms $List_{CT}$, and the results show how importance to recommend at opportune moments.

Furthermore, on answering the second problem mentioned in 4.3, "the more information of the early part, the better Precision we can get on predicting the late part," we check Fig. 8 and find the P@1 is maximized at $t_{20\%}$. So if we are at the time before the boundary $t_{20\%}$, we had better use all information we know because the P@1 is increasing before the boundary $t_{20\%}$. Then again if we are at the time after the boundary $t_{20\%}$, we had better to choose boundary $t_{20\%}$ for our recommender system because the P@1 is decreasing after the boundary $t_{20\%}$.

6   Discussion

Most people are curious about the user amount of our work. How do we know the user amount so to know the boundary $t_{x\%}$? Actually, it is not a big problem. On the one hand, you can use some prediction theories such as Bass Diffusion Model to predict the user amount. On the other hand, here we also propose a solution by the local optimal boundary which is mentioned in 4.4. The concept of the local optimal boundary is to divide all users of a product into two groups – the leader group and the follower group by a dynamic boundary. That is, if the leader groups are similar, the follower group may be similar too. In other words, the local optimal boundary doesn't really use a fixed boundary so it does not need to know the user amount. Table 5 is the results of the different boundary cases of the leave-one-out new movies solved by the local optimal boundary. The results look good at any time.

Table 5. The solution by the local optimal boundary

by Local Optimal Boundary	$t_{25\%}$		$t_{50\%}$		$t_{75\%}$	
Precision	$List_{OUR}$	$List_{CV}$	$List_{OUR}$	$List_{CV}$	$List_{OUR}$	$List_{CV}$
TOP1	0.9194	0.9155	0.8921	0.8751	0.8307	0.7413
TOP5	0.8662	0.8551	0.8421	0.7971	0.7476	0.6365
TOP20	0.7813	0.7584	0.7486	0.6763	0.6260	0.5034

7   Conclusion

In the paper, we propose an approach to adapt the existing item-based (movie-based) collaborative filtering algorithm based on the timestamp of ratings to recommend movies to users at opportune moments. We first present the observation and bring up the main idea of our system architecture. The experimental results show the performance of our system is very promising on MovieLens Data Sets.

More than increasing the precision, another feasible application is to decrease the computational cost. Researchers have been working on the problem of a small group of users to represent whole users, and our method may work well on this problem. That is, if the boundary $t_{x\%}$ is well-chosen, users of the early part of a product may represent whole users of the product.

In the future, to add one more boundary is a good further development in our system. Add one more boundary to reduce the $t_{specific}$ range and then make the recommendation list more accurate. However, a system with two boundaries is more complex than it with only one boundary. For example, the definition of the best case and the similarity functions of different combinations of the vectors become more complicated. We hope we can find out the solution in the near future.

Reference

1. Sarwar, B., Karypis, G., Konstan, J., Riedl, J.: Item-based collaborative filtering recommendation algorithms. In: WWW10 (2001)
2. Kempe, D., Kleinberg, J., Tardos, E.: Maximizing the spread of influence through a social network. In: ACM SIGKDD 2003 (2003)
3. Rogers, E.M.: Diffusion of Innovations. The Free Press, New York (1995)
4. Adomavicius, G., Tuzhilin, A.: Towards the next generation of recommender systems: a survey of the state-of-the-art and possible extensions. IEEE TKDE 17(6), 734–749 (2005)
5. Linden, G., Smith, B., York, J.: Amazon.com recommendations. IEEE Internet Computing 7(1), 76–80 (2003)
6. Konstan, J.A.: Introduction to recommender systems. In: ACM SIGMOD 2008, p. 1373 (2008)
7. Riedl, J., Konstan, J.: Movielens dataset,
   http://www.cs.umn.edu/Research/GroupLens
8. Kawamae, N.: Serendipitous recommendations via innovators. In: ACM SIGIR 2010, pp. 218–225 (2010)
9. Liu, N.N., Yang, Q.: Eignerank: A ranking-oriented approach to collaborative filtering. In: ACM SIGIR 2008, pp. 83–90 (2008)
10. Netflix Prize Dataset, http://www.netflixprize.com/download
11. Domingos, P., Richardson, M.: Mining the network value of customers. In: ACM SIGKDD 2001 (2001)
12. Resnick, P., Iacovou, N., Suchak, M., Bergstrom, P., Riedl, J.: An open architecture for collaborative filtering of netnews. In: ACM CSCW 1994, pp. 175–186 (1994)
13. Rusmevichientong, P., Zhu, S., Selinger, D.: Identifying early buyers from purchase data. In: ACM SIGKDD 2004 (2004)
14. Tyler, S.K., Zhu, S., Chi, Y., Zhang, Y.: Ordering innovators and laggards for product categorization and recommendation. In: ACM RecSys 2009, pp. 29–36 (2009)
15. Koren, Y.: Tutorial on recent progress in collaborative filtering. In: ACM RecSys 2008, p. 333 (2008)
16. Yao, Y.: Hodge decomposition, spectral embedding, and the netflix dataset. In: Workshop on Algorithms for Modern Massive Data Sets, MMDS 2008 (2008)

Emotion Tokens: Bridging the Gap among Multilingual Twitter Sentiment Analysis*

Anqi Cui, Min Zhang, Yiqun Liu, and Shaoping Ma

State Key Laboratory of Intelligent Technology and Systems,
Tsinghua National Laboratory for Information Science and Technology,
Dept. of Computer Science and Technology, Tsinghua Univ., Beijing 100084, China
cuianqi@gmail.com, {z-m,yiqunliu,msp}@tsinghua.edu.cn

Abstract. Twitter is a microblogging service where worldwide users publish their feelings. However, sentiment analysis for Twitter messages (tweets) is regarded as a challenging problem because tweets are short and informal. In this paper, we focus on this problem by the analysis of emotion tokens, including emotion symbols (e.g. emoticons), irregular forms of words and combined punctuations. According to our observation on five million tweets, these emotion tokens are commonly used (0.47 emotion tokens per tweet). They directly express one's emotion regardless of his language; hence become a useful signal for sentiment analysis on multilingual tweets. Firstly, emotion tokens are extracted automatically from tweets. Secondly, a graph propagation algorithm is proposed to label the tokens' polarities. Finally, a multilingual sentiment analysis algorithm is introduced. Comparative evaluations are conducted among semantic lexicon based approach and some state-of-the-art Twitter sentiment analysis Web services, both on English and non-English tweets. Experimental results show effectiveness of the proposed algorithms.

Keywords: Multilingual sentiment analysis, Twitter sentiment analysis, Emotion token, Sentiment lexicon, Network informal language.

1 Introduction

Nowadays millions of users publish short messages on Twitter. It is widely spread all over the world and becomes a rich resource of texts in many different languages. Twitter's messages (*tweets*) are full of opinions and emotions, thus sentiment analysis for tweets is important for information spreading and marketing. However, this is more difficult than traditional text analysis.

Tweets are limited with no more than 140 characters and are usually composed on mobile devices, hence people often use irregular expressions both for convinience and to save room for more words. Emotion tokens (including *emotion symbols*, *irregular forms of words* and *combined punctuations*) are usually seen

---

* Supported by Natural Science Foundation (60736044, 60903107, 61073071) and Research Fund for the Doctoral Program of Higher Education of China (20090002120005). This work has been done at Tsinghua-NUS NExT Search Centre.

M.V.M. Salem et al. (Eds.): AIRS 2011, LNCS 7097, pp. 238–249, 2011.

in tweets, such as "– *Me tienes olvidada :(– (-.- otra vez esta...) Disculpa es que estaba dormido. – AAAAH ok, ¿qué harás ahora? – Dormir mucho más.*" (Spanish: "*– I have forgotten :(– (-.- again this...) Sorry is that he was asleep. – AAAAH ok, what will you do now? – Sleep more.*") Based on our observation (see section 3.2), there are about 0.47 emotion tokens per tweet; about one third tweets contain at least one emotion token. Emotion token is one of the most remarkable features of Internet text. It strongly expresses the feelings of the author and is often utilized across languages. Thus, emotion tokens are helpful for multilingual Twitter sentiment analysis, to determine if a tweet expresses a positive or a negative feeling, no matter what language the author uses.

Although the emotion tokens have been studied previously, they are usually considered as annotation of the texts and are chosen manually [6]. Different from these studies, we automatically extract different types of emotion tokens and use a propagation algorithm to label their polarities by few "seed" tokens. Therefore, many different tokens and their scores are discovered to build a sentiment lexicon which helps multilingual Twitter sentiment analysis.

The highlight of our work is: Different types of emotion tokens are extracted automatically, without considering the semantic information; their sentiment polarities are labeled with an unsupervised propagation algorithm. The sentiment lexicon built based on the tokens works as a bridge over the gap among different languages, while most state-of-the-art Twitter sentiment analysis approaches only deal with English tweets. In addition, the corpus for building the lexicon is independent of time which is practical and feasible for real-world applications.

This paper is organized as follows. Related work on sentiment analysis and Twitter is introduced in Sec. 2. Emotion tokens and their characteristics are analyzed in Sec. 3. In Sec. 4 and 5, algorithms for sentiment lexicon construction and sentiment analysis are proposed respectively. In Sec. 6, the algorithms are evaluated. In the last section, conclusions and future work are addressed.

2   Related Work

Sentiment analysis plays an important role with the growing of user-generated-content services. In traditional studies, most researchers build statistical models for sentiment and affect analyses, where semantic information is highly considered as features [17,20]. These models require annotated corpus, which is often limited for online texts. Alternatively, manually built sentiment lexicons can be used as a useful resource [2,22]. Linguistic information (part-of-speech tags, syntactic information) is used for rule-based approaches [18]. Even though the feature words can be extracted with automatic algorithms [16], the semantic differences among languages limit multilingual analysis.

An intuitive idea for multilingual sentiment analysis is to translate languages into a well-studied language (e.g. English); hence traditional methods can be applied. Previous studies on news and blogs work on sentence level [5] or word level [11]. Cross-language dictionaries work as bridges between different languages [9]. Obviously, without language techniques, these methods do not work. Some machine learning models may be independent of languages, but training requires multilingual annotations [7,19] or is based on machine translation [3].

Emoticons and irregular words, however, are commonly seen in Internet texts of many languages. Emoticons are considered as annotations since they directly express the one's attitude [6]. Unfortunately, they have various forms; most studies manually choose some smileys (e.g. "*:-)*") as labels [19]. They fail to consider many other figures such as "*<3*" (heart, means *love*). In Twitter, we need to discover more possible emoticons with different forms, since they are independent of languages and are helpful for multilingual analysis. On the other hand, the irregular words are usually seen when people wish to save keystrokes, or the length of message is limited. We focus on the emphasized spelling, i.e. the repeating of consecutive letters in a word (e.g. "***nooooo** WTF everyone left me*").

Twitter is a popular research topic nowadays. Besides its network characteristics [15], social impacts reflected by Twitter sentiment are also of interests, such as word-of-mouth branding [13]. Other work follows the trends of sentiment [8]. As mentioned before, smileys are usually considered as annotations [6,19]. Most studies use linguistic rules or supervised learning to help sentiment analysis [14], which is difficult to be generalized into multiple languages.

There are websites that provide sentiment detection of Twitter. However, a study on the comparison of these results concludes that they contain much noise and lack precision [4]. The *twitrratr* site (http://twitrratr.com/) builds lists of positive and negative keywords and classifies the sentiment of tweets based on matching. The *twittersentiment* site (http://twittersentiment.appspot.com/) uses distant supervision to classify the sentiment of Twitter messages [12]. Although some smileys are used to collect training data as labels, emoticons are removed in their classification. We compare our results with these two websites.

To sum up, we notice that the traditional sentiment analysis methods are in shortage on Twitter and are limited to a specific language. To achieve a better multilingual Twitter sentiment analysis, we consider the emotion tokens as a bridge over the gap among languages.

3   Study on Emotion Tokens in Tweets

3.1   Types of Emotion Tokens

The three types of emotion tokens are listed in Table 1. They express emotions and cover most of the emotional informal words on the Internet.

Note that some of the repeating letters words may be relevant to language. However, based on our observation, many of them are onomatopoeic words (Table 2). Therefore, they can be simply considered as another type of *emoticons*.

3.2   Characteristics of Emotion Tokens in Twitter

In this paper, we use Stanford's SNAP data (http://snap.stanford.edu/data/) which contains more than 400 million tweets in over six months.

Table 1. Types of emotion tokens

Type	Definition	Example	Explanation
Emotion symbols	Every combination of symbols with alphabets and numbers. This type is an extended set of the traditional *emoticons*, since they are not limited to "faces".	\o/ *Domingo + Canjica quentinha = tudo de bom! :D* (In English: \o/ *Sunday + warm Canjica = all the best! :D*)	This Portuguese tweet contains "\o/" (man raising arms, cheering) and ":D" (laughing face), telling us the author is happy.
Repeating punctuations	The repeating (or combination) of the exclamation mark (!) and the question mark (?). They reflect if a tweet contains strong emotions.	*@sinceday1 http://twitpic.com/76hen - No!!!! Dnt eat it we got 2 eat healthy remember?!? Smh!! LOL*	The author doesn't want @sinceday1 to eat candies (the URL) for health reason, and he shake his head. The punctuations enhance his negative opinion of the candies.
Repeating letters	One letter or a group of letters[a]repeat within a real word, because of the author's excitement when typing the word.	*YAY. Presentation done. @Brockaldersley is the **besttt**. @BindinDTP thinks he's **soooo** cool.* ***Hummm hummm** hum.*	The author repeats letters in the words "best", "so" and "hum" to emphasize his praise on the user @Brockaldersley.

[a] Most real words contain less than three consecutive same letters [10], so we set the minimal repeated times to be three. The repeating patterns are removed to recover the word's origin; hence "lololol" and "looooool" are all reduced to "lol".

Table 2. Word origins of the most frequent (more than 0.1%) repeating letters words

Rank	Origin	Frequency	POS[a]	Example	Explanation
1	ha	36,110	Excl	haha	Laughing
2	so	17,232	Adv	sooo	For emphasizing
3	ah	10,618	Excl	ahhh	Surprise, pleasure, etc.
4	hm	8,025	Onmt	hmm	Thinking or pondering
5	aw	7,572	Excl	awww	Entreaty, commiseration, etc.
6	oh	5,991	Excl	ohhh	Surprise, anger, etc.
7	m	5,370	Onmt	mmm	Similar as hmm

[a] Part-of-speech: Excl: Exclamation. Adv: Adverb. Onmt: Onomatopoeia.

For a simple classification of English and non-English tweets, we examine the Unicode of the characters in each tweet. If all the characters in one tweet are from the Basic Latin or symbols section, the tweet is called a *Basic Latin* tweet. We find that most of them are in English. If some characters are in the Latin extended section, it is called an *Extended Latin* tweet. These tweets are often in Portuguese, Spanish, German, etc. Tweets containing characters beyond these sections (such as Chinese) are not studied in this paper.

From the SNAP dataset of more than 400 million tweets, we uniformly sample five million tweets. Among them, 1,649,503 (33.0%) tweets contain emotion tokens. Their proportions in each character set are shown in Table 3. Shown in Table 4, each tweet with emotion tokens contains about 1.4 tokens. There are about 0.47 emotion tokens (either of the three types) per tweet.

Table 5 lists how many tweets contain such types of emotion tokens. Many types of emotion tokens co-occur in tweets. This is the basic idea of our propagation algorithm. For example,

> @xClaire_Cullenx LOVE YOU TOO!!! *gives party hat* **xxxxx <3**
> I'm in a slplendid mood right now =**D**

The misspelled word "splendid" may lead to a missing of the emotion in semantic-based methods, but the emotion tokens help us identify its sentiment.

Since the tweets are rich of co-occurred emotion tokens, a propagation algorithm based on the co-occurrence can be applied to label the polarities.

4 Multilingual Sentiment Lexicon Construction Based on Graph Propagation

As mentioned in Section 2, sentiment lexicons are commonly used for sentiment analysis. Previous studies take semantic links (WordNet relations, conjunctions, etc.) to build such lexicons [1]. The emotion tokens, however, do not have semantic links between each other. Considering the frequent emotion tokens in tweets, the co-occurred tokens are likely to have similar sentiment. Thus the co-occurrences between words are links for constructing a graph. Then a few initial seeds are used to propagate and discover new tokens.

4.1 Co-occurrence Graph Construction of Emotion Tokens

An undirected graph $G = (V, E)$ is constructed to represent the links of words. Each node $v \in V$ is a word, while each edge $(v_i, v_j) \in E$ represents a co-

Table 3. Proportion of tweets with emotion tokens in different character sets

	Basic Latin	Extended Latin
Total tweets	4,536,590 (90.7%)	266,831 (5.4%)
Tweets w/ emotion tokens	1,494,499 (29.9%)	115,025 (2.3%)

Table 4. Avg. number of emotion tokens per tweet (tweets with tokens)

Token type	Basic Latin	Extended Latin
Emot.Symb.	0.79	1.00
Rept.Punc.	0.32	0.23
Rept.Ltr.	0.25	0.20

Table 5. Percentage of each type of emotion tokens in tweets (with tokens)

Token type	Basic Latin	Extended Latin
Emot.Symb.	65.9%	78.1%
Rept.Punc.	26.8%	19.4%
Rept.Ltr.	21.0%	16.9%

Tweets	Emotion tokens	Normal words
ˆ_ˆ yes we did ! RT @JetLife24_7 Me and @tinyy_tee had some good times last summer :)	ˆ_ˆ :)	good, ...
@JASMINEVILLEGAS Pretty <3 , How are you ? . I wanna see you in the #MyWorldTour on S.America :) . Love you	<3 :)	love, ...
That's A Good Boyfriend(: RT @CallMeYoshi : I Rather Stay In With My Girlfriend All Night Than Go Out To Party I Love Her To Much <3	(: <3	good, love, ...
I can't wait for high-school (: gonna be back with my friends <3 on top of that my girls @_BRILove and @_THATSLEX are gonna be there too :)	(: <3 :)	...

(a) Example tweets for building the graphs

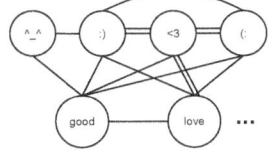

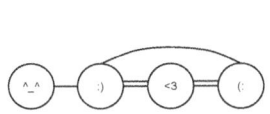

(b) Using emotion tokens only

(c) Using emotion tokens and normal words

Fig. 1. Example of co-occurrence graphs construction

occurrence between the two words v_i and v_j. The weight w_{ij} of edge (v_i, v_j) is the count of co-occurrence between v_i and v_j. This co-occurrence matrix W (i.e. the adjacent matrix of G) is symmetric. Each diagonal element w_{ii} of the matrix is the frequency of the corresponding v_i in the corpus.

A direct idea is to build such a graph with only the emotion tokens. However, the lexicon built with this graph does not contain any normal words; it could not deal with tweets without any emotion tokens. Therefore, we build the graph on both emotion tokens and normal words. Note that the semantic information of normal words is not considered. We show an example for the graph construction (with the four tweets) in Fig. 1. Only two normal words are shown in the figure for simplicity. Fig. 1(c) is the graph we propagate to build the sentiment lexicon.

To illustrate a clear scale of the graph, 100,000 Basic Latin and Extended Latin tweets from August, 2009 are sampled from the SNAP dataset. Both emotion tokens and normal words are extracted from them. The built graph contains 98,924 vertices (with only 10,390 emotion tokens) and 3,353,873 edges (22,515 among emotion tokens themselves, 314,186 among normal words and the rest are bridges). This undirected graph is extremely sparse (edges take only 0.08%).

4.2   The Propagation and Smoothing Algorithm

Similar to the SentiWordNet [1], we assign a positive score and a negative score to each word, which are calculated separately – the propagation starts with one seed for calculating the positive scores and one for negative scores, respectively.

A general algorithm for label propagation is used. Let x_k be the vector of scores of each word after the k-th iteration. The x_{k+1} is calculated by a co-occurrence matrix W and a bias vector $\mathbf{b}$, formally,

$$x_{k+1} = W \cdot x_k + \mathbf{b} \tag{1}$$

Normalizations in each iteration are applied after the $W \cdot x_k$ and $W \cdot x_k + \mathbf{b}$. The convergence of x_k has been proved [23]. This form of graph propagation is used in many algorithms such as Page-Rank and TrustRank. The bias vector $\mathbf{b}$ is set to the seed vector x_0 to keep the superiority of seeds. Also due to this reason, $W \cdot x_k$ is normalized before adding $\mathbf{b}$ to make them in a same scale. Since the initial x_0 is always added in each iteration, the seed token may have a much higher score than the other thousands of tokens. To smooth the scores into a reasonable scale, we add a logarithm transformation on each word followed by a normalization to the $[0,1]$ interval. The positive and negative scores are normalized separately. This method maps the scores into a natural distribution.

We choose only one seed to start the propagation (one for positive and one for negative, respectively). Since the graph contains both emotion tokens and normal words, two types of initial seeds are proposed: (1) smileys: ":)" for positive scores, ":(" for negative. (2) good/bad: "good" for positive, "bad" for negative. We build two *SentiLexicons* based on the two types of initial seeds.

With the graph built in Section 4.1, we find many of the scores (positive score minus negative score) of the emotion tokens are labeled correctly after the propagation. Many of the tokens do not have explicit emotions when judged by humans, but may contain hidden emotions brought by the context. We examine the scores by $P^+@100$ and $P^-@100$, i.e. the precision of the first 100 tokens with the largest absolute scores. Only 25% and 34% tokens have obvious emotions within the positive and negative ones, respectively. Among them, $P^+@100 = 0.92$ and $P^-@100 = 0.53$. Similarly, $P^+@200 = 0.88$, $P^-@200 = 0.56$, while $P^+@300 = 0.83$ and $P^-@300 = 0.59$. This demonstrates that the tokens are usually propagated with larger positive scores.

5   Sentiment Analysis with Emotion Tokens

The sentiment polarity of a tweet t is determined by both its positive score, $\text{score}^+(t)$ and its negative score, $\text{score}^-(t)$, shown in the equation below. Note the scores of the emotion tokens (v_e) and normal words (v_w) of t are looked up from the built *SentiLexicon*. The $\text{score}^-(t)$ is calculated similarly as $\text{score}^+(t)$.

$$\text{polarity}(t) = \begin{cases} \text{neutral} & \max\{\text{score}^+(t), \text{score}^-(t)\} \leq \theta, \text{or score}^+(t) = \text{score}^-(t) \\ \text{positive} & \text{score}^+(t) > \max\{\text{score}^-(t), \theta\} \\ \text{negative} & \text{score}^-(t) > \max\{\text{score}^+(t), \theta\} \end{cases}$$

$$\text{score}^+(t) = \alpha \sum \text{score}^+(v_e) + (1-\alpha) \sum \text{score}^+(v_n) \tag{2}$$

This model is similar as a bag-of-words model. Though simple, it does not involve any linguistic (semantic) information of the sentence. Thus it can be used for multilingual sentiment analysis without much linguistic knowledge.

6   Experiments and Discussions

6.1   Dataset

Tweets for building the *SentiLexicon* and evaluating the algorithms are sampled from the Basic Latin and Extended Latin tweets in the SNAP dataset. Over 99.99% tweets are from June 11th, 2009 to December 31st, 2009, so only the tweets within this period are considered.

During this 204 days period, we pick eight tweets per day for evaluation (no overlap with the tweets building lexicons), including English, Portuguese, Spanish and German tweets (two of each language), which are among the most popular languages in Twitter [21]. Google's Translation API is used to automatically pick out the tweets of a certain language. Each tweet is then given one of the three labels: positive, negative or neutral with two annotators. The third annotator is introduced when there is no majority. If the label is still uncertain, the tweet is discarded (it is difficult even for human judgements). We finally have 449 positive, 211 negative and 553 neutral tweets (total 1,213).

6.2   Strategies of Comparative Evaluations

1. *SentiWordNet*: The baseline method

The SentiWordNet provides positive and negative scores for senses, part-of-speech tags of English words. We use this lexicon with the strategy as referred to its website (http://sentiwordnet.isti.cnr.it/), summing up the scores of each POS tags of a word. With this strategy, the SentiWordNet provides 17,778 positive words (whose positive score is greater than its negative score), 20,350 negative and 1,565 neutral words (with non-zero scores) among 39,693 words. The positive and negative scores of a tweet is the sum of each word. Then a threshold θ is used to classify its sentiment.

2. *twitrratr*

The *twitrratr* provides two lists of positive and negative keywords. We try to match them in a tweet and count their numbers. Similarly, we determine the sentiment of the tweet by the bigger count of positive and negative words. If two numbers are equal, we consider it as a neutral tweet.

3. *twittersentiment*

We retrieve *twittersentiment*'s results of our data from its API. The authors test their method on their own dataset[12]. However, the API they provided makes it comparable for both their and our methods on our dataset.

6.3   Results and Discussions

We examine the lexicon from several aspects. Besides the evaluation on English and non-English tweets, we build the lexicons with different sizes of tweets to see if the size is "the bigger, the better". Moreover, the lexicons are built from different months' tweets in SNAP data, to examine if the lexicon built from a specific month is stable for the analysis on tweets in other months. Our results are all compared with the SentiWordNet baseline and the two websites.

Table 6. Comparative evaluations of algorithms in English tweets

Algorithm	Accuracy	Positive (166) F_1 / P / R	Negative (62) F_1 / P / R	Neutral (111) F_1 / P / R	$\bar{F}_1$
1. *SWN*, $\theta = 0.5$	51.3%	0.591 / 59.9% / 58.4%	0.429 / 38.5% / 48.4%	0.448 / 47.5% / 42.3%	0.489
2. *twitrratr*	49.3%	0.557 / 87.2% / 41.0%	0.276 / 27.9% / 27.4%	0.527 / 41.0% / 73.9%	0.454
3. *twittersentiment*	59.0%	0.648 / 78.0% / 55.4%	0.549 / 70.0% / 45.2%	0.548 / 44.2% / 72.1%	0.582
SentiLexicon	57.8%	0.642 / 65.2% / 63.3%	0.149 / 100.0% / 8.1%	0.606 / 49.7% / 77.5%	0.466

Table 7. Comparative evaluations of algorithms in non-English tweets

Algorithm	Accuracy	Positive (283) F_1 / P / R	Negative (149) F_1 / P / R	Neutral (442) F_1 / P / R	$\bar{F}_1$
1. *SWN*, $\theta = 0.5$	–	–	–	–	–
2. *twitrratr*	52.2%	0.311 / 72.7% / 19.8%	0.130 / 20.9% / 9.4%	0.659 / 52.9% / 87.3%	0.366
3. *twittersentiment*	51.0%	0.218 / 44.1% / 14.5%	0.129 / 52.4% / 7.4%	0.656 / 51.8% / 89.1%	0.334
SentiLexicon	**57.4%**	0.500 / 51.3% / 48.8%	0.123 / 76.9% / 6.7%	0.685 / 59.8% / 80.1%	**0.436**

Parameters and Seeds. The parameter α determines how much the emotion tokens influence the sentiment score, while θ determines the proportion of neutral tweets. We conduct experiments with several combinations of them (both in the $[0, 1]$ interval), based on the *SentiLexicon* built with both smileys and good/bad as seeds from 10,000 tweets in August, 2009 without loss of generality. In general, $\alpha = 1$ is better, i.e. the scores of emotion tokens are weighted with 1 while normal words are weighted with 0. This implies that it is the emotion tokens that affect the sentiment of the tweet. The θ is somehow stable among different α's. Similarly, there are no significant differences between the two types of seeds. For simplicity, we fix $\theta = 0.7$ and smileys as seeds in the following experiments.

Comparative Evaluations on Different Languages. To show our method's efficiency on multilingual tweets, we compare *SentiLexicon* (smileys as seeds, $\alpha = 1$, $\theta = 0.7$) with the three algorithms mentioned above. The lexicons are built with tweets from June to December, respectively. Since the results are similar, we only show the results built with the August tweets. The performances are compared on English tweets and non-English tweets respectively, shown in Table 6 and Table 7. The F_1, P and R under each class stand for F_1-score, Precision and Recall, respectively. The last column $\bar{F}_1$ is the average of three F_1-scores in the three classes. For the performance of *SentiWordNet*, we only list the best one with $\theta = 0.5$. Since this lexicon is for English, it should not be used for non-English sentiment analysis.

These two tables suggest that our method is efficient on multilingual tweets. Most of the F_1-scores, precisions and recalls of the *SentiLexicon* are higher than the current state-of-the-art methods. In English tweets, the recall rate on negative tweets of our method is rather low, which pull down the overall accuracy. We examine that these negative tweets do not contain many strong emotion tokens; hence we classify most of them as neural ones. Another reason is that the tokens are usually have larger positive scores. Therefore, many of the negative tweets are classified as positive ones. We find that in Twitter, there are usually more positive tweets than negative ones (e.g. 449 positive vs. 211 negative ones with our annotation). As a result, the construction of the co-occurrence

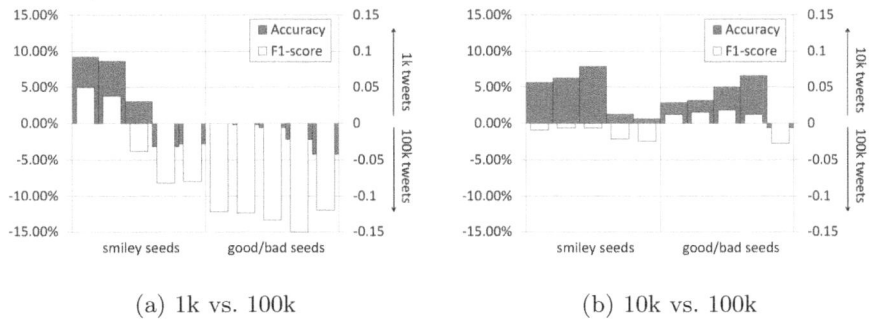

(a) 1k vs. 100k   (b) 10k vs. 100k

Fig. 2. Comparison of different datasizes for building lexicons

graph links many tokens with positive words implicitly; the propagation process assigns larger positive scores for many of the tokens. This indicates that the negative tweets may be processed differently from the positive tweets. However, in non-English tweets, our method outperforms the other methods.

Sizes of Datasets for Building Lexicons. With the tweets from August, 2009, three sizes of datasets are extracted: 1,000 (1k), 10,000 (10k) and 100,000 tweets (100k). We compare their performances on the evaluation tweets with the *SentiLexicon* built from them, and draw the differences between accuracies and average F_1-scores on the same scale to compare 1k vs. 100k (Fig. 2(a)) and 10k vs. 100k (Fig. 2(b)). We see that 10k vs. 100k is less different than 1k vs. 100k. Hence we conclude that 1,000 tweets is not sufficient to build a good lexicon, since many tokens may not even appear in such small amount of tweets. On the other hand, the lexicon built with 100,000 tweets does not perform much better than just 10,000 tweets. This finding is helpful for practical use – we do not have to build a very big lexicon. The tokens covered in 10,000 tweets are enough to build a helpful lexicon for sentiment analysis.

Stability of the Lexicons over Time. The lexicons are built with tweets from only one month, hence we propose to examine whether or not the lexicon from one month can work on future months. One strategy is to build the lexicon with tweets in the first month in the dataset (June 2009), and evaluate it on tweets in the succeeding months in the evaluation set. The other strategy is to build with each month (except the last one) and evaluate it on tweets in just the next month (e.g. use June to evaluate July tweets). We also build a lexicon with the first week (June 11th to June 17th, 2009) and evaluate it on July to December tweets. The performances of each strategy are shown in Fig. 3.

The results show the accuracies are all around 50% to 60% in each month's evaluation tweets, and the average F_1-scores are also within 0.4 to 0.5. The performances of the lexicons do not rely on tweets in a specific month or week.

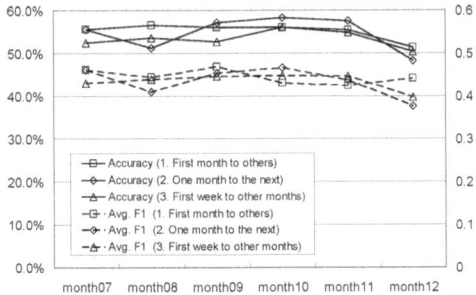

Fig. 3. Stability of *SentiLexicon* with smiley seeds, $\alpha = 1$ and $\theta = 0.7$

This infers that the lexicon is stable along with time. Therefore, we can build a lexicon with the current tweets and use it for future sentiment analysis.

7   Conclusion and Future Work

In this paper, we propose the emotion tokens to help sentiment analysis on multilingual Twitter messages. A graph propagation algorithm with a smoothing method is applied; hence the polarities of the tokens are labeled automatically based on their popular co-occurrences. With this lexicon, we perform a multilingual sentiment analysis for tweets, and achieve a better performance than traditional semantic based approach as well as several Twitter sentiment analysis websites. The comparative evaluations indicate that the emotion tokens are helpful for both English and non-English Twitter sentiment analysis, and are independent with the tweets in different time periods to build the lexicon.

There are also several technical issues we would like to address as future work, such as improving the bag-or-words model for sentiment analysis for higher accuracies, tracking Twitter's sentiment within a longer period and to discover if some tokens have opposite or weak emotions.

Acknowledgments. We would like to thank the anonymous reviewers for their valuable comments. We are deeply grateful to our annotators: Yongfeng Zhang, Qianli Xing, Yin Wang, Kuan Liu, Junwei Miao, Tong Zhu, Qian Wang, Fei Chen, Xiaoguang Wang, Bin Liang and many other participants. The NExT Search Centre is supported by the Singapore National Research Foundation & Interactive Digital Media R&D Program Office, MDA under research grant (WBS:R-252-300-001-490).

References

1. Baccianella, S., Esuli, A., Sebastiani, F.: Sentiwordnet 3.0: An enhanced lexical resource for sentiment analysis and opinion mining. In: LREC, pp. 2200–2204 (2010)

2. Banea, C., Mihalcea, R., Wiebe, J.: A bootstrapping method for building subjectivity lexicons for languages with scarce resources. In: Proc. LREC 2008 (2008)

3. Banea, C., Mihalcea, R., Wiebe, J.: Multilingual subjectivity: are more languages better? In: Proc. 23rd COLING Conference, pp. 28–36 (2010)

4. Barbosa, L., Feng, J.: Robust sentiment detection on twitter from biased and noisy data. In: Coling 2010: Posters, Beijing, China, pp. 36–44 (2010)

5. Bautin, M., Vijayarenu, L., Skiena, S.: International sentiment analysis for news and blogs. In: Proc. International Conference on Weblogs and Social Media (2008)

6. Bifet, A., Frank, E.: Sentiment Knowledge Discovery in Twitter Streaming Data. In: Pfahringer, B., Holmes, G., Hoffmann, A. (eds.) DS 2010. LNCS, vol. 6332, pp. 1–15. Springer, Heidelberg (2010)

7. Boiy, E., Moens, M.F.: A machine learning approach to sentiment analysis in multilingual web texts. Information Retrieval 12, 526–558 (2009)

8. Bollen, J., Pepe, A., Mao, H.: Modeling public mood and emotion: Twitter sentiment and socio-economic phenomena. arXiv:0911.1583 (2009)

9. Boyd-Graber, J., Resnik, P.: Holistic sentiment analysis across languages: multilingual supervised latent Dirichlet allocation. In: EMNLP 2010, pp. 45–55 (2010)

10. Brody, S., Diakopoulos, N.: Cooooooooooooooooollllllllllllllll!!!!!!!!!!!!!! using word lengthening to detect sentiment in microblogs. In: EMNLP 2011, pp. 562–570 (2011)

11. Denecke, K.: Using SentiWordNet for multilingual sentiment analysis. In: IEEE 24th International Conference on Data Engineering Workshop, pp. 507–512 (2008)

12. Go, A., Bhayani, R., Huang, L.: Twitter sentiment classification using distant supervision. Tech. rep., Stanford CS224N Project (2009)

13. Jansen, B.J., Zhang, M., Sobel, K., Chowdury, A.: Micro-blogging as online word of mouth branding. In: CHI 2009, pp. 3859–3864 (2009)

14. Jiang, L., Yu, M., Zhou, M., Liu, X., Zhao, T.: Target-dependent twitter sentiment classification. In: Proc. 49th ACL: HLT, vol. 1, pp. 151–160 (2011)

15. Krishnamurthy, B., Gill, P., Arlitt, M.: A few chirps about twitter. In: Proceedings of the First Workshop on Online Social Networks, pp. 19–24 (2008)

16. Li, Z., Zhang, M., Ma, S., Zhou, B., Sun, Y.: Automatic Extraction for Product Feature Words from Comments on the Web. In: Lee, G.G., Song, D., Lin, C.-Y., Aizawa, A., Kuriyama, K., Yoshioka, M., Sakai, T. (eds.) AIRS 2009. LNCS, vol. 5839, pp. 112–123. Springer, Heidelberg (2009)

17. Liu, B.: Sentiment analysis and subjectivity. In: Handbook of Natural Language Processing, 2nd edn. CRC Press, Taylor and Francis Group (2010)

18. Neviarouskaya, A., Prendinger, H., Ishizuka, M.: Sentiful: A lexicon for sentiment analysis. IEEE Transactions on Affective Computing 2(1), 22–36 (2011)

19. Pak, A., Paroubek, P.: Twitter as a corpus for sentiment analysis and opinion mining. In: LREC 2010 (2010)

20. Pang, B., Lee, L.: Opinion mining and sentiment analysis. Foundations and Trends in Information Retrieval 2(1-2), 1–135 (2008)

21. Semiocast: Half of messages on twitter are not in english. Tech. rep. (2010)

22. Strapparava, C., Mihalcea, R.: Learning to identify emotions in text. In: Proceedings of the 2008 ACM Symposium on Applied Computing, pp. 1556–1560 (2008)

23. Zhu, X., Ghahramani, Z.: Learning from labeled and unlabeled data with label propagation. Tech. rep., CMU-CALD-02-107 (2002)

Identifying Popular Search Goals behind Search Queries to Improve Web Search Ranking

Wang Ting-Xuan and Lu Wen-Hsiang

Computer Science and Information Engineering,
No.1, University Road, Tainan City 701, Taiwan
{p78981320,whlu}@mail.ncku.edu.tw

Abstract. Web users usually have a certain search goal before they submit a search query. However, many laypersons can't transform their search goals into suitable queries. Thus, understanding original search goals behind a query is very important for search engines. In the past decade, many researches focus on classifying search goals behind a query into different search-goal categories. In fact, there may be more than one search goal behind a certain query. We thus propose a novel Popular-Search-Goal-based Search Model to effectively identify search goals by the features extracted from search-result snippets and click-through data. Furthermore, we proposed a Search-Goal-based Ranking Model which exploits the identified search goals to re-rank the search result. The experimental result shows our proposed model can effectively identify the search goals behind a search query (achieve precision of 0.94) and enhance the search result ranking (achieve precision of 0.72 for top-1 returned snippet).

Keywords: Web search, information retrieval, user need, search goals, short query, language model.

1   Introduction

Different from classic Information Retrieval (IR) works, search engines help users to retrieve not only textual (informational) data, but also various resources from the web. However, conventional search engines usually relied on classic IR techniques like keyword matching mechanisms, and they only considered that search queries are driven by informational need. In fact, many users' search goals behind the queries are not always informational. Perhaps their search goals are navigational or transactional in the viewpoints of Broder [1]. The definitions of search goal types are given as follows:

- Informational goal is that users want to learn/know something about the query topic.
- Navigational goal is that users want to go to specific websites that users already have in mind.
- Transactional goal is that users want to obtain a resource available on web pages or to transact online.

M.V.M. Salem et al. (Eds.): AIRS 2011, LNCS 7097, pp. 250–262, 2011.

In recent years, some researchers have studied for the research area of identifying user search goal. Kang and Kim [3] try to classify the search queries based on search goals by utilizing content, link, and URL information. Lee et al. [4] proposed to use past user-click behavior and anchor-link distribution as potential features for predicting informational and navigational goal. The above two works only consider automatic classification of the two broad categories, navigational and informational goal. In this paper, we propose a novel Popular-Search-Goal-based Search Model to identify a variety of search goals behind search queries based on features extracted from search-result snippets and click-through data.

A few researchers [5,6] argued that accurate understanding and modeling of user search goals has a great benefit in applications to web search ranking, click spam detection, web search personalization, and other tasks. In this paper, we focus on improving the web search ranking, thus we proposed a Search-Goal-based Ranking Model which exploits the identified search goals to re-rank the search result. The experimental result shows our proposed model can effectively identify the search goals behind a search query and enhance the search result ranking.

There are three main contributions in this paper. First, we addressed the problem of search-result snippet classification for three search-goal categories. Second, we proposed a generative framework to identify the search goals behind a search query. Finally, we further re-rank search results based on identified search goals. The following sections consist of our idea and observation (in Section 2), the main method (in Section 3), experimental results (in Section 4), and the conclusion (in Section 5).

2   Observation and Main Idea

In general, web users often have certain implicit search goals in their minds to drive their searching behaviors before submitting queries to a search engine. Therefore, we assume that such searching behaviors can be implicitly expressed in users' minds with certain natural sentences like "I want to do something ..." Generally, a typical sentence in Chinese/English consists of a subject, a verb, and an object (SVO syntactic structure). Obviously, the subject of the implicit sentences in the user's mind is the user himself/herself, and the remaining part of verb-object (VO) pair can be used to represent the user's search goal.

Figure 1 shows an example, if a user submits a query "Google Map" to a search engine, he/she may have a certain search goal among the goal-categories including informational, navigational, and transactional. For informational goal example (Figure 1a), the user wants to learn how to use Google Map API, and thus has an informational goal "參考原始檔(refer to source code)". For navigational goal example (Figure 1b), users want to visit Google Map official web sites, and thus has a navigational goal "前往首頁(go to home page)". For transactional goal example (figure 1c), users want to download map from the web and thus has a transactional goal "下載離線地圖(download offline map)".

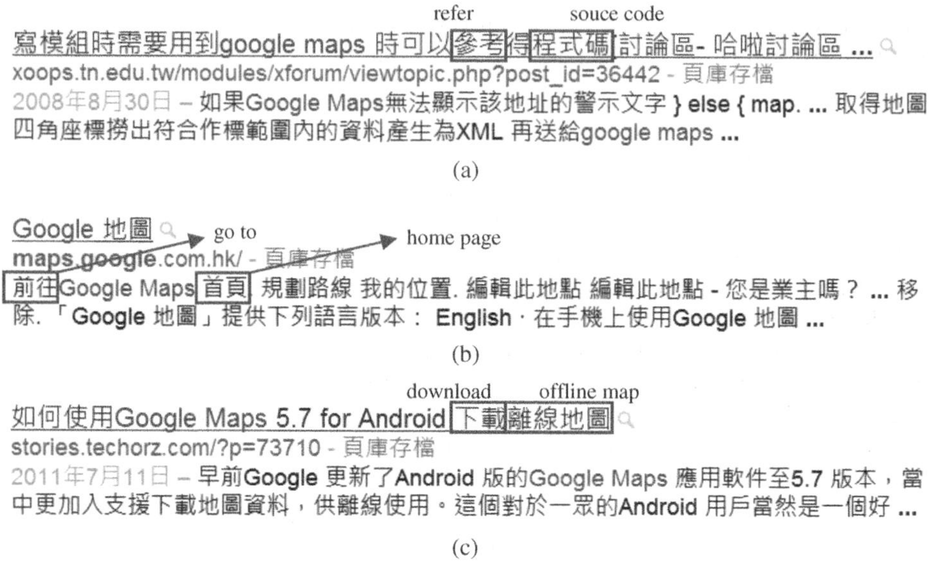

refer   souce code

寫模組時需要用到google maps 時可以 參考得程式碼 討論區- 哈啦討論區 ...

xoops.tn.edu.tw/modules/xforum/viewtopic.php?post_id=36442 - 頁庫存檔

2008年8月30日 – 如果Google Maps無法顯示該地址的警示文字 } else { map. ... 取得地圖
四角座標撈出符合作標範圍內的資料產生為XML 再送給google maps ...

(a)

Google 地圖
maps.google.com.hk/ - 頁庫存檔
go to   home page

前往Google Maps 首頁 規劃路線 我的位置. 編輯此地點 編輯此地點 - 您是業主嗎？ ... 移
除.「Google 地圖」提供下列語言版本： English · 在手機上使用Google 地圖 ...

(b)

download   offline map

如何使用Google Maps 5.7 for Android 下載離線地圖

stories.techorz.com/?p=73710 - 頁庫存檔

2011年7月11日 – 早前Google 更新了Android 版的Google Maps 應用軟件至5.7 版本，當
中更加入支援下載地圖資料，供離線使用。這個對於一眾的Android 用戶當然是一個好 ...

(c)

Fig. 1. Example of research-result snippets returned by Google with the query "Google Map" including (a) a snippet contained informational goal, (b) a snippet contained navigational goal, and (c) a snippet contained transactional goal

3 Popular-Search-Goal-Based Search Model

Figure 2 shows the framework of our proposed Popular-Search-Goal-based Search Model (PSGSM), which is proposed to automatically identify user search goals from search-result snippets and further improve the ranking of search results. First, we collect a set S of search-result snippets returned by search engines and then separate the snippets into three categories. Second, the search goal candidates will be generated from search-result snippets, and then we identify each type of search goals based on features from search-result snippet and click-through data. Finally, the identified search goals are used in re-ranking the original search result snippets.

3.1 Search-Result Snippet Classification

In this paper, we investigate whether user goal identification can be boosted by classifying search-result snippets into three types including transactional, informational, and navigational. Classifying search-result snippets can effectively increase the precision of identifying each type of popular search goals. Considering computational cost, we use a simple but efficient approach to classifying search-result snippets.

First, for navigational snippets, we observed that the URL of navigational snippets is usually the host name of a website. For example, Figure 3 illustrated a snippet about the website of "Apple Computer Inc.", and the page is in the root (e.g., in the first level directory) of the website. Moreover, we also observed that it is a good clue to determine a navigational snippet if the query terms usually appear in the title of the snippet. Figure 3 shows the query "Apple Computer" appears in the title of Apple Computer website. Therefore, a page's title and URL are considered as useful features to determine whether a snippet belongs to navigational category. The scoring function *Navig_Snippet(s)* for navigational snippets is given as follows:

$$Navig\_Snippet(s) = 2^{1-d} + Title(q, s),\qquad(1)$$

where d is the directory level (e.g., for a root page, d is 1), and $Title(q, s)$ returns 1 if the query q occurred in the title of a snippet s, otherwise, returns 0. We select the snippets which the scoring function achieved more than a threshold t_n as navigational snippets, and leave the remaining snippets in further classification.

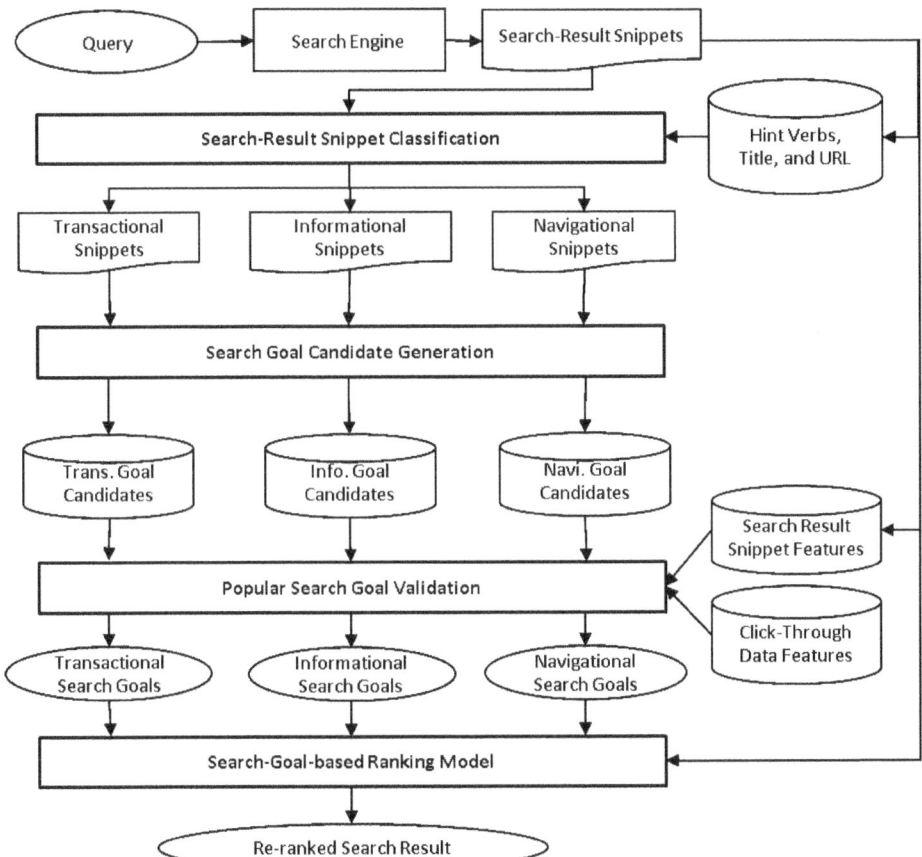

Fig. 2. The framework of Popular-Search-Goal-based Search Model

Apple Computer Inc.
Official site of **Apple Computer**, Inc.
www.**apple**.com/ - 31k - 5 Nov 2006 - Cached - Similar pages

Fig. 3. The Snippet with the URL www.apple.com of the website "Apple Computer Inc."

Second, to identify transactional snippets, we find some verbs occurred in search queries are usually refer to the transactional goals, like 下載(download), 試聽 (listen), 購買(buy), etc. We manually select 394 hint verbs H_v from 3000 high frequent queries within one-month query-log of Sogou. The scoring function $Trans\_Snippet(s)$ for transactional snippets is given as follows:

$$Trans\_Snippet(s) = |V_s \cap H_v|, \tag{2}$$

where H_v is the set of high frequent hint verbs, V_s is the set of verbs occurred in the search-result snippet s, and $|X|$ is the number of elements in set X. We select the snippets which the scoring function achieved more than a threshold t_t as transactional snippets.

Finally, for informational category, it is hard to choose any unique feature to determine an informational snippet since this category of snippets cover any topic contents in the web. Fortunately, the above two categories of identification have strong distinguishability among the three different categories of snippets. Therefore, we simply treat the remaining snippets as informational snippets.

To efficiently classify search-result snippets, we develop an algorithm which is described in Figure 4. The input is a single search-result snippet, and "I", "N", "T" are the output labels representing categories of informational, navigational, and transactional, respectively. For each snippet, we calculate the navigational snippet score ns and transactional snippet score ts. If ns is greater than threshold t_n the snippet is considered as a navigational snippet. If ts is greater than threshold t_t the snippet is considered as a transactional snippet. Otherwise, the snippet is considered as a informational snippet. To estimate t_n and t_t, we manually labeled k snippets with categories and set the values of t_n and t_t according to the percentage of snippets per category.

Algorithm 1: Search_Result_Snippet_Classification(s, q, H_v)

Input: search-result snippet s, search query q, and hint verb vocabulary set H_v.

Output: search-goal category $c \in \{I, N, T\}$

1. Compute navigational snippet score ns of s based on Equation (1).

2. If ns is greater than threshold t_n then return "N"

3. Compute transactional snippet score ts of s based on Equation (2).

4. If ts is greater than threshold t_t then return "T"

5. Return "I"

Fig. 4. Algorithm for search-result snippet classification

3.2   Search Goal Candidate Generation

To extract popular search goals from the classified search-result snippets, we need to generate goal candidates. We extract all verb-noun combinations (i.e., VO-pairs) from each search-result snippet (where nouns include unigram and bigram). We only adopt the VO-pairs whose verb-noun distance is less than w terms (the window size w is set to 3 in this work). In this paper, we only select three types of verb and three types of noun. The types of verb include active intransitive verb, active transitive verb, and active transitive verb with sentential object. The percentage of selected verbs is 92.1%. The types of noun include common noun, proper noun, and location noun. The percentage of selected nouns is 99.8%. We only extracted those three types of verb and three types of noun from search-result.

3.3   Popular Search Goal Validation

Once goal candidates are generated, then we need to determine which goal candidates are correct search goals. The correct search goals defined in this paper are not only relevant to the query, but also should be semantically suitable. To automatically detect correct search goals, we select various useful features which can be extracted from search-result snippets effectively, and then we compare three binary classifiers to classify the search-goal candidates into two categories including correct search goals and incorrect search goals.

3.3.1   The Features Used in Popular Search Goal Model

According to our observation, we found there are several useful properties in the search-result snippets and click-through data for identifying the correctness of search goal candidates. In the following, we will introduce several effective features, and the usage of features for different search-goal categories is described in Section 3.3.2.

(1)   Search-Result Snippet Features

Noun Phrase Length: According to our preliminary experiments, we found that some noun phrases can describe the complete meaning of the objects. For example, "下載手機遊戲(download cellphone game)" is more clear than "下載手機(download cellphone)" or "下載遊戲(download game)." To deal with noun phrase extraction problem, we simply extract all unigrams and bigrams as noun part in the VO-pair, and assume that longer noun phrases are probably more meaningful than shorter nouns.

$$f_{NPLength}(g) = Length(g_n), \tag{3}$$

where $Length(g_n)$ is the number of terms in g_n, and $g=(g_v, g_n)$.

VO-Distance: In general, a smaller distance between verb and noun indicates its VO combination is more semantically suitable. We found that the probability of distance d approximately obey normal distribution. Therefore, we employ normal distribution model as follows:

$$f_{VODiatance}(g) = \frac{1}{\sqrt{2\pi}\sigma} \exp\left[-\frac{1}{2\sigma^2}(d - \mu)^2\right], \tag{4}$$

where $g=(g_v, g_n)$ and d is distance between g_v and g_n, μ is the distance mean, and σ is the standard deviation of distance.

TFIDF: This feature is used to calculate the importance of search goals by using weighting mechanism Term Frequency- Inverted Document Frequency.

$$f_{TFIDF}(g) = fre(g)log\frac{N}{|D(g)|},$$ (5)

where $fre(g)$ represents frequency of search goal g in a search-result snippet, $|D(w)|$ the number of set of search-result snippet that contains search goal g, and N is the number of search-result snippets in set S. Intuitively, more frequent terms are more likely to be better candidates of search goals; while terms with higher document frequency might be less informative.

Search-Result Snippet Rank: For some search engines, top ranked-search result snippets would possibly contain popular search goals. We employ a power-law distribution to estimate the snippet rank feature function.

$$f_{SRSnippetRank}(g) = \eta_s \times r_s^{-p_s},$$ (6)

where η_s is a constant, r_s is search-result snippet rank, and p_s is the scaling parameter.

Title: If a search goal occurred in the title of a snippet, it would have more attraction for users to click the snippet. Thus, the title feature function is

$$f_{Title}(g) = \begin{cases} 1, & \text{if search goal } g \text{ appears in a snippet title} \\ 0, & \text{otherwise} \end{cases}$$ (7)

URL: For navigational search goals, we assume users intend to browse a specific web site (home page, especially). Thus, we use the directory level of URL as a feature. Here is the URL feature function

$$f_{URL}(g) = 2^{1-l},$$ (8)

(2) Click-through Data Features
Hint Verb and Hint Noun: According to our observations, we found some verbs and nouns in click-through data are more suitable to be the verb or noun parts of search goals. The feature function is given as follows:

$$f_{HintWord}(g) = \begin{cases} 1, & \text{if } g_v \text{ is a hint verb or } g_n \text{ is a hint noun} \\ 0, & \text{otherwise} \end{cases},$$ (9)

where g_v and g_n are the verb or noun parts in the VO-pair $g=(g_v, g_n)$.

User Click Order: As our observation, user would like to click the snippets which contain their search goals. We assume that the more early a user clicks a snippet, the more probable the snippet contains search goals. Thus we exploit a power-law distribution to estimate the user-click-order feature function.

$$f_{UserClickOrder}(g) = \eta_u \times r_u^{-p_u},$$ (10)

where η_u is a constant, r_u is search-result snippet rank, and p_u is the scaling parameter.

Snippet Click Count: A snippet which has been clicked many times is probably containing search goals. We use a log function to alleviate the domination of click count of popular web site.

$$f_{SnippetClickCount}(g) = \log(count(g)), \qquad (11)$$

where $count(g)$ is the click count of the snippets which contains the search goal g.

3.3.2   Search Goal Classifiers

Originally, our goal is to identify correct search goals G from search goal candidates G_c, in fact the above problem we addressed can be simplified to classify search goal candidates into two categories, including correct search goal, and incorrect search goal. That is, each search goal candidate belongs to one of two categories $C = \{TRUE, FALSE\}$. Since the properties of specific search-goals in different categories of snippets are different, we select different feature sets for each search-goal category. Table 1 shows the features used in each search-goal category. User Click Order and Snippet Click Count are used in three search-goal categories. For informational category, using TFIDF can effectively identify the query topics from search result snippets, VO-Distance can improve the semantic correctness of identified search goals, and using Noun Phrase Length to favor the noun phrases (bigram nouns). For navigational goal, URL & Title are the most important features used in this category. For transactional goals, Hint Verb & Hint Noun are the major features used in this category, and using TF (Term Frequency) to favor high-frequency terms occurred in search-result snippets.

Table 1. The features used in three different categories of search goals

	Search-Result Snippet Features	Click-Through Data Features
Informational Goals	TFIDF VO-Distance Noun Phrase Length	User Click Order Snippet Click Count
Navigational Goals	URL Title Search-Result Snippet Rank	User Click Order Snippet Click Count
Transactional Goals	TF Noun Phrase Length	Hint Verb & Hint Noun User Click Order Snippet Click Count

To deal with the problem of binary classification, we employ three different classifiers including log-linear model, support vector machine, and adaptive boost, and compare their performance among the three classifiers.

Support Vector Machine

Support vector machine, SVM, is the state-of-the-art classifier, which maps samples into a higher dimensional space [Cortes and Vapnik, 1995], and is very effective to handle the case when the relation between categories and features is nonlinear. A well-known implementation of SVM naming LIBSVM is adopted in this paper. We use the radial basis function as the kernel function

$$K(x_i, x_j) = \exp\left(-\gamma \|x_i - x_j\|^2\right) \tag{12}$$

where the capacity constant and γ are set to 0.03 and 0.125 respectively.

Log-Linear Model

Log-linear model is a far-reaching extension of logistic regression [Christensen, 1997], which maps each of features into a feature function and is modeled by conditional probability $P(c|g)$ as follows:

$$P(c|g; w) = \frac{\exp\left(\sum_{i=1}^{|F|} w_i f_i(g, c)\right)}{\sum_{c \in C} \exp\left(\sum_{j=1}^{|F|} w_j f_j(g, c)\right)} \tag{13}$$

where c is the category of candidate search goal g (i.e., $c \in C$), F is the set of features, $|F|$ is the number of features, w_i is a feature weight parameter, and $f_i(g, c)$ can be estimated by the feature functions mentioned in Section 3.3.1 when given the category c.

AdaBoost

AdaBoost (short for Adaptive Boosting) is a machine learning algorithm formulated by Freund and Schapire [1995]. AdaBoost can deal with the binary classification task by adapting in the sense that subsequent classifiers built are tweaked in favor of those instances misclassified by previous classifiers. Although, AdaBoost is sensitive to noisy data and outliers, it can be less susceptible to the overfitting problem. In this paper, we use decision stumps as the weak learners. The simple algorithm is as follows:

(1) Computes classification score of weak learner f_t (in this paper, we use the feature functions mentioned in Section 3.3.1).

(2) Multiplies by learned confidence value α_t.

(3) Sums over T rounds.

(4) Compares sum to zero (return the discrete classification value +1 or -1).

Thus the decision function is given as follows:

$$H(g) = sign\left(\sum_{t=1}^{T} \alpha_t f_t(g)\right).$$

3.4   Search-Goal-Based Ranking Model

One of most important application of detecting search goal is to improve the search-result snippet ranking. We propose a Search-Goal-based Ranking Model (SGRM) to rank the search-result snippet s for a given query q as follows:

$$P(s|q) = \sum_{g \in G} P(g|q) P(s|g, q) \tag{14}$$

$P(g|q)$ is the probability of search goals in top t search result snippets with the query q.

$$P(g|q) = \sum_{c \in \{I,N,T\}} P(c|q) P(g|c,q), \qquad (15)$$

where c is the category of search goals, including informational I, navigational N transactional T. $P(c|q)$ is the probability of the query q belongs to the search goal category c. The equation is given as follows:

$$P(c|q) = \frac{|S_c|}{|S|}, \qquad (16)$$

where $|S|$ is the count of search-result snippets we crawled from search engines with the search query q, $|S_c|$ is the count of search-result snippets in the goal-category c and $S_c \in S$.

$$P(g|c,q) = \frac{count(g_i)}{\sum_{g_i \in G_c} count(g_i)}, \qquad (17)$$

where $count(g)$ is the occurrence count of the search goal g in all search result snippets with the query q, and G_c is the set of correct search goals of the category c.

$P(s|g,q)$ represents the probability that a user will click on the snippet s in the set S of returned search results given a query q with a user goal g. It is estimated as follows.

$$P(s|g,q) = \frac{gcount(s,g)}{\sum_{s_i \in S} gcount(s_i,g)}, \qquad (18)$$

where $gcount(s,g)$ is the occurrence count of the search goal g in the search result snippet s.

4   Performance Evaluation

To realize the performance of our proposed Search Goal Detection Model, we conducted a series of experiments. First, we need to verify the accuracy of search-result snippet classification. Second, three classifiers were compared against three different search goal types. Finally, we compared our Search-Goal-based Ranking Model with the Query-Expansion-based Ranking Model.

4.1   Experimental Setup

4.1.1   Dataset
We selected 1200 top-frequency queries (named top-1200 query set) from the one-month query log (which contains 21,422,773 records and 3,167,170 distinct queries) of Sogou search engine in 2006. For each query, we crawl top 100 snippets from Sogou search engine, and generate about 200-300 search goal candidates for each query in average. We employed three judges to label "TRUE" or "FALSE" for each search goal candidate. For each judge, only the queries and search goal candidates are observable, and they label the search goal candidate based on the guideline "Label a search goal candidate with "TRUE" only when the search goal is relevant to the query

and semantically suitable." For example, given the query "Youtube," a judge labeled search goals with "TRUE" such as "看影片(watch video)." On the other hand, a judge labeled search goals with "FALSE" such as "購買手機(buy mobile phone)" (since it is not relevant to the query) and "聽電影(listen movie)" (since it is not semantically suitable). We only consider a search goal candidate as a "correct search goal" only if that there are at least two judges label the search goal with "TRUE". In average, there are 23.2 correct search goals for each query.

4.1.2   Evaluation Metrics

We use overall precision, recall, and F1 as the metrics to evaluate the performance of three compared classifiers. Recall evaluation is very difficult in this work since we do not really understand the total number of relevant search goals behind a search query. Therefore, we consider all correct search goals occurred in crawled search-result snippets as the total relevant search goals. To evaluating the performance of two compared search result ranking models, we employ the metrics of top-t NDCG, precision, and recall.

4.2   Experimental Result

4.2.1   Performance of Snippet Classification

To determine the effectiveness of search-result snippet classification, we employed three judges to manually classify the search-result snippets into three categories, including informational, navigational, and transactional. For this preliminary experiment, we use 5000 search-result snippets returned by 50 queries which were randomly selected from our top-1200 query set (100 snippets for each query) as our testing data. The 5000 snippets consist of 2612 informational snippets, 1493 navigational snippets, and 895 transactional snippets. Table 2 shows the results of search-result snippet classification, and each column shows the numbers of snippets classified to a certain category by our proposed algorithm (see Figure 4). The overall accuracy is 0.94 ((2543+1322+837)/5000), which is adequate to be exploited in the following experiments.

Table 2. The overall results of classified categories over real (human labeled) categories

	Informational	Navigational	Transactional
Classified Info.	**2543**	45	12
Classified Navi.	7	**1322**	46
Classified Trans.	62	117	**837**

4.2.2   Performance of Popular Search Goal Validation

To enhance the quality of search goals, we use binary classifiers to validate the correctness of our extracted search goal candidates. We compare the performance of three state-of the-art classifiers for each goal-category over the top-1200 query set under the measure of 5-fold cross validation. The classifiers we evaluated including

Support Vector Machine (SVM), Log-Linear Model (LLM), and AdaBoost (AB). Table 3 shows the performance comparison of identified popular search goals of three classifiers over three different search goal categories. The SVM achieved best F1 measure of 0.74 in average, but LLM achieved best recall of 0.654 in average. Therefore, we use the SVM classifier to detect correct search goals for Search-Goal-based Ranking Model in next experiment.

Table 3. The overall results of three classifiers over different search goal types

	SVM			LLM			AB		
	Prec.	**Rec.**	**F1**	**Prec.**	**Rec.**	**F1**	**Prec.**	**Rec.**	**F1**
Info.	0.927	0.613	**0.734**	0.901	0.611	0.728	0.865	0.607	0.713
Navi.	0.832	0.667	**0.736**	0.808	0.668	0.731	0.782	0.581	0.667
Tran.	0.847	0.674	**0.749**	0.831	0.682	0.749	0.812	0.623	0.705
Aver.	0.865	0.648	**0.740**	0.847	**0.654**	0.736	0.820	0.604	0.695

4.2.3   Performance of Search-Goal-Based Ranking Model

In order to evaluate the performance of our Search-Goal-based Ranking Model (SGRM), we employ a pseudo-relevance feedback ranking method based on Okapi BM25 Query Expansion (QE) as the baseline. We selected top 10 search result snippets as "relevant snippets", and utilized top 2 expanded terms with original query as a new expanded query to re-rank the original search results.

Table 4 shows the top-k NDCG and precision of SGRM and QE. Obviously, SGRM outperforms QE. The reason is that, although they are both based on non-explicit relevance feedback approach, the SGRM adopts more useful keywords (i.e., search goals) contained in search-result snippets instead of considering query expansion based weighting.

Table 4. The top-k NDCG and precision of SGRM and QE

Top-t snippets	SGRM		QE	
	NDCG	**Precision**	**NDCG**	**Precision**
1	**72.81%**	**72.81%**	45.44%	45.44%
3	**78.72%**	**55.62%**	53.42%	27.86%
5	**82.64%**	**43.86%**	59.34%	22.78%
10	**85.75%**	**29.01%**	63.48%	12.02%

5   Conclusion

We have presented a novel approach to effectively extract popular search goals based on features of search-result snippet features and click-through data features. The experimental results show that our proposed approach is effective for suitable search goal identification and further improve the search-result ranking.

References

1. Broder, A.: A taxonomy of web search. SIGIR Forum 36(2) (2002)
2. Rose, D.E., Levinson, D.: Understanding User Goals in Web Search. In: Proceedings of the 13th International Conference on World Wide Web, pp. 13–19 (2004)
3. Kang, I.H., Kim, G.: Query type classification for web document retrieval. In: Proceedings of the 26th Annual International ACM SIGIR Conference on Research and Development in Information Retrieval (2003)
4. Lee, U., Liu, Z., Cho, J.: Automatic Identification of User Goals in Web Search. In: Proceedings of the 14th International Conference on World Wide Web, pp. 391–400 (2005)
5. Joachims, T., Granka, L., Pang, B., Hembrooke, H., Gay, G.: Accurately Interpreting Click-through Data as Implicit Feedback. In: Proceeding of the ACM SIGIR Conference on Research and Development on Information Retrieval (2005)
6. Agichtein, E., Brill, E., Dumais, S.: Improving Web Search Ranking by Incorporating User Behavior. In: Proceedings of the ACM Conference on Research and Development on Information Retrieval (2006)

A Novel Crawling Algorithm for Web Pages

Mohammad Amin Golshani, Vali Derhami, and AliMohammad ZarehBidoki

Department of Electrical and Computer Engineering, Yazd University, Yazd, Iran
Golshani.ma@stu.yazduni.ac.ir,
{Vderhami,AliZareh}@yazduni.ac.ir

Abstract. Crawler is a main component of search engines. In search engines, crawler part is responsible for discovering and downloading web pages. No search engine can cover whole of the web, thus it has to focus on the most valuable web pages. Several Crawling algorithms like PageRank, OPIC and FICA have been proposed, but they have low throughput. To overcome the problem, we propose a new crawling algorithm, called FICA+ which is easy to implement. In FICA+, importances of pages are determined based on the logarithmic distance and weight of the incoming links. To evaluate FICA+ we use web graph of university of California, Berkeley. Experimental result shows that our algorithm outperforms other crawling algorithms in discovering highly important pages.

Keywords: World Wide Web, Search engines, Web crawling, Web Graph, Hot pages.

1 Introduction

Nowadays, World Wide Web is the best environment for producing information, publication and accessing to the required data. However, the web is a highly dynamic environment which is growing fast in content and developing fast in the structure [18]. New pages are built, old pages are deleted and links are changed, all at a high rate. Each week, 320 million new pages are added to the web graph. Every year, 20 percent of the web pages of today will disappear and 50 percent of all contents will be changed. The web graph (link structure) will change even faster: about 80 percent of all links (web graph) will have changed or will be new within a year [12]. The results show how important it is for a search engine to find important pages faster (earlier) in this vast dynamic environment.

One of the main components of the search engines is crawler. The crawler is a program for the bulk downloading of the web pages. The aim of the crawling process is to retrieve whole of the web content. Over the time frame of crawler development, the web has been growing rapidly, and so crawlers need to operate efficiently and effectively. Most crawlers will not be able to visit every possible page for three main reasons:

- The network bandwidth is expensive [2].
- The crawlers may have limited storage capacity, so it is reasonable to expect that most crawlers will not be able to cope with all data [5].

M.V.M. Salem et al. (Eds.): AIRS 2011, LNCS 7097, pp. 263–272, 2011.

- Crawling takes time, so at some point the crawler should revisit previously scanned pages to prevent index inconsistency [5].

The Crawling algorithm usually uses a ranking mechanism to calculate the importance of pages as a crawling priority. In this paper, we propose a new method based on FICA (Fast Intelligent Crawling Algorithm) [19], called FICA+, which has higher performance than the previous algorithms. It acts based on the connection links between web pages.

The remainder of this paper is structured as follows: The next section reviews the background and related work. In Section 3, we introduce our algorithm, FICA+. Experimental analysis and comparison to some of the well-known algorithms are given in Section 4, and finally our conclusion and future work of research are presented in section 5.

2 Background and Related Work

Web crawlers have been studied since the advent of the web. Nowadays, crawling algorithms are the subject of extensive research. These studies can be categorized into one of the following topics [5]: Crawler architecture, page selection, page update (freshness), and change frequency estimation for web pages. This paper is placed in the page selection category.

By retrieving important pages earlier, a crawler can improve the quality of the downloaded pages. Methods based on link analysis have been widely used to calculate the page importance such as HITS [11] and PageRank [15]. In the following some of the well-known algorithms are considered.

PageRank is a popular ranking algorithm used by Google which measures the importance of web pages. PageRank weights each link based on the importance of the document from which it originates and the number of outlinks in the origin document. It models the users' browsing behaviors as a random surfer model [3][16]. In this model a person who surfs the web by randomly clicking links on the visited pages. When she (PageRank) reaches to a web page that does not have any outward link, she will randomly jump to another page. PageRank assumes that a user either follows a link from the current page or jumps to a random page on the web graph. The rank of page j is then computed by following equation:

$$r(j) = \frac{1-d}{n} + d * \sum_{i \in B(j)} r(i)/o(i) \qquad (1)$$

where n is the number of web pages and O(i) denotes the number of outgoing links from page i, and B(j) shows the set of pages that point to page j. Parameter d, damping factor, is used to guarantee the convergence of PageRank and remove the effects of sink pages-pages with no outputs.

There is a similar work that a new metric called RankMass has been proposed to find highly important pages [7]. The RankMass metric is based on commonly used variations of PageRank such as Personalized PageRank [10] and TrustRank [9] which assumes that users' random jumps are limited to a set of specified pages.

In [14], breadth-first algorithm is used as a crawling algorithm. They examined the average quality of downloaded pages during a web crawling of 328 million unique pages and connectivity-based metric PageRank was used to measures the quality of downloaded pages.

In [6], a comparison between some crawling algorithms including PageRank, Backlink count and breadth-first has been done. It was found that the crawling based on PageRank finds the hot (important) pages earlier than others.

Abiteboul, Preda and Cobena [1] proposed an algorithm called OPIC, to find the importance of pages online in the crawling process. In their method, each page has a value called cash. Initially all pages have the same cash equal to 1/n (n is the number of web pages). The crawler will download web pages with the higher cash and when a page is downloaded its cash will be distributed among the pages it points to. In this method, each page will be downloaded many times leading to increasing crawling time. Unfortunately, the experiments were done on a synthetic web graph including at most 600,000 nodes with the power law distribution. There is no comparison between OPIC and other crawling strategies.

A site-based method named largest site first has been proposed [4]. In this method the sites with the larger number of pending pages have higher priority for crawling. It is found that this algorithm is better than the breadth-first method.

A crawling algorithm has been proposed to schedule web pages for (re)downloading into a search engine repository [16]. The objective of the algorithm is to maintain the freshness of the search engine's index based on a quality metric using users' experiences.

Dasgupta, Ghosh and Kumar et al. [8] proposed a new crawling algorithm in order to discover newly-arrived content on the web. They measured the overhead of discovering new content, defined as the average number of fetches required to discover a new page. They showed that with perfect foreknowledge of where to explore for links to new content, it is possible to discover 90 percent of all new content with under 3 percent overhead and 100 percent of new content with 9 percent overhead.

ZarehBidoki and Yazdani proposed an intelligent crawling algorithm based on reinforcement learning, called FICA [19]. Our algorithm is based on FICA, so we explain it in more details in the following subsection.

2.1   FICA

FICA is an intelligent crawling algorithm that models a random surfer user. FICA acts like breadth-first method, but with a new definition of distance between web pages [13].

FICA models a user browsing the web. In the initial stages of the crawling process, she (FICA) does not have any background (knowledge) about the web pages and selects them only by current status and over time her knowledge gradually increases. Over time, she learns more and more, with accumulating knowledge from the web environment she learns which page is more important than others and improves her selections.

FICA defines two new metrics, the first is the link weight and the second is the logarithmic distance:

- Definition 1. Link weight: if page i points to page j then the weight of the link between i and j equals to $\log_{10}O(i)$ where $O(i)$ denotes i's out-degree.

- Definition 2. Logarithmic distance: the distance between pages i and j is the weight of shortest path (the path with the minimum value) or sum of link weights in the shortest path from i to j. They denoted it with d_{ij}. Also they denoted the logarithmic distance between the root and page i with d_i.

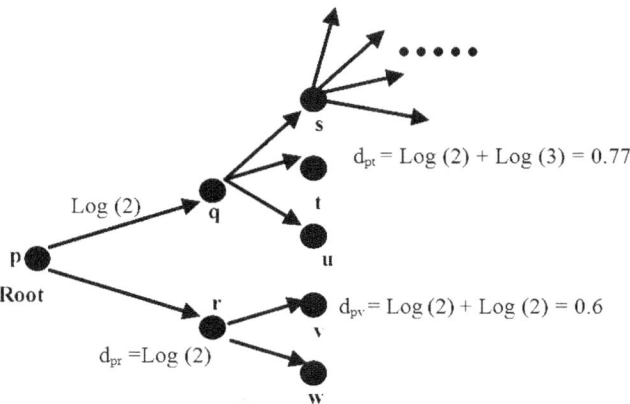

d_{pt} = Log (2) + Log (3) = 0.77

d_{pv} = Log (2) + Log (2) = 0.6

d_{pr} =Log (2)

Fig. 1. Logarithmic distance in the crawling tree [19]

For example, in Fig. 1, the weight of outward links in pages p, q and s are equal to log2, log3 and log4, respectively. The distance between p and t is log2+log3 and between p and v is log2+log2. Thus, whereas both t and v are the same number of links away from p (two clicks), v is closer to p in terms of logarithmic distance ($d_{pv}<d_{pt}$) [19].

If a crawled page i has distance d_i from root page, by using Definition 2 the distance of each of its child nodes is computed as follows:

$$d_j = \log(O(i)) + d_i \qquad (2)$$

If Eq. (2) was used as the selection criteria, after passing several iteration, the values of $\log(O(i))$ and d_i are not comparable and almost the effect of current link's weight will be lost. So they proposed the following formula which is similar to the reinforcement learning algorithm [17].

$$d_j = \log(O(i)) + \gamma * d_i, 0 \le \gamma \le 1 \qquad (3)$$

The distance factor γ is used to regulate the effects of parent nodes. For example, if there is a path like $i \rightarrow k \rightarrow l \rightarrow j$, then the effect of distance of i on j is γ^3. Eq. (4) shows the main formula of FICA which is based on reinforcement learning [17].

$$d_{j_{t+1}} = (1-\alpha) * d_{j_t} + \alpha * (\log(O(i)) + \gamma * d_{i_t}) \qquad (4)$$

$$i \in B(j), \ 0 < \alpha \le 1, \ 0 \le \gamma \le 1$$

Where α is the learning rate that is modelled in Eq. (5) and $\log(O(i))$ is the instantaneous punishment that the crawler receives in transition from i to j, the old distances, d_{j_t} and d_{i_t} show the distance values of pages j and i in time t respectively and $d_{j_{t+1}}$ is the new distance of page j at time t+1. The aim of FICA is to decrease the sum of the punishments (distances) received from the web environment.

$$\alpha = e^{-\beta * t} \tag{5}$$

In Eq. (5) t shows time and β is a static value to control learning rate, α. Initially, the agent has little knowledge about the web pages (environment), hence $\alpha = 1$, as she visits more pages, she slowly learns from environment (α decreases).

Eq. (4) used when we reach page j for the first time. In situations that a page has different parents, the parent which produces the least distance has been chosen. Thus Eq. (4) changes to Eq. (6) which is based on Q-learning [19].

$$d_{j_{t+1}} = (1 - \alpha) * d_{j_t} + \alpha * \min_i (\log(O(i)) + \gamma * d_{i_t}), \; i \in B(j) \tag{6}$$

where β and γ were set to 0.1 and 0.5, respectively. Although FICA proposes a well-defined method for crawling, but it has two major weaknesses.

FICA weaknesses. FICA has two main problems:

- Problem 1. FICA is only dependent on out-degree of web pages. As Eq. (6) shows the main operand in FICA is O(i).
- Problem 2. In initial stages of the crawling process the crawler agent has little knowledge about web environment, and it is possible it makes wrong choices in her decisions.

Suppose page j currently has the distance 2.3 and has been reached directly through page i (Fig. 2), the crawler then finds page j for the second time through page k.

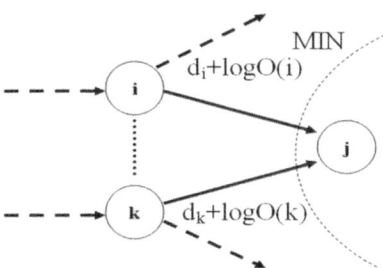

Fig. 2. Logarithmic distance for page j [19]

Suppose page k has the distance 0.4 and an out-degree of 10. Then the distance of page j through page k is calculated as 1.4. Since this number is smaller than 2.3, it will be chosen as the new distance value for page j [19]. In fact, FICA chooses a parent which produces the least distance. This paper is the extended version of FICA [19], called FICA+. In this version we eliminate the defects of previous version and tried to cover all of the weaknesses of the old version.

3   Proposed Algorithm

At first we explain an interesting feature of the breadth-first algorithm which has an important role in developing our new algorithm. The breadth-first crawling algorithm traverses the graph by following its links. The distance of each crawled page from the root (seed) is always less than or equal to that of the uncrawled pages. The breadth-first ordering is not the best method for crawling [6], but breadth-first has an interesting feature. It can discover pages with high PageRank in the initial stages of the crawling process. Because important pages have many backlinks from different web sites, so in the breadth-first algorithm high-quality pages have more chance to be crawled, in [14], this issue has been proven by Najork and Wiener. They examined the average page quality during a web crawl of 328 million unique pages. They used the PageRank metric as benchmark. Fig. 3 shows the average PageRank (unnormalized) of all downloaded pages on each day of the crawl.

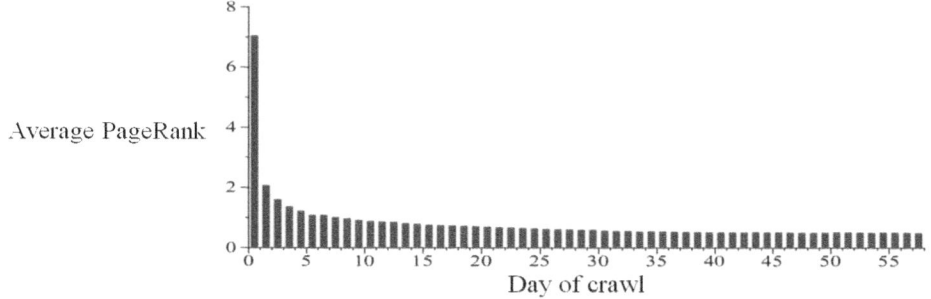

Fig. 3. Average PageRank score by day of crawl [14]

The average score of crawled pages on the first day is 7.04 which is more than three times the average score of 2.07 for crawled pages on the second day. The average score of pages decrease to 1.08 on the first week, then to 0.84 after the second week, and to 0.59 after the fourth week.

According to Figure 3, we can say that the breadth-first method downloads the hot pages in the first day of the crawling process. It happens due to the large number of links from others pages to these pages, then the average quality of pages decrease gradually. We use this feature of breadth-first in our algorithm.

3.1   FICA+

As was mentioned, in the initial stages of the crawling process the crawler agent has little knowledge about web environment, and it is possible it makes wrong choices in her decision. Furthermore, it is based on outward links. To solve these problems, we propose a new crawling algorithm, called FICA+ which is based on FICA, Backlink count, and the feature of the Breadth-first method. We introduce Eq. (7) as a new distance formula.

$$d_{j_{t+1}} = (1-\delta_t)*[(1-\alpha)*d_{j_t} + \alpha*\min_i(\log(O(i)) + \gamma*d_{i_t})] - \delta_t \qquad (7)$$

In other words:

$$d_{j_{t+1}} = (1-\delta_t)*(\text{ the gained knowledge by the crawler })\qquad (8)$$

$$-(\text{ the weight of each incoming link which}$$

$$\text{depends on the percentage of crawled pages })$$

Where $i \in B(j)$ and δ_t is balancing factor in time t. We use δ_t to create balancing between knowledge gained by the crawler agent and the web structure feature. Balancing factor would help her (crawler agent) in her page selection policy, especially in the initial stages of the crawling process. So, in the early stages of the crawling process that crawler agent does not have any background about web structure, $\alpha = 1$, and δ_t has its maximum value. Over time as it accumulates more knowledge about environment, δ_t will reduce linearly. In fact, FICA+ in initial stages of the crawling process tends to the Backlink count method- to utilize the feature of the Breadth-first method. In the final stages of the crawling process, balancing factor approaches to zero- because in the final stages she has almost compelete knowledge about web environment. Experimentally, we found if the initial value of δ_t was in the range of [0.35, 0.45], the algorithm has high throughput, and at the end of crawling process it will reach $\varepsilon \approx 0.01$. In our experiments, we model the balancing factor, δ_t as in Eq. (9):

$$\delta_t = -0.39 * \text{Percentage of crawled web pages} + 0.4 \qquad (9)$$

4 Experimental Result

In this section, we report the result of our algorithm evaluation. For evaluation of FICA+, we used web graph of university of California, Berkeley. Our goal was to compare FICA+ with other crawling algorithms, and see which one of them finds more important pages (high PageRank) faster. We compare our algorithm with the following Crawling algorithms:

- Breadth-first: The crawling process is done in the breadth-first order. Initially, the algorithm starts with some starting URLs as the roots of the crawling tree.
- Backlink count: In this algorithm, pages with more input links are crawled first [13], that is, pages with more input links have higher ranks.
- Partial PageRank: This method uses the PageRank algorithm [7] on the web pages seen so far and crawls the pages with higher PageRank first.

- OPIC: In this algorithm, all pages start with the same amount of cash [14]. Every time a page is crawled, its cash is distributed to its outward links. In each step the next page for crawling is the one with the highest amount of cash up to now.
- FICA: It is an intelligent crawling algorithm based on reinforcement learning that models a random surfer user. The priority for crawling pages is based on a new concept, called logarithmic distance, web pages with low logarithmic distance are more important than others.

Initially, we start crawling the web with every algorithm with some starting URLs (5000 URLs as seed URLs). Every time by crawling k new web pages (k is set to 125,000 web pages), we run one of the above ranking algorithms. Afterward, we sort the web pages in the queue according to the produced ranking. This process continues until a specified portion of the web is crawled. All methods will be run in this way with their own ranking criteria [6]. Unlike other ranking algorithms, FICA+, FICA and OPIC are scheduling algorithms and they do not require an additional ranking stage.

The aim of the crawling is to find hot pages. To do this, we choose the PageRank algorithm as benchmark. First, the ranks of all web pages are computed using PageRank algorithm on the entire graph. In a set of K pages, gathered from a running algorithm, a page is hot if it exists in the first K hot pages of the benchmark ranking. Clearly, the algorithm that retrieves the most hot pages will be better than others. We define throughput at each step as a fraction of crawled hot pages to all hot pages that can be discovered.

We compared the aforementioned algorithms with FICA+ in Figure 4 on the web graph of Berkeley University. The damping factor of PageRank, β, γ were set to 0.85, 0.1, and 0.5, respectively. As the following figure shows, FICA+ outperforms all other algorithms in the tested web graph. For example, in Figure 4, when 25 percent of pages are crawled, FICA+ finds about 54 percent of hot pages whereas partial PageRank, FICA and OPIC find 49 percent, 45 percent, and 50 percent of hot pages, respectively. In comparison to PageRank, FICA and OPIC, FICA+ exhibits an *6.0 percent* increase in the average throughput.

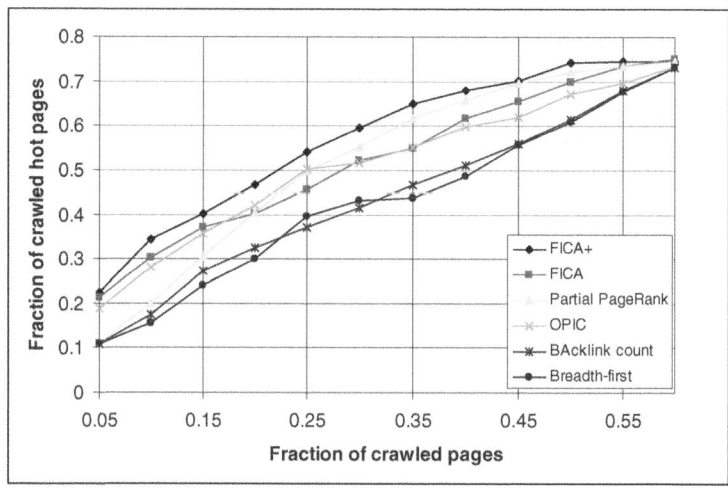

Fig. 4. University of California, Berkeley- 2,500,000 web pages

5   Conclusion and Future Work

In this paper we proposed a new crawling algorithm called "FICA+". This algorithm selects each page based on its background knowledge from visited pages and the feature of web environment. In fact, it makes a relationship between the obtained knowledge by the crawler agent and the web structure feature. For evaluation of FICA+ we used the web graph of Berkeley university.

The result shows FICA+ is an efficient crawling algorithm for web pages. The main contribution of the paper is an efficient crawling algorithm that finds hot pages faster (earlier) than previous algorithms. FICA+ does not need to save the matrix of web graph and only a vector of web graph nodes for saving the distances of pages is enough.

There are two directions in which we would like to extend this work. One direction is to execute FICA+ as a crawling algorithm on a dynamic web graph and the second direction is to evaluate FICA+ as a ranking algorithm.

Acknowledgments. This work has been partially supported by Iran Telecommunication Research Center (ITRC).

References

1. Abiteboul, S., Preda, M., Cobena, G.: Adaptive On-Line Page Importance Computation. In: 12th International Conference on World Wide Web, pp. 280–290. ACM Press (2003)
2. Ali, H.: Effective Web Crawlers. Phd thesis, School of Computer Science and Information Technology, Science, Engineering, and Technology Portfolio, Melbourne, Victoria, Aus
3. Brin, S., Page, L.: The anatomy of a large-scale hypertextual Web search engine. In: 7th International Conference on World Wide Web. Elsevier Science (1998)
4. Castillo, C., Marin, M., Rodríguez, A., Baeza-Yates, R.: Scheduling algorithms for Web crawling. In: Latin American Web Conference (WebMedia/LA-WEB), pp. 10–17. IEEE CS Press, Ribeirao Preto (2004)
5. Cho, J.: Crawling the web: discovery and maintenance of large-scale web data. Phd thesis, Department of computer science. Stanford University (2001)
6. Cho, J., Garcia-Molina, H., Page, L.: Effcient Crawling Through URL Ordering. In: 7th International Conference on World Wide Web, Brisbane, Australia (1998)
7. Cho, J., Schonfeld, U.: RankMass crawler: a crawler with high personalized pagerank coverage guarantee. In: 33th International Conference on Very Large Data Bases, pp. 375–386 (2007)
8. Dasgupta, A., Ghosh, A., Kumar, R., Olston, C., Pandey, S., Tomkins, A.: The Discoverability of the Web. In: 16th International Conference on World Wide Web, pp. 412–430 (2007)
9. Gyongyi, Z., Garcia-Molina, H., Pedersen, J.: Combating web spam with trustrank. In: 13th International Conference on Very Large Data Basess (2004)
10. Haveliwala, T.H.: Topic-Sensitive PageRank. In: 11th International Conference on World Wide Web, ACM, Honolulu (2002)
11. Kleinberg, J.M.: Authoritative sources in a hyperlinked environment. Journal of the ACM (JACM) 46 (1999)

12. Lewandowski, D.: A three-year study on the freshness of web search engine databases. Journal of Information Science 34 (2008)
13. Matsuo, Y., Ohsawa, Y., Ishizuka, M.: Average-clicks: A New Measure of Distance on the World Wide Web. Journal of Intelligent Information Systems 20, 51–62 (2003)
14. Najork, M., Wiener, J.L.: Breadth-first crawling yields high-quality pages. In: 10th International Conference on World Wide Web (2001)
15. Page, L., Brin, S.: The PageRank citation ranking: bring order to the web. Technical Report, Computer science department, Stanford University (1998)
16. Pandey, S., Olston, C.: User-Centric Web Crawling. In: 14th International Conference on World Wide Web, pp. 401–411 (2005)
17. Sutton, R.S., Barto, G.A.: Reinforcement learning an introduction. MIT Press, Cambridge (1998)
18. ZarehBidoki, A.M.: Effective web ranking & Crawling. Phd thesis, School of Electrical and Computer Engineering. University of Tehran, Iran (2009)
19. ZarehBidoki, A.M., Yazdani, N., Ghodsnia, P.: FICA: a novel intelligent crawling algorithm based on reinforcement learning. Web Intelligence and Agent Systems (WIAS) Journal 7, 363–373 (2009)

Extraction of Web Texts
Using Content-Density Distribution

Saori Kitahara[1], Koya Tamura[2], and Kenji Hatano[3]

[1] Graduate School of Culture and Information Science, Doshisha University
1-3 Tatara Miyakodani, Kyotanabe, Kyoto 610-0394, Japan
`kitahara@ilab.doshisha.ac.jp`
[2] UX Department, Mixi Inc.,
1-2-20 Higashi, Shibuya, Tokyo 150-0011, Japan
`koya.tamura@mixi.co.jp`
[3] Faculty of Culture and Information Science, Doshisha University
1-3 Tatara Miyakodani, Kyotanabe, Kyoto 610-0394, Japan
`khatano@mail.doshisha.ac.jp`

Abstract. We propose a method for grasping the content of each Web page and extracting a part of the Web page related to query keywords, in order to make more effective snippets of a Web search engine. We regard the content as a set of words in the text of a Web page, and we generate the content-density distribution by using both the position and the influence of the word. In our experiments, we found that the proposed method facilitated the recognition of the content of Web pages, as compared to conventional methods based on snippets.

Keywords: Web page recognition, content-density distribution.

1 Introduction

Currently, the Internet has witnessed a proliferation of Web pages. Although users access Web pages in order to obtain useful information, it is difficult to search for the Web pages that contain such information. When users perform a search, they get a ranked list of Web pages with their summaries, which are called result snippets [1]. Using the result snippets, users can select the pages that contain relevant information. However, users are sometimes unable to recognize the contents of Web pages because the snippets consist of a combination of text strings and query keywords; these strings are not sufficiently long to determine whether the corresponding Web page contains relevant information.

If the contents of Web pages are represented using summaries that are more comprehensible than the snippets, users will be able to identify the Web pages that contain relevant information. Thus, we propose a method for extracting a relevant text string based on the content-density distribution of a Web page; this method provides the user with an alternative snippet of the Web page, thereby enabling him/her to grasp its content. Then, the content can be regarded as a set of words in a text string. The content-density distribution helps users to understand the position and influence of the content of the Web page.

M.V.M. Salem et al. (Eds.): AIRS 2011, LNCS 7097, pp. 273–282, 2011.

2   Related Work

Ercan et al. extracted nouns from a Web page by using lexical chains for its summarization [3]; this approach is a type of text segmentation [4,5]. Text segmentation is a method for separating text into blocks; it is adopted in several research fields including query extraction [6,10]. A lexical chain is a sequence of words in each sentence of a Web page; each lexical chain has a lexical chain occurrence vector. When we regard a lexical chain as a type of content, the its occurrence vector is regarded as the position and the influence of the content. However, it is difficult to grasp the content if there is a sudden in-text content change; this is because a lexical chain is defined for each sentence.

Another related concept is passage retrieval [7], which involves the extraction of parts of a Web text related to query keywords. Consequently, users can obtain these parts or the entire Web pages. In other words, they can search for a part of a Web text in the vicinity of the query keywords. In contrast, our objective is to determine the position and influence of the content of a Web text; we calculate the content related to the keywords.

Lv et al. constructed the positional language model by extending the information retrieval model on the basis of the language model [8]. They carried out passage retrieval without determining the size of passages, and they determined the position and proximity of query keywords. They described the construction of the positional language model and its application to the passage retrieval score. In addition, they calculated the values representing the estimated word count at each position in a Web text. Thus, they did not use the model to determine the position and influence of the content of the Web text. They evaluate their study not on the basis of the content but passage retrieval. Moreover, if the estimated word count is regarded as the influence of the word in a Web text, the influence does not take the value of zero at any position in their method. In our method, zero value of the influence of the word indicates no content of the word in that position. Therefore, we cannot compare their method to our method.

3   Our Method

In this section, we describe the construction of the content-density distribution, and we extract a text as a summary of the Web page. As described in Section 1, the content-density distribution denotes the positions of the words and their influence in a Web text. Thus, the content-density distribution reflects both the position and the influence of the content. We construct it in the following steps:

1. Calculation of word-density distribution in a Web text
2. Construction of content-density distribution in a Web text and extraction of a text

Before calculating the word-density distribution, we should extract words in a text string of a Web page (Web text, in short). Therefore, we search for Web pages using a search engine, extract the Web text, and specify the POS (part of speech) of each word in the Web text. If the word is not in dictionary form, we convert the word into its dictionary form.

3.1   Calculation of Word-Density Distribution in a Web Text

We calculate the word-density distribution of each word from the extracted Web text. The Web text is composed of a set of words, and we can recognize the appearance position of each word in it. In addition, if we define the influence of the word to be the highest at the appearance position, we can briefly survey both the position and the influence of the word, and calculate the word-density distribution. We explain its calculation using Fig. 1.

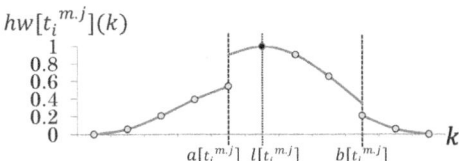

Fig. 1. Weighted Hanning window function

If we extract the Web text s_m, t_i^m denotes the word t_i in s_m. When we count the words in the Web text s_m, $hw[t_i^{m.j}](k)$ denotes a value of the word-density distribution, which is taken by the j-th word of t_i^m at the k-th word in the Web text s_m ($j = 1, 2, \cdots$) (See Fig. 1). The j-th word t_i^m appears at the position $l[t_i^{m.j}]$, so that $hw[t_i^{m.j}](k)$ becomes the highest. The greater the distance of the position from $l[t_i^{m.j}]$, the smaller is $hw[t_i^{m.j}](k)$. In addition, the influence of the content may suddenly change at the end of a sentence; this is known as a sentence separator[1]. We define $a[t_i^{m.j}]$ as the sentence separator just before $l[t_i^{m.j}]$, and $b[t_i^{m.j}]$ as the sentence separator just after $l[t_i^{m.j}]$. In this case, the Web text from $a[t_i^{m.j}]$ to $b[t_i^{m.j}]$ is a sentence. Therefore, if k is far from $l[t_i^{m.j}]$, $hw[t_i^{m.j}](k)$ would be smaller. $hw[t_i^{m.j}](k)$ is given by Equation (1), which we call the weighted Hanning window function; it has been suggested that this function is the most useful function for extracting the influence of words [9][2]:

$$
hw[t_i^{m.j}](k) = \begin{cases} \dfrac{1}{2}(1 + \cos 2\pi \dfrac{k - l[t_i^{m.j}]}{W}) & (a[t_i^{m.j}] < k < b[t_i^{m.j}]) \\ \dfrac{1}{2}S(1 + \cos 2\pi \dfrac{k - l[t_i^{m.j}]}{W}) & (a[t_i^{m.j}] \geq k, \, b[t_i^{m.j}] \leq k) \end{cases} \tag{1}
$$

$$
(\mid k - l[t_i^{m.j}] \mid \leq \tfrac{W}{2}, \, 0 \leq S \leq 1)
$$

We use the window function because we assume that the word has an effect on other words that precede and follow it. Therefore, we assume its effect in the range $\mid k - l[t_i^{m.j}] \mid \leq \frac{W}{2}$. S is a weighting parameter whose value lies between

---

[1] For example, periods, exclamation marks, and question marks.

[2] In this related work, the authors discuss the effective function to extract the influence of the words. Then, they use this function to generate links between documents.

zero and one, and we define this parameter in order to consider the change in the content of a Web text. Although the Hamming window function was proposed as an improved variant of the Hanning window function, the Hamming window function does not take the value of zero at the tail. The Hamming window function is not suitable for treating this influence because we assume that the influence of the word gradually tends to zero.

We present the following example in order to provide readers with a better understanding of word-density distribution. Fig. 2 shows the calculation of the word-density distribution of t_1^m. We assume three t_1 in a Web text, and the words are denoted by $t_1^{m.1}$, $t_1^{m.2}$, and $t_1^{m.3}$. Then, we calculate $hw[t_1^{m.1}](k)$, $hw[t_1^{m.2}](k)$, and $hw[t_1^{m.3}](k)$ by using Equation (1). We also calculate $hw[t_1^m](k)$ by summing up and normalizing $hw[t_1^{m.1}](k)$, $hw[t_1^{m.2}](k)$, and $hw[t_1^{m.3}](k)$.

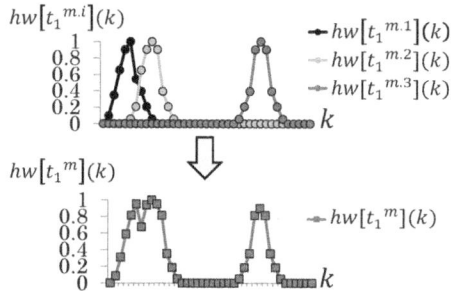

Fig. 2. Calculation of word-density distribution

There are several methods for integrating multiple values into one value [2]. We select the summation of all $hw[t_i^{m.j}](k)$ at each position k because summation is an intuitive approach. Here, the word t_i^m may appear many times, in which case we have to combine $hw[t_i^{m.j}](k)$ to construct the entire content-density distribution. However, if we treat the summation as the word-density distribution of t_i^m, the highest value of this distribution depends on the occurrence of t_i^m in the Web text s_m. To solve this problem, we normalize $hw[t_i^{m.j}](k)$ by dividing each $\sum_j hw[t_i^{m.j}](k)$ by the highest value among these values. $hw[t_i^m](k)$ is defined as

$$hw[t_i^m](k) = \frac{\sum_j hw[t_i^{m.j}](k)}{\max_k \sum_j hw[t_i^{m.j}](k)} \tag{2}$$

3.2 Construction of Content-Density Distribution in a Web Text and Extraction of a Text

After calculating the word-density distribution, we use it to construct the content-density distribution. In this paper, we denote the content-density distribution of

s_m as $hw[Q^m](k)$, where Q^m is a query composed of word t_i^m ($i = 1, 2, \cdots, n$) in s_m. If the query has only one word, the word-density distribution and the content-density distribution are equivalent.

Because the query is usually composed of two or more keywords, we need to judge whether a set of query keywords forms content. We believe that a set of keywords forms content if each keyword is closely located in a Web text. We judge the existence of the content as the overlap between the word-density distributions of keywords. In this study, we assume the presence of a range in which the content exists if the word-density distributions overlap with each other. Fig. 3 shows the construction of the content-density distribution of s_m. The two-headed arrows in Fig. 3 represent content related with the query Q^m, comprising t_1^m and t_2^m, because two word-density distributions overlap in s_m.

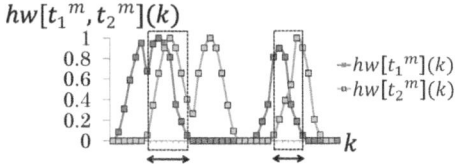

$$hw[t_1{}^m, t_2{}^m](k)$$

Fig. 3. Definition of content in the Web text s_m

Here, we construct the content-density distribution from the word-density distributions. In much the same fashion as the construction of word-density distributions, we use the summation of each word-density distribution related to each query keyword. If we calculate the content-density distribution related to query keywords Q^m to sum up $hw[t_i^m](k)$ at each k, the content-density distribution relies on the number of query keywords at the same position; thus, we divide their summation by the number of query keywords to get $hw[Q^m](k)$ as:

$$hw[Q^m](k) = \begin{cases} \dfrac{1}{n} \sum_i hw[t_i^m](k) & (hw[t_i^m](k) > 0, i = 1, 2, \cdots, n) \\ 0 & (others) \end{cases} \tag{3}$$

We illustrate the following example to facilitate the understanding of word density. In Fig. 4, if we judge whether t_1^m and t_2^m form content, we check whether two word-density distributions overlap in s_m. According to Fig. 4, these words in the query Q^m have the range of content. Then, we can construct the content-density distribution related to Q^m. We can regard these word-density distributions as content-density distributions because t_1^m and t_2^m overlap with each other. Finally, we can extract three contents related to the query keywords in the Web text s_m because we can construct three content-density distributions; $hw[t_1^m](k)$, $hw[t_2^m](k)$, and $hw[Q^m](k)$.

Finally, we extract a part of a Web text including the content by using the content-density distribution. If the value of the content-density distribution at a position is larger than a threshold, we can say that the content related to the query Q^m exists at this position. In this paper, we regard this threshold as zero because we believe that the content will exist, no matter how small the value of the content-density distribution is, unless the value is zero.

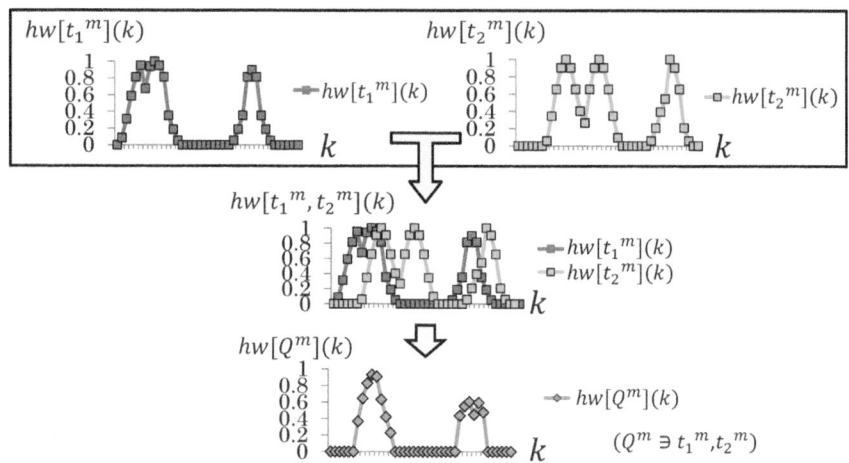

Fig. 4. Construction of content-density distribution

4   Experiment

The content extracted by the content-density distribution has two aspects. One is the range of the content; the other is the influence of the content. If we evaluate these aspects precisely, we must confirm whether our method is appropriate with regard to these aspects. In this section, we evaluate a Web text extracted by the content-density distribution and the position containing the highest value of the content-density distribution in each Web text. Next, we assume that these aspects are suitable for the content users consider useful. Then, they use the Google AJAX Search API to obtain Japanese Web texts [12]. We describe the evaluation of our method and the objectives of these experiments as follows:

- Evaluating the extracted Web text
  We evaluate the extracted Web text, that is, whether the extracted Web text is suitable for the content of the Web page.
- Evaluating the influence of the extracted Web text
  We evaluate the position of the highest value of the content-density distribution.

In order to evaluate the aspects mentioned above, we create two indicators, and in order to conduct the experiments from the viewpoint of these aspects, we have to create a data set for evaluating our method. We request an external individual (collaborator) to create it in the following steps:

1. The collaborators of the experiments generate queries composed of two familiar words to each collaborators, and they issue these queries to the Google AJAX Search API for extracting Web texts. As described in previous sections, we define content as a set of several words; however, we treat it as two words for the purpose of simplicity. By issuing queries to the API, we can get eight Web texts per query[3]. Thus, we can store numerous Web texts for constructing the data set.
2. In order to construct the data set, the extracted Web texts should be parsed to divide into words. We employ a part-of-speech and morphological analyzer [13] to carry out the partition[4]. The collaborators also select the answer part of each Web text.
3. The collaborators choose a word that they believe to be the most important part of the Web text. This part is equivalent to the answer part described above; in this experiment, this position expresses the content of the Web text, so that we can evaluate a method, i.e., whether this position expresses the content or not. We call this position the answer position c_{ans}. We also define the position with the highest value of the content-density distribution as the extracted position c_{max}.

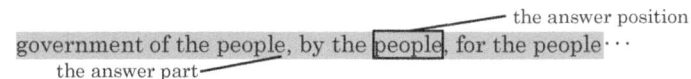

the answer position
government of the people, by the people, for the people···
the answer part

Fig. 5. Answer part and the answer position

Fig. 5 shows the answer data in the Web text s_m. The shaded positions in Fig. 5 represent the answer part of the word "people." In addition, the square shows the answer position.

In this experiment, we used 89 Web texts for constructing the data set because some Web pages do not include the Web text. Moreover, we conduct preliminary experiments in order to obtain good parameters. As a result, we set some parameters as W and S. W equals 0.6, and S is three times the average length of sentences in each Web text.

4.1   Evaluating the Extracted Web Text

In this experiment, we evaluate whether the part of the Web text, which is extracted on the basis of a threshold of the content-density distribution, coincides with the answer part of the Web text. Thus, we compare the Web text extracted by our method with that extracted by the Google Snippet using the answer parts in the data set. For this comparison, we use an indicator called the concordance

---

[3] Google AJAX Search API can return only eight Web texts as a search result.
[4] MeCab is the most famous morphological analyzer for Japanese texts. This morphological analyzer processes texts faster than the conventional analyzer [11].

rate of an answer. Equation (4) represents the indicator used for the comparison. This indicator shows the concordance rates of an answer, the part calculated by the numbers of words in both the answer part and the positions of the values of the content-density distribution.

$$IT_{s_m} = \frac{N_{a \cdot e}}{N_a} \qquad (4)$$

This indicator shows how much of the answer part can be extracted for evaluating all extracted Web texts by using each method. We utilize the answer part for the evaluation. In this equation, N_a indicates the number of words in the range of the answer extracted by all Web texts, $N_{a \cdot e}$ indicates the number of words in the answer part and the part of all Web texts extracted by our method.

When we utilize the number of words to evaluate the extracted Web text, the number of words in the result snippet is smaller than the number of words in our method. At this time, we are unable to compare these methods directly. To solve this problem, we adopt the proposed method, which can extract the Web text containing the same number of words as the proposed method.

We calculate the concordance rate of an answer to compare this rate, based on each method. According to the results, this value based on the proposed method is 0.1801. In contrast, this value based on the conventional method is 0.1645. Therefore, the proposed method coincides with the answer more than the conventional method in this evaluation. This is because we consider not only the proximity of the query keywords but also the position where the content may change. As a result, we can precisely extract the position where the value of the influence is large. Thus, we can say that the Web text extracted by the content-density distribution is more appropriate for a comprehensible summary of a Web page than a result snippet.

4.2   Evaluating the Influence of the Extracted Web Text

In the other experiment, we determine whether the influence of the content-density distribution is suitable for c_{ans}. In the paper, we denote the collaborators believe that the influence of the content is the highest at c_{ans}, and the value of the content-density distribution at c_{max} is the highest. When c_{ans} coincides with c_{max}, we can say that we can extract the influence of the content using the content-density distribution. Even though the gap between c_{ans} and c_{max} is small, the value of the content-density distribution at c_{ans} is not always small. Consequently, we cannot evaluate it using only the distance between c_{ans} and c_{max}. Thus, we also define another indicator to evaluate the proximity between the values of the content-density distribution at c_{ans} and c_{max}.

First, we evaluate our method on the based on the gap between c_{ans} and c_{max}. When we define the number of words in each Web text s_m as N_{s_m}, we utilize these positions to define this indicator as

$$IP_{s_m} = \frac{|c_{ans} - c_{max}|}{N_{s_m}} \qquad (5)$$

Equation (5) represents the normalized distance between c_{ans} and c_{max}. Therefore, if the value calculated by Equation (5) is smaller, c_{ans} is closer to c_{max}, we can assume that the content-density distribution grasps c_{ans}.

We also evaluate our method based on the proximity between the value of the content-density distribution at c_{ans} and the highest value of the content-density distribution. We define the equation to calculate the indicator as

$$IV_{s_m} = \frac{hw[Q^m](c_{ans})}{hw[Q^m](c_{max})} \qquad (6)$$

In Equation (6), $hw[Q^m](c_{ans})$ describes the value of the content-density distribution at c_{ans}, and $hw[Q^m](c_{max})$ describes the value of the content-density distribution at c_{max} for a query Q^m in the Web text s_m. If IV_{s_m} is large, we can say that our method can precisely extract c_{ans} from the Web text. As a result, we can assume that the gap between $hw[Q^m](c_{max})$ and $hw[Q^m](c_{max})$ is small.

We evaluate the influence of extracted Web text, whether c_{max} can grasp the answer position of the content by combining these two indicators, and plot the data for evaluations on a scatter diagram. When IP_{s_m} is small, the content-density distribution of the Web text s_m can grasp the answer position c_{ans}. On the other hand, when IV_{s_m} is large, the value of the content-density distribution at c_{ans} is similar to the highest value of the content-density distribution. Thus, if the Web text produces a good result by the two indicators, the content-density distribution of the Web text can grasp the content by evaluation of the position as well as by evaluation of the influence.

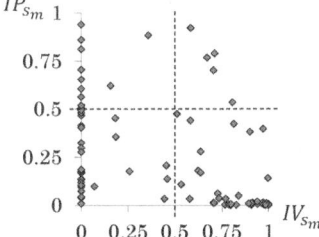

Fig. 6. Evaluating the influence of extracted Web text

Fig. 6 shows the result calculated for each Web text. The vertical axis in Fig. 6 denotes the indicator IP_{s_m} and the horizontal axis in Fig. 6 denotes the other one IV_{s_m} In this experiment, the plots on the lower-right indicate the Web text precisely extracted the center of the content by the content-density distribution. This is because the result of the evaluation will be better, when IP_{s_m} is smaller and IV_{s_m} is larger. According to Fig. 6, 43 plot appear on the lower-right; we can precisely extract the important part of 48% texts related to the content from the viewpoint of both the position and the influence. When we use the summary extracted by the content-density distribution, we can recognize whether the content exists in one of the two Web pages. Therefore, we can show that the range and influence of the content extracted by the content-density distribution are appropriate.

5 Conclusion

In this paper, we proposed a method for grasping the content of text on a
Web page in order to construct the content-density distribution related to query
keywords. For the sake of evaluation, we compared the range of the content-
density distributions and result snippets on the basis of the average concordance
rate of an answer. Consequently, the proposed method produces much better
results than conventional one. According to discussions on the influence of the
content-density distributions, users believe that the center of content in a Web
text has a larger value of the content-density distribution; moreover, we also
have to evaluate our method from the the viewpoint of time complexity.

Acknowledgments. This work is partly supported by JSPS Grant-in-aid for
Scientific Research (A) #22240005 and for Young Scientists (B) #22700248.

References

1. Manning, C.D., Raghavan, P., Schuetze, H.: Introduction to Information Retrieval.
   Cambridge University Press (2008)
2. Montague, M., Aslam, A.J.: Relevance score normalization for metasearch. In:
   CIKM 2001, pp. 427–433. ACM (2001)
3. Ercan, G., Cicekli, I.: Lexical Cohesion Based Topic Modeling for Summarization.
   In: Gelbukh, A. (ed.) CICLing 2008. LNCS, vol. 4919, pp. 582–592. Springer, Heidelberg (2008)
4. Hearst, M.A.: Multi-paragraph segmentation of expository text. In: ACL 1994, pp.
   9–16. ACL (1994)
5. Kojima, H., Teiji, F.: Segmenting Narrative Text into Coherent Scenes. Literary
   and Linguistic Computing 9(1), 13–19 (1994)
6. Li, Q., Candan, K.S., Qi, Y.: Extracting Relevant Snippets from Web Documents
   through Language Model based Text Segmentation. In: WI 2007, pp. 287–290.
   IEEE Computer Society (2007)
7. Salton, G., Allan, J.M., Buckley, C.: Approaches to Passage Retrieval in Full Text
   Information System. In: ACM SIGIR 1993, pp. 49–58. ACM (1993)
8. Lv, Y., Zhai, C.X.: Positional Language Models for Information Retrieval. In: ACM
   SIGIR 2009, pp. 299–306. ACM (2009)
9. Kise, K., Mizuno, H., Yamaguchi, M., Matsumoto, K.: On the Use of Density Distribution of Keywords for Automated Generation of Hypertext Links from Arbitrary
   Parts of Documents. In: ICDAR 1999, pp. 301–304. ACM (1999)
10. Tur, G., Mori, R.D.: Spoken Language Understanding: Systems for Extracting
    Semantic Information from Speech. Wiley (2011)
11. Kudo, T., Yamamoto, K., Matsumoto, Y.: Applying Conditional Random Fields
    to Japanese Morphological Analysis. In: EMNLP 2004, pp. 230–237. ACL (2004)
12. Google Code, http://code.google.com/more/
13. MeCab: Yet Another Part-of-Speech and Morphological Analyzer,
    http://mecab.sourceforge.net/

A New Approach
to Search Result Clustering and Labeling

Anil Turel and Fazli Can

Bilkent Information Retrieval Group
Computer Engineering Department, Bilkent University
Bilkent, Ankara 06800, Turkey
aturel@cs.bilkent.edu.tr,
canf@cs.bilkent.edu.tr

Abstract. Search engines present query results as a long ordered list of web snippets divided into several pages. Post-processing of retrieval results for easier access of desired information is an important research problem. In this paper, we present a novel search result clustering approach to split the long list of documents returned by search engines into meaningfully grouped and labeled clusters. Our method emphasizes clustering quality by using cover coefficient-based and sequential k-means clustering algorithms. A cluster labeling method based on term weighting is also introduced for reflecting cluster contents. In addition, we present a new metric that employs precision and recall to assess the success of cluster labeling. We adopt a comparative strategy to derive the relative performance of the proposed method with respect to two prominent search result clustering methods: Suffix Tree Clustering and Lingo. Experimental results in the publicly available AMBIENT and ODP-239 datasets show that our method can successfully achieve both clustering and labeling tasks.

Keywords: Cluster labeling, search result clustering, web information retrieval.

1 Introduction

The utility of search result clustering (SRC) and associated cluster labeling algorithms for easy access to the query results has been widely investigated [6]. Without a proper arrangement of search results, finding the desired query result among ranked list of document snippets is usually difficult for most users. This problem is further aggravated when the query belongs to a general topic which contains documents from a variety of subtopics. At this point, the burden of solving inter-relations among documents and extracting the relevant ones are left to the user. More recently; however, there are continuous research and commercial efforts for developing online search result clustering and labeling methods [6].

Even though there exists some search result clustering algorithms, embedding these methods in search engines is not a common practice. There are three main

M.V.M. Salem et al. (Eds.): AIRS 2011, LNCS 7097, pp. 283–292, 2011.
© Springer-Verlag Berlin Heidelberg 2011

reasons behind this problem: (1) existing algorithms are not able to capture the relationships among documents since the snippets are too short to convey enough information about query subtopics; (2) finding descriptive and meaningful labels for clusters is a difficult problem; (3) the evaluation of SRC task is not well-defined. Motivated by these observations, we present a new search result clustering method based on cover coefficient (C^3M) [3] and sequential k-means clustering algorithms [14].

Early works on the SRC problem include the Scatter-Gather system [13], Suffix Tree Clustering (STC) [22], and Lingo [16]. Apart from those, MSEEC [12] and SHOC [9] also contribute to the use of words proximity in the input documents. Clustering web results is also essential for mobile devices since it decreases the amount of information transmitted, provides a more effective and informative user interface that require less interactions in terms of page scroll or query reformulation [5] [6]. Search result diversification is another approach to postprocessing of search results. Related studies re-rank search results for presenting documents from different subtopics at the beginning of search results list [4] which is similar to but different from the SRC problem. Although SRC seems as a subset of document clustering, it has distinguishing constraints coming from efficiency, effectiveness and labeling quality requirements [6]. While both keyword extraction and labeling task of SRC are based on frequent phrases, labeling differentiates from keyword extraction with efficiency requirement it possesses.

Note that, among all SRC methods, for comparison we study two prominent algorithms; Lingo [16] and STC [22]. Lingo uses singular value decomposition to generate cluster descriptions that are crucial for user-friendly search engines. The Lingo method is currently being used in Carrot^2 open source search result clustering engine [21]. Besides, STC introduced in [22] is based on suffix tree data structure that enables the usage of phrases instead of single words as cluster labels. In this method, clustering and labeling steps are accomplished using suffix tree.

Our search result clustering method, C^3M+K-means is based on C^3M and sequential k-means algorithms. The adaptation of these two methods to the search result clustering problem is one of the contributions of this paper. Additionally, a new labeling approach "labeling via term weighting" is introduced. The key contribution of this paper is the labeling evaluation strategy. To assess the effectiveness of cluster labeling, we introduce a new metric called $sim_{F-measure}$, by employing precision and recall. We provide experimental results by systematically evaluating the performance of our method in the AMBIENT [7] and ODP-239 [8] test collections. We show that our method can successfully achieve both clustering and labeling tasks [19].

2   An Approach to Search Result Clustering and Labeling

The methodology we use in this study is to extract the relationships among documents with C^3M method and to construct the final clusters through feeding the results of C^3M to the sequential k-means algorithm. We then use our term weighting-based approach to label the generated clusters.

2.1   Preprocessing

The first step is to clean the document text from non-letter characters and to convert all characters to lower case. Afterwards, stopwords are eliminated and stemming is applied by the Porter Stemmer [18]. Finally, the terms appearing in the 3-30% of the snippets constitute the term list (that is used for document description).

In order to generate meaningful cluster labels, phrase discovery is a crucial phase in SRC problem. Most of the time, a combination of words, namely, phrases are needed to reflect the cluster content. In this study, we use suffix tree structure [22] to extract phrases from the document snippets. Suffix tree indexes sequence of words in the nodes and stores number of occurrences. Then, the inner nodes with sufficient occurrences are considered as a phrase (in our experiments nodes that occur in more than %2 of the documents are selected as phrases) and they are added to the term list.

Before passing to the clustering phase, we index each document using its terms that appear in the term list. The term weights are computed by using the log entropy formula [10] [19]. Entropy based term weighting considers the distribution of term over documents. Finally, we reduce the weights of single-word terms by multiplying them with a constant value, in our experiments 0.3, and to increase the importance of phrases, they are multiplied with 0.7. Then, we normalize the term weights of documents and the collection becomes ready for clustering [19].

2.2   Clustering

Cover coefficient-based clustering. It is a seed oriented, partitioning, single-pass, linear-time clustering algorithm introduced in [3]. The main goal of C^3M is to convey the relationships among documents using a two-stage probability experiment. The efficiency and effectiveness of C^3M for information retrieval in texts has been experimentally demonstrated in [1]. To accomplish clustering task, briefly, ten documents are selected as seed documents and for each non-seed document we check the coverage of the document with the seed documents and select the seed that has the highest coverage over the non-seed. If none of the seeds covers the non-seed document, then, it is directly added to the *Others* cluster. Detailed information about C^3M can be found in [3].

Modified sequential k-means algorithm. K-means is a linear-time and widely used clustering algorithm which groups given documents after the initial centroids are provided.The success rate of the k-means algorithm highly depends on the initial cluster centroids. Therefore, we use the results of C^3M clustering to derive the centroids as accurately as possible. The input centroids are the vectorial averages of the documents in each C^3M cluster.

Sequential k-means algorithm [14] updates the cluster centroid after each document assignment to the cluster instead of after all documents distributed in original k-means. We use a modified version of the sequential k-means algorithm where we assign documents to the centroids as in k-means in the first pass. Then, the centroids are re-calculated according to the new distribution of documents. At the beginning of each following pass, we empty the cluster contents. Then, we assign each document to the nearest cluster and update that cluster's centroid again as:

$$centroid_i = \frac{\sum_{j \in cluster_i} doc_j + centroid_i}{|cluster_i| + 1} \tag{1}$$

where $|cluster_i|$ is the number of documents in the cluster and 1 is added to the denominator for the centroid vector in numerator.

2.3   Labeling via Term Weighting

The final step of our method is the labeling phase. We aim to assign descriptive labels to clusters that reflect their contents. This step is very important because meaningless or confusing labels may mislead users to check the wrong clusters for the query and lose extra time. We present a novel labeling strategy called *labeling via term weighting* that assesses significance of terms for clusters. Firstly, the terms of documents in a cluster are merged, then term weighting is applied to the clusters (by assuming them as documents). We use the same term weighting formula as in Section 2.1 [10] [19]. A single-word label generally lacks expressiveness, so we give more weight to phrases than single-word terms during cluster labeling as in Section 2.1. For each cluster, we select the highest weighted terms into the candidate labels list. In our experiments, we add topmost five terms to the list. While we are assigning the final labels of the clusters from these lists, we follow the criteria below:

- Clusters are labeled in descending order of cluster size,
- Label should not be one of the previously given labels to another cluster,
- Phrase label candidate with less than five words is preferred (if exists),
- Term with a higher weight is preferred.

3   Performance Measures

3.1   Clustering Evaluation

To be able to quantify clustering performance, we first need to define a success measure which reflects the actual performance of clustering results as fairly as possible, regardless of the clustering method we choose. In this paper, we use weighted average F-measure ($w_{F\text{-measure}}$) [20] which is the average of total weighted F-measure of each class. Intuitively, precision reflects to what extent presented cluster includes documents of ground truth class and recall reflects to what extent ground truth class is presented to the user. The necessary equations to measure the similarity between a ground truth class i and represented cluster j are given.

$$precision(i,j) = \frac{class_i \cap cluster_j}{|cluster_j|} \qquad (2)$$

$$recall(i,j) = \frac{class_i \cap cluster_j}{|class_i|} \qquad (3)$$

$$F\text{-}measure(i,j) = \frac{2 \times recall(i,j) \times precision(i,j)}{recall(i,j) + precision(i,j)} \qquad (4)$$

For each class in the ground truth we find the best matching cluster (that has the maximum F-measure among all clusters). We are interested in weighted F-measure to better evaluate the contribution of each class to the overall performance. Clustering performance is computed as follows.

$$w_{F\text{-}measure} = \frac{1}{\sum_{i=1}^{n_{class}} |class_i|} \sum_{i=1}^{n_{class}} \left(\max_{j} \{F\text{-}measure(i,j)\} \, |class_i| \right) \qquad (5)$$

where n_{class} represents the number of classes.

3.2   Labeling Evaluation

Although human judgment is preferred to evaluate the labeling performance of most of the SRC methods, this approach is very expensive and difficult to repeat for different parameters. It is also difficult to compare distinct labeling methods based on human judgment. Due to such drawbacks, we propose a new labeling evaluation measure called $\text{sim}_{F\text{-measure}}$ based on the assessment of similarity between two labels (ground truth and generated label).

3.2.1   Comparison of Ground Truth and Generated Label

We use four similarity metrics to automatically find similarity between generated label and ground truth label and they are semantic similarity, exact, partial and overlap match. Each metric reflects the labeling performance of SRC methods from different aspects. While exact match is strict to the ground truth, partial match requires the ground truth structure (also human readability) is preserved partially. Overlap match considers how close suggested labels are to the ground truth. Lastly, semantic similarity finds the indirect relationship between labels. Before applying these metrics, stopwords are eliminated and stemming is applied. If the ground truth class is *Others* cluster, and algorithm cluster is not, or vice versa, the similarity score between labels is set to 0. Similarity metrics give Boolean output; 1 for similarity and 0 for dissimilarity, except semantic similarity.

Semantic similarity. It is a research field in artificial intelligence, that aims to determine the similarity between concepts by mapping them into an ontology and investigating their relationship within the ontology. In this paper, we use semantic similarity to detect the similarity between the ground truth and proposed labels. For the experiments, we use Java WordNet Similarity Library [17] that exploits WordNet [11] as the ontology source. The semantic similarity metric outputs a similarity value within the range of 0 and 1 to quantify the measure of similarity between two labels. Although there are different formulations of this metric, we are using the approach presented in [15] that uses the information content concept of information theory. For example, in our experiments, ground truth and generated label pairs "News" - "Broadcasts" and "Sound Files" - "Streaming Audio" are found to share respectively 0.90 and 0.78 similarity according to the semantic similarity metric of [15].

Exact match. It suggests similarity if the generated label is the same as the ground truth or the generated label covers the other. To exemplify, when ground truth and generated label pair is "Instruments" - "Musical Instrument," exact match is ensured.

Partial match. It suggests similarity if the cluster label covers the ground truth label or vice versa. For instance, the ground truth - extracted label pair "USS Coral Sea, disambiguation" - "USS Coral Sea" is accepted. The partial and exact match do not cover the case when the words in ground truth change order in generated label.

Overlap match. It aims to catch the slightest similarity between labels. If the intersection between the label and ground truth label is not empty, then the overlap match accepts the label. As an example; if the ground truth label is "Editorial Illustration," the overlap match accepts the generated label "Digital Illustrations."

3.2.2 Labeling Evaluation Measure: $sim_{\text{F-measure}}$

In order to obtain a robust labeling evaluation metric for the entire clustering structure, we introduce a new measure, $sim_{\text{F-measure}}$, based on precision and recall. It is inspired by [20]. In this formulation, similarity precision ($sim_{\text{precision}}$) represents to what extent labels presented to the user resemble ground truth labels and similarity recall (sim_{recall}) defines to what extent ground truth labels are reflected to the user. The methodology for computing the overall similarity can be summarized as follows. For each class in the ground truth, we find the matching cluster that gives the highest F-measure with the class. Then, we compute the similarity between the labels by using one of the similarity metrics (represented as *similarity* function in equation 6). After that, we sum up the similarity scores for all classes and normalize by the number of classes to find the sim_{recall}. We find the $sim_{\text{precision}}$ by applying the same procedure to the clusters. Finally, $sim_{\text{F-measure}}$ is computed as the harmonic mean of sim_{recall} and $sim_{\text{precision}}$. The necessary formulation for this procedure can be derived as follows (note that all of them have a value between 0 and 1).

$$sim_i = similarity \ \{label(class_i), \ label(cluster_{\max F\text{-}measure(i,j)})\} \qquad (6)$$

$$sim_j = similarity \ \{label(cluster_j), \ label(class_{\max F\text{-}measure(i,j)})\} \qquad (7)$$

$$sim_{precision} = \frac{\sum_{j=1}^{n_c} sim_j}{n_c} \qquad sim_{recall} = \frac{\sum_{i=1}^{n_{class}} sim_i}{n_{class}} \tag{8}$$

$$sim_{F\text{-}measure} = \frac{2 \times sim_{recall} \times sim_{precision}}{sim_{recall} + sim_{precision}} \tag{9}$$

where n_{class} and n_c are respectively the number of classes and clusters.

3.3   Experimental Results

In order to assess the the performance of clustering and cluster labeling algorithms, we perform experiments in two publicly available datasets specific to SRC task: the AMBIENT Dataset [7] and ODP-239 Dataset [8]. They consist of 44 and 239 queries, respectively and 100 snippets for each query. We present both the results of C^{3}M and C^{3}M+K-means methods to discuss the effect of using sequential k-means clustering. We use a comparative strategy to derive the relative performance of our algorithm with respect to the two state-of-the-art algorithms: Lingo and Suffix Tree Clustering (STC). Implementation of these methods are available in Carrot^2 API [21].

Clustering results. The first step of the clustering evaluation is to prove that the algorithm shows significant difference from random clustering according to the Monte Carlo method [14]. If the cluster sizes are preserved and documents are added to the clusters randomly, we obtain random clustering. A target cluster of a class contains at least one relevant document of the class. As a rule, the average number of target clusters of the clustering method should be significantly less than the average number of target clusters of random clustering [3]. The random clustering is performed 1000 times and as a result, on the average, the proposed method outperforms %97.3 (in AMBIENT) and %98.8 (in ODP-239) of the 1000 random clusterings. So we conclude that the proposed method performs significantly different from random.

Table 1. Clustering results in terms of w$_{\text{F-measure}}$

Algorithm	AMBIENT	ODP-239
C^{3}M	0.444	0.386
C^{3}M+K-means	**0.603**	0.464
STC	0.413	**0.510**
Lingo	0.370	0.420

Afterwards, we test our algorithm in the AMBIENT and ODP-239 datasets by using w$_{\text{F-measure}}$ success measure. Table 1 details the average results for all queries in both datasets including the results for STC and Lingo. As seen in this table, the proposed C^{3}M+K-means algorithm performs the best among all methods in the AMBIENT dataset when we look at the w$_{\text{F-measure}}$ results. To

prove that our results are statistically significantly different from those of the
other algorithms, we also run a paired t-test over $w_{F\text{-measure}}$ scores of all queries
in AMBIENT. With a threshold level of 0.01, we achieve statistical significance
in our results. The proposed method ranks second in the ODP-239 dataset after
STC, but the difference between the proposed method and STC is not statistically significant. Therefore, we conclude that the proposed method is successful
at clustering search results. Notice that, the usage of sequential k-means as a
secondary clustering mechanism after the C^3M method increases the clustering
performance significantly.

Labeling results. Labeling performances of the proposed method are provided
in Table 2. Success rates are shown based on the previously mentioned semantic
similarity, exact, partial and overlap match similarity metrics applied on *similarity F-measure* ($sim_{F\text{-measure}}$) label evaluation measure. In contrast to the smaller
exact match scores by all methods in AMBIENT relative to ODP-239, we observe higher scores in the other measures. The reason behind is that the ground
truth labels, which define the meaning of ambiguous words, are too long in the
AMBIENT dataset (on average 8.6, 1.63 words in AMBIENT and ODP-239
datasets, respectively). Note that scores are low by all methods because according to the labeling evaluation strategy, success of labeling depends on how good
clusters are obtained.

For the AMBIENT dataset, our algorithm performs best with overlap match,
while ranking second in other measures following the STC algorithm. We show
the significance of these results using a t-test as described previously. In contrast,
our method outperforms the other methods in all the success metrics in the
ODP-239 dataset (with one exception and in that case there is a tie with STC).
However, statistical significance is not observed due to the close results of the
proposed method and STC. In the light of these results, it can be concluded
that, the proposed method shows comparable performance on labeling clusters.

In fact, the automatically computed similarity metrics are more strict than
human judgment and they produce smaller similarity scores since they only compare with ground truth label, while human can also consider cluster content. In
addition, automatic evaluation finds similarity between labels if they share words

Table 2. Labeling results in terms of $sim_{F\text{-measure}}$. Similarity between labels are decided
by exact (E), partial (P), overlap (O) match and semantic similarity (S) metrics.

Dataset	Algorithm	E	P	O	S
AMBIENT	C^3M	0.002	0.151	0.481	0.214
	C^3M+K-means	0.005	0.235	**0.488**	0.261
	STC	**0.086**	**0.335**	0.455	**0.331**
	Lingo	0.049	0.209	0.406	0.225
ODP-239	C^3M	0.091	0.112	0.149	0.108
	C^3M+K-means	**0.151**	**0.185**	**0.221**	**0.172**
	STC	0.119	0.176	0.195	**0.172**
	Lingo	0.112	0.144	0.168	0.137

or have a relationship in the ontology, but human infer similarity intuitively, even such an association does not exists. However, the disadvantage of such an evaluation method is that the results may vary from person to person. Therefore, we can say that, using an automatic similarity metric simplifies the comparison of search result labeling methods. Inserting F-measure constraint into the computation of $sim_{F\text{-measure}}$ provides that the cluster content should match with the class content. This ensures that not only the label similarity is enough but also the documents in the cluster should be common with the ground truth subtopic.

4 Conclusion

In this paper, we propose methods for solving two key information retrieval problems; search result clustering and cluster labeling. Our study addresses the difficulty of clustering and labeling search results. Our contribution on SRC can be summarized as taking document relationships into account by using cover coefficient-based clustering method and using its results as an initial clustering structure for the sequential k-means clustering algorithm to improve the SRC performance. We experimentally show that our approach generates meaningful clustering structures.

A novel cluster labeling approach called "labeling via term weighting" is introduced. This labeling method observes both the behavior of terms within the documents of cluster and in the document collection. The key contribution of this study is the proposed labeling evaluation strategy. We introduce a new metric, similarity F-measure, by employing precision and recall, to assess the effectiveness of cluster labeling. The resemblance between the generated and ground truth labels is determined by semantic similarity, exact, partial, and overlap match metrics.

Extensive experimental results for both clustering and labeling show that the proposed method successfully cluster and label search results while maintaining a performance competitive with the two state-of-the-art methods Lingo and Suffix Tree Clustering. In our future research we plan to embed the proposed method to the information retrieval interface of Bilkent News Portal [2].

References

1. Can, F., Altingovde, I.S., Demir, E.: Efficiency and Effectiveness of Query Processing in Cluster-based Retrieval. Information Systems 29(8), 697–717 (2004)
2. Can, F., Kocberber, S., Baglioglu, O., Kardas, S., Ocalan, H.C., Uyar, E.: Bilkent News Portal: A personalizable system with new event detection and tracking capabilities. In: Proceedings of the 31st International ACM SIGIR Conference on Research and Development in Information Retrieval, p. 885. ACM Press, Singapore (2008)
3. Can, F., Ozkarahan, E.A.: Concepts and Effectiveness of the Cover-Coefficient-based Clustering Methodology for Text Databases. ACM Transactions on Database Systems 15(4), 483–517 (1990)

4. Carbonell, J., Goldstein, J.: The Use of MMR, Diversity-based Reranking for Reordering Documents and Producing Summaries. In: Proceedings of the 21st International ACM SIGIR Conference on Research and Development in Information Retrieval, pp. 335–336. ACM Press, Melbourne (1998)
5. Carpineto, C., Mizzaro, S., Romano, G., Snidero, M.: Mobile Information Retrieval with Search Results Clustering: Prototypes and evaluations. Journal of the American Society for Information Science and Technology 60(5), 877–895 (2009)
6. Carpineto, C., Osinski, S., Romano, R., Weiss, D.: A Survey of Web Clustering Engines. ACM Computing Surveys 41(3(17)) (2009) ISSN:0360-0300
7. Carpineto, C., Romano, G.: AMBIENT Dataset (2008),
   http://credo.fub.it/ambient/
8. Carpineto, C., Romano, G.: ODP239 Dataset (2009),
   http://credo.fub.it/odp239/
9. Dong, Z.: Towards Web Information Clustering. PhD thesis, Southeast University, Nanjing, China (2002)
10. Dumais, S.: Improving the Retrieval of Information from External Sources. Behavior Research Methods, Instruments, and Computers 23(2), 229–236 (1991)
11. Fellbaum, C.: WordNet: An Electronic Lexical Database. MIT Press, Cambridge (1998)
12. Hannappel, P., Klapsing, R., Neumann, G.: MSEEC a Multi Search Engine with Multiple Clustering. In: Proceedings of the 1999 Information Resources Management Association Conference (1999)
13. Hearst, M.A., Pedersen, J.O.: Reexamining the Cluster Hypothesis: Scatter/Gather on Retrieval Results. In: Proceedings of the 19th International ACM SIGIR Conference on Research and Development in Information Retrieval, pp. 76–84 (1996)
14. Jain, A.K., Dubes, R.C.: Algorithms for Clustering Data. Prentice Hall (1988)
15. Jiang, J., Conrath, D.: Semantic Similarity Based on Corpus Statistics and Lexical Taxonomy. In: Proceedings of the International Conference Research on Computational Linguistics, ROCLING (1997)
16. Osinski, S., Stefanowski, J., Weiss, D.: Lingo: Search Results Clustering Algorithm Based on Singular Value Decomposition. In: Proceedings of the International Conference on Intelligent Information Systems (2004)
17. Pirro, G.: A Semantic Similarity Metric Combining Features and Intrinsic Information Content. Data and Knowledge Engineering 68(11), 1289–1308 (2009)
18. Porter, M.F.: An Algorithm for Suffix Stripping. Program 14(3), 130–137 (1980)
19. Turel, A.: A New Approach to Search Result Clustering and Labeling. M.Sc. thesis, Bilkent University, Ankara, Turkey (2011)
20. Weiss, D.: Descriptive Clustering as a Method for Exploring Text Collections. PhD thesis, Poznań University of Technology, Poznań, Poland (2006)
21. Weiss, D., Osinski, S.: Carrot2 Open Source Search Results Clustering Engine (2002), http://project.carrot2.org/
22. Zamir, O., Etzioni, O.: Web Document Clustering: A Feasibility Demonstration. In: Proceedings of the 19th International ACM SIGIR Conference on Research and Development in Information Retrieval, pp. 46–54 (1998)

Efficient Top-k Document Retrieval Using a Term-Document Binary Matrix

Etsuro Fujita[2] and Keizo Oyama[1,2]

[1] National Institute of Informatics, Tokyo, Japan
[2] The Graduate University for Advanced Studies (SOKENDAI), Tokyo, Japan
{ffujita,oyamag}@nii.ac.jp

Abstract. Current web search engines perform well for "navigational queries." However, due to their use of simple conjunctive Boolean filters, such engines perform poorly for "informational queries." Informational queries would be better handled by a web search engine using an informational retrieval model along with a combination of enhancement techniques such as query expansion and relevance feedback, and the realization of such a engine requires a method to prosess the model efficiently. In this paper, we describe a novel extension of an existing top-k query processing technique. We add a simple data structure called a "term-document binary matrix," resulting in more efficient evaluation of top-k queries even when the queries have been expanded. We show on the basis of experimental evaluation using the TREC GOV2 data set and expanded versions of the evaluation queries attached to this data set that the expanded technique achieves significant performance gains over existing techniques.

Keywords: web search engine, top-k query processing, early pruning, early termination, term-document binary matrix.

1 Introduction

Current web search engines perform well for navigational queries which are used to acquire particular web pages that a user has in mind. However, due to their use of simple conjunctive Boolean filters [4], such engines perform poorly for informational queries, which are used to acquire information about a certain topic where the information may be on one or more web pages. To better support informational queries, such engines should incorporate an information retrieval (IR) model [8], such as the term frequency-inverse document frequency (TF-IDF) or BM25, along with a combination of enhancement techniques [8], such as query expansion and relevance feedback. Moreover, highly optimized techniques are needed for efficiently evaluating expanded queries because the sizes of the expanded queries may often result in fifty or more.

There has been a large amount of work on optimization techniques including index compression and caching [12], result caching [7], and top-k query processing [1], [2], [3] [5], [6], [9], [10]. In this paper, we focus on top-k query processing

M.V.M. Salem et al. (Eds.): AIRS 2011, LNCS 7097, pp. 293–302, 2011.
© Springer-Verlag Berlin Heidelberg 2011

techniques, which are used to find the correct top-k documents without processing the entire posting list for each query term. This approach is especially efficient in the case of large-scale IR systems such as web search engines, where k is small and the posting lists can be overwhelmingly long. Much of the previous work on this approach mainly dealt with relatively short queries, not longer queries. We have developed a novel extension of an existing top-k query processing technique that enables it to efficiently evaluate informational queries using an IR model even when the queries have been expanded. It uses a simple data structure called a "term-document binary matrix," which indicates which document contains which query term. To the best of our knowledge, there have been no reports on integrating such a data structure into top-k query processing to increase the efficiency.

1. We describe the integration of the term-document binary matrix into the best known top-k query processing technique, the Combined Algorithm (CA), enabling more efficient evaluation of top-k queries even when the queries have been expanded using enhancement techniques such as query expansion and relevance feedback.
2. We describe our experimental evaluation of the extended technique using the TREC GOV2 data set and expanded versions of the evaluation queries attached to this data set and show that it performs significantly better than CA.

The paper is organized as follows. Section 2 discusses related work on top-k query processing. Section 3 presents the model and algorithm that are used. Section 4 describes our approach. Section 5 explains the experimental evaluation done using the TREC GOV2 collection, describes the key results, and presents our conclusions. The key points are summarized and future work is mentioned in Section 6.

2   Related Work

Top-k query processing is an efficient method to retrieve top-k documents by combining the values from sorted posting lists for query terms without processing the entire lists. There has been considerable work on top-k query processing in the IR and database communities. The IR community has a long history of research on efficient evaluation of vector space queries. Earlier work includes [3] and [9]. Although Buckley and Lewit [3] and Persin et al. [9] dealt with longer queries, their techniques are intended for relatively small collections. There has been recent work, e.g., Anh and Moffat [1], aimed at efficient evaluation for top-k queries for large collections, but the techniques reported mainly focus on relatively short queries consisting of at most ten or so terms. In the database community, seminal work has been done by Fagin [5] who introduced a family of threshold algorithms for top-k query processing. He introduced the notion of instance optimality and showed that his family of threshold algorithms satisfy this notion. Inspired by his work, many researchers in both the IR and database communities extended Fagin's threshold algorithms. Among them, those most relevant to our work are those using upper level information of sorted posting lists such as intersections of such lists [6], and those using lower level information of sorted posting lists, such as histograms of value distributions in such lists and/or co-occurrence statistics between such lists [2]. For the former type, Kumar et al. [6] generalized Fagin's threshold algorithms to the case in which pre-aggregated intersection lists of different sorted posting lists are available in addition to the original sorted posting lists. Although these ideas are relevant to our approach, which uses upper level information in a term-document binary matrix, they deal with conjunctive queries. In contrast, we focus on disjunctive queries, which are traditionally studied in the IR community. Schenkel et al. [10] developed an efficient top-k query processing technique using upper level information such as indexes for pairs of terms in each document. However, their work aims at the case in which term proximity, i.e., the distance between term occurrences in a document, is integrated into the IR model used. Therefore their work is orthogonal and complementary to our work. Among the techniques using lower level information of sorted posting lists, Bast et al. [2] proposed integrating into Fagin's threshold algorithms a novel access scheduling based on statistics on such lists, such as histograms of per-term score distributions in such lists and co-occurrence statistics between such lists. While they did not report the instance optimality of the resulting algorithms, they demonstrated significant performance improvements in evaluating shorter disjunctive queries. The approach of Bast et al. [2] is also orthogonal and complementary to our approach.

3   Preliminaries

3.1   Model

We describe the underlying model that is used.

Queries. We assume that a query q contains m terms $t_1, \ldots, t_m$ and that the score of a document d for q is of the form $score(d) = score(w(d, t_1), \ldots, w(d, t_m))$ where *score* is a given scoring function and $w(d, t_i)$ is the per-term score of d for a term t_i.

Indexes. We assume that a posting list L_i is maintained for each term t_i, that each posting in each L_i has a unique document ID and a per-term score, i.e., each posting is of the form $\langle d, w(d, t_i) \rangle$, and that the postings in each L_i are sorted in descending order by the per-term score, $w(d, t_i)$, of d for t_i. As in Fagin [5], we consider two modes of access to the sorted posting lists. The first mode is sorted access in which the query processor obtains the per-term score of a document for t_i in L_i by proceeding through L_i sequentially from the top. The second mode is random access in which the query processor obtains the per-term score of a document for t_i in L_i in one random access. In addition, we assume that we have a list for each term which contains only the IDs of documents containing that term. These document-ID-only lists are used for creating the term-document binary matrices presented in the next section.

3.2 Combined Algorithm

We extended CA, the best known top-k query processing technique and one of the threshold algorithms of Fagin [5]. CA combines sorted access with random access and is appropriate when random access is expensive relative to sorted access as in the case of IR systems. The outline of CA is following:

1. CA begins by doing sorted access in parallel for each of the sorted posting lists, and every h steps (that is, every time the depth of the sorted access increases by h), random access is performed to compute the total score of a viable document seen so far but for which not all per-term scores for query terms are known.
2. At each step of the execution of the algorithm, the query processor maintains the set of current top-k documents denoted as T_k, based on the lower bound score of each document seen so far, i.e., the score computed from the per-term scores of already known query terms in that document. Then the query processor halts when it observes that the upper bound score of each document left outside T_k, i.e., the score computed from the per-term scores of already known query terms in that document and the last obtained per-term scores through sorted accesses in the other query terms' lists, is not larger than the lower bound score of the rank-k document in T_k.

The characteristics of CA include early pruning and early termination, which are described next.

Early pruning. Let $score_{\mathrm{LB}}(d)$ be the lower bound score of document d seen so far:

$$score_{\mathrm{LB}}(d) = \sum_{t_i \in a(d)} w(d, t_i) \ , \tag{1}$$

and let $score_{\mathrm{UB}}(d)$ be the upper bound score of document d seen so far:

$$score_{\mathrm{UB}}(d) = score_{\mathrm{LB}}(d) + \sum_{t_i \in q \setminus a(d)} \underline{w}_i \ , \tag{2}$$

where $a(d) = \{t_{i_1}, \ldots, t_{i_l}\} \subseteq q = \{t_1, \ldots, t_m\}$ contains already known query terms in d, with per-term scores $w(d, t_{i_1}), \ldots, w(d, t_{i_l})$, and $\underline{w}_i$ is the last (smallest) per-term score obtained through sorted access in sorted list L_i. Let $mink$ be the lower bound score of the rank-k document in T_k. The query processor safely prunes off d when the $score_{\mathrm{UB}}(d)$ is no longer larger than $mink$. Thus, CA can keep bookkeeping cost small. Early pruning is very important for the efficient execution of the algorithm.

Early termination. Let S be the set of documents seen so far left outside T_k and U be the set of documents which have not been seen so far. Let $score_{\mathrm{UB}}(S)$ be the maximum upper bound score of documents in S:

$$score_{\mathrm{UB}}(S) = \max_{d \in S} score_{\mathrm{UB}}(d) \ . \tag{3}$$

And let $score_{\mathrm{UB}}(U)$ be the maximum upper bound score of documents in U:

$$score_{\mathrm{UB}}(U) = \sum_{i=1}^{m} \underline{w}_i \ . \tag{4}$$

The query processor halts, yielding the correct top-k documents, when both $score_{\mathrm{UB}}(S)$ and $score_{\mathrm{UB}}(U)$ are no longer larger than $mink$. Thus, CA can stop without processing the entire sorted posting list for each query term. Early termination is especially efficient in the case of large-scale IR systems such as web search engines where k is small and the posting lists can be overwhelmingly long.

4   Our Approach

4.1   The Key Idea

As reported by Bast et al. [2], CA shows poor efficiency when used for retrieving top-k documents for longer queries for a number of reasons. First, the upper bound score, $score_{\mathrm{UB}}(d)$, computed using Equation (2), becomes looser (larger) when the number of query terms is increased because of the increase in unknown query terms. Looser upper bound scores restrict early pruning, which significantly increases bookkeeping overhead. Second, the maximum upper bound scores, $score_{\mathrm{UB}}(S)$ and $score_{\mathrm{UB}}(U)$, computed using Equation (3) and (4) respectively, also become looser (larger) when the number of query terms is increased. This restricts the possibility of early termination. Thus, CA is not effective in this scenario. To overcome these shortcomings of CA for longer queries, we propose integrating a simple data structure B_q, which depends on query q, into CA. The (i, j) entry of B_q is 1 if query term t_i is contained in document d_j, and 0 if not. We call this matrix a "term-document binary matrix." We re-estimate the upper bound score, $score_{\mathrm{UB}}(d)$, and the maximum upper bound scores, $score_{\mathrm{UB}}(S)$ and $score_{\mathrm{UB}}(U)$, more tightly by using the term-document binary matrix. When $d = d_j$, referring to the j-th column of the term-document binary matrix,

$$score_{\mathrm{UB}}(d_j) = score_{\mathrm{LB}}(d_j) + \sum_{\substack{t_i \in q \backslash a(d_j) \\ B_q(i,j)=1}} \underline{w}_i \ . \tag{5}$$

Thus, $score_{\mathrm{UB}}(S)$ is re-estimated using Equation (3) and (5). $score_{\mathrm{UB}}(U)$ is re-estimated using

$$score_{\mathrm{UB}}(U) = \sum_{i=1}^{m'} \underline{w}'_i \ , \tag{6}$$

where $\underline{w}'_i$ is ith largest value of $\underline{w}_1, \ldots, \underline{w}_m$ and m' is the maximum co-occurrence of query terms in documents in U, i.e., $m' = \max_{d_j \in U} \sum_{i=1}^{m} B_q(i, j)$. The m' can be efficiently computed by caluculating the co-occurrence statistics for all

documents in a collection based on B_q at the begining of the algorithm and
updating the statistics each time an unknown document is obtained during the
execution of the algorithm. CA using these tighter upper bounds leads to better
early pruning and early termination.

4.2   Term-Document-Binary-Matrix-Based Combined Algorithm

We call our version of the CA algorithm "BMCA," short for "term-document-
binary-matrix-based Combined Algorithm."

1. Create B_q for $q = \{t_1, \ldots, t_m\}$ from the document-ID-only lists of $t_1, \ldots, t_m$.
   Initialize T_k, S, U, $\underline{w}_i$ and $mink$; i.e., $T_k \leftarrow \{\}$, $S \leftarrow \{\}$, $U \leftarrow$ all documents,
   $\underline{w}_i \leftarrow$ a sufficiently large value, and $mink \leftarrow 0$.
2. Choose sorted posting lists and perform sorted accesses:
   (a) Compute $score_{\mathrm{UB}}(U)$. If $score_{\mathrm{UB}}(U) > mink$, choose the m' sorted post-
       ing lists corresponding to the largest m' $\underline{w}'_i$. Otherwise, choose the fol-
       lowing sorted posting lists: $\{L_i \,|\, \exists$ not-yet-pruned $d_j \in T_k \cup S$ s.t. $t_i \in q \setminus a(d_j)$ and $B_q(i,j) = 1\}$.
   (b) Perform sorted access in parallel for each of the chosen lists. When the
       sorted accesses are completed,
       – Maintain the last obtained $\underline{w}_1, \ldots, \underline{w}_m$ in the sorted posting lists.
       – Maintain the set of current top-k documents, T_k, based on the lower
         bound score of each document seen so far; if two documents have
         the same lower bound score, the tie is broken using the upper bound
         scores, such that the document with the largest upper bound score
         wins. A tie among documents with the largest upper bound score
         is arbitrarily broken. Update $mink$ to the lower bound score of the
         new rank-k document in T_k.
       – Maintain S and U.
3. Call document d viable if $score_{\mathrm{UB}}(d) > mink$. Every μ sorted accesses (that
   is, every time sorted accesses for μ sorted posting lists are performed), do the
   following: pick ν viable documents (if any) seen so far which have the largest
   upper bound scores but in each of which not all per-term nonzero scores for
   $t_1, \ldots, t_m$ are known (again, ties are broken arbitrarily). Perform random
   accesses for all of its (at most $m - 1$) missing per-term nonzero scores for
   unknown query terms in each of those ν documents. Update $mink$. If there
   is no such documents, do not perform random accesses in this step.
4. Compute $score_{\mathrm{UB}}(S)$ and $score_{\mathrm{UB}}(U)$. If the following conditions are satisfied, halt and return documents in T_k. Otherwise return to step 2.
   – T_k contains k distinct documents.
   – There are no viable documents left outside T_k; that is, $score_{\mathrm{UB}}(S) \leq mink$ and $score_{\mathrm{UB}}(U) \leq mink$.

This algorithm consists of two phases. The first phase of this algorithm (that is,
when $score_{\mathrm{UB}}(U) > mink$, which means that some document in U might make
it into the final top-k documents.), is to find all the documents which could

qualify for the final top-k documents, and the second phase of this algorithm (that is, when $score_{\mathrm{UB}}(U) \leq mink$, which means that no document in U could make it into the final top-k documents.), is to choose the final top-k documents from documents found in the first phase. The algorithm optimizes the execution of the sorted and random access, which is described next.

Optimizing Sorted Access. In the first phase, the algorithm decreases $score_{\mathrm{UB}}(U)$ by performing as few sorted accesses enough to decrease $score_{\mathrm{UB}}(U)$ as possible based on Equation (6), not performing sorted accesses for all the sorted posting lists as in CA. This reduces sorted accesses. In the second phase, the algorithm identifies and ignores sorted posting lists based on B_q, from each of which any missing per-term nonzero score for any query term in any not-yet-pruned document in T_k or S can not be obtained. This avoids useless sorted accesses.

Optimizing Random Access. By Equation (5), which estimates $score_{\mathrm{UB}}(d)$ more tightly than Equation (2) of CA, the algorithm chooses more viable documents for random accesses than does CA. This leads to an increase in $mink$ and thereby better early pruning and early termination.

5   Experimental Evaluation

5.1   Setup

We used the TREC GOV2 collection. This collection contains about 25 million web documents crawled from the gov domain during early 2004. The uncompressed size of this collection is 426GB. To evaluate the performance for longer queries, we created expanded queries from the title fields of TREC topics 701–850 using the query expansion feature of the Indri search engine.[1] We used 32- and 64-term expanded queries. In the experiments in described here, no stopwords were removed. We implemented CA and BMCA on the Zettair search engine.[2] Because the Zettair search engine provides a standard technique for processing the entire posting list for each query term for the pivoted cosine model [11], we implemented this model both in CA and BMCA. For handling sorted access, both CA and BMCA were implemented such that they obtained a partition of 64KB in each sorted posting list per sorted access. This partition contained about 13,000 postings. For handling random access, we integrated the additional data structure which assigned each document to a document vector containing all terms in that document with nonzero per-term scores along with their scores. Thus, both CA and BMCA were implemented such that they obtained a document vector for each random access.[3] For a term-document binary matrix, we

---

[1] http://www.lemurproject.org/indri/
[2] http://www.seg.rmit.edu.au/zettair/
[3] μ and ν were set as follows: $\mu = 32$ for 32-term queries, and $\mu = 64$ for 64-term queries for the first iteration, and $\mu = 100$ for k = 5 and 10, and $\mu = 200$ for k = 50 and 100 after the first iteration for both query sizes. $\nu = 50$ for k = 5 and 10, and $\nu = 100$ for k = 50 and 100 for both query sizes.

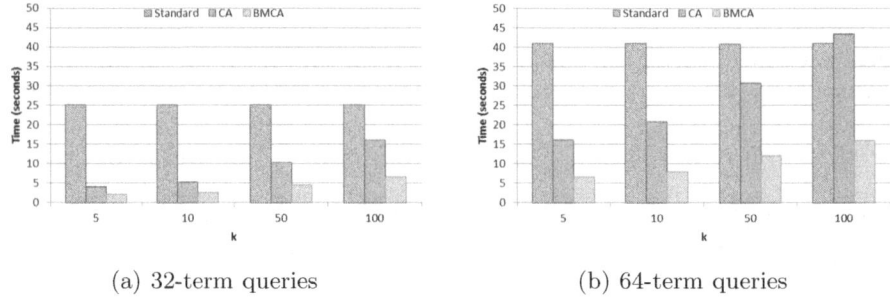

(a) 32-term queries   (b) 64-term queries

Fig. 1. Average execution times for standard technique, CA, and BMCA for 32- and 64-term queries for various values of k for TREC GOV2 data set

created a matrix as an array of 32- or 64- bit integers for a 32- or 64-term expanded query from document-ID-only lists corresponding query terms. All runs were performed on a single core of a 2.50-GHz Intel(R) Xeon CPU. For BMCA, the runs were performed on condition that the term-document binary matrices were created beforehand and cached.

5.2 Results

We compared the performance of our algorithm, BMCA, with that of a standard technique, in which the entire posting list for each query term is processed, and that of the original algorithm, CA. We measured the average execution time per query. Figure 1 shows the average execution times for 32- and 64-term queries for various values of k for the TREC GOV2 data set. BMCA significantly outperformed the standard technique and CA for every query size and every k. In particular, while the performance of both CA and BMCA deteriorated as k increased for both query sizes, the CA degration was much greater, and its performance actually became worse than that of the standard technique for query size = 64 and k = 100. BMCA showed high performance, yielding performance gains of up to a factor of 2.7 over CA. This is because our term-document-binary-matrix-based algorithm leads to better early pruning and early termination even for expanded queries. Figure 2 illustrates the effects of the better early pruning and early termination property of BMCA. It shows the number of documents retained at each iteration during the execution of both algorithms for a 32-term expanded query from the title field of TREC TOPIC 741 and k = 100. BMCA pruned more documents than CA by an order of magnitude and terminated after about three fifths of the number of iterations performed by CA. That is, BMCA performed about three fifths of sorted and random accesses performed by CA. We statistically analyzed this property of BMCA. Figure 3 shows the average total number of sorted and random accesses performed by CA and BMCA, and the average maximum number of documents retained by both algorithms, for

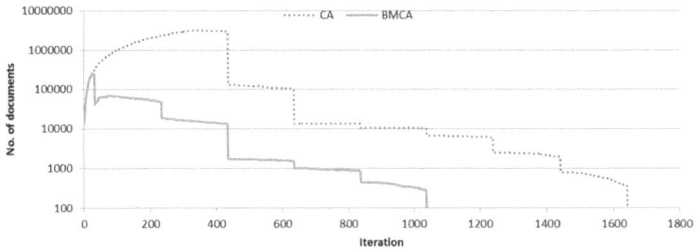

Fig. 2. Number of documents retained at each iteration during execution for a 32-term expanded query from the title field of TREC TOPIC 741 for k = 100 for CA and BMCA

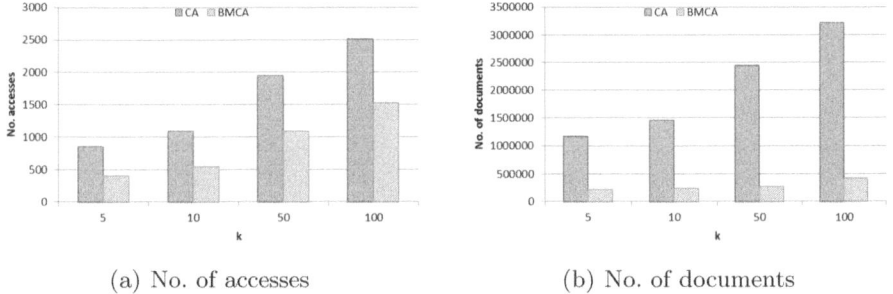

(a) No. of accesses   (b) No. of documents

Fig. 3. Average total number of sorted and random accesses performed by CA and BMCA and average maximum number of documents retained by both algorithms for 32-term queries for various values of k for TREC GOV2 data set

32-term queries for various values of k for TREC GOV2 data set.[4] The results shown in these figures indicate that, for longer queries, CA incurs substantial I/O overhead due to more sorted and random accesses along with considerable CPU overhead due to the explosive increase of retained documents, resulting in performance degradation for longer queries. In contrast, BMCA incurs much less I/O overhead due to efficient early pruning and early termination based on the term-document binary matrices along with much less CPU overhead due to a reduced number of retained documents, even for longer queries. As a result, BMCA has significantly better performance than CA.

6 Conclusion

Our integration of a simple data structure, the "term-document binary matrix," into the CA algorithm resulted in efficient evaluation of top-k disjunctive queries,

---

[4] Similar results are obtained for 64-term queries.

even ones expanded using enhancement techniques such as query expansion and relevance feedback. Experimental evaluation using the TREC GOV2 data set and expanded versions of the evaluation queries attached to this data set showed that the resulting algorithm performed significantly better than CA. In addition, we have studied the instance optimality of the resulting algorithm, but we will discuss on this issue at the next chance.

The evaluation was done under the assumtion that the term-document binary matrices were created beforehand and cached because the creation from the document-ID-only lists imposes relatively large overhead due to our farily simple implementation of the functionality. However, the term-document binary matrices should be created at runtime in the case of large-scale IR systems such as web search engines because it is not realistic for such systems to create beforehand and cache the term-document binary matrices for a large number of queries. We plan to re-implement the functionality using state-of-the-art techniques, e.g., [12], which should greately reduce the overhead. We also plan to investigate the combination of our approach with that of Bast et al. [2] to achive a more sophisticated random access scheduling.

References

1. Anh, V.N., Moffat, A.: Pruned query evaluation using pre-computed impacts. In: Proc. of SIGIR, pp. 372–379 (2006)
2. Bast, H., Majumdar, D., Schenkel, R., Theobald, M., Weikum, G.: IO-Top-k: index-access optimized top-k query processing. In: Proc. of VLDB, pp. 475–486 (2006)
3. Buckley, C., Lewit, A.: Optimization of inverted vector searches. In: Proc. of SIGIR, pp. 97–110 (1985)
4. Downey, D., Dumais, S., Liebling, D., Horvitz, E.: Understanding the relationship between searchers' queries and information goals. In: Proc. of CIKM, pp. 449–458 (2008)
5. Fagin, R.: Combining Fuzzy Information: an Overview. SIGMOD Record 31(2), 109–118 (2002)
6. Kumar, R., Punera, K., Suel, T., Vassilvitskii, S.: Top-k aggregation using intersections of ranked inputs. In: Proc. of WSDM, pp. 222–231 (2009)
7. Long, X., Suel, T.: Three-level caching for efficient query processing in large Web search engines. In: Proceedings of WWW, pp. 257–266 (2005)
8. Manning, C.D., Raghavan, P., Schtze, H.: Introduction to Information Retrieval. Cambridge University Press (2008)
9. Persin, M., Zobel, J., Sacks-Davis, R.: Filtered document retrieval with frequency-sorted indexes. Journal of the American Society for Information Science 47(10), 749–764 (1996)
10. Schenkel, R., Broschart, A., Hwang, S.-w., Theobald, M., Weikum, G.: Efficient Text Proximity Search. In: Ziviani, N., Baeza-Yates, R. (eds.) SPIRE 2007. LNCS, vol. 4726, pp. 287–299. Springer, Heidelberg (2007)
11. Singhal, A., Buckley, C., Mitra, M.: Pivoted document length normalization. In: Proc. of SIGIR, pp. 21–29 (1996)
12. Zhang, J., Long, X., Suel, T.: Performance of compressed inverted list caching in search engines. In: Proc. of WWW, pp. 387–396 (2008)

Topic Analysis for Online Reviews
with an Author-Experience-Object-Topic Model

Yong Zhang[1,2], Dong-Hong Ji[1], Ying Su[3], and Po Hu[1,2]

[1] Computer School, Wuhan University, Wuhan, P.R. China
[2] Department of Computer Science, Huazhong Normal University, Wuhan, P.R. China
[3] Department of Computer Science, Wuchang Branch,
Huazhong University of Science and Technology, Wuhan, P.R. China
ychang.cn@gmail.com, donghong_ji2000@yahoo.com.cn
suying929@163.com, phu@mail.ccnu.edu.cn

Abstract. In this paper, we propose a new probabilistic generative model for topic analysis of online reviews, called Author-Experience-Object-Topic Model (AEOT). This model is to capture the relationship between the authors, objects and reviews in order to improve the performance of topic analysis. The model, as a general one, can be transformed to six simpler models, and can produce topic-word, author-topic and object-topic distributions. Experimental results show that the model is suitable for topic analysis of online reviews, and outperforms other existing methods.

Keywords: Latent Dirichlet Allocation, Author-Experience-Object-Topic Model, Social Review Network, Topic Model.

1 Introduction

Statistical topical modeling [1, 2] has attracted much attention recently due to its broad applications in text mining and information retrieval. Particularly, the Latent Dirichlet Allocation (LDA) model has become popular due to its solid theoretical foundation and promising performance, and several strategies have been proposed to extend the model in order to simulate the contexts for different purposes [3, 4, 5, 8].

On the other hand, social data available in the internet has drawn attention to improve traditional topic analysis. For instances, the Topic-Perspective model [9] simulates the generation process of social annotations by modeling the tag generation and word generation process separately and incorporating the user information into the process. The Author-Topic Model (AT) [3] extends the basic topical model to include author information in which topics and authors are jointly modeled. Each author is a multinomial distribution over topics and each topic is a multinomial distribution over words. The Author-Recipient-Topic model (ART) [4] is proposed for social network analysis, which learns topic distributions based on the direction sensitive messages sent between entities.

M.V.M. Salem et al. (Eds.): AIRS 2011, LNCS 7097, pp. 303–314, 2011.
© Springer-Verlag Berlin Heidelberg 2011

The online reviews, free text generated by the authors to comment on target objects (i.e. services or products), is one popular type of social media in the current internet. The authors, target objects as well as the reviews form a heterogeneous network, which we refer to as social review network. Although much work has been done for online reviews, few of them can model the review content alongside with the social review network [7, 12]. The AT model adds an author layer to LDA, and the ART model is conditioned distinctly on both the sender and recipient for social network analysis. But for online reviews, the author and object is heterogeneous with the reviews. Previous work in recommendation systems and collaborative filtering has modeled the existence of such links from the authors to objects, but no text content exist on those links [10, 11].

In order to utilize the social network information of online reviews, we propose an Author-Experience-Object-Topic Model (AEOT), a new probabilistic generative model, which models the relationships among the different entities involved in the social review network, including the authors, objects and content of reviews.

The rest of this paper is organized as follows. Section 2 reviews related work on topic modeling and various models for social media proposed in previous research. Section 3 presents the proposed AEOT Model for online reviews and introduces the parameter estimation process. In section 4, we evaluate the performance of the proposed model based on an online movie dataset. In section 5, we conclude the paper and present the future work.

2   Related Work

For social media, several strategies extending LDA have been proposed, which generally employ information other than document words for topic learning. For instance, the APT model assumes that each author write under one or more personas, which are represented as independent distributions over hidden topics for matching papers with reviewers [6]. The Pairwise-Link-LDA and the Link-PLSA-LDA models jointly model both text and link for Academic Social Networks [5]. The author-topic model [3] uses the authorship information together with the text to learn topics. Among these models only the authorship information is incorporated into the model.

In recommendation system and collaborative filtering domain, the users, items and the links between them form a heterogeneous network. Generally the user and the item are modeled together [17]. The Topic-Perspective model [9] simulates the generation process of social annotations, by representing all related entities (users, documents, words, and tags) as well as latent variables (topics, user perspectives) in a unified model. FolkRank [11] is proposed to exploit the structure of the users, resources, and tags, and the user-based assignment of tags to resources called folksonomies for social bookmark. To our knowledge, few of work utilize the social review network for topic analysis.

3   Author-Experience-Object-Topic Model

In this section, we introduce the proposed Author-Experience-Object-Topic Model for topic analysis of online reviews. This model depicts the social commenting process and the generation process of content terms in a unified framework. This model is to represent and connect all the observed and hidden variables of social review network in a unified framework. By estimating this model, we can learn the topical structures of the reviews relevant with the authors and objects.

3.1   Motivation

In web communities for online reviews, one author may write multiple reviews towards different objects and one object may be commented by multiple authors. Thus the authors, objects and reviews form a heterogeneous social review network.

In social review network, we assume that the content of each review entails the hidden relationship between the author and the object. The assumption is motivated by three reasons. Firstly each author has a different preference and style of commenting. For the same target object and the nearly same experience, different authors may give different review. Secondly, each object has multiple aspects which may have different qualities. Thirdly, the author's experience on the object should affect the content of review. For example an author might feel well after he went to see a film because he had a nice seat. As shown in Table 1, the snippet 1 is about the author, the snippet 2 discusses the experience, and the snippet 3 talks about the movie.

Table 1. Three textual snippets in lowercase from review18645 in IMDB

1: they recently sent me a vhs copy of their down with america trilogy and i decided to spend an hour of my day watching it.
2: sure , the risky use of vhs instead of super 8mm or 16mm was a pain , and the natural light was one of the most annoying things about public access films,
3: but the movie itself was fairly enjoyable. down with america concerns a government agent , needless murder , and a book containing everything from the unabomber's manifesto to the 1995 apple computer profit report .

3.2   The Author-Experience-Object-Topic Model

The model is designed based on the real social commenting process, in which a term may be generated from the author, the object or the experience. In order to reflect this nature of online review in the generative model, we adopt a switch variable x to control the influence of the author, the object and the experience. The proposed model is illustrated in Fig. 1 (c), where D denotes the number of reviews, N_d denotes the number of terms in review d, K denotes the number of topics, V denotes the size of vocabulary, U denotes the number of authors and O denotes the number of objects in the dataset. The α^e, α^u, α^o, η and β are hyper parameters and priors of Dirichlet distributions.

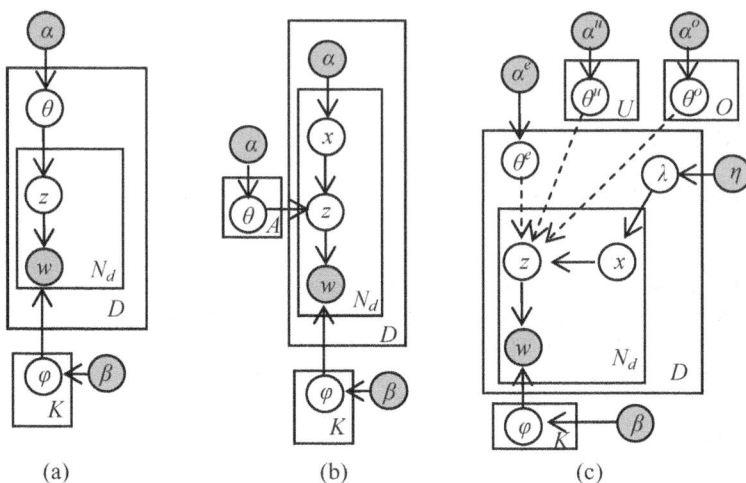

Fig. 1. Generative models for documents. (a) Latent Dirichlet Allocation (LDA). (b) The author model. (c) The Author-Experience-Object-Topic model.

As shown in Fig. 1, the generative process of review in the model can be described as follows:

- For each document d, sample $\theta^{(e)}_d \sim Dir(\alpha^e)$
- For each author u, sample $\theta^{(u)}_u \sim Dir(\alpha^u)$
- For each object o, sample $\theta^{(o)}_o \sim Dir(\alpha^o)$
- For each document d, sample $\lambda_d \sim Dir(\eta)$
- For each of the K topics k, sample $\varphi_k \sim Dir(\beta)$
- For each of the N_d word tokens w_i in document d:

  - Choose $x_i \sim Multinomial(\lambda_d)$
    - If $x_i = exp$, Choose a topic $z_i \sim Multinomial(\theta^{(e)}_d)$
    - If $x_i = user$, Choose a topic $z_i \sim Multinomial(\theta^{(u)}_u)$
    - If $x_i = obj$, Choose a topic $z_i \sim Multinomial(\theta^{(o)}_o)$
  - Choose a word $w_i \sim Multinomial(\varphi_{zi})$

As shown in Fig. 1 (c), our model adds an author layer, an object layer based on LDA. $Dir(\eta)$ is the prior Dirichlet distribution for choosing x among experience, author and object for each term. The latent variable x indicates that the term is associated with author, object or experience. Through the latent variable x, we try to distinguish which term is associated with the author, which term is associated with the object and which term is associated with the experience, as shown in Table 1. In fact our model is very similar to author-topic model [3] except the generation of variable x as shown in Section 3.4.

3.3   Gibbs Sampling Algorithms

Several methods have been developed for estimating the latent parameters in LDA model, such as the variational expectation maximization [1], expectation propagation

[13], and Gibbs sampling [2]. Compared to the other two methods, Gibbs sampling often yields relatively simple algorithms for approximate inference in high-dimensional models such as LDA. Therefore we select this approach for parameter estimation. For our model, we derive the sampling equations for our model.

In the AEOT model, we have two sets of latent variables: z and x. The joint probability of the z, x assignments and the words can be factored into the following terms:

$$
\begin{aligned}
p(w, z, x) &= p(w \mid z) p(z \mid x) = p(w \mid z)(p(z \mid \theta^e) p(x = exp) \\
&+ p(z \mid \theta^u) p(x = user) + p(z \mid \theta^o) p(x = obj))
\end{aligned} \tag{1}
$$

We draw each $(z_i; x_i)$ pair as a block, conditioned on all other variables. When $x=exp$, the sampling equation is:

$$
p(z_i = k, x_i = exp \mid w, z_{-i}, x_{-i}) \propto \frac{n_{w_i}^k + \beta}{n_{\cdot}^k + V\beta} \cdot \frac{n_k^e + \alpha^e}{n_d^e + K\alpha^e} \cdot \frac{n_d^e + \eta_{exp}}{N_d + \sum_{x \in \{exp,user,obj\}} \eta_x} \tag{2}
$$

where n_d^e is the number of times that words are generated from experience in review d; n_k^e is the number of times that words assigned topic k are from experience in the review d; n^k is the number of times that words are assigned topic k. $n_{w_i}^k$ is the number of times that word in the position i of review d are assigned topic k; V is the number of terms in the vocabulary.

When $x=user$, the equation is:

$$
p(z_i = k, x_i = exp \mid w, z_{-i}, x_{-i}) \propto \frac{n_{w_i}^k + \beta}{n_{\cdot}^k + V\beta} \cdot \frac{n_k^u + \alpha^u}{n^u + K\alpha^u} \cdot \frac{n_d^u + \eta_{user}}{N_d + \sum_{x \in \{exp,user,obj\}} \eta_x} \tag{3}
$$

where n^u is the number of times that words are generated from author u; n_k^u is the number of times that words assigned topic k are from author u; n_d^u is the number of times that words are generated from author u in the review d;.

When $x=obj$, the equation is:

$$
p(z_i = k, x_i = exp \mid w, z_{-i}, x_{-i}) \propto \frac{n_{w_i}^k + \beta}{n_{\cdot}^k + V\beta} \cdot \frac{n_k^o + \alpha^o}{n^o + K\alpha^o} \cdot \frac{n_d^o + \eta_{obj}}{N_d + \sum_{x \in \{exp,user,obj\}} \eta_x} \tag{4}
$$

where n^o is the number of times that words are generated from object o; n_k^o is the number of times that words assigned topic k are from object o; n_d^o is the number of times that words are generated from object o in the review d;.

After a set of sampling processes based on the posterior distributions calculated with above equations, we can estimate the parameters for any single sample using the following equations:

$$\theta_k^e \approx \frac{n_k^e + \alpha^e}{n_d^e + K\alpha^e} \qquad (5) \qquad \theta_k^u \approx \frac{n_k^u + \alpha^u}{n^u + K\alpha^u} \qquad (6)$$

$$\theta_k^o \approx \frac{n_k^o + \alpha^o}{n^o + K\alpha^o} \qquad (7) \qquad \varphi_w^k \approx \frac{n_w^k + \beta}{n^k + V\beta} \qquad (8)$$

3.4   Generation of Variable x

Note that AEOT, AT and ART models all have two latent variables: z and x for each word, where x is to control the generation of z and w. In AT [3] for each word w, an author, x is chosen at uniform from a_d, all authors of the documents. And in ART [4] for each word w, a recipient, x, is chosen uniformly from r_d, all recipients of the email. The ACT model in [14] is similar to AT model for the generation of latent author x_d.

But in our model, the generation of x is not at random, but is chosen from a multinomial distribution λ, which means that x is not distributed evenly in the document.   In fact the author's and the object's roles are different in the review content. Modeling the object can achieve better performance, which suggests that the object plays a more important role in the model than the author. It seems that the generation of x in this paper is more reasonable than in AT and ART. For comparison, another model is proposed which we refer to as AEOT-R. In AEOT-R model, the x is chosen at random. Experiments show the strategy for the generation of x in AEOT model is better than AEOT-R for online reviews.

3.5   Variants of AEOT Model

In our model, we use an additional multinomial distribution λ (λ_{exp}, λ_{user} and λ_{obj}, subject to $\lambda_{exp}+\lambda_{user}+\lambda_{obj}=1$) to record the probability that each word is generated from the experience, the author or the object. Greater value of λ_{exp} indicates a higher probability that the word is generated from the experience and vice versa. Note that the words with $\lambda_{exp}=1$ are completely generated from the experience and not affected by the author and the object, and thus our model is degraded and equivalent with LDA. Contrarily, the words with $\lambda_{user} = 1$ are totally generated from the author. Thus the AEOT model is similar to the AT model. Similarly the words with $\lambda_{obj}=1$ are totally generated from the object.

Especially if only λ_{user} is set to 0, the words are generated from the experience or the object, not from the author. If only λ_{obj} is set to 0, the words are generated from the experience or the author, not from the object. If only λ_{exp} is set to 0, the words are generated from the author or the object.

Summarily our model can be transformed to six simpler models, as listed below.

- LDA, when $\lambda_{exp}=1$, $\lambda_{user}=0$, $\lambda_{obj}=0$;
- AT, when $\lambda_{exp}=0$, $\lambda_{user}=1$, $\lambda_{obj}=0$;
- OT, when $\lambda_{exp}=0$, $\lambda_{user}=0$, $\lambda_{obj}=1$;
- EOT, when $\lambda_{exp}>0$, $\lambda_{user}=0$, $\lambda_{obj}>0$;
- EAT, when $\lambda_{exp}>0$, $\lambda_{user}>0$, $\lambda_{obj}=0$;
- AOT, when $\lambda_{exp}=0$, $\lambda_{user}>0$, $\lambda_{obj}>0$;

Note that for document with single author the AT model in this paper is equivalent with the AT model in [3]. And each review has only one author.

4   Experiments

In this section, we investigate the performance of the proposed AEOT model based on a movie review dataset [19]. We also compare our model to other six simpler models for online reviews.

4.1   Dataset

The dataset used in the experiments is from a social movie review dataset collected from [19] by Bo Pang[1]. The original dataset contains 27886 html pages. From the 27886 html pages we extracted 27857 reviews contents, and their authors and movie ids. And the information of the left 29 html pages is incomplete. For the experiments, we selected 17657 reviews as the dataset. To further clean the dataset, stop words and words with term frequency less than 5 or greater than 7000 are filtered out. The final dataset used for experimentation contains 17657 reviews, 155 authors, 1540 movies, 46661 unique words and 4,970,358 word tokens. Then we randomly selected 1601, around 9% of the reviews as a held-out test data and trained the model on the remaining 90%. The data is available online[2].

4.2   Experimental Setup

The AEOT model has five Dirichlet prior parameters. Previous research also found that these parameters only affect the convergence of Gibbs sampling but not much the output results [15]. So we set $\alpha^e=0.5$, $\alpha^u=0.5$, $\alpha^o=0.5$, $\beta=0.05$ and $\eta=0.5$ for all experiments. For the number of topics K, we set to 50, 100, 150 and 200 respectively in the experiments. For the training process, the number of iteration of the Gibbs sampler is set to 1000, and the iteration number is set to 400 for the held-out test data in all experiments empirically.

4.3   Document Modeling

In the experiments, we use perplexity as the criterion for model evaluation. Perplexity is a standard measure for evaluating the generalization performance of a probabilistic

---

[1] http://www.cs.cornell.edu/People/pabo/movie-review-data/
[2] http://www.clr.org.cn/ychang/social-review/

model. The value of perplexity reflects the ability of a model to generalize to unseen data. A lower perplexity score indicates better generalization performance. Formally, the perplexity for a test set of D_{test} documents is calculated as formulation 9, where $\theta_{(train)_k}^u$, $\theta_{(train)_k}^o$ and $\varphi_{(train)_{w_d^i}}^k$ are learned from the training process, and $p(x)$ and $\theta_{(test)_k}^e$ are estimated through a Gibbs Sampling process on the test data based on the parameters learned from the training data. M_{test} denotes the number of reviews in test data.

$$\text{perplexity}(D_{test} \mid D_{train}) = \exp\left(-\frac{\sum_d^{M_{test}} \log(p(\mathbf{w}_d \mid D_{train}))}{\sum_d^{M_{test}} N_d}\right)$$

$$\log(p(\mathbf{w}_d \mid D_{train})) = \sum_i^{N_d} \log(p(w_d^i \mid D_{train})) \tag{9}$$

$$p(w_d^i \mid D_{train}) = p(x = exp) \sum_{k=1}^K \theta_{(test)_k}^e \varphi_{(train)_{w_d^i}}^k$$

$$+ p(x = user) \sum_{k=1}^K \theta_{(train)_k}^u \varphi_{(train)_{w_d^i}}^k + p(x = obj) \sum_{k=1}^K \theta_{(train)_k}^o \varphi_{(train)_{w_d^i}}^k$$

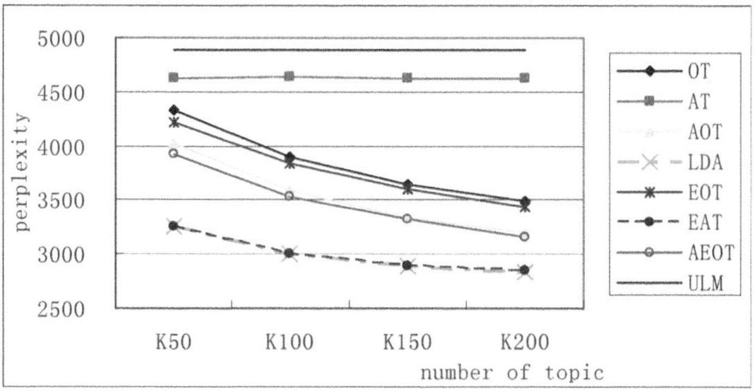

Fig. 2. The perplexity results of eight models with topic number K=50, 100, 150 and 200

Fig. 2 plots the perplexities of eight models being compared, introduced in the section III-E on the test data for topic number set to 50, 100, 150 and 200 respectively. The ULM is the abbreviation of smoothed unigram language model, where the smoothing parameter is set to 0.05.

From Fig. 2, we can see that the perplexity values of EAT model and LDA perform similarly, and are the best of all the eight models. It's very strange that the performance of EAT is so good. Further studies uncover the secret that the most of words are assigned to the experience, and little words are assigned to the author for each review in EAT model. That is to say, the EAT is degraded and similar to the LDA.

However the AT performs very badly, just better than the ULM. But OT and EOT perform similarly and better than AT. It seams that the object plays a more important role in social review network. In fact the most of words in the movie review are related with the target movie, and the results are consistent with our intuitive.

The AEOT works better than AOT slightly. At the same time, we note that the result of EAT is better than AT, and EOT is better than OT. It suggests that the experience can improve the performance and plays an important role in the AEOT model for online reviews. In fact the role of the experience is similar to the fictitious authors in AT Model [3], which could improve the performance of AT. However the role of experience is limited for the AOT and AEOT perform similarly.

However the performances of AOT and AEOT are worse than the results of EAT and LDA, but the AOT and AEOT can produce the distributions over topic of authors and objects simultaneously from the social review network.

4.4   Two Strategies about the Generation of x

As mentioned in section 3.4, the variable x can be drawn at uniform, such as AT [3], ART [4] and ACT [14]. We implemented the AEOT-R and AOT-R, where the x is drawn uniformly.

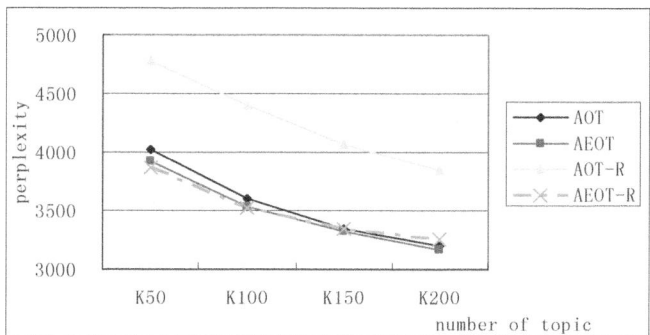

Fig. 3. The perplexity results of AEOT, AOT, AEOT-R and AOT-R models with topic number K=50, 100, 150 and 200

For perplexity, the AEOT-R is a little worse than AEOT except for K=50 and AOT-R is much worse than AOT, as shown in Fig. 3. Especially the performance of AOT-R is much worse than other models. The experimental results suggest that the drawing strategy of variable x in this paper is more efficient than that in AT, ART and ACT for online reviews. The results provide a good suggestion for design of the switch variable in probabilistic generative model.

4.5   Discovered Topics

Table 2 shows 9 topics discovered by LDA and AEOT respectively in the training dataset, where the topic number K is set to 200. Each topic is shown with the top 10 words. The symmetrized Kullback Leibler (KL) distance [2] is used to find the most similar topic discovered by AEOT (as shown in the right of Table 2) for each topic obtained by LDA (as shown in the left of Table 2). In machine learning there has been less effort on qualitative understanding of the semantic nature of the learned topics. Some work has been done on the evaluation of topic coherence [16], but there is no standard measure for it yet. We leave the problem as future work.

In our experiments, we remained the words "the, their, them, then" in the vocabulary, not contained in the stop word list. In most of topics discovered by LDA, the word "the" is the top 1 word out of the top 50 words. As shown in the left column of Table 2, there are 8 topics in which the word "the" takes the first place. But in the topics discovered by AEOT, there is only one topic, in which the list of top 50 words contains the word "the", as shown as topic No. 9 in Table 2. It seems that the AEOT model could produce more coherent topics than LDA intuitively. This is a very interesting result that we will explore how to evaluate in the future work.

Table 2. A subset of discovered topics by LDA and AEOT

No.	Topics by LDA	Topics by AEOT
1	the, lawyer, trial, law, court, courtroom, legal, attorney, firm, civil	trial, lawyer, court, courtroom, attorney, judge, accused, richard, defense, murder
2	the, horse, annie, pleasantville, color, grace, redford, jennifer, tom, ross	horse, grace, redford, whisperer, annie, andrew, thomas, tom, sam, golf
3	the, eye, usual, todd, apt, evil, suspects, stephen, ian, singer	todd, singer, apt, thomas, pupil, angus, kurt, sally, russell, celebration
4	the, scream, horror, killer, halloween, slasher, genre, urban, scary, sequel	scream, horror, scary, halloween, slasher, killer, genre, summer, sequel, williamson
5	the, musical, song, band, rock, songs, singing, soundtrack, singer, sing	dance, musical, dancing, numbers, songs, dancer, singing, song, sing, tango
6	boy, parents, child, children, kid, dad, age, mom, older, brother	child, boy, children, parents, emotional, heart, drama, living, tale, sister
7	the, space, armageddon, summer, bay, team, bruce, rock, deep, impact	space, mission, mars, armageddon, crew, nasa, earth, apollo, astronauts, team
8	jokes, laughs, laugh, gags, joke, comic, hilarious, comedies, amusing, funniest	jokes, laughs, comic, hilarious, laugh, gags, joke, comedies, their, amusing
9	the, girls, their, godfather, coppola, francis, them, boys, ford, ii	the, their, them, then, side, called, hand, full, including, early

5   Conclusion and Future Work

In this paper, we presented a social review network, and proposed the Author-Experience-Object-Topic Model for topic analysis of online reviews. This model provides a probabilistic view to explore the relationships between authors, objects, reviews, topics, and words in the social review network. The proposed model can not only produce the topical distribution of each review, but also generates the topical distributions of each user and each object simultaneously in an unsupervised manner. The proposed AEOT is general and can be transformed to simpler models including LDA. Experiments on the movie review dataset that we extracted from the original 27886 review html files show that the method is promising. Furthermore, this model also generates topical structures for authors and objects, which may be helpful for social recommendation, review classification, author's community analysis, target object clustering, information retrieval for reviews, review sentiment analysis and other review's text mining tasks.

How to evaluate the coherence of topics produced by topic models is intractable. In future work, we will focus on the evaluation of coherence of topics and study how to produce more coherence topics for online reviews. At the same time, we will try to apply the model to recommendation system. Recommender systems are usually classified into three categories based on their approach: content-based, collaborative, and hybrid approaches combining content-based and collaborative methods [18]. Recommendation system based on social review network may be another promising future task. However other social datasets of different characteristics will be chosen for experimentation to test our model's applicability in the future.

Acknowledgments. We thank anonymous reviewers for their useful comments. The paper is supported in part by the National Nature Science Foundation of China (Grant No. 90820005 and 61070082). Ying Su is the corresponding author.

References

1. Blei, D.M., Ng, A., Jordan, M.: Latent Dirichlet allocation. J. Machine Learning Research 3, 993–1022 (2003)
2. Steyvers, M., Griffiths, T.: Probabilistic Topic Models. In: Handbook of Latent Semantic Analysis (2007)
3. Zvi, M.R., Chemudugunta, C., Griffiths, T., Smyth, P., Steyvers, M.: Learning author-topic models from text corpora. ACM Transactions on Information Systems 28(1), 1–38 (2010)
4. McCallum, A., Wang, X., Corrada-Emmanual, A.: Topic and role discovery in social networks and experiments in enron and academic e-mail. J. Artificial Intelligence Research 30, 249–272 (2007)
5. Tang, J., Zhang, J., Yao, L., Li, J., Zhang, L., Su, Z.: ArnetMiner: Extraction and Mining of Academic Social Networks. In: Proceeding of the 14th ACM SIGKDD International Conference on Knowledge Discovery and Data Mining, pp. 990–998 (2008)
6. Mimno, D., McCallum, A.: Expertise modeling for matching papers with reviewers. In: Proceedings of the 13th ACM SIGKDD International Conference on Knowledge Discovery and Data Mining, pp. 500–509 (2007)
7. Lu, Y., Zhai, C., Sundaresan, N.: Rated Aspect Summarization of Short Comments. In: Proceedings of the 18th International Conference on World Wide Web (2009)
8. Newman, D., Chemudugunta, C., Smyth, P.: Statistical entity-topic models. In: Proceedings of the 12th ACM SIGKDD international Conference on Knowledge Discovery and Data Mining, Philadelphia, PA, pp. 680–686 (2006)

9. Lu, C., Hu, X., Chen, X., Park, J.-R.: The topic-perspective model for social tagging systems. In: Proceedings of the 16th ACM SIGKDD International Conference on Knowledge Discovery and Data Mining, Washington D.C., USA, July 25-28, pp. 683–692 (2010)
10. Lu, C., Hu, X., Park, J.-R., Huang, J.: Post-based collaborative filtering for personalized tag recommendation. In: iConference, pp. 561–568 (2011)
11. Hotho, A., Jäschke, R., Schmitz, C., Stumme, G.: FolkRank: A Ranking Algorithm for Folksonomies. In: Proceeding of Fachgruppe Information Retrieval, FGIR (2006)
12. Titov, I., McDonald, R.T.: Modeling online reviews with multi-grain topic models. In: Proceeding of the 17th International Conference on World Wide Web, Beijing, China, pp. 111–120 (2008)
13. Minka, T., Lafferty, J.: Expectation-propagation for the generative aspect model. In: The 18th Conference in Uncertainty in Artificial Intelligence, pp. 352–359. Morgan Kaufmann, Edmonton (2002)
14. Celikyilmaz, A., Hakkani-Tur, D., He, H., Kondrak, G., Barbosa, D.: The Actor-Topic Model for Extracting Social Networks in Literary Narrative. In: Proceedings of the NIPS 2010 Workshop Machine Learning for Social Computing, p. 7 (2010)
15. Zhou, D., Bian, J., Zheng, S., Zha, H., Giles, C.L.: Exploring Social Annotations for Information Retrieval. In: Proceeding of the 17th International Conference on World Wide Web, Beijing, China, pp. 715–724 (2008)
16. Newman, D., Lau, J.H., Grieser, K., Baldwin, T.: Automatic evaluation of topic coherence. In: HLT 2010 Human Language Technologies (2010)
17. Ovsjanikov, M., Chen, Y.: Topic Modeling for Personalized Recommendation of Volatile Items. In: Balcázar, J.L., Bonchi, F., Gionis, A., Sebag, M. (eds.) ECML PKDD 2010. LNCS, vol. 6322, pp. 483–498. Springer, Heidelberg (2010)
18. Adomavicius, G., Kwon, Y.: Improving Aggregate Recommendation Diversity Using Ranking-Based Techniques. IEEE Transactions on Knowledge and Data Engineering 99(PrePrints) (2011)
19. The Internet Movie Database (IMDB), http://reviews.imdb.com/reviews/

Predicting Query Performance Directly from Score Distributions

Ronan Cummins

Department of Information Technology
National University of Ireland, Galway
ronan.cummins@nuigalway.ie

Abstract. The task of predicting query performance has received much attention over the past decade. However, many of the frameworks and approaches to predicting query performance are more heuristic than not. In this paper, we develop a principled framework based on modelling the document score distribution to predict query performance directly.

In particular, we (1) show how a standard performance measure (e.g. average precision) can be inferred from a document score distribution. We (2) develop techniques for query performance prediction (QPP) by automatically estimating the parameters of the document score distribution (i.e. mixture model) when relevance information is unknown. Therefore, the QPP approaches developed herein aim to estimate average precision directly. Finally, we (3) provide a detailed analysis of one of the QPP approaches that shows that only two parameters of the five-parameter mixture distribution are of practical importance.

1 Introduction

Query performance prediction (QPP) has become an important problem in the area of information retrieval (IR). These predictors aim to automatically estimate the performance of queries so that different strategies (e.g. query expansion or reduction) can be applied based on their estimated performance. The performance of these predictors are usually compared by measuring the correlation between the output of the predictor and query performance (e.g. average precision). However, many approaches to QPP are unprincipled, and it is unclear how to improve their performance, or if their performance can even be improved.

In this paper, we develop a principled framework based on modelling document score distributions that aims to predict query performance directly. Fig. 1 shows an example of a document score distribution returned for a query (when relevance information is known). We (1) develop formulae that directly infer average precision from a document score distribution, (2) develop simple heuristics that can estimate the *important* parameters of the score distribution when relevance information is unknown, and (3) provide an analysis that informs us of the most important parameters in the distributional model. This analysis helps in narrowing the focus of future research.

M.V.M. Salem et al. (Eds.): AIRS 2011, LNCS 7097, pp. 315–326, 2011.

The remainder of the paper is organised as follows: Section 2 reviews related work on score distributions and query performance prediction (QPP). Section 3 outlines our principled model, before the formulae for calculating the average precision from a mixture are introduced. In section 4, we outline three approaches to automatically predict the performance of a query from the score distribution when relevance information is unknown. Furthermore, we present an analysis that shows that only two parameters of the model are crucial in the estimation of average precision. Section 5 presents comparative results of the newly developed QPP approaches versus existing predictors. Finally, section 6 outlines our conclusions and future work.

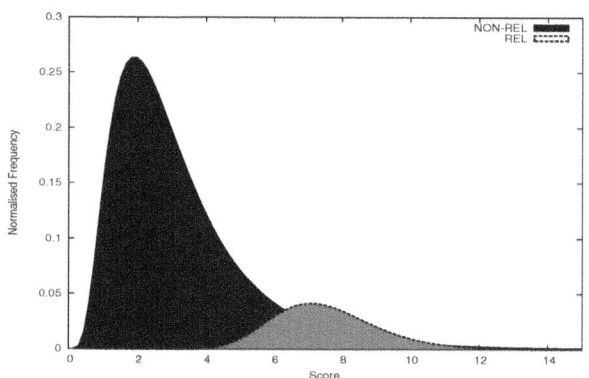

Fig. 1. A Typical Distribution of Scores Returned from a Classical IR System

2   Related Work

Modelling the distribution of document scores returned from IR systems has been studied from a theoretical perspective since the early days of IR [13]. More recently renewed interest has led to research that uses score distributions for data fusion [14]. Other researchers have modelled document score distributions for threshold filtering [1]. Others [9] have studied the generation process of the score distribution and have provided reasons for the typical shape (Fig. 1) of the distribution.

Automatically predicting query performance can aid information retrieval systems by enabling these systems to apply different strategies (e.g. query expansion) to queries of varying difficulty. One of the earliest approaches to QPP has been that of the clarity score [4], which measures the KL-divergence between the query and collection model in a language modelling framework. Some approaches [15] have measured the robustness of a ranking to perturbations and have developed novel predictors from this, while others [7] have investigated the clustering ability of similarly ranked documents to develop predictors.

Recent research has shown that the standard deviation (σ) (i.e. dispersion) of scores in a ranked-list is a good predictor of query performance [10,12,5]. These approaches are more heuristic based and lack a deeper theoretical understanding. The performance of predictors are usually measured by calculating the correlation (i.e. linear and/or non-parametric) between the output of the predictor and the performance of the query (i.e. usually average precision) over a set of queries.

However, to the authors knowledge, to date there has been no research that has directed aimed to estimate the performance of a query (either using score distributions or other methods). While some predictors use document scores returned from a system, and use various measures of the dispersion of such scores to develop their predictors, the methods are unprincipled and do not aim to directly predict performance, rather some surrogate of performance.

3   Explicitly Modelling Query Performance

In this section, we present a mixture distribution that is used in this paper to model the scores of both relevant and non-relevant documents.

3.1   Assumptions and Mixture Model

Consider an IR system that retrieves a returned set of N documents, and thus N scores given a query (Q). We assume that a system ranks documents independently of each other, in accordance with the probability ranking principle (PRP) [11] and that the relevance judgments are binary.

The log-normal distribution has been used successfully [14] to model scores for fusion tasks in IR, and therefore, we adopt this distribution[1]. The probability density function (pdf) of the log-normal distribution is as follows:

$$P(x) = \frac{1}{x\sigma\sqrt{2\pi}}e^{-\frac{(ln(x)-\mu)^2}{2\sigma^2}} \tag{1}$$

where μ and σ are the parameters. This distribution is supported from 0 to ∞ and the cumulative density function (cdf) is again simply the integral of this function from 0 to ∞. The mean of the distribution is $e^{\mu+\sigma^2/2}$, while the variance is $(e^{\sigma^2} - 1) \cdot (e^{2\mu+\sigma^2})$. Therefore, by rewriting these equations the method-of-moments estimates (MME) are as follows:

$$\hat{\mu} = ln(m) - \frac{1}{2}(1 + \frac{v}{m^2}) \qquad\qquad \hat{\sigma}^2 = ln(1 + \frac{v}{m^2}) \tag{2}$$

where m and v are the sample mean and variance respectively. Therefore, similar to previous approaches, the document score distribution can be thought of as a mixture of relevant and non-relevant documents as follows:

$$P(s) = (\lambda) \cdot P(s|1) + (1 - \lambda) \cdot P(s|0) \tag{3}$$

where $P(s|1)$ is the probability density function (pdf) for the scores (s) of relevant documents, $P(s|0)$ is the pdf for the scores of non-relevant documents, and where $\lambda = \frac{R}{N}$ is the proportion of relevant documents R in the entire returned set N.

---

[1] Noting that any reasonable choice of distribution can be substituted into the mixture.

3.2 Inferring Average Precision

We will now show how average precision (a standard metric for the effectiveness of a query) can be calculated directly from the mixture of continuous distributions. As recall is the proportion of relevant returned documents compared to the entire number of relevant documents, the recall at score s can be defined as follows:

$$recall(s) = \int_s^\infty \frac{\lambda \cdot P(s|1) \cdot ds}{\lambda} = \int_s^\infty P(s|1) \cdot ds \qquad (4)$$

which is the cumulative density function (cdf) of the distribution of relevant documents (viewed from ∞). Under the distributions outlined earlier for our model, we know that $recall(s)$ will vary between 0 and 1, (i.e. when $s = 0$, $recall(s) = 1$ as ensured by the cdf). Similarly, the precision at s (the proportion of relevant returned documents over the number of returned documents) can be defined as follows:

$$precision(s) = \frac{\int_s^\infty \lambda \cdot P(s|1)}{\int_s^\infty (\lambda) \cdot P(s|1) + (1 - \lambda) \cdot P(s|0)} \qquad (5)$$

Now that we can calculate the precision and recall at any score s in the range $[0 : \infty]$, we can create a precision-recall curve. Furthermore, as average precision can be estimated geometrically by the area under the precision-recall curve [2], the average precision ($avg.prec$) of a query can be calculated as follows:

$$avg.prec() = \int_0^1 precision(s) \cdot dz(s) \qquad (6)$$

where $z(s) = recall(s)$ which is in the range [0:1]. As these expressions are not closed-form, they can be calculated using relatively simple geometric numerical integration methods. It is worth noting that the formulae given for calculating average precision can over-estimate the actual average precision value calculated from TREC runs. This is due to the fact that recall is calculated as the number of relevant documents in the returned set, rather than the total number of relevant documents in the collection.

4   Estimating Parameters without Relevance Information

In this section, we develop approaches to automatically estimate (i.e. when no relevant information is known) the five parameters of the mixture model (i.e. λ, μ_1, σ_1, μ_0, σ_0) using a number of different methods. The section is comprised of three approaches to estimating the parameters of the mixture models. The first two approaches are based on heuristics and the MME of parameters. The third approach makes use of the standard EM algorithm for mixture models. We perform an analysis to find the most important parameters in one of the new parameter estimation approaches. Firstly Table 1 outlines the TREC[2] datasets used in this paper.

---

[2] http://trec.nist.gov/

Table 1. Test Collection Details

Collection		# docs	# topics	range
Tuning	LATIMES	131,896	144	301-450
	AP	242,918	149	051-200
Test	FT	210,158	188	251-450
	WT2G	221,066	50	401-450
	WT10G	1,692,096	100	451-550

4.1 Estimating Moments and Mixture

In this section we aim to estimate the sample moments so that the parameters of the model can, in turn, be automatically calculated using method of moment estimates (MME) from equations (2) (i.e. Section 3.1). Therefore, to estimate the five parameters of the log-normal model using MME, we must estimate the sample mean (m_1 and m_0) and variances (v_1 and v_0) for the relevant and non-relevant document scores and the mixture parameter (λ).

Firstly, as the number of non-relevant documents (NR) is usually much larger than the number of relevant documents (R) in the entire returned set (i.e. $NR \gg R$), we can estimate the sample mean (m_0) and sample variance (v_0) of the non-relevant documents by using the mean and variance of the scores in the entire returned set (i.e. $N \simeq NR$), as this seems a rather sound heuristic. However, the estimation of the mean and variance (m_1 and v_1) of relevant documents is more problematic.

Recent research has posited that a theoretically valid distribution should be able to approach Dirac's delta function under the strong SD hypothesis [1]. Fundamentally, as IR systems are striving to separate the set of relevant documents (R) from the set of non-relevant documents (NR), we estimate the mean (m_1) and variance (v_1) of the relevant set by assuming that all documents over a certain threshold score (min-max normalised for convenience) are relevant. Fig. 2 shows the tuning[3] of this threshold on a separate tuning collection collection (i.e. the LATIMES for both title and desc queries) averaged over five different IR systems (i.e BM25, LM, Pivoted Normalisation, F2EXP [8] and ES [6]). We can see that a common stable performance (i.e. average Spearman correlation with average precision) for both title and desc queries, can be achieved at a min-max normalised score of around 0.5 (i.e. midway between the minimum and maximum score of a ranked list). Therefore, the sample mean (m_1) and variance (v_1) of the relevant document scores are estimated by calculating the mean and variance of all scores that lie in the top half of a min-max normalised score range.

At this stage, we have estimates of the mean and variance of both relevant and non-relevant document scores, and consequently, from these we can calculate four parameters of the mixture model using MME. However, the final parameter that needs to be estimated is the mixture parameter λ. We apply a similar approach as before and assume that all documents over a certain threshold normalised score

---

[3] During this tuning process, the actual mixture value (λ) is assumed to be known.

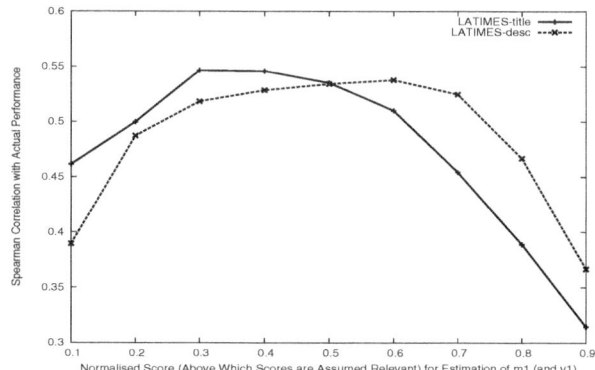

Fig. 2. Estimating m_1 (and v_1) threshold by tuning on LA Times collection averaged over five IR systems

are deemed relevant[4]. Similar to the previous tuning experiment, Fig. 3 shows the performance of the mixture model at various normalised threshold scores for estimating the mixture parameter λ. We can see that the best performance (i.e correlation with actual average precision) occurs when assuming that very few documents are relevant (i.e. only those scores that are at or above a normalised score of 0.95).

Now we have estimated, albeit heuristically, all the information needed to infer average precision without relying on relevance information. Furthermore, for all future experiments using this MMP1 (method of moments predictor) approach, the threshold for estimating m_1 and v_1 remain at normalised score of 0.5, and the threshold used for estimating λ is a normalised score of 0.95. As we will see in the next section, the estimation of the mean and variance of non-relevant documents (m_0 and v_0) is based on a rather sound heuristic. However, the approach to estimating the mean and variance of the relevance document scores, and the mixture parameter, is where loss in predictive performance can be attributed. We shall see later in the results section (section 5.1) that the estimation from these heuristics yields very good performance compared to other predictors. However, in the next section, we analyse the most important parameters (i.e. m_1, v_1, m_0, v_0, λ) for the MMP1 approach outlined in this section.

4.2 Analysing Moments and Mixture

In this section, we aim to identify the parameters (i.e. m_1, v_1, m_0, v_0, λ) that contribute most to the performance of the model outlined in the previous section (i.e. MMP1). It would be beneficial to know which parameters are more

---

[4] Preliminary experiments informed us that the number of documents above the normalised score of 0.5 grossly overestimated the number of actual relevant documents, although the estimates of m_1 and v_1 are suitable. Therefore, a separate and more stringent threshold is needed.

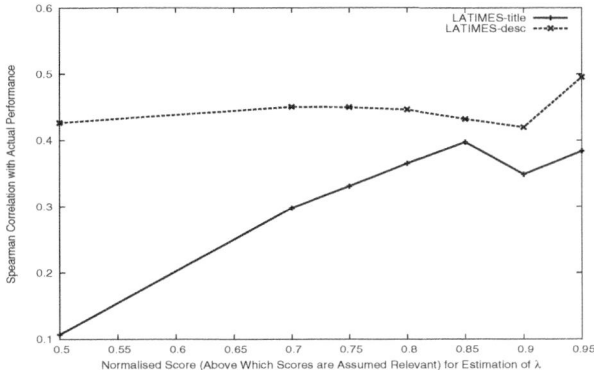

Fig. 3. Estimating λ threshold by tuning on LA Times collection averaged over five IR systems

important in terms of performance (i.e. correlation with average precision). We measure the amount of information contained in each parameter by initially assuming that all parameters (i.e. moments and mixture parameters) are as accurate as possible (i.e. using MME when relevance labels are known). We then substitute an estimated version of each parameter (i.e. estimated without relevance labels) and recalculate the performance of the model. At each stage of the process, a parameter is estimated using the heuristics in the previous section (section 4.1). Therefore, when the process is complete, all of the parameters of the model have been estimated without use of relevance information.

Fig. 4 show the results of such a process[5]. As we view the Figure from left to right, an estimate (i.e. without using labelled data) of each parameter is substituted into the model. It is clear from Fig. 4 that the estimation of the mean and variance of non-relevant documents (m_0 and v_0), and the variance of relevant documents (v_1) can be accurately estimated using the approach previously outlined (section 4.1), as the correlation coefficient does not decrease. However, when the m_1 and λ parameters are estimated without using relevance information, the correlation coefficient decreases a significant amount. Therefore, the two most important parameters in the model are the mean of relevant document scores (m_1) and the mixture parameter (λ), the former being the most important. These results are averages across five different IR systems. We can report that all of the systems tested behaved very similarly.

4.3   Motivation and Improvement

While Fig. 4 (and the results later in Section 5.1) show that the first approach (MMP1) to estimating the mean and mixture of the set of relevant documents seems to be effective to some degree, there is little motivation as to why this

---

[5] The results of all other collections show similar trends.

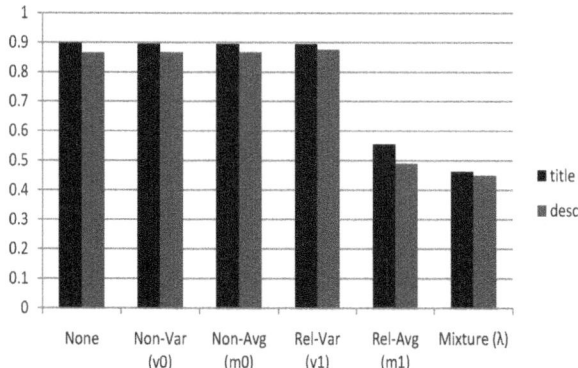

Fig. 4. Decrease in Spearman correlation with actual average precision as moments and mixture are estimated (FT Collection) from unlabelled data

may be so. The MMP1 approach estimated the mean and variance of relevant documents by using all of the scores above a normalised score of 0.5. Consider a system which returns N documents and where $K = \#(\|S(d)\| > 0.5)$ is the number of documents that are above a min-max normalised score of 0.5 (i.e. they have a score in the top half of the distribution). If K is small it implies that the system has also succeeded in promoting a relatively small number of documents, compared to the returned set N. Given the view of score distributions in Fig. 1, we can see that if the relevant and non-relevant scores are separated to a higher degree, the performance of the query will also be higher. Given that the distribution of document scores from systems is positively skewed, a smaller number of documents in this set of K documents will lead to a higher mean for the relevant documents (m_1). This in turn is an indicator that there is a good separation between relevant and non-relevant documents (and subsequently an indiction of a good query).

The initial estimate of m_1 was calculated by averaging all the scores above a normalised score of 0.5. A subsequent analysis on the LATIMES tuning collection has informed us that for 80% of the queries, the mean of the set of relevant document scores lies in the top half of the score distribution. However, our initial method of estimating the mean score of relevant documents (m_1) cannot return an estimate below a normalised score of 0.5. Therefore, we now propose a small modification to the initial estimate of m_1 so that a score of below 0.5 can be achieved when it is detected that the distribution of relevant and non-relevant document have not been seperated to a sufficient degree. Given that a small value of K explicitly indicates good separation, the following formula give us an updated measure of the normalised mean score of relevant documents (m_1') using a simple linear combination with the original normalised m_1 estimate:

$$\|m_1'\| = \alpha \cdot (1 - \frac{log(K)}{log(N)}) + (1 - \alpha) \cdot \|m_1\| \tag{7}$$

where K is the number of documents above a normalised score of 0.5, N is the returned set, and α is a parameter we set to 0.5 for all subsequent experiments. The left-hand side of this equation will reduce the estimate of the normalised mean score ($\|m_1'\|$) when K is relatively large. Consider a query which returns $N = 10,000$ documents, for which a relatively large proportion $K = 4,000$ lie in the top half of the distribution. The left-hand side of the equation $(1 - \frac{log(4,000)}{log(10,000)} = 0.099)$ will return a low value which can reduce the initial normalised estimate of $\|m_1\|$ below 0.5. The new estimate can be unnormalised to recover a new updated mean m_1'. This updated mean m_1' can be used in place of m_1 in the initial MMP1 approach to yield a second approach (MMP2). A further discussion of the comparative results of these approaches is undertaken in the results section.

4.4   Expectation Maximisation Approach

The EM algorithm is a popular unsupervised learning algorithm for estimating the parameters in mixture models [3]. We initialised the EM algorithm with the parameter estimates from the first MME approach (Section 4.1) that were generated using heuristics. We ran the EM algorithm for 50 iterations. Our initial experiments showed that the parameters converged prior to the 50th iteration.

5   Results and Discussion

In this section we present comparative results of the two QPP approaches based on heuristics that estimate the model parameters via moments (MMP1 and MMP2), and the approach based on the EM algorithm (EM). We then discuss the contributions and limitations of the research undertaken. In the subsequent results we focus on the two most popular IR systems (i.e. BM25 and LM).

5.1   Comparative Results

In this section, we compare the performance of the new QPP approaches developed in Section 4 (labelled MMP1, MMP2 and EM) against other state-of-the-art post-retrieval approaches. The state-of-the-art baseline approaches that we use are the clarity score [4] (a principled approach using KL-divergence), the standard deviation of document scores at 100 ($\sigma(100)$) [10], and NQC [12] also at 100 documents. We also tested the automatically tuned version of the standard deviation [10], and the maximum retrieval score of a ranked list, and found that the baselines presented in Tables 2 and 3 are stronger.

Tables 2 and 3 show the Spearman correlation[6] of the output of each predictor and average precision, for the approaches on four test collections for two prominent IR systems (BM25 and LM). The column labelled 'OPT' is the theoretically

---

[6] Best results are in bold. Due to the sizes of the differences and the number of queries in some of the test collections, statistical tests tend not to find significant differences between most of the correlations. However, for all but the EM approach the individual correlations are significant.

maximum correlation of the mixture model, if the parameters could be predicted using the MME from labelled relevance data. We can see that the new MMP approaches outperform the clarity score on most of the collections and, in general, are comparable in performance to that of the best baselines for longer queries. In general, on short queries, the new MMP1 and MMP2 approaches outperform the baselines, with MMP2 noted as the best predictor. We performed statistical tests[7] on the correlation coefficients of the new MMP approaches against both baseline approaches for each collection, and found that on most of the collections, the correlation coefficients were not significantly higher. We can report that when any of the baselines outperformed the MMP approaches the result was not significant, but on some collections, the MMP approaches significantly outperformed one (always the lower) of the baselines (denoted by †). The MMP2 approach tends to outperform the MMP1 approach especially for longer queries. It should be noted that we have not tuned the linear combination (i..e $\alpha = 0.5$) parameter in this approach.

The results of the unsupervised EM learning approach are particularly poor. We analysed the parameters returned from the approach and determined that the EM algorithm tends to grossly over-estimate the mixture parameter (λ), while not estimating values that are close to the actual values for $\hat{\mu}_1$, $\hat{\sigma}_1$, $\hat{\mu}_0$, or $\hat{\sigma}_0$.

Table 2. Spearman correlation of output of various predictors vs average precision for title (top half of table) and desc (bottom half of table) queries for BM25

Collection		BM25					
	clarity	$\sigma(100)$	NQC	EM	MMP1	MMP2	OPT
AP	0.393	0.280	0.265	0.037	**0.511** †	0.495 †	0.87
FT	0.426	0.492	0.513	0.173	**0.596** †	0.590 †	0.88
WT2G	0.352	0.445	0.411	-0.125	0.423	**0.473**	0.82
WT10G	**0.357**	0.328	0.342	-0.031	0.298	0.344	0.74
Avg(title)	0.382	0.386	0.382	0.013	0.457	**0.475**	0.83
AP	0.508	**0.591**	0.543	0.060	0.513	0.571	0.84
FT	0.382	0.431	0.518	-0.025	0.519	**0.543** †	0.86
WT2G	0.321	**0.584**	0.592	-0.129	0.507	0.552	0.81
WT10G	0.400	**0.501**	0.491	-0.042	0.411	0.456	0.72
Avg(desc)	0.402	0.526	**0.536**	-0.034	0.487	0.530	0.81

It is true that the methods for estimating the parameters of the distributions are heuristic, but these can be removed when more theoretically sound methods for estimating these are discovered. There are many approaches to query performance prediction that have not been evaluated against the new approaches developed here, but comparative studies [10] would tend to suggest that our approach is highly competitive. Furthermore, other approaches to QPP have not

---

[7] We transformed both coefficients to z-scores and tested whether the 0.95 confidence interval levels overlapped.

Table 3. Spearman correlation of output of various predictors vs average precision for title (top half of table) and desc (bottom half of table) queries on for a Jelinek-Mercer Language Model

Collection	clarity	$\sigma(100)$	NQC	EM	MMP1	MMP2	OPT
					LM		
AP	0.387	0.170	0.205	0.184	**0.389** †	0.378 †	0.89
FT	0.467	0.432	0.467	0.105	0.442	**0.469**	0.89
WT2G	0.335	0.467	0.428	-0.158	0.453	**0.514**	0.80
WT10G	0.246	0.276	0.253	0.040	0.523 †	**0.537** †	0.76
Avg(title)	0.358	0.336	0.338	0.042	0.451	**0.474**	0.83
AP	**0.525**	0.519	0.456	-0.038	0.430	0.499	0.86
FT	**0.414**	0.296	0.368	0.002	0.347	0.388	0.87
WT2G	0.249	0.533	0.517	-0.139	0.513	**0.577** †	0.82
WT10G	0.333	**0.567**	0.455	0.017	0.381	0.482	0.75
Avg(desc)	0.380	0.478	0.449	-0.039	0.417	0.486	**0.83**

aimed to explicitly estimate the performance measure in question. One interesting practical advantage of the predictors developed here is that they can be easily modified to predict other performance measures.

6   Conclusion

In this work, we have developed new query performance predictors that explicitly aim to predict average precision. The new predictors (MMP1 and MMP2) based on estimating the moments and mixture parameter are comparable to state-of-the-art predictors. Furthermore, an analysis of the parameters of the predictor has determined that only two parameters (m_1 and λ) are of crucial importance to the performance of the predictor. This analysis aids in narrowing the focus of future work. In a broader IR sense, it follows that only these two parameters are of importance to any IR application using score distributions. Future work, involves researching other unsupervised learning approaches to parameter estimation in the hope that they may yield higher performance predictors.

References

1. Arampatzis, A., Robertson, S.: Modeling score distributions in information retrieval. Inf. Retr. 14(1), 26–46 (2011)
2. Aslam, J.A., Yilmaz, E.: A geometric interpretation and analysis of r-precision. In: CIKM, pp. 664–671 (2005)
3. Bishop, C.M.: Pattern Recognition and Machine Learning (Information Science and Statistics). Springer-Verlag New York, Inc., Secaucus (2006)
4. Cronen-Townsend, S., Zhou, Y., Bruce Croft, W.: Predicting query performance. In: SIGIR 2002, pp. 299–306. ACM, New York (2002)

5. Cummins, R., Jose, J., O'Riordan, C.: Improved query performance prediction using standard deviations. In: SIGIR, pp. 524–531 (2011)
6. Cummins, R., O'Riordan, C.: Learning in a pairwise term-term proximity framework for information retrieval. In: SIGIR, pp. 251–258 (2009)
7. Diaz, F.: Performance prediction using spatial autocorrelation. In: SIGIR, pp. 583–590 (2007)
8. Fang, H., Zhai, C.: An exploration of axiomatic approaches to information retrieval. In: SIGIR, pp. 480–487 (2005)
9. Kanoulas, E., Dai, K., Pavlu, V., Aslam, J.A.: Score distribution models: assumptions, intuition, and robustness to score manipulation. In: SIGIR, pp. 242–249 (2010)
10. Pérez-Iglesias, J., Araujo, L.: Standard Deviation as a Query Hardness Estimator. In: Chavez, E., Lonardi, S. (eds.) SPIRE 2010. LNCS, vol. 6393, pp. 207–212. Springer, Heidelberg (2010)
11. Van Rijsbergen, C.J.: Information Retrieval, 2nd edn. Butterworth-Heinemann, Newton (1979)
12. Shtok, A., Kurland, O., Carmel, D.: Predicting Query Performance by Query-Drift Estimation. In: Azzopardi, L., Kazai, G., Robertson, S., Rüger, S., Shokouhi, M., Song, D., Yilmaz, E. (eds.) ICTIR 2009. LNCS, vol. 5766, pp. 305–312. Springer, Heidelberg (2009)
13. Swets, J.A.: Information retrieval systems. Science 141(3577), 245–250 (1963)
14. Wilkins, P., Smeaton, A.F., Ferguson, P.: Properties of optimally weighted data fusion in cbmir. In: SIGIR 2010, pp. 643–650 (2010)
15. Zhou, Y., Bruce Croft, W.: Ranking robustness: a novel framework to predict query performance. In: CIKM 2006, pp. 567–574. ACM, New York (2006)

Wikipedia-Based Smoothing for Enhancing Text Clustering

Elahe Rahimtoroghi and Azadeh Shakery

School of Electrical and Computer Engineering, College of Engineering,
University of Tehran, Tehran, Iran
e.rahimtoroghi@ece.ut.ac.ir, shakery@ut.ac.ir

Abstract. The conventional algorithms for text clustering that are based on the bag of words model, fail to fully capture the semantic relations between the words. As a result, documents describing an identical topic may not be categorized into same clusters if they use different sets of words. A generic solution for this issue is to utilize background knowledge to enrich the document contents. In this research, we adopt a language modeling approach for text clustering and propose to smooth the document language models using Wikipedia articles in order to enhance text clustering performance. The contents of Wikipedia articles as well as their assigned categories are used in three different ways to smooth the document language models with the goal of enriching the document contents. Clustering is then performed on a document similarity graph constructed on the enhanced document collection. Experiment results confirm the effectiveness of the proposed methods.

Keywords: Text clustering, Wikipedia, Smoothing, Language models.

1 Introduction

With the fast growth of text data in management systems and on the web, it is essential to develop efficient methods to organize the content of text documents. Text clustering is a fundamental method for organizing large number of documents, and can also be used to enhance the performance of information retrieval systems.

However, most of the previous text clustering algorithms are based on the "bag of words" (BOW) model in which the semantic relations between the words are not considered. In BOW approach, a document is represented by its individual terms and their frequencies in the document's content. However, two documents describing the same topic may falsely be categorized into different clusters if they use different sets of words that are not necessarily identical but have semantic relations (e.g., synonymy). This issue negatively affects the reliability of the text clustering results. One way to overcome this drawback is to use background knowledge to enrich the documents.

WordNet, ODP, and Wikipedia have already been used to improve document clustering. WordNet has limited coverage. It mostly contains common words but not scarce ones. In that sense, using WordNet may cause information loss [10]. In

M.V.M. Salem et al. (Eds.): AIRS 2011, LNCS 7097, pp. 327–339, 2011.
© Springer-Verlag Berlin Heidelberg 2011

contrast, Wikipedia is a large online repository which contains millions of articles and is regularly updated. It has a widespread coverage over different concepts and is more complete than other resources. In this paper we propose to utilize Wikipedia to enhance the results of text clustering. There have been few attempts to use Wikipedia to improve text clustering [9]. These methods mostly extract features from Wikipedia for documents to enrich their contents [10].

In this research, we adopt a language modeling approach for text clustering and propose to smooth the language models of text documents using Wikipedia articles in order to enhance the content of the documents. Language modeling has recently been successfully applied to many different information retrieval and text mining problems [13] and smoothing has been shown to be a critical part for many of these solutions. A short document pertaining to a particular topic may not contain all the words related to that topic. As a result, it becomes unlikely to have all related words in the language model, a problem addressed by smoothing in order to enrich the documents. Smoothing refers to the adjustment of the maximum likelihood estimator of a language model so that it will be more accurate [14]. When estimating a language model based on a limited amount of text, smoothing is extremely important.

In this paper, we propose two smoothing methods and for each method, we define a neighborhood of the Wikipedia articles, the target neighborhood, for each text document to smooth the content of documents. In the first method we extract the top features of each document, which are the terms, and define the target neighborhood in Wikipedia using these features. Intuitively, since top features are appropriate representatives of the document, document expansion based on these features can enhance the document content. In the second approach the content similarity of the documents and Wikipedia articles are used to define the neighborhoods. We also propose to combine the two methods and use both neighborhood sets for smoothing to benefit from the advantages of both methods.

After smoothing, we model the smoothed documents by a graph constructed upon the language models, and perform link-based clustering on the resultant graph. For link-based clustering, we use a link-based algorithm proposed in [12]. For baseline we use the K-Means algorithm on the BOW model. Our experiments on Reuters-21578 [15] and 20-newsgroup (20NG) [16] show that both proposed smoothing methods improve the clustering results, and the combination of them provides the best performance, confirming the power of proposed methods in enriching document contents using Wikipedia articles. Link-based clustering of the documents has also shown to be superior to the traditional content based clustering method, K-Means.

The rest of this paper is organized as follows. In section 2 we review some related works. Section 3 introduces the proposed smoothing methods based on Wikipedia articles and presents construction of the document similarity graph and link-based clustering algorithm. Experimental results are described in section 4. Finally we discuss our main results and conclude the paper in section 5.

2   Related Works

A number of different methods have been applied to improve basic text clustering algorithms. Recently the research on utilizing background knowledge and ontology to enhance text mining tasks such as text classification and clustering is increasing.

In [1], WordNet is used to improve text clustering. They enrich the term vector of each document using three different strategies based on WordNet concepts. Another algorithm that represents the vector of text documents using WordNet is introduced in [2]. Both these researches somehow resolve the problem of BOW model. But they have some limitations. WordNet covers a limited number of words and is focused on common words not rare ones. Another problem is that most of the descriptions presented by WordNet for each term are too short.

Another resource studied in recent researches is Wikipedia. Though using Wikipedia is more difficult because it is not as structured as WordNet, but researches on utilizing Wikipedia to enhance text mining are growing recently because it is widespreading and rectifies the limitations of WordNet. In [3] and [4], a method was proposed to incorporate encyclopedia knowledge with text classification systems. Wikipedia was also used in [5] to improve text categorization. Other text mining tasks have been enhanced by Wikipedia, such as cluster labeling [6]. Wikipedia can also improve information retrieval, for example in [7] Wikipedia was used for query expansion and relevance feedback.

In addition, Wikipedia has been used for text clustering. The representation of short text documents are improved using Wikipedia features in [8]. In [9] an algorithm is presented which maps the terms in the text documents into Wikipedia concepts and the term vector of the document is reduced to a vector of Wikipedia concepts. They also compare Wikipedia with WordNet and show that most of the terms in a document that are appropriate features for the document are not covered in WordNet but they are included in Wikipedia concepts.

Beside the concepts of Wikipedia, the categories of the articles have also been utilized to improve text clustering [10]. Although these methods show improvement in text clustering using Wikipedia, they have some limitations. They use Wikipedia to add new features to the document mostly using concepts and categories of the articles. But Wikipedia has a rich content for each article that can also be used to improve text clustering. In this paper we use the whole content and the categories of Wikipedia articles to smooth the language models of text document to enrich the documents and enhance text clustering.

3   Wikipedia-Based Smoothing

In this paper we propose to utilize Wikipedia to improve text clustering. Our approach is to smooth the language models of the text documents using Wikipedia articles in order to enrich the documents. We then construct a document similarity graph using the method proposed in [11] and based on the contents of the enriched documents. The document similarity graph is a graph of documents citing each other, where the weighted links are induced from the new, improved contents of documents. We finally cluster the documents using the result graph.

For the first step we propose two smoothing methods and for each method, we define a neighborhood of the Wikipedia articles for any of the text documents. Next, we smooth each document using the given set. Wikipedia articles have different parts; in both proposed methods we choose to use the content and the categories of the articles for smoothing and discuss the effects of each part. The contents of Wikipedia articles are very rich as they are encyclopedia knowledge on the subject of the articles; and thus can be used for expansion and enriching text documents. Wikipedia contains a hierarchical categorization system, and each article belongs to at least one category. The category information of each article is an appropriate summarized narration of its subject and therefore is suitable for enriching documents.

In the first proposed method, "Top-Feature Smoothing", we extract the top features of each document and define the target neighborhood in Wikipedia using these features. In the second approach, "Similarity Smoothing", the content similarity of the documents and Wikipedia articles are used to define the neighborhoods. We also propose to combine the two methods and use both neighborhood sets for smoothing to benefit from advantages of both methods. The details of these algorithms are described below.

3.1   Top-Feature Smoothing Algorithm

For this algorithm we first extract the top m features of each document. We rank the document features using:

$$\text{Score}(w, d) = \text{TF}(w, d) * \log(\text{IDF}(w)). \tag{1}$$

where, w is a word in the document d and $Score(w, d)$ is the score of w as a feature of document d indicating the importance of w in d. $TF(w, d)$ is the raw frequency of w in document d, and $IDF(w)$ is the inverse of the document frequency of the word w in the whole collection. For each document, we rank its words based on this score and select the top m words as top features of the document: $f_1, f_2 \ldots f_m$.

Next, we use these features to define a neighborhood of Wikipedia articles for each document d. For this purpose, we match each feature f_i with Wikipedia concepts, considering the titles of Wikipedia articles as Wikipedia concepts. We bring in all Wikipedia articles whose titles match f_i and construct a set of Wikipedia articles for each feature. The neighborhood set of d, named S, is defined as the union of all these sets and is used to smooth the document d. Smoothing is done in two steps. In the first step, the raw content of the document is smoothed using Dirichlet prior smoothing method [14]:

$$p_S(w \mid d) = (\frac{|d|}{|d|+\mu}) * p_{ML}(w \mid d) + (\frac{\mu}{|d|+\mu}) * p(w \mid S). \tag{2}$$

In the second step, the smoothed content of the previous step is smoothed with the contents of the whole collection:

$$p(w \mid d) = (1-\beta) * p_S(w \mid d) + \beta * p_{ML}(w \mid C). \tag{3}$$

In the above equations, C is the whole collection of the documents, p_{ML} is the maximum likelihood estimation, and $|d|$ is the length of the document. Here, β and μ are smoothing parameters; and $p(w \mid S)$ is computed in two ways:

1. TopF-Content: In this method, we only use the content of the Wikipedia articles for smoothing, and $p(w \mid S)$ is:

$$p_1(w|S) = \sum_{i=1}^{m} \alpha_i \, p_{ML}(w|D_{fi}) \,.$$ (4)

where, D_{fi} is the set of Wikipedia articles corresponding to feature f_i. Since the top features of the document have different importance compared to each other, we combine their language models in a weighted manner, and define α_i as:

$$\alpha_i = \frac{2 * (m-i+1)}{m * (m+1)} \quad , \quad \sum_{i=1}^{m} \alpha_i = 1 \,.$$ (5)

2. TopF-Content&Cat: As described before, beside the content of the Wikipedia articles, we also use the categories of each article for smoothing. We combine the language model of the articles with the language model of categories. For this purpose, we define $p_{ML}(w \mid Cat_{fi})$ as the probability of w occurring in the categories of the articles in D_{fi}; and we linearly combine this probability with $p_{ML}(w \mid D_{fi})$. In this method, $p(w \mid S)$ is calculated as follows:

$$p_2(w|S) = \sum_{i=1}^{m} \alpha_i \, [(1 - \gamma) * p_{ML}(w|D_{fi}) + \gamma * p_{ML}(w|Cat_{fi})] \,.$$ (6)

Here, weighting scheme is similar to equation (4); and $\boldsymbol{\gamma}$ is a parameter for combining the probabilities.

3.2   Similarity Smoothing Algorithm

In our second smoothing method, the language model similarities are used. We define the distance between two text documents based on the Kullback-Leibler divergence (KL-divergence) measure between their language models:

$$\text{Dist } (d_i, d_j) = \text{KL-divergence } (\theta(d_i), \theta(d_j)) \,.$$ (7)

Where, $\theta(d_i)$ is the language model of the document d_i. We compute the distance measure defined above between each document and all Wikipedia articles. Then we smooth the documents using the two stage smoothing method described in section 3.1, equations (2) and (3); only the parameter $p(w \mid S)$ is different in this method; it is computed in two ways:

1. Sim-Content: In this method we use the contents of Wikipedia articles weighted based on the similarity between the articles and each document; so $p(w \mid S)$ is:

$$p_3(w|S) = \sum_{a_i \text{ is a Wikipedia article}} \beta_i * p_{ML}(w|a_i) \,.$$ (8)

where a_i is a Wikipedia article and the summation is performed on all Wikipedia articles for the document d. This summation is weighted by the KL-divergence distance between the document and Wikipedia articles, so the articles which are more similar to d will have more effect on its language model:

$$\beta_i = \frac{\left(\frac{1}{Dist\,(d,\ a_i)}\right)}{D} \quad , \quad D = \sum_{a_j \text{ is a Wikipedia article}} \frac{1}{Dist\,(d,\,a_j)} \cdot \qquad (9)$$

2. Sim-Content&Cat: In this approach, both content and categories of the Wikipedia articles are used for smoothing. Therefore $p(\,w\,/\,S)$ is:

$$p_4(w|S) = \sum_{a_i \text{ is a Wikipedia article}} \beta_i * [(1-\gamma)*p_{ML}(w|a_i) + \gamma*p_{ML}(w|Cat_i)] \,. \qquad (10)$$

Here, Cat_i is the set of categories of the article a_i and γ is the parameter for combining the language models.

3.3   Combination of the Two Methods

"Top-Feature Smoothing" and "Similarity Smoothing" algorithms can be used together for smoothing a document taking advantage of both methods. *Similarity* smoothing benefit from all Wikipedia articles, while *Top-Feature* smoothing increases the effect of the words resembling top features of the document. We again use the two stage smoothing formula in section 3.1; except in this method $p(w\,/\,S)$ is:

$$p_5(w\mid S) = \alpha * p_2(w\mid S) + (1-\alpha) * p_4(w\mid S) \,. \qquad (11)$$

Here, α is a parameter to control the effect of each part.

3.4   Construction of the Document Similarity Graph and Link-Based Clustering

After smoothing, we model the smoothed documents with a directed graph by using the method proposed in [11]. This method considers each document as a query q; and then defines D_{init} as a set of top documents returned by some initial retrieval algorithm in response to the query q. As defined in [11], the top n generators of a document $o \in D_{init}$, denoted $TopGen(o)$, is the set of n documents $g \in D_{init}$- $\{d\}$ that yield the highest $p_g(o)$, where, $p_g(.)$ is the unigram language model induced from g. Then the graph is defined as:

$$wt_W(o \rightarrow g) = \begin{cases} p_g(o) & \text{if } g \in TopGen(o) \\ 0 & \text{otherwise;} \end{cases} \qquad (12)$$

So the links created between document o and other documents are weighted based on the probability which their induced language model assigns to o. Because the links are weighted, we set $n = |D_{init}|$ and consider all generators instead of selecting top n. Then we cluster the resultant graph to achieve a clustering for the documents. In all parts of this paper when we talk about modeling documents with a graph, this algorithm is used. To cluster the graph of the documents, we applied a link-based clustering method [12]. This algorithm needs an initial clustering of all nodes of the graph as its input which can be a random clustering. In this algorithm, for each node n_i, we examine all its adjacent nodes, and assign n_i to the cluster which includes more neighbors of n_i. This algorithm can be extended for using in graphs with weighted edges, as in our research. In this case, we use the summation of the weights of the edges between n_i and its neighbors, instead of the number of the adjacent nodes in clusters. Given K clusters, we define the assignment of n_i to cluster c_m as $x_{i,m}$ where $x_{i,m} = 1$ if n_i is in cluster c_m and 0 otherwise. We represent our graph with a matrix M, where $m_{i,j}$ indicates the weight of the edge from n_i to n_j. Then, we rank all clusters for n_i based on the score defined as: $\text{Score}(n_i, c_m) = \Sigma_j [(m_{i,j} + m_{j,i}) * x_{j,m}]$.

Then n_i will be assigned to the cluster which has the highest score. All nodes are reassigned iteratively until no reassignment is possible or a specific number of iterations have been performed. After smoothing the documents and modeling them with a graph as described before, we use this algorithm to cluster the resultant graph. We have given different random clusterings of nodes as input to this algorithm, but the best results were achieved when the results of K-Means were given as input to the algorithm. Thus, in all experiments on graph clustering we used the result of K-Means as input of the graph clustering algorithm.

4   Experiments and Results

The Wikipedia articles are available in form of database dumps and are updated periodically. They can be downloaded from http://download.wikipedia.org. The version we have used in this research was released on May, 2009. It contains about 3,000,000 articles. The experiments are performed on two datasets: Reuters-21578 [15] which contains 21,578 documents and 135 categories and 20-newagroup (20-NG) [16] with 19,997 documents and 20 classes.

4.1   Evaluation Metrics

Three metrics are used in this research to evaluate clustering quality: purity [17], f-score [18] and normalized mutual information (NMI) [19]. Given a dataset with N documents and M classes in which documents are labeled as $C = (c_1, c_2, \dots c_M)$, if our algorithm generates K clusters $Q = (q_1, q_2, \dots q_K)$, then purity measure of a cluster is defined as:

$$\text{purity}(Q, C) = \frac{1}{N} \sum_i \max_j |q_i \cap c_j|. \tag{13}$$

Purity is used extensively in clustering tasks but it increases as the number of clusters increases and equals to 1 when each document is in an individual cluster. Thus, we also use other metrics for evaluating the clustering. The f-score measure combines precision and recall: F-score = [2 (precision * recall)] / (precision + recall); and NMI is defined as the mutual information between the clustering results (Q) and the class labels (C):

$$\text{NMI}(X, Y) = \frac{I(X;Y)}{(\log M + \log K)/2} \tag{14}$$

where X is a random variable for cluster assignments and Y is a random variable for class labels. Unlike purity, when the number of the clusters increases, NMI does not necessarily grow. The values of the three measures described in this section range between 0 and 1, and the higher values indicate better clustering results.

4.2   Baselines

In text clustering, the number of documents is high, and the feature space will be very large. Thus we need a clustering algorithm which can address these two issues. K-Means [18] is a partitioning algorithm which iteratively calculates the cluster centroids and reassigns each document to the closest cluster until no document can be reassigned or a specific number of iterations has been performed. The time complexity of K-Means is linear in the number of documents and it is the most widely used algorithm for text clustering [18]; therefore we apply it as the baseline method in this research: **K-Means**, is K-Means clustering method with cosine similarity measure. We also perform another experiment as baseline: **Link-based**, is the link-based clustering algorithm described in section 3.4 performed on the similarity graph constructed upon initial documents. The results achieved on baseline experiments are shown in Table 1.

Table 1. Baseline experiments on Reuters-21578 dataset

Experiment	Purity	F-score	NMI
K-Means	0.538	0.323	0.431
Link-based	0.562	0.330	0.438

We conducted several experiments to evaluate our proposed algorithms. The details of the experiments are described in the remainder of this section.

4.3   K-Means Clustering of the Smoothed Documents

After smoothing documents using the three methods described in section 3, and before construction of the similarity graph, we clustered documents using K-Means algorithm. In the first set of experiments, **K-Means-TopF-Content**, we used K-Means to cluster the documents smoothed using the contents of Wikipedia articles and specifically the proposed *TopF-Content* smoothing algorithm. In the second set of experiments, **K-Means-TopF-Content&Cat,** we applied K-Means to documents smoothed with the contents as well as categories of Wikipedia articles, the proposed *TopF-Content&Cat* algorithm. The results on the Reuters-21578 dataset are shown in Table 2 (Numbers in parentheses show the percentage of improvement compared with the baseline experiment presented in each table).

Table 2. Comparison of K-Means-TopF-Content and K-Means-TopF-Content&Cat with baseline on Reuters-21578

Experiment	Purity	F-score	NMI
K-Means (baseline)	0.538	0.323	0.431
K-Means-TopF-Content	0.572 (6.3%)	0.332 (2.8%)	0.455 (5.6%)
K-Means-TopF-Content&Cat	0.585 (8.7%)	0.338 (4.6%)	0.463 (7.4%)

In *Top-Feature* smoothing we tested different values for parameter m and the best result was achieved for $m = 8$. Thus in all experiments with this smoothing algorithm, we set $m = 8$. In Dirichlet prior smoothing we set $\mu = 2000$ due to the research in [14]. In the second smoothing step, we set $\beta = 0.5$. Parameter γ for combining content and categories is set to 0.1; and $\alpha = 0.5$ in the combination of two smoothing methods so that both algorithms have same effect in the combination.

From the performance results in Table 2, it is observed that clustering results are enhanced after applying *Top-Feature* smoothing. The same experiments are performed for *Similarity* smoothing algorithm and the combination of the two smoothing methods: **K-Means-Sim-Content:** applying *Sim-Content* smoothing and cluster documents by K-Means algorithm. **K-Means-Sim-Content&Cat:** applying *Sim-Content&Cat* smoothing and cluster documents by K-Means algorithm. **K-Means-Combination:** applying combination of *TopF-Content&Cat* smoothing method with *Sim-Content&Cat* smoothing algorithm. The results of the experiments are shown in Table 3.

Table 3. Comparison of K-Means-Sim-Content, K-Means-Sim-Content&Cat and K-Means-Combination with baseline on Reuters-21578

Experiment	Purity	F-score	NMI
K-Means (baseline)	0.538	0.323	0.431
K-Means-Sim-Content	0.561 (4.3%)	0.331 (2.5%)	0.451 (4.6%)
K-Means-Sim-Content&Cat	0.569 (5.8%)	0.335 (3.7%)	0.458 (6.3%)
K-Means-Combination	0.597 (**11 %**)	0.346 (**7.1%**)	0.470 (**9%**)

As indicated in Table 3, *Similarity* smoothing improves text clustering results. The combination of the two smoothing methods improves clustering purity by the factor of 11% compared to the baseline which is the best performance among these experiments. From the results in Tables 2 and 3, in both smoothing algorithms, adding categories to the content for smoothing works better than using content only; we can also see that *Top-Feature* smoothing achieves more improvement than *Similarity* smoothing algorithm, on the Reuters-21578 dataset.

4.4   Link-Based Clustering of the Smoothed Documents

Further, we construct the document similarity graph and apply link-based clustering as described in section 3.4. At this point, we performed the following experiments: **Link-based-TopF-Content:** applying *TopF-Content* algorithm for smoothing. **Link-based-TopF-Content&Cat:** applying *TopF-Content&Cat* algorithm for smoothing.

Table 4. Comparison of Link-based-TopF-Content and Link-based-TopF-Content&Cat with Link-based on Reuters-21578

Experiment	Purity	F-score	NMI
Link-based (baseline)	0.562	0.330	0.438
Link-based-TopF-Content	0.595 (5.8%)	0.348 (5.4%)	0.464 (5.9%)
Link-based-TopF-Content&Cat	0.608 (8.2%)	0.355 (7.6%)	0.473 (8%)

The results in Table 4 indicate that not only *Top-Feature* smoothing enhances clustering performance, but also link-based clustering of documents has shown improvement, compared to the results in Table 2.

We performed similar experiments for *Similarity* smoothing algorithm and the combination of the two smoothing methods: **Link-based-Sim-Content:** applying *Sim-Content* algorithm for smoothing. **Link-based-Sim-Content&Cat:** applying *Sim-Content&Cat* algorithm for smoothing. **Link-based-Combination:** applying combination of *TopF-Content&Cat* method with *Sim-Content&Cat* algorithm for smoothing. The results are shown in Table 5.

Table 5. Comparison of Link-based-Sim-Content, Link-based-Sim-Content&Cat, and Link-based-Combination with Link-based on Reuters-21578

Experiment	Purity	F-score	NMI
Link-based (baseline)	0.562	0.330	0.438
Link-based-Sim-Content	0.579 (3%)	0.340 (3%)	0.457 (4.3%)
Link-based-Sim-Content&Cat	0.591 (5.1%)	0.347 (5.2%)	0.466 (6.4%)
Link-based-Combination	0.629 (**11.9 %**)	0.365 (**10.3%**)	0.481 (**9.8%**)

From the performance results in Table 5, we observe that the combination of the two smoothing algorithms improves clustering results more than individual smoothing algorithms. Though *Similarity* smoothing algorithm has improved clustering results but compared to *Top-Feature* smoothing it achieved less improvement. Among all proposed methods in this research, *Link-based-Combination* has the best results with the three metrics and improved purity of the K-Means clustering results by the factor of 16.9%. The comparison of *Link-Com* results with K-Means is shown in Table 6.

Table 6. Link-based-Combination method compared with K-Means, on Reuters-21578

Experiment	Purity	F-score	NMI
K-Means (baseline)	0.538	0.323	0.431
Link-based-Combination	0.629 (**16.9%**)	0.365 (**13%**)	0.481 (**11.6%**)

4.5   Experiments on 20-NG Dataset

To compare our results with previous clustering methods which utilized Wikipedia, we performed the experiments on 20-NG dataset and compared the results f our proposed methods with the algorithms presented in [10]. The results are summarized in Table 7. The **Word-Category** and **Word-Concept-Category** methods are proposed in [10] and had the best results in that research. Improvements are computed compared to the K-Means baseline.

Table 7. Comparison of the experiments with K-Means, on 20-NG dataset

Experiment	Purity	F-score	NMI
K-Means (baseline)	0.411	0.381	0.388
Link-based	0.420 (2.2%)	0.387 (1.6%)	0.393 (1.3%)
K-Means-TopF-Content	0.415 (1%)	0.384 (0.8%)	0.391 (0.8%)
K-Means-TopF-Content&Cat	0.419 (1.9%)	0.390 (2.4%)	0.398 (2.6%)
K-Means-Sim-Content	0.417 (1.5%)	0.388 (1.8%)	0.394 (1.5%)
K-Means-Sim-Content&Cat	0.422 (2.7%)	0.392 (2.9%)	0.401 (3.4%)
K-Means-Combination	0.425 (3.4%)	0.397 (4.2%)	0.413 (6.4%)
Link-based-TopF-Content	0.428 (4.1%)	0.393 (3.1%)	0.408 (5.2%)
Link-based-TopF-Content&Cat	0.439 (6.8%)	0.403 (5.8%)	0.419 (8%)
Link-based-Sim-Content	0.433 (5.3%)	0.401 (5.2%)	0.412 (6.2%)
Link-based-Sim-Content&Cat	0.442 (7.5%)	0.411 (7.9%)	0.425 (9.5%)
Link-based-Combination	**0.446 (8.5%)**	**0.421 (10.5%)**	**0.432 (11.3%)**
Word-Category	0.442 (7.5%)	0.412 (8.1%)	0.429 (10.5%)
Word-Concept-Category	0.442 (7.5%)	0.418 (9.7%)	0.412 (6.2%)

As indicated in the Table 7, *Similarity* smoothing algorithm outperforms *Top-Feature* smoothing algorithm on 20-NG dataset, unlike Reuters-21578. Since the performance of *Top-Feature* smoothing depends on the top features extracted from the documents, this can be because the feature extraction method did not get acceptable results on 20-NG dataset. This issue is to be examined in the future works. Using categories in addition to the content of the Wikipedia articles improves clustering more than using content only. The combination of both smoothing algorithms (*Link-based-Combination*) gives the best improvement (8.5% in purity) among all other experiments. Compared to *Word-Category* and *Word-Concept-Category*, that map text documents to Wikipedia concepts and categories, our best algorithm (*Link-based-Combination*) achieved more improvement based on three measures. Besides, in [10] the authors did not obtain any improvement without using categories, but our smoothing algorithms which used content only, had improvements based on all three metrics; because we utilize the whole content of Wikipedia articles that are very rich in content compared with concepts of Wikipedia (only titles of the articles).

5   Conclusions and Future Works

In this paper, we adopt a language modeling approach for text clustering and propose to smooth the language models of the text documents using Wikipedia articles in order to enhance the content of the documents. We propose two smoothing methods and for each method, we define a neighborhood of the Wikipedia articles for each text document to smooth the content of documents. In the first method we extract the top features of each document and define the target neighborhood in Wikipedia using these features. In the second approach the content similarity of the documents and Wikipedia articles are used to define the neighborhoods. We also propose to combine the two methods and use both neighborhood sets for smoothing to benefit from the advantages of both methods. After smoothing, we construct a document similarity graph using the enriched contents of documents, and perform link-based clustering on the resultant graph.

The proposed algorithms are tested on two datasets: Reuters-21578 and 20NG. In order to evaluate the effectiveness of proposed smoothing methods on clustering we perform different experiments and compare them with K-Means clustering algorithm as the baseline. Based on the empirical results, both proposed smoothing algorithms improve text clustering performance on both collections. *Top-Feature* smoothing algorithm achieves more improvement than *Similarity* smoothing algorithm, on Reuters-21578 dataset; and, *Similarity* smoothing algorithm outperforms *Top-Feature* smoothing algorithm on 20NG dataset. Since the performance of *Top-Feature* smoothing depends on the top features extracted from the documents, this can be because the feature extraction method did not get acceptable results on 20-NG dataset. This issue is to be examined in the future works. Another conclusion obtained from experimental results is that using categories in addition to the content of the Wikipedia articles for smoothing, improves text clustering performance. The combination of both smoothing algorithms achieves the best improvement, taking advantage of both methods. Similarity smoothing benefit from all Wikipedia articles, while Top Feature smoothing increases the effect of the words resembling top features of the document.

Link-based clustering of the documents has also shown to be superior to the traditional content based clustering method, K-Means. Based on the experimental results, creating content similarity graph and performing link-based clustering generates better text clustering results. Our best algorithm (Link-based-Combination) achieves 16.9% and 8.5% improvement in purity compared with K-Means on Reuters-21578 and 20NG, respectively. Comparison with existing text clustering methods using Wikipedia has shown that our proposed smoothing approach is more effective in text clustering results; because we utilize the whole content of Wikipedia articles that are very rich in content compared with concepts of Wikipedia (only titles of the articles). The results achieved in this research can be extended to other applications such as text classification and information retrieval. For future works we will use parameter tuning methods to make the proposed algorithms more accurate. Moreover, we will study how to utilize other parts of Wikipedia such as hyperlinks, to improve text clustering.

Acknowledgments. This research is partially supported by Iran Telecommunication Research Center (ITRC).

References

1. Hotho, A., Staab, S., Stumme, G.: Wordnet improves text document clustering. In: Proceedings of the Semantic Web Workshop at SIGIR 2003, 26th Annual International ACM SIGIR Conference on Research and Development in Information Retrieval, Toronto, Canada. ACM Press, New York (2003)

2. Sedding, J., Kazakov, D.: WordNet-based text document clustering. In: Proceedings of COLING 2004 Workshop on Robust Methods in Analysis of Natural Language Data (2004)
3. Gabrilovich, E., Markovitch, S.: Feature generation for text categorization using world knowledge. In: Proceedings of the 19th International Joint Conference on Artificial Intelligence, Edinburgh, Scotland, July 30-August 05, pp. 1048–1053 (2005)
4. Gabrilovich, E., Markovitch, S.: Overcoming the brittleness bottleneck using Wikipedia: enhancing text categorization with encyclopedic knowledge. In: Proceedings of the 21st National Conference on Artificial Intelligence, Boston, Massachusetts, July 16-20, pp. 1301–1306 (2006)
5. Wang, P., Hu, J., et al.: Improving text categorization by using encyclopedia knowledge. In: ICDM 2007 (2007)
6. Carmel, D., Roitman, H., Zwerdling, N.: Enhancing cluster labeling using Wikipedia. In: Proceedings of the 32nd International ACM SIGIR Conference on Research and Development in Information Retrieval, Boston, MA, USA, July 19-23 (2009)
7. Xu, Y., Jones, G.J.F., Wang, B.: Query dependent pseudo-relevance feedback based on Wikipedia. In: Proceedings of the 32nd International ACM SIGIR Conference on Research and Development in Information Retrieval, Boston, MA, USA, July 19-23 (2009)
8. Banerjee, S., Ramanathan, K., Gupta, A.: Clustering short texts using Wikipedia. In: Proceedings of the 30th Annual International ACM SIGIR Conference on Research and Development in Information Retrieval, Amsterdam, The Netherlands, July 23-27 (2007)
9. Hu, J., Fang, L., Cao, Y., et al.: Enhancing text clustering by leveraging Wikipedia semantics. In: Proceedings of the 31st Annual International ACM SIGIR Conference on Research and Development in Information Retrieval, Singapore, July 20-24 (2008)
10. Hu, X., Zhang, X., Lu, C., et al.: Exploiting Wikipedia as external knowledge for document clustering. In: Proceedings of the 15th ACM SIGKDD International Conference on Knowledge Discovery and Data Mining, Paris, France (June 2009)
11. Kurland, O., Lee, L.: Page Rank without hyperlinks: Structural re-ranking using links induced by language models. In: Proceedings of SIGIR 2005, pp. 306–313 (2005)
12. He, X., Zha, H., Ding, C., Simon, H.D.: Web document clustering using hyperlink structure, TechReport CSE-01-006 (April 2001)
13. Zhai, C.: Statistical language models for information retrieval a critical review. Found. Trends Inf. Retr. 2(3), 137–213 (2008)
14. Zhai, C., Lafferty, J.: A study of smoothing methods for language models applied to ad hoc information retrieval. In: Proceedings of SIGIR 2001 (September 2001)
15. Reuters-21578 text categorization test collection, Distribution 1.0. Reuters, http://www.daviddlewis.com/resources/testcollections/reuters 21578/
16. 20-newsgroup text categorization dataset, http://people.csail.mit.edu/jrennie/20Newsgroups
17. Zhao, Y., Karypis, G.: Criterion functions for document clustering: experiments and analysis, Technical Report. Department of Computer Science, University of Minnesota (2001)
18. Steinbach, M., Karypis, G., Kumar, V.: A Comparison of document clustering techniques. Technical Report. Department of Computer Science and Engineering, University of Minnesota (2000)
19. Zhong, S., Ghosh, J.: Generative model-based document clustering: a comparative study. Knowledge and Information Systems 8(3), 374–384 (2005)

ASVMFC: Adaptive Support Vector Machine Based Fuzzy Classifier

Hamed Ganji and Shahram Khadivi

Amirkabir University of Technology, Computer Engineering & IT Department,
No. 424, Hafez Ave, Tehran 15914, Iran

Abstract. SVM[1] and FNN[2] are popular techniques for pattern classification. SVM has excellent generalization performance, but this performance is dependent on appropriate determining its kernel function. FNN is equipped with human-like reasoning, but the learning algorithms used in most FNN classifiers only focus on minimizing empirical risk. In this paper, a new classifier called ASVMFC has offered uses capabilities of SVM and FNN together and does not have the mentioned disadvantages. In fact, ASVMFC is a fuzzy neural network that its parameters is adjusted using a SVM with an adaptive kernel function. ASVMFC uses a new clustering algorithm to make up its fuzzy rules. Moreover, an efficient sampling method has been introduced in this paper that drastically reduces the number of training samples with very slight impact on the performance of ASVMFC. The experimental results illustrate ASVMFC can achieve very good classification accuracy with generating only a few fuzzy rules.

Keywords: Fuzzy neural network (FNN), support vector machine (SVM), adaptive kernel, pattern classification, clustering.

1   Introduction

Pattern classification is one of the most important issues in machine learning and data mining that encompasses a wide domain of problems. Support vector machine (SVM) and fuzzy neural network (FNN) are considered as important tools for pattern classification; among them, SVM is a kernel based machine learning method. The theory of SVM is based on statistical learning theory and had been proposed by Vapnik in 1992 [1]. Although SVM due to reliance on the principle of structural risk minimization, has high generalization power theoretically; but this power is heavily dependent on appropriate determining its kernel function according to problem. This means that using common kernel functions for SVM, due to lack of compliance with the conditions of different problems, is caused SVM has good performance only in

---

[1] Support Vector Machine.
[2] Fuzzy Neural Network.

M.V.M. Salem et al. (Eds.): AIRS 2011, LNCS 7097, pp. 340–351, 2011.

special problems; experimental results given in this paper confirm this fact. Up to now, some efforts have been made to determine the SVM kernel function or kernel matrix adaptively, but the results of these efforts have not been very satisfactory [2, 3]. On the other hand, FNN has human-like reasoning benefit for handling uncertainty of information [4], but the learning algorithms, which so far have been used in most FNN classifiers, suffer from the weakness of focusing on minimizing empirical risk (training error) regardless of expected risk (testing error). In this paper a new classifier called ASVMFC is presented that uses capabilities of SVM and FNN together and does not have the mentioned disadvantages.

In fact, ASVMFC is a fuzzy neural network that its rules are produced by a new fuzzy clustering method and its parameters are adjusted by training a SVM equipped with an adaptive kernel. ASVMFC tries to achieve generalization power of SVM and human-like reasoning benefit of FNN together. The clustering method, which is used in ASVMFC to determine fuzzy rules, generates clusters that appropriately approximate data distribution of the problem and counteract the effects of outliers in creating ASVMFC. In addition, using an adaptive kernel for training SVM in ASVMFC has been caused ASVMFC achieves good classification performance for different problems and also generates a few number of support vectors. Moreover, a method for reducing number of training samples of ASVMFC is introduced in this paper. This method can identify samples, which are effective in adjusting ASVMFC, according to fuzzy membership values of samples and reduces number of training samples drastically. Empirical results show that the ASVMFC classification accuracy is more than classification accuracy of the three types of SVM which use RBF kernel, polynomial kernel, and sigmoid kernel respectively.

This paper is organized as follows. In section 2 the theory of SVM is presented briefly. Section 3 describes structure of ASVMFC and its learning algorithms in details. The results of experiments are expressed in section 4. Finally, the conclusions are summarized in section 5.

2   Support Vector Machine

The major purpose of SVM in binary classification problems is finding optimal discriminator hyper-plane in n dimensional problem space. To do that, SVM find a discriminator hyper-plane that has maximum margin to boundaries of two classes. Suppose we are given a set S of labeled training set, $S = (X_1, y_1), (X_2, y_2), ..., (X_N, y_N)$, where $X_i \in \mathbb{R}^m$, and $y_i \in \{+1, -1\}$. The goal of SVM is to find an optimal hyper-plane such that

$$y_i(W^T X_i + b) \geq 1 - \xi_i, \qquad i = 1, ..., N \tag{1}$$

where $W \in \mathbb{R}^m$, $b \in \mathbb{R}$, and $\xi_i \geq 0$ is a slack variable. For $\xi_i > 1$, the data are misclassified. To find an optimal hyper-plane is to solve the following constrained optimization problem:

$$\text{Minimize} \frac{1}{2} \|W\|^2 + C \sum_{i=1}^{N} \xi_i$$

$$\text{Subject to } y_i(W^T X_i + b) \geq 1 - \xi_i, \qquad i = 1, \dots, N \tag{2}$$

where C is a user defined positive cost parameter and $\sum \xi_i$ is an upper bound on the number of training errors. In fact, (2) is a quadratic programing problem and we can write (2) as (3) by using the Lagrange technique. Values of Lagrange multipliers are obtained by solving (3) and we can use them to compute values of W and b [5].

$$L(\alpha) = \sum_{i=1}^{N} \alpha_i - \frac{1}{2} \sum_{i,j=1}^{N} \alpha_i \alpha_j y_i y_j X_i^T X_j$$

$$\sum_{i=1}^{N} \alpha_i y_i = 0, \qquad \alpha_i \geq 0, \quad i = 1, \dots, N \tag{3}$$

To extend the above linear SVM to a nonlinear classifier we can use the kernel trick for realizing the idea of "in order to achieve better separability mapping the input data into a higher dimensional feature space". To this end, we apply a kernel function with form of $K(X_i, X_j)$ instead of $X_i^T X_j$ term in (3). A standard kernel is a function that is symmetric and positive-definite and satisfies the Mercer conditions [6].

3   Adaptive Support Vector Machine Based Fuzzy Classifier (ASVMFC)

ASVMFC is a classifier in a form of FNN, which uses a new fuzzy clustering method to generate its fuzzy rules and a SVM which equipped with an adaptive kernel to adjust its parameters. The ASVMFC learning algorithm consists of three phases as follow: 1) Clustering phase, 2) SVM training phase, 3) Creating and adjusting FNN phase. In this section, these phases will be described in details.

3.1   Clustering Phase

The basis of the clustering phase is to approximate the probability distribution of either classes of problem by a set of clusters with Gaussian distribution. To this end, we run a new clustering algorithm that its pseudo code is given in follow for either class individually. The clusters obtained in this phase will be considered as fuzzy rules of ASVMFC.

Pseudocode 1. The algorithm of clustering method of ASVMFC.

```
Clusters Clustering (Samples, InitCovariance, Thershold)
{
   FOR i ← 1 TO number of the samples
      FOR j ← 1 TO number of the samples
         IF i = j THEN
            InfluenceMatrix[i,j] ← infinite
         ELSE
            InfluenceMatrix[i,j] ← CalculateInfluence(
                  Samples[i], Samples[i], InitCovariance)
         ENDIF
      ENDFOR
   ENDFOR
   Influences ← sum of the rows of InfluenceMatrix
   Influences ← Influences/max(Influences)
   WHILE Influences is not empty
      I ← argmax(Influences);
      IF Influences[I] != -infinite THEN
         L ← InfluenceMatrix[I, all];
         I_th ← indexes of entries of L that their values
                  are greater than Thershold and are not
                  infinite
         WHILE I_th is not empty
            L ← L + sum of the rows of
                  InfluenceMatrix[I_th, all]
            I_th ← indexes of entries of L that their
                     values are greater than Thershold and
                     are not infinite
         ENDWHILE
         I_inf ← indexes of entries of L that their
                  values are infinite
         IF I_inf has more than one entity THEN
            InfeluenceCoeff ← max(Influences[I_inf])
            AddNewCluster(Clusters, Samples[I_inf],
                     InfeluenceCoeff)
         ENDIF
         Influences[I_inf] ← -infinite
      ENDIF
   ENDWHILE
}
```

According to this algorithm, first we consider one small Gaussian distribution for each sample and calculate sum of influences of these distributions on each sample and indicate it as Influence value of that sample (δ_i). In fact, the Influence value of each sample is calculated as follows,

$$\delta_i = \sum_{j}^{N} \exp\{-\frac{1}{2}(X_j - X_i)^T \Sigma_{init}^{-1}(X_j - X_i)\} \tag{4}$$

where N is the number of the samples of the class which the desired sample is belong to it and Σ_{init} is the predetermined covariance matrix of the small Gaussian distributions. We select the sample, which has maximum influence value, as the initial member of a new cluster. We calculate effect of small distribution of selected samples on the rest samples and select the samples that the acquired value for them are greater than a special threshold (δ_th) and add these samples to members of the new cluster. Then we continue with calculating effect of the small distributions of the selected samples on the rest samples until no sample is exist that its calculated influence value is greater than δ_th. Finally, if the new cluster has more than one member then we compute its parameters by using its elements.

In above algorithm, `CalculateInfluence()` function takes two samples and calculates effect of the small Gaussian distribution of the first sample on the second sample. Each cluster obtained in this phase, forms a hyper-ellipsoid in n dimensional problem space and probability distribution of the data within it, is considered as a Gaussian distribution. So, each of these clusters are determined by a center vector and a covariance matrix that are computed by (5). Also each cluster has an influence coefficient which determines amount of cluster influence in the next learning steps. The value of this parameter is equal to maximum Influence values of members of desired cluster. Indeed, `AddNewCluster()` function calculates center and covariance matrix of the new cluster and add it to list of clusters.

$$\mu_{C_i} = \frac{\sum_{X \in C_i} \delta_X X}{\sum_{X \in C_i} \delta_X}, \qquad \Sigma_{C_i} = \frac{\sum_{X \in C_i} \delta_X (X - \mu_{C_i})(X - \mu_{C_i})^T}{\sum_{X \in C_i} \delta_X} \tag{5}$$

Σ_{init} and δ_th are actually the internal parameters of the clustering algorithm and they can be calculated by the algorithm automatically. We can use (6) to determine value of Σ_{init} automatically,

$$\Sigma_{init} = \frac{1}{N_C}\sum_{i=1}^{N_C}|X_i - \mu_{init}|, \qquad \mu_{init} = \frac{1}{N_C}\sum_{i=1}^{N_C}X_i \tag{6}$$

where N_C is number of the samples of the class. Also, value of δ_th can be calculated by (7),

$$\delta\_th = (\mu_\delta - S_\delta)^2 \tag{7}$$

$$\mu_\delta = \frac{1}{N_C}\sum_{i=1}^{N_C}\delta_i, \qquad S_\delta = \frac{1}{N_C}\sum_{i=1}^{N_C}|\delta_i - \mu_\delta| \tag{8}$$

where μ_δ is mean of δ_is, and S_δ is standard deviation of δ_is.

3.2   SVM Training Phase

As noted earlier, SVM uses the kernel trick on to deal with problems which their samples are not linearly separable. Although, some functions as common standard kernel functions are available, but due to the diversity and complexity of various problems, these kernel functions cannot have reasonable performance in all problems. Hence, requirement to have an adaptive kernel for SVM is strongly felt in this domain. Because of this, in the learning algorithm of ASVMFC we try to realize the goal of SVM kernel being adaptive. For this purpose, in this phase, the kernel function of SVM is made up automatically according to the data distribution of training set which approximated in previous phase. (9) defines the adaptive kernel function which has been used in ASVMFC,

$$K(X_i, X_j) = X_i \cdot X_j + \delta_A(X_i) \cdot \delta_A(X_j) + \delta_B(X_i) \cdot \delta_B(X_j) \tag{9}$$

$$\delta_A(X_i) = \frac{\sum_j^{N_{CA}} \beta_{CA_j} \exp\{-\frac{1}{2}\left(\mu_{CA_j} - X_i\right)^T \Sigma_{CA_j}^{-1} \left(\mu_{CA_j} - X_i\right)\}}{\sum_j^{N_C} \beta_{C_j} \exp\{-\frac{1}{2}\left(\mu_{C_j} - X_i\right)^T \Sigma_{C_j}^{-1} \left(\mu_{C_j} - X_i\right)\}} \tag{10}$$

$$\delta_B(X_i) = \frac{\sum_j^{N_{CB}} \beta_{CB_j} \exp\{-\frac{1}{2}\left(\mu_{CB_j} - X_i\right)^T \Sigma_{CB_j}^{-1} \left(\mu_{CB_j} - X_i\right)\}}{\sum_j^{N_C} \beta_{C_j} \exp\{-\frac{1}{2}\left(\mu_{C_j} - X_i\right)^T \Sigma_{C_j}^{-1} \left(\mu_{C_j} - X_i\right)\}} \tag{11}$$

with $\beta_{CA_j} > 0$ and $\beta_{CB_j} > 0$, where β_{CA_j} is the influence coefficients of the jth cluster of class A and β_{CB_j} is the influence coefficients of the jth cluster of class B, N_{CA} indicates the number of the clusters of class A, N_{CB} indicates the number of the clusters of class B and , N_C is equal to $N_{CA} + N_{CB}$. In (9), $\delta_A(X_i)$ returns a value, which approximated data distribution of class A assigns to X_i. In other word, the output of $\delta_A(X_i)$ is equal to normalized sum of effects of the Gaussian distributions of the obtained clusters for class A on X_i. Also $\delta_B(X_i)$ is similar to $\delta_A(X_i)$ except that is related to class B.

In order that the kernel function defined by (9) is suitable for application in SVM, we must prove that the kernel function has the Mercer conditions. According to Mercer's work [6, 7], we know that if K is the symmetrical and continuous kernel of an integral operator $O_k: L^2 \to L^2$, such that

$$(O_{Kg})(X) = \int K(X, Y)g(Y)dY \tag{12}$$

is positive, i.e.,

$$\int K(X, Y)g(X)g(Y)dXdY \geq 0 \qquad \forall g \in L^2, \tag{13}$$

then K can be expanded into a uniformly convergent series

$$K(X, Y) = \sum_{i=1}^{\infty} \lambda_i \phi_i(X) \phi_i(Y) \tag{14}$$

with $\lambda_i \geq 0$. In this case, the mapping from input space to feature space produced by the kernel is expressed as

$$\phi: X \rightarrow (\sqrt{\lambda_1}\phi_1(X), \sqrt{\lambda_1}\phi_1(X), \dots) \tag{15}$$

such that K acts as the given dot product, i.e.,

$$\left(\phi(X), \phi(Y)\right) = \phi(X)^T \phi(X) = K(X, Y) \tag{16}$$

Lemma 1. *A nonnegative linear combination of Mercer kernels is also a Mercer kernel.*

According to Lemma1, whereas the kernel function of ASVMFC is linear combination of three terms, to prove the kernel function of ASVMFC is a standard kernel, it is sufficient to prove that each of these terms is a standard kernel. Like linear kernels, the first term is product of two vectors and satisfies conditions of a standard kernel.

Lemma 2. *The product of Mercer kernels is also a Mercer kernel.*

Also according to Lemma 2, whereas the second and third terms of the kernel function of ASVMFC are product of two functions, it is sufficient to prove that each of these functions satisfies Mercer conditions.

Lemma 3. *For any function $\psi(x)$ that can be expanded as uniformly convergent power series of x with nonnegative coefficients; that is, $\psi(x) = \sum_{i=0}^{\infty} a_i x^i$, $a_i \geq 0$, If we define the kernel function, then it is a Mercer kernel.*

According to Lemma3, many common functions such as$\exp(x)$, $\cosh(x)$, $\sinh(x)$, etc., can be used as possible kernel. Hence, $\delta_A(X)$ and $\delta_B(X)$ which are linear compositions of some exponential functions with positive coefficients, satisfy Mercer conditions according to Lemma1 and Lemma2. Therefore, we can claim that the kernel function of ASVMFC is a standard kernel.

Finally, in this phase we train a SVM with the kernel function defined in (9) and use its Lagrange multipliers for adjusting ASVMFC parameters in the next phase.

3.3 Creating and Adjusting ASVMFC Phase

In this phase a FNN is created using the results of previous phases. This FNN, which is shown in Fig. 1, is final classifier system of ASVMFC and consists three layers. In the first layer called input layer no computation is done. Each node in the input layer only transmits input values to the next layer directly. The second layer is called rule layer. A node in this layer represents one fuzzy rule and performs precondition matching of a rule. In fact, these rules are corresponding to the obtained clusters. The fuzzy rules of this FNN are divided into two groups and each of these groups consist the rules which are related to one of two classes. A and B indices show class label of the rules. The output weights of the rules are influence coefficients of the clusters and are computed by (17),

$$w_{A_i} = \beta_{CA_i} \sum_{j=1}^{N} \alpha_j y_j \delta_A(X_j) \ , \qquad w_{B_i} = \beta_{CB_i} \sum_{j=1}^{N} \alpha_j y_j \delta_B(X_j) \qquad (17)$$

where N is the size of training set and X_j is the jth sample of training set. Also, β_{CA_i} is the influence coefficient of the ith cluster of class A and β_{CB_i} is the influence coefficient of the ith cluster of class B. The parameters α_j and y_j are Lagrange multiplier (acquired in the previous phase) and class label of the jth sample of training set respectively.

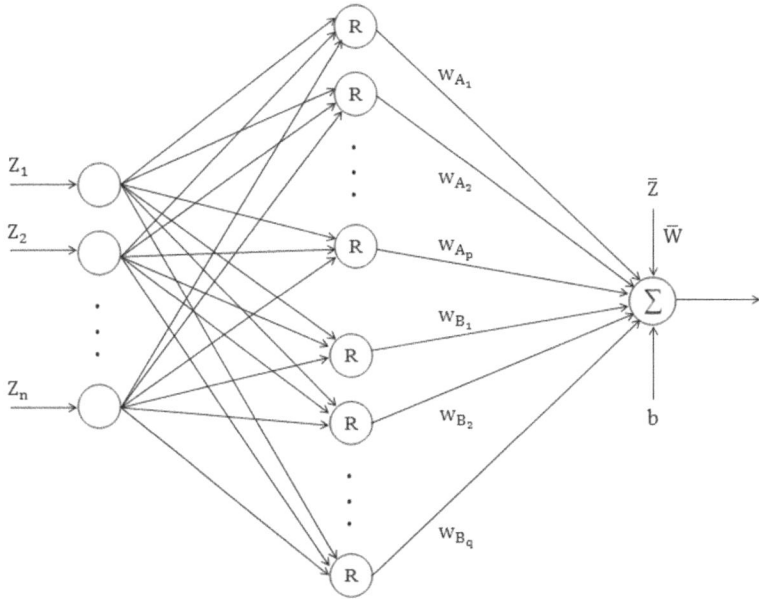

Fig. 1. The structure of the fuzzy neural network of ASVMFC

In Fig. 1 p represents number of data samples of class A and q represents number of data samples of class B. The third layer is output layer and has a single neuron unit. This unit, which is labeled with Σ, computes the overall output as the summation of second layer outputs and two additional inputs. Also, W and b are computed by (18).

$$W = \sum_{j=1}^{N} \alpha_j y_j X_j, \qquad b = \frac{1}{N} \sum_{j=1}^{N} (y_j - W^T X_j). \qquad (18)$$

3.4   Sample Reduction

One of the main problems of SVM is high time consumption of its training phase in the case of large training sets. Due to this problem, using the standard SVM is not affordable for cases with very large training set. A solution to overcome this problem is reducing number of samples by sampling from training set [8]-[10]. This means that we can remove samples which are not very impressive to form separator hyper plane. In other word, if we can identify the samples which probably will be chosen as support vectors, then we can use only these samples for training SVM.

The samples, which are lied in the area between two classes, have more potential to be selected as support vectors; so the problem of sample reduction can be converted to find samples in the area between two classes. This seems as a good solution but it is impractical for a SVM which operates based on a kernel. Because, in this kind of SVM the separator hyper plane is chosen based on the samples which have been mapped by kernel and there is no assurance that samples in the area between two classes in the original space are samples in the area between two classes in the mapped space. However, because in ASVMFC the feature space is scaled form of the original space, we can apply that solution here.

Another interesting point that should be noted is that finding the samples in the area between two classes is done easily based on fuzzy belonging degrees of samples to each of two classes and this process does not require much processing. We can say the samples which have high belonging degrees to both classes are in the area between two classes. Also, it is better that we add some samples from outer boundary of each class to training samples to determine class areas for preventing deviation of hyper plane. To this end, we apply (19) on the all samples and select some of smallest ones for training.

$$\theta(X_i) = |\delta_A(X_i) - \delta_B(X_i)| \tag{19}$$

The absolute of difference between two belonging degrees for each sample of the classes' boundaries is small value. So by choosing an appropriate threshold we can select efficient samples for training SVM. Interesting point here is that $\theta(.)$ assigns large values to samples located in area of opposite class; due to this, these samples are not selected for training and separator hyper plane will not be deviated by them.

4   Experiments

We used the six Benchmark datasets from UCI Repository [11] and LIBSVM software [12] to assess performance of ASVMFC. Table 1 describes these datasets in details. In these experiments we apply 10-fold cross validation to estimate the generalized accuracy of ASVMFC, RBF-kernel-based SVM, polynomial-kernel-based SVM and sigmoid-kernel-based SVM For each data set. In each round of cross validation we run the four classifiers on all the "one class against rest class" cases of each dataset and average all the results for it. The value of cost parameter of SVM is $C = 2^{12}$ for all cases. Table 2 shows the classification accuracy and support vector (SV) count of (the numbers of SVs have been rounded).

Table 1. Descriptions of the benchmark datasets

Dataset name	Number of samples	Number of Features	Number of classes
Iris	150	4	3
Ecoli	336	8	8
Glass	214	10	6
Wine	178	14	3
Heart Disease	270	14	2
Breast Cancer	683	10	2

Table 2. Comparison between the accuracies and number of support vectors

Dataset name	ASVMFC		RBF SVM		Polyno. SVM		Sigmoid SVM	
	%Acc	#SVs	%Acc	#SVs	%Acc	#SVs	%Acc	#SVs
Iris	**98.75**	**8**	95.29	13	96.58	9	66.76	90
Ecoli	**96.05**	**33**	95.27	34	95.17	37	92.78	38
Glass	**93.06**	**26**	91.70	33	91.01	27	87.40	50
Wine	**98.18**	**10**	70.34	160	95.98	11	66.87	107
Heart Disease	**84.53**	**60**	56.65	243	75.41	64	55.58	216
Breast Cancer	**97.22**	**42**	96.48	259	93.96	56	30.61	430

As shown in Table 2, ASVMFC has the best classification accuracy for all cases. The accuracies of polynomial-kernel-based SVM on Heart data set and RBF-Kernel-based SVM on Wine and Heart data sets are very low. Also, sigmoid-kernel-based SVM has unpromising accuracy for majority cases. Better performance of ASVMFC in all cases is because of being adaptive. In other word, using the common kernel functions for SVM, due to lack of compliance with the conditions of the problem, is caused SVM does not has good performance in any type of problems. Moreover corresponding to results shown in Table 1, numbers of fuzzy rules that ASVMFC has generated for Iris, Ecoli, Glass, Wine, Heart, and Breast datasets are 3, 2, 3, 6, 3, and 3 respectively. Also, as shown in Table 2 the number of generated support vectors in ASVMFC is lowest for all data sets. These results show that the adaptive kernel function used in ASVMFC increase the capabilities of linear separability of the data and so find the separator hyper plane with lower support vectors.

Fig. 2 shows how the proposed sample reduction method reduces the number of training samples for ASVMFC. Fig. 2 illustrates that by decreasing the value of threshold the number of samples is reduced significantly while the changes of the classification accuracy of ASVMFC has been minor for all datasets. For example, when the value of threshold is 0.2, 250 samples are almost reduced from 336 samples of Ecoli dataset while the classification accuracy of ASVNFC is still near 96%. These results actually confirm the good performance of the sample reduction method.

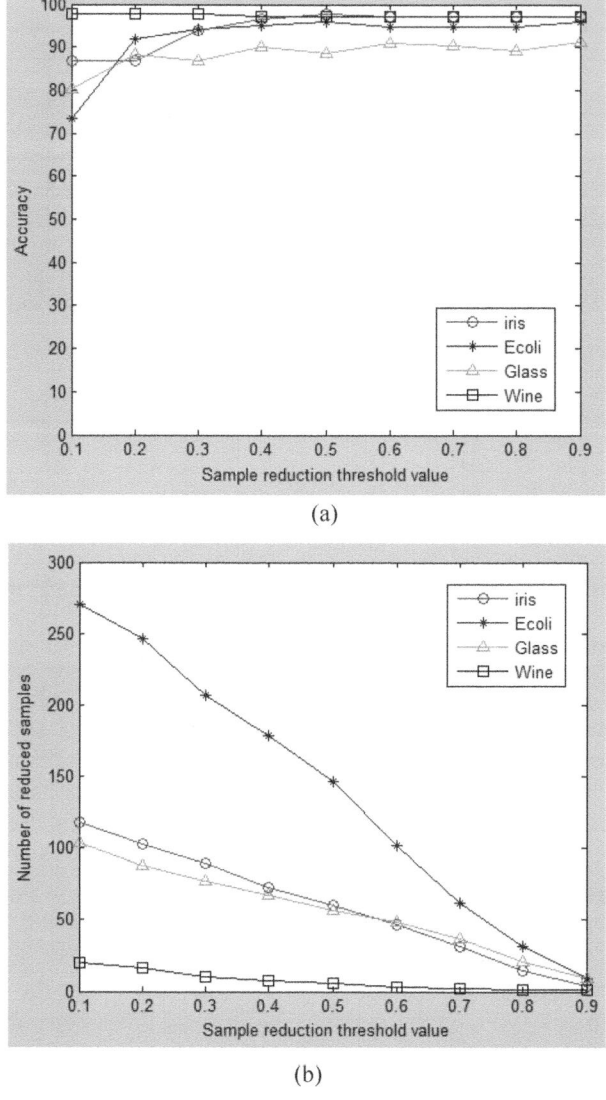

Fig. 2. (a) The effect of sample reduction on the accuracy of ASVMFC; (b) The effect of sample reduction threshold value on the number of reduced samples for ASVMFC

5   Conclusion

In this paper a new classifier called ASVMFC is introduced. ASVMFC makes up a FNN and adjusts the parameters of this FNN by using a new fuzzy clustering method and training a SVM equipped with an adaptive kernel. ASVMFC actually takes advantages of the SVM generalization and human-like reasoning of FNN. The fuzzy rules of ASVMFC are generated automatically by a new clustering method according to data distribution of problem. This enhances the accuracy of ASVMFC against the common standard kernel functions. Moreover, because ASVMFC uses an adaptive kernel for training SVM, it has a good performance with a few fuzzy rules in most problems while creates a few support vectors in SVM training phase of its learning algorithm. Also, we can easily reduce number of training samples for ASVMFC by using belonging degrees of samples to both classes.

References

1. Vapnik, V.: Statistical Learning Theory. Wiley, New York (1998)
2. Pontil, M., Poggio, T.: Regularization networks and support vector machines. Advances in Computational Mathematics 13(1), 1–50 (2000)
3. Lin, C.T., Yeh, C.M., Liang, S.F., Chung, J.F., Kumar, N.: Support-Vector-Based Fuzzy Neural Network for Pattern Classifications. Fuzzy Sets and Systems 142, 243–265 (2004)
4. Chen, M.Y., Linkens, D.A.: Rule-base self-generation and simplification for data-driven fuzzy models. IEEE Transaction on Fuzzy Systems 14(1), 31–40 (2006)
5. Abe, S.: Support Vector Machines for Pattern Classification. Springer, London (2005)
6. Mercer, J.: Functions of positive and negative type and their connection with the theory of integral equations. Philo. Trans. Royal Soc. 209, 415–446 (1909)
7. Williamson, R.C., Smola, A.J., Scholkopf, B.: Generalization Performance of Regularization Networks and Support Vector Machines via Entropy Numbers of Compact Operators. Technical Report 19, NeuroCOLT (1998)
8. Lin, K.M., Lin, C.J.: A study on reduced support vector machines. IEEE Transaction on Neural Networks 14(6), 1449–1459 (2003)
9. Pothin, J.B., Richard, C.: Incorporating prior information into support vector machine in the form of ellipsoidal knowledge sets. In: 14th European Signal Processing Conference (EUSIPCO), Florence, Italy (2006)
10. Guo, G., Zhang, J.S.: Reducing examples to accelerate support vector regression. Pattern Recognition Letters 28, 2173–2183 (2007)
11. Blake, C.L., Merz, C.J.: UCI repository of machine learning databases. University of California, http://www.ics.uci.edu/mlearn/MLSummary.html
12. Chang, C.C., Lin, C.J.: A library for support vector machines LIBSVM, http://www.csie.ntu.edu.tw/~cjlin/libsvm/index.html

Ensemble Pruning for Text Categorization Based on Data Partitioning

Cagri Toraman and Fazli Can

Bilkent Information Retrieval Group,
Computer Engineering Department,
Bilkent University, 06800, Ankara, Turkey
{ctoraman,canf}@cs.bilkent.edu.tr

Abstract. Ensemble methods can improve the effectiveness in text categorization. Due to computation cost of ensemble approaches there is a need for pruning ensembles. In this work we study ensemble pruning based on data partitioning. We use a ranked-based pruning approach. For this purpose base classifiers are ranked and pruned according to their accuracies in a separate validation set. We employ four data partitioning methods with four machine learning categorization algorithms. We mainly aim to examine ensemble pruning in text categorization. We conduct experiments on two text collections: Reuters-21578 and BilCat-TRT. We show that we can prune 90% of ensemble members with almost no decrease in accuracy. We demonstrate that it is possible to increase accuracy of traditional ensembling with ensemble pruning.

Keywords: Data partitioning, ensemble pruning, text categorization.

1 Introduction

Ensemble of classifiers are known to perform better than individual classifiers when they are accurate and diverse [5], [19]. In text categorization, they are proven to perform better in some cases [6]. However, they are not efficient due to computational workload. For instance, in news portals, it is a burden to train a new ensemble model or test new documents. Various ensemble selection methods are proposed to overcome this problem [3]. The main idea is to increase the efficiency by reducing the size of ensemble without hurting the effectiveness. Besides, it can increase the effectiveness if selected classifiers are more accurate and diverse than base classifiers. In this work, we study these two aspects of ensemble selection by giving the accuracy (i.e. effectiveness) results [16].

Ensemble selection mainly consists of three stages: constructing base classifiers (ensemble members), selecting target classifiers among base classifiers, and combining their predictions. Base classifiers are constructed homogeneously or heterogeneously. Homogeneous classifiers are trained by the same algorithm and constructed by data partitioning methods in which training documents are manipulated [5], [6]. Heteregeneous classifiers are usually created by training different algorithms on the training set [3]. There are also mixed constructions in

M.V.M. Salem et al. (Eds.): AIRS 2011, LNCS 7097, pp. 352–361, 2011.

which data is partitioned and different algorithms are applied separately. There are various ensemble selection approaches [17]. In general, they search for an optimal subset of ensemble members. Searching evaluation is done with a validation (hillclimbing or hold-out) set, which can be used either in training or as a separate part of training set. Lastly, ensemble predictions are combined by simple/weighted voting, mixture of experts or stacking [17]. Voting is the most popular approach. It combines predictions of ensemble based on sum, production or other rules. It is called weighted when each prediction is multiplied by a coefficient.

Construction of base classifiers, training them and getting predictions from each of them require a considerable amount of time in text categorization when there are huge numbers of text documents. This becomes crucial when text documents become longer as experienced in news portals. There is a need for pruning as many base classifiers as possible. Therefore, in this study, we examine ensemble pruning in text categorization by applying different data partitioning methods for construction of base classifiers and popular classification algorithms to train them. We select a simple ranked-based ensemble pruning method in which base classifiers are ranked (ordered) according to their accuracy in a separate validation set and then pruned pre-defined amounts. We choose to use weighted voting to combine predictions of pruned ensemble.

Our answers to the following questions are the contributions of this study:

1. How much data can we prune without hurting the effectiveness using data partitioning?
2. Which partitioning and categorization methods are more suitable for ensemble pruning in the text categorization domain?
3. How do English and Turkish differ in ensemble pruning?
4. Can we increase effectiveness with ensemble pruning in the text categorization domain and which combination of partitioning method and categorization algorithm gives the highest accuracy?

The rest of the paper is organized as follows. Section 2 gives the related work on ensemble selection. Section 3 explains the experimental design and the datasets used in our study. Section 4 gives the experimental results. Finally, Section 5 concludes the paper.

2   Related Work

There are several ensemble selection studies. Tsoumakas et al. [17] give a taxonomy and short review on ensemble selection. Their taxonomy divides ensemble selection methods into search-based, clustering-based, ranked-based, and other methods. Search-based methods apply greedy search algorithms (forward or backward) to get the optimal ensemble. Clustering-based methods employ a clustering algorithm and then prune clusters. Ranked-based methods rank ensemble members once, and then prune a pre-defined amount of members. Our approach also uses a ranked-based selection approach that examines different pruning levels.

Margineantu and Dietterich [11] study search-based ensemble pruning considering memory requirements. Classifiers constructed by the AdaBoost algorithm are pruned according to five different measures for greedy search based on accuracy or diversity. Their results show that it is possible to prune 60-80% ensemble members in some domains with good effectiveness performance.

Prodromidis et al. [14] define pre-pruning and post-pruning for ensemble selection in fraud detection domain using meta-learning. In our study, their pre-pruning corresponds to forward greedy search and post-pruning means backward greedy search. They produce their base classifiers in a mixed way such that they divide the train data into data partitions by time divisions and then apply different classification algorithms including decision trees to these partitions. They get up to 90% pruning with 60-80% of the original performance.

Caruana et al. [3] employ forward greedy search with heteregeneous ensembles on binary machine learning problems. They show that their selection approach outperforms traditional ensembling methods such as bagging and boosting. Caruana et al. [2] then examine some unexplored aspects of ensemble selection. They indicate that increasing validation set size improves performance. They also show that pruning up to 80-90% ensemble members rarely hurts the performance.

Martínez-Muñoz and Suárez [12] examine search-based ensemble pruning with bagging. They use CART trees and three different measures for forward greedy search. They show that 80% members can be removed with Margin Distance Minimization (MDM). Hernández-lobato et al. [7] study search-based ensemble pruning with bagging on regression problems. They decide to use 20% of ensemble members by looking regression errors. Martínez-Muñoz and Suárez [13] use training error defined in boosting in order to use in greedy search of ensemble pruning. Results are similar with the work by Hernández-lobato et al. [7].

In a recent work, Lu et al. [10] introduce ensemble selection by ordering according to a heuristic measure based on accuracy and diversity. Similar to our study, they prune the ordered (ranked) ensemble members using a pre-defined number of ensemble sizes. They compare their results with bagging and the approach used by Martínez-Muñoz and Suárez [12]. Their method usually outperforms the others when 15% and 30% of ensemble members arc selected.

Our study is different from the above studies in terms of the production method of ensemble members, the way of ensemble selection, and the domain to which ensemble selection applied. We introduce a novel approach that examines data partitioning ensembles in ensemble selection. We also examine different classification algorithms that are popular in text categorization for ensemble selection. Our ensemble selection method is also simple such that we do not use greedy search or a genetic algorithm.

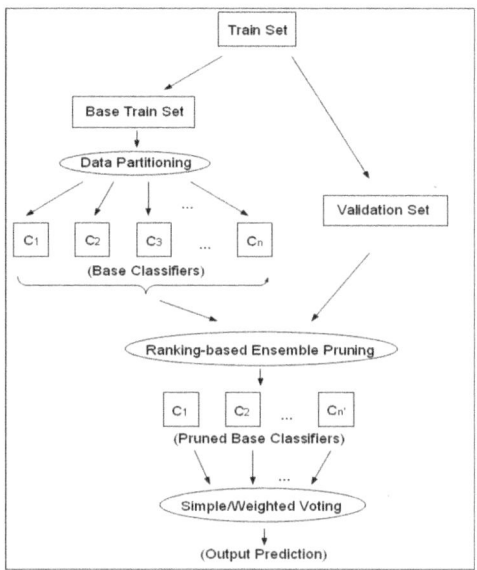

Fig. 1. Ensemble selection process used in this study (adapted from [16])

3   Experimental Environment

3.1   Experimental Design

Figure 1 represents the ensemble selection process used in this study. Firstly, the training set is divided into two separate parts. The base training set is used for training the base classifiers (i.e the ensemble). We construct the ensemble by dividing the base training set with homogeneous (in which base classifiers are trained by the same algorithm) data partitioning methods.

We apply four different partitioning methods: bagging, random-size sampling, disjunct, and fold partitioning [6].

- *Bagging* [1] creates ensemble members each of size N documents by randomly selecting documents with replacement where N is the size of the training set.
- *Disjunct partitioning* divides the training set into k equal-size partitions randomly and each k partition is trained separately.
- *Fold partitioning* divides the training set into k equal-size partitions and k-1 partitions are trained for each partitions.
- *Random-size sampling* is similar to bagging, but the size of each ensemble member is chosen randomly.

The base classifiers are then trained with four popular machine learning algorithms: C4.5 decision tree [15], KNN (k-nearest Neighbor) [4] , NB (Naive Bayes) [8], and SVM (Support Vector Machines) [18]. KNN's k value is set as 1 and the default parameters are used for other classifiers.

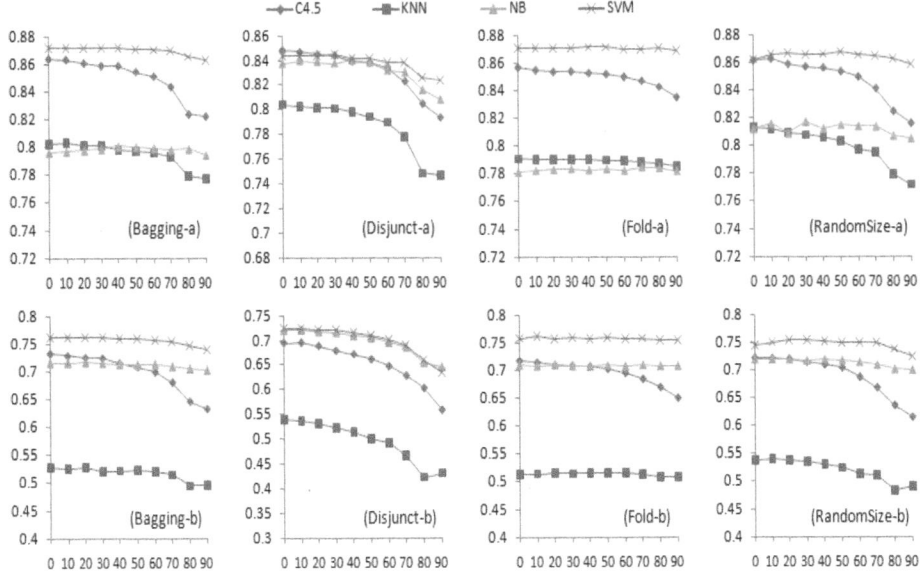

Fig. 2. Accuracy vs. pruning level: experimental results of different data partitioning and categorization methods on two datasets: (a) Reuters-21578 (b) BilCat-TRT. (Figures are not drawn to the same scale.)

After constructing the ensemble we decide to use simple solutions for ensemble selection since constructing data partitioning ensembles is a time-consuming process for large text collections. We choose ranking-based ensemble pruning that does not use complex search algorithms of other ensemble selection methods. Each ensemble member is ranked according to their accuracy on the validation set. We use a distinct part of the training set for the validation. The size of the validation set is set as 5% of the training set since we observe reasonable effectiveness and efficiency and accordingly, 5% of each category's documents are chosen randomly without replacement. After ranking, we prune the ranked-list 10% to 90% by 10% increments.

For the combination of the pruned base classifiers, we choose weighted voting that avoids the computational overload of stacking, mixture of experts etc. Class weight of each ensemble member is taken as its accuracy performance on the validation set. If the validation set of a class is empty (when number of documents in a class is not enough), then simple voting is applied.

Considering each four data partitioning methods with four classification algorithms, we use a thorough experimental approach and repeat the above ensemble pruning procedure for 16 different scenarios. All experiments are repeated 30 times and results are averaged. Documents are represented with term frequency vectors. Ensemble size (i.e data partitioning parameter) is set as 10 and the most frequent 100 unique words per category are used to increase efficiency. We use the classification accuracy for effectiveness measurement.

Table 1. The highest ensemble pruning degrees(%) obtained by unpaired t-test for each partitioning and categorization method on both datasets*

	Reuters-21578					BilCat-TRT				
	C4.5	KNN	NB	SVM	Avg.	C4.5	KNN	NB	SVM	Avg.
Bagging	10	10	90	60	42.5	10	20	60	50	35
Disjunct	10	30	60	40	35	10	0	20	30	15
Fold	0	0	90	50	35	10	60	90	90	62.5
Random-size	10	10	90	90	50	0	20	50	70	35
Avg.	7.5	12.5	82.5	60	40,6	7.5	25	55	60	36,8

* All accuracy differences between traditional ensemble and ensemble pruning approaches are statistically insignificant ($p > 0.05$) up to the pruning degrees given above. This means that, for example, with Reuters-21578, NB, and Bagging we can prune 90% of ensemble members with no statistically significant decrease in accuracy with respect to traditional ensemble approach.

3.2 Datasets

We use the following two datasets in the experiments: Reuters-21578 and BilCat-TRT (http://cs.bilkent.edu.tr/~ctoraman/datasets/ensemblePruning). The former one is a well-known benchmark dataset [9]. After splitting it with ModApte, eliminating multi-class documents and choosing the 10 most frequent topics, we get 5,753 training and 2,254 test news articles. The latter consists of 3,184 training and 1,660 test Turkish news articles. We choose these two datasets to observe the performance in both English and Turkish.

4 Experimental Results

4.1 Pruning Results

The four questions given in introduction are answered in this section. Firstly, Figure 2 gives the results of how much ensemble member we can prune with different data partitioning and categorization methods. These figures can be interpreted either heuristically or statistically. In heuristic way, one can look at Figure 2 and choose an appropriate pruning degree regarding some accuracy reduction. In general, fold partitioning seems to be more robust to accuracy reduction while disjunct partitioning is the weakest one. Similarly, NB and SVM are more suitable for ensemble pruning while C4.5 prunes the least number of base classifiers.

One can also apply some statistical methods to obtain a pruning degree regarding no accuracy reduction. We apply unpaired two-tail t-test between each pruning degree and traditional ensembling to check whether accuracy reduction is statistically significant. We apply unpaired t-test until difference becomes statistically significant. Pruning degrees regarding no accuracy reduction with unpaired t-test are listed in Table 1. We can prune up to %90 ensemble members using fold partitioning and NB on both datasets. Disjunct partitioning seems to

Table 2. Traditional ensembling accuracy and pruning's highest accuracy for each data partitioning and categorization method on Reuters-21578

	Traditional / Pruning's Highest (Pruning Degree)			
	C4.5	KNN	NB	SVM
Bagging	0.8646/-	0.8044/-	0.7928/0.8007(40%)**	0.8714/**0.8722(10%)**
Disjunct	0.8490/-	0.8024/-	0.8351/0.8404(40%)*	0.8414/0.8452(30%)**
Fold	0.8576/-	0.7921/-	0.7780/0.7846(60%)**	0.8718/-
Random-size	0.8624/0.8629(10%)	0.8139/-	0.8092/0.8174(30%)**	0.8565/0.8682(40%)**

Table 3. Traditional ensembling accuracy and pruning's highest accuracy for each data partitioning and categorization method on BilCat-TRT

	Traditional / Pruning's Highest (Pruning Degree)			
	C4.5	KNN	NB	SVM
Bagging	0.7325/-	0.5277/0.5282	0.7128/0.7163(20%)*	0.7605/**0.7620(20%)**
Disjunct	0.6987/-	0.5529/-	0.7209/0.7220(10%)	0.7206/-
Fold	0.7159/0.7171	0.5180/-	0.7076/0.7101(20%)*	0.7554/0.7612(10%)**
Random-size	0.7290/-	0.5423/-	0.7186/0.7205(10%)	0.7479/0.7549(30%)*

* Difference between traditional and pruning's highest is highly statistically significant when $p < 0.05$

** Difference between traditional and pruning's highest is extremely statistically significant when $p < 0.01$

be the worst method for ensemble pruning with no accuracy reduction. Similar to heuristic observations, we get better pruning degrees when either NB or SVM is used. Small amount of ensemble members are pruned using C4.5 and KNN with no accuracy reduction.

Table 1 also suggests that all partitioning and categorization methods prune similar number of ensemble members in both English and Turkish when no accuracy reduction is considered. However, NB prunes more or equal number of ensemble members with all partitioning methods in English than those of Turkish.

In some pruning degrees, we observe that ensemble pruning even increases accuracy of traditional ensembling. Table 2 and 3 list accuracies of traditional ensembling and highest increased accuracy that we can obtain by ensemble pruning using Reuters-21578 and BilCat-TRT respectively. If any degree of ensemble pruning makes no increase in accuracy, then we only give its traditional ensembling accuracy. We also give the pruning degree in which we get the highest accuracy within parentheses. Note that these pruning degrees are not the same as those in Table 1. Unpaired t-test is applied for all comparisons between traditional ensemble and pruning's highest increased accuracy. Results show that, in general, it is possible to increase accuracy with NB and SVM when ensemble pruning is applied. The combination when highest accuracies are seen is bagging with SVM on both datasets. Fold with SVM and random-size with SVM are almost as good as bagging with SVM.

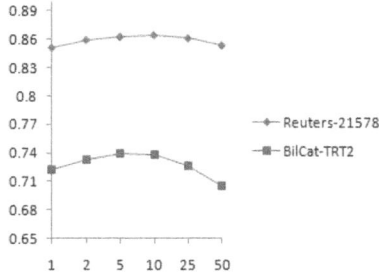

Fig. 3. Accuracy vs. validation set size: effect of different validation set size between 1% and 50% of original training set

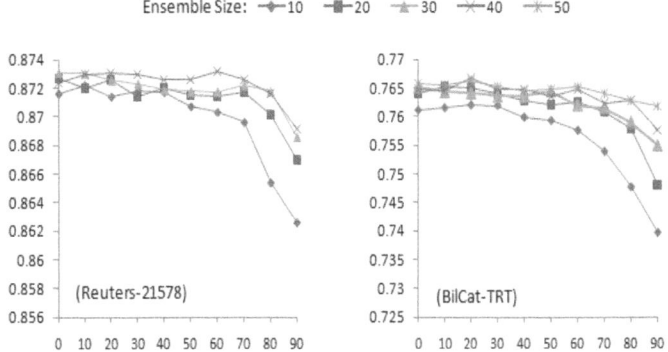

Fig. 4. Accuracy vs. pruning level: effect of different ensemble set size between 10 and 50 base classifiers. (Figures are not drawn to the same scale.)

4.2   Pruning-Related Parameters

Ranked-based ensemble pruning explained in Section 3.1 is a simple strategy that depends on choosing an appropriate validation set and ensemble size. In the previous experiments, validation set is chosen as 5% of the original training set and ensemble size is set as 10. These decisions are chosen for simplicity. However, other decisions may affect the accuracy result of ranked-based ensemble pruning. The following experiments are conducted on only bagging with SVM.

Different validation set sizes on both datasets are examined in Figure 3. Validation size experiments are conducted by 90% pruning of 10 base classifiers. We randomly select news documents for each category between 1% and 50% of the original training set and set this separate part as validation set. Figure 3 shows that if validation set size is either too small or too big, accuracy becomes reduced. Optimal validation set size is somewhere between 5% and 10% of the original training set.

Ensemble size is another parameter for ensemble pruning. Figure 4 displays pruning accuracies of different number of base classifiers between 10 and 50.

In ensemble set size experiments, validation set size is selected as 5% of the original training set. Accuracy is slightly increased with increasing number of base classifiers as expected. Moreover, accuracy reduction due to pruning becomes lower as ensemble size increases. However, efficiency is reduced due to the additional workload of training base classifiers. Thus, one should consider the trade-off between reduction in efficiency and increase in accuracy.

5   Conclusion

In this work we study ensemble pruning in text categorization. Ensembles are created with different data partitioning methods and trained by four different popular text categorization algorithms. The controlled experiments are conducted on English and Turkish datasets. We plan to perform further experiments with additional datasets. However, our statistical tests results provide strong evidence about the generalizability of our results. The main goals are to find how many ensemble members we can prune in text categorization without hurting accuracy, which data partitioning methods and categorization algorithms are more suitable for ensemble pruning, how English and Turkish differ in ensemble pruning, and lastly whether we can increase accuracy with ensemble pruning.

This study employs data partitioning methods with several classification algorithms in ensemble pruning. The main results of this study are:

1. Up to 90% of ensemble members can be pruned with almost no decrease in accuracy (See Table 1).
2. NB and SVM prune more ensemble members than C4.5 and KNN. Using disjunct partitioning prunes less members than other methods.
3. Pruning results are similar for both English and Turkish.
4. It is possible to increase accuracy with ensemble pruning (See Table 2 and 3). But pruning degrees are decreased in comparison to degree values without accuracy decrease (See Table 1). The best accuracy results are obtained by bagging with SVM on both datasets.

We also examine the effect of different ensemble and validation set sizes. It is seen that using 5-10% of the training set for validation is an appropriate decision for both datasets. We also find that accuracy reduction becomes smaller as ensemble size increases.

In future work different ensemble selection methods and validation measures can be studied. Additional test collections in other languages can be used in further experiments.

References

1. Breiman, L.: Bagging predictors. Mach. Learn. 24, 123–140 (1996)
2. Caruana, R., Munson, A., Niculescu-Mizil, A.: Getting the most out of ensemble selection. In: ICDM 2006, pp. 828–833. IEEE Computer Society, Washington, DC (2006)

3. Caruana, R., Niculescu-Mizil, A., Crew, G., Ksikes, A.: Ensemble selection from libraries of models. In: Proceedings of The Twenty-First Int. Conf. on ML, ICML 2004, p. 18 (2004)
4. Cover, T., Hart, P.: Nearest neighbor pattern classification. IEEE Transactions on Information Theory 13, 21–27 (1967)
5. Dietterich, T.G.: Ensemble Methods in Machine Learning. In: Kittler, J., Roli, F. (eds.) MCS 2000. LNCS, vol. 1857, pp. 1–15. Springer, Heidelberg (2000)
6. Dong, Y.S., Han, K.S.: Text classification based on data partitioning and parameter varying ensembles. In: Proceedings of the 2005 ACM Symposium on Applied Computing, SAC 2005, pp. 1044–1048 (2005)
7. Hernández-lobato, D., Martínez-Muñoz, G., Suárez, A.: Pruning in ordered regression bagging ensembles. In: Proceedings of IJCNN 2006, IEEE WCCI 2006, Vancouver, BC, pp. 1266–1273 (2006)
8. John, G.H., Langley, P.: Estimating continuous distributions in bayesian classifiers. In: UAI 1995, pp. 338–345 (1995)
9. Lewis, D.D., Ringuette, M.: A comparison of two learning algorithms for text categorization. In: Symposium on Document Analysis and Information Retrieval, pp. 81–93. ISRI, Univ. of Nevada, Las Vegas (1994)
10. Lu, Z., Wu, X., Zhu, X., Bongard, J.: Ensemble pruning via individual contribution ordering. In: Proceedings of the 16th ACM SIGKDD, KDD 2010, pp. 871–880 (2010)
11. Margineantu, D.D., Dietterich, T.G.: Pruning adaptive boosting. In: Proceedings of the Fourteenth International Conference on ML, ICML 1997, pp. 211–218 (1997)
12. Martínez-Muñoz, G., Suárez, A.: Aggregation ordering in bagging. In: Proc. of the IASTED, pp. 258–263. Acta Press (2004)
13. Martínez-Muñoz, G., Suárez, A.: Using boosting to prune bagging ensembles. Pattern Recognition Letters 28, 156–165 (2007)
14. Prodromidis, A.L., Stolfo, S.J., Chan, P.K.: Effective and efficient pruning of metaclassifiers in a distributed data mining system. Tech. rep. (1999)
15. Quinlan, J.R.: C4.5: Programs for machine learning. Morgan Kaufmann Publishers Inc., San Francisco (1993)
16. Toraman, C.: Text Categorization and Ensemble Pruning in Turkish News Portals. M.Sc. Thesis. Bilkent University, Ankara, Turkey (2011)
17. Tsoumakas, G., Partalas, I., Vlahavas, I.: A taxonomy and short review of ensemble selection. In: ECAI 2008, Workshop on Supervised and Unsupervised Ensemble Methods and Their Applications (2008)
18. Vapnik, V.: Estimation of Dependences Based on Empirical Data. Springer Inc., Secaucus (1982)
19. Witten, I.H., Frank, E.: Data Mining: Practical Machine Learning Tools and Techniques, 2nd edn. Morgan Kaufmann Series in Data Management Systems. Morgan Kaufmann Publishers Inc., San Francisco (2005)

Sentiment Analysis for Online Reviews Using an Author-Review-Object Model

Yong Zhang[1,2], Dong-Hong Ji[1], Ying Su[3], and Cheng Sun[1]

[1] Computer School, Wuhan University, Wuhan, P.R. China
[2] Department of Computer Science, Huazhong Normal University, Wuhan, P.R. China
[3] Department of Computer Science, Wuchang Branch, Huazhong University of Science and Technology, Wuhan, P.R. China
{Ychang,gensun.cn}@gmail.com,
donghong_ji2000@yahoo.com.cn,
suying929@163.com,

Abstract. In this paper, we propose a probabilistic generative model for online review sentiment analysis, called joint Author-Review-Object Model (ARO). The users, objects and reviews form a heterogeneous graph in online reviews. The ARO model focuses on utilizing the user-review-object graph to improve the traditional sentiment analysis. It detects the sentiment based on not only the review content but also the author and object information. Preliminary experimental results on three datasets show that the proposed model is an effective strategy for jointly considering the various factors for the sentiment analysis.

Keywords: Sentiment Analysis, Topic Model, Author-Review-Object Model.

1 Introduction

In business sectors, lots of research work has been done to find out customers' sentiments and opinions from online reviews [1, 4, 5, 6, 7, 8, 9, 10, 11]. Usually the sentiment polarities are dependent on topics or domains [14]. It is more suitable to analyze the topic and sentiment simultaneously. At the same time, lots of research works have been done focusing on utilizing the social data to improve the traditional data mining [12, 13]. The online reviews are one type of such social data in Web 2.0, which are free text generated by the users to comment on target objects (i.e. services or product). The users, target objects and reviews form a heterogeneous graph.

Although much work has addressed the problem of sentiment detection on such social reviews in various levels [6, 10] and various joint sentiment and topic models are proposed, none of them can model joint sentiment and topic analysis alongside with such user-review-object graph.

For joint topic and sentiment analysis, the sentiment is analyzed in more detailed topic or domain level. Multi-grain LDA (MG-LDA) and Multi-Aspect Sentiment model (MAS) proposed by Titov [10] are based on LDA and focus on extracting text for sentiment summaries of each aspect ratings for online reviews. Joint Sentiment/Topic Model (JST) [1, 8] based on LDA draws a word from the distribution over words jointly

M.V.M. Salem et al. (Eds.): AIRS 2011, LNCS 7097, pp. 362–371, 2011.

defined by topic and sentiment label. Li [9] presents an extension of JST and forces on dependency of the sentiment of words. The model proposed in [5] tries to learn a set of properties of a product and captures aggregate user sentiments towards these properties by supervised. But these models do not exploit the structure of user-review-object graph.

For social media, many models extending LDA have been proposed and explored information other than document words for topic learning. For instance, the author-topic model proposed in [11] uses the authorship information together with the words to learn topics. To our knowledge, few of work utilize the user-review-object graph for sentiment analysis.

In this paper, we propose a new probabilistic generative model on sentiment analysis for online reviews using user-review-object graph. The rest of this paper is organized as follows. Section 2 presents the proposed ARO Model and formulation. In section 3, we evaluate the performance of ARO model, and in section 4, we conclude the paper and present the future work.

2   Social Review Sentiment-Topic Model

In this section, we introduce the joint Author-Review-Object model (ARO), which integrates the user preference and object quality to the estimation of review sentiment at topic level in a unified framework.

2.1   Motivation

In general, a review is about some topics of an object, and reflects some sentiments or opinions of the author about the topics. In a web community with reviews, one user may contribute multiple reviews towards different objects, and one object may be commented by multiple users. Thus the users, objects and reviews form an inter-connected network as shown in Fig. 1.

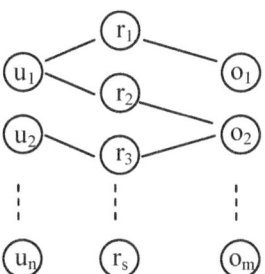

Fig. 1. The user-review-object graph for online reviews

For users, they have different preference and styles of commenting. Some users tend to be very strict, and give more negative reviews in average, while some users may be the opposite. On the other hand, some users may focus on certain aspects of an object, but some users may be concerned with the overall performance. So for one same object, different users may give different opinions. As to objects, each object has multiple aspects or topics for users to comment on. If an aspect of an object performs well, most of users tend to give positive comments, and the vise versa. On the other hand, an ambiguous review may be clarified resolved for an object with good quality. Generally the user preference can be mined from reviews commented by him, and the object quality can be mined from reviews commenting it. At the same time, the user preference and object quality can contribute to the analysis of the reviews associated with them.

2.2   Model Formulation

In our model, we adopt a switch variable to control the influence of user preference, object quality and the experience between the user and object. The proposed model is illustrated in Fig. 2, where D denotes the number of reviews, N_d denotes the number of terms in review d, K denotes the number of topics, V denotes the size of vocabulary, S denotes the number of sentiment label, U denotes the number of users and O denotes the number of objects in the dataset. The α^e, α^u, α^o, γ^e, γ^u, γ^o, η and β are hyper parameters and priors of Dirichlet distributions. As shown in Fig. 2 (c), the generative process of review in the model can be described as follows:

- For each document d, sample $\pi^{(e)}_d \sim Dir(\gamma^e)$
- For each sentiment label s under each documents d, sample $\theta^{(e)}_{d,s} \sim Dir(\alpha^e)$
- For each user u, sample $\pi^{(u)}_u \sim Dir(\gamma^u)$
- For each sentiment label s under each user u, sample $\theta^{(u)}_{u,s} \sim Dir(\alpha^u)$
- For each object o, sample $\pi^{(o)}_o \sim Dir(\gamma^o)$
- For each sentiment label s under each object o, sample $\theta^{(o)}_{o,s} \sim Dir(\alpha^o)$
- For each document d, sample $\lambda_d \sim Dir(\eta)$
- For each of the topics k and sentiment labels s, sample $\varphi^s_k \sim Dir(\beta)$
- For each of the N_d word tokens w_i in document d:
    - Choose $x_i \sim Multinomial(\lambda_d)$

        - If $x_i = exp$, choose a sentiment label $s_i \sim \pi^{(e)}_d$, Choose a topic $z_i \sim \theta^{(e)}_{d,s}$
        - If $x_i = user$, choose a sentiment label $s_i \sim \pi^{(u)}_u$, Choose a topic $z_i \sim \theta^{(u)}_{u,s}$
        - If $x_i = obj$, choose a sentiment label $s_i \sim \pi^{(o)}_o$, Choose a topic $z_i \sim \theta^{(o)}_{o,s}$
        - Choose a word w_i from the distribution over words defined by topic z_i and sentiment label s_i, $\varphi^{s_i}_{z_i}$

As shown in Fig. 2 (c), our model adds a user layer, an object layer based on JST, and it integrates the user preference and object quality to the estimation of review sentiment at topic level. The ARO model seems much more complex than both LDA and JST. Actually its running time of parameter estimation is a little more than the JST because the time complexity is mainly estimated by the number of tokens, K, S and the number of iteration.

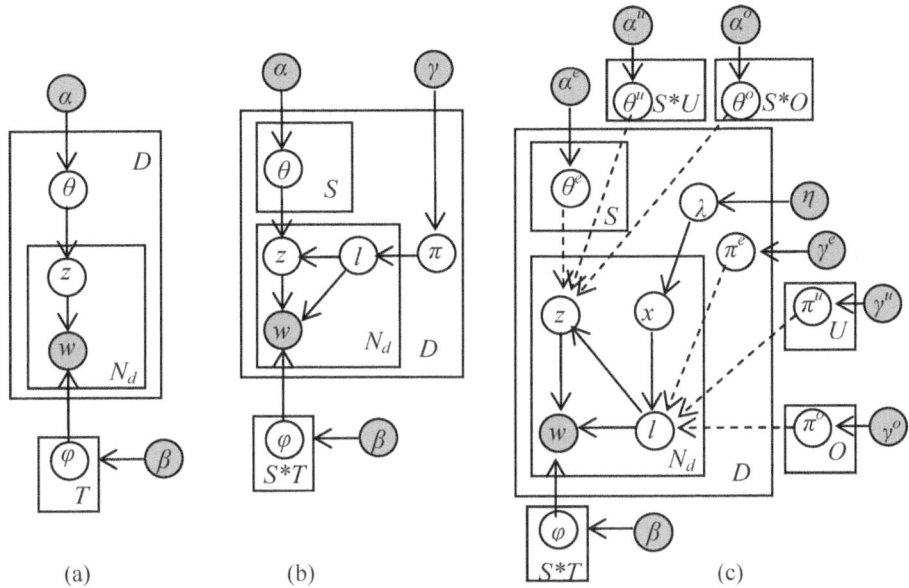

Fig. 2. Generative models for documents. (a) Latent Dirichlet Allocation (LDA). (b) The JST model. (c) The ARO model.

2.3   Parameter Estimation

In ARO model, we select Gibbs sampling [3] to derive the equations for our model, and we have three sets of latent variables: l, z and x. The joint probability of the l, z, x assignments and the words can be factored into the following terms:

$$
\begin{aligned}
p(w,z,l,x) &= p(w\,|\,z,l)\,p(z,l\,|\,x) \\
&= p(w\,|\,z,l)(\,p(z\,|\,l,\theta^e)\,p(s\,|\,x=exp)\,p(x=exp) \\
&+ p(z\,|\,l,\theta^u)\,p(l\,|\,x=user)\,p(x=user) \\
&+ p(z\,|\,l,\theta^o)\,p(l\,|\,x=obj)\,p(x=obj))
\end{aligned} \tag{1}
$$

We draw each (l_i; z_i; x_i) pair as a block, conditioned on all other variables. When $x=exp$, the sampling equation is:

$$
\begin{aligned}
&p(z_i = k, l_i = s, x = exp\,|\,w, z_{-i}, l_{-i}, x_{-i}) \\
&\propto \frac{n_{w_i}^{k,s} + \beta}{n_{.}^{k,s} + V\beta} \cdot \frac{n_{s,k}^e + \alpha^e}{n_{s}^e + K\alpha^e} \cdot \frac{n_{s}^e + \gamma^e}{n_{.}^e + S\gamma^e} \cdot \frac{n_d^e + \eta_{exp}}{N_d + \displaystyle\sum_{x\in\{exp,user,obj\}} \eta_x}
\end{aligned} \tag{2}
$$

where n_d^e is the number of times that words are generated from experience in review d; n_s^e is the number of times that words assigned sentiment label s are generated from experience in the review d; $n_{s,k}^e$ is the number of times that words assigned sentiment label s and topic k are from experience in the review d; $n^{k,s}$ is the number of times that words are assigned sentiment label s and topic k. $n_{w_i}^{k,s}$ is the number of times that word in the position i are assigned sentiment label s and topic k.

When x=*user*, the conditional posterior for z_i, l_i is:

$$p(z_i = k, l_i = s, x = user \mid w, z_{-i}, l_{-i}, x_{-i})$$

$$\propto \frac{n_{w_i}^{k,s} + \beta}{n_.^{k,s} + V\beta} \cdot \frac{n_{s,k}^u + \alpha^u}{n_s^u + K\alpha^u} \cdot \frac{n_s^u + \gamma^u}{n_.^u + S\gamma^u} \cdot \frac{n_d^u + \eta_{user}}{N_d + \sum_{x \in \{exp, user, obj\}} \eta_x} \tag{3}$$

where n^u is the number of times that words are generated from user u; n_d^u is the number of times that words in review d are generated from user u; n_s^u is the number of times that words assigned sentiment label s are generated from user u; $n_{s,k}^u$ is the number of times that words assigned sentiment label s and topic k are generated from user u;

When x=*obj*, the conditional posterior for z_i, l_i is:

$$p(z_i = k, l_i = s, x = user \mid w, z_{-i}, l_{-i}, x_{-i})$$

$$\propto \frac{n_{w_i}^{k,s} + \beta}{n_.^{k,s} + V\beta} \cdot \frac{n_{s,k}^o + \alpha^o}{n_s^o + K\alpha^o} \cdot \frac{n_s^o + \gamma^o}{n_.^o + S\gamma^o} \cdot \frac{n_d^o + \eta_{obj}}{N_d + \sum_{x \in \{exp, user, obj\}} \eta_x} \tag{4}$$

where n^o is the number of times that words are generated from object o; n_d^o is the number of times that words are generated from object o in review d; n_s^o is the number of times that words assigned sentiment label s are generated from object o; $n_{s,k}^o$ is the number of times that words assigned sentiment label s and topic k are generated from object o.

After a set of sampling processes based on the posterior distributions calculated with above equations, the overall approximated predictive distribution over sentiment for review d can be estimated by equation 5:

$$\pi_s^d \propto \frac{n_s^e + \gamma^e}{n_.^e + S\gamma^e} \cdot \frac{n^e + \eta_{exp}}{N_d + \sum_{x \in \{exp, user, obj\}} \eta_x} + \frac{n_s^u + \gamma^u}{n_.^u + S\gamma^u} \cdot \frac{n_d^u + \eta_{user}}{N_d + \sum_{x \in \{exp, user, obj\}} \eta_x}$$

$$+ \frac{n_s^o + \gamma^o}{n_.^o + S\gamma^o} \cdot \frac{n_d^o + \eta_{obj}}{N_d + \sum_{x \in \{exp, user, obj\}} \eta_x} \tag{5}$$

As shown in Equation 5, the distribution over sentiment for review is estimated based on three factors: the experience, the user preference and the object quality.

The distribution of words in topics and sentiment labels can be approximated by

$$\varphi_{i,k,s} \propto \frac{n_{i,k,s} + \beta}{n_{k,s} + V\beta} \tag{6}$$

2.4   Sub Models of ARO

In ARO model, we use three additional variables λ (λ^{exp}, λ^{user} and λ^{obj}, $\lambda^{exp}+\lambda^{user}+\lambda^{obj}=1$) to record the probability that each word is generated from experience, user or object. Especially if only λ^{user} is set to 0, the words are generated from experience or object, not from user. Our model is degraded to another model, which we call **Object-Review Model** (OR) which is based on the object-review graph, as shown as Figure 1. Similarly if only λ^{obj} is set to 0, our model is transformed to a simpler model called **User-Review Model** (UR) which is based on the user-review graph, as shown as Fig. 1.

3   Experiment

3.1   Experimental Setup

The review datasets with social information are very rare. We exacted the user's name and movie's id for each review of movie review polarity dataset 2.0[1] based on the pool of 27886 unprocessed html files. And we found 312 users and 1107 movies in the 2000 reviews. Fig. 3 (a) depicts the number of reviews for each user, and Fig. 3 (b) depicts the number of reviews for each movie sorted in descending order. The data now is available online[2].

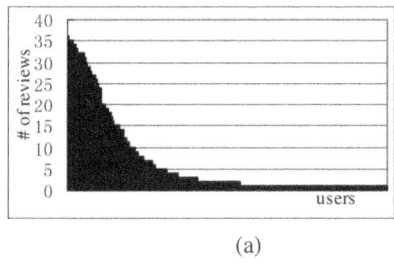 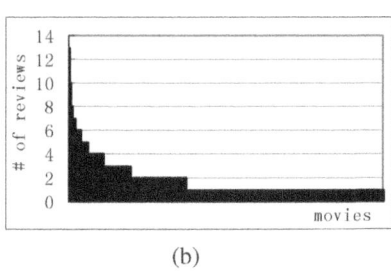

(a)   (b)

Fig. 3. The number of reviews for each user (a) and each movie (b)

Before experiments, preprocessing was performed. The punctuation, numbers, stop words and other non-alphabet characters were removed, and stemming was performed.

---

[1] http://www.cs.cornell.edu/people/pabo/movie-review-data/
[2] http://www.clr.org.cn/ychang/social-review/

The ARO has eight Dirichlet prior parameters, which affect the convergence of Gibbs sampling but not much the output results [13]. Generally we set $\alpha^e=0.1$, $\alpha^u=0.1$, $\alpha^o=0.1$, $\beta=0.05$, $S=3$ and $\eta=0.5$ for all experiments. For the rest hyper parameters γ^e, γ^u and γ^o, we set to 0.01 for positive sentiment label and 5 for negative sentiment label.

3.2   Define and Incorporate Sentiment Prior

In the experiments, we chose two lexicons: MPQA subjectivity lexicon and filtered MPQA subjectivity lexicon, which were both used in JST [1].

3.3   Sentiment Classification

The document sentiment is classified based on $p(l|d)$, which is approximated using Equation 5. In the experiments, we only consider the probability of positive and negative label given document. At the same time, we followed the approach in (B. Pang, 2004) and performed subjectivity detection prior to sentiment classification, which we named as "subjective MR".

Table 1. Results of experiments using different models and prior sentiment knowledge

Prior Sentiment	OR(%)	UR(%)	ARO(%)	JST(%)
MPQA	69.8	71.0	71.5	70.4
Filtered MPQA	81.5	82.3	83.5	82.8
Filtered MPQA(subjective MR)	84.8	85.2	86.3	84.6
Supervised Classification (Yessenalina et al. (2010)) [7]	The best accuracy: 93.22%			

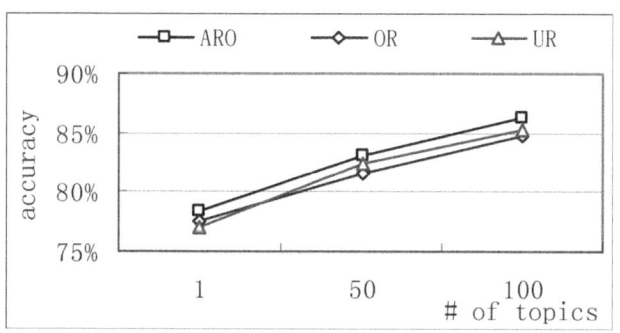

Fig. 4. Sentiment classification accuracy with different topic numbers

Results with Different λ. As mentioned in Section 2.4, we evaluated the performances of OR and UR models. From Table 1, we see that the performances of OR and UR are comparable to the result of JST, and are not better than ARO. Actually OR and UR models just utilize the user-review-object graph partly.

At the same time, we can see that the performance of UR is a little better than OR. Intuitively the object should play a more important role than the user. We guess that the results are dependent with dataset. In the movie dataset, there are 312 users, but 1107 movies. Because the object-review graph is sparser than user-review graph.

Results with Different Topic Numbers. We also analyze the influence of topic numbers. We used Filtered MPQA as the prior sentiment knowledge on the subjective MR Dataset. Fig. 4 shows the sentiment classification accuracy of our model ARO, OR and UR with the number of topic set to 1, 50 and 100. From the results, we can conclude that topic information indeed helps in sentiment classification.

3.4   Topic Extraction

The top words (most probable word) for each distribution could approximately reflect the meaning of the topic. Table 2 shows the selected 4 examples of global topics extracted with our model. The 2 topics on the left columns of Table 2 were generated under the positive sentiment label and the remaining topics on the right were generated under the negative sentiment label, each of which is represented by the top 20 words (all words were stemmed). As shown in Table 2, the 4 extracted topics are quite informative and coherent. For example, topic 1 is likely to be very positive review comments for a movie; topic 2 is apparently about the movies "Lost World: Jurassic Park, The (1997)" and "Jurassic Park III (2001)"; topic 3 is likely to be negative and topic 4 is probably about gangster movies.

Table 2. Extracted topic examples

Positive sentiment label		Negative sentiment label	
Topic 1	Topic 2	Topic 3	Topic 4
like just	dinosaur park	hollow sleepi	action cop
time good	jurass sky	depp susan	fight car
come love	octob grant	gaug granger	polic crime
best want	homer island	burton ichabod	gui chase
great star	rocket cooper	crane gilbert	crimin thriller
effect live	fund taylor	town johnni	plot gun
better real	laura kirbi	horseman murder	partner stunt
funni right	dern dino	particularly kevin	explos villain
origin moment	uplift adventur	tim headless	tough big
friend person	jake skillfulli	andrew wood	littl laugh

3.5   Experiments for Chinese Reviews

In another set of experiments, we chose to evaluate with Chinese movie and restaurant reviews. More details about the datasets are described in Table 3. The datasets is available online[3].

---

[3] http://www.clr.org.cn/ychang/social-review/

Table 3. The details about Chinese review datasets

Domains	# of pos. reviews	# of neg. reviews	Average length	# of users	# of objects
restaurant	3000	3000	363	336	4514
movie	3000	3000	284	5797	396

Table 4 shows the experimental results with different methods using HowNet sentiment lexicon[4]. The results of SVM method are obtained using SVM-light[5] with 3-fold cross-validation. The results show that our ARO model performs better than the JST. However the accuracies of ARO are about 10% lower than the supervised SVM method.

Table 4. Accuracies for Chinese reviews

	OR(%)	UR(%)	ARO(%)	JST(%)	SVM(%)
restaurant	70.1	71.5	72.7	69.3	84.6
movie	73.7	67.3	73.4	67.5	82.8

4   Conclusion and Future work

In this paper, we present a joint Author-Review-Object model for sentiment analysis of online reviews. The proposed model can not only produce the topical and sentiment distributions for each review, but also generate the topical and sentiment distributions for each user and each object simultaneously in an unsupervised manner. Consequently, the opinion polarity is jointly determined based on the review content including topics and sentiments as well as the author and object information. Experiments demonstrated that our model give competitive performance in document level sentiment classification compared with existing approaches, which means that both author and object information is meaningful for sentiment classification.

One of the limitations of our model and JST is that it represents each document as a bag of words and ignores syntax. How to model sentiment at sentence level or phrase level is a promising work. At the same time, the user preference and object quality at topic level could be captured by ARO model, and we will carry more experiments in social recommendation and collaborative filtering domain to analysis them in future work.

Acknowledgments. We thank anonymous reviewers for their comments. The paper is supported by the National Nature Science Foundation of China (Grant No. 90820005 and 61070082). Ying Su is the corresponding author.

References

1. Lin, C., He, Y.: Joint Sentiment/Topic Model for Sentiment Analysis. In: The 18th ACM Conference on Information and Knowledge Management, CIKM (2009)

---

[4] http://www.keenage.com/html/c_index.html
[5] http://svmlight.joachims.org

2. Blei, D.M., Ng, A.Y., Jordan, M.I.: Latent Dirichlet Allocation. J. Mach. Learn. Res. 3, 993–1022 (2003)
3. Griffiths, T., Steyvers, M.: Finding scientific topics. Proceedings of the National Academy of Sciences 101(90001), 5228–5235 (2004)
4. Pang, B., Lee, L.: A sentimental education: sentiment analysis using subjectivity summarization based on minimum cuts. In: Proceedings of the 42nd Annual Meeting on Association for Computational Linguistics, Morristown, NJ, USA, p. 271 (2004)
5. Sauper, C., Haghighi, A., Barzilay, R.: Content Models with Attitude. In: Proceedings of the 49nd Annual Meeting on Association for Computational Linguistics, pp. 350–358 (2011)
6. Zhai, Z., Liu, B., Xu, H., Jia, P., Lei, Z.: Identifying Evaluative Opinions in Online Discussions. In: Proceedings of AAAI-2011, San Francisco, USA, August 7-11 (2011)
7. Yessenalina, A., Yue, Y., Cardie, C.: Multi-level Structured Models for Document-level Sentiment Classification. In: Proceedings of the 2010 Conference on Empirical Methods in Natural Language Processing, pp. 1046–1056 (2010)
8. He, Y., Lin, C., Alani, H.: Automatically Extracting Polarity-Bearing Topics for Cross-Domain Sentiment Classification. In: Proceedings of the 49th Annual Meeting of the Association for Computational Linguistics, Portland, Oregon, June 19-24, pp. 123–131 (2011)
9. Li, F., Huang, M., Zhu, X.: Sentiment Analysis with Global Topics and Local Dependency. In: AAAI 2010, Atlanta, Georgia, USA, July 11-15 (2010)
10. Titov, I., McDonald, R.: A Joint Model of Text and Aspect Ratings for Sentiment Summarization. In: 46th Meeting of Association for Computational Linguistics, Columbus, OH, USA (2008)
11. Michal, R.: Zvi, Chaitanya Chemudugunta, Thomas Griffiths, Padhraic Smyth and Mark Steyvers: Learning author-topic models from text corpora. ACM Transactions on Information Systems 28(1), 1–38 (2010)
12. Lu, C., Hu, X., Park, J.-R., Huang, J.: Post-based collaborative filtering for personalized tag recommendation. In: iConference, pp. 561–568 (2011)
13. Zhou, D., Bian, J., Zheng, S., Zha, H., Giles, C.L.: Exploring Social Annotations for Information Retrieval. In: Proceedings of the 17th International Conference on World Wide Web, Beijing, China, pp. 715–724 (2008)
14. Mei, Q., Ling, X., Wondra, M., Su, H., Zhai, C.: Topic sentiment mixture: modeling facets and opinions in weblogs. In: Proceedings of the 16th International Conference on World Wide Web, pp. 171–180. ACM, New York (2007)

Semantic-Based Opinion Retrieval
Using Predicate-Argument Structures
and Subjective Adjectives

Sylvester Olubolu Orimaye, Saadat M. Alhashmi, and Siew Eu-Gene

Monash University, Sunway Campus, Malaysia
{sylvester.orimaye,alhashmi,siew.eu-gene}@monash.edu

Abstract. We present the results of our experiment on the use of predicate-argument structures containing subjective adjectives for semantic-based opinion retrieval. The approach exploits the grammatical tree derivation of sentences to show the underlying meanings through the respective predicate-argument structures. The underlying meaning of each subjective sentence is then semantically compared with the underlying meaning of the query topic given in natural language sentence. Rather than using frequency of opinion words or their proximity to query words, our solution is based on frequency of semantically related subjective sentences. We formed a *linear relevance model* that uses explicit and implicit semantic similarities between predicate-argument structures of subjective sentences and the given query topic. Thus, the technique ensures that opinionated documents retrieved are not only subjective but have semantic relevance to the given query topic. Experimental results show that the technique improves performance of topical opinion retrieval task.

Keywords: semantic-based opinion retrieval, predicate-argument structures, subjective adjectives, linear relevance model.

1   Introduction

Opinionated documents (documents that contain opinion) in opinion-driven sources such as collection of blogs called *blogosphere* (e.g. Google Blogs Search[1], Technorati Blog Directory[2]) and review websites (e.g. *reviewcenter*[3]) often contain opinion words used in different semantic contexts. However, current opinion retrieval techniques are only concerned with retrieving opinionated documents as long as they contain some opinion words. Usually, the opinion words may be within a proximity window [1] (a certain distance of opinion words to a query word) or randomly located within the document. For example, consider an opinion finding scenario below:

---

[1] http://blogsearch.google.com/
[2] http://technorati.com/
[3] http://www.reviewcentre.com/

M.V.M. Salem et al. (Eds.): AIRS 2011, LNCS 7097, pp. 372–385, 2011.
© Springer-Verlag Berlin Heidelberg 2011

User query topic: *Why Apple made huge sales on iPhones?*

A relevant blog: *"One [interesting] piece of information revealed by the report is that [Apple] may be its own [worst] [enemy]. The numbers show that iPod sales [fell] by about 20 percent – a trend that has been in place since the [iPhone] 3G hit the market. While [Apple] noted that most of those iPod [sales] are iPod touches, which run on [Apple's] iOS platform and therefore spread its mobile software's influence and gather revenue through the iTunes App Store"*

If we assume a unified opinion retrieval model that combines topical relevance and opinion relevance to retrieve the above document [1], it is quite obvious that the result has been favored by only the frequency or co-occurrence of query words (i.e. Apple, iPhone, sales) used in the *topical relevance* and then the proximity or presence of opinion words (i.e. interesting, worst, enemy, fell) used in the *opinion relevance*. However, if we consider the semantic context of the query topic based on human judgment, it could be observed that the query topic and the retrieved document are not semantically relevant. Consequentially, opinions retrieved for the qualitative analysis of products, performance measurement for companies, and public reactions to political decisions are more likely to be largely biased.

Few approaches have made attempts to perform sentence-level opinion retrieval, for example, by merging opinionated nature of individual sentences (sentiment flow) with the degree of relatedness of individual sentences to the given query (topic relevance flow). The result forms a *sentiment-relevance flow* (SRF) [2] that is still largely based on frequency of opinion words within sentences.

Considering these challenges, our work introduces a semantic-based approach for identifying semantic and subjective relationships between the given query topic and the sentences in opinionated documents. The proposed technique employs the semantic similarities between predicate-argument structures of a given query topic and the sentences in a given opinionated document. The predicate-argument structures are derived from the output parse tree of a syntactic parser. In addition, each predicate-argument structure is checked to contain a subjective component identified by subjective adjectives. This ensures semantically relevant sentences are also subjective. To overcome the efficiency problem due to syntactic parsing, predicate-argument structures are derived offline and used to index the opinionated documents. We summarize our contributions as follows:

- Rather than proximity-based or keyword-based opinion retrieval approaches, we employ frequency of semantically relevant subjective sentences.
- Our technique employs the predicate-argument structures output of syntactic parse trees for subjective sentences to detect semantic similarities.

The rest of this paper is organized as follows. Section 2 presents related work. Section 3 presents the overview of the proposed technique. Section 4 explains the approaches used in our model. Section 5 explains model formations. Section 6 discusses experiments and results. Finally, section 7 presents conclusions and future work.

2   Related Work

Opinion polarity detection techniques have recorded some level of success [3]. In opinion polarity detection, specific keywords within a document are labeled with a particular polarity (e.g. positive or negative). However, determining an effective way to differentiate between what is positive and what is negative is still an unsolved problem[4][10][16]. For example in [4], the presence of *ironic* phrases and inverted polarity in opinionated documents led to lower precision for positive opinions with just 77% accuracy. As a result of this, the choice of individual words for polarity detection in opinionated documents is still a big challenge.

Subjectivity detection shows if a sentence contains opinion or not [5]. Words or sentences that contain opinions are systematically categorized according to the degree of subjectivity shown by their lexicons. However, an automatic and effective way to detect hidden and inverted polarities in phrases and sentences is still a huge research challenge [4]. According to Pang and Lee [5], subjectivity is a two-state task and can be interpreted differently in some cases. It is still very challenging to effectively determine appropriate subjectivity state in many documents that contain opinion.

Lexicon-based approaches consider domain-specific evidences to form lexicons for opinion retrieval [6]. For generating lexicons, individual opinionated keywords are selected from each sentence in a document. However, we believe opinionated words alone cannot completely and independently express the overall opinion contained in a document, without taking into consideration the grammatical dependencies between words. We suggest opinion should be syntactically retrieved from a complete sentence rather than individual words. Individual keywords in the lexicon might have been selected from varying grammatical contexts.

Probabilistic approaches are commonly used to evaluate and explain theoretical foundations for information retrieval models [7-8]. In many cases, *estimates* or *assumptions* made in probabilistic approaches may not be practically applicable to real scenarios. It is only effective where there are high chances of frequency of observations as applicable to *BM25*[9] and *Divergence from Randomness* probabilistic models. Some probabilistic models use proximity density information to calculate probabilistic opinion score for individual query terms [10]. However, proximity of words to some of the query terms may not reflect the semantic context at which opinion is required.

Language model approach combines prediction of occurrence for natural language words and then shows a probabilistic interpretation of such occurrences. A common limitation is the estimation of model parameters. It is often difficult to effectively model a document to give a higher probability of relevance to user query. For example, [11] addressed the parameter tuning or optimization problem in higher order language models by incorporating different component models as features. A common approach is to perform smoothing at varying levels to ensure high chance of document retrieval. However, this approach usually leads to having higher number of model parameters, where all of such parameters would also require optimization at different levels.

3   Overview of Semantic-Based Opinion Retrieval

As shown in Figure 1, given a document from a collection, sentences are transformed to grammatical trees. Predicate-argument structures are derived from the trees and then annotated semantically for indexing. In the online process, the given query topic is semantically annotated as done for each sentence in the offline process. The index is then searched for relevant documents based on a scoring scheme that compensates each relevant annotation containing a subjective adjective with an empirical score.

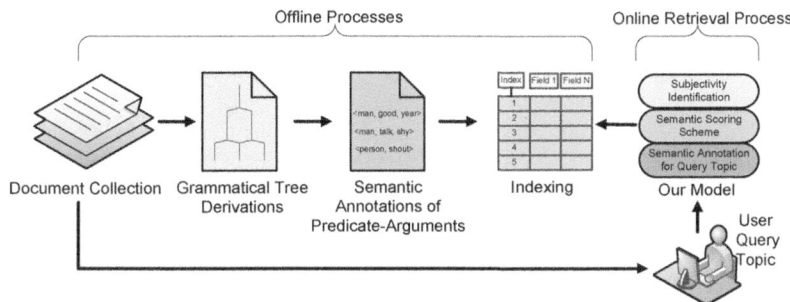

Fig. 1. Overview of the semantic-based opinion retrieval

4   Approaches Used for Semantic-Based Opinion Retrieval

4.1   Grammatical Tree Derivations

We use a context-dependent syntactic parser, specifically, Categorial Combinatory Grammar (CCG) [12]. This does one-time complete natural language processing (NLP) tasks to show the underlying meaning of each query topic or well formed sentence. CCG has a relatively straightforward way of providing a compositional semantics for the intending grammar by providing completely transparent interface between syntax and semantics [13]. We chose CCG because of its distinct ability to integrate syntactic and semantic information in its single parse process even for long-range dependencies in natural language sentences. It also has speed advantage over other parsing models such as Collins parser and Charniak parser [13]. Derivation examples are shown below:

Sentence I: *China proposed regulations on Tobacco.*   Sentence II: *regulations that China proposed*

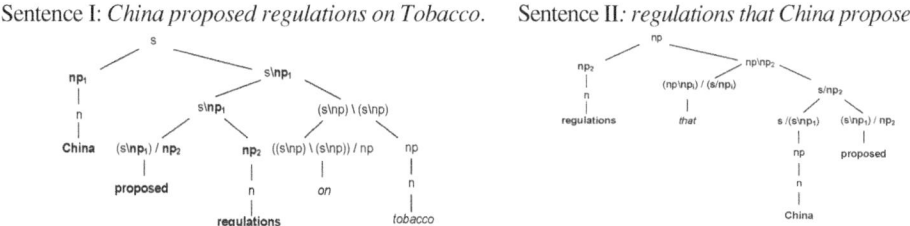

Fig. 2. Grammatical tree derivation using CCG notations

The focus of this paper is beyond the deep understanding of the CCG parsing processes as it has been thoroughly discoursed in existing literatures [12-13]. For the purpose of our experiment, we use a log-linear CCG syntactic parser by Clark and Curran [13]. The model is described in [14] as part of the popular C&C tools[4].

4.2   Predicate-Argument Structures

Predicate-argument structures are derived from the output parse tree of a syntactic parser as words that fill the argument slots of predicates. They essentially show the local and long-range word-word dependencies structures for a sentence. These structures can be retrieved with F-measure accuracy of about 83% labeled recovery and 91% unlabeled recovery [15]. Usually, a transitive verb category denoted as $(S\backslash NP)/NP$, contains two predicate-argument relations. One is the object NP argument and the other is the subject NP argument. To differentiate between argument slots, the arguments are indexed from left to right. Indices can be used to number the argument slots of the categories. For example, the transitive verb can be represented as $(S\backslash NP_1)/NP_2$. Thus, the predicate-argument structure can be presented as tuples $\langle w_H, c_{LH}, i_{args}, w_i \rangle$, where w_H is the *head word*, c_{LH} is the *lexical category* of the head word w_H, i_{args} is the *index* of the argument slot, and w_i is the head word that shows the *semantic* and *dependency relation* of c_{LH}. A detailed illustration for representing predicate-argument semantic structure as tuples can be found in [15,13].

Table 1. Representation of predicate-argument structure for the sentence *"China proposed regulations on Tobacco"*

w_H	c_{LH}	i_{args}	w_i
proposed	$(S\backslash NP_1)/NP_2$	1	China
proposed	$(S\backslash NP_1)/NP_2$	2	regulations
on	$((S\backslash NP_1)\backslash(S\backslash NP)_2)/NP_3$	2	proposed
on	$((S\backslash NP_1)\backslash(S\backslash NP)_2)/NP_3$	3	Tobacco

Table 2. Representation of predicate-argument structure for the sentence *"Regulations that China proposed"*

w_H	c_{LH}	i_{args}	w_i
proposed	$(S\backslash NP_1)/NP_2$	1	China
proposed	$(S\backslash NP_1)/NP_2$	2	regulations
that	$((S\backslash NP_1)\backslash(S\backslash NP)_2)$	2	proposed

In Table 1 and Table 2, both sentences have the same predicate-argument *semantic relation* between *China* and *proposed* and *regulations* and *proposed*. In both sentences, the subject *China* holds index 1 of the argument slots of the predicate *proposed,* that is, NP_1 in $(S\backslash NP_1)/NP_2$, and the object *regulations* holds index 2 of the argument slots of the predicate *proposed,* that is, NP_2 in $(S\backslash NP_1)/NP_2$. For the purpose of our experiment, we are interested in the w_i column above. Thus we use the

---

[4] http://svn.ask.it.usyd.edu.au/trac/candc/wiki

Boxer[5] component of the C&C tools [14], to get the predicate-argument semantic structure from the CCG output parse trees.

5   Constructing a Semantic-Based Model for Opinion Retrieval

5.1   Subjective Component Identification

Adjectives play significant role in *expression of feelings* in English language. We use a list of 1336 subjective adjectives reported by Hatzivassiloglou and Mckeown [24] to identify subjectivity from the predicate-argument structure of each sentence. Given a structure, the subjectivity score $score_{subj}$ of a sentence is computed as follows:

$$score_{subj} = (args_{struct} \ni (a_n \in A_l), n \geq 1 \leq l) = 0.5 \tag{1}$$

where $args_{struct}$ is the predicate-argument structure, a is an adjective element in list of subjective adjectives A_l, n is the number of adjectives in $args_{struct}$, l is the length of $args_{struct}$, and the constant **0.5** can vary empirically while **0** is assigned to a non-subjective structure. For example, if the predicate-argument structure of a sentence contains five terms, at least one or more of these terms must be present in the list of subjective adjectives.

5.2   Semantic Similarity between Structures

Having derived predicate-argument structures that represent underlying meaning of sentences, we propose to identify semantic similarities between predicate-argument structures. Given that $D = \{s_1, \dots, s_m\}$ is a set of well formed sentences in document d, the semantic similarity function sim_s is used to estimate the *similarity score* between a given natural language query q and each well formed sentence s, i.e., $sim_s(q, s)$. Given that $Q_{predargs} = \{t_1, \dots, t_m\}$ is a set of predicate-argument terms derived from q, $S_{predargs} = \{w_1, \dots, w_m\}$ is a set of predicate-argument terms derived from s. Since the semantic similarity between q and s can be shown using the predicate-argument *relation*, for each s, $sim_s(q, s)$ can either be 1 (q similar to s) or 0 (q not similar to s). However, it is often difficult to determine when $sim_s(q, s)$ must be 1 or 0. One straightforward way is to model similarity between each term in $Q_{predargs}$ and $S_{predargs}$. For example, by using topic models such as LSA [16] and PLSA [17] to measure term co-occurrences (t,s) between $Q_{predargs}$ and $S_{predargs}$. We use PLSA because of its advantages over the conventional LSA.

$$sim(s, t) = P(s)P(t|s), \qquad P(t|s) = \sum_{z \in Z} P(t|z)P(z|s) \tag{2}$$

where sim is the *similarity* score between $Q_{predargs}$ and $S_{predargs}$, s is the derived predicate-argument *structure* for sentences, t is each *term* in the derived predicate-argument *structure*, and z is an un-observed class variable which acts as a bottleneck variable in predicting terms for each observation. Intuitively, similar predicate-

---

[5] http://svn.ask.it.usyd.edu.au/trac/candc/wiki/boxer

argument structures would show a *similarity score* that tends towards 1. However, this process suffered a particular set back. We observed direct *word overlap* between q and s since the *similarity score* in equation 2 is designed to give values between 0 and 1. Whereas, a *similarity score* of 0.85 does not necessarily indicate similarity. For example, a single dissimilarity among terms in the predicate-argument structures may change the semantic meaning of the concerned sentence. Consider q and s in the following *word overlap* problem as shown by their predicate-argument structures.

q: Amazing clothes are designed for Oscar awards.
$Q_{predargs}$: designed\amazing/clothes/Oscar/awards
s_1: The amazing fashion show at Oscar awards.
$S1_{predargs}$: show\amazing/fashion/Oscar/awards

In the above example, the predicate-argument structure for query q (i.e. $Q_{predargs}$) has *word overlap* (i.e. *designed/show*) with the predicate-argument structure for s_1 (i.e. $S1_{predargs}$). Although, the two predicate-argument structures share the same words like "amazing" and "awards", but q and s_1 turned out to have different semantic meaning. Thus, we propose a more intuitive similarity function that can help model semantic relevance such that the *relations score* of $Q_{predargs}$ to $S_{predargs}$ can be equal to 1. For each structure pair of $Q_{predargs}$ and $S_{predargs}$, we propose a *transformed term-term similarity matrix* on which Jaccard Similarity Coefficient (JSC) value can be easily calculated for each pair. We chose JSC since it is largely used to measure word overlaps between the attributes of paired sets. Ideally, the best JSC value is 1. For each document, we then consider *linear combinations* of the JSCs and the *relevance scores* given by a popular *relevance model*. The chosen relevance model was proposed by Lavrenko and Croft [18], and it has shown good performance for effectively detecting topical relevance in sentences [19]. We give the reason for using this relevant model in section 5.4.

5.3   Transformed Terms Similarity (TTS)

Since we can no longer apply PLSA because of its word overlap problem, our model may be prone to implicit relevance problem resulting from *synonymy* and *polysemy*. These problems are without doubt solved in LSA and PLSA respectively. We understand that two terms with the same implicit meaning (e.g. *clothes and dresses)* can exist among $Q_{predargs}$ and $S_{predargs}$. Thus, we propose a straightforward transformation mechanism that uses *synonyms* and *hyponyms* of individual terms in both $Q_{predargs}$ and $S_{predargs}$. Arguably, we use the method to determine implicit similarity between terms, such that two paired terms i and j have implicit meaning if either term is the synonym or hyponym of the other. Thus, we form a *similarity matrix* distribution over terms to measure implicit relationship between $Q_{predargs}$ and $S_{predargs}$. This approach is used within LSA, termed as LSA term-term matrix.

Table 3. Term-term similarity matrix with semantic word overlap problem

		$Q_{predargs}$				
		~~designed~~	amazing	clothes	Oscar	awards
$S_{predargs}$	~~show~~	**0**	0	0	0	0
	amazing	0	**1**	0	0	0
	fashion	0	0	**1**	0	0
	Oscar	0	0	0	**1**	1
	awards	0	0	0	1	**1**

We form a similarity matrix where each $sim_{ij} (0 \leq sim_{ij} \leq 1)$ is the similarity between i and j. The similarity sim_{ij} is calculated as the absolute observation of $i \equiv j$, or the presence of i in sets j_{sym} or j_{hyp}, where j_{sym} and j_{hyp} is the synonyms and hyponyms of term j respectively. To calculate the JSC, we use the Binary Independent Model (BIM) [20], to construct the term-term similarity matrix. The BIM method avoids using frequency of terms but takes each term as binary vector over the vocabulary space of j_{sym} and j_{hyp}. That is, sim_{ij} shows the similarity between term i and j as a binary attribute 1 (term i is similar to j) if i is j or i is present in j_{sym} or i is present in j_{hyp}. Conversely, sim_{ij} can also show dissimilarity between term i and j as a binary attribute 0 (term i not similar to j) if i is not j or i is not present in j_{sym} or i is not present in j_{hyp}. Synonyms in j_{sym} and hyponyms in j_{hyp} were derived using Wiktionary [22]. More importantly, the binary attributes help in calculating the JSC for the term-term similarity matrix which can only contain asymmetric binary attributes of each pair predicate-argument structures. Let $S_{predargs} = (i_1, ..., i_n)$ and $Q_{predargs} = (j_1, ..., j_m)$ be the pair predicate-argument structures respectively. Thus, using BIM, we calculate sim_{ij} as the probability of term incidence vectors $P(R|i,j)$.

$$P(R = 1|i,j) = \frac{P(i|R=1,j)P(R=1|j)}{P(i|j)} \qquad (3)$$

$$P(R = 0|i,j) = \frac{P(i|R=0,j)P(R=0|j)}{P(i|j)} \qquad (4)$$

In the above, $P(i|R = 1, j)$ in equation 3 and $P(i|R = 0, j)$ in equation 4 are the probabilities that term i is j and term i is not j respectively. However, since the actual probabilities are not known, the prior probability of i is j or i is not j is defined as $P(R = 1|j)$ and $P(R = 0|j)$ respectively. Since i can either be present or not present in $Q_{predargs}$, then:

$$P(R = 1|i,j) + P(R = 0|i,j) = 1 \qquad (5)$$

Consider the following term-term similarity matrix with asymmetric binary attributes:

q: Amazing clothes are designed for Oscar awards.
$Q_{predargs}$: designed\amazing/clothes/Oscar/awards
s_1: The amazing clothes were made for Oscar awards.
$S1_{predargs}$: made\amazing/clothes/Oscar/awards

Table 4. Term-term similarity matrix without semantic word overlap problem

		$Q_{predargs}$				
		designed	amazing	clothes	Oscar	awards
$S_{predargs}$	made	**1**	0	0	0	0
	amazing	0	**1**	0	0	0
	clothes	0	0	**1**	0	0
	Oscar	0	0	0	**1**	1
	awards	0	0	0	1	**1**

The term-term similarity matrix shown in Table 4 is constructed for a sample $Q_{predargs}$ and $S_{predargs}$ by using BIM to compute the binary attributes. It is therefore straightforward to compute the JSC value for the term-term similarity matrix. Intuitively, the best JSC would be 1 provided all diagonal elements of the similarity matrix is filled with 1 (i.e. all ij elements must be identical to all ji elements).

$$JSC = \frac{M_{11}}{M_{01} + M_{10} + M_{11}} \tag{6}$$

where M_{11} is the total number of attributes where i and j both have a value of 1, M_{01} is the total number of attributes where the attribute of i is 0 and the attribute of j is 1, M_{10} is the total number of attributes where the attribute of i is 1 and the attribute of j is 0. Using equation 6, the JSC for Table 4 is equal to 1, which means query q is semantically similar to sentence s unlike Table 3 which has a JSC value of 0 because of the word overlap problem.

5.4   Linear Relevance Model (LRM)

The Transformed Term Similarity (TTS) approach that is described above is efficient provided $Q_{predargs}$ and $S_{predargs}$ are squared (i.e. equal rows and columns). This is based on the fundamental principle of term-term similarity matrix which we adopted. However, in some cases, $Q_{predargs}$ and $S_{predargs}$ may not have the same number of rows and columns since there may be variations in the number of terms in each predicate-argument structure. For example, a given query may have just three terms in its predicate-argument structure, while another sentence with much longer grammatical dependencies may have five or more terms in its predicate-argument structure. Which is why we use the relevance model proposed by [18] to estimate $P(i \mid Q_{predargs} \cap S_{predargs})$. The relevance model $P(i \mid Q_{predargs} \cap S_{predargs})$ is estimated in terms of the joint probability of observing a term i together with terms from $Q_{predargs} \cap S_{predargs}$ [18]. Consider, for example, the following scenario:

q: important regulations that China proposed
$Q_{predargs}$: proposed\important/regulations/China
s_1: China proposed some important regulations on tobacco.
$S1_{predargs}$: proposed\important/regulations/China/tobacco

s_2: The proposed regulation on tobacco is a new and important development in China.
$S2_{predargs}$: proposed\important/regulation/China/tobacco/development

Each predicate-argument structure above has different number of terms. The chosen relevance model solves this problem with the fact that i and all terms from $Q_{predargs} \cap S_{predargs}$ are sampled independently and identically to each other. By this way, the relevance model would be appropriate to determine when the predicate-argument structure $Q_{predargs}$ is independently present as part of the predicate-argument structure of some long range sentences such as shown in $S1_{predargs}$ and $S2_{predargs}$ above. The relevance model is computed below:

$$RM_{score}(i, j_1 \ldots j_k) = P(i) \prod_{i=1}^{k} \sum_{M_i \in Z} P(M_i|i) \, P(j_1|M_i) \qquad (7)$$

where M_i is a unigram distribution from which i and j are sampled identically and independently, Z is the finite set of terms in each predicate-argument structure. For each document, we do a *linear combination* of the absolute sum of subjectivity scores in equation (1) and the results of TTS in equation (6) and the relevant model in equation (7). Thus, the predicate-argument structure of a given query can be compared with the predicate-argument structures of sentences with either short-range or long-range grammatical dependencies. While the subjectivity score ensures a document is subjective, we believe the TTS model and the relevance model would solve the short-range and long-range dependencies efficiently and respectively. The *linear relevance score* for ranking a document is computed as follows:

$$R_{\mathcal{L}}(q,d) = \frac{k}{N} \sum_{i=1}^{k} TTS_{score} + \frac{v}{N} \sum_{i=1}^{v} RM_{score} + \sum_{i=1}^{k+v} Score_{subj} \qquad (8)$$

where $R_{\mathcal{L}}(q,d)$ is the linear relevance model that takes input as query q and document d satisfying a linear combination expression $C = aX + bY + Z$, where a and b can be empirical constants, and X, Y and Z are the values to be linearly combined. k is the number of sentences with short-range dependencies that satisfy TTS, N is the total number of sentences in document d, TTS_{score} is derived using equation 6, v is the number of sentences with long-range dependencies that satisfy RM_{score} in equation 7.

6   Experimental Results on Opinion Retrieval Task

We perform experiments on TREC Blog 08 [21] and compared our results with Blog 08 best run and one of the TREC Blog 08 best runs that used title and description fields for opinion finding task (KLEDocOpinTD) [23]. 50,000 English TREC Blog posts were extracted using our ad-hoc blog posts extraction algorithm with optimized sentence model (BlogTEX)[6]. At least 90% of the extracted posts have well formed sentences and are transformed to their equivalent syntactic parse trees and then predicate-argument structures. Again, we use the log-linear CCG parsing model and the Boxer tool[14]. The description fields of 50 query topics (TREC 1001-1050) are

---

[6] https://sourceforge.net/projects/blogtex

also transformed to their equivalent syntactic parse trees and predicate-argument structures. We use Lucene[7] for indexing with focus on high precision at top-ranked documents. Thus, we use a re-ranking technique by initially retrieving top 20 documents using BM25 popular ranking model with empirical parameters $k=1.2$ and $b=0.75$. These top 20 documents are then re-ranked by our model. The IR evaluation metrics used include Mean Average Precision (MAP), R-Prec, and precision at ten (P@10). A comparison of opinion MAPs with the increased number of top-K blog documents is shown in Figure 3. Our model shows improved performance over Blog 08 best run and significantly outperforms Blog 08's KLEDocOpinTD. Performance improvement in precision and recall curves upon the re-ranking technique is shown in Figure 4. We observed improved performance *with* and *without* our re-ranking technique with marginal difference between the two performances. This shows the effectiveness of our model and the possibility of good performance without a re-ranking mechanism. In Table 5 below, the best significant improvements over Blog 08 best run is indicated with *.

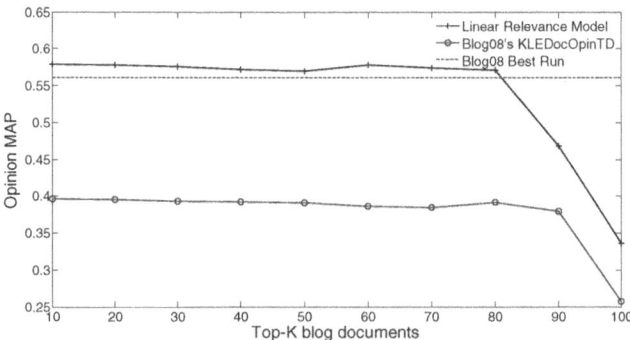

Fig. 3. MAP comparisons for different models at Top-K documents

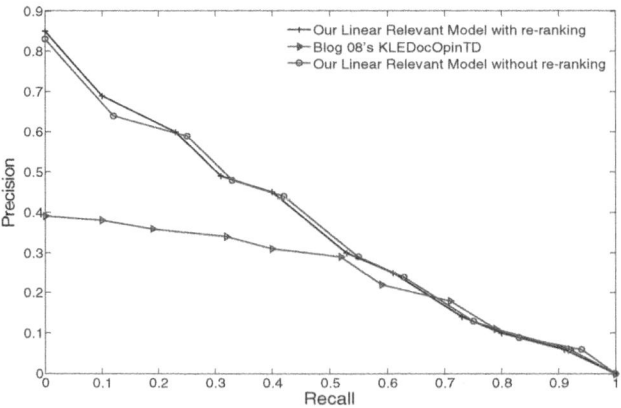

Fig. 4. Precision and Recall curves with and without re-ranking mechanism

---

[7] http://lucene.apache.org/java/docs/index.html

Table 5. Performance and improvements on Opinion Retrieval task

Model	MAP	R-Prec	P@10
Best run at Blog 08	0.5610	0.6140	0.8980
Linear Relevance Model re-ranked	**0.5937** *	**0.6446** *	0.9051
Improvement over Best run at Blog 08	5.8%	5.0%	0.8%
Linear Relevance Model without re-ranking	**0.5882** *	**0.6364** *	0.9036
Improvement over Best run at Blog 08	4.8%	3.6%	0.6%
Blog 08's KLEDocOpinTD	0.3937	0.4231	0.6273
Improvement with Linear Relevance Model re-ranked	50.8%	52.4%	44.3%

In terms of MAP, our model has significant performance more than Blog 08 best run and Blog 08's KLEDocOpinTD. Our model also shows significant improvement in terms of R-prec in all cases. This shows the effectiveness of our model in terms of retrieving semantically relevant opinionated documents.

7   Conclusions and Future Work

We proposed semantic-based opinion retrieval using predicate-argument structures and subjective adjectives. The predicate-argument structures were derived from the output of a syntactic parser. We developed a *linear relevance model* that is based on semantic similarity between a pair of structures for a query topic and each subjective sentence. We performed experiments on TREC Blog 08 and compared the results with Blog 08 best run and Blog 08's KLEDocOpinTD. Our model outperforms the baselines in the results shown above. We observed that relevant document to not only contain opinionated query words but opinionated sentences that lexically combine synonyms and hyponyms of query words. This implies that frequency of keywords may not be helpful to context-dependent opinion retrieval afterall. In future, we will consider semantic and opinion dependencies between multiples sentences in the same document. We will also consider using sentence-level subjective component to detect and assign weight to opinionated sentences rather than using only subjective adjectives.

References

1. Zhang, M., Ye, X.: A generation model to unify topic relevance and lexicon-based sentiment for opinion retrieval. In: The 31st Annual International ACM SIGIR Conference on Research and Development in Information Retrieval, Singapore (2008)
2. Lee, S.-W., Lee, J.-T., Song, Y.-I., Rim, H.-C.: High precision opinion retrieval using sentiment-relevance flows. In: The 33rd International ACM SIGIR Conference on Research and Development in Information Retrieval, Geneva, Switzerland (2010)

3. Stefan Siersdorfer, S.C., Pedro, J.S.: How Useful are Your Comments? Analyzing and Predicting YouTube Comments and Comment Ratings. In: International World Wide Web Conference (WWW), North Carolina, USA, pp. 891–900 (2010)

4. Luís Sarmento, P.C., Silva, M.J., de Oliveira, E.: Automatic creation of a reference corpus for political opinion mining in user-generated content. In: Proc. of the 1st International CIKM Workshop on Topic-Sentiment Analysis for Mass Opinion, Hong Kong, China (2009)

5. Pang, B., Lee, L.: Opinion Mining and Sentiment Analysis. Found. Trends Inf. Retr. 2, 1–135 (2008)

6. Ding, X., Liu, B., Yu, P.S.: A holistic lexicon-based approach to opinion mining. In: Proc. of the International Conference on Web Search and Web Data Mining, California, USA (2008)

7. Hiemstra, D.: Using language models for information retrieval. PhD Thesis, Centre for Telematics and Information Technology, The Netherlands (2000)

8. Amati, G., Amodeo, G., Bianchi, M., Gaibisso, C., Gambosi, G.: A Uniform Theoretic Approach to Opinion and Information Retrieval. In: Armano, G., de Gemmis, M., Semeraro, G., Vargiu, E. (eds.) Intelligent Information Access. SCI, vol. 301, pp. 83–108. Springer, Heidelberg (2010)

9. Robertson, S., Zaragoza, H.: The Probabilistic Relevance Framework: BM25 and Beyond. Found. and Trends in Information Retrieval 3, 333–389 (2009)

10. Shima Gerani, M.J.C., Crestani, F.: Proximity-Based Opinion Retrieval. In: SIGIR ACM, Geneva, Switzerland, p. 978 (2010)

11. Javanmardi, S., Gao, J., Wang, K.: Optimizing two stage bigram language models for IR. In: Proc. of the 19th International Conference on World Wide Web (WWW), North Carolina, USA (2010)

12. Steedman, M.: The Syntactic Process (Language, Speech, and Communication). The MIT Press (2000)

13. Clark, S., Curran, J.R.: Wide-coverage efficient statistical parsing with ccg and log-linear models. Comput. Linguist. 33, 493–552 (2007)

14. Curran, J.R., Clark, S., Bos, J.: Linguistically motivated large-scale NLP with C & C and boxer. In: Proc. of the 45th Annual Meeting of the ACL on Interactive Poster and Demonstration Sessions, Prague, Czech Republic (2007)

15. Gildea, D., Hockenmaier, J.: Identifying semantic roles using Combinatory Categorial Grammar. In: Proc. of the 2003 Conference on Empirical Methods in Natural Language Processing, vol. 10 (2003)

16. Dumais, S.T., Furnas, G.W., Landauer, T.K., Deerwester, S., Harshman, R.: Using latent semantic analysis to improve access to textual information. In: Proc. of the SIGCHI Conference on Human Factors in Computing Systems, Washington, D.C., United States (1988)

17. Hofmann, T.: Probabilistic latent semantic indexing. In: Proc. of the 22nd Annual International ACM SIGIR Conference on Research and Development in Information Retrieval, USA (1999)

18. Lavrenko, V., Croft, W.B.: Relevance based language models. In: Proc. of the 24th Annual International ACM SIGIR Conference on Research and Development in Information Retrieval, New Orleans, Louisiana, USA (2001)

19. Lv, Y., Zhai, C.: A comparative study of methods for estimating query language models with pseudo feedback. In: Proc. of the 18th ACM Conference on Information and Knowledge Management (CIKM), Hong Kong, China (2009)

20. Rijsbergen, C.J.V.: A Theoretical Basis for the use of Co-Occurrence Data in Information Retrieval. Journal of Documentation 33, 106–119 (1977)

21. Ounis, I., Macdonald, C., Soboroff, I.: Overview of the TREC 2008 Blog Track. In: TREC 2008 (2008)
22. Torsten, Z., Christof, M., Iryna, G.: Extracting Lexical Semantic Knowledge from Wikipedia and Wiktionary. In: LREC (2009)
23. Yeha Lee, S.-H.N., Kim, J., Nam, S.-H., Jung, H.-Y., Lee, J.-H.: KLE at TREC 2008 Blog Track: Blog Post and Feed Retrieval. In: TREC 2008 (2008)
24. Hatzivassiloglou, V., McKeown, K.R.: Predicting the semantic orientation of adjectives. In: Proc. of the 8th Conference on European Chapter of the Association for Computational Linguistics, EACL, Madrid, Spain, pp. 174–181 (1997)

An Aspect-Driven Random Walk Model for Topic-Focused Multi-document Summarization

Yllias Chali, Sadid A. Hasan, and Kaisar Imam

University of Lethbridge
Lethbridge, AB, Canada
{chali,hasan,imam}@cs.uleth.ca

Abstract. Recently, there has been increased interest in topic-focused multi-document summarization where the task is to produce automatic summaries in response to a given topic or specific information requested by the user. In this paper, we incorporate a deeper semantic analysis of the source documents to select important concepts by using a predefined list of important aspects that act as a guide for selecting the most relevant sentences into the summaries. We exploit these aspects and build a novel methodology for topic-focused multi-document summarization that operates on a Markov chain tuned to extract the most important sentences by following a random walk paradigm. Our evaluations suggest that the augmentation of important aspects with the random walk model can raise the summary quality over the random walk model up to 19.22%.

Keywords: Topic-focused summarization, multi-document summarization, aspects, random walk model.

1 Introduction

The main goal of topic-focused multi-document summarization is to create a summary from the given documents that can answer the need for information expressed in the topic. We consider the problem of producing extraction-based[1] topic-focused multi-document summaries given a collection of relevant documents. To generate the summaries, we focus on a deeper semantic analysis of the source documents instead of relying only on document word frequencies to select important concepts. We use a predefined list of important aspects to direct our search for the most relevant sentences, and generate topic-focused summaries that cover all these aspects. For example, a topic about *Natural Disasters* might consider the aspects: *what happened; date; location; reasons for the disaster; casualties; damages; rescue efforts etc.* while generating the summary.

---

[1] Extract summaries contain original sentences extracted from the documents whereas abstract summaries can employ paraphrasing [8].

M.V.M. Salem et al. (Eds.): AIRS 2011, LNCS 7097, pp. 386–397, 2011.
© Springer-Verlag Berlin Heidelberg 2011

In this paper, we propose a novel topic-focused multi-document summarization framework that operates on a Markov chain model and follows a random walk paradigm in order to generate possible summary sentences. We build three alternative systems for summary generation that are based on important aspects, random walk model, and a combination of both. We run our experiments on the TAC[2]-2010, and DUC[3]-2006 data and based on the evaluation results we argue that augmenting important aspects with a random walk model often outperforms the other two alternatives. Contributions of this work are: a) constructing an aspect-based summarization model that generates summaries based on given important aspects about the topics, b) building a novel summarization model based on a random walk paradigm that operates on a Markov chain exploiting topic signature [6] and Rhetorical Structure Theory (RST) [9] as node weights and WordNet[4]-based sentence similarities as edge weights, and c) generating a hybrid summarization model combining the aspect-based model with the random walk approach. Extensive automatic evaluations suggest that the combined model can raise the performance up to 19.22% while manual evaluations further confirm this improvement. The rest of the paper is organized as follows: Section 2 presents the related work, Section 3 describes our three alternative summary generation models, Section 4 shows the evaluation results, and finally, Section 5 concludes the paper.

2   Related Works and Motivation

Although the task of topic-focused summarization has got a lot of attention recently (TAC-2010), the task is not new. A topic-sensitive LexRank is proposed in [12], where the set of sentences in a document cluster is represented as a graph. In this graph, the nodes are sentences and links between the nodes are induced by a similarity relation between the sentences. A substantial body of work on summarization using Information Extraction (IE) templates have been accomplished over the years in the Message Understanding Conferences (MUC[5]), DUC-2004 biography-related summarization task[6], as well as TREC[7]. In [15], they discuss the use of MUC templates for summarization. In [16], the authors define several biographical facts that should be included into a good biography. Filatova et al. [3] automatically create templates for several domains and use summarization-like task to evaluate the quality of the created templates. All the templates and facts are used in these researches to generate more focused summaries. Nastase [11] expands the query by using encyclopedic knowledge in Wikipedia and use the topic expanded words with activated nodes in the graph to produce an extractive summary. New features such as topic signature

---

[2] http://www.nist.gov/tac/2010/

[3] http://duc.nist.gov/

[4] http://wordnet.princeton.edu/

[5] http://www-nlpir.nist.gov/related_projects/muc/proceedings/ie_task.html

[6] http://duc.nist.gov/duc2004/

[7] http://trec.nist.gov/

are used in the NeATS system by Lin and Hovy [7] to select important content from a set of documents about some topic to present them in coherent order. An enhanced discourse-based summarization framework by rhetorical parsing tuning is proposed by Marcu [10]. In our work, we exploit topic signature and rhetorical structure theory [9] to weight the sentences. In [4], they introduced a paradigm for producing summary-length answers to complex questions. Their method operates on a Markov chain, by following a random walk with mixture model on a bipartite graph of relations established between concepts related to the topic of a complex question and subquestions derived from topic-relevant passages. Motivated by all these related researches, we propose to augment a predefined list of important aspects (that provides a better coverage of the topic on the entire document collection) into a random walk framework that no other study has used before to the best of our knowledge.

3   Our Approaches

In this section, we give a detailed description of all the three models that we build for the task of topic-focused multi-document summarization. Our first model is solely based on aspect information, while the second follows a novel random walk framework, and the third model is the aspect-driven random walk approach that combines the intuitions of the first two models. We get a candidate summary from each of the model at the end of the summary generation procedure. Therefore, three models give us three candidate summaries for the same given topic. Figure 1 presents the overall architecture of our systems.

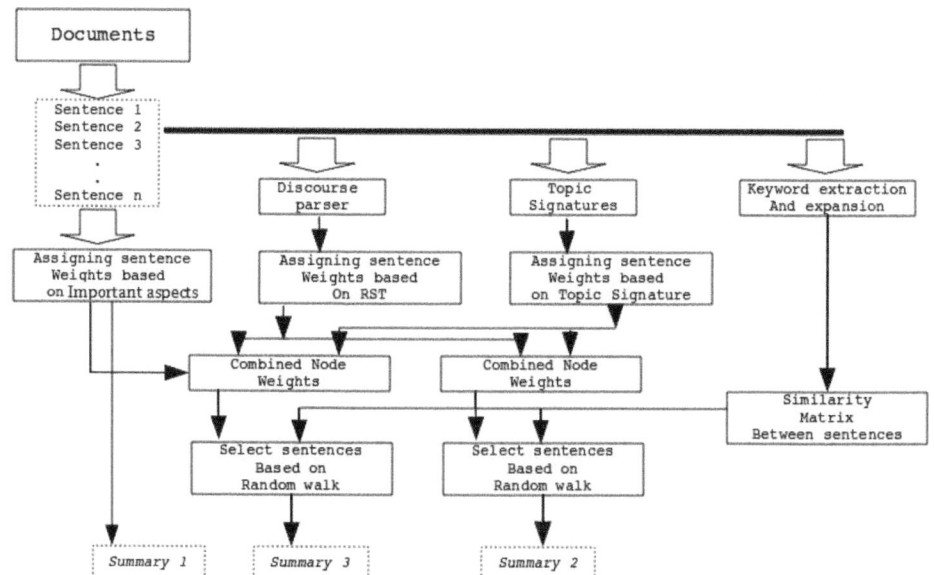

Fig. 1. The overall architecture of our approaches

3.1   Aspect–Based Model

Our first approach exploits the predefined list of important aspects to find the most relevant sentences from the document collection for creating the summaries. For each question (i.e. aspect) of a topic, we did keyword expansion using Word-Net[8] [2]. For example, the word "happen" being a keyword in the given *aspect*: "What happened?" returns the words: *occur, pass, fall out, come about, take place* from WordNet. On the other hand, for each document sentence in the collection we perform Named Entity (NE) tagging using the *OAK* system [13]. Named Entities (NE) are defined as terms that refer to a certain entity. For instance, *USA* refers to a certain *country*, and *$200* refers to a certain *quantity of money*. OAK system has 150 named entities (such as PERSON, LOCATION, ORGANIZATION, GPE (Geo-Political Entity), FACILITY, DATE, MONEY, PERCENT, TIME etc.) that can be tagged. They are included in a hierarchy. We weight each sentence based on the presence of one or more Named Entity classes. We rank the document sentences based on the following two criteria:

1. Similarity of each sentence with the expanded aspect (in terms of word matching), and
2. weight assigned to each sentence by the NE tagging procedure[9].

Finally, we select the top-ranked sentences to be included in the candidate summary (Summary 1 in Figure 1).

3.2   Random Walk Model

To include into our second candidate summary, we select the most relevant sentences by following a random walk on a graph where each node is a document sentence and the edges represent similarity between sentences. The whole procedure operates on a Markov chain (MC). A Markov chain is a process that consists of a finite number of states and some known probabilities p_{ij}, where p_{ij} is the probability of moving from state j to state i. For each node (i.e. sentence) and each edge in the graph, we calculate *"node weight"* and *"edge weight"*, respectively. Once we find all the node weights and edge weights, we perform a random walk on the graph following a Markov chain model in order to select the most important sentences. The initial sentence is chosen simply based on the node (sentence) weights using the following formula:

$$InitialSentence = \arg \max_{i=1}^{N} \left(weight\left(S_i \right) \right) \qquad (1)$$

where N is the total number of nodes in the graph. After finding the initial best sentence, in each step of the random walk we calculate the probability (transition probability) of choosing the next relevant sentence based on the following equation:

---

[8] For simplicity, we consider the synsets up to level 1 in this research.
[9] For example, for an aspect like *"When did the accident happened?"*, we search for $< Time >$ tag in the NE tagged sentences and give them higher weights if found.

$$P(S_j|S_i) = \frac{1}{\alpha} \arg \max_{j=1}^{Z} \left(weight\,(S_j) * similarity\,(S_i, S_j)\right) \qquad (2)$$

where S_i is the sentence chosen early, S_j is the next sentence to be chosen, Z is the set of sentence indexes that does not contain i, the $similarity(S_i, S_j)$ function returns a similarity score between the already selected sentence and a new sentence under consideration, and α is the normalization factor that is determined as follows:

$$\alpha = \sum_{j=1}^{Z} \left(weight\,(S_j) * similarity\,(S_i, S_j)\right) \qquad (3)$$

Node Weight. We associate each node (sentence) in the graph a weight that indicates the importance of the node with respect to the document collection. Node weights are calculated based on a Topic Signature (TS) model [6] and Rhetorical Structure Theory (RST) [9]. We combine the weights of TS and RST, and normalize it to get the final weights of the sentences/nodes.

Using Topic Signature. Topic signatures are typically used to identify the presence of a complex concept–a concept that consists of several related components in fixed relationships [6]. Inspired by the idea presented in [6], for each topic present in the data set, we calculate its topic signature defined as below:

$$TS = \{topic, signature\}$$
$$= \{topic, \langle(t_1, w_1), \cdots, (t_n, w_n)\rangle\} \qquad (4)$$

where *topic* is the target concept and signature is a vector of related terms. Each t_i is a term highly correlated to the *topic* with association weight, w_i. We use the following log-likelihood ratio to calculate the weights associated with each term (i.e. word) of a sentence:

$$w_i = log \frac{occurrences\ of\ t_i\ in\ topic\ j\ sentences}{occurrences\ of\ t_i\ in\ all\ topics'\ sentences} \qquad (5)$$

To calculate the topic signature weight for each sentence, we sum up the weights of the words in that sentence and then, normalized the weights. Thus, a sentence gets a high score if it has a set of terms that are highly correlated with a target concept (topic).

Exploiting Rhetorical Structure Theory (RST). Rhetorical Structure Theory provides a framework to analyze text coherence by defining a set of structural relations to composing units ("spans") of text. The most frequent structural pattern in RST is that two spans of text are related such that one of them has a specific role relative to the other. A paradigm case is a claim followed by evidence for the claim. RST assumes an "Evidence" relation between the two spans that is denoted by calling the claim span a *nucleus* and the evidence span a *satellite*[10]. In this paper, we parse each document sentence within the framework of Rhetorical Structure Theory (RST) using a Support Vector Machine (SVM)-based discourse parser described in [1] that was shown 5% to 12% more accurate than current state-of-the-art parsers. We observe that in a relation the nucleus often contains the main information while the satellite provides some additional information. Therefore, we assign a weight to each sentence that is a nucleus of a relation and normalize the weights at the end.

Edge Weight. Edge weight is determined by measuring similarity between the sentences. Initially, we remove the stopwords from the sentences using a stopword list. Then, we use the *OAK* system [13] to get the stemmed words of a sentence. We expand the remaining keywords of the sentence using WordNet. Finally, we find the similar words between each pair of sentences that denotes the edge weight between the two sentences. We build a similarity matrix by populating into it the edge weights between sentences.

3.3   Aspect-Driven Random Walk Model

The third model that we construct to generate a candidate summary is based on augmentation of the predefined important aspects into the random walk framework. Motivated by Harabagiu et al. [4], where they describe how a random walk can be used to populate a network with potential decompositions of a complex question, we propose to use the list of aspects (given in TAC-2010) in the random walk model as a guided way to provide a better coverage to satisfy a wide range of information need on a given topic. Through out the rest of the paper, we term this model as a *Combined Model* since it combines the important aspects with the random walk paradigm. The whole procedure can be again formulated according to a Markov Chain principle described in Section 3.2 except the fact that the node(sentence) weights will also include the weights obtained by using the list of aspects' information as defined in Section 3.1. Figure 2 shows a part of the graph with node and edge weights (after applying the combined model) for the top ranking sentences that were chosen by the random walk. This is an example of a DUC-2006 topic outlined below.

```
<topic id = "D0626H"
category = "2">
<title> bombing of US embassies
in Africa  </title>
```

S1: Among them is Saudi dissident Osama bin Laden, who allegedly runs al Qaida, a radical Islamic network accused of planning the bombings.

---
[10] http://www.sfu.ca/rst/01intro/intro.html

S2: In an interview Tuesday, Home Affairs Minister Ali Ameir Mohamed likened Ahmed to a chameleon.

S3: It said Khalid, who can not speak English or Kiswahili but only Arabic, was identified by a guard and a civilian worker at the embassy and a third witness.

S4: Although no details were released in court, local media said traces of chemicals that could have been used to make the bomb had been found in Saleh's home and car.

S5: The action contrasted markedly to a decision by Kenya, where the American Embassy was bombed on the same day.

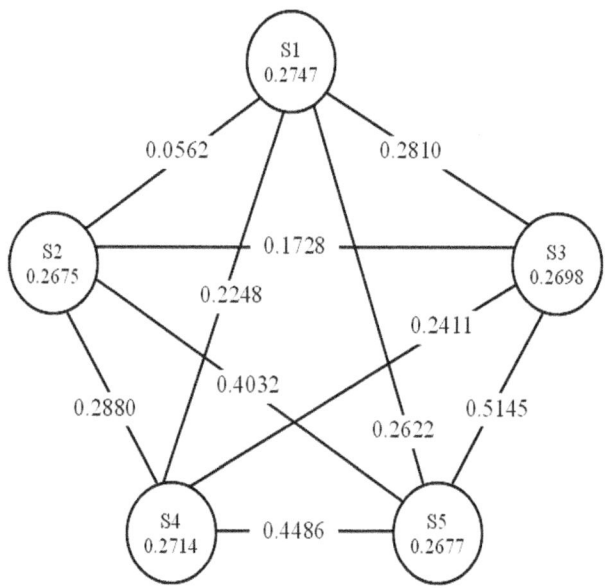

Fig. 2. Important aspects with random walk model

From Figure 2, we get to the fact that initially, sentence *S1* is chosen into the candidate summary as it has the highest node (sentence) weight, then, by performing a random walk based on the transition probabilities of the Markov chain model, we find *S2* as the next candidate sentence, then, *S3*, *S4*, *S5* and so on. The random walk stops after the k steps which is related to reaching the summary-length of 250 words.

4   Evaluation Results and Analyses

4.1   Task Description

TAC-2010 provides a new research direction for multi-document summarization by the means of predefined supervision or guide (the category and its aspects) that defines what information the reader is actually looking for. The task of DUC-2006 models the real-world complex question answering in terms of multi-document summarization. That is: *"Given a complex question (topic description) and a collection of relevant documents, the task is to synthesize a fluent, well-organized 250-word summary of the documents that answers the question(s) in the topic"*. In this paper, we consider a modified task description that induces the guided concept of TAC-2010 in order to automatically generate 250-word summaries like DUC-2006. Our summarization task can be defined as:

"To write a 250-word summary of a set of given newswire articles for a given topic, where the topic falls into a predefined category."

4.2   Data

In this research, we run our experiments using the TAC-2010 and DUC-2006 data applying three different models to generate three candidate summaries for each topic. The test dataset in TAC-2010 is composed of 44 topics, divided into five categories: Accidents and Natural Disasters, Attacks, Health and Safety, Endangered Resources, Investigations and Trials. In this paper, we consider only the first two categories[11]. As DUC-2006 data were not categorized, we manually categorize them to put into our chosen categories: Accidents and Natural Disasters, and Attacks. Since human-generated abstract summaries are not publicly available, we perform an extensive manual evaluation on the TAC-2010 data to report comparisons based on linguistic quality and responsiveness of the summaries. For DUC-2006 data, we obtain both an automatic[12] and a manual evaluation.

4.3   Automatic Evaluation

For the DUC-2006 data, we carried out automatic evaluation of our candidate summaries using ROUGE [5] toolkit, which has been widely adopted for automatic summarization evaluation. For all our systems, we report the widely accepted important metrics: ROUGE-2 and ROUGE-SU. We also present the ROUGE-1 scores since they provide a better correlation with the human judgement. We show the 95% confidence intervals of the important evaluation metrics for our systems to report significance for doing meaningful comparison. ROUGE uses a randomized method named bootstrap resampling to compute the confidence interval. We used 1000 sampling points in the bootstrap resampling. Table 1 to Table 3 show the ROUGE-1, ROUGE-2, and ROUGE-SU scores of our three different summary generation models.

For all the three systems, Table 4 shows the F-scores of the reported ROUGE measures while Table 5 reports the 95% confidence intervals of the important ROUGE measures. Table 4 clearly shows that the **Combined** system improves the ROUGE-1, ROUGE-2, and ROUGE-SU scores over the **Random walk** system by **3.67%**, **19.22%**, and **8.21%**, respectively, whereas, it could not beat

---

[11] TAC provides already categorized data.

[12] Abstract summaries are available for comparisons.

Table 1. ROUGE-1 measures

Scores	Aspects	Random Walk	Combined
Recall	0.3488	0.3344	0.3624
Precision	0.3415	0.3604	0.3556
F-score	0.3444	0.3460	0.3587

Table 2. ROUGE-2 measures

Scores	Aspects	Random Walk	Combined
Recall	0.0711	0.0500	0.0633
Precision	0.0693	0.0545	0.0609
F-score	0.0701	0.0520	0.0620

the ROUGE-2 score of **Aspect**–based system but improves the ROUGE-1, and ROUGE-SU scores by **4.15%**, and **4.97%**, respectively. These results suggest that augmenting important aspects with the random walk model provides a better content coverage to satisfy the information need. The proposed methods are also compared with a *Baseline* system. The *Baseline* is the official baseline system established in DUC-2006 that generated the summaries by returning all the leading sentences (up to 250 words) in the ⟨*TEXT*⟩ field of the most recent document(s). We also list the average ROUGE scores of all the participating systems of DUC-2006 (i.e. *DUC-Average*). Table 6 presents this comparison which denotes that the **Combined** system improves the ROUGE-1, and ROUGE-2 scores over the **Baseline** system by **11.77%**, and **17.78%**, respectively, whereas, it performs closely to the average DUC-2006 systems.

4.4   Manual Evaluation

Even if the ROUGE scores come up promising, it might be possible to generate bad summaries that get state-of-the-art ROUGE scores [14]. So, we conduct an

Table 3. ROUGE-SU measures

Scores	Aspects	Random Walk	Combined
Recall	0.1159	0.1029	0.1211
Precision	0.1109	0.1182	0.1156
F-score	0.1123	0.1090	0.1179

extensive manual evaluation in order to analyze the effectiveness of our approach. We judged the summaries for linguistic quality and overall responsiveness according to the DUC evaluation guidelines[13]. Table 7 and Table 8 presents the average linguistic quality and overall responsive scores of all the systems on TAC-2010

---

[13] http://www-nlpir.nist.gov/projects/duc/duc2007/quality-questions.txt

Table 4. ROUGE F-scores for different systems

Systems	ROUGE-1	ROUGE-2	ROUGE-SU
Aspects	0.3444	0.0701	0.1123
Random walk	0.3460	0.0520	0.1090
Combined	0.3587	0.0620	0.1179

Table 5. 95% confidence intervals for different systems

Systems	ROUGE-2	ROUGE-SU
Aspects	0.0569 - 0.0844	0.1053 - 0.1190
Random walk	0.0373 - 0.0682	0.0894 - 0.1262
Combined	0.0364 - 0.0879	0.0989 - 0.1363

Table 6. Comparison with DUC-2006 systems

Systems	ROUGE-1	ROUGE-2
Aspects	0.3444	0.0701
Random walk	0.3460	0.0520
Combined	0.3587	0.0620
Baseline	0.3209	0.0526
DUC-Average	0.3778	0.0748

Table 7. Linguistic quality and responsiveness scores (TAC-2010 data)

Systems	Lin. Quality	Responsiveness
Aspects	4.00	4.00
Random walk	3.60	3.00
Combined	4.00	3.00

data and DUC-2006 data, respectively. To compare the proposed models' performance with the state-of-the-art systems, in Table 8 we also list the scores of *LCC's GISTexter*[14] system that participated in the DUC-2006 competition and was ranked as one of the best systems. Analyzing these results yields the fact that augmenting important aspects with the random walk model often outperforms the random walk model alone in terms of linguistic quality and responsiveness scores. Table 8 shows that the proposed aspect-driven random walk model (i.e. *Combined*) performs very close to LCC's system in terms of linguistic quality while considerably outperforming it in terms of overall responsiveness scores. This confirms that the use of the aspect information enhances the coverage of the information that is necessary to satisfy the quest of the users.

---

[14] http://duc.nist.gov/pubs/2006papers/lcc2006.pdf

Table 8. Linguistic quality and responsiveness scores (DUC-2006 data)

Systems	Lin. Quality	Responsiveness
Aspects	3.72	3.00
Random walk	3.52	3.00
Combined	3.76	3.20
LCC	4.10	2.84

5   Conclusion

In this paper, we present a novel methodology to solve the topic-focused multi-document summarization task that uses a predefined list of important aspects in a random walk framework by performing a deeper semantic analysis of the source documents instead of relying only on document word frequencies to select important concepts. Experiments on the DUC-2006 and TAC-2010 data indicate that augmenting the important aspects into the random walk model considerably outperforms the random walk model alone. This suggests the fact that the aspects can provide a certain amount of supervision to cover all the relevant perspectives of a topic and hence, the use of it with any sophisticated model such as random walk can enhance the model's performance substantially in comparison to the model if used alone.

Acknowledgments. The research reported in this paper was supported by the Natural Sciences and Engineering Research Council (NSERC) of Canada – discovery grant and the University of Lethbridge.

References

1. duVerle, D.A., Prendinger, H.: A Novel Discourse Parser Based on Support Vector Machine Classification. In: Proceedings of the Joint Conference of the 47th Annual Meeting of the ACL and the 4th International Joint Conference on Natural Language Processing of the AFNLP (ACL 2009), vol. 2, pp. 665–673 (2009)
2. Fellbaum, C.: WordNet - An Electronic Lexical Database. MIT Press, Cambridge (1998)
3. Filatova, E., Hatzivassiloglou, V., McKeown, K.: Automatic Creation of Domain Templates. In: Proceedings of the COLING/ACL on Main Conference Poster Sessions, COLING-ACL 2006, pp. 207–214 (2006)
4. Harabagiu, S., Lacatusu, F., Hickl, A.: Answering Complex Questions with Random Walk Models. In: Proceedings of the 29th Annual International ACM SIGIR Conference on Research and Development in Information Retrieval, pp. 220–227. ACM (2006)
5. Lin, C.Y.: ROUGE: A Package for Automatic Evaluation of Summaries. In: Proceedings of Workshop on Text Summarization Branches Out, Post-Conference Workshop of Association for Computational Linguistics, Barcelona, Spain, pp. 74–81 (2004)

6. Lin, C.Y., Hovy, E.H.: The Automated Acquisition of Topic Signatures for Text Summarization. In: Proceedings of the 18th Conference on Computational Linguistics, pp. 495–501 (2000)
7. Lin, C.Y., Hovy, E.H.: From Single to Multi-Document Summarization: A Prototype System and Its Evaluation. In: Proceedings of the 40th Annual Meeting of the Association for Computational Linguistics (ACL), Philadelphia, pp. 457–464 (2002)
8. Mani, I., Maybury, M.T.: Advances in Automatic Text Summarization. MIT Press (1999)
9. Mann, W.C., Thompson, S.A.: Rhetorical Structure Theory: Toward a Functional Theory of Text Organization. Text 8(3), 243–281 (1988)
10. Marcu, D.: Improving Summarization Through Rhetorical Parsing Tuning. In: The Sixth Workshop on Very Large Corpora, Montreal, Canada, pp. 206–215 (1998)
11. Nastase, V.: Topic-Driven Multi-Document Summarization with Encyclopedic Knowledge and Spreading Activation. In: Proceedings of the Conference on Empirical Methods in Natural Language Processing (EMNLP 2008), pp. 763–772 (2008)
12. Otterbacher, J., Erkan, G., Radev, D.R.: Using Random Walks for Question-focused Sentence Retrieval. In: Proceedings of Human Language Technology Conference and Conference on Empirical Methods in Natural Language Processing, Vancouver, Canada, pp. 915–922 (2005)
13. Sekine, S.: Proteus Project OAK System (English Sentence Analyzer) (2002), http://nlp.nyu.edu/oak
14. Sjöbergh, J.: Older Versions of the ROUGEeval Summarization Evaluation System Were Easier to Fool. Information Processing and Management 43, 1500–1505 (2007)
15. White, M., Korelsky, T., Cardie, C., Ng, V., Pierce, D., Wagstaff, K.: Multidocument Summarization via Information Extraction. In: Proceedings of the First International Conference on Human Language Technology Research, HLT 2001, pp. 1–7 (2001)
16. Zhou, L., Ticrea, M., Hovy, E.H.: Multi-document Biography Summarization. CoRR abs/cs/0501078 (2005)

An Effective Approach for Topic-Specific Opinion Summarization

Binyang Li, Lanjun Zhou, Wei Gao, Kam-Fai Wong, and Zhongyu Wei

Department of Systems Engineering and Engineering Management
The Chinese University of Hong Kong Shatin, NT, Hong Kong, China
Key Laboratory of High Confidence Software Technologies Ministry of Education, China
{byli,ljzhou,wgao,zywei,kfwong}@se.cuhk.edu.hk

Abstract. Topic-specific opinion summarization (TOS) plays an important role in helping users digest online opinions, which targets to extract a summary of opinion expressions specified by a query, i.e. topic-specific opinionated information (TOI). A fundamental problem in TOS is how to effectively represent the TOI of an opinion so that salient opinions can be summarized to meet user's preference. Existing approaches for TOS are either limited by the mismatch between topic-specific information and its corresponding opinionated information or lack of ability to measure opinionated information associated with different topics, which in turn affect the performance seriously. In this paper, we represent TOI by word pair and propose a weighting scheme to measure word pair. Then, we integrate word pair into a random walk model for opinionated sentence ranking and adopt MMR method for summarization. Experimental results showed that salient opinion expressions were effectively weighted and significant improvement achieved for TOS.

Keywords: Topic-specific opinion summarization, topic-specific opinionated information, word pair, MMR.

1 Introduction

With the development of Web 2.0, people have become interested in expressing their personal opinions through online tools. There are a great amount of opinions widely spread from the comments on the health reform to the evaluations on a feature of a consumer product. In practice, people would like to focus on the summary of opinion expressions with their own preference to make decision [1]. For instance, users would like to give a query, "what are the opinions on X?" to express their preference of X, where X could not only be a feature of a product, but also be a target of a general domain opinion. Therefore, it is significant to study topic-specific opinion summarization(TOS) to meet the user's personal preference.

In Example 1, there are three opinion expressions tagged in bold in sentence (a). If a query is given about a *computer game*, traditional summarization regards only topical relevance, such as "*game*", "*operation*", "*screen*" to be the information need. For TOS, however, it is supposed to take both topic-specific information and

M.V.M. Salem et al. (Eds.): AIRS 2011, LNCS 7097, pp. 398–409, 2011.
© Springer-Verlag Berlin Heidelberg 2011

opinionated information into consideration, e.g., "*game is very small*". In this paper, we define user's information need of TOS as topic-specific opinionated information (TOI), i.e. the opinion expressions about the user's query.

> Example 1:
> (a) *The* [**game is very small**], [**operation is very flexible**], *and* [**screen is beautifully smooth**].
> (b) *The* [**screen is small but adequate**].

One of the fundamental problems in TOS is how to effectively represent the user's information need so as to evaluate and summarize salient opinion expressions.

Previous methods using KL-divergence [2] or feedback-style learning [3] have the limitation that TOI is represented by one single word. In practice, one single word can hardly represent both topic-specific information and opinionated information at the same time, especially for those domain-independent sentiment words that barely represent topic-specific information, e.g., "*small*" in Example 1. [4, 5] proposed to express topic-specific information and opinionated information by topic-specific words and sentiment words, respectively. However, they regarded the document as *bag-of-word*, and neglected the contextual information, which means word-based representation cannot hold the associative information between topic-specific information and opinionated information in individual opinion expression and lead to a mismatch. In Example 1, although topic-specific information and opinionated information of three opinion expressions can be represented separately, the associative information between them is lost due to lack of contextual information. In an extreme situation, the fake opinion expression "*small screen*" will be selected as the salient opinion expression.

Li et al. proposed to adopt word pair to represent TOI [6]. A word pair is constructed by a sentiment word together with its corresponding topic-specific word. The sentiment word represents opinionated information, i.e. the opinion, and the corresponding topic-specific word represents topic-specific information, i.e. target (also refer to the topic-specific word in this paper). With the help of pairwise representation, the contextual information between the opinion and its corresponding target could be maintained. Nonetheless, Li et al. neglected to measure the variant associations between the sentiment word and the topic-specific word within different word pair. In [6], a unified trade-off parameter was introduced to balance the topic-specific information and opinionated information for all word pairs. In practice, it is inadequate to describe the distinct association of individual word pair, because one sentiment word is supposed to modify different targets in different opinion expressions. In Example 1, both sentence (a) and sentence (b) include the sentiment word "*small*" which associated with different topic-specific words ("*game*" in (a) and "*screen*" in (b)). Intuitively, "*small*" is a domain-independent sentiment word, and it should be dynamically assigned different weights when modifying different targets. Therefore, we argue that the trade-off parameter should be estimated for each word pair to describe the distinct association.

In this paper, we also follow the TOI representation by word pair. According to the above analysis of word pair, we first propose an effective weighting scheme from the perspective of information gain by selecting the word pair to measure both topic-specific words and sentiment words, and then provide an individual associative score for each word pair. Thus, the TOI of individual opinion expression is able to be measured. Finally, we integrate word pair into a random walk model for sentence ranking and adopt maximal marginal relevance (MMR) method to generate the topic-specific opinion summary.

To investigate the effectiveness of our approach, experiments were made based on the TAC2008 and OpQA benchmark datasets. Significant improvements over the best run in TAC 2008 and those models with word-based representation were shown in this paper.

In the remainder of this paper, we first present pairwise representation of topic-specific opinionated information together with a new weighting scheme in Section 2. We then integrate word pair into a random walk model for sentence ranking and generate opinion summary by using MMR method in Section 3. In Section 4, we will show the experimental results. We review the related work in Section 5. Finally, this paper is concluded and the future work is suggested in Section 6.

2   Representation of Topic-Specific Opinionated Information

Topic-specific opinion summarization (TOS) was first proposed in the Text Analysis Conference (TAC) 2008, and the objective is to extract an informative summary of opinion expressions about a given query, as found in a document collection [7]. Different from traditional topic-specific summarization that concentrates only on the topic-specific information, TOS concerns on the topic-specific opinionated information (TOI). More precisely, the TOI of an opinion expression is supposed to contain the following attributes: opinion (i.e. opinionated information), holder, target (i.e. topic-specific information) and polarity [8]. In this section, we will first describe how to express TOI by pairwise representation, word pair. Then a new weighting scheme is introduced for measuring the pairwise representation.

2.1   Pairwise Representation

Without loss of generality, we assume that there is a document set $\mathcal{D}$ ($\mathcal{D} = \{d_1, d_2, d_3, \cdots, d_n\}$) that includes a set of sentences $\mathcal{S} = \{s_1, s_2, s_3, \cdots, s_N\}$, and a user generated query $Q = \{q_1, q_2, q_3, \cdots, q_z\}$, where $q_1, q_2, q_3, \cdots, q_z$ are the key words. TOS aims at extracting an informative summary of $\mathcal{S}'$ ($\mathcal{S}' \subseteq \mathcal{S}$) with opinion expressions from $\mathcal{D}$ about the Q.

In order to represent TOI, we need to consider all the attributes of an opinion expression together with the associations between these attributes. We also utilize topic-sentiment word pair [6] in this paper. The notion of word pair was first proposed for opinion retrieval to capture the contextual information between the opinion and its corresponding target [6]. Since most opinion holders are implicit to be the author in the blogosphere, we do not take opinion holder attribute into consideration.

Definition 1: topic-sentiment word pair p_{ij} consists of two elements, one represents the opinion, and the other one represents the modified target, $p_{ij} = \{< t_i, o_j > | t_i \in V_t, o_j \in V_o\}$.

V_t is the topic-specific words collection with all candidate targets, and the target reflects the preference of user by the query matching. V_o is the sentiment word lexicon which is used to express opinions. We maintain the semantic information between the topic-specific word and the sentiment word by pairwise representing. Thus, TOI of an opinion expression is represented by a word pair. We assume that for one query the candidate targets and opinions are in V_t, and V_o, respectively. The total number of the word pair is $|p_{ij}| = m \times M$, $(|V_t| = m, |V_o| = M)$.

In practice, the weight of a sentiment word may differ from variant targets. In the following section, we will describe a weighting scheme to measure word pair from the perspective of information gain, and introduce a method by normalizing Point-wise Mutual Information (PMI) between the sentiment word and the corresponding topic-specific word to compute the associative score for individual opinion expression.

2.2   Weighting Scheme for Word Pair

Previous weighting schema would like to capture both topic-specific information and opinionated information by one single word. It either assigns a *relevance* weight to the sentiment word, such as using the distribution divergence from the query words [9], or integrates the sentiment weight into query word, e.g., computes the Point-wise Mutual Information (PMI) score between sentiment word and the target combined with *tf-idf* value of query words [10]. In this work, we regard the topic-specific information and opinionated information of an opinion expression to be represented by topic-specific word and sentiment word of a word pair, respectively. Therefore, we measure both topic-specific words and sentiment words by computing the information gain in selecting the word pair.

In TOS, both sentiment words and topic-specific words are considered as informative content words, and described as *"term"* (denoted by w). Additionally, we would like to concentrate on the granularity of sentence rather than document. We can compute term weight of w, $TW(w)$ by Equation (1):

$$TW(w) = -\log_2 P(w)/P'(w) \qquad (1)$$

For simplicity, we assume that any term w_i follows *Poisson* distribution. (discuss other distributions in Section 4) $P(w)$ on the whole set of words with the parameter $\lambda_w = |w|/|S|$ ($|S|$ is the total number of words in S), while it also follows another *Poisson* distribution $P_1'(w)$ on the set of sentences including w with the parameter $\lambda_w' = |w|/|S(w)|$ ($|S(w)|$ is the total number of words in the sentences including w). Obviously $\lambda_w \le \lambda_w'$.

We measure each term according to its distributions between the sentence set it occurs and the whole sentence set. In other words, we weigh sentiment word by computing the gain in selecting a sentence containing the sentiment word.

Recall Example 1, one sentiment word (resp. target) may be assigned different weights when associated with different targets (resp. sentiment words). Therefore, there is a must to embody the different associations between sentiment words and topic-specific words.

Previous works [4, 5, 6] focus on using a unified parameter to express variant combinations between topic-specific information and opinionated information. It is inadequate to express the variant of associations even to one specific domain. We propose to compute an associative score (also referred as trade-off parameter in this paper) for each individual association between the sentiment word and the topic-specific word.

Inspired by the fact that Mutual Information is a measurement to assess how two words are associated, and achieves better performance in [12], we therefore utilize mutual information to estimate the trade-off parameter for both sentiment words and the target words.

In our method, for each word t_i in V_t, we compute its mutual information scores for all words in V_o and normalize the scores. Informally, mutual information compares the probability of observing t_i and o_j together (the joint probability) with the probabilities of observing t_i and o_j independently. The mutual information between words t_i and o_j are calculated as follows:

$$I\left(X_{t_i}; X_{o_j}\right) = \Sigma_{X_{t_i}=0,1} \Sigma_{X_{o_j}=0,1} \log \frac{p(X_{t_i}, X_{o_j})}{p(X_{t_i})p(X_{o_j})} \tag{2}$$

where X_{t_i} and X_{o_j} are binary variables indicating whether t_i or o_j is present or absent.

The parameters are estimated as follows:

$$p\left(X_{t_i} = 1\right) = c\left(X_{t_i} = 1\right)/N$$
$$p\left(X_{t_i} = 0\right) = 1 - p\left(X_{t_i} = 1\right)$$
$$p\left(X_{o_j} = 1\right) = c\left(X_{o_j} = 1\right)/N$$
$$p\left(X_{o_j} = 0\right) = 1 - p\left(X_{o_j} = 1\right)$$
$$p\left(X_{t_i} = 1, X_{o_j} = 1\right) = \frac{c\left(X_{t_i} = 1, X_{o_j} = 1\right)}{N}$$
$$p\left(X_{t_i} = 1, X_{o_j} = 0\right) = \frac{c(X_{t_i} = 1) - c\left(X_{t_i} = 1, X_{o_j} = 1\right)}{N}$$
$$p\left(X_{t_i} = 0, X_{o_j} = 1\right) = \frac{c\left(X_{o_j} = 1\right) - c\left(X_{t_i} = 1, X_{o_j} = 1\right)}{N}$$
$$p\left(X_{t_i} = 0, X_{o_j} = 0\right) = 1 - p\left(X_{t_i} = 1, X_{o_j} = 1\right) - p\left(X_{t_i} = 0, X_{o_j} = 1\right) - p\left(X_{t_i} = 1, X_{o_j} = 0\right)$$

where $c(X_{t_i} = 1)$ and $c\left(X_{o_j} = 1\right)$ are the numbers of sentences containing word t_i and o_j, respectively, $c\left(X_{t_i} = 1, X_{o_j} = 1\right)$ is the number of sentences that contain both t_i and o_j, and N is the total number of sentences in the collection. We then normalize the mutual information score to obtain a trade-off parameter $\mu_{mi}(o_j|t_i)$ to balance the weight of t_i when associated with o_j:

$$\mu_{mi}(o_j|t_i) = \frac{I(X_{t_i}; X_{o_j})}{\Sigma_{o \in V_o} I(X_{t_i}; X_o)}$$

$$\mu_{mi}(t_i|o_j) = \frac{I(X_{t_i}; X_{o_j})}{\Sigma_{t \in V_t} I(X_{o_j}; X_t)}$$

$\mu_{mi}(t_i|o_j)$ is computed in the same way to balance the weight of o_j when associated with t_i. The probability would be higher if the two words co-occur with each other more frequently.

After estimating the associative score between the two elements of a word pair, we can assign the weight to individual word pair p_{ij}.

$$w_{p_{ij}} = \mu_{mi}(t_i|o_j) \times TW(t_i) + \mu_{mi}(o_j|t_i) \times TW(o_j) \qquad (3)$$

where $TW(t_i)$ and $TW(o_j)$ are the term weights of t_i and o_j, which can be computed from Equation (1).

As to those word pairs with a negation operator around, an alternative would be to rewrite the individual word pair p_{ij} as $\neg p_{ij}$. Since the negation operator only shifts the polarity of the word pair, we assign $\neg p_{ij}$ with the same weight as p_{ij}.

3   Word Pair Based TOS

In this section, we first integrate word pair into a random walk model [14], for sentence ranking. Then, we utilize MMR method to generate a summary.

3.1   PageRank Based on Word Pair

In Section 2, we introduce a weighting scheme for measuring individual word pair. According to the definition of word pair, it is intuitively that the sentence with word pair representing salient opinion expression should be assigned a relatively high weight. We, therefore, consider the global information of word pair for sentence ranking by the recursive procedure in the random walk model, PageRank.

One of our objectives is to investigate the effectiveness of our proposed weighting scheme for word pair, so we compute the similarity between sentences according to the weighted word pair. Moreover, in our approach, we do not explicit divide sentiment words into domain-dependent and domain-independent, but use the corresponding target as an indicator. This will weaken the effect of domain-independent sentiment words. In order to correct the opinionated information of a domain-independent sentiment of a word pair, we utilize synonym dictionary SentiWordNet [13]. We choose the sentiment word with the highest PMI score over a topic-specific word in Equation (2) as the cue word and consider all synonyms of the cue word together with the corresponding target to be the same word pair.

We define a PageRank model that has sentences to be summarized as nodes and edges placed between two sentences that are similar to each other.

We can then score all the sentences based on the expected probability of a random walker visiting each sentence. We use the short-hand $P(s_u|s_v)$ to denote the probability of being at node s_u at a time t given that the walker was at s_v at time $t-1$. The jumping probability from node s_v to node s_u is given by:

$$P(s_u|s_v) = \frac{sim(s_v, s_u)}{\sum_{s_{u'} \in S \setminus s_v} sim(s_v, s_{u'})} \qquad (4)$$

where *sim* is a similarity function defined on two sentence/excerpt nodes based on the word pair they contain.

$$sim(s_u, s_v) = \frac{\sum_{p_{ij} \in s_u, s_v} w_{p_{ij}}^{su} \cdot w_{p_{ij}}^{sv}}{\sqrt{\sum_{p_{ij} \in s_u}(w_{p_{ij}}^{su})^2} \times \sqrt{\sum_{p_{ij} \in s_v}(w_{p_{ij}}^{sv})^2}} \tag{5}$$

The saliency score $Score(s_u)$ for sentence s_u can be calculated by mixing query similar score and scores of all other sentences linked with it as follows:

$$Score(s_u) = \gamma \sum_{v \neq u} Score(s_v) \cdot P(s_u|s_v) + (1 - \gamma) sim'(s_u|Q) \tag{6}$$

where $sim'(s_u|Q) = \frac{sim(s_u|Q)}{\sum_{k=1}^{N} sim(s_k|Q)}$

$$sim(s_u|Q) = \frac{\sum_{w_{t_i} \in s_u, Q} w_{t_i}^{su} \cdot w_{t_i}^{Q}}{\sqrt{\sum_{w_{t_i} \in s_u}(w_{t_i}^{su})^2} \times \sqrt{\sum_{w_i \in Q}(w_{t_i}^{Q})^2}} \tag{7}$$

where $w_{t_i}^{su}$ $w_{t_i}^{Q}$ are the weights of t_i in s_u, and Q, respectively.

Finally, all the sentences will rank according to the saliency score. As for each query, we choose a number of sentences with weights higher than a threshold as candidate set R for TOS.

3.2   Summary Generation

In order to generate a summary, we adopt maximal marginal relevance (MMR) method and incrementally add the top ranked sentences from R into the answer set.

$$MMR = Arg \max_{s_u \in R \setminus S'} \left[\theta \big(sim(s_u|Q) \big) - (1 - \theta) \max_{s_v \in S'} sim(s_u, s_v) \right]$$

R is the ranked list of sentences retrieved by the PageRank model in Section 3.1, given the document set D and a query Q. We set the relevance threshold δ, below which it will not be regarded as candidate sentences. S' is the subset of sentences in R already selected; $R \setminus S'$ is the set difference, i.e., the set of as yet unselected sentences in R. We compute $sim(s_u, s_v)$ and $sim(s_u|Q)$ and by Equation (5) and Equation (7), respectively in Section 3.1. The parameter θ lying in [0,1] controls the relative importance given to "relevance" versus redundancy. As different users with different information needs may require a totally different summary, especially for TOS, one of the attractive points of MMR is by setting the value of parameter θ, it can particularly generate summaries according to a user's need. In the experiment, we set $\theta = 0.5$ to balance the novelty and the relevance.

4   Evaluation

4.1   Experiment Setting

4.1.1   Benchmark Datasets

Our experiments are based on two benchmark datasets for topic-specific opinion summarization, TAC2008 and OpQA.

TAC2008 dataset is the benchmark data set for the topic-specific opinion summarization track in the Text Analysis Conference 2008 (TAC2008), which contains 87 squishy opinion questions. The initial topic words for each question are also provided. Summarizations for all queries must be retrieved from the TREC Blog06 collection [15], which consists of review and blog data. The top 50 documents were retrieved for each query.

The Opinion Question Answering (OpQA) corpus consists of 98 documents appeared in the world press between June 2001 and May 2002. The documents covered four general topics, and 30 questions were given. [16]

4.1.2   Sentimental Lexicon and Topic Collection

In our experiment, we use SentiWordNet as the sentiment lexicon. SentiWordNet is a popular lexical resource for opinion mining, which consists of 4800 negative sentiment words and 2290 positive sentiment words. For each sentiment word, SentiWordNet also provides its synonyms.

In order to acquire the collection of topic terms, we adopt two expansion methods, dictionary- based method and pseudo relevance feedback method [6].

4.1.3   TOS Approaches for Comparison

To demonstrate the effectiveness of pairwise representation for TOS, we compared it with the following models:

(1) Baseline 1: This model [18] achieved the best run in TAC2008 opinion summarization task. We treated it as Baseline 1 in the experiment.

(2) OPM-1: This model was proposed for opinion question and answering, which achieved 2% improvement over the best run in TAC2008 Opinion QA track [19].

(3) OPM-2: This model was similar with OPM-1, but use PageRank model for sentence ranking instead.

(4) GOSM: This model was originally designed for opinion retrieval, and it adopted pairwise representation of TOI. GOSM adopted *"relevance"* measurement for sentiment word and utilized a uniform parameter to balance topic-specific information and opinionated information. We re-designed GOSM to deal with TOS by using Pair-based HITS model so that we could compare the effectiveness of different weighting schema for word pair [6].

(5) PPM: our proposed approaches.

Additionally, in our experiments, we will also investigate the performance of sentence retrieval with different probability models. We used the metrics in the Text Retrieval Conference (TREC), which are average precision (AvPr), R-precision (R-Pre) and precision at 10 sentences (P@10).

4.2   Performance Evaluation

4.2.1   Parameter Tuning

In our proposed approach, there are two parameters θ and γ. θ is a user-defined parameter according to the specific need. In our experiment, we set the parameter $\theta = 0.5$ to balance the novelty and the relevance.

We studied how the parameter γ (in Equation (6)) influenced the performance of sentence ranking in both TAC2008 and OpQA datasets. The results are given in Fig. 1.

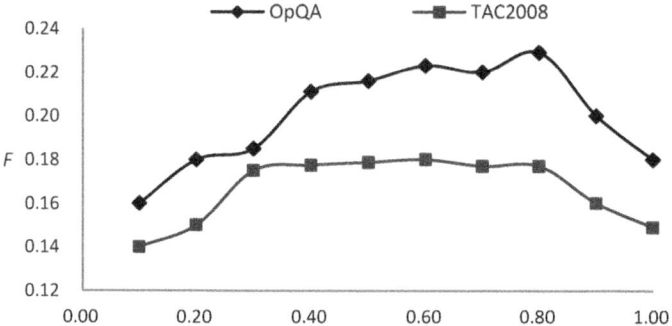

Fig. 1. Pair-based PageRank Performance with varying parameter γ on TAC2008 and OpQA

Best *F* value was achieved, when γ was set around 0.8 in both TAC2008 and OpQA datasets. Therefore, in the following experiments, we set γ = 0.8.

4.2.2   Comparisons on Sentence Ranking

In our evaluation, we also tested the performance of sentence ranking of other probability models, including *tf-idf* model and Bose-Einstein model. We used the metrics in the Text Retrieval Conference 10 (TREC), which are average precision (AvPr), R-precision (R-Prec) and precision at 10 sentences (P@10). In our experiment, we created the judgment through pooling method.

The experimental results based on these metrics are shown in Table 1.

Table 1. Comparison of sentence ranking on OpQA and TAC2008 datasets

Dataset	Probability	Metrics		
		AvPr	R-Prec	P@10
OpQA	Poisson	0.212	0.233	0.408
	tf-idf	**0.229**	0.230	0.397
	Bose-Einstein	0.208	**0.245**	**0.421**
TAC 2008	Poisson	**0.180**	0.206	0.361
	tf-idf	0.177	0.198	0.354
	Bose-Einstein	0.175	**0.212**	**0.369**

Table 1 showed that Bose-Einstein model achieved best R-Prec and P@10 on OpQA and TAC2008 datasets. Thus, we chose Bose-Einstein model for further evaluation.

4.2.3   Comparisons on TOS

We were also interested in the performance comparison with the other models for TOS.

Table 2. Comparison of TOS on OpQA and TAC2008 datasets

Data set	Approaches	Measurements		
		Precision	Recall	F(3)
OpQA	Baseline 1	0.280	0.356	0.325
	OPM-1	0.274	0.368	0.343
	OPM-2	0.281	0.354	0.362
	GOSM	0.286	0.360	0.379
	PPM	0.276	0.375	**0.385**
TAC 2008	Baseline 1	0.101	0.217	0.186
	OPM-1	0.102	0.256	0.195
	OPM-2	0.113	0.245	0.208
	GOSM	0.102	0.241	0.216
	PPM	0.103	0.268	**0.231**

Table 2 showed that PPM achieved around 6% and 5% improvement in F value compared with Baseline 1 in OpQA and TAC2008, respectively.

5   Related Work

Research on opinion summarization started mostly on review-type data, and much progress has been made in automatic sentiment summarization in the review domain [20]. These summarizations referred to as feature-based summarization or aspect summarizations are extracted from a collection of reviews on some specific product. Benefited from the limited topics and fixed sentiment words in specific domain, technologies such as LDA, LSA, pLSA, have been utilized and they achieved good performance in product review [21, 22, 23, 24]. In this paper, we focus on TOS, which is about general domain summarization, and the above works are out of the scope of TOS due to the constraints of limited targets and fixed sentiment words.

For TOS, lots of work concentrates on term weighting to improve the precision of sentence ranking. A weighted sentiment dictionary was generated from previous Text Retrieval Conference (TREC) relevance data [11]. This dictionary was submitted as a query to a search engine to get an initial query-independent opinion score of all retrieved documents. Similarly, a pseudo opinionated word composed of all opinion words was first created, and then used to estimate the opinion score of a document [3]. This method was shown to be very effective in TREC evaluations.

Ernsting et al. applied the KL divergence to weigh opinionated word [25]. However, the weights of the terms in the sentiment word dictionary were biased towards the terms with high values. Experimental results showed that this method had detrimental effect on the performance. [9] followed the KL divergence measurement and made a positive experimental result by taking term frequency into consideration.

Li et al. proposed a new representation based on word pair [6] for TOI. With the help of word pair, the associative information between the opinion and its corresponding target could be uniformly represented. However, [6] didn't give an explicit approach to weigh word pair but utilize the relationship between word pair and document instead, which is in accordance with "*relevance*".

In this paper, we also utilize word pair to represent TOI. Different from previous work, we present a weighting scheme, which regards both topic-specific words and sentiment words as informative content to represent topic-specific information and opinionated information. Moreover, regarding the specialty of TOS, we propose a method to estimate individual associative score for each word pair to measure the association of topic-specific information and opinionated information, and take *negation* into consideration and integrate it into word pair for TOS.

6   Conclusion and Future Work

In this work, we present a method for topic-specific opinion summarization inspired by the representation of word pair. Based on word pair, we further propose a weighting scheme so that both topic-specific words and sentiment words are weighed. We also provide a method by normalizing PMI between sentiment word and topic-specific word to compute the associative score for individual word pair. Thus, the topic-specific opinionated information of individual opinion expression is able to be well expressed and measured. We integrate word pair into the PageRank model for sentence ranking and adopt maximal marginal relevance method to extract salient sentences as the result of TOS.

In the future, more research is required in the following directions:

(1) Deeper NLP techniques e.g., discourse analysis [26], dependency parser, collocation identification[17] may help to extract word pair and understand the meaning of opinion so as to improve the accuracy.
(2) Opinion holder is another important attribute of TOI [27]. It would be interesting to study opinion holders for QOS.
(3) Since the new weighting scheme and the trade-off parameter indicate topic-specific opinionated information effectively, it is worth further study on other opinion oriented applications.

Acknowledgments. This work is partially supported by National 863 program of China (Grant No. 2009AA01Z150), the Innovation and Technology Fund of Hong Kong SAR (Project No. GHP/036/09SZ) and 2010/11 CUHK Direct Grants (Project No. EE09743).

References

[1] Hu, M., Liu, B.: Mining and summarizing customer reviews. In: Proceedings of the Tenth ACM SIGKDD 2004 (2004)
[2] Huang, X., Croft, W.: A Unified Relevance Model for Opinion Retrieval. In: Proceedings of CIKM (2009)
[3] Na, S.-H., Lee, Y., Nam, S.-H., Lee, J.-H.: Improving Opinion Retrieval Based on Query-Specific Sentiment Lexicon. In: Boughanem, M., Berrut, C., Mothe, J., Soule-Dupuy, C. (eds.) ECIR 2009. LNCS, vol. 5478, pp. 734–738. Springer, Heidelberg (2009)
[4] Zhang, M., Ye, X.: A generation model to unify topic relevance and lexicon-based sentiment for opinion retrieval. In: SIGIR 2008 (2008)

[5] Yang, K., Yu, N., Valerio, A., Zhang, H.: WIDIT in TREC-2006 Blog track. In: Proceedings of TREC (2006)

[6] Li, B., Zhou, L., Feng, S., Wong, K.F.: A unified graph model for sentence-based opinion retrieval. In: ACL 2010 (2010)

[7] Dang, H.: Overview of the TAC 2008 opinion question answering and summarization tasks. In: TAC 2008 (2008)

[8] Stoyanov, V., Cardie, C.: Toward opinion summarization: linking the sources. In: Proceedings of the Workshop on Sentiment and Subjectivity in Text (2006)

[9] He, B., Macdonald, C., He, J., Ounis, I.H.: An effective statistical approach to blog post opinion retrieval. In: CIKM 2008 (2008)

[10] Turney, P.: Thumbs up or thumbs down? semantic orientation applied to unsupervised classification of reviews. In: ACL 2002 (2002)

[11] Amati, G.: Probabilistic models for information retrieval based on Divergence from Randomness. PhD thesis, University of Glasgow (2003)

[12] Kim, J., Li, J., Lee, J.: Discovering the discriminative views: Measuring term weights for sentiment analysis. In: ACL-IJCNLP 2009 (2009)

[13] Esuli, A., Sebastiani, F.: Sentiwordnet: A publicly available lexical resource for opinion mining. In: LREC (2006)

[14] Erkan, G., Radev, D.R.: Lexpagerank: Prestige in multi-document text summarization. In: EMNLP 2004 (2004)

[15] Macdonald, C., Ounis, I.: Overview of the TREC-2006 Blog Track. In: TREC 2006 (2006)

[16] Wilson, T., Wiebe, J., Hoffmann, P.: Recognizing contextual polarity in phrase-level sentiment analysis. In: EMNLP 2005 (2005)

[17] Xia, Y., Xu, R., Wong, K.F., Feng: The unified collocation framework for opinion mining. In: International Conference on Machine Learning and Cybernetics 2007 (2007)

[18] Varma, V., Pingali, P., Katragadda, R., Krishna, S., Ganesh, S., Sarvabhotla, K., Garapati, H., Gopisetty, H., Reddy, V.B., Reddy, K., Bysani, P., Bharadwaj, R.: IIIT Hyderabad at TAC 2008. In: TAC 2008 (2008)

[19] Li, F., Tang, Y., Huang, M., Zhu, X.: Answering opinion questions with random walks on graphs. In: ACL 2009 (2009)

[20] Lerman, K., Godensohn, S., McDonald, R.: Sentiment Summarization: Evaluating and learning User Preferences. In: EACL 2009 (2009)

[21] Paul, M., Zhai, C., Girju, R.: Summarizing contrastive viewpoints in opinionated text. In: EMNLP 2010 (2010)

[22] Zhao, X., Jiang, J., Yan, H., Li, X.: Jointly modeling aspects and opinions with a MaxEnt-LDA hybrid. In: EMNLP 2010 (2010)

[23] Lin, C., He, Y., Everson, R.: A comparative study of Bayesian models for unsupervised sentiment detection. In: ACL 2010 (2010)

[24] Blitzer, J., Dredze, M., Pereira, F.: Biographies, bollywood, boomboxes and blenders: Domain adaptationfor sentiment classification. In: ACL (2007)

[25] Ernsting, B., Weerkamp, W., de Rijke, M.: Language modeling approaches to blog post and feed finding. In: TREC 2007 (2007)

[26] Zhou, L., Li, B., Gao, W., Wei, Z., Wong, K.F.: Unsupervised Discovery of Discourse Relations for Eliminating Intra-sentence Polarity Ambiguities. In: Proceedings of EMNLP 2011 (Oral presentation), Edinburgh, Scotland, July 27-31 (2011)

[27] Xu, R., Wong, K.F., Xia, Y.: Opinmine - Opinion Analysis System by CUHK for NTCIR-6 Pilot Task. In: Proceedings of NTCIR-6

A Model-Based EM Method for Topic Person Name Multi-polarization

Chien Chin Chen and Zhong-Yong Chen

Department of Information Management, National Taiwan University
No.1, Sec. 4, Roosevelt Rd., Taipei City 10617, Taiwan (R.O.C.)
`paton@im.ntu.edu.tw, d98725003@ntu.edu.tw`

Abstract. In this paper, we propose an unsupervised approach for multi-polarization of topic person names. We employ a model-based EM method to polarize individuals into positively correlated groups. In addition, we present off-topic block elimination and weighted correlation coefficient techniques to eliminate the off-topic blocks and reduce the text sparseness problem respectively. Our experiment results demonstrate that the proposed method can identify multi-polar person groups of topics correctly.

Keywords: Sentiment Analysis, Person Name Clustering, Text Mining.

1   Introduction

Topics that involve competing viewpoints are interesting and may be reported by hundreds of documents. Generally, topic documents cover every detail of different polarities (stances) and provide a balanced view of the topics. However, readers often have difficulty assimilating the information in numerous topic documents that cover various perspectives. Thus, several topic mining techniques have been developed to help readers. Existing topic mining approaches focus on extracting important themes from documents, and summaries of the themes are provided to help readers digest topics quickly. For instance, Mei and Zhai [13] developed a mixture of unigram models to extract salient themes from topic documents; while Nallapati et al. [16] and Feng and Allan [4] grouped topic documents into clusters, each of which represents a unique theme in a topic. Even though the extracted themes and summaries distill the topic contents clearly, readers must still expend a great deal of time in comprehending the extracted information if they are not familiar with the topics.

Basically, a topic is associated with specific times, places, and persons [16]. Identifying the polarity of important persons in topics that involve competing viewpoints can help readers construct the background of the topics and facilitate reading. For example, if readers know the important persons of the competing political parties in a presidential campaign, they can learn about the campaign more easily. Chen and Wu [1, 2] proposed a PCA-based method to identify the polarities of topic persons, and showed that the signs of the entries in the principal eigenvector of PCA are effective in partitioning topic persons into different polarities; however, the

M.V.M. Salem et al. (Eds.): AIRS 2011, LNCS 7097, pp. 410–421, 2011.

method is only useful for bipolar topics. In practice, a topic that involves competing viewpoints may contain more than two polarities, i.e., multi-polarities. We define the person name multi-polarization problem in this paper and propose an effective method for identifying the multi-polarities of person names in topic documents. The identified person polarities function as character descriptions of novels and fictional stories to help readers construct the background of the topics and facilitate reading. A challenging issue in the multi-polarization of topic person names is that the polarity of the individuals is context-dependent. For instance, politicians often change their policies for the sake of expediency, so their polarities change accordingly. In such cases, no external knowledge source is available for person name multi-polarization. To solve the problem, we propose a model-based EM method that polarizes person names in an unsupervised manner. As the method only considers the word usage patterns of person names in topic documents, it does not require external knowledge sources and it can capture the polarity dynamics effectively. We also present off-topic block elimination and weighted correlation coefficient techniques to eliminate off-topic blocks and reduce the text sparseness problem respectively. Our experiment results demonstrate that the proposed method can identify multi-polar person groups of topics effectively.

The remainder of this paper is organized as follows. Section 2 contains a review of related works. We describe the proposed method in Section 3, and evaluate its performance in Section 4. Then, in Section 5, we present our conclusions.

2   Related Work

Although person name multi-polarization is closely related to sentiment analysis [17], which focuses on discovering textual units with bipolar orientations, it differs from sentiment analysis in a number of respects. First, most sentiment analysis approaches identify the polarity of adjectives, adverbs, and verbs because the syntactic constructs generally convey sentimental semantics. For instance, Hatzivassiloglou and McKeown [6] employed language conjunctions, such as *and*, *or*, and *but*, to judge the polarity of conjoined adjectives. Ganapathibhotla and Liu [5] investigated the polarity of comparative adjectives (e.g., quick) or adverbs (e.g., quickly) combined with product features (e.g., run time) to identify the pros and cons of products discussed in product reviews. Ding et al. [3] also considered sentiment verbs, such as *like* and *hate*, to extract further sentiment comments about a product. In contrast, person name multi-polarization considers the polarity of person names, which are nouns that rarely express sentiment information. Second, sentiment analysis generally classifies textual units in terms of a positive or negative orientation, but a person's polarity may not have a positive or negative meaning. Specifically, people with different polarities take opposing stances regarding a certain topic, while people in the same polarity group reach a consensus or have the same goal. Finally, sentiment analysis usually requires external knowledge sources or human-composed sentiment lexicons. For example, Turney and Littman [21] manually selected seven positive and seven negative words as a sentiment lexicon and used pointwise mutual information (PMI) to calculate the polarity of a word. Kim and Hovy [10] and Hu and Liu [8] determined a word's polarity by classifying the synonyms and antonyms of the word in WordNet [14];

while Ku et al. [11] dealt with Chinese sentiment analysis by considering the sentiment words in the General Inquirer lexicon [20]. However, as a person's polarity is dynamic and context-dependent, no external knowledge source is available for person name multi-polarization research. The property of context-dependence and the lack of knowledge sources make the person name multi-polarization task a challenging research issue.

3   Method

3.1   Model-Based Person Name Multi-polarization

Given a set of documents about a topic that involves competing viewpoints with K polarities, we first decompose the documents into a set of non-overlapping blocks $B = \{\underline{b_1},...,\underline{b_L}\}$. A block is a content coherent unit, i.e., a document or a paragraph, which we represent as an M-dimensional frequency vector $\underline{b_l}$ whose i'th entry, denoted as $b_{l,i}$, is the frequency of person name i in block l. Let $N = \{\underline{n_1},...,\underline{n_M}\}$ represent a set of topic person names. A person name $\underline{n_i}$ is an L-dimensional frequency vector whose l'th entry, denoted as $n_{i,l}$, is the frequency of person name i in block l. We employ a model-based approach to identify the polarities of person names. Let $\theta = \{\underline{\omega_1}, ..., \underline{\omega_K}, \alpha_1, ..., \alpha_K\}$ represent a polarity model, where $\underline{\omega_k}$ is an L-dimensional representation vector of polarity k whose l'th entry, denoted by $\omega_{k,l}$, is the weight of block l in polarity k, and α_k is polarity k's weight, s.t., $\Sigma\alpha_k=1$. Given the person name occurrence data N, the topic person name multi-polarization task searches for an appropriate polarity model. The task can be formulated as the following model search problem:

$$\hat{\theta} = \underset{\theta \in \, search\, space}{\arg\max} \ P(\theta \,|\, N). \tag{1}$$

As the number of polarity models is infinite, we assume that all models have the same prior probability [15]; then, using Bayes theorem, Eq. (1) can be converted into the following form:

$$\hat{\theta} = \underset{\theta \in \, search\, space}{\arg\max} \ P(\theta) * P(N \,|\, \theta)$$

$$= \underset{\theta \in \, search\, space}{\arg\max} \ P(N \,|\, \theta)$$

$$= \underset{\theta \in \, search\, space}{\arg\max} \ \prod_{i=1}^{M} \sum_{k=1}^{K} \alpha_k P(\underline{n_i} \,|\, \underline{\omega_k}). \tag{2}$$

To search for an appropriate polarity model, we need to define $P(\underline{n_i}|\underline{\omega_k})$. Kanayama and Nasukawa [9] validated that polar text units tend to occur (not occur) jointly to make contexts coherent. Consequently, if the occurrence of a person name are coincident with those of a polarity's members, the person should be a member of that polarity and have a high $P(\underline{n}_i|\underline{\omega}_k)$. In this research, we learn $P(\underline{n}_i|\underline{\omega}_k)$ from topic blocks and use a correlation coefficient to discover the joint behavior of person names. However, as the range of the correlation coefficient is [-1,1], to avoid negative probabilities, we convert the range of the correlation coefficient by using the following equation:

$$corr^{\wedge}(\underline{n}_i,\underline{\omega}_k) = (corr(\underline{n}_i,\underline{\omega}_k)+1)/2, \tag{3}$$

where $corr(.)$ denotes the correlation coefficient between the frequency vectors of person i and polarity k. The range of the converted correlation coefficient $corr^{\wedge}(.)$ is within [0,1]. It returns 1 when person i and polarity k are positively correlated, and 0 when they are negatively correlated. Next, we define $P(\underline{n}_i|\underline{\omega}_k)$ as follows:

$$P(\underline{n}_i \mid \underline{\omega}_k) = \frac{corr^{\wedge}(\underline{n}_i,\underline{\omega}_k)}{\sum_{i=1}^{M} corr^{\wedge}(\underline{n}_i,\underline{\omega}_k)}. \tag{4}$$

The denominator in Eq. (4) is a normalization factor; hence, persons positively correlated with polarity k would be members of $\underline{\omega}_k$. Then, our objective, i.e., Eq. (2), is to cluster persons into positively correlated groups.

Let $<z_{i,1}, ..., z_{i,K}>$ denote a sequence of binary variables of person i. $z_{i,k} = 1$ if person i belongs to polarity k; otherwise, $z_{i,k} = 0$. As person name multi-polarization is an unsupervised problem, the values of the variables are unobserved. Here, we employ an EM method to search for appropriate person polarities. First, we randomly initialize the model parameters, and execute the following EM steps iteratively until convergence.

$$E\text{ - }step : E[z_{i,k}] = \alpha_k * P(\underline{n}_i \mid \underline{\omega}_k) \Big/ \sum_{j=1}^{K} \alpha_j * P(\underline{n}_i \mid \underline{\omega}_j). \tag{5}$$

$$M\text{ - }step : \alpha_k = \sum_{i=1}^{M} E[z_{i,k}]/M \quad and$$
$$\underline{\omega}_k = \sum_{i=1}^{M} E[z_{i,k}]\underline{n}_i \Big/ \sum_{i=1}^{M} E[z_{i,k}]. \tag{6}$$

The E-step computes the expectation of an unobserved variable $z_{i,k}$, by using the current polarity model. The M-step re-computes the polarity model as the maximum likelihood estimates given all the calculated expectations. Wu [22] proved the convergence of model-based EM methods. When converged, $E[z_{i,k}]$ indicates the probability that person i belongs to polarity k. We then assign person i to the polarity with the maximum probability.

3.2   Off-Topic Block Elimination

While collecting the experimental data, we observed that topic blocks are sometimes off-topic. Since persons tend to be absent from the blocks jointly, including the blocks in the dataset would cause the EM method to overestimate the correlation between opposing persons and polarities. To reduce the influence of off-topic blocks, we propose an off-topic block elimination (OBE) procedure. For each topic, we construct a topic representation vector by averaging $\underline{b_i}$'s. Then, the blocks whose cosine similarity [18] to the representation vector is lower than a predefined threshold γ are considered off-topic blocks and excluded from the process of the EM method.

3.3   Weighted Correlation Coefficient

Although OBE eliminates off-topic blocks, the proposed EM method is still affected by the text sparseness problem. Based on the principle of least effort [23], document authors tend to use a small vocabulary of common words to reduce the reading (writing) effort of readers (the authors). Consequently, the frequency distribution of person names follows Zipf's law [19] that is, there are only a few frequent (important) person names, and most person names rarely occur in topic blocks. Consequently, many frequency vectors of topic person names contain a lot of zeros, which affect the calculation of the correlation coefficient. The absence of person names from topic blocks could cause overestimation (or underestimation) of the correlation between persons and polarities. We therefore propose the following weighted correlation coefficient, called $wcorr(.)$, to weight absent blocks:

$$
wcorr(\underline{n}_i, \underline{\omega}_k) =
$$

$$
\frac{\left(\begin{array}{l} (1-\beta) \sum\limits_{b \in co(i,k)} (n_{i,b} - n_i^{\tilde{}}) * (\omega_{k,b} - \omega_k^{\tilde{}}) + \\ \beta \sum\limits_{b \in B - co(i,k)} (n_{i,b} - n_i^{\tilde{}}) * (\omega_{k,b} - \omega_k^{\tilde{}}) \end{array} \right)}{\left(\begin{array}{l} \sqrt{(1-\beta) \sum\limits_{b \in co(i,k)} (n_{i,b} - n_i^{\tilde{}})^2 + \beta \sum\limits_{b \in B - co(i,k)} (n_{i,b} - n_i^{\tilde{}})^2} * \\ \sqrt{(1-\beta) \sum\limits_{b \in co(i,k)} (\omega_{k,b} - \omega_k^{\tilde{}})^2 + \beta \sum\limits_{b \in B - co(i,k)} (\omega_{k,b} - \omega_k^{\tilde{}})^2} \end{array} \right)},
\tag{7}
$$

where $co(i,k)$ denotes the set of blocks whose frequencies in both $\underline{n}_i$ and $\underline{\omega}_k$ are non-zero. In other words, if we treat $\underline{\omega}_k$ as the representative person of polarity k, $co(i,k)$ denotes the set of blocks in which person i and polarity k co-occur. Here, $n_i^{\tilde{}} = 1/L \Sigma_{l=1}^{L} n_{i,l}$ and $\omega_k^{\tilde{}} = 1/L \Sigma_{l=1}^{L} \omega_{k,l}$ are the average frequencies of person i and polarity k respectively. Parameter β, whose range is within [0,1], weights the influence of non-co-occurring blocks when we calculate the correlation coefficient. A large β makes non-co-occurring blocks important for multi-polarization. Like correlation coefficient, the range of $wcorr(.)$ is within [-1,1]. We therefore employ Eq. (3) to avoid negative probabilities when calculating $P(\underline{n}_i | \underline{\omega}_k)$. In the experiment section, we examine the effect of β on person name multi-polarization.

4   Performance Evaluations

In this section, we first introduce the data corpus and the evaluation metric. Then, we examine the effect of the system components. Next, we compare our model-based person name multi-polarization method with well-known clustering algorithms. Finally, we demonstrate multi-polarization results via case studies.

4.1   Data Corpus and Evaluation Metric

In natural language processing, evaluations are normally based on official benchmarks. However, to the best of our knowledge, there are no official corpora for person name multi-polarization because the research subject is relatively new. We therefore compiled a data corpus for evaluations. As shown in Table 1, the corpus comprises eight topics with competing viewpoints about 4 sports tournaments and 4 global business events. All the topic documents were downloaded from Google News[1]. We selected the topics for evaluation because they are all related to global news events, so readers can comprehend the multi-polarization example presented in Sec. 4.4 without specific cultural or background knowledge. To compare our method with Chen and Wu's bipolarization approach, we prepared four bipolar topics (i.e., Topics *A~D*). Topics *E~H* are multi-polar topics, each of which involves four basketball teams competing in a tournament. For each of the eight topics, we extracted all the person names mentioned in the topic documents by using the Stanford Named Entity Recognizer[2]. Given an input text, the Stanford Named Entity Recognizer extracts all possible named entities from the text. The recognizer also tags an extracted entity as a person name, a location name, or an organization name. We used the extracted person names for evaluation. Since there is no perfect named entity recognition approach, the recognizer identified false person name entities, e.g., person name typos. To evaluate the true multi-polarization performance, we removed the false entities comprised of the name of a person and the name of an organization (or a location) because they were ambiguous. We did not remove any typo entities because they refer to specific (unambiguous) persons and retaining them for the evaluations helps us test the robustness of our method. We counted the frequency of each extracted person name. We observe that many of the extracted person names rarely occur in the topic documents and the rank-frequency distribution of person names follows Zipf's law, as shown in Figure 1. The low frequency names are usually persons that are irrelevant to the topic (e.g., journalists), so they are excluded from the evaluation. For the evaluation, we selected the first frequent person names whose accumulated frequencies reached λ percent of the total person name frequency count. In other words, the evaluated person names accounted for λ percent of the person name occurrences in the examined topic. In the following experiments, we assess the system performance under $\lambda = 60\%$. All the evaluated person names represent important topic persons.

---

[1] http://news.google.com
[2] http://nlp.stanford.edu/software/CRF-NER.shtm

Table 1. The statistics of the evaluation corpus

	Date	Topic description # of topic documents	# of extracted persons	# of evaluated persons
A	2010/7/18- 2010/7/22	Smartphone manufacturers deny Apple reception claims 123	74	5
B	2010/8/4- 2010/8/6	Google-Verizon denies tiered-web deal report 74	53	7
C	2010/6/1- 2010/6/3	Prudential's shareholders opposed buying AIG's Asian Unit 154	93	3
D	2010/1/13- 2010/1/15	Google ends four years of censoring the Web for China. 48	103	13
E	2008/5/20- 2008/5/30	The 2008 NBA Conference Finals 119	77	12
F	2009/5/19- 2008/5/30	The 2009 NBA Conference Finals 78	147	17
G	2010/5/16- 2010/5/30	The 2010 NBA Conference Finals 166	162	17
H	2011/5/14- 2011/5/27	The 2011 NBA Conference Finals 292	135	13

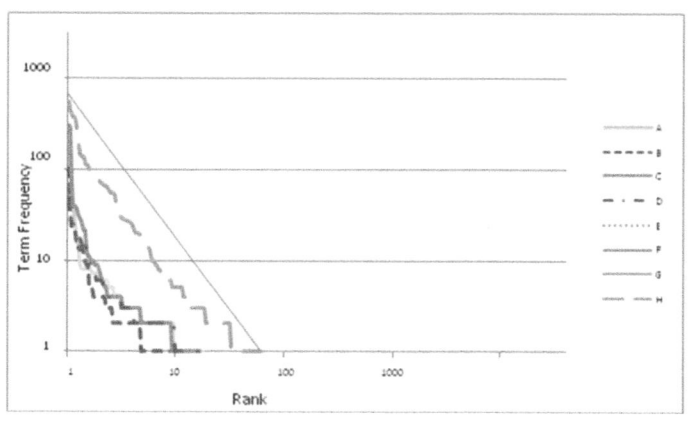

Fig. 1. The rank-frequency distribution of person names

We asked two experts to annotate the person polarity and to establish a reliable ground-truth for evaluations. We utilized the rand index [12], a conventional evaluation metric frequently used to compare clustering algorithms, to evaluate the performance of our multi-polarization method. Specifically, the rand index is based on name pairs. After a set of person names are partitioned into K polarities (clusters), the index measures the percentage of clustering decisions that are correct (e.g., placing a name pair with the same polarity in the same polarity). As large topics generally dominate small topics in micro-averaging [19], we use the macro-averaging to average the multi-polarization rand indices of the evaluated topics in order to compare the global performance. Paragraph tags are not provided in the evaluated topic documents, so a block is a topic document in our evaluation.

4.2   Effect of System Components

In this section, we examine the effect of system components. As the performance of our method depends on EM initialization, we run the method twenty times with random initializations. For each EM experiment run, we examine the performance difference with and without off-topic block elimination. The OBE threshold γ is set at 0.3 as the setting produces superior performances under a validation process. The parameter β is set between 0.1 and 1[3], and increased in increments of 0.1 to examine the effectiveness of the weighted correlation coefficient. To prevent overestimation by our method, we take the average rand indices of the twenty EM runs under various settings, as shown in Figures 2.

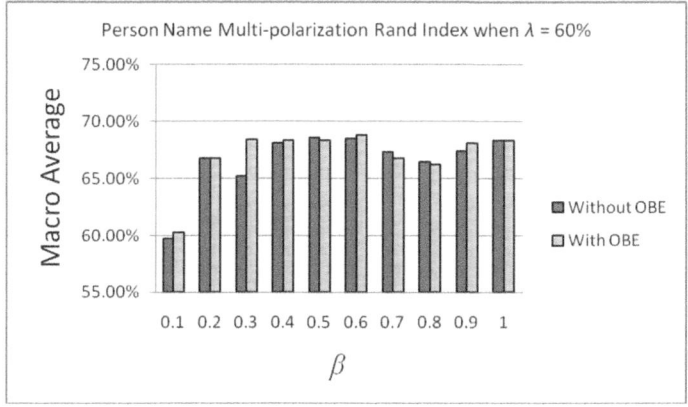

Fig. 2. Person name multi-polarization rand index under $\lambda = 60\%$

Although eliminating off-topic blocks improves the multi-polarization performance overall, the improvement is small under some parameter settings. We observe that the blocks eliminated by OBE only account for 8.6% of the topic content. As OBE just excludes a small portion of the topic content, it only corrects the correlation between persons and polarities slightly. Consequently, the improvement derived by OBE is small. We observe that some of the evaluated person names seldom occur in topic blocks. Since the purpose of the weighted correlation coefficient is to reduce text sparseness, the weighted correlation coefficient is effective in identifying the polarity of infrequent person names. Figure 2 shows that a small β generally yields inferior multi-polarization performances. We observe that many topic blocks, especially in the sports topics, are recaps of polarity competitions, which tend to mention persons of different polarities together. As a small β makes such co-occurring blocks important, the corresponding rand index is inferior. It is noteworthy that a large β produces superior multi-polarization performances. This is because a large β makes non-co-

---

[3] We did not test $\beta = 0$ because person names sometimes have no co-occurring block (i.e., $co(i,k)$ is empty). The sparseness phenomenon leads to a zero denominator in Eq. (7) and makes the weighted correlation coefficient non-calculable.

occurring blocks important for multi-polarization and non-co-occurring blocks between persons and polarities provide useful information in determining polarity relationships. To summarize, by eliminating off-topic blocks, a large β generally produces superior multi-polarization performances. Hence, the proposed off-topic block elimination and weighted correlation coefficient polarize topic person names effectively.

4.3   Comparison with other Methods

The person name multi-polarization task is a special clustering problem that groups persons into polarity-coherent clusters. Here, we compare our method with the following four clustering algorithms: 1) the K-means algorithm [19]; 2) the PLSI algorithm [7]; 3) the HAC algorithm [18]; and 4) Chen and Wu's PCA-based method, which polarizes persons by examining the sign of the entries in the principal eigenvector of PCA. In addition, a naive method, which considers all the person names as a single polarity, serves as a baseline to evaluate the efficiency of the clustering-based multi-polarization approaches. Both the K-means method and the HAC method represent a person as a high-dimensional frequency vector (i.e., $\underline{n}_j$) and employ a cosine similarity measure to group similar person names into clusters. Hofmann treats each latent variable z of PLSI as a concept and groups the terms (or documents) of a text corpus into clusters according to $P(z|w)$ (or $P(z|d)$). In our experiment, a term w is a person name. For our method, the parameter β is set at 0.6 based on the experiment results in Section 4.2. As the clustering performance of PLSI, K-means, and our method depends on cluster initialization, we randomly initialize the algorithms twenty times and examine their best, worst, and average results. For each of the algorithms, we also show the rand indices of the runs (denoted as obj-best and obj-worst) that produce the best and worst objective function values for a fair comparison.

As the PCA-based method cannot polarize multi-polar topics (i.e., Topics A-D), Table 2 compares the multi-polarization rand index and the bipolarization rand index. As shown in the table, the proposed method outperforms the compared algorithms in terms of the rand index. Although the PCA-based method yields comparable rand indices, it cannot deal with multi-polar topics. We observe that the performance of the K-means approach is inferior when popular persons are selected as the centroids of initial clusters. The frequency vector of a popular person usually contains a lot of non-zero entries, which tend to produce a high similarity score because the cosine similarity is the inner product of the normalized vectors. Under K-means, selecting such a person as the centroid of the initial cluster would merge cosine-similar but opposing persons together, and therefore impacts the multi-polarization performance. The inferior performance of the HAC single-link strategy also reflects the shortcomings of the cosine similarity measure. This is because the strategy determines the similarity between two clusters by examining the cosine similarity of the most similar persons in the clusters. Consequently, a high cosine similarity score between popular but opposing persons would result in opposing person groups being merged together. The PLSI algorithm also groups popular but opposing person names together. This is because its objective function tends to compute a high $P(z|w)$ to persons that co-occur frequently in topic blocks. Consequently, popular but opposing persons would be grouped together. By contrast, the proposed method determines person-polarity relationships in terms of the correlation coefficient, which indicates how the occurrences of persons and polarities vary jointly. Therefore, it can identify the relationships between popular persons correctly and achieve a better multi-polarization performance.

Table 2. The multi-polarization results of the compared methods

	Topics $A\sim D$	Topics $E\sim H$	All topics
PAC-based method	55.27%	n.a.	n.a.
Our method (best)	**84.49%**	**83.77%**	**84.13%**
Our method (avg.)	62.11%	75.52%	68.81%
Our method (worst)	38.28%	63.57%	50.92%
Our method (obj-best)	80.64%	78.16%	79.40%
Our method (obj-worst)	38.28%	65.49%	51.88%
K-means (best)	77.35%	77.27%	77.31%
K-means (avg.)	59.54%	64.76%	62.15%
K-means (worst)	31.36%	50.77%	41.06%
K-means (obj-best)	72.86%	70.94%	71.90%
K-means (obj-worst)	42.13%	56.69%	49.41%
PLSI (best)	77.35%	74.40%	75.87%
PLSI (avg.)	54.84%	66.25%	60.54%
PLSI (worst)	42.77%	57.88%	50.32%
PLSI (obj-best)	69.78%	71.17%	70.47%
PLSI (obj-worst)	44.05%	62.39%	53.22%
HAC (single-link)	55.40%	72.00%	63.70%
HAC (complete-link)	62.51%	74.21%	68.36%
HAC (average-link)	63.10%	73.56%	68.33%
HAC (centroid-link)	64.89%	74.21%	69.55%
Baseline	25.07%	19.01%	22.04%

4.4   Multi-polarization Example

Due to space limitations, we only provide a multi-polarization example in Table 3. The table shows the multi-polarization results of Topic E. The first column of the table list the evaluated person names of topic E and the remaining columns list the expectation $E[z_{i,k}]$ generated by our method. A person belongs to the polarity with the maximum expectation. The results show that the proposed method polarizes the important players of the conference finals perfectly. It is noteworthy that, in Topic E, our method correctly polarizes Chauncey Billups as a member of the Detroit Pistons, as shown in Table 3. However, in Topic F, Billups was traded to the Denver Nuggets which also played in the 2009 conference finals. As our method polarizes person names in terms of word usage patterns in topic documents, it captured the polarity dynamics and polarized him correctly as a member of the Denver Nuggets. The examples show that our unsupervised method is context-oriented and can identify polarity dynamics without using any external knowledge source.

Table 3. The multi-polarization results for Topic E ($\lambda = 60\%$, $\beta = 0.6$ with OBE)

	$E[z_{i,Pistons}]$	$E[z_{i,Celtics}]$	$E[z_{i,Spurs}]$	$E[z_{i,Lakers}]$
Chauncey Billups	**0.41**	0.28	0.14	0.17
Richard Hamilton	**0.36**	0.29	0.16	0.19
Antonio McDyess	**0.47**	0.14	0.15	0.24
Kevin Garnett	0.30	**0.36**	0.16	0.18
Paul Pierce	0.32	**0.33**	0.16	0.19
Ray Allen	0.30	**0.39**	0.14	0.17
Tony Allen	0.25	**0.37**	0.18	0.20
Manu Ginobili	0.19	0.17	**0.37**	0.27
Tim Duncan	0.17	0.16	**0.38**	0.29
Kobe Bryant	0.16	0.15	0.33	**0.36**
Lamar Odom	0.17	0.16	0.23	**0.44**
Pau Gasol	0.19	0.17	0.31	**0.33**

5   Conclusions

A topic that involves competing viewpoints usually creates a lot of interest and generates a large number of documents. Identifying the polarities of the important persons mentioned in the topic documents can help readers understand the topic quickly. In this paper, we have defined a topic person name multi-polarization problem and proposed an effective EM method to identify person polarities without using external knowledge sources. The number of topic polarities is pre-defined. In our future work, we will incorporate the number of polarities into a multi-polarization objective function to determine the appropriate number of polarities and polarity members automatically. In addition, effective model initialization techniques will be investigated to produce high quality multi-polarization results.

Acknowledgements. The authors would like to thank the anonymous reviewers for their valuable comments and suggestions. This research was supported in part by NSC 100-2628-E 002-037-MY3 from the National Science Council, Republic of China.

References

1. Chen, C.C., Wu, C.-Y.: Bipolar person name identification of topic documents using principal component analysis. In: Proceedings of the 23rd International Conference on Computational Linguistics, pp. 170–178. Association for Computational Linguistics, Beijing (2010)
2. Chen, C.C., Chen, Z.-Y., Wu, C.-Y.: An Unsupervised Approach for Person Name Bipolarization Using Principal Component Analysis. IEEE Transactions on Knowledge and Data Engineering (to appear, 2012)
3. Ding, X., Liu, B., Yu, P.S.: A holistic lexicon-based approach to opinion mining. In: Proceedings of the International Conference on Web Search and Web Data Mining, pp. 231–240. ACM, Palo Alto (2008)
4. Feng, A., Allan, J.: Finding and linking incidents in news. In: Proceedings of the Sixteenth ACM Conference on Information and Knowledge Management, pp. 821–830. ACM, Lisbon (2007)

5. Ganapathibhotla, M., Liu, B.: Mining opinions in comparative sentences. In: Proceedings of the 22nd International Conference on Computational Linguistics, vol. 1, pp. 241–248. Association for Computational Linguistics, Manchester (2008)
6. Hatzivassiloglou, V., McKeown, K.R.: Predicting the semantic orientation of adjectives. In: Proceedings of the Eighth Conference on European Chapter of the Association for Computational Linguistics, pp. 174–181. Association for Computational Linguistics, Madrid (1997)
7. Hofmann, T.: Probabilistic latent semantic indexing. In: Proceedings of the 22nd Annual International ACM SIGIR Conference on Research and Development in Information Retrieval, pp. 50–57. ACM, Berkeley (1999)
8. Hu, M., Liu, B.: Mining opinion features in customer reviews. In: Proceedings of the 19th National Conference on Artifical Intelligence, pp. 755–760. AAAI Press, San Jose (2004)
9. Kanayama, H., Nasukawa, T.: Fully automatic lexicon expansion for domain-oriented sentiment analysis. In: Proceedings of the 2006 Conference on Empirical Methods in Natural Language Processing, pp. 355–363. Association for Computational Linguistics, Sydney (2006)
10. Kim, S.-M., Hovy, E.: Determining the sentiment of opinions. In: Proceedings of the 20th International Conference on Computational Linguistics, p. 1367. Association for Computational Linguistics, Geneva (2004)
11. Ku, L.W., Liang, Y.T., Chen, H.H.: Opinion extraction, summarization and tracking in news and blog corpora. In: Proceedings of AAAI 2006 Spring Symposium on Computational Approaches to Analyzing Weblogs (2006)
12. Manning, C.D., Raghavan, P., Schütze, H.: Introduction to information retrieval. Cambridge University Press (2008)
13. Mei, Q., Zhai, C.: Discovering evolutionary theme patterns from text: an exploration of temporal text mining. In: Proceedings of the Eleventh ACM SIGKDD International Conference on Knowledge Discovery in Data Mining, pp. 198–207. ACM, Chicago (2005)
14. Miller, G.A., Beckwith, R., Fellbaum, C., Gross, D., Miller, K.J.: Introduction to WordNet: An On-line Lexical Database*. International Journal of Lexicography 3, 235–244 (1990)
15. Mitchell, T.: Machine learning. MacGraw-Hill (1997)
16. Nallapati, R., Feng, A., Peng, F., Allan, J.: Event threading within news topics. In: Proceedings of the Thirteenth ACM International Conference on Information and Knowledge Management, pp. 446–453. ACM, Washington, D.C (2004)
17. Pang, B., Lee, L.: Opinion Mining and Sentiment Analysis. Found. Trends Inf. Retr. 2, 1–135 (2008)
18. Salton, G.: Automatic text processing: the transformation, analysis and retrieval of information by computer (1989)
19. Schütze, H.: Foundations of statistical natural language processing. The MIT Press (1999)
20. Stone, P., Dunphy, D., Smith, M., Ogilvie, D.: The General Inquirer: A Computer Approach to Content Analysis. MIT Press (1966)
21. Turney, P.D., Littman, M.L.: Measuring praise and criticism: Inference of semantic orientation from association. ACM Trans. Inf. Syst. 21, 315–346 (2003)
22. Wu, C.F.J.: On the Convergence Properties of the EM Algorithm. The Annals of Statistics 11, 95–103 (1983)
23. Zipf, G.K.: Human behavior and the principle of least effort: an introduction to human ecology. Addison-Wesley Press (1949)

Using Key Sentence to Improve Sentiment Classification

Zheng Lin[1,2] , Songbo Tan[1], and Xueqi Cheng[1]

[1] Institute of Computing Technology, Chinese Academy of Sciences,
100190 Beijing, China
[2] Graduate School of the Chinese Academy of Sciences,
100049 Beijing, China
{linzheng,cxq}@ict.ac.cn, tansongbo@software.ict.ac.cn

Abstract. When predicting the polarity of a review, not all sentences are equally informative. In this paper, we divide a document into key sentence and trivial sentences. The key sentence expresses the author's overall view while trivial sentences describe the details. To take full advantage of the differences and complementarity between the two kinds of sentences, we incorporate them in supervised and semi-supervised learning respectively. In supervised sentiment classification, a classifier combination approach is adopted; in semi-supervised sentiment classification, a co-training algorithm is proposed. Experiments carried out on eight domains show that our approach performs better than the baseline method and the key sentence extraction is effective.

Keywords: key sentence, sentiment classification, classifier combination, co-training algorithm.

1   Introduction

Sentiment classification [1][2][3][18-20] has gradually become a research hotspot with broad application prospects. A key problem of sentiment analysis is to determine the polarity of a review is positive (thumbs up) or negative (thumbs down). Unlike topic-based text classification [4][21-22], where a high accuracy can be achieved because the topic clusters are typically well-separated from each other, resulting from the fact that word usage differs considerably between two topically-different documents [5]. However, many reviews are sentimentally ambiguous for a variety reasons. Objective statements interleaved with the subjective statements can be confusing for learning methods, and subjective statements with conflicting sentiment further complicate the classification task [6].

One of the main challenges for document-level sentiment classification is that not every part of the document is equally informative for inferring the polarity of the whole document. Thus, we think that making a distinction between key sentence and trivial sentences will be helpful to improve the sentiment classification performance. In this paper, we first propose an approach for key sentence extraction, and then incorporate key sentences in supervised and semi-supervised sentiment classification respectively.

M.V.M. Salem et al. (Eds.): AIRS 2011, LNCS 7097, pp. 422–433, 2011.

For key sentence extraction, our approach takes three attributes into account: sentiment attribute, position attribute and special words attribute. Sentiment attribute discriminates whether a sentence is accompanied by subjective feelings. Position attribute guarantees that sentences at the beginning and end have higher probability than the middle part. Special words attribute increases the weight for those sentences that contain special words such as "overall". Finally, a weighted sum model is used to extract the key sentence from each document.

In supervised learning, a classifier combination approach is adopted. This kind of method requires individual classifier as independent as possible for ensemble methods to work well. In our work, key sentences are usually summative and brief while trivial sentences are descriptive and multifarious, so key sentence and trivial sentences are different and complementary. As a result, key sentence classifier and trivial sentence classifier have different feature space and they provide different knowledge and benefit, that is why classifier combination approach could work well in our approach.

In semi-supervised learning, a co-training algorithm is proposed to incorporate unlabeled data for sentiment classification. The main idea behind this method is: Key sentences and trivial sentences follow different distributions. When an example can be labeled confidently by the key sentence classifier, it may be not easy to be classified by the trivial sentence classifier, so the trivial sentence classifier will get useful information to improve itself and vice versa. It is the diversity of each classifier that makes co-training algorithm applicable.

The rest of this paper is organized as follows: Section 2 introduces the related work of sentiment classification. Section 3 presents the key sentence extraction approach. Section 4 and Section 5 show the supervised and semi-supervised methods on sentiment classification respectively. Experimental results are presented and analyzed in Section 6. Lastly we conclude this paper in Section 7.

2   Related Work

According to granularity, sentiment classification can be conducted on: words, sentences and documents. According to training mode, sentiment classification can be categorized into: supervised, unsupervised and semi-supervised. In this paper we focus on supervised and semi-supervised document sentiment classification.

Supervised methods usually regard the polarity predicting task as a classification task and use the labeled corpus to train a sentiment classifier. Pang et al. [1] conducted early polarity classification of reviews using supervised approaches. They employed Support Vector Machines, Naïve Bayes and Maximum Entropy classifiers using a diverse set of features and concluded that sentiment classification is more difficult than standard topic-based classification. Mullen and Collier [7] used SVM and expanded the feature set for representing documents with favorability measures from a variety of diverse sources. Gamon [8] demonstrated that using large feature vectors in combination with feature reduction, high accuracy can be achieved in the very noisy domain of customer feedback data. Koppel and Schler [9] used neutral reviews to help improve the classification of positive and negative reviews.

McDonald et al. [10] investigated a structured model for jointly classifying the sentiment of text at varying levels of granularity. Li et al. [11] employed two-view model for sentiment classification based on personal/impersonal views.

Semi-supervised methods make use of both labeled and unlabeled data for training, typically a small amount of labeled data with a large amount of unlabeled data [12]. Goldberg and Zhu [13] presented a graph-based algorithm that addresses the rating inference problem in the semi-supervised learning setting, where a closed-form solution to the underlying optimization problem is found through computation on a matrix induced by a graph representing inter-document similarity relationships, and the loss function encodes the desire for similar items to receive similar labels. Li et al. [14] employed lexical prior knowledge for semi-supervised sentiment classification based on non-negative matrix tri-factorization, where the domain-independent prior knowledge was incorporated in conjunction with domain-dependent unlabelled data and a few labeled documents. Dasgupta and Ng [5] firstly mined the unambiguous reviews using spectral techniques, and then exploit them to classify the ambiguous reviews via a novel combination of active learning, transductive learning, and ensemble learning.

3   Key Sentence Extraction

The polarity of a review mainly depends on the author's overall evaluation rather than the details of the specific product or service. We think that a key sentence is the statement that expresses the author's overall attitude or opinion, which is more discriminative than the trivial sentences. Here, we aim to extract the key sentence from the document automatically. There're three attributes to be considered and three functions to be constructed respectively. The final score of each sentence is the weighted sum of three attributes score and sentence with the highest score is considered as key sentence.

Each document d is represented as a sequence of sentences: $d=\{s_1, s_2, \ldots, s_m\}$ where m is the number of sentences and each sentence s_i is represented as a sequence of words: $s_i=\{w_{i1}, w_{i2}, \ldots, w_{in}\}$ where n is the number of the words.

3.1   Sentiment Attribute

The opinion and preference of the product reviews are usually reflected by the opinion words, thus we take opinion words into account and introduce sentiment attribute. Here, sentiment attribute measures the sentiment importance of a sentence and its function is defined as:

$$f\_sentiment(s_i) = \frac{|\sum_{j=1}^{n} positive(w_{ij}) - \sum_{j=1}^{n} negative(w_{ij})|}{n} \tag{1}$$

where positive(w_{ij}) and negative(w_{ij}) are defined as :

$$positive(w_{ij}) = \begin{cases} 1 & w_{ij} \in positive \\ 0 & w_{ij} \notin positive \end{cases} \tag{2}$$

$$negative(w_{ij}) = \begin{cases} 1 & w_{ij} \in negative \\ 0 & w_{ij} \notin negative \end{cases} \tag{3}$$

Here, a sentiment lexicon (http://www.keenage.com/ html/e_index.html) is used to identify whether a word is positive or negative. Denominator n of $f\_sentimen\ (s_i)$ is set to avoid extracting overlong sentence. Numerator of $f\_sentiment(s_i)$ in the form of absolute value is to solve the sentiment ambiguity when a sentence has opposite polarity words. For example, "The computer looks clumsy but works well", the word "clumsy" and "well" are in the same sentence but have opposite polarity which may confuse the polarity predicting. Therefore, only the unipolar sentiment sentence is regarded as key sentence. Overall, function of sentiment attribute reveals characteristics of the key sentence from two aspects. On one hand, a key sentence is accompanied with emotion and preference and the more the better. On the other hand, the emotion and preference of a key sentence is unipolar and confident.

3.2   Position Attribute

Two psycholinguistic and psychophysical experiments showed that in order to efficiently extract polarity of written texts such as customer reviews on the Internet, one should concentrate computational efforts on messages in the final position of the text [15]. However, the first sentence of the text usually comes straight to the point, so we think that both beginning and ending sentences have higher probability to be a key sentence than the middle part. Function of position attribute is defined as:

$$f\_position(s_i) = \alpha * pos(s_i)^2 + \beta * pos(s_i) + c \tag{4}$$

subject to

$$\frac{-\beta}{2\alpha} = \frac{m}{2}$$

$$pos(s_i) = i$$

$$\alpha > 0 \; ; \; \beta > 0 \; ;$$

where m is the number of sentences in document d; $pos(s_i)$ is an integer indicating the position of s_i in d; α, β and c are polynomial coefficients. As a matter of fact, $f\_position(s_i)$ is a parabola concave up and its peak represents the position of middle sentence whose score is the lowest. According to function of position attribute, sentences near the beginning and end will have higher score than the middle part.

3.3   Special Words Attribute

Some conclusive words such as "overall" often appeared in a key sentence, which offers good heuristic information to key sentence extraction. In this paper, we collected ten common special words that usually occur in key sentences. The special words are collected as follows: Firstly, we suppose that the last sentence of a review is key sentence. Secondly, we compute the frequency of the front unit (segmented by comma) of all pseudo key sentences. Finally, we regard the units with high frequency as special words.

Function of keyword attribute is defined as:

$$f\_keyword(s_i) = \sum_{j=1}^{n} keyword(w_{ij}) \tag{5}$$

$$keyword(w_{ij}) = \begin{cases} 1 & w_{ij} \in keywords \\ 0 & w_{ij} \notin keywords \end{cases} \tag{6}$$

4   Classifier Combination with Key Sentences

After key sentence extraction, the training data is divided into two parts: key sentences and trivial sentences. Key sentences and trivial sentences follow different word distributions. The feature space of key sentences is usually smaller than that of trivial sentences because key sentence is just one sentence while trivial sentences consist of many sentences. Besides, key sentence is usually summative and trivial sentences are descriptive, so expressions in key sentences are less diverse but more discriminative than trivial sentences.

In supervised sentiment classification, we adopt classifier combination method. Firstly, three base classifiers are trained: f1 and f2 are trained on datasets of key sentences and trivial sentences; f3 is trained on the whole training data. Each base classifier outputs not only the class label but also some kinds of confidence measurements, e.g. posterior probabilities of the testing sample belonging to each class. Secondly, the class label of a testing sample is assigned by the combination of f1, f2 and f3. There're many combining methods, and we choose the simple sum rule as combining rule without loss of generality.

The chosen sum rule combines base classifiers by adding the posterior possibilities together and using the sum possibility for decision. Given a testing example d, the final class label cj is assigned by Equation 7.

$$j = \arg\max \sum_{i=1}^{3} p(c_j \mid d) \tag{7}$$

5   Co-training with Key Sentences

Classifier learned on key sentences and classifier learned on trivial sentences provide different knowledge and benefits when predicting the polarity of a document. If the document has a key sentence, the key sentence classifier will be more confident about its decision. Furthermore, key sentence classifier is appropriate for ambiguity existing case because it can ignore the complex content and focus on limited features. However, not every document has a key sentence, so trivial sentence classifier is still necessary.

As mentioned before, key sentence classifier and trivial sentence classifier are from different views and sometimes complementary. If one classifier can confidently predict the class of an example, which is very similar to some labeled ones, it can provide one more training example for the other classifier. But, of course, if this example happens to be easy to be classified by the first classifier, it does not mean that this example will be easy to be classified by the second classifier, so the second classifier will get useful information to improve itself and vice versa.

In semi-supervised sentiment classification, a co-training algorithm is adopted. There're three views in our co-training algorithm: key sentence view, trivial sentence view and full-text view. The algorithm of co-training is described as follows:

Given:

- F_{key}, $F_{trivial}$, F_{full} are sets of features, where F_{key} represents the key sentence features, $F_{trivial}$ represents the trivial sentence features, F_{full} represents the full-text features;
- L is a set of labeled training reviews;
- U is a set of unlabeled reviews;

Loop for I iterations:

(1) Learn the first classifier f_{key} from L based on F_{key};
(2) Use f_{key} to label reviews from U based on F_{key};
(3) Choose n_1 positive and n_1 negative the most confidently predicted reviews S_{key} from U;
(4) Learn the second classifier $f_{trivial}$ from L based on $F_{trivial}$;
(5) Use $f_{trivial}$ to label reviews from U based on $F_{trivial}$;
(6) Choose n_2 positive and n_2 negative the most confidently predicted reviews $S_{trivial}$ from U;
(7) Learn the third classifier f_{full} from L based on F_{full};
(8) Use f_{full} to label reviews from U based on F_{full};
(9) Choose n_3 positive and n_3 negative the most confidently predicted reviews S_{full} from U;
(10) Removes reviews $S_{key} \square S_{trivial} \square S_{full}$ from U;
(11) Add reviews $S_{key} \square S_{trivial} \square S_{full}$ with the corresponding labels to L;

---

6   Evaluation

6.1   Experimental Setup

To validate the effectiveness and robustness of proposed method, we conduct experiments on product reviews of eight different domains. The product reviews on the first four domains (book, DVD, electronic, and kitchen appliances) are collected from http://www.amazon.com/ by Blitzer et al. [16]. The product reviews on the other four domains (network, software, pet, health) are collected from http://www.amazon.com/ by Li et al. [11]. Each of the eight domains contains 1000 positive and 1000 negative reviews. In the experiments, we choose Naïve Bayesian as base classifier and use all words as features without reduction and selection. In supervised sentiment classification, the baseline classifier is trained on full-text. Besides, if a document has only one sentence, it is regarded as key sentence.

In supervised sentiment classification, we choose 50% labeled data as training data and the rest 50% labeled for testing for each domain. The classifier combination approach is compared with the baseline classifier which is trained on full-text.

In semi-supervised sentiment classification, we choose 10% labeled data as training set , 20% labeled data as testing set and 70% data as unlabeled set for each domain. The co-training approach is compared with the following baseline methods:

(1)Self-learning: uses the unlabeled data in a bootstrapping way and only the baseline classifier is used to select most confident unlabeled samples in each iteration.

(2)Transductive SVM: seeks the largest separation between labeled and unlabeled data through regularization [17].

(3)Co-training with full-text classifier: samples are selected by three base classifiers and labeled by full-text classifier in each iteration.

(4)Co-training with combined classifier based on random key sentence: samples are selected by three base classifiers and labeled by combined classifier based on random key sentence.

6.2   Experimental Results

Table 1 presents ten random examples of key sentences extracted from book domain by our method, and the original texts with bold key sentences are shown in appendix. The first five are extracted from positive reviews and the last five are extracted from negative reviews. The evaluation of key sentence is too subjective, thus we don't test the accuracy of extracted key sentences. Seen from the examples in appendix, all key sentences are appropriate and discriminative, which justifies the effectiveness of our key sentences extraction method.

Table 1. Examples of extracted key sentences

Polarity	Key Sentence
Positive	I would strongly recommend this book.
Positive	I really like the book and I think you will too.

Table 1. (*continued*)

Positive	Overall, the book is a captive read and strongly recommended.
Positive	Do yourself a favor and buy this book.
Positive	I love this book.
Negative	I found this book very boring.
Negative	This book waste my life.
Negative	In short, please give this book a pass.
Negative	I was really disappointed with this book.
Negative	Overall, I thought this was mediocre and a bit of a waste of time.

To make use of the discrepancy and complementarity of key sentences and trivial sentences, we incorporate the key sentences both in supervised sentiment classification and semi-supervised classification. In the experiments, we first compare the proposed classifier combination approach with the baseline method. Table 2 shows the comparison results in supervised sentiment classification.

Table 2. Comparison results in supervised sentiment classification

Domain	Key Sentences Classifier	Trivial Sentences Classifier	Full-text Classifier (baseline)	Combined Classifier with key sentences
Book	0.699	0.691	0.714	0.742
DVD	0.727	0.729	0.74	0.773
Electronic	0.766	0.741	0.762	0.780
Kitchen	0.778	0.792	0.815	0.837
Network	0.812	0.684	0.739	0.788
Software	0.640	0.645	0.686	0.715
Pet	0.640	0.623	0.636	0.650
Health	0.659	0.547	0.568	0.602
Average	0.7151	0.6815	0.7075	0.7359

From Table 2, we can see that combined classifier based on the extracted key sentences consistently outperforms the baseline across eight domains with an average performance improvement of 2.84%, which justifies that using key sentences can improve performance of supervised sentiment classification. Moreover, before classifier combination, single classifier trained on key sentences outperforms the baseline (classifier trained on full-text) on average, not to mention that the scale of key sentences is much smaller than that of full-text. In some case, key sentence classifier is more confident about its decision because of its compact features distribution, such as in "Network" domain and in "Health" domain, the classifier trained on key sentences even outperforms combined classifier.

In the experiments of semi-supervised sentiment classification, we set the classifier trained on initial 10% labeled data as baseline and conduct four bootstrapping like methods to compare with our co-training approach. Table 3 shows the comparison results in semi-supervised sentiment classification after 100 iterations. In each iteration, two top-confident samples are chosen in each category, i.e. n1=n2=n3=2.

Table 3. Comparison results in semi-supervised sentiment classification

Domain	Baseline	Self-learning	Co-training with full-text classifier	Co-training with random view	Transductive SVM	Co-training with combined classifier
Book	0.66	0.6875	0.6975	0.6875	0.6575	0.7225
DVD	0.575	0.7075	0.665	0.7025	0.6525	0.6975
Electronic	0.6875	0.685	0.7275	0.6975	0.6825	0.735
Kitchen	0.7025	0.775	0.7825	0.790	0.750	0.7975
Network	0.6825	0.7575	0.7475	0.750	0.755	0.7775
Software	0.6575	0.660	0.695	0.6825	0.7475	0.7075
Pet	0.465	0.5875	0.5675	0.5775	0.500	0.605
Health	0.4125	0.405	0.520	0.4775	0.410	0.5525
Average	0.6053	0.6581	0.6753	0.6706	0.6444	0.6994

Seen from Table 3, co-training with combined classifier based on key sentences significantly outperforms the baseline with 9.41% accuracy improvement across all eight domains on average. On "Pet" domain and "Health" domain, the accuracy even improves 14%.

Co-training with random views performs worse than co-training with full-text classifier, which proves that the impressive improvements are mainly due to the key sentences extraction rather than the combination strategy again. Of all the five semi-supervised learning methods, co-training with combined classifier based on extracted key sentences performs best.

The main reason for the effectiveness of our approach in co-training algorithm is that we don't threat sentences of a document equally. Key sentences and trivial sentences have different feature space and different ways of expressing feelings and opinions. Key sentence classifier can confidently predict the class of an example, which may be not easy to be classified by the trivial sentence classifier, so the trivial sentence classifier will get useful information to improve itself and vice versa.

7   Conclusion

In this paper, we propose an approach for key sentence extraction, which takes three attributes into account: sentiment attribute, position attribute and special words attribute. After key sentences extraction, we incorporate key sentences and trivial sentences in supervised and semi-supervised sentiment classification. In supervised learning, a classifier combination approach is adopted to take advantage of the discrepancy of key sentences and trivial sentences. In semi-supervised learning, a co-training algorithm is proposed to make each classifier learn from each other. The experiments carried out on eight domains show our classifier combination approach and co-training algorithm significantly improve the performance.

Acknowledgments. This work was mainly supported by two funds, i.e., 60933005 & 60803085, and one another project, i.e., 2012CB316303.

References

1. Pang, B., Lee, L., Vaithyanathan, S.: Thumbs up? Sentiment Classification using Machine Learning Techniques. In: Proceedings of EMNLP (2002)
2. Turney, P.: Thumbs up or thumbs down? Semantic orientation applied to unsupervised classification of reviews. In: Proceedings of ACL (2002)
3. Tan, S., Zhang, J.: An empirical study of sentiment analysis for chinese documents. Expert Systems with Applications 34(4), 2622–2629 (2008)
4. Tan, S., Cheng, X., et al.: A novel refinement approach for text categorization. In: Proceedings of the ACM CIKM (2005)
5. Dasgupta, S., Ng, V.: Mine the Easy and Classify the Hard: Experiments with Automatic Sentiment Classification. In: Proceedings of ACL (2009)
6. Yessenalina, A., Yue, Y., Cardie, C.: Multi-level Structured Models for Document Sentiment Classification. In: Proceedings of EMNLP (2010)
7. Mullen, T., Collier, N.: Sentiment analysis using support vector machines with diverse information sources. In: Proceedings of EMNLP (2004)
8. Michael, G.: Sentiment classification on customer feedback data: noisy data, large feature vectors, and the role of linguistic analysis. In: International Conference on Computational Linguistics (2004)
9. Koppel, M., Schler, J.: The importance of neutral examples for learning sentiment. Computational Intelligence 22(2), 100–109 (2006)
10. McDonald, R., Hannan, K., Neylon, T., Wells, M., Reynar, J.: Structured Models for Fine-to-coarse Sentiment Analysis. In: Proceedings of ACL (2007)
11. Shoushan, L., et al.: Employing Personal/Impersonal Views in Supervised and Semi-supervised Sentiment Classification. In: Proceedings of ACL (2010)
12. Zhu, X.: Semi-supervised learning literature survey. University of Wisconsin, Madison (2007)
13. Goldberg, A.B., Zhu, X.: Seeing stars when there aren't many stars: graph-based semi-supervised learning for sentiment categorization. In: Proceedings of TextGraphs: the First Workshop on Graph Based Methods for Natural Language (2006)
14. Li, T., Zhang, Y., et al.: A Non-negative Matrix Tri-factorization Approach to Sentiment Classification with Lexical Prior Knowledge. In: Proceedings of ACL (2009)
15. Becker, I., Aharonson, V.: Last but Definitely not Least: On the Role of the Last Sentence in Automatic Polarity-Classification. In: Proceedings of ACL (2010)
16. Blitzer, J., Dredze, M., Pereira, F.: Biographies, bollywood, boom-boxes and blenders: domain adaptation for sentiment classification. In: Proceedings of ACL (2007)
17. Joachims, T.: Transductive Inference for Text Classification using Support Vector Machines. In: ICML (1999)
18. Tan, S., Cheng, X., Wang, Y., Xu, H.: Adapting Naive Bayes to Domain Adaptation for Sentiment Analysis. In: Boughanem, M., Berrut, C., Mothe, J., Soule-Dupuy, C. (eds.) ECIR 2009. LNCS, vol. 5478, pp. 337–349. Springer, Heidelberg (2009)
19. Tan, S., Wang, Y., Wu, G., Cheng, X.: A novel scheme for domain-transfer problem in the context of sentiment analysis. In: CIKM 2007 (2007)
20. Tang, H., Tan, S., Cheng, X.: A survey on sentiment detection of reviews. Expert Syst. Appl. 36(7), 10760–10773 (2009)

21. Tan, S.: An effective refinement strategy for KNN text classifier. Expert Syst. Appl. 30(2), 290–298 (2006)
22. Tan, S., Wang, Y., Wu, G.: Adapting centroid classifier for document categorization. Expert Syst. Appl. 38(8), 10264–10273 (2011)

Appendix: Examples of Original Text and Key Sentence

Sphere by Michael Crichton is an excellant novel. This was certainly the hardest to put down of all of the Crichton novels that I have read. The story revolves around a man named Norman Johnson. Johnson is a phycologist. He travels with 4 other civilans to a remote location in the Pacific Ocean to help the Navy in a top secret misssion. They quickly learn that under the ocean is a half mile long spaceship. The civilans travel to a center 1000 feet under the ocean to live while researching the spacecraft. They are joined by 5 Navy personel to help them run operations. However on the surface a typhoon comes and the support ships on the surface must leave. The team of ten is stuck 1000 feet under the surface of the ocean. After a day under the sea they find out that the spacecraft is actually an American ship that has explored black holes and has brought back some strange things back to earth. This novel does not have the research that some of the other Crichton novels have, but it still has a lot of information on random things from the lawes of partial pressure to behavior analysis. **I would strongly recommend this book.**
I think you would like this book if you read it. It was a little bit wordy at the beginnig but each page gets better than the one before. You will be glad you got past the first few pages because it gets a lot more interesting as you go along. And the ending of this book is really something else, it makes the whole book well worth reading. **I really liked the book and I think you will too**
This is a simple, yet heartfelt and elegant tale of the lives and struggles of women in rural China of the 1920s and 1930s. The story centers on Pei, who as a young girl, is sold by her parents to work in a silk factory. The story evolves as Pei befriends other women in the silk factory, participates in a strike for fair working conditions, reconnects with her family, loses loved ones, and escapes from advancing Japanese soldiers. The plot is succinct and the characters are interesting, varied, and believable, if perhaps lacking somewhat in psychological depth. Socio-cultural aspects of life in 1920s-30s China are elegantly woven into the plot. Themes of friendship, love, and courage are convincingly presented. **Overall, the book is a captivating read, and strongly recommended.**
This is the best Rich Dad book to date. The book is filled with priceless information that 90% of people in the U.S.(particularly baby-boomers) are clueless about today. Most Rich Dad fans will know that Kiyosaki's books tend to talk about the same general principles and don't touch on many new and groundbreaking concepts. This book, however, is the exception. I learned so many eye-opening things from this masterpiece. It has truly changed my life and the way I look at money. **Do yourself a favor and BUY THIS BOOK.**
Wow..I LOVE this book! The recipes are creative and the presentations beautiful. I love the simplicity of each recipe, which give the warmth and richness of traditional old fashioned taste with a twist of the class & sophistication of today's refined culinary experience! Sara's food philosophy and the atmosphere she has created in her market has been my dream. I'm living it through her cookbook right now
I found this book very boring. Tyler does a good job on detailing the charictors but I kept waiting for something to happen. Poor Maryam never developed into a person she was so caught up in trying to be a proper American. I will read another of her books since she is so well thought of

THis book was horrible. If it was possible to rate it lower than one star i would have. I am an avid reader and picked this book up after my mom had gotten it from a friend. I read half of it, suffering from a headache the entire time, and then got to the part about the relationship the 13 year old boy had with a 33 year old man and i lit this book on fire. One less copy in the world...don't waste your money. I wish i had the time spent reading this book back so i could use it for better purposes. **THis book wasted my life**

Simple arithmetic and you know it's stupid to expect much from a 227 content book that attempts to tell the success stories and forumlas of 55 business leaders. Definitely everyone of them deserves its own biography instead of an average 4.13 pages. So I had lowered my standard before I read it. Still I had been quite disappointed. All passages were columns published previously on IBD written by different reporters that not only the writing style but the focus on individual leaders fluctuated much between hard data/history and success formula, primarily on the former. I am sorry that I could gain little knowledge/insight (that I really wanted to learn "how" they succeeded) during the reading. I felt even worse after reading from the previous reviewer that the content had been available on the net. **In short, please give this book a pass!**

I was really disappointed with this book.   A lot of the recipes are identical to one another, with just one ingredient - eg a different type of vegetable - substituted. With no pictures and not much more content, I really regret buying it.

I agree with one the comments posted below. The main problem with this book is Mitch is a totally one-dimensional character and it's impossible to find any depth in him. Over the course of the book he doesn't change, he's unsympathetic, and by the time we've reached page 500, we don't know anything more about him than we did on page 10. I found myself dying to get to the end of this book, just so I could start another book with some substance and some well-drawn characters.   Another major problem is that most of the characters in this book use the same wise-cracking speaking style so there is no sense of the characters being different from each other in any way.   **Overall, I thought this was mediocre and a bit of a waste of time**

Using Concept-Level Random Walk Model and Global Inference Algorithm for Answer Summarization

Xiaoying Liu[1,2], Zhoujun Li[1,3,4], Xiaojian Zhao[1,3], and Zhenggan Zhou[2]

[1] State Key Laboratory of Software Development Environment, Beihang University,
Beijing 100191, China
[2] School of Mechanical Engineering and Automation, Beihang University,
Beijing 100191, China
[3] School of Computer Science and Engineering, Beihang University,
Beijing 100191, China
[4] Beijing Key Laboratory of Network Technology, Beihang University,
Beijing 100191, China
{liuxiaoying,lizj}@buaa.edu.cn, zhaoxj01@gmail.com,
zzhenggan@buaa.edu.cn

Abstract. Community Question Answer (cQA) archives contain rich sources of knowledge on extensive topics, in which the quality of the submitted answer is uneven, ranging from excellent detailed answers to completely unrelated content. We propose a framework to generate complete, relevant, and trustful answer summaries. The framework discusses answer summarization in terms of maximum coverage problem with knapsack constraint on conceptual level. Global inference algorithm is employed to extract sentences according to the saliency scores of concepts. The saliency score of each concept is assigned through a two-layer graph-based random walk model incorporating the user social features and text content from answers. The experiments are implemented on a data set from Yahoo! Answer. The results show that our method generates satisfying summaries and is superior to the state-of-the-art approaches in performance.

1 Introduction

Community Question Answer (cQA) archives are successful instances of Social Media, in which users pose and reply questions in natural languages as well as evaluate and select the best answers, e.g., Naver (www.naver.com), Yahoo! Answer (answers.yahoo.com), Baidu Zhidao (zhidao.baidu. com) and AnswerBag (answerbag.com). The user-generated question-answer pairs in cQA websites have inherent advantages: humongous amounts, rapid growth rate, wide range of subject matter, and multiple languages support [2]. Nevertheless, the quality of the content is with high variance, ranging from excellent detailed answers to completely irrelevant or commercial advertisements, and even abusive language. This reduces the efficiency and user satisfaction of question-answer-pair reuse, if answers from similar questions are used to tailor users' information needs.

M.V.M. Salem et al. (Eds.): AIRS 2011, LNCS 7097, pp. 434–445, 2011.

The approach to resolve the uneven quality in answer content is to composite multiple answers from a question and generate a complete, succinct summary. The task approximates to the query-biased multi-document summarization task, in which questions are mapped into queries and answers into documents.

We present a framework for answer summarization, which employs the graph-based random walk model to weight each concept in answers and extract a set of sentences which maximize the value of the objective function by exploiting a global inference algorithm based on dynamic programming. The summaries with the maximal values are chosen as the final answers.

We focus on several essential questions to fulfill our objective:

(1) What granularity of concepts distilled from answers is suitable in the graph-based random walk model for reinforcing diversity and relevance of summaries?

Generally, sentences are chosen as nodes to construct the graphs in most of literatures about document or answer summarization [8,1]. Under limitation of the amount of words in summaries, these reduplicative information between sentences might subduct diversity of summaries. In this paper, we use concepts with a finer granularity than that of sentences to represent nodes in the graph-based random walk model. Concept-level representation helps to recognize and decrease redundancy in sentences when abstracting sentences from sets of answers. In addition, measuring the similarity between concepts is more accurate and easier than between sentences as fewer words are included in concepts.

We employ a topic-sensitive model to improve the relevance of generated summaries. The model assigns each concept a saliency score according to its relevance to questions as well as its similarity with other concepts in answers. Hence concepts related with questions are weighed relatively high saliency scores.

(2) What roles does the authority of users play in enhancing trustworthiness of answer summaries?

In most cases, the answers responded by more authoritative users are more trustworthy. Hence, the concepts from authoritative users should have greater saliency scores than those from other users. For such purpose, a two-layer link graph is constructed to redistribute concept saliency scores through incorporating user social features and text content from answers.

(3) How to improve the coverage and conciseness of answer summaries through global inference algorithms?

With concepts weighted by the saliency scores from graph-based random walk model, summary generation is cast to the problem of finding the maximum coverage with knapsack constraint on all answers. This problem can be resolved by exploiting a global inference algorithm. The objective function of the algorithm is the key of boosting the coverage and conciseness of answer summaries. A good objective function should reduce redundancy and retain relevant information in summaries as much as possible.

The remainder of this paper is organized as follows: Section 2 describes how to employ graph-based random walk model to assign the saliency scores of concepts according to the information of text contents and user social features in detail.

Section 3 illustrates how to apply the global inference algorithm to generate the complete and succinct summaries. Section 4 exhibits the conduct of several experiments and discusses the results. Section 5 contains an overview about the document and answer summarization. Finally, conclusions and orientations for future work are presented.

2   Ranking Concepts Using Graph-Based Random Walk Model

2.1   Graph-Based Random Walk Model on Text Content

A graph-based random walk model is employed to rank the answer content through assigning a saliency score to each nodes in the graph. In this paper, we suppose each answer consist of many concepts and represent each node in graphs by a concept from answers.

Given a question q and a set of concepts from its answers $C=\{c_1, c_2, \ldots, c_n\}$, we first define a graph $G=<V, E>$ as an undirected graph, where V is a set of vertices representing n concepts, and E is a set of edges representing the similarity between vertices, a subset of $V \times V$. The graph G is regarded as $n \times n$ weighted matrix A, where a_{ij} is a similarity score from the node i and j. Then, we normalize the matrix A into a similarity matrix S such that each element s_{ij} in S is equal to $\frac{a_{ij}}{\sum_{k=1}^{n} a_{ik}}$ and all rows in S sum to 1.

According to a given question q, a relevant vector B is constructed, where b_i is defined as the maximum relevant score between the concept c_i from answers and the concepts from the question. We derive an normalized vector D from B such that $d_i = \frac{b_i}{\sum_{k=1}^{n} b_k}$ and the sum of all values in d is 1. Then, we transform D into a square relevant matrix R, where each element in the ith column is assigned to d_i.

We define a transition matrix M as follows:

$$M = [d \cdot R + (1 - d) \cdot S] \tag{1}$$

where d is a damping factor with a real value of $[0, 1]$. Since all rows in the matrix M have non-zero values and sum to 1, the transition matrix M is a stochastic matrix. Each element m_{ij} in M means the transition probability from i to j in Markov chain. Thus, M has a unique stationary distribution $\mathbf{p}=M^T\mathbf{p}$. The stationary distribution $\mathbf{p}$ can be used to rank the concepts in answers. We write the model (denoted as Rank-1) into the equation (2) in matrix notation.

$$\mathbf{p} = [d \cdot R + (1 - d) \cdot S]^T \cdot \mathbf{p} \tag{2}$$

The simplified equation of the model Rank-1 is rewritten as follows:

$$p(c_i|q) = d \cdot \frac{rel(c_i|q)}{\sum_{c_j \in C} rel(c_j|q)} + (1 - d) \cdot \sum_{c_k \in C} \frac{sim(c_i, c_k)}{\sum_{c_j \in C} sim(c_i, c_j)} \cdot p(c_k|q) \tag{3}$$

where C is the set of all concepts in answers, and d is a damping factor with a real value of $[0, 1]$ indicating the "question bias". The topic-sensitive saliency score $p(c_i|q)$ of a concept c_i is determined as the sum of the concept's relevance to the question q and the similarity with other concepts in the answer set. The damping factor d is used to control which parts is more important in answer summaries: relevance to the question or the similarity with other concepts. $sim(c_i, c_j)$ is the function to measure the similarity in c_i and c_j. We adopt a similarity measure proposed in [12] to calculate the concept similarity. $rel(c_i|q)$ is the function for calculating the relevance between c_i and q. The relevance of a concept c_i is defined as the maximum similarity between c_i and the concepts from questions.

For improving the diversity and generating compendious summaries, we employ concepts with finer granularity than sentences to represent the nodes in graphs. The concept representations are listed as following:

Phrase: A phrase is a non-overlapping span in a sentence and can be obtained through partitioning a sentence based on the syntactic structure. We employ a natural language processing tool SST [4] to obtain phrases in sentences.

N-gram: N-gram is a subsequence containing the continuous n words in a sentence. In this paper, unigram, bigram and trigram are chosen as the graph nodes, respectively.

Phrase and n-gram are two kinds of common approaches of partitioning sentences and widely used in natural language processing. Both phrase and n-gram contain fewer words than sentences, which contributes to accurately reflect the relationship of nodes and identify redundancy information. Noted that stop words, such as articles, pronouns, prepositions, and conjunctions, are removed before calculating the similarity.

2.2   Two-Layer Link Graph for User Social Features

In cQA websites, the authoritative users tend to provide answers with high quality. Hence, user social features can facilitate the generation of a reliable summary. A two-layer link graph is utilized to incorporate the text content and user social features in a unified framework. The two-layer graph-based random walk model is shown in Figure 1. The first layer denotes the concept relationship in the topic-sensitive random walk model. The second layer represents all users which reply questions. The connection between two layers indicates the influence for ranking concepts under condition of user authority.

Given a question q, a set of concepts from its answers $C=\{c_1, c_2, \ldots, c_n\}$, and the set of the users $U=\{u_1, u_2, \ldots, u_m\}$, in which u_i is the corresponding author of the answers of q, the two layer graph is denoted as $G^*=<C, U, E_{c-c}, E_{u-c}>$. E_{c-c} represents the set of the edges between concepts, and E_{u-c} denotes the set of edges between concepts and users. If the user u_i is the author of the answer including the concept c_j, there is an edge e_{ij} in E_{u-c}, otherwise no connection.

Taking user authority into account, the similarity matrix S is rewritten into a new matrix S^*, where each element $s_{ij}^* = \frac{sim(c_i, c_j | u_i, u_j)}{\sum_{c_k \in C} sim(c_i, c_k | u_i, u_k)} \cdot sim(c_i, c_j | u_i, u_j)$

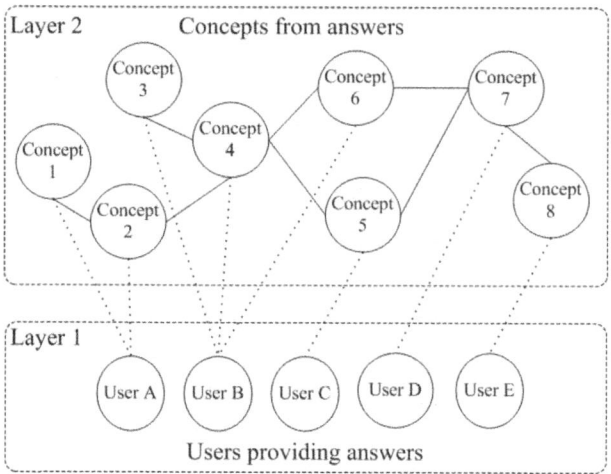

Fig. 1. Two-layer link graph

is a new similarity measure under the condition of u_i and u_j and is calculated according to the following formula:

$$sim(c_i, c_j | u_i, u_j) = sim(c_i, c_j | u_i) + sim(c_i, c_j | u_j)$$
$$= sim(c_i, c_j) \cdot auth(u_i) + sim(c_i, c_j) \cdot auth(u_j) \qquad (4)$$

where u_i and u_j are the authors of the original answers including c_i and c_j, respectively. $auth(u_i)$ denotes the user authority in community. $auth(u_i)$ is calculated through a HITS-based method from [6], which regarded users who pose questions as "hubs" and users who provide answers as "authorities". An algorithm based on HITS is employed to compute the hub and authority value of users on the link graph according to three relationship: user-question, question-answer and answer-user. The hub and authority values of users are calculated by the following equation:

$$H(i) = \sum_{j \in U_A} A(j)$$
$$A(j) = \sum_{i \in U_Q} H(i) \qquad (5)$$

where the set U_A contains all the users who provide answers, and U_Q includes all the users who pose questions. $H(i)$ is the hub value of the user i and $A(i)$ is its authority value.

The final saliency score for the two-layer graph-based random walk model is denoted by:

$$p(c_i|q) = d \cdot \frac{rel(c_i|q)}{\sum_{c_j \in C} rel(c_j|q)} + (1-d) \cdot \sum_{c_k \in C} \frac{sim(c_i, c_k | u_i, u_k)}{\sum_{c_j \in C} sim(c_i, c_j | u_i, u_j)} \cdot p(c_k|q) \quad (6)$$

The matrix form of the model is: $\mathbf{p} = [d \cdot R + (1-d) \cdot S^*]^T \cdot \mathbf{p}$. The model is denoted as Rank-2.

The model Rank-2 can provide more reliable answers than the model Rank-1. This is because the model Rank-2 tends to extract sentences from authoritative user into answer summaries compared with the model Rank-1 through assigning the concepts from answers of authoritative user a higher saliency score than the concepts from other users.

3   Global Inference Algorithm for Answer Summarization

After a sentence is represented as a set of concepts with saliency scores deduced based on a graph-based random walk model, the aim of answer summarization is transformed to maximize the covered concepts and minimize the number of sentences included in the summary. Namely, we require to find a subset of sentences which satisfy two conditions: the length (the number of words or bytes) of summaries must be at most L (cardinality constraint), and summaries should cover as many concepts as possible. The first condition guarantees the conciseness of generated summaries, and the second one ensures the diversity and coverage. Hence, the answer summarization is mapped into a maximum coverage problem with knapsack constraint (MCKP). MCKP is an NP-hard problem [7]. For alleviating this problem, a global inference algorithm based on dynamic programming is employed.

We set the following objective function (denoted as Obj-1) to maximize the sum of the weights of concepts included in the generated summary:

$$\text{Maximize:} \quad \sum_i w_i \cdot x_i$$

$$\text{Subject to:} \quad \sum_j l_j \cdot y_j \leq L$$

$$\sum_j occ_{ij} \cdot y_j \geq x_i \qquad \forall i$$

$$x_i, y_j, occ_{ij} \in \{0, 1\} \quad \forall i, j \qquad (7)$$

In the above program, s_j is a sentence from answers and its length is l_j. c_i is a concept from sentences and its weight obtained from a graph-based random walk model is w_i. S denotes the generated summary and is a subset of sentences included in multiple answers. The variable x_i and y_j indicate whether the concept c_i and the sentence s_j occur in the summary S, respectively. If appearing, the corresponding variable x_i or y_j is equal to 1, otherwise 0. The variable occ_{ij} represents whether the concept c_i exists in the sentence s_j. If existing, the value of occ_{ij} is 1, otherwise 0.

The objective function accumulates the weight of concepts only once regardless of total times of occurrence in S, which benefits to redundancy control of summaries. There are two constrains in the program: one is the length constrain, which limits the number of words in summaries, and the other is the consistency constrain, which demonstrates that S must contain at least one sentence s_j such that occ_{ij} is equal to 1 if the concept c_i occurs in the summary S.

For further eliminating redundancy, we introduce the concept groups, in which concepts are partitioned into the same group when the similarity between two concepts is above a threshold δ. All of concepts in the set of answers are divided into concept groups under the threshold δ. The new objective function (denoted as Obj-2) is described as follows:

$$\text{Maximize: } \sum_i w_i^g \cdot x_i^g$$

$$\text{Subject to: } \sum_j l_j \cdot y_j \leq L$$

$$\sum_j occ_{ij}^g \cdot y_j \geq x_i^g \qquad \forall i$$

$$x_i^g, y_j, occ_{ij}^g \in \{0, 1\} \quad \forall i, j \qquad (8)$$

In the above program, s_j is a sentence from answers and its length is l_j. g_i is a concept group and its weight w_i^g is defined as the maximum saliency score in the concept group. S denotes the generated summary and is a subset of sentences included in multiple answers. The variable x_i^g denotes whether the concept group contains the concepts in S. If containing, the variable x_i^g is equal to 1, otherwise 0. y_j indicates whether s_j occurs in the summary S. If appearing, the corresponding variable y_j is equal to 1, otherwise 0. The variable occ_{ij}^g represents whether a mutual concept exists in both the sentence s_j and g_i. If existing, the value of occ_{ij}^g is 1, otherwise 0.

The objective function Obj-2 takes the similar concepts above the threshold δ only once into account when summaries are generated. Hence, it is inclined to contain more information into answer summaries than the objective function Obj-1, thus improving in the coverage and diversity of summaries.

4   Experiments and Discussion

4.1   Experimental Setup

Data Sets. We conducted experiments on the data sets from Yahoo! Answers portal, which were compiled by Tomasoni and Huang [14]. The original dataset contained $216, 563$ questions and $1, 982, 006$ answers created by $171, 676$ users in 100 categories. A filtered version were picked out which reserved $89, 814$ question-answer pairs with adequate statistical and linguistic characteristic. Trivial, factoid and encyclopedia-answerable questions were eliminated. They also considered some factors related to the task of summarization, such as the number of answers, length of the longest answer and length of the sum of all answers, and produced a smaller dataset for summarization. The new dataset was composed of 100 questions and 358 answers manually selected according to subjective and human interest from $89, 814$ question-answer pairs. Three human annotators were asked to generate extractive summaries of 100 questions with the limitation of 250 words.

Table 1. Performances under different concepts

Method	ROUGE-1	ROUGE-2	ROUGE-L
Unigram+Rank-1+Obj-1	**0.6213**	0.4944	**0.6041**
Bigram+Rank-1+Obj-1	0.6201	**0.4953**	0.6032
Trigram+Rank-1+Obj-1	0.6180	0.4935	0.6010
Phrase+Rank-1+Obj-1	0.6133	0.4837	0.5951
LexRank	0.6022	0.4623	0.5838
NegativeRank	0.6012	0.4604	0.5819
Bigram-GIM	0.6174	0.4911	0.6005
Baseline			
Best Answer	0.5632	0.4676	0.5515

In the following experiments, the 100 questions with 358 answers were used for evaluation of summarization. The 300 manual summaries were regarded as the gold standard for 100 questions. The data set including 89, 814 question-answer pairs was exploited to measure the user authority which is referred to in section 2.2.

Compared Algorithm. We compared our method with the topic-sensitive LexRank [13] and NegativeRank [1], which were the sentence-level methods based on random walk model. The semantic similarity between sentences was measured by the approach proposed in [12]. We also carried out the concept-level global inference algorithm (denoted by Bigram-GIM) proposed in [5] based on a dynamic programming, which weighted concepts only by the number of bigrams appearing in documents. Bigrams were chosen as concepts because they obtain better performance than unigrams or trigrams under a variety of ROUGE measure. We treated the best answer chosen by the users as the evaluation baseline.

Evaluation Metrics. We employed the standard ROUGE (version 1.5.5) [9] for evaluation. ROUGE has been widely used in automatic summarization evaluation through counting the number of overlapping units, such as n-gram, word sequences and word pairs between the summaries generated by machines and the gold standard by humans. We calculated three metrics against human annotation on data set: ROUGE-1, ROUGE-2 and ROUGE-L. ROUGE-1 depends on the co-occurring unigrams between a candiate summary and a set of reference summaries, ROUGE-2 depends on the co-occurring bigrams, and ROUGE-L relies on the Longest Common Subsequence (LCS).

4.2 Results

Experiment 1: Performances under different concepts. We analyzed the impact of different concepts through the task of answer summarization. We chose four kinds of concept representations: unigrams, bigrams, trigrams, and phrases. The graph-based random walk model only on text content (Rank-1) was employed to assign concept saliency scores. The best answers in Yahoo!

Answer were regarded as the baseline. We compared with two sentence-level methods: topic-sensitive LexRank [13] and NegativeRank [1]. We also implemented a concept-level approach Bigram-GIM [5]. The experiment results of methods under different concepts were listed in Table 1.

All of four methods using concept-level representations excel the best answer in performance. Especially, unigrams obtains the best performance in four concept representations. "Unigram+Rank-1+Obj-1" climbs up the performance 10.3% in ROUGE-1, 5.7% in ROUGE-2 and 9.5% in ROUGE-L than best answers, respectively.

Four concept-level methods also are superior to LexRank and NegativeRank. "Unigram+ Rank-1+Obj-1" improves the performance 3.2% in ROUGE-1, 6.9% in ROUGE-2 and 3.5% in ROUGE-L than LexRank, and 3.3% in ROUGE-1, 7.4% in ROUGE-2 and 3.8% in ROUGE-L than NegativeRank, respectively. The increase derives from the different node representations in link graph. Concept-level representations are beneficial to more accurately capture the relations between nodes than sentence-level ones.

"Unigram+Rank-1+Obj-1" also slightly outperforms "Bigram-GIM", improving 0.6% in ROUGE-1, 0.6% in ROUGE-2 and 1.7% in ROUGE-L respectively. This shows that the saliency scores of concepts assigned according to the graph-based random walk model performs better than the weights from the number of concept appearance in documents.

Experiment 2: Performances under the graph-based random walk model Rank-1 and Rank-2. We evaluated the effect of the user information in generating summaries. Table 2 shows the experiment results employing Rank-1 and Rank-2, respectively. In generally, the methods using Rank-2 obtained better performance than the ones using Rank-1 under all of concept representations. "Trigram + Rank 2" had the greatest increase in all concept representations.

The model Rank-1 is a graph-based random walk model only on document content, which prefers to pick out the sentences containing important concepts. The model Rank-2 is a two-layer random walk model on document content and user links, which tends to extract the sentences from authoritative users with vital information. Rank-2 is based on the assumption that authoritative users tend to provide reliable answers. The experiment results in Table 2 verify this assumption is correct.

Experiment 3: Performances under the objective function Obj-1 and Obj-2. We also compared the objective function Obj-1 and Obj-2 to evaluate the influence of grouping concepts in improving diversity and coverage of answer summarization. The model Rank-2 was used to rank concepts under different concept representations and objective functions. The results were listed in Table 3.

Grouping concepts under a given threshold is effective for the methods employing bigrams and trigrams as graph nodes. "Trigram+Rank-2+Obj-2" obtained the optimal result when varying concepts, ranking models and objective functions.

Table 2. Performances under the graph-based random walk model Rank-1 and Rank-2

Method	ROUGE-1	ROUGE-2	ROUGE-L
Unigram+Rank-1+Obj-1	0.6213	0.4944	0.6041
Unigram+Rank-2+Obj-1	0.6214	0.4988	0.6055
Bigram+Rank-1+Obj-1	0.6201	0.4953	0.6032
Bigram+Rank-2+Obj-1	0.6193	0.4970	0.6037
Trigram+Rank-1+Obj-1	0.6180	0.4935	0.6010
Trigram+Rank-2+Obj-1	**0.6224**	**0.5006**	**0.6061**
Phrase+Rank-1+Obj-1	0.6133	0.4837	0.5951
Phrase+Rank-2+Obj-1	0.6187	0.4959	0.6027

Table 3. Performances under the objective function Obj-1 and Obj-2

Method	ROUGE-1	ROUGE-2	ROUGE-L
Unigram+Rank-2+Obj-1	0.6214	0.4988	0.6055
Unigram+Rank-2+Obj-2	0.6215	0.4989	0.6056
Bigram+Rank-2+Obj-1	0.6193	0.4970	0.6037
Bigram+Rank-2+Obj-2	0.6224	0.5013	0.6069
Trigram+Rank-2+Obj-1	0.6224	0.5006	0.6061
Trigram+Rank-2+Obj-2	**0.6234**	**0.5023**	**0.6072**
Phrase+Rank-2+Obj-1	0.6187	0.4959	0.6027
Phrase+Rank-2+Obj-2	0.6193	0.4964	0.6031

5   Related Works

The appearance of Community Question Answer portals has appealed to numerous researchers, and their work involves searching for similar questions already answered [15], ranking content quality using social features such as the authority of users [6,2], predicting asker satisfaction [10] and other application.

A graph-based random walk model is widely applied on the document and answer summarization, most of which chooses sentences as the nodes in graphs. Otterbacher et al. [13] introduced a topic-sensitive random walk model LexRank. Chall and Joty [3] improved LexRank through substituting the sentence similarity based on syntactic and semantic tree kernels for the sentence similarity based on tf-idf. Achananuparp et al. [1] proposed the algorithm NegativeRank for increasing answer diversity through assigning negative weights to edges in the graph-based random walk model.

McDonald [11] applied the global inference model in multi-document summarization, which consider relevance and redundancy at the sentence level. The objective function is designed into the form of the sum of relevance scores of sentences minus the sum of the redundancy scores of each sentence pair in a summary. Gillick and Favre [5] enhanced McDonald's algorithm and proposed a concept-level global inference model, which assumed the concepts are independent. Tomasoni and Huang [14] proposed an answer summarization on cQA, which utilized four metadata-aware measures on the concept level to evaluate answers: Quality, Coverage, Relevance and Novelty. The final summaries about questions were generated through a global inference algorithms similar with a maximum coverage model.

Our method has two main differences compared with those above algorithms. First, we adopt the graph-based link analysis to distribute the node weights under a finer level than sentences. This contributes to the analysis of dependency relationship between nodes and redundancy reduction in generated summaries. Second, we construct a two-layer link graph for strengthening the impact of the user authority in answer summarization.

6   Conclusion

We propose a framework to automatically generate answer summarization for questions posed by user in cQA services. In the framework, a series of methods are employed to improve diversity, coverage, relevance, trustworthiness, and conciseness of answer summaries: (1) We choose different concept representations with finer granularity than sentences on a graph-based random walk to enhance the diversity. (2) We incorporate user authority and document content from answers into a two-layer link graph to strengthen the trustworthiness of summaries. (3) We cast the summary generation into a maximum coverage problem with knapsack constraint, boosting the coverage and conciseness through the optimization process. The experimental results demonstrate our approach outperforms current methods.

In future work, we plan to distill syntactic or shallow semantic structures to represent concepts. We also plan to implement the evaluations on the larger-scale data set.

Acknowledgments. This research is supported by State Key Laboratory of Software Development Environment (under grant no. SKLSDE-2011ZX-03) and the National Natural Science Foundation of China (under grant no. 61170189).

References

1. Achananuparp, P., Yang, C.C., Chen, X.: Using negative voting to diversify answers in non-factoid question answering. In: 18th ACM Conference on Information and Knowledge Management, CIKM 2009, pp. 1681–1684 (2009)
2. Chali, Y., Joty, S.R., Surdeanu, M., Ciaramita, M., Zaragoza, H.: Learning to rank answers on large online qa collections. In: 46th Annual Meeting for the Association for Computational Linguistics: Human Language Technologies, ACL 2008: HLT, pp. 719–727 (2008)
3. Chali, Y., Joty, S.R.: Improving the performance of the random walk model for answering complex questions. In: 46th Annual Meeting of the Association for Computational Linguistics on Human Language Technologies: Short Papers, HLT-Short 2008, pp. 9–12 (2008)

4. Ciaramita, M., Altun, Y.: Broad-coverage sense disambiguation and information extraction with a supersense sequence tagger. In: The 2006 Conference on Empirical Methods in Natural Language Processing, EMNLP 2006, pp. 594–602 (2006)
5. Gillick, D., Favre, B.: A scalable global model for summarization. In: Workshop on Integer Linear Programming for Natural Langauge Processing, ILP 2009, pp. 10–18 (2009)
6. Jurczyk, P., Agichtein, E.: Discovering authorities in question answer communities by using link analysis. In: The Sixteenth ACM Conference on Information and Knowledge Management, CIKM 2007, pp. 919–922 (2007)
7. Khuller, S., Moss, A., Naor, J.S.: The budgeted maximum coverage problem. Inf. Process. Lett. 70, 39–45 (1999)
8. Li, L., Zhou, K., Xue, G.R., Zha, H., Yu, Y.: Enhancing diversity, coverage and balance for summarization through structure learning. In: 18th International Conference on World Wide Web, WWW 2009, pp. 71–80 (2009)
9. Lin, C.Y.: Rouge: a package for automatic evaluation of summaries. In: ACL Workshop on Text Summarization Branches Out
10. Liu, Y., Bian, J., Agichtein, E.: Predicting information seeker satisfaction in community question answering. In: 31st Annual International ACM SIGIR Conference on Research and Development in Information Retrieval, SIGIR 2008, pp. 483–490 (2008)
11. McDonald, R.: A Study of Global Inference Algorithms in Multi-Document Summarization. In: Amati, G., Carpineto, C., Romano, G. (eds.) ECiR 2007. LNCS, vol. 4425, pp. 557–564. Springer, Heidelberg (2007)
12. Mihalcea, R., Corley, C., Strapparava, C.: Corpus-based and knowledge-based measures of text semantic similarity. In: 21st National Conference on Artificial Intelligence, vol. 1, pp. 775–780 (2006)
13. Otterbacher, J., Erkan, G., Radev, D.R.: Using random walks for question-focused sentence retrieval. In: The conference on Human Language Technology and Empirical Methods in Natural Language Processing, HLT 2005, pp. 915–922 (2005)
14. Tomasoni, M., Huang, M.: Metadata-aware measures for answer summarization in community question answering. In: 48th Annual Meeting of the Association for Computational Linguistics, ACL 2010, pp. 760–769 (2010)
15. Wang, K., Ming, Z.Y., Hu, X., Chua, T.S.: Segmentation of multi-sentence questions: towards effective question retrieval in cqa services. In: 33rd International ACM SIGIR Conference on Research and Development in Information Retrieval, SIGIR 2010, pp. 387–394 (2010)

Acquisition of Know-How Information from Web

Shunsuke Kozawa[1], Kiyotaka Uchimoto[2], and Shigeki Matsubara[1]

[1]Nagoya University, Furo-cho, Chikusa-ku, Nagoya, 464-8601, Japan
[2]National Institute of Information and Communications Technology
3-5 Hikari-dai, Seika-cho, Soraku-gun, Kyoto, 619-0289, Japan
kozawa@el.itc.nagoya-u.ac.jp, uchimoto@nict.go.jp, matubara@nagoya-u.jp

Abstract. A variety of know-how such as recipes and solutions for troubles have been stored on the Web. However, it is not so easy to appropriately find certain know-how information. If know-how could be appropriately detected, it would be much easier for us to know how to tackle unforeseen situations such as accidents and disasters. This paper proposes a promising method for acquiring know-how information from the Web. First, we extract passages containing at least one target object and then extract candidates for know-how from them. Then, passages containing the know-how are discriminated from non-know-how information considering each object and its typical usage.

Keywords: know-how, how-to type question answering, object, usage information, procedural question.

1 Introduction

A variety of know-how such as recipes and solutions for troubles have been stored on the Web and they are often referred to by using web search. It has been reported that about 40% of all non-factoid questions on Q&A sites are how-to questions [8], and the know-how information has potential to give us the answers to the questions. However, it is not so easy to appropriately find know-how information related to a particular how-to question. If know-how could be appropriately detected, it would become easy to find answers to how-to questions. Also, it would become possible to let us know how to tackle unforeseen situations if know-how information could be stored beforehand and classified into several classes according to their topics.

In this research, we assume that know-how is a procedure or an advice. Figure 1 shows two examples of know-how. In previous researches, know-how information was acquired by using typical words such as "how to" and clue words such as "only if" and "first" [10]. However, it is difficult to efficiently acquire know-how information based on the conventional methods because a variety of clue expressions are required to appropriately detect know-how information which is often described without using clue words or the typical words "procedure", "how to" and "first" as you can see in Figure 1.

Well then, what is a clue to efficiently obtaining know-how information? Our observation for this question is as follows. Firstly, know-how information often

M.V.M. Salem et al. (Eds.): AIRS 2011, LNCS 7097, pp. 446–457, 2011.

ID sentence
9   A very easy way to remove the stickers without damaging them is
    to heat them up with a hair dryer.
10 It sounds kind of odd, but it works well, and doesn't discolor, or melt them.
11 Just heat them up to soften up the adhesive and carefully peel them off.
12 I usually use a razor blade to start peeling it off because you lessen the chance of
    wrinkling the sticker. (exacto knives work well becaue they have a pointed end)
13 This works well on those unwanted bumber stickers also

ID sentence
1   How do you treat a heat stroke victim?
2   Victims of heat stroke must receive immediate treatment to avoid permanent organ damage.
3   First and foremost, cool the victim.
4   Get the victim to a shady area, remove clothing, apply cool or tepid water to the skin
    (for example you may spray the victim with cool water from a garden hose), fan the victim
    to promote sweating and evaporation, and place ice packs under armpits and groins.
5   Monitor body temperature with a thermometer and continue cooling efforts
    until the body temperature drops to 101-102°F (38.3-38.8°C).
6   Always notify emergency services (911) immediately.
7   If their arrival is delayed, they can give you further instructions for treatment of the victim.

Fig. 1. Examples of know-how

includes at least one object that plays an important role; for instance, "hair dryer" in the first example and "thermometer" in the second one in Figure 1. Secondly, the typical usage of the object is often described in each know-how; for instance, "to heat something" in the first example, and "to monitor a temperature" in the second one. As seen in the examples, know-how is often characterized by an object name and the description of its usage. We examined 100 lists of know-how information randomly sampled from a web site[1] and found that 75 out of the 100 lists included at least one object and the description of its usage. This supports our intuition.

In this paper, we propose a method for acquiring know-how information by focusing on each object and how it is used. First, we extract passages containing at least one target object and then extract candidates for know-how from them. Then, lists of know-how information are acquired based on the description of the object and its typical usage.

2   Related Works

In previous works, a few studies have addressed the acquisition of procedural texts. Takechi et al. proposed a method to categorize HTML texts tagged with ⟨OL⟩ or ⟨UL⟩ as either procedural or non-procedural by using word N-grams [7]. Aouladomar proposed a method to estimate *questionability* of web texts by using tag information (title, advice, warning, etc.) annotated by rule-base methods [2,3] and clue expressions [1]. Yin et al. proposed a method for measuring the

---

[1] http://know-how.fc2.com/ (in Japanese)

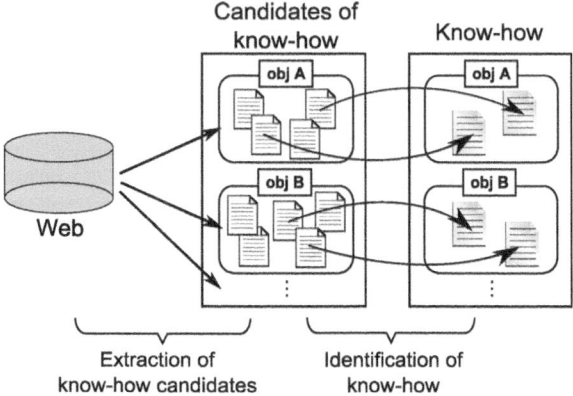

Fig. 2. Flow for acquiring know-how information

degree of procedurality of texts containing the phrase "how to" by using syntactic tags, morphological tags and cue phrases [10].

In this study, we acquire procedures and advices as know-how information. Our proposed method uses not only clue expressions which were also used in previous works but also an object and how it is used.

3   Acquisition of Know-How

Aouladomar classified procedural texts into three categories [1]; procedures (e.g. recipes, maintenance and construction manuals, etc.), injunctions (e.g. orders, regulations, etc.), and advices (e.g. beauty advices, health management methods, etc.). In this research, we acquire procedures and advices as know-how information since they are frequently asked on Q&A sites, namely there are many demands to obtain information on procedures and advices.

We acquire know-how information by focusing on an object and how it is used. In this research, we assume that objects are noun phrases which are hyponym of "physical entity" in Japanese WordNet 1.1[2]. We use expressions representing utilization of an object (utilization roles) to capture how the object is used. For example, the utilization role of "hair dryer" is to "heat" and that of "thermometer" is to "monitor." Figure 2 illustrates the flow for acquiring know-how information. We assume that a unit of know-how is a passage since know-how is very often composed of multiple sentences, but it is not always composed of all sentences in a document. We also assume that lists of know-how information are acquired for each object. First, given a target object, the candidate passages for know-how are extracted by using several methods since most passages would not contain know-how information. Then, the candidates are classified as either they contain know-how information or not by using clue patterns, an object and its utilization roles.

---

[2] http://nlpwww.nict.go.jp/wn-ja

In the following sections, we assume that the target language is Japanese because know-how information is not well organized yet, although the method can be expanded to any languages. We used the Web corpus consisting of 500M Japanese parsed sentences extracted automatically from the Web [5]. The sentences in the corpus have been automatically annotated with morphological and syntactic information. The syntactic information in a sentence is represented as a dependency structure between Japanese phrasal units, *bunsetsu*.

3.1   Extraction of Know-How Candidates

First, passages are extracted in the following way. Every segment that contains at least one target object and is enclosed by a pair of the following HTML tags is extracted:

body, div, table, span, p, blockquote, h1, h2, h3, h4, h5, h6

If the number of sentences contained in a segment is less than or equal to a threshold α, the segment is extracted as a passage. Otherwise, we split the segment into one or more passages with the TextTiling algorithm [4] using α as the window size. This is because there exist Web sites where HTML tags are used incorrectly.

Next, the candidate passages that might contain know-how information are extracted. They are extracted by using the following four conventional methods (A,B,C,D) and our proposed method (E). Multiple methods are used since the acquired know-how information might be limited when only one method is used. We take the union of the candidate passages as the candidates for know-how.

(A) Extraction of the passages containing the term *houhou* (how to) [10][3].
(B) Extraction of the passages tagged with ⟨OL⟩ or ⟨UL⟩ tags [7].
(C) Extraction of the passages containing any of 47 expressions (e.g. "in order not to", "prefer", "as long as", etc.) which have been manually generated by referring to the policy of generating patterns in the previous works [3,10].
(D) Extraction of the passages containing any of 638 expressions such as "get well" and "feel good" found in the semantic lexicon constructed by Kobayashi et al. and tagged with *keiken* (experience) tags in the lexicon [6]. This is because we assume that know-how information contains expressions representing experience.
(E) Extraction of the passages containing any of 3-tuples which are composed of an object and its utilization roles. Utilization roles of a given object o are defined as paraphrases of such expressions as "using o" or "enjoying o" and expressed by a pair of a postposition p and a verb v. For example, the utilization role of a hair dryer is ⟨*de*(by), *atatameru*(heat)⟩. The 3-tuples ⟨o, p, v⟩ (ex. ⟨ *doraiya*(hair dryer), *de*(by), *atatameru*(heat) ⟩) are assumed to appear in the dependent *bunsetsus*; o and p (*doraiya+de* (by using hair

---

[3] The term was originally proposed in English. We manually translated them into Japanese and used them.

dryer)) appear in a *bunsetsu* which depends on another *bunsetsu* containing *v* (*atatameru* (heat)). The method for acquiring utilization roles is described in detail in Section 3.2.

3.2   Identification of Know-How

The extracted candidates are classified as either they contain know-how information or not by using a machine learning model. Clue patterns, part-of-speech information, target objects and their utilization roles (3-tuple) are used as features[4].

- clue patterns
  Clue patterns are manually generated by referring to know-how and non-know-how information in a development data. The patterns are applied to a sentence or *bunsetsu* (Japanese phrasal unit) sequences in a sentence in a target passage. The frequency of the sentences matched with each clue pattern, the total frequency of the sentences matched with the patterns and the number of types of the matched patterns are used as features.
- part-of-speech information
  The appearance frequency of each part-of-speech normalized by the number of the sentences is used as a feature.
- 3-tuples
  The total number of the frequency of the 3-tuples $\langle o, p, v \rangle$ is used as a feature.

The target objects and their utilization roles are our newly added features. Henceforth, we call a set of clue patterns and part-of-speech information **PT** features. The following subsections describe the method for acquiring utilization.

Acquisition of Utilization Roles. Torisawa made the following three assumptions about the characteristics of utilization role $\langle p, v \rangle$ for a given noun n: 1) An n marked by p often appears with v. 2) First-person pronouns such as "watashi (I)" often occupy the agent role of v. 3) The postposition "*de*" is a good candidate of a postposition in a utilization role. Utilization roles were acquired by using the following formula reflecting these assumptions [9].

$$U(n) = argmax_{(v',p') \in V \times A}\{Uscore(n, p', v')\}$$

V is a set of verbs, which can be a verb in possible utilization roles. As verbs in V, we used 6,485 verbs which appeared in the Web corpus with the verbal suffix "*tai*," which can be translated to "want." We also manually removed 20 verbs that can never take utilization roles such as "*naru* (become)" or mean literally "using" and "enjoying" such as "*tsukau* (use)" from V. A is a set of postpositions.

---

[4] Word n-grams were not used because they did not work well in our preliminary experiments.

Table 1. Sizes of training and testing data

objects	# of passages containing know-how						# of passages
	A	B	C	D	E	total	
air conditioner	14	16	33	5	6	70	492
cell-phone	11	5	11	0	0	27	495
digital camera	8	14	19	5	9	53	487
electric fan	17	29	37	5	11	92	476
iron	34	43	31	9	37	144	468
microwave	24	61	27	13	69	183	482
oven	28	79	32	7	70	208	476
refrigerator	28	56	22	1	60	165	494
vacuum cleaner	14	34	31	5	45	117	478
washer	13	26	14	3	27	78	479
total	201	363	258	53	334	1137	4827

$$Uscore(n, p', v') = \frac{P(n, p', v')P(S|AP, v')Bias(p')}{P(n)}$$

$P(n, p', v')$ is the co-occurrence probability between the verb v' and n marked by the postposition p'. As for $P(S|AP, v')$, S denotes a set of first-person pronouns, and 17 pronouns were used. AP is a set of postpositions which can mark agent roles, and "ga" and "ha" were used. $P(S|AP, v')$ is the probability that the first-person pronouns occupy the agent role of v'. $Bias(p')$ denotes the bias concerning the postposition "de." If p' is "de," the bias is 25, otherwise the bias is 1 by referring to Torisawa's method [9].

4 Evaluation

4.1 Preparation

Construction of Training and Testing Data. For our experiment, ten objects were selected from electric products that had appeared 10,000 times or more in the web corpus [5] and also appear in Japanese WordNet 1.1. We constructed training and testing data from the web corpus using the methods in Section 3.1. Passages were extracted using 20^5 as the window size α. We chose two sets of 100 passages at random from each group of passages extracted by using the methods A and B, and three sets of 100 passages in decreasing order of frequency from the groups of passages extracted based on the C, D and E methods, and judged manually whether they contained know-how information or not. As utilization roles, we used top 25 pairs of a proposition and a verb automatically produced by the method described in Section 3.2. In this research, we assume that the passage is judged to be correct if a given passage contains

---

[5] The same window size was used throughout all of the experiments mentioned in this paper, and it was not well tuned to particular data.

Table 2. Size of development data

# of passages containing know-how						# of passages
A	B	C	D	E	total	
43	44	41	16	34	148	444

know-how information. Note that if a given passage contained only a fragment of know-how information, the passage was judged as incorrect. The sizes of the data are shown in Table 1.

Generation of Clue Patterns. We constructed the development data in the same way as training and testing data by extracting candidates for know-how from the web corpus with the methods described in Section 3.1 and judging them manually whether they contained know-how information or not. The breakdown of the development data is shown in Table 2.

We manually generated 79 types of patterns by referring to the development data. Some examples of the patterns are shown in Table 3. The symbols "|", "+" and ".∗" in column 2 represent a disjunction, a word boundary and any word sequences, respectively. Column 3 represents the targets that the patterns are applied to. S, L, R1, R2 and R3 represent a sentence, the leftmost *bunsetsu* in a sentence, the rightmost *bunsetsu* in a sentence, the rightmost two *bunsetsus* in a sentence and the rightmost three *bunsetsus* in a sentence, respectively.

In our experiments, the patterns were generated based on the development data constructed focusing on the object "hair dryer." However, we found that the patterns tended to depend not on an object but on the types of know-how information such as procedures and advices according to our preliminary investigation.

4.2 Settings

In the experiment, the 3-tuples and PT features (clue patterns and part-of-speech information) were used to train machine learning models. We used the following two types of 3-tuples:

3T$_{auto}$ As utilization roles, the top 25 pairs of a postposition and a verb produced by the method described in Section 3.2, which were the same pairs used for constructing the training and testing data, were used.

3T$_{man}$ As utilization roles, pairs of a postposition and a verb were manually selected from among top 100 pairs produced by the method described in Section 3.2, and were used. The average number of the manually selected pairs was 25.

Support Vector Machines were used as machine learning models and they were trained using LibSVM[6]. For the experiments, we prepared the following baseline method:

---

[6] http://www.csie.ntu.edu.tw/~cjlin/libsvm/

Table 3. Clue patterns

No.	patterns	target
1	*mazu|hajimeni|hajime+ha|saisho+ha* (first, primarily, to begin with)	L
2	*sore+kara|tsugini|konoato|sonoato|soshite* (then, next, secondly)	L
3	*dekiagari|kansei|shuryo|kanryo* (finish, end, complete, accomplish)	R1
4	verb+*hou+ga.**yoi*|verb+*no+ga.**anshinda* (prefer, preferable)	S
5	verb+*yasui* (easier to + verb)	R2
6	*kinmotsu|dameda|genkin|kiken* (danger, caution, prohibition)	S
7	verb+*nai+youda.**verb|verb+*nu+ni.**verb (never, avoid + verb)	S
8	*ki+wo+tsukeru|te+wo+nuku+nai|tyuui* (see to, warning)	R3
9	*wo+taishou|ni+gentei|ni+kagiru* (limit, target, restrict)	R3
10	*hitsuyouda|youi|kakaseru+nai|hissuda* (necessary, need, essential, vital)	R3
11	*yakudatsu|katsuyaku|benrida|kouritsu* (useful, helpful, efficient)	R2
12	*teineida|shintyouda|kinnitsu|shikkari* (carefully, advisedly, fastly)	S

baseline (conventional method) Identify whether a given passage contains know-how information or not by using the model based only on **PT** features.

In the next section, the effectiveness of using both an object and its utilization roles is shown by comparing models with and without the 3-tuples.

4.3   Experimental Results

We split the data for each object in five and carried out 5-fold cross validation by using data composed of ten objects. The experimental results are shown in Table 4. In comparison with the model using only PT features (baseline), the models using both PT and 3T features significantly improved in both the precision and the recall. This indicates that lists of know-how can be efficiently acquired by focusing on an object and its utilization roles. In addition, the model using both PT and $3T_{auto}$ features achieved good results although the rates were a little lower than ones using both PT and $3T_{man}$ features.

In the above experiment, both the training and testing sets contained the same 10 objects. Thus, the results do not show that the proposed models would achieve good results for the data containing objects which are not contained in the training set since the patterns and the 3-tuples appearing in know-how information might depend on an object. Therefore, we carried out 10-fold cross validation by training the data containing nine objects out of ten and testing the data containing the rest (one object). The experimental results are shown in Table 5. The results show that both the precision and the recall were significantly improved by using the 3-tuples. This indicates that using an object and its utilization roles is useful even though the training set does not contain data for the target object. The differences between the proposed methods and the baseline method are statistically significant according to Mcnemar's test (p < 0.01).

454   S. Kozawa, K. Uchimoto, and S. Matsubara

Table 4. Experimental results by 5-fold cross validation

Feature	Precision	Recall	F-measure
PT (baseline)	73.0% (588/805)	51.7% (588/1137)	60.6
PT + 3T$_{auto}$	74.6% (660/885)	58.1% (660/1137)	65.3
PT + 3T$_{man}$	75.6% (666/881)	58.6% (666/1137)	66.0

Table 5. Experimental results by 10-fold cross validation

Feature	Precision	Recall	F-measure
PT (baseline)	71.9% (550/765)	52.8% (550/1137)	57.8
PT + 3T$_{auto}$	72.8% (600/824)	52.8% (600/1137)	61.2
PT + 3T$_{man}$	73.4% (614/836)	54.0% (614/1137)	62.2

The 3-tuples feature of our proposed method is a template or a type of features while the patterns used as baseline features are surface strings or tokens. Therefore, the training data do not have to contain all of the 3-tuples appearing in the testing data. This indicates that the proposed method works without creating training data for all objects.

4.4   Discussion

Error Analysis. In order to detect the causes of errors, we investigated the results by 10-fold cross validation with the model that uses both **PT** and **3T**$_{man}$.

We investigated 222 passages acquired incorrectly and found following three main causes:

- Passages containing only a fragment of know-how information
  There were 93 (41.9%) passages. This is because passages were extracted inaccurately from the documents. If these passages had been extracted accurately, the passages would have been judged as correct. There were 190 passages containing only a fragment of know-how information in the training and testing data. It would be important to develop a method for accurately extracting passages if the acquired passages are separated from documents and then used. In the case that people refer to the acquired know-how information, it is a matter of no importance that the boundaries of passages are incorrect since they can understand know-how by referring to the know-how information in conjunction with surrounding contexts.
- Passages containing utilization roles
  There were 69 (31.1%) passages. They were either diaries or commercial articles which contained typical usage of the target object and did not contain know-how information. This type of errors would be reduced by taking into account co-occurrences and sequences of the patterns, the target object and its utilization roles.

Table 6. Experimental results using arbitrary pairs and results combining models

Feature	Prec	Rec	F1
PT (baseline)	71.9%	48.4%	57.8
PT+3T$_{all}$	71.5%	49.3%	58.3
***PT+3T**$_{auto}$	72.8%	52.8%	61.2
***PT+3T**$_{man}$	73.4%	54.0%	62.2
PT (without the method E)	71.5%	46.0%	56.0

– Passages containing injunctions
  There were 21 (9.5%) passages. We did not target them in this research. However, it is a type of procedural texts, and it could be a type of know-how.

We investigated 523 passages that are know-how but could not be acquired. Objects besides the target object were used in 450 (86.0%) passages. That is to say, these know-how information might be acquired by targeting other objects. Furthermore, we expect that the recall will be more improved by simultaneously considering two or more objects and their utilization roles.

Effect of an Object and Its Utilization Roles. In order to show the contribution of our 3-tuples compared with that of arbitrary 3-tuples to acquisition of know-how information, we carried out 10-fold cross validation using the following feature:

3T$_{all}$ The total number of the frequency of the 3-tuples $\langle o, p, v \rangle$, where o is a target object and $\langle p, v \rangle$ is the pairs of any postposition and any verb, was used.

In addition, we carried out experiments by using only conventional methods in both extraction of know-how candidates and identification of know-how. That is, know-how candidates were extracted by using the method A through D in Section 3.1 (without the method E considering 3-tuples) and know-how information was identified by using the model based on PT features. Experimental results are shown in Table 6. The models attached with * in Table 6 are statistically significant than the baseline method according to Mcnemar's test (p < 0.01). The model based on PT and 3T$_{all}$ features is not statistically significant than the baseline method. Moreover, the models attached with * in Table 6 are statistically significant than the model based on PT and 3T$_{all}$ according to Mcnemar's test (p < 0.01). These results indicate that the 3-tuples should be selected for efficiently acquiring know-how information and one of the promising selection method is based on an object and its utilization roles. The last row in Table 6 shows the results obtained by using only conventional methods. The results show that the 3-tuples play important roles in both extraction of know-how candidates and identification of know-how although we can not simply compare them since the size of the data are different.

Table 7. Experimental results of open-domain experiments

Feature	Precision
PT (baseline)	68.4% (171/250)
PT + 3T$_{auto}$	70.5% (172/244)
PT + 3T$_{man}$	69.0% (176/255)

To our knowledge, this study is the first trial to show that an object and its usage play an important role in acquiring know-how information.

4.5   Open-Domain Tests

We performed experiments on data for different domains. For the experiments, we selected eight objects (curtain, hunger, ladder, lighter, mirror perfume, scissors, stove) which are different types of objects from electric products. They were selected almost randomly except that frequent nouns were preferred to others. The passages were automatically extracted in the same way as the construction of the training and testing data and classified whether they contained know-how information by the models trained using the training and testing data.

Table 7 shows the precisions for acquiring know-how information. The model using automatically acquired 3-tuples improved the precision and the recall (the number of lists of successfully acquired know-how information).

Thus, we expect that know-how information can be acquired by focusing on an object and how it is used without depending on domain.

5   Conclusion

This paper presented a method for acquiring know-how information by focusing on an object and how it is used. First, we extracted know-how candidates for each object. Then, by using both an object and its utilization roles, passages containing know-how information were efficiently acquired from them. Various lists of know-how information could be acquired by expanding the number of target objects although they were restricted in our experiments as a first trial.

In this paper, we used manually extracted patterns to compare our method with those in the previous works. We would like to acquire know-how information using automatically extracted patterns. In the future, we would like to consider (grammatical) objects of verbs for utilization roles since we expect that they are more useful information for identifying know-how information. For example, to "monitor a temperature with thermometer" is more important than to "monitor something with thermometer" in the second example in Figure 1. In addition, we would like to focus on multiple objects and how they are used since it is often the case that multiple objects are used in know-how. Know-how information would be more efficiently acquired by simultaneously considering two or more objects and their utilization roles.

In this study, we assume that know-how information could be mostly covered by the physical entities of WordNet. However, the coverage should be improved to get statistically valid data for obtaining utilization roles. We are planning to generalize nouns not found in WordNet.

References

1. Aouladomar, F.: Towards answering procedural questions. In: Proceedings of the IJCAI Workshop on Knowledge and Reasoning for Answering Questions, pp. 21–31 (2005)
2. Delpech, E., Saint-Dizier, P.: Investigating the structure of procedural texts for answering how-to questions. In: Proceedings of the 6th International Conference on Language Resources and Evaluation, pp. 46–51 (2008)
3. Fontan, L., Saint-Dizier, P.: Analyzing the explanation structure of procedural texts: dealing with advice and warnings. In: Proceedings of the 2008 Conference on Semantics in Text Processing, pp. 115–127 (2008)
4. Hearst, M.A.: TextTiling: Segmenting text into multi-paragraph subtopic passages. Computational Linguistics 23(1), 33–64 (1997)
5. Kawahara, D., Kurohashi, S.: A fully-lexicalized probabilistic model for Japanese syntactic and case structure analysis. In: Proceedings of the 7th Human Language Technology Conference of the North American Chapter of the Association for Computational Linguistics, pp. 176–183 (2006)
6. Kobayashi, N., Inui, K., Matsumoto, Y., Tateishi, K., Fukushima, T.: Collecting evaluative expressions for opinion extraction. In: Proceedings of the 2nd International Joint Conference on Natural Language Processing, pp. 584–589 (2004)
7. Takechi, M., Tokunaga, T., Matsumoto, Y., Tanaka, H.: Feature selection in categorizing procedural expressions. In: Proceedings of the 6th International Workshop on Information Retrieval with Asian Languages, pp. 49–56 (2003)
8. Tamura, A., Takamura, H., Okumura, M.: Classification of multiple-sentence questions. In: Proceedings of the 2nd International Joint Conference on Natural Language Processing, pp. 426–437 (2005)
9. Torisawa, K.: Automatic acquisition of expressions representing preparation and utilization of an object. In: Proceedings of the 5th Recent Advances in Natural Language Processing, pp. 556–560 (2005)
10. Yin, L., Power, R.: Adapting the Naive Bayes Classifier to Rank Procedural Texts. In: Lalmas, M., MacFarlane, A., Rüger, S.M., Tombros, A., Tsikrika, T., Yavlinsky, A. (eds.) ECIR 2006. LNCS, vol. 3936, pp. 179–190. Springer, Heidelberg (2006)

Topic Based Creation of a Persian-English Comparable Corpus

Zahra Rahimi and Azadeh Shakery

School of Electrical and Computer Engineering,
College of Engineering,
University of Tehran, Tehran, Iran
z.rahimi@ece.ut.ac.ir, shakery@ut.ac.ir

Abstract. One of the most important issues in cross language information retrieval (CLIR) is where to obtain the translation knowledge. Multilingual corpora are valuable resources for this purpose, but few studies have been done on constructing multilingual corpora in Persian language. In this study, we propose a method to construct a Persian-English comparable corpus using two independent news collections and based on date and topic criteria. Unlike most existing methods which use publication dates as the main basis for aligning documents, we also consider date-independent alignments: alignments based only on topics and concept similarities. In order to avoid low quality alignments, we cluster the collections based on their topics prior to alignments which allows us to align similar documents whose publication dates are distant. Evaluation results show the high quality of constructed corpus and the possibility of extracting high quality association knowledge from the corpus for the task of CLIR.

Keywords: Clustering, Comparable Corpora, Cross Language Information Retrieval, Topic-based Alignment.

1 Introduction

Because of the fast growth of the World Wide Web and the amount of information existing in different languages, cross language information retrieval (CLIR) has become a very important task. CLIR is the task of retrieving documents where queries and documents are in different languages. One of the most important issues in CLIR is where to obtain translation knowledge. Different translation resources have been used for this purpose. Comparable corpora are one of the useful resources widely used in different languages. However, few studies have been done on constructing and using comparable corpora in Persian [8,6]. Karimi's Persian-English comparable corpus [8] consists of 1100 loosely translated BBC News documents and UTPECC Persian-English comparable corpus [6] is constructed using Persian articles of Hamshahri newspaper[1] and English articles of BBC News[2].

---

[1] www.hamshahrionline.ir

[2] www.bbc.co.uk

M.V.M. Salem et al. (Eds.): AIRS 2011, LNCS 7097, pp. 458–469, 2011.
© Springer-Verlag Berlin Heidelberg 2011

Different resources have been used to construct comparable corpora. Many researches use news articles to obtain comparable corpora [10,4,16,1,6], some methods crawl the web [18,13,20,7], and some others use the research corpora like CLEF and TREC collections [17,3,14] to obtain comparable corpora. Many studies to construct comparable corpora, easily align news articles in different languages by date [19,20,1]. Although this approach may work in some languages, this is not the case in Persian because of the lack of news agencies that publish appropriate articles in both Persian and English. Some studies try to align documents in comparable corpora based on both date criteria and content similarities [3,17,6]. These methods can align news articles which are related to the same event, however, are not able to identify all possible alignments with distant dates. Intuitively, some news articles which are published in distant dates, may have similar concepts, or be about related events or discuss the same topic even though are not related to any specific event, for example two documents that discuss "skin cancer prevention". These kinds of alignments can also be useful for the CLIR task. However these alignments should be considered very carefully to avoid aligning unrelated documents with common keywords.

In this paper, we propose a method for aligning related documents that cannot be identified using date criteria. On top of the previous methods for aligning the documents, we propose to align documents based on "concept similarities" and "topics" and without the date criteria. For this purpose, we propose to first cluster the collections based on their topics and then align the documents in the corresponding clusters in the two languages based on their concept similarities. This enables us to align similar documents whose publication dates are distant, like scientific articles. These documents with similar concepts not only have common keywords, but also are about the same topic, thus we can avoid aligning unrelated documents in different topics that share common keywords.

For topical clustering of the documents, we propose a method based on edge betweenness measure [5] and extract major features of each cluster using the mutual information (MI) graph. These features represent the topics of the collection and are used to cluster the documents in the collection. To construct the comparable corpus, we follow the general procedure proposed in [6]. We first apply the proposed alignment method on the whole collection to find the alignments based on concept similarities and date criteria. We then apply a similar method on each cluster to align the documents based on concept similarities and topics. Our experiment results show that the proposed approach improves both the quality and size of the existing Persian-English comparable corpora. The rest of the paper is organized as follows. We explain the details of constructing the comparable corpus in section 2. We discuss the experiment results in section 3 and finally bring the conclusions and future work of our study in section 4.

2   Constructing the Comparable Corpus

We use two independent news collections, one in English and another in Persian, to construct the comparable corpus. We construct the comparable corpus considering two kinds of alignments: (1) alignments based on the concept similarity of the documents and the publication dates, similar to the method proposed in [6], and (2) alignments based on topics and concept similarities.

In the first phase, two different news articles are aligned if their publication dates are not distant and they have common describing keywords. Using date criteria, we cannot align news with similar concepts which are not published in same period of time. Intuitively, some news articles which are published in distant dates may have similar concepts. For these reasons we try to further align related news that cannot be aligned with date criteria. We cannot align such news simply by omitting date criteria and just based on concept similarity because a lot of noise will be added to the corpus this way. For example we may mistakenly align news articles that talk about different topics, but share some keywords. In order to avoid such alignments, we propose to first cluster the collections based on their topics and then align the documents in the corresponding clusters based on their concept similarities.

We follow these steps in the proposed method: (1) extract best representative features of the English collection, (2) construct the mutual information graph of the features (the MI graph describes the relatedness of the features and by analyzing these relations, we can recognize different topics of the collection), (3) cluster the MI graph and extract independent components that represent the topics of the collection, (4) enrich the features of each component using Chi-Square method and cluster the documents based on these features, (5) extract the corresponding clusters in the Persian collection based on translation of the features and Persian MI graph, (6) and finally align the documents in each cluster. In the rest of this section, we present the details of the method for constructing and evaluating the comparable corpus.

2.1 Base Features Selection and Construction of the MI Graph

In these steps, we extract the best representative features of the English collection and construct a pruned mutual information graph using these features. Extraction of the features is necessary because using all unique words of the collection for constructing the MI graph can considerably decrease the quality of clusters and the performance of the clustering method. In order to select the words which best represent the news collection, we apply the RATF formula [11] and extract the main features of each document of the collection. In this formula, we set the parameters to their best values reported in [17].

We then sort all these extracted features based on their collection frequencies. Each feature is representative of at least one document. If the feature occurs rarely in the collection, it cannot be a good representative of a topic in the collection and thus we omit it. In this way, we come up with the list of features that are expected to be the representatives of major domains of the collection. We use these features to construct the Mutual Information (MI) Graph. The features are vertices of this directional graph and each vertex is connected to 100 top related vertices based on their MI score. We decrease the noise in the graph and the size of the graph by pruning unimportant edges and vertices in the following steps:

1. Omit the edge from vertex X to vertex Y if the Mutual Information between the two features X and Y is less than threshold MI_thr.
2. Omit the edge between two vertices if it is not bidirectional. If the edge is not bidirectional it shows that one of the two words is not in the top MI related words of the other one so these two words are not tightly related to each other.
3. Omit the vertices with few edges because such words are loosely connected to the graph and they are not informative.
4. Repeat the steps 2 and 3 until the convergence, i.e., to the point that no weak edges or vertices can be further removed.

This final graph is less noisy than the MI graph of all features of the collection because it doesn't contain high frequency and low frequency words. Besides, the size of the graph is much smaller which increases the performance of our further computations.

2.2   Extraction of Major Topics Using MI Graph

The goal in this step is to recognize major topics of the collection by clustering of the MI graph. Our intuition is that the features related to each topic mostly co-occur, so in the MI graph they construct tightly connected components that are loosely connected to the other components. In order to cluster the graph, we propose to use the edge betweenness measure [5]. Edge betweenness measures the betweenness centrality of the edges by summing up the number of all shortest paths that run along them. If the graph contains tightly connected components that are loosely connected to the other ones by some inter-group edges, the shortest paths between different components must go along these few inter-group edges and these edges will have high edge betweenness. By removing the edges with the highest betweenness in a hierarchical algorithm, we can separate different components that represent different topics of the collection.

In [2], the authors propose a fast algorithm for calculating betweenness centrality of vertices in $O(mn)$ for unweigthed graphs where m is the number of edges and n is the number of nodes. We make a minor change in this algorithm for generalizing the node betweenness centrality to edge betweenness centrality.

Starting from the constructed MI graph, we remove edges to the point that the components have acceptable number of vertices and the size of all the components are in an equal range. It is important to note that by removing the edges, we will come up with some vertices with zero degree which are ignored. Applying this algorithm to the MI graph, we achieve the major clusters of the graph. Each resulting component will be representative of a major domain of the collection.

2.3   Construction of Document Clusters

In the previous step, we obtained features that are representative of major topics in the English collection. In order to improve the quality of clusters, we enrich the features. We extract representative features of each topic of the collection using the Chi-Square feature selection. If each new feature extracted by chi-square method has relation with more than 10% of the representative features of the topic on the MI graph, we add it to the feature word list.

We use these extracted features to cluster the documents in the collection. To this end, we construct queries using the MI graph of extracted features and retrieve documents using these queries. In each topic, we first use the node centrality algorithm [2] to select the most central words. The high centrality score shows the vertices are on considerably large number of shortest paths and that they can reach to the other vertices on relatively short paths. By removing the central vertices, we divide the graph to a number of subcomponents. We then construct the queries based on these subcomponents. The words of each subcomponent are added to one query. In order to prevent constructing large queries, we divide large subcomponents to more than one query and add specific number of words to each query. Central words are added to the query if their corresponding vertices in the graph are connected to one or more query words.

After constructing the queries, we use a retrieval model to rank the top 1000 documents based on their similarities to the queries. The more times a document appears in the results, and the higher it is ranked, the higher score it gets. The documents are then sorted based on their scores. In this model documents can appear in more than one topic, we assign a document to a cluster if its score in that topic is higher than its score in the other topics. Major topics of the collection are extracted and at next step, we derive the same topics from the Persian collection.

2.4 Extraction of Related Topics in the Persian Collection

Now that we have clustered the English collection to major topics, we extract corresponding clusters in the Persian collection. Having the English features of a topic, we use an English-Persian dictionary to translate the features to Persian. We use the top 3 translations for each word in the dictionary. Some English words are translated to phrases in Persian (containing verbs, prepositions, and stop words), and some of the translations are not related to the considered topic, so we need to clean up the translated features and remove the noise as much as possible. To this end, we first break down all translated phrases to corresponding words, and remove the stop words. We then use the Persian mutual information graph extracted from the Persian collection to recognize words related to the topic: we first remove the words that are not connected to the other words and then we construct Persian queries using the final feature words for the cluster. We extract central features and subcomponents from the MI graph and construct the queries using discussed methods in section 2.3. We finally rank and cluster the Persian documents in the same way as we did for English documents. At the end of this step, we come up with an English and Persian cluster for each topic.

2.5   Document Alignment

As discussed before, we align the documents in the comparable corpus either based on concept similarity and publication date or based on concept similarity and topic. In order to align the documents in the whole collections based on the concept similarities and publication dates, we use the method proposed in [6]. We first construct a query for each English document by applying the RATF formula. We then translate the query words to Persian using a dictionary, Google machine translation system for the words not found in the dictionary, and Google transliteration system for the words which are still not translated. The idea for this kind of alignment is that if an English document and a Persian document have a high similarity score and are published at close dates, they are likely talking about a related event. We apply a combination of three different score thresholds ($\theta_1 < \theta_2 < \theta_3$) to search for suitable document alignments. The threshold becomes tighter for more distant publication dates.

In the second phase, in order to align the documents based on the concept similarities and topics, we extract topics as explained in section 2.2, construct the corresponding English and Persian clusters and align the documents in these two clusters. We align two documents in the corresponding clusters of a topic if their similarity score is higher than $\theta\%$ of all the similarity scores on the topic. We should set the θ threshold to a high value in order to prevent low quality alignments. The combination of all obtained alignments comprises our final comparable corpus.

3   Experiments and Results

In this section, we explain our experiments on creating the Persian-English comparable corpus as well as extracting word associations from the corpus and using them in cross-lingual IR. In our experiments, we used two independent news collections, the English news articles of BBC News dated from Jan. 2002 to Dec. 2006, and the Persian news articles of Hamshahri newspaper published between 1996 and 2007. We used the Lemur toolkit[3] as our retrieval system and Porter stemmer to stem the English words. All the parameters are tuned to their best values.

3.1   Creating and Evaluating the Comparable Corpus

To construct the comparable corpus, first we select major representative features of the collection and construct their mutual information graph using the method discussed in section 2.1. We further omit the features that rarely occur in the collection: we remove the words with collection frequency less than 80 and we set $MI\_thr = 0.001$. In this way, we come up with 53142 extracted features. We then cluster the MI graph to its subcomponents by removing central edges. In our experiments, we force the size of each cluster to be greater than 100 words.

---

[3] http://www.lemurproject.org/

Table 1. Statistics on the proposed comparable corpus

# of alignments	14979
# of unique English (BBC) documents	10724
# of unique Persian (Hamshahri) documents	5544
# of alignments using concept similarity and date	8041
# of alignments using concept similarity and topic	6938

By removing 6000 edges, the termination conditions are met and 6 different clusters are obtained in the English collection. To cluster the documents in the collection using the method discussed in 2.3, we select 20 central words of each topic and add them to the queries. We also divide large subcomponents to more than one query, adding 10 words to each query.

Table 2. Evaluation of the proposed comparable corpus (a) The coverage percentage of BBC classes (b) Quality of alignments for BBC articles published between 1st of Dec. 2006 and 13th of Dec. 2006

(a)

BBC Class	% of Coverage in UTPECC	% of Coverage in Proposed CC
Health	3.3	19.4
Technology	3.8	20.3
Business	6.1	20
Nature	8.1	24

(b)

	# of Alignments	% of Alignments
Class1	7	11.67
Class2	18	30.0
Class3	18	30.0
Class4	15	25.0
Class5	2	3.33
Total	60	100

In the experiments we set the score thresholds $\theta_1, \theta_2, \theta_3$ to the best values reported in [6] which are $\theta_1 = 60$, $\theta_2 = 80$ and $\theta_3 = 85$. In the second phase in which we align documents in clusters based on concept similarities and topics, we set the threshold to a tighter value, $\theta = 98$, to prevent adding low quality alignments. Table 1 shows some statistics about the proposed comparable corpus.

To assess the coverage of the comparable corpus on different topics, we leverage the BBC News classes. Each article in BBC News is tagged based on its topic. We have not used these tags for our topical clustering for two reasons. First these classes have different granulites. For example, the topic of one class is 'health' which is very general and the topic of another class is 'boxing'. Second the topics of some of the classes are not very meaningful. For example some articles are classified based on the region that the event occurred in like Europe, Africa, and pacific Asia. Although these tags are not good topic indicators for our clustering purposes, we can use them to compare the coverage of our comparable corpus with the existing ones. We compute the coverage percentage of the most meaningful classes for the UTPECC. We also measure the coverage percentage of these classes in the proposed corpus. The results are shown in Table 2 (a). As can be seen, the proposed method increases the coverage percentage of the classes considerably.

We further show that although we increase size of the corpus, the proposed corpus has proper quality. We evaluate the proposed corpus by measuring the quality of alignments. We manually assess the quality of alignments whose BBC news publication dates are between 1st of Dec. 2006 and 13th of Dec. 2006, on a five-level relevance scale [3]. The five levels of relevance are: (Class 1) Same story, (Class 2) Related story, (Class 3) Shared aspect, (Class 4) Common terminology, and (Class 5) Unrelated. The results in Table 2 (b) show that most of the alignments belong to classes (1) through (4) and only few unrelated documents are aligned in the proposed comparable corpus. As discussed in [3], for extracting terms for CLIR applications, classes (1) through (4) are helpful and a corpus with such alignments is considered a high quality corpus.

Table 3. Word associations for eight English words and their associations extracted from UTPECC

English Word	Persian Word (Proposed)	Google Translation of Persian Word	Persian Word (UTPECC)	Google Translation of Persian Word
wimbledon	تنیس	tennis	فوتبال	football
corrupt	فساد	corruption	انتخابات	election
pollut	آلاینده	pollutants	کیوتو	kyoto
south	جنوبی	south	کشور	country
anti	ویروس	virus	کشور	country
kidnie	کلیه	kidnei	دیابت	diabetes
blood	خون	blood	بیماری	illness
industri	تولید	production	سال	year

3.2 Extracting Word Associations

As the second criterion to examine the proposed comparable corpus, we try to extract word association knowledge from the corpus. Using the method proposed in [17] with a minor change. We replace the $Maxtf_k$, maximum term frequency in document d_k, in the formula [17] by $Avgtf_k$. As discussed in [15], and confirmed in our experiments, $Avgtf$ normalization performs better than $Maxtf$.

Intuitively, a high quality corpus should lead to high quality associations. Table 3 shows some sample Persian-English word associations extracted from the proposed corpus and its comparison with the associations extracted from UTPECC. For each English word, we show the top associated Persian words. The English words are stemmed and suffixes are omitted. We also report the

Persian word Google translations for the readers not familiar with Persian. As can be seen, most of the extracted associations from the proposed corpus have very high quality and they show considerable improvement compared to the associations extracted from UTPECC corpus.

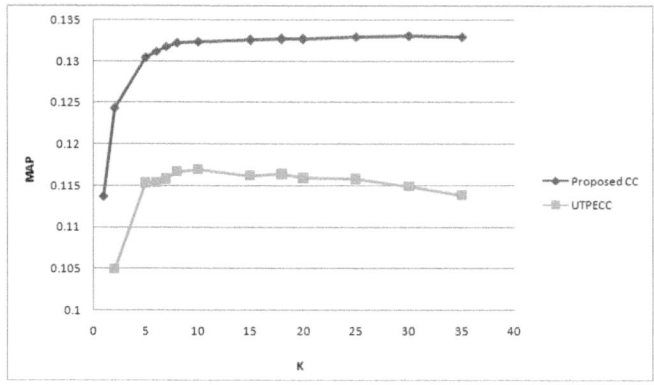

Fig. 1. Mean average precision with different number of word associations

3.3   Cross-Language Information Retrieval Experiments

In this step, we use the extracted word associations to do cross language information retrieval. As the cross-language information retrieval task, we focus on the CLIR task of CLEF-2008. We also repeat some of the CLIR experiments on the INFILE collection (CLEF INFILE track). In our experiments, we use the top K extracted word associations to translate English queries to Persian. We do an exponential transformation on the scores of the associations and then normalize the scores to obtain word pair probabilities. We use the obtained probabilities to construct the Persian query language model for each English query and retrieve documents using the KL-divergence retrieval model [9]. We run a monolingual retrieval as baseline. We did not stem the Persian queries because of the lack of high performance Persian stemmer and we only use the topic field of each query. For evaluating our experiments we use mean average precision (MAP), precision at top 5, precision at top 10 and recall of 1000 top documents.

We repeat our experiments using different values for K. Figure 1 depicts the MAP-K graph which shows resulting MAP for different values of K using the associations extracted from both the proposed corpus and UTPECC. The result shows that using this weighting system and by increasing in the amount of K, the resulting MAP does not change and thus the selection of best K is not an issue in our retrieval system. It also shows the improvement in the quality of extracted associations compared to the associations extracted from UTPECC.

In our experiments on CLEF-2008 CLIR task, the Persian document collection and CLIR are both on the same Hamshahri collection. In order to have a fair set of experiments and to make sure that the extracted associations are

Table 4. The performance of cross language retrieval

Method (Collection)	MAP	% of Mono	Prec @5	% of Mono	Prec @10	% of Mono	Recall	% of Mono
Mono (Hamshahri)	0.4081		0.648		0.624		0.868	
UTPECC	0.1169	28.64	0.224	34.5	0.216	34.6	0.468	53.9
The Proposed CC	0.1331	32.61	0.252	38.8	0.218	34.9	0.546	62.9
Mono (INFILE)	0.211		0.344		0.322		0.634	
UTPECC	0.085	39.8	0.1125	32.7	0.10	31.05	0.481	75.15
The Proposed CC	0.1272	60.28	0.1917	55.72	0.1854	57.57	0.615	97

not biased toward the Hamshahri corpus, we ran the set of experiments on the INFILE data set as well. Table 4 shows the results for both collections. The results on Hamshahri collection show that using the extracted associations from the proposed comparable corpus, compared to the monolingual retrieval, we can achieve up to 32.61% of MAP and also compared to the experiments using the UTPECC, we can achieve 13.85% improvement in MAP. The results on INFILE collection also show that using the extracted associations from the proposed comparable corpus we can achieve up to 60.28% of monolingual MAP and compared to the experiments using the UTPECC, we can achieve 49.64% improvement in MAP. This improvement shows the improvement in quality of the associations and quality of the proposed comparable corpus.

Table 5. Retrieval results using the extracted knowledge from Wikipedia, dictionary, and the proposed comparable corpus

Method	MAP	% of Mono	Prec @5	% of Mono	Prec @10	% of Mono	Recall	% of Mono
Dic&Wiki	0.209	51.2	0.392	60.49	0.360	57.6	0.534	61.5
CC+Wiki(k=2)	0.227	55.6	0.428	66	0.371	59.4	0.596	68.6
CC+Dic(k=1)	0.226	55.3	0.476	73.4	0.411	65.8	0.636	73.2
CC+Dic+wiki(k=1)	0.24	58.8	0.476	73.4	0.419	67.2	0.636	73.2

We set up another experiment to combine two other translation resources with our extracted word associations to improve the performance of cross language IR. The first resource is an English-Persian dictionary which contains more than 50,000 entries. The second resource is the Persian-English translation knowledge extracted from Wikipedia using the method proposed in [12]. For each English query word, we compute the intersection of Persian translations extracted from the proposed comparable corpus and those extracted from Wikipedia and/or the dictionary. We sort the result Persian word list, I, based on the association scores. If the size of I is smaller than K, the number of translations for the query word, we extend the list with translations extracted from the comparable corpus. In this approach we do not blindly add translations from dictionary and translation knowledge of Wikipedia to the queries, but instead using these resources, we emphasize the importance of common translations in different resources.

In this experiment, we retrieve documents using the simple-KL-divergence model and we set up K to different values and we report the best results. Table 5 shows the results of CLIR using different combinations of translation resources. The results show that using the dictionary or the translation knowledge extracted from Wikipedia improves the quality of the extracted associations from the corpus by rearranging the associations. Results in Table 5 show this method improves the MAP of the proposed method using the associations as the only resource by 80.4% and the MAP of the method using the dictionary and translation knowledge of Wikipedia by 14.8%.

4   Conclusions and Future Work

In this work, we constructed a Persian-English comparable corpus from two independent news collections. We proposed a method to construct a comparable corpus based on concept similarities and topics. This approach enabled us to align similar documents whose publication dates are distant, avoiding alignments of unrelated documents with common words. We assessed the quality of the proposed corpus using a five-level relevance scale. We also extracted the word associations from the corpus and used these associations as translation knowledge in cross language information retrieval. Experiment results show considerable improvement in performance of CLIR compared to the same method extracting translation knowledge from the existing Persian-English comparable corpus [6]. We also set up an experiment to combine a dictionary and the translation knowledge extracted from Wikipedia [12] with extracted associations from the corpus. Using the three resources, we achieved up to 58.8% of MAP compared to the monolingual baseline and were able to improve the MAP of the method using the same dictionary and extracted knowledge from Wikipedia [12] by 14.8% that shows the high quality of the extracted associations.

In future works, we are going to construct a comparable corpus using other resources like web pages instead of the news collections. We will also try to construct a bigger corpus that covers more topics. We are going to focus on extraction of translation knowledge from the corpus and improving the quality of the translations using other useful resources like the web pages and search engine results. We will also focus on CLIR task to improve the performance of Persian-English cross language IR.

Acknowledgments. This research is partially supported by Iran Telecommunication Research Center (ITRC).

References

1. Bekavac, B., Osenova, P., Simov, K., Tadić, M.: Making monolingual corpora comparable: a case study of Bulgarian and Croatian. In: LREC 2004, pp. 1187–1190 (2004)

2. Brandes, U.: A faster algorithm for betweenness centrality. Journal of Mathematical Sociology 25, 163–177 (2001)
3. Braschler, M., Schäuble, P.: Multilingual Information Retrieval Based on Document Alignment Techniques. In: Nikolaou, C., Stephanidis, C. (eds.) ECDL 1998. LNCS, vol. 1513, pp. 183–197. Springer, Heidelberg (1998)
4. Collier, N., Kumano, A., Hirakawa, H.: An application of local relevance feedback for building comparable corpora from news article matching. NII J. (Natl. Inst. Inform.) 5, 9–23 (2003)
5. Girvan, M., Newman, M.E.J.: Community structure in social and biological networks. Proceedings of the National Academy of Sciences of the United States of America 99, 404–409 (2001)
6. Baradaran Hashemi, H., Shakery, A., Faili, H.: Creating a Persian-English Comparable Corpus. In: Agosti, M., Ferro, N., Peters, C., de Rijke, M., Smeaton, A. (eds.) CLEF 2010. LNCS, vol. 6360, pp. 27–39. Springer, Heidelberg (2010)
7. Huang, D., Zhao, L., Li, L., Yu, H.: Mining large-scale comparable corpora from chinese-english news collections. In: COLING 2010, pp. 472–480 (2010)
8. Karimi, S.: Machine Transliteration of Proper Names between English and Persian. Ph.D. thesis, RMIT University, Melbourne, Victoria, Australia (2008)
9. Lafferty, J., Zhai, C.: Document language models, query models, and risk minimization for information retrieval. In: SIGIR 2001, pp. 111–119 (2001)
10. Munteanu, D.S., Marcu, D.: Improving machine translation performance by exploiting non-parallel corpora. Comput. Linguist. 31(4), 477–504 (2005)
11. Pirkola, A., Leppanen, E., Järvelin, K.: The RATF formula (Kwoks formula): exploiting average term frequency in cross-language retrieval. Information Research 7(2) (2002)
12. Rahimi, Z., Shakery, A.: Creating a Wikipedia-based Persian-English word association dictionary. In: IST 2010, pp. 562–567 (2011)
13. Sharoff, S.: Creating general-purpose corpora using automated search engine queries. In: WaCky! Working Papers on the Web as Corpus (2006)
14. Sheridan, P., Ballerini, J.P.: Experiments in multilingual information retrieval using the spider system. In: SIGIR 1996, pp. 58–65 (1996)
15. Singhal, A., Buckley, C., Mitra, M.: Pivoted document length normalization. In: SIGIR 1996, pp. 21–29 (1996)
16. Steinberger, R., Pouliquen, B., Ignat, C.: Navigating multilingual news collections using automatically extracted information. CIT 13(4), 257–264 (2005)
17. Talvensaari, T., Laurikkala, J., Järvelin, K., Juhola, M.: Creating and exploiting a comparable corpus in cross-language information retrieval. ACM TOIS 25(4) (2007)
18. Talvensaari, T., Pirkola, A., Järvelin, K., Juhola, M., Laurikkala, J.: Focused web crawling in the acquisition of comparable corpora. Information Retrieval 11, 427–445 (2008)
19. Tao, T., Zhai, C.X.: Mining comparable bilingual text corpora for cross-language information integration. In: SIGKDD 2005, pp. 691–696 (2005)
20. Utsuro, T., Horiuchi, T., Chiba, Y., Hamamoto, T.: Semi-Automatic Compilation of Bilingual Lexicon Entries from Cross-Lingually Relevant News Articles on WWW News Sites. In: Richardson, S.D. (ed.) AMTA 2002. LNCS (LNAI), vol. 2499, pp. 165–176. Springer, Heidelberg (2002)

A Web Knowledge Based Approach
for Complex Question Answering

Han Ren[1], Donghong Ji[1,2], Chong Teng[1], and Jing Wan[2]

[1] School of Computer, Wuhan University, Wuhan, China
{hanren,dhji,tengchong}@whu.edu.cn
[2] Center for Study of Language and Information, Wuhan University, Wuhan, China
Jennifer_wanj@yahoo.com.cn

Abstract. Current researches on Question Answering concern more complex questions than factoid ones. Since complex questions are investigated by many researches, how to acquire accurate answers still becomes a core problem for complex QA. In this paper, we propose an approach that estimates the similarity by topic model. After summarizing relevant texts from web knowledge bases, an answer sentence acquisition model based on Probabilistic Latent Semantic Analysis is introduced to seek sentences, in which the topic is similar to those in definition set. Then, an answer ranking model is employed to select both statistically and semantically similar sentences between sentences retrieved and sentences in the relevant text set. Finally, sentences are ranked as answer candidates according to their scores. Experiments show that our approach achieves an increase of 5.19% F-score than the baseline system.

Keywords: Question Answering, Answer Sentence Acquisition, Probabilistic latent semantic analysis, Topic Model.

1 Introduction

Current researches on Open-Domain Question Answering (QA) mainly target more complex questions than factoid ones. The ciQA track in TREC2007 [1] focused on 'Relationship' questions, which is defined as the ability of one entity to influence another, including both the means to influence and the motivation for doing so. In NTCIR-8, complex questions [2] are taxonomically divided into five types(Event, Definition, Biography, Relationship and Why). Sentences that contain correct responses are extracted as answer candidates.

Complex questions refer to complex relations between semantic concepts, or synthesizing processes of deep knowledge as well as rich information need. Take a question-answer pair in Table 1 as an example. The question is a definition one, in which the information need is supposed to be the characteristic of some people who have much concern about the security of computer and network, and their activities of breaking computer security as well. Therefore, Answer 1 and Answer 2 are both correct answers for the question.

M.V.M. Salem et al. (Eds.): AIRS 2011, LNCS 7097, pp. 470–478, 2011.
© Springer-Verlag Berlin Heidelberg 2011

Table 1. A question answer pair

Question	*What is the hacker?*
Answer 1	*In computer security, a hacker is someone who focus on security mechanisms of computer and network systems.*
Answer 2	*Hacking in the sense of breaking computer security had already been in use as computer jargon, but there was no public awareness about such activities.*

Essentially, the complexity of a complex question lies in two folds: (1) it is difficult to acquire user requirements from the question and; (2) the answer to the complex question is a mixture of complex semantic relations and should be accurate as while as non-redundant. Most current researches combined the two problems as one core problem, namely how to acquire accurate answers. To this end, many systems aim at investigating the sentence patterns of the complex questions, i.e., leverage the syntactic styles of complex sentences and convert them to patterns for answer sentence retrieval. Wu et al. [3] extracted definitional patterns from the Wikipedia data using regular expressions. Cui et al. [4] employed soft patterns (also known as probabilistic lexico-syntactic patterns), which were produced by unsupervised learning. These approaches are supported either by manual work or by annotated corpus more or less. On the other hand, some approaches consider the two problems independently. Harabagiu [5] decomposed the complex questions to the factoid ones using WordNet, and answers to the complex questions are appropriately combined as the answer to the original question. Such approaches cast the problem of acquiring information need of the complex questions to that of question decomposition, by which the unspecific information need is alternated to the specific concepts in terms of hyponymy. However, general lexico-semantic resources against some specific resources, i.e., paraphrases or manually collected entailment rules, may result in low performance because lexical alternation doesn't take the semantically related context into account.

For a better performance, most systems employ specific knowledge bases such as Wikipedia or Encarta; and the essential reason is that the specific knowledge involves almost entire information need for complex questions. Hickl et al. [6] searched the original complex questions into Wikipedia and calculate similarity between sentences in Wikipedia and the corpus. Zhang et al. [7] utilized multiple web knowledge bases to improve EAT acquisition. However, most of these systems achieve low performances, since complex questions imply complex(i.e., deep semantic) relations between terms, whereas those systems just statistically retrieve and rank documents or sentences by centroid-based (or bag-of-words) methods.

In this paper, we treat the two problems independently. For acquiring user requirements in questions, we summarize texts from web knowledge bases, i.e., Wikipedia, as the answer references. At the answer acquisition stage, we employ a topic model for answer sentence acquisition to extract sentences that may contain the answers. By mapping into a latent semantic space, sentences that are semantically related with the questions are selected and ranked as the answer candidates. We also consider linguistic information in answer ranking and employ the web knowledge bases to expand the questions.

The rest of the paper is organized as follows. In Section 2, we give methods of answer acquisition model in detail. Section 3 discuss the experimental results. Finally, the conclusion and the future work are given in Section 4.

2   Complex Question Answering Based on Web Knowledge Bases

2.1   Summarization from Web Knowledge Bases

Although web knowledge bases are cleaned up and organized by manual work, they still contain insignificant information that may decrease the performance of document retrieval or answer ranking. In order to get significant information for a question, some systems make use of summarization methods to acquire important portions from web knowledge bases. In this case, the performance of the systems depends on the quality of summarized texts. In other words, when the contents related to a question in web knowledge bases are not rich, the initial query generated is insufficient to retrieve semantically related documents. Since some web knowledges such as Wikipedia provide the links to combine the relevant concepts of documents, they can be leveraged to obtain more reliable texts. Ye et al. [8] proposed a novel approach that produces summaries with various lengths. By building an extended document concept lattice model, concepts and non-textual features such as wiki article, infobox and outline are combined. Experiments showed that system performance outperformed not only traditional summarizing methods but also some soft-pattern approaches. In this paper, we utilize the approach to summarize contents in Wikipedia. Sentences summarized for each question are put into a text set.

Although most relevant texts can be found in Wikipedia, we still utilize other two web knowledges as a supplement. Following an unsupervised summarization approach proposed by Ji et al. [9] to rank sentences, text summarized are put into the text set without sentence redundancy.

The summarization method carries out with the following steps. First, extract the words and named entities from a question as well as eliminate the stop words or clue words, i.e., interrogative words, from the questions. Note that a longer word in the question gives the more reliable indication for the summarization method. For example, the question "What is the greenhouse effect" contains two words "greenhouse" and "greenhouse effect" that are both indexed by Wikipedia, whereas "greenhouse effect" is more reliable than "greenhouse" for the question. More specifically, summarized texts of the search result of "greenhouse effect" are more semantically relevant than that of "greenhouse" for the question. Then, make use of the words and the named entities to search from web knowledge bases. All texts summarized from the searched results are finally combined without redundancy.

2.2   Answer Sentence Acquisition Using Topic Model

For a complex question, the accurate answer sentences mostly indicate the complex semantic relations within it, whereas the bag-of-word models that compute the similarity between the answers and the questions are clearly insufficient. To this end, some systems [11,12] employ Latent Semantic Analysis(LSA) model to build a semantic layer, in which the similarity between documents and words, or documents and documents are capable to compute by using the semantically latent relations. However, the meaning of the decomposition algorithm is indefinite so that the model could lead to an uncontrolled   performance for retrieval. For a clear decomposition meaning, Probabilistic Latent Semantic Analysis(PLSA) model is introduced to build a semantic space of underlying topics, in which words and documents can be mapped as vectors. Some QA systems utilize PLSA to improve answer validation, i.e., modeling languages for document relevance estimation. In our system, we employ PLSA model as the answer sentence retrieval model. Following is the description of the PLSA model in our system.

Given a sentence set S, a term set W and a topic set Z, the conditional probability of sentence-term P(s, w) can be described as follows:

$$P(s, w) = P(s) \sum_{z \in Z} P(w|z)P(z|s) \tag{1}$$

In (1), P(w | z) represents the conditional probability of words in latent semantic layers(or topics), P(z | s) represents the conditional probability of topics in sentences. Here the count for topic set Z is between 20 and 100. Then the model fits with the EM algorithm and export the optimal P(Z), P(W | Z) and P(Z | S). When a new query is coming, it is projected to the topic space Z by using EM algorithm. The similarity of the query and each sentence can be acquired by computing the similarity of the probabilistic distribution between them in the topic space.

Our algorithm is described as follows. First, initialize the P(s, w) for each sentence $s \in \{s_1, s_2, ..., s_n\}$ in the retrieved documents(here we make use of a general retrieval model Lucene to search words and named entities extracted in section 2.1 and get the retrieved documents) by using the ratio of the frequency of w in s and in the sentence set; and P(w | z) for each word and P(z | s) for each sentence are iteratively computed by EM algorithm. Then the query built from the question is mapped into the topic space to compute   by using EM algorithm, keeping P(w | z) invariably. After that, the conditional probability $P(w|q)$ and $P(w|s)$ for each sentence in the retrieved documents is computed according to the formula (2)-(3).

$$P(w|q) = \sum_{z \in Z} P(w|z)P(z|q) \tag{2}$$

$$P(w|s) = \sum_{z \in Z} P(w|z)P(z|s) \tag{3}$$

Finally, a cosine similarity method is utilized to compute the similarity between the query q and each sentence in retrieved documents, and the sentences that the weighting values are above a threshold are selected as the answer sentences.

$$Sim(s,q) = \frac{\sum_{w \in W} P(w|s)P(w|q)}{\sqrt{\sum_{w \in W} P(w|s)^2}\sqrt{\sum_{w \in W} P(w|q)^2}} \qquad (4)$$

For a better performance, we submit each question to Wikipedia and summarizing the results as the additional queries. Then we compute the similarity between each sentence in retrieved documents and query sentence set. Sentences that their similarities are above the threshold are selected into the answer set.

2.3   Answer Ranking

For a better performance of answer validation and ranking, some approaches employ various resources, whereas most of them are language-dependent and less helpful for the complex questions. To implement a general system, our answer ranking model considers two main factors, namely semantically structural and bag-of-word features of sentences. The motivation is, a sentence can be a potential answer candidate if 1) the sentence and the question are semantically relevant and; 2) most of the words in the sentence are also appear in questions or sentences derived from web knowledge bases. More specifically, although the retrieved sentences have the latent semantic similarity with questions, we should still consider the long distance dependency relations that could probably result in a low score for an answer candidate. On the other hand, a statistical similarity that treats the sentences as a bag of words could probably balance the impact of the semantic bias.

For semantic similarity, we consider the similarity of the main semantic roles because they cover most of the meaning of a sentence and easy to acquire. We choose the verb based labeling architecture derived from PropBank, in which 'predicate', 'subject', 'object' and some modifiers are the core roles for a sentence. For each sentence in answer candidate set we only label PRED, A0-A4, AM-LOC and AM-TMP and combine each predicate and the corresponding argument as well as its type of semantic dependency relation to a pair. The structure of a pair is described as follows:

$$\{w_1|Predicate, w_2|Argument, Role\}$$

If a pair lies in both a sentence and a question, it is viewed as a matched pair of within them. More specifically, every term and its semantic dependency relation should be matched if the pair is matched. Following is the weighting formula that we compute the semantic similarity:

$$Sim(s,q_i) = \frac{\sum_{p_s \in s, p_{q_i} \in q_i} \{sim(p_s, p_{q_i})\}}{max\{\sum p_s, \sum p_{q_i}\}} \qquad (5)$$

p_s denotes the pair in sentence s, while p_{q_i} denotes the pair in query q_i. $sim(p_s, p_{q_i})$ denotes the similarity between p_s and p_{q_i}. Here the similarity defines as a boolean value, representing whether the two pairs are full matched.

For labeling of semantic roles, we utilize a system that we proposed in [10] to extract pairs. The system handles syntactic dependency parsing with a transition-based approach and utilizes MaltParser[1] as the base model. The system also utilizes a Maximum Entropy model to identify predicate senses and classifies arguments.

For statistical similarity, we simply utilize a cosine similarity to compute it. The weighting formula of our method for answering ranking is as follows:

$$w_s = max\{\alpha \cdot Sim(s, q_i) + (1 - \alpha) \cdot cos(s, q_i)\} \tag{6}$$

α is an adjusting parameter, and q denotes each sentence in the query sentence set described in the previous subsection. According to (5), the sentence that mostly similar to a query is acquired. Finally, 30 answer sentences is extracted and ranked by their ranking scores.

3   Experimental Results

We select NTCIR-8 CCLQA CS dataset as experimental document collection. The collection contains Chinese questions of five types: Definition(10), Biography(10), Event(20), Relationship(20) and Why(20). The baseline system in our experiments makes use of the same architecture with JAVELIN [13], whereas it does not take answer types into account and simply extracts noun phrases from questions as key terms. For answer extraction and ranking, the system selects sentences that contain key terms from high ranked documents. Table 2 shows the result of the experiment.

Table 2. Overall Performance (F-score)

	Definition	Biography	Relationship	Event	Why	Average
this paper	0.1933	0.1514	0.2065	0.1697	0.1082	0.1658
baseline	0.1359	0.1178	0.1505	0.1146	0.0508	0.1139

The results indicate that our approach based on web knowledge bases can improve the overall performance of the complex QA system. We can also see that, results for definition and relationship questions are better than those for other type of questions. It is mainly because the information need of the questions of these two types are simplex, while the questions of other types concern more complex relations. Take the biography type as an example, temporal terms almost appear at every biographic text whereas they can not represent a rich semantic concept; the PLSA retrieval model can not make a relationship with them and a topic in layer Z. Thus those sentences may

---

[1] http://w3.msi.vxu.se/~jha/maltparser/

not have a tight relationship with definitions from web knowledge bases although they could be answers. We also notice that, the system achieves a very low performance for the question of the type 'Why'. It is mainly because that the semantic relations in these questions are more complex than others. Actually, most answers logically related with the question rather than synonyms or shallow semantic ones. For example, when asking 'Why do the Shenzhou spacecrafts always launched in cold season', the answer is probably like a reasoning chain:

The spacecraft is recovered by the survey vessels. →
The survey vessels are mainly located in the southern hemisphere. →
The recovery task befits in summer. →
The climate is different between the southern and the northern hemisphere. →
For the sake of recovery, Shenzhou spacecraft always launched in cold season.

However, most keywords, i.e., southern hemisphere, survey vessels and recovery task, do not appear at the question. Alternately, sentences in answers have logical(more specifically, casual) relations that should be inferred from one to another. The answer to the question can be acquired only if all the relative events are chained through inference. Hence the inference models should be considered for a better performance of 'Why' questions.

Table 3. Comparsion of system performances based on different parts (F-score)

	Definition	Biography	Relationship	Event	Why	Average
VSM+key term	0.1167	0.0723	0.1290	0.0923	0.0406	0.0902
PLSA+key term	0.1564	0.1187	0.1602	0.1272	0.0690	0.1263
PLSA+sem	0.1859	0.1422	0.1896	0.1533	0.0905	0.1523
ts+PLSA+sem	0.1933	0.1514	0.2065	0.1697	0.1082	0.1658

We also investigate performance of each part in our system. *ts* means the summarization method in our approach mentioned in Section 2.1. For sentence retrieval, two models, which are based on VSM and PLSA mentioned in Section 2.2, are also involved in the experiments. For answer ranking, as a comparison, we invoke the approach of the baseline system that rank sentences by key terms to replace our method proposed in Section 2.3(sem). Table 3 shows the result of the experiments.

From Table 3 we can see that when using VSM model to replace PLSA, the average performance decreases 3.61%. Noticeably, the average performance decreases 2.6% by using key term method to replace the method of computing semantic plus statistical similarity. Data show the facts that: (1) effective summarizing method from Wikipedia improves ordinary QA systems to a better performance; (2) a large number of synonyms as well as data sparse phenomenon exist in texts, thus topic based models can be more fit than centroid-based models for definition QA systems; (3) a sentence that just contains key terms could not be considered as an answer unless it has semantic pattern features that the question bears.

4   Conclusion

In this paper, we propose an approach to acquire answer sentences from documents effectively. The main contribution to our approach lies in two folds: 1) a PLSA model based answer acquisition approach is introduced for complex questions; and 2) semantic and statistical information are combined in the answer ranking model for a better performance. Although a general performance is achieved according to the official evaluation results, we still face two important problems: 1) with the improvement of the complexity of questions, deep knowledge in them should be acquired by some inference mechanisms; 2) more refined nugget extraction model are needed to improve the precision of the system.

Acknowledgements. This research is supported by Natural Science Foundation of China(Grant Nos. 90820005, 61070082) and the Post-70s Scholars Academic Development Program of Wuhan University.

References

1. Dang, H.T., Kelly, D., Lin, J.: Overview of the TREC 2007 Question Answering Track. In: Proceedings of the Sixteenth Text Retrieval Conference. NIST, Gathersburg (2007)
2. Mitamura, T., Shima, H., Sakai, T., Kando, N., Mori, T., Takeda, K., Lin, C.-Y., Song, R., Lin, C.-J., Lee, C.-W.: Overview of the NTCIR-8 ACLIA Tasks: Advanced Cross-Lingual Information Access. In: Proceedings of the 8th NTCIR Workshop, Tokyo, Japan (2010)
3. Wu, Y., Chen, W., Kashioka, H.: NiCT/ATR in NTCIR-7 CCLQATrack: Answering Complex Cross-lingual Questions. In: Proceedings of the Seventh NTCIR Workshop Meeting, Tokyo, Japan (2008)
4. Cui, H., Kan, M.-Y., Chua, T.-S.: Unsupervised Learning of Soft Patterns for Generating Definitions from Online News. In: Proceedings of Thirteenth World Wide Web Conference (2004)
5. Harabagiu, S., Lacatusu, F., Hickl, A.: Answering Complex Questions with Random Walk Models. In: Proceedings of SIGIR 2006, Seattle, Washington (2006)
6. Hickl, A., Roberts, K., Rink, B., Bensley, J., Jungen, T., Shi, Y., Williams, J.: Question Answering with LCC's CHAUCER-2 at TREC 2007. In: Proceedings of the Sixteenth Text Retrieval Conference. NIST, Gathersburg (2007)
7. Zhang, Z., Zhou, Y., Huang, X., Wu, L.: Answering Definition Questions Using Web Knowledge Bases. In: Dale, R., Wong, K.-F., Su, J., Kwong, O.Y. (eds.) IJCNLP 2005. LNCS (LNAI), vol. 3651, pp. 498–506. Springer, Heidelberg (2005)
8. Ye, S., Chua, T.-S., Lu, J.: Summarizing Definition from Wikipedia. In: Proceedings of the Fourty-Seventh Annual Meeting of the Association for Computational Linguistics, Singapore (2009)
9. Ji, P.: Multi-Document Summarization Based on Unsupervised Clustering. In: Ng, H.T., Leong, M.-K., Kan, M.-Y., Ji, D. (eds.) AIRS 2006. LNCS, vol. 4182, pp. 560–566. Springer, Heidelberg (2006)
10. Ren, H., Ji, D., Wan, J., Zhang, M.: Parsing Syntactic and Semantic Dependencies for Multiple Languages with A Pipeline Approach. In: Proceedings of the Thirteenth Conference on Computational Natural Language Learning, Boulder, Colorado (2009)

11. Tomás, D., Vicedo, J.L.: Re-ranking Passages with LSA in a Question Answering System. In: Peters, C., Clough, P., Gey, F.C., Karlgren, J., Magnini, B., Oard, D.W., de Rijke, M., Stempfhuber, M. (eds.) CLEF 2006. LNCS, vol. 4730, pp. 275–279. Springer, Heidelberg (2007)
12. Yu, Z., Fan, X., Guo, J., Geng, Z.: Answer Extracting for Chinese Question-Answering System Based on Latent Semantic Analysis. Chinese Journal of Computers 10, 1889–1893 (2006)
13. Mitamura, T., et al.: JAVELIN III: Cross-Lingual Question Answering from Japanese and Chinese Documents. In: Proceedings of the Sixth NTCIR Workshop Meeting, Tokyo, Japan (2007)

Learning to Extract Coherent Keyphrases from Online News

Zhuoye Ding, Qi Zhang, and Xuanjing Huang

Fudan University
School of Computer Science
{09110240024,qz,xjhuang}@fudan.edu.cn

Abstract. Keyphrases extracted from news articles can be used to concisely represent the main content of news events. In this paper, we first present several criteria of high-quality news keyphrases. After that, in order to integrate those criteria into the keyphrase extraction task, we propose a novel formulation which coverts the task to a learning to rank problem. Our approach involves two phases: selecting candidate keyphrases and ranking all possible sub-permutations among the candidates. Three kinds of feature sets: lexical feature set, locality feature set and coherence feature set are introduced to rank the candidates, and then the best sub-permutation provides the keyphrases. The proposed method is evaluated on a multi-news collection and experimental results verify that our proposed method is effective to extract coherent news keyphrases.

Keywords: keyphrase extraction, learning to rank, Support Vector Machine.

1   Introduction

Keyphrase extraction is a long studied topic in natural language processing. A keyphrase, which consists a word or a group of words, is defined as a precise and concise expression of one or more documents. In recent years, keyphrase extraction has received much attention [2,11,3,6,7].

Keyphrases are usually manually chosen by authors, for scientific publications, magazine articles, books, et al. Because the manual effort of assigning keyphrase is expensive and time consuming, web pages and online news rarely contain keyphrases. It should be useful to automatically extract keyphrases from online news to represent their main contents. There are already a number of studies which focus on extracting keyphrases from scientific publications or single news article [13,8,10,4]. We also notice that, currently, many websites provide the service which group related news together to facilitate users browsing. In this paper, we focus on extracting keyphrases from a group of news articles which describe the same news event by different publishers.

Most of the current methods focus on judging the importance of each phrase, and individually extract phrases with the highest scores. In this paper, we regard the keyphrases as a sequence and aim to extract syntactically coherent

M.V.M. Salem et al. (Eds.): AIRS 2011, LNCS 7097, pp. 479–488, 2011.

keyphrases for summarizing a collection of news articles about the same subject or event. After analyzing the human assigned keyphrases, we observe that the keyphrases of news should satisfy the following properties:

1. **Relevance.** The keyphrases should be semantically relevant to the news theme. The most important ones should be selected as keyphrases.
2. **Indication.** The keyphrases should be indicative of the whole news event.
3. **Coherence.** Here, coherence means keyphrases should be not only semantically related but also syntactically coherent.
4. **Conciseness.** We should keep least redundancy in keyphrases.

In order to automatically select keyphrases which can satisfy the above properties, we propose a keyphrase extraction approach under a learning to rank framework. This approach involves two phases: selecting candidate keyphrases and ranking all sub-permutations among candidates. Three feature sets: lexical feature set, locality feature set and coherence feature set are introduced to rank the candidates, and then the best sub-permutation is extracted as keyphrases.

The major contributions of this work can be summarized as follows: 1) We present a novel approach based on learning to rank framework to extract coherent keyphrases from news. 2) Keyphrases are extracted as a sequence, and the relationships among them are considered. 3) Some novel features are introduced to improve the quality of keyphrases. 4) Experimental results on the dataset consisting of 150 groups of news articles with human annotated keyphrases demonstrate that our method is effective to extract coherent news keyphrases.

The remaining of the paper is organized as follows: In section 2, we review the related work and the state-of-the-art approaches in related areas. Section 3 presents the proposed method. Experimental results in the test collection of "NewsKEX" are shown in section 4. Section 5 concludes this paper.

2   Related Work

Existing methods can be categorized into supervised and unsupervised approaches.

Unsupervised approaches usually select general sets of candidates and use a ranking step to select the most important candidates. For example, Mihalcea and Tarau propose a graph-based approach called TextRank, where the graph nodes are tokens and the edges reflect co-occurrence relations between tokens in the document [9]. Wan and Xiao expand TextRank by using a small number of topic-related document to provide more knowledge, which improves results compared with standard TextRank and a tf.idf baseline [13]. Liu and Sun argued that the extracted keyphrases for documents should be understandable, relevant, and good coverage and an unsupervised method based on cluster was proposed to extract high-quality keyphrases [6].

Supervised approaches use a corpus of training data to learn a keyphrase extraction model that is able to classify candidates as keyphrases. A well known supervised system is Kea that uses all n-grams of a certain length as candidates, and ranks them based on a Naive Bayes classifier using tf.idf and position as its features [2]. Extractor is another supervised system that uses stems and stemmed n-grams as candidates [11]. Turney introduces a feature set based on statistical word association to ensure that the returned keyphrase set is coherent [12]. Hulth argues that some linguistic knowledge, such as NP chunks and POS tag, can improve the performance when learning model is applied [3].Nguyen and Kan present a keyphrase extraction algorithm for scientific publications [10].

3   Learning to Extract Keyphrases

3.1   Learning to Rank

Recently, learning to rank technologies have been intensively studied. Different from classification and regression, the goal of learning to rank is to learn a function that can rank objects according to their degree of preference, importance, or relevance[14,1].

In this paper, we employed a kind of learning to rank method to perform our two-phase task. We will take the phase 2 as an example to explain why it is suitable to employ a pairwise learning to rank method. The reason is that, it is more natural to consider the likelihood of a sub-permutation's being keyphrase sequence in a relative sense than in an absolute sense. For example, given two phrase sequences p1 and p2, suppose, p1 is more suitable to be keyphrase sequence than p2, but the absolute scores of them can not be easily evaluated. So we should consider the problem in a relative sense and a pairwise method can be employed. Specifically, in pairwise learning to rank framework, the optimization goal is p1's priority to p2 in the ranking list. And their absolute scores are not our concern.

3.2   Two-Phase Ranking Approach to Keyphrase Extraction

In this section, learning to rank model is employed for our two-phase task.

Candidate Keyphrase Selection. In this step, we are aiming to select suitable phrases as candidate keyphrases according to their importance. First, we regard each word w as a ranking instance, create a feature vector for each w and train a ranking model to sort the words by employing a pairwise learning to rank framework described in section 3.1. Then, top-n phrases are selected as candidate keyphrases(n is assigned to 6 after tuning the parameter).

Keyphrase Extraction from Candidates. This step is to find the indicative, coherent and concise keyphrases from the candidate keyphrases. So we need to find, not only the appropriate keyphrases from the candidates, but also the appropriate order. To accomplish this goal, we generate all the possible sub-permutations of candidate keyphrases. However, there will be exponential number of possible sub-permutations. To reduce the number of candidate sub-permutations, we use the following strategies. A sub-permutation will be selected if both conditions are satisfied.

1. Select only sub-permutations composed of two to four phrases. As observed by our annotators, most news has keyphrases in this range.
2. Select only sub-permutations with POS tag sequence which exists in the training data.

Then each sub-permutation is regarded as a ranking instance. We employ a pairwise learning to rank framework to sort all sub-permutations and select the top one as keyphrase sequence.

3.3   Features in Two Phases

In this section, we will describe the features we used in two phases. The two phases of keyphrase extraction share some similar features, especially, lexical features and locality features. The difference between them is that features for candidate keyphrases selection are in word-level, while features for keyphrases extraction are in phrase sequence level.

Single Word Features for Candidate Keyphrase Selection. In this section, features are all in word-level, including TF-IDF score, position of a word's first occurrence, a word's appearance in news title [4] and keyword prior. Here, TF-IDF score and keyword prior are lexical features, and the other two are locality features. Actually, the features in this paper are all based on multi-news.

Phrase Sequence Features for Keyphrase Extraction. Similar to phase 1, the lexical and locality feature set are also used to represent a keyphrase sequence, but their meaning is slightly different - this time, they are defined on phrase sequence level. Additionally, a new set of features for measuring coherence are introduced in this section. Next we will explain each feature set in detail.

Lexical Feature Set

1) TF.IDF
Here we calculate the average TF.IDF value for the keyphrase sequence of K.

$$TF.IDF(K) = \sum_{w \in K} \frac{TF.IDF(w)}{n}$$

Where n is the total number of words in the keyphrase sequence K.

Locality Feature Set

1) Appearance in Title
The case of word existence in title will offer evidence to keyphrases extraction. So we also calculate the average value for keyphrase sequences.

$$InTitle(K) = \sum_{w \in K} \frac{InTitle(w)}{n}$$

n is the number of words in the keyphrase sequence.

2) First Occurrence in the Article

When given sub-permutation K, the feature can be calculated as:

$$FirstOccur(K) = \sum_{w \in K} \frac{FirstOccur(w)}{n}$$

Where w is a word in the keyphrase sequence K, and n is the number of words which appears in K.

Coherence Feature Set

1) Mutual Information(MI)

Turney argued that keyphrases should be semantically related [12]. So a well-known phrase association measure mutual information is applied. To tackle the situation of arbitrary number of variables, the two-variable case can be extended to the multivariate case. The extension called multivariate mutual information (MVMI) can be generalized to:

$$I(p_1, p_2, ...p_k) = \frac{1}{n} \sum_{i,j=\{1,2...k; i \neq j\}} I(p_i; p_j)$$

where n is the number of pair $I(p_i, p_j)$. $i, j = \{1, 2...k; i \neq j\}$

$$I(p_i; p_j) = p(p_i, p_j) \log \frac{p(p_i, p_j)}{p(p_i)p(p_j)}$$

and $p(p_i, p_j)$ is co-occur times of p_i and p_j within a window of 10 words[5].

2) Sequence-Based

Sequence-based feature aims to guarantee a suitable sequence for keyphrases by ensure the consistence between keyphrase sequence and news article. Specifically, we obtain the sequence-based feature by calculating average of pairwise values for keyphrase sequence.

$$Sequence(K) = \frac{2}{m(m-1)} \sum_{i<j} (t(p_i, p_j) - t(p_j, p_i))$$

Where m is the number of terms in keyphrase sequence, and $t(p_i, p_j)$ is the frequency that phrase p_i exists in front of phrase p_j in a sentence in the news article.

3) Syntactic Feature

Hulth argued that linguistic features can help when extracting keyphrases[3]. So in this paper, syntactic feature are applied to ensure the coherence of keyphrases. Specifically, dependency relation of keyphrase sequence is employed.

In this step, keyphrase sequence are first parsed by a dependency parser, in which 24 dependency relations (e.g. SBV, VOB, ATT...) are defined.

In a dependency tree of keyphrase sequence, their dependency relation sequence are applied to represent the keyphrase sequence. For instance, (SBV,VOB) can represent the keyphrase sequence "Toyota" "Recall" "Prius". Dependency relations can provide evidence to the coherence of keyphrases to some extent. We can acquire possible dependency relation sequences from the training data, and score them according to their occurrence probabilities. The value can be calculated as:

$$DependencyValue(K) = \frac{num(Dependency(K))}{\sum_{K_i} num(Dependency(K_i))}$$

Where $num(Dependency(K))$ is the number of dependency relation sequence of K the in training set.

4   Experiment

4.1   Experimental Data

There are almost no publicly available datasets with manually annotated gold standard keyphrases for news, due to the high expense of labor and time for manual annotation. In this experiment, we randomly selected 150 groups of online news articles from Goolge News. We divided the dataset into two subsets: 100 for training and 50 for test. The detail of our dataset is shown in table 1.

Table 1. Details of the "NewsKEX" dataset

Description	value
News Articles	1103
Words	345K
News Articles per group	7.35
Labeled Keyphrases per group	4.6

In the phase of candidate keyphrases selecting, human annotators are asked to label the words with two levels of relevance, which represent whether the word is a part of keyphrases or not. In the second phase, human annotators are asked to label the each sub-permutation of candidate keyphrases with three levels. We assign level 2 to sub-permutations which are the most suitable as keyphrase sequences and level 1 to those probable as keyphrase sequences, others are assigned with level 0.

4.2   Evaluation Measures

In order to evaluate news keyphrases clearly, we introduce two kinds of evaluation measures: quantitative measures and qualitative measures.

Quantitative Measures

For evaluate the obtained keyphrases, we first report the in quantitative measures: Precision, Recall, and F-value. Precision means the percentage of "keyphrases truly extracted" among "keyphrases extracted by system". Recall means that "keyphrases truly extracted" among "keyphrases manually assigned". F1 is the average of Precision and Recall.

Qualitative Measures

We also ask 6 human annotators to evaluate the quality of extracted keyphrases, meaning that whether they can meet the properties we have presented: indication, coherence and conciseness. The average Kappa statistic among them is around 85%, which shows good agreement. Then we respectively report Indication, Coherence and Conciseness to measure these three performances. The measure is calculated by the number of news which obtains indicative (coherent or concise) keyphrases divided by the total number of news.

4.3   Performance of Keyphrase Extraction

We applied the RankSVM implementation available in the SVM light package[1]. Table 2 shows the different performance based on seven different feature sets. We report the quantitative results in Precision, Recall, F-value, as well as qualitative results in Indication, Coherence, Conciseness. The results show that F4 feature set obtains the best performance in Precision and F6 gets the best in Recall. These confirm that locality feature and lexical feature are both effective to extract keyphrases. The table also illustrates that coherence feature set can help to improve the overall quality of keyphrases. When combined with coherence feature set, the performances are significantly improved, especially, in measure Indication, Coherence and Conciseness. And F7 feature set obtains the best performance in measure F-value, Indication and Coherence.

Table 2. Comparison of news keyphrase extraction in Precision, Recall, F-value, Indication(Ind), Coherence(coh), Conciseness(con) for different feature sets. The best in each column is highlighted.

Feature set	Precision	Recall	F-value	Ind	Coh	Con
F1: Lexical	82.00%	67.13%	73.83%	32%	40%	68%
F2: Locality	89.00%	76.07%	82.03%	40%	70%	86%
F3: Coherence	84.00%	84.90%	84.45%	68%	78%	80%
F4: Lexical+locality	**93.3%**	77.3%	84.56%	42%	68%	**92%**
F5: Lexical+Coherence	85.33%	84.23%	84.78%	70%	84%	88%
F6: Locality+Coherence	83.50%	**86.07%**	84.76%	**72%**	78%	82%
F7: Lexical+Locality+Coherence	87.00%	85.40%	**86.19%**	**72%**	**86%**	86%

---

[1] http://kodiak.as.cornell.edu/svm_light

4.4 Comparing with other Algorithms

In this section, we implement two baseline methods on the same dataset for comparison and here our approach is based on F7 feature set applied due to the best performance.

Baseline-1: Baseline-1 is based on a classification method, SVM. We classify phrases in the multi-news documents as keyphrases and non-keyphrases by applying SVM method. The features include TF-IDF, InTitle, FirstOccur, Keyphrases prior.

Baseline-2: The title of news articles provides a good enough summary or keyphrase sequence. So baseline-2 is performed based on the titles of news articles. We sort the phrases in multi-news titles according to the TF*IDF scores and select top-k as keyphrases. We assign k to 4 after tuning the parameter.

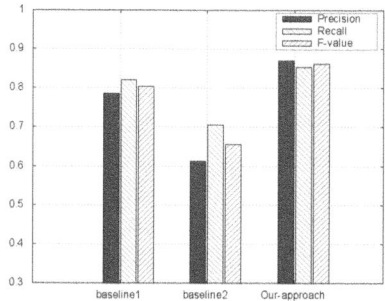
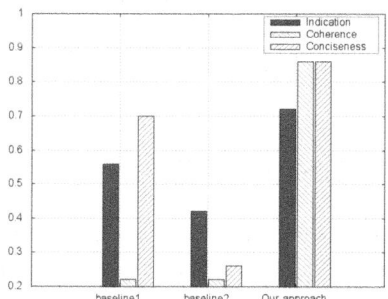

Fig. 1. Comparison of news keyphrases extraction in quantitative measures(Precision, Recall, F-value) for two baseline methods and our approach

Fig. 2. Comparison of news keyphrases extraction in quantitative measures(Indication, Coherence, Conciseness) for two baseline methods and our approach

Figure 1 shows the results in quantitative measures. The results illustrate that our approach perform better than both two baseline approaches in quantitative measures. As shown in Figure 2, our approach obtain significant improvement in qualitative measures compared with two baseline methods, since our method introduces some coherence features in phrase sequence level to ensure the quality of keyphrases. These confirm that our approach can significantly improve the quality of news keyphrases with no loss in quantitative measure. The results also show that we can't obtain high-quality keyphrases simply based on titles of news article.

4.5 Discussion

Contribution of Individual Feature. To determine the contribution of individual feature, we employ our approach, omitting one feature each time and evaluation measurements including quantitative measures, as well as qualitative measures. Table 3 shows our results in quantitative measures. We can find that each feature is effective. Specifically, when eliminating locality features "InTitle" it results in a decrease in quantitative measures. The observations imply that the locality features are discriminative in predicting the keyphrases, because the information of occurrence of title in the articles are really helpful. Table 3 also shows the lexical feature TF.IDF score have strong connection to quantitative measures, which support that lexical information is helpful in prediction the importance of phrases in news articles. We also find when eliminating coherence features "sequence-based" it results in a notable decrease in all quantitative measures. That confirms that consistence between keyphrase sequence and news articles is helpful in quantitative measures.

Table 3. Contribution of each feature in quantitative measures Precision, Recall, F-value. The best in each column is highlighted.

Features	Precision	Recall	F-value
NoTFIDF	83.5%	86.07%	84.76%
NoInTitle	85.33%	84.23%	84.77%
NoFirOcc	87.00%	85.0%	85.99%
NoMI	82.67%	**86.97%**	84.76%
NoSeq	77.67%	63.73%	70.01%
NoSyn	86.00%	84.90%	85.45%
All	**87.00%**	85.40%	**86.19%**

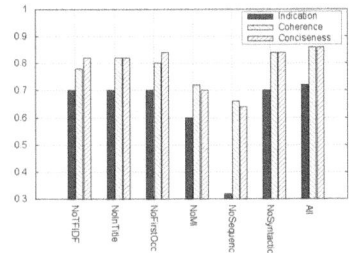

Fig. 3. Contribution of news keyphrases extraction in measure Indication, coherence, conciseness for different features

As shown in figure 3, when omitting the coherence features such MI and Sequence-based, the qualitative measure: Indication, Coherence and Conciseness significantly decreases. That implies coherence feature are effective when extracting high-quality keyphrases. Additionally, the feature of "InTitle" and "TFIDF" are also related to qualitative measure.

5   Conclusion

In this paper, we try to extract coherence keyphrases from online group news instead of single news. we first present several criteria of high-quality news keyphrases. After that, in order to integrate those criteria into the keyphrase extraction task, we propose a novel formulation which coverts the task to a learning to rank problem. The proposed model has been verified on multi-news collections and the results show that our approach is effective to extract coherence keyphrases. Additionally, the effect of different feature sets and even contributions of individual feature are analyzed. We find that all of three kinds of feature set can help to extract higher quality keyphrases. Future works include developing ranking algorithm and trying other measures representative of performance for keyphrases.

Acknowledgements. The author wishes to thank the anonymous reviewers for their helpful comments. This work was partially funded by National Natural Science Foundation of China (61003092, 61073069), Shanghai Science and Technology Development Funds(10dz1500104), Doctoral Fund of Ministry of Education of China (200802460066), and Key Projects in the National Science & Technology Pillar Program(2009BAH40B04).

References

1. Cao, Z., Qin, T., Liu, T.Y., Tsai, M.F., Li, H.: Learning to rank: From pairwise approach to listwise approach. In: Proceedings of the International Conference on Machine Learning (2007)
2. Frank, E., Paynter, G.W., Witten, I.H., Gutwin, C., Nevill-Manning, C.G.: Domain-specific keyphrase extraction. In: Proceedings of IJCAI, pp. 668–673 (1999)
3. Hulth, A.: Improved automatic keyword extraction given more linguistic knowledge. In: Proceedings of EMNLP, pp. 216–223 (2003)
4. Jiang, X., Hu, Y.H., Li, H.: A aanking approach to keyphase extraction. In: Proceedings of SIGIR (2009)
5. Kumaran, G., Carvalho, V.R.: Reducing long queries using query quality predictors. In: Proceedings of SIGIR (2009)
6. Liu, Z.Y., Li, P., Zheng, Y.B., Sun, M.S.: Clustering to find exemplar terms for keyphrase extraction. In: Proceedings of EMNLP, pp. 257–266 (2009)
7. Liu, Z., Huang, W., Zheng, Y., Sun, M.: Automatic keyphrase extraction via topic decomposition. In: Proceedings of EMNLP, pp. 366–376 (2010)
8. Matsuo, Y., Ishizuka, M.: Keyword extraction from a single document using word co-occurrence statistical information. International Journal on Artificial Intelligence Tools (2004)
9. Mihalcea, R., Tarau, P.: Textrank: Bringing order into texts. In: Proceedings of EMNLP, pp. 404–411 (2004)
10. Nguyen, T.D., Kan, M.Y.: Keyphrase Extraction in Scientific Publications. In: Goh, D.H.-L., Cao, T.H., Sølvberg, I.T., Rasmussen, E. (eds.) ICADL 2007. LNCS, vol. 4822, pp. 317–326. Springer, Heidelberg (2007)
11. Turney, P.D.: Learning algorithms for keyphrase extraction. Information Retrieval, 303–336 (2000)
12. Turney, P.D.: Coherent keyphrase extraction via web mining. In: Proceedings of IJCAI, pp. 434–439 (2003)
13. Wan, X., Xiao, J.: Single document keyphrase extraction using neighborhood knowledge. In: Proceedings of AAAI, pp. 855–860 (2008)
14. Xia, F., Liu, T.Y., Wang, J., Zhang, W., Li, H.: Listwise approach to learning to rank - theory and algorithm. In: Proceedings of ICML (2008)

Maintaining Passage Retrieval Information Need Using Analogical Reasoning in a Question Answering Task

Hapnes Toba, Mirna Adriani, and Ruli Manurung

Faculty of Computer Science, University of Indonesia
Depok 16424, Indonesia
hapnes.toba@ui.ac.id, {mirna,maruli}@cs.ui.ac.id

Abstract. In this paper we study whether a question and its answer can be related using analogical reasoning by using various kinds of textual occurrences in a question answering (QA) task. We argue that in a QA passage retrieval context, low cost language features can contribute some positive influence in the representation of the information need that also appears in other passages, which have some analogical features. We attempt to leverage this through query expansion and query stopwords exchange strategies among analogical question answer pairs, which are modeled by a Bayesian Analogical Reasoning framework. Our study by using ResPubliQA 2009 and 2010 dataset shows that the predicted analogical relation between question answer pairs can be used to maintain the information need of the QA passage retrieval task, but has a poor performance in determining the question type. Our best accuracy score was achieved by using *'bigram occurrences by using stemmer and TF-IDF weighting completed with named-entity'* feature set for the query expansion approach, and *'bigram occurrences by using stemmer and TF-IDF weighting'* feature set for the stopwords exchanged approach.

Keywords: Bayesian Analogical Reasoning, Question Answering System, Passage Retrieval, Query Expansion, ResPubliQA.

1 Introduction

Question Answering System (QAS) is a form information retrieval that tries to produce an exact answer given a natural language question. Despite its natural task to find a single answer, a QAS needs supporting textual context from one or more document collections, in the size of a sentence, a passage, a paragraph or even the whole document [1]. This is one of the reasons why question answering also considered as a challenging field in information retrieval.

Most typical QAS pipeline architectures consist of four main components, i.e.: question analyzer, query formulation, information retrieval and answer validation. The task of a question analyzer is to classify a question into one or more question types, which will be used as the expected answer type during the answer validation phase. In the query formulation component, a question will be formulated into a

M.V.M. Salem et al. (Eds.): AIRS 2011, LNCS 7097, pp. 489–498, 2011.

specific keyword-based query, for instance by using bag-of-words (BOW) approach after removing stopwords, or by using WordNet for term expansion [2]. In the information retrieval component, usually by using third-party search engines, such as: Indri, or Lucene, appropriate top-n textual candidates will be retrieved. Finally, the answer validation component needs to validate whether a retrieved answer candidate reflects some information need, with respect to the expected answer type, and produce a final single answer. The difficult task of constructing a final answer will be made easier if the final answer is already included in a limited set of passage retrieval results [3, 4, 5]. In this context, the performance of an underlying information retrieval system is important to retrieve relevant passages.

Recent works in information retrieval strategies that are specific to the question answering (QA) task are mostly focused on: linguistic and semantic constraints [4, 6], relevance feedback [7], semantic role labeling [8] or by topic indexing [9]. Despite these recent approaches, performing QA passage retrieval in a more conventional information retrieval way, i.e. by using textual features consisting of appropriate question terms, could be preferable if important search terms are already stated in the question. Recently, a new approach has been developed that focuses in the relational data between existing questions answer pairs [10]. By assuming that answers are related to their questions through certain types of implicit links, it is theoretically possible to learn these links from existing data, and to apply the learned model for relating unseen questions to their appropriate answers.

Table 1. Example of overlapping information need between QA pairs collections

QA Pairs Collections	Question	Passage Gold Standard
ResPubliQA 2010 (#91, question type: Factoid)	In which country will the **2010 FIFA World Cup** be held?	Repeats its demand that the Mugabe regime … value from either the run-up to the **2010 World Cup** or the tournament itself; in this regard, calls on [South Africa], the host nation, and on **FIFA** to exclude Zimbabwe … in pre-World Cup matches, … national teams involved in the event;
ResPubliQA 2009 (#7, question type: Factoid, feature: 'unigram occurrences')	**In** which **areas** will objective information be provided on drugs and drugs addiction?	The Centre's **objective** is to provide, [in the areas referred to in Article 4], the Community and … with **objective**, reliable and comparable information … … **drugs and drug addiction** and their consequences.
Overlapping unigram between QA pairs	in, which, will, be, on, and, the	

Inspired by this work, we study whether a question and its answer can be related using low cost language features in a QA passage retrieval scenario. We argue that in a QA passage retrieval context, low cost language features, such as n-gram, can actually contribute some positive influence to represent the information need that also appear in other passages, which have some analogical or related features. Table 1 gives an example of such a case, the words in **bold** show the overlapping words between the question and the answer, <u>underlined</u> words show the overlapping words between both question answer pairs, and the text surrounded by '[ ]' are the exact answers.

In Table 1, seems that the two questions have different question types. The first question could be classified into *'COUNTRY'* question type and the second one into *'LOCATION'* question type. But if we consider the QA pairs as a relation, the answers of both question are directing into something in common, i.e. a kind of named-entity, by using the question word *'WHICH'*, either it is about 'location of an event' or 'location of a section in a regulation document'. In this way, we could define an analogy as measure of similarity between structures of related objects.

2   Question Answering Approach

As stated in [2, 3], most typical QA architectures consider questions and answers as independent elements. The consequence of this kind of architecture is that question type and the related expected answer type cannot be learned in a single learning mechanism framework. To compensate for the independence of a question and its answer pair, we exclude the question type component in our approach, and use a single learning mechanism framework to learn the relations of a question and its answer pair. We propose to use the related features of a question and its answer as a means to recognize the information need of a question, which at the same time could also give an indication of how a question should be answered. The related features are learnt from a collection of question answer pairs, in which the answer is given in the form of a passage. The complete approach can be seen in Fig. 1.

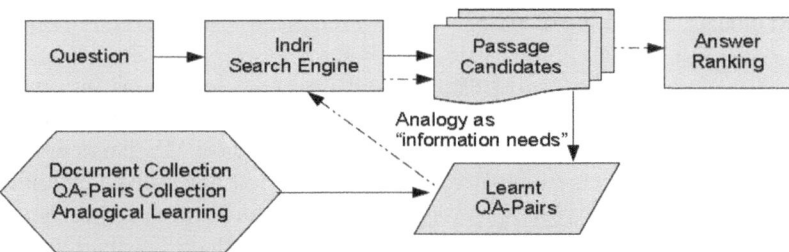

Fig. 1. Proposed QAS Approach. Question analysis component is excluded and the learnt question answer pairs are used as a means to recognize the information need.

In our approach, we assume that all words that appear in a question are important and have influences during the information retrieval phase. Each passage candidate and its related question will be compared to a set of question answer pairs which relation has been learnt by using an analogical learning mechanism. This comparison value will produce a new score that shows how information need from a passage candidate is related to the set of learnt question answer pairs, which also depends on the feature set during the analogical learning. We hypothesized that it is theoretically possible to exchange those related features or to enhance the original question, and use them to re-run the query in a second phase (shows as dotted-line in Fig. 1), to form the final passage answer.

3   Bayesian Analogical Reasoning

Bayesian Analogical Reasoning (BAR) was originally introduced in [11, 12], which basic idea is to learn model parameters and priors from related objects, and update it during the comparison process of a query to obtain marginal probability that relates the query with the objects that have been learnt.

Assume there is a space of unseen functions $Q \times A \rightarrow \{0,1\}$. If two objects, a question Q and an answer A are members of a set S, which are related by an unknown function $f(Q,A) = 1$, what needs to be quantified is how similar the function $f(Q,A)$ is to another unseen function $g(.\,,\,.)$, that classifies all pairs of $(Q^i, A^j) \in S$ as being linked where $g(Q^i, A^j) = 1$. The functions $f(.,\,.)$ and $g(.\,,\,.)$ are unseen, and thus we need a set of priors that will be used to integrate them over the function space.

Suppose for each pair $(Q^i \in Q, A^j \in A)$, there exists a feature vector:

$$X^{ij} = \left[\phi_1(Q^i, A^j)...\phi_k(Q^i, A^j)\right]^T \text{ , defined by the mapping: } \Phi : Q \times A \rightarrow \Re^K \text{ , as a single}$$

point of link representation on the feature space Φ.

This feature space mapping computes a K-dimensional vector of features of the question answer pairs, which is hoped to have a relevant link prediction between the objects in the pairs. The feature vector X^{ij} , for each pair of question and answer consists of the same number of features, and thus we can define a measure as the link representation between such pair. In this case we use the cosine distance.

If there is an unseen label L^{ij} , with $L^{ij} \in \{0,1\}$ as a predicted indicator of the existence of a relation between Q^i and A^j, then we will have a model parameters vector $\Theta = [\Theta_1 ... \Theta_K]^T$ which models the presence or absence of interaction between objects, and could be learnt by performing the logistic regression model:

$$P\left(L^{ij} = 1 / X^{ij}, \Theta\right) = logistic\left(\Theta^T X^{ij}\right) \tag{1}$$

where $logistic(x)$ is defined as: $1/(1 + e^{-x})$.

The priors are learnt by using:

$$P(\Theta) = N(\tilde{\Theta}, (c\tilde{T})^{-1}) \tag{2}$$

where $\tilde{\Theta}$ is the Maximum Likelihood Estimator (MLE) of Θ $N(m, v)$ is a normal of mean m and variance V. Matrix $\tilde{T}$ is the empirical second moment's matrix of the link object features, and c is a smoothing parameter, which is set to the number of links that exist in the trained set.

During the retrieval process of linked pairs, a query is compared by the functions for links prediction by marginalizing over the parameters of the functions. If we have L^S as the vector of link predictions for S, then each $L \in S$ has the value $L = 1$, indicating that every pair of objects in S is linked. The final score of a retrieval process indicating the order of predicted links between the query and the related objects that has been learnt, and is compute as follows:

$$score\left(Q^i, A^j\right) = \log P\left(L^{ij} = 1|X^{ij}, S, L^S = 1\right) - \log P\left(L^{ij} = 1|X^{ij}\right) \tag{3}$$

Silva et al. in [11, 12] use the variational logistic regression to compute this scoring function.

4   Experimental Setting

To evaluate the influence of analogical reasoning in our question answering scenario, we use Indri [13, 14], to retrieve the top-5 passage candidates of the testing question set, in a bag-of-words approach. We evaluate the performance in terms of accuracy at the top-1 answer against the gold standard. The accuracy is computed according to the formula:

$$C@1 = tp/(tp + fp) \tag{4}$$

where tp is the number of 'true' answer at the top-1, and fp is the number of 'wrong' answer.

We use the question answer pairs from ResPubliQA 2009 paragraph selection gold standard as our training set, and ResPubliQA 2010 as our testing data. To maintain the question type's equality between the two question collection sets, we use the: *Definition* (95 questions), *Factoid* (139), *Reason-Purpose* (187), and *Procedure* (79) question types during the experiments, and exclude the Opinion and Other question types. In total, we have 500 question answer pairs from ResPubliQA 2009 collection, and 133 questions from ResPubliQA 2010, which consist of: *Definition* (32 questions), *Factoid* (35), *Reason-Purpose* (33), and *Procedure* (33) .

The document collections that were used during this study are JRC-ACQUIS and EUROPARL [15]. We created an index that was based on paragraph segmentation with Indri indexing tools. In total, we have about 1.5 million paragraphs indexed that were considered as documents. Indri is a search engine that is specially designed for passage retrieval, thus will be fitted to the retrieval task in this study [14]. The works in [15] showed that paragraph selection is a challenging task, and one of the successful methods is to improve paragraph retrieval by using overlapping uni- and bigram occurrences' as contextual information. This is the main motivation that we explored the following textual feature sets during the experiments:

- unigram occurrences;
- unigram occurrences after using (Porter) stemmer;
- unigram occurrences by removing stopwords and using a stemmer;
- unigram occurrences by using TF-IDF weighting;
- unigram occurrences by using stemmer and TF-IDF weighting;
- unigram occurrences completed with named-entity;
- bigram occurrences;
- bigram occurrences by using a stemmer and TF-IDF weighting;
- bigram occurrences by using stemmer and TF-IDF weighting completed with named-entity;
- named-entity occurrences, by using Stanford NER and dictionary-based NER [2].

Finally, we decomposed the feature sets into SVD 25-dimension, as the main feature dataset, in order to reduce the word features dimensionality.

During the passage retrieval phase, we made two types of query enhancement. The first one is to add overlapping non-stopwords word occurrences, which appear in the top-5 retrieved analogous question answer pairs, to the original question, as a kind of query expansion method. The second one is to use the stopwords that appear in the best analogous question answer pair, to complete the stopwords that was removed from the original question. Each query is considered as a BOW model. To evaluate our expansion method, we also run some experiments of the original question by using Indri pseudo-relevance feedback, which were set to some configuration of smoothing parameters and weights for the original query [14, 16].

5   Results and Discussions

We present our results in the following aspects: the accuracy performance of the query expansion strategies, the comparison of the performance with respect to Indri relevance feedback, and the influence of the retrieved analogous pairs to the passage retrieval ranking performance.

5.1   Performance of Query Expansion

The result of the re-run scenario by using overlapping of non-stopwords which occur in the top-5 of analogical pairs is given in Table 2. The best accuracy score was achieved by the *'bigram occurrences by using stemmer and TF-IDF weighting completed with named-entity'* feature set, i.e. 0.31. For our example in Table 1, by using the *'bigram-stem-TFIDF-ne'* feature set, the question will be enriched by some other non-stopwords term(s) that occur in the top-5 analogical pairs, as follows : *"In which country will the 2010 FIFA World Cup be held + European"*.

The low accuracy performance is mainly influenced by out-of-topic terms. Such cases are mostly occurred when the analogical pairs come from different topic or when the semantics of the question and answer are too far to be captured in the analogical model. For example the question: *"what is maladministration?"* (Q#6-2010: Definition type), the analogical model only considered the word *"what is"*, as related important features, and thus fail to enrich the query with something that is related to *"maladministration"*.

To create another view of retrieval performance, a running result of the original questions that were expanded using WordNet is included in Table 2. This expansion strategy is performed for every verb and noun from the original questions [2, 17]. The result is mostly below the accuracy of our approach. The original questions were mostly enriched with out-of-topic terms which decreased the retrieval accuracy. For our example in Table 1, the query would be expanded as follows: *"In which (country OR commonwealth OR state OR land OR nation OR "res publica" OR "body politic") will the 2010 FIFA World Cup be (held OR maintained OR kept)"*.

Table 3 shows the result of the re-run scenario by using the stopwords that appear in the best analogous question answer pair. Again, the retrieval result for the WordNet query expansion is also included, which performance is lower than our approach. The best performed feature set is the *'bigram occurrences by using stemmer and TF-IDF weighting'*, with 0.34 accuracy. For our example in Table 1, the question, by using the *'bigram occurrences by using stemmer and TF-IDF weighting'*, will be reformulated as the following bag-of-words query: *"country FIFA World Cup held + What is the of A top side to of which is"*.

The stopwords exchanged have the same failure analysis as the non-stopwords enhancement. In overall the stopwords exchanged performance is slightly better in terms of accuracy than the non-stopwords expansion.

Table 2. Overlapping non-stopwords performance in decreasing order

"Overlap-of-Terms-Top5" Features	C@1
Indri BOW	0.35
Bigram-stem-TFIDF-NE	0.31
Bigram-stem-TFIDF	0.30
Named- entity	0.30
Bigram	0.29
Unigram	0.26
Unigram-stem-remove stopwords	0.26
Unigram-stem-remove stopwords-NE	0.24
Unigram-TFIDF	0.24
Unigram-stem	0.23
WordNet	0.23
Unigram-TFIDF-stem	0.21

Table 3. Stopwords exchanged performance in decreasing order

"Stopwords-Exchanged" Features	C@1
Indri BOW	0.35
Bigram-stem-TFIDF	0.34
Bigram-stem-TFIDF-NE	0.33
Unigram-TFIDF-stem	0.33
Unigram-stem	0.31
Bigram	0.29
Unigram	0.29
Unigram-stem-remove stopwords	0.28
Unigram-stem-remove stopwords-NE	0.28
Unigram-TFIDF	0.27
Named-entity	0.26
WordNet	0.23

5.2   Indri Pseudo-relevance Feedback

To evaluate the performance of our best feature set expansion approach, we compare our results to the Indri pseudo-relevance feedback of the original questions, with various parameter settings. The first parameter setting is regarding the document smoothing to overcome data sparseness problem [14]. We use Dirichlet smoothing, and experimenting with: $\mu = 2500$ *(default)*, and $\mu = 2000$ *(optimum for the query and document length)*. Those values were chosen based on the work in [16]. Another parameter setting is the Indri feedback smoothing *(fbMu = 0.0 (default), and 0.5)*, the query word weighting *(fbOrigWeight = 0.5, 0.8 and 1.0)*, the number of terms for the feedback *(fbTerms=10)*, and the number of documents for the feedback *(fbDocs=5)*. The comparison is presented in Table 4.

Table 4. Best BAR feature set (*Bigram-stem-TFIDF-NE*) of non-stopwords expansion vs. Indri pseudo-relevance feedback in decreasing order

Parameter Setting	C@1	MRR@5
Indri BOW (mu = 2500, no relevance feedback)	0.35	0.45
Indri BOW (mu = 2500, fbOrigWeight = 1, fbMu = 0)	0.35	0.45
Indri BOW (mu = 2000, fbOrigWeight = 0.8, fbMu = 0.5)	0.32	0.43
BAR expansion with '*Bigram-stem-TFIDF-NE*' (mu=2500; no rel. feedb.)	0.31	0.43
Indri BOW (mu = 2000, fbOrigWeight = 0.5, fbMu = 0.5)	0.31	0.43
Indri BOW (mu = 2500, fbOrigWeight = 0.5, fbMu = 0.5)	0.29	0.40

From Table 4, we can observe that the accuracy of our expansion approach (0.31) is quite similar to the accuracy of Indri pseudo relevance feedback (0.32). This indicates that the expanded terms of the analogical question answer pairs can maintain the information need of the original query. Further analysis on the top-5 retrieval results, in terms of Mean Reciprocal Rank (MRR) performance; give us promising results for answer validation strategy, which is beyond this study.

5.3   Question Type and Retrieval Performance Issues

Table 5 gives a number of analogous pairs examples, relating to the question type, of the *'Bigram-TFIDF-NE'* feature set in Table 2. The 'question type' classification accuracy from this feature set is 0.31. In our opinion, one of the problems is due to the term variations in the training and testing sets. The ResPubliQA collection [15] is characterized by its wide scope of questions and documents coverage in parliamentary domains. On the other hand, the BAR framework assumes that the feature space should provides a reasonable classifier to predict the existence of links. Such case is not in general decomposable as similarities between only the textual features in the question part, but also the presence or absence of the features in the answer part of related pair.

Table 5. Examples of question analogical pairs with respect to the question type (QT)

No.	2010	Question	QT	2009	Best Analogous Question	QT
1.	#73	What actions does the competent authority for maritime security of a port carry out?	PR	#338	What should be done in the case of epizootic?	PR
2.	#91	In which country will the 2010 FIFA World Cup be held?	F	#216	Who will be involved in radiotherapeutic practices?	F
3.	#105	What is the WTO Agreement?	D	#58	Why is the increase of the weight of the 50 cent coin from 7 g to 7,8 g necessary?	R
4.	#188	What was the purpose of EU states in establishing new permanent political and military bodies?	R	#418	What is the main objective of producing electricity in public thermal plants?	R

If we inspect further into each question type, for instance, from the *'Bigram-TFIDF-NE'* feature set in Table 2, this will gives us a distribution of question type accuracy that can be seen in Table 6. Such typical distribution also occurs in other feature sets in Table 2 and Table 3. The *Reason-Purpose* question type always has the best accuracy, and *Definition* question type always has the lowest one.

Table 6. Question Type Accuracy Distribution *'Bigram-TFIDF-NE'* Feature Set in Table 2

Question Type	Accuracy
Reason-Purpose	0.45
Factoid	0.37
Procedure	0.21
Definition	0.19

This result gives us an indication that the analogical relations among common bigram terms, such as: *'in order'*, *'order to'* or *'objective to'* in the *Reason-Purpose* type, could provide us much better expanded terms, in contrast to the relations of quite specific terms in the *Definition* question types, such as: *'define as'*, or *'the meaning*.

Table 7 presents some cases of expanded queries with their influence to the retrieval ranking performance.

Table 7. Some examples (as in Table 5), of expanded queries with their influence to the retrieval ranking (we only consider the top-5 retrieval)

No.	2010	Question	Expanded Q. Terms	Baseline Retrieval	After Expansion
1.	#73	What actions does the competent authority for maritime security of a port carry out?	application, competent	5	2
2.	#91	In which country will the 2010 FIFA World Cup be held?	european	1	1
3.	#105	What is the WTO Agreement?	order	> 5	> 5
4.	#188	What was the purpose of EU states in establishing new permanent political and military bodies?	account, competent, decide, order	> 5	4

The result presented in Table 7 indicates that a simple question has in fact more term variations in the answer, as for example in the *Definition* type. In contrast, a more complex question with numerous term occurrences' in the answer part, has the tendency to be more related to their analogous pair, and hence could achieved a better retrieval performance, as in the *Reason-Purpose* type.

6   Conclusions and Future Work

In general we conclude that the predicted analogical relation between question answer pairs can be used to maintain the information need of the QA passage retrieval task (c.f. section 5.2), but in the case of our experiments, analogical reasoning does a very poor job of classifying the expected answer type of a question (c.f. section 5.3). The overall passage accuracy in this study is much below the best performance of the ResPubliQA 2010 baseline [15], which is 0.73. It seems that the feature sets which were explored during the experiments are not enough to bridge the semantic gap between question and answer pairs. The choice of feature set is a crucial step in our study, which give significant influence to the retrieval results. In our study the best performed feature set is *'bigram occurrences by using stemmer and TF-IDF weighting completed with named-entity'* for the query expansion approach, and *'bigram occurrences by using stemmer and TF-IDF weighting'* for the stopwords exchanged approach (c.f. section 5.1).

Considering that we cannot always have all possibilities of question answer pairs during the training, it might valuable to aggregate patterns from *n*-most analogous question answer pairs, as recurring patterns, would seem to specify an indicative feature of the information need. Such automatic pattern generation strategy will be useful to expose question type analysis and its expected answer type in a question answering system. To address these issues we plan to conduct study in feature selection mechanism to be fitted in the analogical model as our future work.

References

1. Lin, J., Quan, D., Sinha, V., Bakshi, K., Huynh, D., Katz, B., Karger, D.R.: The Role of Context in Question Answering Systems. In: Extended Abstracts on Human Factors in Computing Systems, Fort Lauderdale, Florida, pp. 1006–1007 (2003)
2. Schlaefer, N., Gieselmann, P., Schaaf, T., Waibel, A.: A Pattern Learning Approach to Question Answering Within the Ephyra Framework. In: Sojka, P., Kopeček, I., Pala, K. (eds.) TSD 2006. LNCS (LNAI), vol. 4188, pp. 687–694. Springer, Heidelberg (2006)
3. Bilotti, M.W.: Linguistic and Semantic Passage Retrieval Strategies for Question Answering. Dissertation Thesis. Language Technologies Institute. School of Computer Science, Carnegie Mellon University (2009)
4. Aktolga, E., Allan, J., Smith, D.A.: Passage Reranking for Question Answering Using Syntactic Structures and Answer Types. In: Clough, P., Foley, C., Gurrin, C., Jones, G.J.F., Kraaij, W., Lee, H., Mudoch, V. (eds.) ECIR 2011. LNCS, vol. 6611, pp. 617–628. Springer, Heidelberg (2011)
5. Cui, H., Sun, R., Li, K., Kan, M.-Y., Chua, T.-S.: Question answering passage retrieval using dependency relations. In: SIGIR 2005, pp. 400–407. ACM, New York (2005)
6. Bilotti, M.W., Elsas, J., Carbonell, J., Nyberg, E.: Rank Learning for Factoid Question Answering with Linguistic and Semantic Constraints. In: Proceedings of CIKM (2010)
7. Pizzato, L.A., Molla, D., Paris, C.: Pseudo-Relevance Feedback using Named Entities for Question Answering. In: Australasian Language Technology Workshop, ALTW (2006)
8. Pizzato, L.A., Molla, D.: Indexing on Semantic Roles for Question Answering. In: Proceedings of the 2nd Workshop on Information Retrieval for Question Answering (2008)
9. Ahn, K., Webber, B.: Topic Indexing and Information Retrieval for Factoid QA. In: Proceedings of the 2nd ACL Workshop on Information Retrieval for Question Answering (2008)
10. Wang, X.-J., Tu, X., Feng, D., Zhang, L.: Ranking Community Answers by Modeling Question-Answer Relationship via Analogical Reasoning. In: Proceedings of SIGIR Conference (2009)
11. Silva, R., Heller, K., Ghahramani, Z.: Analogical Reasoning with Relational Bayesian-sets. In: Proceedings of AISTATS (2007)
12. Silva, R., Heller, K., Ghahramani, Z., Airoldi, E.: Ranking Relations Using Analogies in Biological and Information Networks. The Annals of Applied Statistics 4(2), 615–644 (2010)
13. Strohman, T., Metzler, D., Turtle, H., Croft, W.B.: Indri: A language-model based search engine for complex queries, http://ciir.cs.umass.edu
14. Metzler, D.: Indri Retrieval Model Overview, http://ciir.cs.umass.edu/~metzler/indriretmodel.html
15. Peñas, A., Forner, P., Rodrigo, A., Sutcliffe, R., Forascu, C., Mota, C.: Overview of ResPubliQA 2010: Question Answering Evaluation over European Legislation. In: Working Notes of CLEF ResPubliQA (2010)
16. Fang, H., Tao, T., Zhai, C.X.: A Formal Study of information Retrieval Heuristics. In: Proceeding of SIGIR. ACM, New York (2005)
17. Sun, R., Jiang, J., Tan, Y.F., Cui, H., Chua, T.S., Kan, M.Y.: Using syntactic and semantic relation analysis in question answering. In: Proceedings of the 14th Text REtrieval Conference (2005)

Improving Document Summarization by Incorporating Social Contextual Information

Po Hu[1,2], Donghong Ji[1], Cheng Sun[1], Chong Teng[1], and Yong Zhang[1]

[1] Computer School, Wuhan University, China
[2] Department of Computer Science, Huazhong Normal University, China
geminihupo@163.com, donghong_ji2000@yahoo.com.cn,
{gensun.cc,ychang.cn}@gmail.com, tchong616@126.com

Abstract. We propose a collaborative approach to improve document summarization by incorporating social contextual information into the sentence ranking process. Both the relationships between sentences from document context and the preference information from user context are investigated in the approach. We validate our method on a social tagging dataset and experimentally demonstrate that by incorporating social contextual information it obtains significant improvement over several baseline methods.

Keywords: Document summarization; social context; sentence ranking.

1 Introduction

Document summarization has been widely explored for a long time, which has shown the particular potential to simplify information consumption in many applications such as snippet creation for search engine, contextual advertising [1] [2], etc.

A variety of summarization methods have been developed, which can be roughly divided into two classes: extractive approach and abstractive approach [3]. An extract summary preserves its content in the original form, whereas to generate an abstract summary, the content from the document may be paraphrased, compressed or regenerated. In this study, we focus on extractive approach. Many extractive approaches have been proposed by making use of the local information in a document or incorporating document contextual information into the sentence ranking process [4], in which a number of unsupervised, semi-supervised and supervised algorithms are adopted [5-16].

Recently, based on the assumption that similar documents can provide more useful clues to help summarize a specified document, document expansion methods have been proposed to build an appropriate document context for summarization [17][18]. To the best of our knowledge, most methods usually only consider document or document context while ignore social contextual information (i.e. the combination of document context and user context) when summarizing a single document in an online social environment.

In this study, we propose a collaborative summarization approach called *SocialContextSum* to make use of social contextual information in the summarization

M.V.M. Salem et al. (Eds.): AIRS 2011, LNCS 7097, pp. 499–508, 2011.

process. Firstly, document context and user context are recognized from online documents through clustering algorithm. Secondly, the relationships between sentences from document context and the preference information from user context are employed to rank sentences respectively. Finally, a few most important sentences are extracted from the specified document to generate a summary by fusing document-context based score and user-context based score.

Experiments have been performed on a social tagging dataset, and the results demonstrate that the proposed approach leveraging social contextual information outperforms several baseline methods.

The rest of paper is organized as follows: Section 2 presents the details of the proposed approach. Section 3 evaluates the performance of *SocialContextSum* by comparing with other baselines. Section 4 concludes the paper with future work.

2   The Proposed Approach

SocialContextSum mainly consists of three phases: social context recognition, sentence ranking and sentence extraction. The first phase aims to discover both document context and user context for a specified document from online documents. The second phase aims to use the social contextual information to evaluate the importance of each sentence in the document. The third phase aims to extract a number of sentences with highest overall scores into the summary.

2.1   Social Context Recognition

Given multiple documents with social tagging information, in which each document needs to be summarized respectively, *SocialContextSum* incorporates both document context and user context into the collaborative summarization process.

In our study, each document d_i in the documents is represented by a combined vector D_i consisting of three component vectors (i.e. content vector $D_i^{(c)}$, user vector $D_i^{(u)}$, and tag vector $D_i^{(t)}$), which can be demonstrated as $D_i = (D_i^{(c)}, D_i^{(u)}, D_i^{(t)})$, $D_i^{(c)} = (x_{il} \mid l=1,2,...,|W|)$, $D_i^{(u)} = (y_{im} \mid m=1,2,...,|U|)$, $D_i^{(t)} = (z_{in} \mid n=1,2,...,|T|)$. Here x_{il} denotes the weight of term w_l in document d_i, which can be calculated by the *TF*IDF* scheme. $|W|$ is the total number of words in the documents. y_{im} denotes the times that document d_i is annotated by user u_m, $|U|$ is the total number of users who tagged the documents. z_{in} denotes the times that document d_i has been tagged with tag t_n, and $|T|$ denotes the total number of tags associated with the documents. Similarly, each user can be modeled with all the documents that have been tagged by the user and all the tags that have been used by the user, in our study, each user u_i is represented by a combined vector U_i consisting of two components (i.e. document vector $U_i^{(d)}$ and tag vector $U_i^{(t)}$. $U_i^{(d)} = (a_{ip} \mid p=1,2,...,|D|)$, $U_i^{(t)} = (b_{iq} \mid q=1,2,...,|T|)$, where a_{ip} is the times that document d_p has been annotated by user u_i and b_{iq} is the frequency that tag t_q has been used by user u_i. $|D|$ and $|T|$ are the total number of documents and tags respectively. Based on this representation, we can measure the similarity between any two documents (users) by their corresponding vectors.

Then the social contextual information can be recognized by clustering document vectors and user vectors respectively, in which k-means clustering algorithm is adopted in this study. Since the latent numbers of document clusters and user clusters are hard to predict in advance, in this study we simply set them to the square root of the total number of documents and users respectively.

In the clustering results, each document cluster includes topic-related documents and acts as a document context for any document in the cluster. Each user cluster includes the users with similar information preference and different user cluster shows different preferences. In most social tagging websites, a user's tagging behavior can reflect the user's information preference to certain extent. For a user cluster and two documents x and y, if the number of users who tags document x is greater than that of users who tags document y, we may assume that the user cluster is more interested in the content of document x, so we choose a number of documents that have been tagged by most of the users in each user cluster as the user context for each cluster.

2.2   Sentence Ranking Based on Social Context

For collaboratively summarizing a single document, after the social context is recognized, all the documents in the document context which the specified document belongs to are segmented into sentences and each sentence is ranked by document context and user context respectively.

2.2.1   Sentence Ranking Based on Document Context

To rank sentences based on its document context, a graph-based ranking algorithm is applied by making use of the inter-sentence relationships [19]. The inter-sentence relationships are described by a sentence affinity graph G_s with each vertex s_i representing a sentence in the document context and each edge e_{ij} representing the affinity relationship between sentence s_i and s_j ($i \neq j$) whose weight is associated with the similarity between the pair of sentences. G_s can be encoded by either the matrix M_{intra} or the matrix M_{inter} with each entry corresponding to G_s's sub-graphs G_{intra} and G_{inter}, which describe either the within-document sentence relationships or the cross-document sentence relationships. Then M_{intra} and M_{inter} are normalized to $\widetilde{M}_{intra}$ and $\widetilde{M}_{inter}$ by making the sum of each row equal to 1.

The ranking score of sentence s_i based on document context can be denoted as $Score^{dc}(s_i)$, which is calculated by the following formula:

$$Score^{dc}(s_i) = \lambda * Score^{dc}_{intra}(s_i) + (1-\lambda) * Score^{dc}_{inter}(s_i) \tag{1}$$

where λ is a weight adjusting parameter specifying the relative contribution from the within-document relationships and the cross-document relationships, which is set to 0.4 in the study. $Score^{dc}_{intra}$ and $Score^{dc}_{inter}$ are the scores of the corresponding sentence by considering either the within-document relationship or the cross-document relationship, which can be computed respectively as follows.

$$Score^{dc}_{intra}(s_i) = \delta * \sum_{all\ j \neq i} Score^{dc}_{intra}(s_j) * (\widetilde{M}_{intra})_{j,i} + \frac{(1-\delta)}{n} \tag{2}$$

$$Score^{dc}_{inter}(s_i) = \delta * \sum_{all\ j \neq i} Score^{dc}_{inter}(s_j) * (\widetilde{M}_{inter})_{j,i} + \frac{(1-\delta)}{n} \tag{3}$$

where n is the number of sentences in the document context of the specified document d, s_j is any other sentence linked with s_i, and δ is a damping factor usually set to 0.85. The convergence of the above iteration computing process is achieved when the difference between the scores calculated at two successive iterations falls below 0.0001.

2.2.2 Sentence Ranking Based on User Context

To rank sentences by considering all the user context's influences, each user context UC_k is firstly transformed into a pseudo-query-sentence q_k that is represented by the centroid vector of all the sentences in the user context. q_k can be processed in the same way as other sentences. Then, the graph-based manifold ranking algorithm is adopted to rank sentences based on each user context UC_k by naturally making use of the sentence-to-sentence relationships and the sentence-to-user-context relationships in a manifold-ranking process [16].

In this study, an affinity graph G_{uk} can be constructed in which the vertex set includes all the sentences in the specified document's document context and the kth pseudo-query associated with the kth user context, and the edges encode both the relationships among the sentences and the relationship between the kth pseudo-query and the sentences.

The ranking score of sentence s_i based on the kth user context can be denoted as $Score_k^{uc}(s_i)$, which is calculated by the Table 1.

Table 1. User-context based sentence ranking via the manifold ranking algorithm

---

Input:

G_{uk}: The affinity graph.

N: The total sentence number in the sentence set S, which includes all the sentences in the specified document's document context.

Output:

The limit value $V_{S_i}^*$ of the sentence ranking function V_S: $S \rightarrow \Re$, which refers to $Score_k^{uc}(s_i)$ in this study and can be represented as a vector $V_S = [V_{q_k}, V_{S_1}, ..., V_{S_N}]^T$ with each element V_{S_i} denoting the user-context based score of the corresponding sentence.

Process:

Step 1: Define a prior sentence vector $Y_S = [Y_{q_k}, Y_{S_1}, ..., Y_{S_N}]^T$, in which Y_{q_k} is set to 1 since it corresponds to the kth pseudo-query associated with the kth user context, which can be regarded as the only labeled seed on the affinity graph, and other vector elements in Y_S are set to 0.

Step 2: Define the affinity matrix $W_S = (w_{S_{i,j}})_{(N+1)\times(N+1)}$ with each element $w_{S_{i,j}}$ denoting the affinity weight $w(S_i, S_j)$ between the sentences S_i and S_j.

Step 3: Normalize W_S by $N_S = D_S^{-1/2} \cdot W_S \cdot D_S^{-1/2}$, where D_S is the diagonal matrix whose entry (i, i) equals to the sum of the i-th row of W_S.

Step 4: Iterate according to the following equations until convergence.

$$V_S(t+1) = \beta N_S V_S(t) + (1-\beta) Y_S$$

Where the parameter $\beta \in [0, 1]$ specifies the relative contribution to the ranking scores from the neighborhood sentences and the initial scores, which is set to 0.6 in our study.

Step 5: Let $V_{S_i}^*$ denotes the limit of the sequence $\{V_{S_i}(t)\}$, each sentence S_i gets its ranking score $V_{S_i}^*$.

In the above algorithm, all sentences spread their ranking scores to their neighbors via the corresponding affinity graph, and the whole spreading process is repeated until a stable state is achieved and each sentence gets the user-context-based score. In this way, the ranking score for sentence s_i on the rest of user contexts can be deduced similarly.

The final score of sentence s_i assigned by all the user contexts can be denoted as $Score^{uc}(s_i)$, which is calculated by the combination of all scores from different user contexts as follows:

$$Score^{uc}(s_i) = \frac{\sum_{k=1}^{N_{uc}} \text{Confidence}(UC_k, DC_{s_i}) * Score_k^{uc}(s_i)}{\sum_{k=1}^{N_{uc}} \text{Confidence}(UC_k, DC_{s_i})} \qquad (4)$$

where N_{uc} is the number of user contexts, and $\text{Confidence}(UC_k, DC_{s_i})$ denotes the recommendation confidence of the user context UC_k to the document context DC_{s_i} of sentence s_i, which is calculated by the similarity between UC_k and DC_{s_i}.

2.3   Sentence Extraction

In order to evaluate the importance of each sentence s_i within the specified document based on social context, we fuse both the document-context based score $Score^{dc}(s_i)$ and the user-context based score $Score^{uc}(s_i)$ in a unified way as follows:

$$Score(s_i) = \eta * Score^{dc}(s_i) + (1-\eta) * Score^{uc}(s_i) \qquad (5)$$

where $\eta \in [0,1]$ is a weight adjusting parameter, specifying the relative contribution to the sentence's final score from the impact of document context and user context. If $\eta = 1$, only the document context's impact is considered and the final score equals to $Score^{dc}(s_i)$; if $\eta = 0$, only the user context's impact is considered and the final score equals to $Score^{uc}(s_i)$; if $\eta = 0.5$, the two context's impacts are considered equally.

Finally, the overall score for each sentence within document is obtained and the sentences with highest scores and least redundancy are chosen into the summary.

3   Experiments

3.1   Data Set

Since there is no existing evaluation benchmark dataset for social contextual summarization, in this study, we construct a dataset to evaluate the proposed method by downloading 200 bookmarked CNN news web documents from delicious website[1] on diverse topics. The "Story Highlights" are extracted from each news document to form the gold-standard summary.

3.2   Evaluation Methods

The ROUGE toolkit is adopted for evaluation because it is an automatic evaluation method that was officially adopted by DUC[2] [20]. ROUGE measures a summary's content quality by counting overlapping units such as the n-gram, word sequences, and word pairs between the automatically generated summary and the gold-standard reference summaries. Formally, ROUGE-N is an n-gram recall based measurement between a candidate summary and a set of reference summaries, which is computed as follows.

$$ROUGE-N = \frac{\sum\limits_{S \in \{reference\ summaries\}} \sum\limits_{gram_n \in S} Count_{match}(gram_n)}{\sum\limits_{S \in \{reference\ summaries\}} \sum\limits_{gram_n \in S} Count(gram_n)} \qquad (6)$$

where n stands for the length of the n-gram, $gram_n$, and $Count_{match}(gram_n)$ is the maximum number of n-grams co-occurring in a candidate summary and a set of reference summaries.

In the following, we will show three ROUGE scores in the experimental results: ROUGE-1 (unigram based metric), ROUGE-2 (bigram based metric), and ROUGE-SU4 (skip bigram and unigram based metric with maximum skip distance 4).

---

[1] http://www.delicious.com/
[2] http://duc.nist.gov/

3.3   Experimental Results

As a preprocessing step, in the experiments, all the documents were segmented into sentences, stop-words were removed and the remaining words were stemmed by the Porter Stemmer. All the sentences were represented as the term vectors. The process of redundancy removing and the setup of the corresponding parameters of the following baselines are also the same as that of the proposed approach.

Given a document set, we implement the following baselines for comparison based on how to use the contextual information when computing the importance scores of sentences.

Baseline-R: Baseline-R is implemented by extracting sentences randomly from the specified document to generate a summary.

Baseline-DCI: Baseline-DCI computes the importance score of a sentence in the specified document based on only intra-document links between sentences, which can be computed by formula (2).

Baseline-DC: Baseline-DC computes the importance score of a sentence in the document context based on intra-document and inter-document links between sentences, which can be computed by formula (1).

All the above baselines employ either the specified document's internal information or the document contextual information, yet the user contextual information is neglected. The proposed summarization approach *SocialContextSum* considers the social contextual information in a unified framework, which includes both document context and user context.

We show the evaluation results of different approaches in Table 2.

Table 2. Evaluation results of different approaches

Approach	ROUGE-1	ROUGE-2	ROUGE-SU4
SocialContextSum	0.35267	0.05190	0.09974
Baseline-DC	0.34249	0.05076	0.08308
Baseline-DCI	0.33837	0.05002	0.08228
Baseline-R	0.24575	0.02859	0.05989

In our study, the best result of *SocialContextSum* is achieved when the weight adjusting parameter η is set to 0.5.

Seen from Table 2, the proposed approach (i.e. *SocialContextSum*) using the social contextual information achieves the best performance on all ROUGE metrics comparable to that of the baselines, which demonstrates that both document context and user context are very important for improving the performance of single document summarization.

We also observe that the Baseline-DC performs better than the Baseline-DCI and Baseline-R. It shows that document context can benefit the sentence's evaluation by proving more external clues related to the specified document.

506   P. Hu et al.

3.4 Impact of the Adjusting Parameter η

To discover how the relative contribution from document context and user context influences the summarization performance, we investigate tuning the weight adjusting parameter η.

Figure 1 shows the ROUGE-1 evaluation results of the proposed approach with respect to the parameter η.

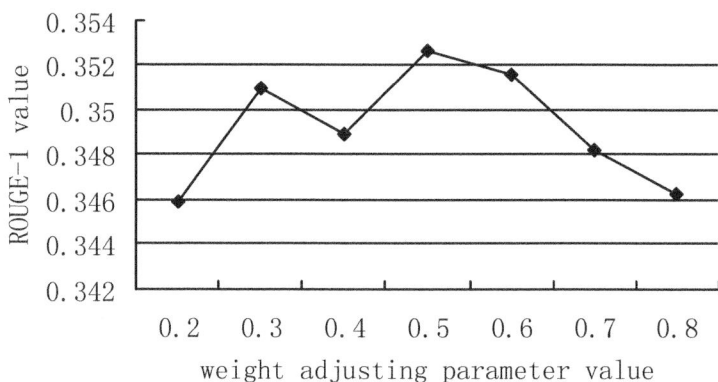

Fig. 1. The ROUGE-1 evaluation results of the proposed approach with respect to the weight adjusting parameter

Seen from Figure 1, it is clear that the summarization performance on ROUGE-1 firstly increases with η, when η is larger than 0.5, the performance tends to decrease or at least stop increasing. It shows that considering appropriate user contextual information is important for improving document summarization performance.

The reason underlying the above observations that the proposed social contextual summarization approach can improve the performance of document summarization is that there are many different online documents to discuss the same topic from various perspectives, and the recognized appropriate social context would guarantee that the influences through the different documents and different user clusters are more reliable.

4   Conclusion and Future Work

Although document summarization has been comprehensively analyzed in the past decade, the study of social-context based summarization just started. In this paper, we propose a collaborative summarization approach based on social contextual information. Both document context and user context are incorporated into the sentence ranking process, and the preliminary experimental results have demonstrated the significant improvement of the proposed approach over several baselines and clearly highlight the potential benefits of the incorporation of social contextual information.

In future work, we would like to investigate the performance of the approach on other data sets with richer social contextual information such as blog and micro-blog entries, comments on social networks, and other social bookmarking services, etc. In addition, we also plan to investigate how to generate community-focused summaries by making use of more implicit or explicit social media information from community user's feedback to further improve summarization system's performance.

Acknowledgments. This work was supported by the National Natural Science Foundation of China (No. 90820005 and No. 61070082).

References

1. Dragomir, R., Weiguo, F.: Automatic Summarization of Search Engine Hit Lists. In: 2000 ACL Workshop on Recent Advances in NLP and IR (2000)
2. Armano, G., Giuliani, A., Vargiu, E.: Studying the Impact of Text Summarization on Contextual Advertising. In: 8th International Workshop on Text-Based Information Retrieval (2011)
3. Dragomir, R., Eduard, H., Kathleen, M.: Introduction to the Special Issue on Text Summarization. Computational Linguistics 28(4) (2002)
4. Delort, J., Bernadette, B., Maria, R.: Web Document Summarization by Context. In: 12th World Wide Web Conference, WWW 12 (2003)
5. You, O.Y., Li, W.J., Li, S.J., Lu, Q.: Applying Regression Models to Query-Focused Multi-Document Summarization. Information Processing and Management (2010)
6. Conroy, J.M., O'Leary, D.P.: Text Summarization via Hidden Markov Models. In: 24th Annual International ACM SIGIR Conference on Research and Development in Information Retrieval (SIGIR 2001), pp. 406–407 (2001)
7. Shen, D., Sun, J.T., Li, H., Yang, Q., Chen, Z.: Document Summarization Using Conditional Random Fields. In: 20th International Joint Conference on Artificial Intelligence, IJCAI 2007 (2007)
8. Toutanova, K.: The PYTHY Summarization System: Microsoft Research at DUC 2007. In: Document Understanding Conference 2007 (2007)
9. Gong, Y., Liu, X.: Generic Text Summarization Using Relevance Measure and Latent Semantic Analysis. In: 24th Annual International ACM SIGIR Conference on Research and Development in Information Retrieval (SIGIR 2001), pp. 19–25 (2001)
10. Lee, J.H., Sun, P., Ahn, C.M., Daeho, K.: Automatic Generic Document Summarization Based on Non-Negative Matrix Factorization. Information Processing and Management, 20–34 (2009)
11. Carbonell, J., Goldstein, J.: The use of MMR, Diversity-Based Reranking for Reordering Documents and Producing Summaries. In: 21st Annual International ACM SIGIR Conference on Research and Development in Information Retrieval (SIGIR 1998), pp. 335–336 (1998)
12. Nomoto, T., Matsumoto, Y.: A new Approach to Unsupervised Text Summarization. In: 24th Annual International ACM SIGIR Conference on Research and Development in Information Retrieval (SIGIR 2001), pp. 26–34 (2001)
13. ErKan, G., Radev, D.R.: LexPageRank: Prestige in Multi-Document Text Summarization. In: 2004 Conference on Empirical Methods in Natural Language Processing, EMNLP 2004 (2004)

14. Mihalcea, R., Tarau, P.: TextRank: Bringing Order into Texts. In: 2004 Conference on Empirical Methods in Natural Language Processing, EMNLP 2004 (2004)
15. Zha, H.Y.: Generic Summarization and Keyphrase Extraction Using Mutual Reinforcement Principle and Sentence Clustering. In: 25th Annual International ACM SIGIR Conference on Research and Development in Information Retrieval (SIGIR 2002), pp. 113–120 (2002)
16. Wan, X., Yang, J., Xiao, J.: Manifold-Ranking Based Topic-Focused Multi-Document Summarization. In: 20th International Joint Conference on Artificial Intelligence (IJCAI 2007), pp. 2903–2908 (2007)
17. Wan, X., Yang, J.: Single Document Summarization with Document Expansion. In: 22nd AAAI Conference on Artificial Intelligence (AAAI 2007), pp. 931–936 (2007)
18. Wan, X., Yang, J.: CollabSum: Exploiting Multiple Document Clustering for Collaborative Single Document Summarizations. In: 20th International Joint Conference on Artificial Intelligence (SIGIR 2007), pp. 143–150 (2007)
19. Wan, X.: Using Only Cross-Document Relationships for Both Generic and Topic-Focused Multi-Document Summarizations. Information Retrieval 11, 25–49 (2008)
20. Lin, C.Y., Hovy, E.: Automatic Evaluation of Summaries Using N-gram Co-occurrence Statistics. In: 2003 Human Language Technology Conference of the North American Chapter of the Association for Computational Linguistics, HLT-NAACL 2003 (2003)

Automatic Classification of Link Polarity in Blog Entries

Aya Ishino, Hidetsugu Nanba, and Toshiyuki Takezawa

Graduate School of Information Sciences, Hiroshima City University
3-4-1 Ozuka-higashi, Asaminami-ku, Hiroshima 731-3194, Japan
{ishino,nanba,takezawa}@ls.info.hiroshima-cu.ac.jp

Abstract. In this paper, we propose a method for classification of an author's sentiment for a linked blog (we call this sentiment link polarity), as a first step for finding authoritative blogs in the blogosphere. Generally, blogs that are linked positively from many other blogs are considered more reliable. In citing a blog entry, there are passages where the author describes his/her sentiments about a linked blog (which we call citing areas). We extract citing areas in a Japanese blog entry automatically, and then classify a link polarity using the information in the citing areas. To investigate the effectiveness of our method, we conducted experiments. For classification of link polarity, we obtained a high precision and recall than baseline methods. For the extraction of the citing areas, we obtained the same Precision and Recall as manual extraction. From our experimental results, we confirmed the effectiveness of our methods.

Keywords: Blog, Link polarity, Sentiment analysis.

1   Introduction

Recently, with the explosive spread of blogs, users can express their private ideas or opinions on the internet easily and actively. The importance of this information is recognized widely; however, the information in blogs contains a mixture of wheat and chaff. Therefore, identifying reliable information efficiently has become an important issue. Many researchers have been trying to collect individual opinions from blogs and analyze them. We have been studying the automatic identification of authoritative blogs in the blogosphere.

At present, methods based on the number of links are used to find authoritative information, such as the PageRank algorithm used by Google [1]. However, the algorithm does not reflect author sentiment about the site being linked. Therefore, blogs on the receiving end of abuse are sometimes highly ranked on by search engines.

To solve this problem, we focus on the author's sentiment about a linked blog. We call this sentiment link polarity. In a citing blog entry, there are passages where the author describes their sentiments about a cited blog, as shown in Figure 1 (which we call citing areas). In the case of Figure 1, both cited blogs A and B have three citing blogs. Based on the number of links, these cited blogs have the same authority. However, we identify cited blog A as more authoritative using link polarity.

M.V.M. Salem et al. (Eds.): AIRS 2011, LNCS 7097, pp. 509–518, 2011.
© Springer-Verlag Berlin Heidelberg 2011

In this paper, we automatically extract the citing areas in a blog entry, and classify the link polarity using the information within the citing areas. This information is useful for identifying authoritative blogs in the blogosphere efficiently, because blogs that are linked positively from many other blogs are considered more reliable.

The remainder of this paper is organized as follows. Section 2 describes related work. Section 3 explains our methods for classifying link polarity. To investigate the effectiveness of our methods, we conducted some experiments, and Section 4 reports the experimental results. We present conclusions in Section 5.

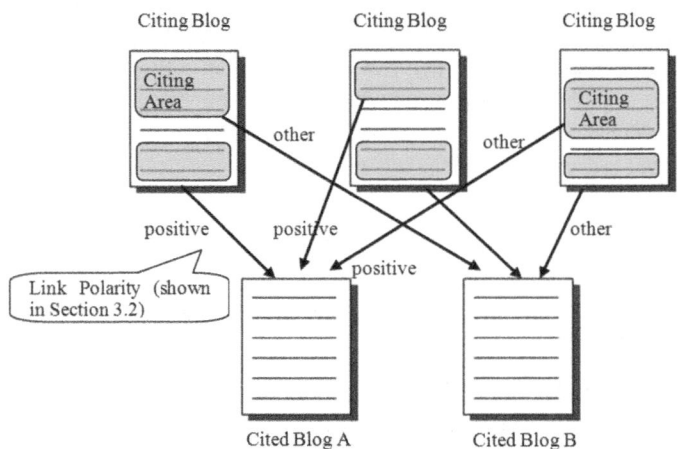

Fig. 1. Citation relationships between blogs

2   Related Work

In this section, we describe some related studies on sentiment analysis, lexicon for sentiment analysis, and classification using citing area.

2.1   Sentiment Analysis

In this paper, we focus on the author's sentiment for a linked blog, classifying sentiment as a link polarity. We can regard the classification of link polarity as a kind of sentiment analysis. Nanno *et al.* [2] presented a system called "blogWatcher", which collects Japanese blogs, performs searches on them, and classifies the sentiment for a search query as positive, negative or other, using a sentiment analysis technique. If a user uses the name of a commercial product, such as "iPhone", as a search term, the system extracts and classifies the sentiment for "iPhone" as a polarity for the commercial product.

There have been several reports of research that automatically classify online movie reviews [3, 4]. Turney applied an unsupervised learning technique based on polar words, such as "excellent" and "poor" [3]. Manual information is computed using statistics gathered by search engines. Pang *et al.* classified reviews using the three machine learning methods (Naive Bayes, maximum entropy classification, and SVM) [4]. They used uni-gram and bi-gram as features for machine learning.

Nanno *et al.* classify the polarity of commercial products. Turney and Pang *et al.* classify movie reviews. In our work, we aim to classify the polarity of author sentiment of a cited blog.

2.2   Lexicon for Sentiment Analysis

For sentiment analysis, researchers use a lexicon that contains polar phrases and their polarity values. Kaji and Kitsuregawa built a lexicon for sentiment analysis from Japanese HTML documents [5]. We call the lexicon the sent lexicon. The sent lexicon contains approximately 10,000 Japanese polar phrases with their polarity values. The polarity value is a numerical value indicating the strength of polarity, which is referred to as a polarity value. To disambiguate orthography, all polar phrases were transferred into their original form with Juman[1] software for morphological analysis. For example, the sent lexicon contains 綺麗だ ("beautiful"), but does not contain either きれいだ or キレイだ, which translate to the same word. When blog authors mention the link, they sometimes use お勧め ("recommend") and参考 ("reference"). As the sent lexicon contains adjectives and adjective phrases, these words are not included. In this paper, we build a lexicon for classification of link polarity. We call this lexicon the link lexicon.

2.3   Classification of Link Polarity

There have been several reports on research to classify links in blog entries automatically [6, 7]. Kale *et al.* devised a method that classifies links in blog entries as either positive or negative [6]. They used a lexicon of positive and negative oriented words and matched the token words from x characters before and after the links to determine the polarity. The link polarity was calculated as follows.

$$Polarity=(N_p-N_n)/ (N_p+N_n) \tag{1}$$

Here, N_p is the number of positively oriented words and N_n is the number of negatively oriented words. We manually created rules for the automatic extraction of any surrounding sentences that mention the link (citing areas). Alternatively, Martineau proposed a machine-learning approach for link classification from several viewpoints using words that appear in the context of citations of URLs as features [7].

Several researchers focused on links in a social net [8, 9]. Guha and Kumar studied data sets from a general consumer review site, Epinions [8]. Leskovec *et al.* studied data sets from Epinions, Slashdot, and Wikipedia [9]. They predicted positive and negative links in data sets using a method based on machine learning, and compared these with theories of balance and status from social psychology. They use the

---

[1] http://nlp.ist.i.kyoto-u.ac.jp/EN/index.php?JUMAN

number of links as a feature for machine learning. In this paper, we focus on links in blog entries, and classify the link polarities using the sentiment in the text surrounding the link.

In a research paper, there are passages where the author describes the essence of a citing paper and the differences between the current paper and the cited paper (we call them citing areas). These passages can be considered as a kind of summary of the citing paper from the current author's viewpoint. Nanba *et al.* [10] automatically classified the types of citation relationships that indicate the reason for citation, using information in the citing areas. They classified the reason for citation into three categories.

To construct collections of hyperlinks for a tourist spot, Ishino *et al.* [11] classified the link in travel blog entries [12] into four categories with citing areas. They manually created rules using cue phrases for the automatic extraction of citing areas. For the classification of links, they obtained a high precision with the information in the citing areas.

In this paper, we automatically extract citing areas in the same way as Nanba *et al.* and Ishino *et al.*, and classify link polarity with the information in the citing areas.

3   Classification of Link Polarity

In this section, we define link polarity in Subsection 3.1, and then explain our approach for classification of link polarity in blog entries in Subsection 3.2.

3.1   Link Polarity

Authors of blogs often link to other blogs with a sentiment for a linked blog. We call the sentiment link polarity. In general, a link polarity can be classified into three categories: positive, negative and other. However, there were only five negative links in 840 links that we collected. Because the link polarities were overwhelmingly positive, we classify link polarities in this paper as either positive or other.

3.2   Classification of Link Polarity

The procedure for the classification of link polarity is as follows.

1.   Input a blog entry.
2.   Extract a hyperlink and any surrounding sentences that mention the link (a citing area).
3.   Classify the link by taking account of the information in the citing area.

In the following, we will explain Steps 2 and 3.

Extraction of Citing Areas. We manually created rules for the automatic extraction of citing areas. These rules use cue phrases. When authors of a blog mention links, they use particular words, such as "blog" (ブログ), "entry" (記事) , or author's name/title of the linked blog. Therefore, we manually selected the cues, and used them for citing area extraction. For extraction of citing areas, we take account of the HTML anchor types and classify the anchor into the following three categories.

- Anchor type 1: Anchor is a URL of linked blog.
- Anchor type 2: Anchor contains blog author's name of linked blog.
- Anchor type 3: Other than anchor types 1 and 2.

We manually created rules 1, 2, and 3, for Anchor types 1, 2, and 3, respectively.

- Rule 1 for anchor type 1

  1. Extract a sentence that includes the link and extract X sentences that appear before or after a web hyperlink, and add them to the candidate. Here, we used the value of $X = 2$, which was determined via a pilot study.
  2. Extract the author's name or title of the linked site from one sentence that appears before or after a web hyperlink. When blog authors introduce web sites, quotation marks or brackets are often used immediately before and after the title of the site. We extract character strings within quotation marks or brackets as keywords. We also extract a word with "Mr." (君) and "Ms./Mrs." (さん) as an author name of the linked blog.
  3. Extract all sentences including the title of the linked blog in the blog entry as citing areas.
  4. Extract all sentences including the author's name of the linked blog in the blog entry as citing areas. If cues appear in a sentence that appears before or after the sentence including cues, we extract them as citing areas.

- Rule 2 for anchor type 2

  1. Extract a word with "Mr." (君) and "Ms./Mrs." (さん) as the author's name of the linked blog.
  2. Extract all sentences including the author's name of the linked blog in the blog entry as citing areas. If cues appear in a sentence that appears before or after the sentence including the title of the linked blog, we extract them as citing areas.

- Rule 3 for anchor type 3

  1. Extract a sentence that includes the link and extract X sentences that appear before or after a web hyperlink, and add them to the candidate. Here, we used the value of $X = 2$, which was determined via a pilot study.
  2. Extract sentences include the same character strings as the anchor and character strings just before particular cues, such as "blog of" (のブログ), or "entry of" (の記事), as the author's name of the linked blog.
  3. Extract all sentences including the author's name of the linked site in the blog entry as citing areas. If cues appear in a sentence that appears before or after the sentence including cues, we extract them as citing areas.

Building the Link Lexicon for Classification of Link Polarity. We build a lexicon containing positively oriented words for classification of link polarity. We call the lexicon the link lexicon. If the citing area contains positively oriented words, we classify the link polarity as positive. If not, we classify the link polarity as other. The procedure for building the link lexicon is as follows.

First, we collect sentences containing the word "this blog" (このブログ) as candidates for the polar sentence from blog entries and a "Web Japanese N-gram" database[2] provided by Google, and manually classify the polarity of the sentences. Examples of polar sentences are shown in Table 1. Second, we manually extract positively oriented words from sentences classified as positive. The double-underlined parts are the positively oriented words. We extract positively oriented words, such as "recommended" (オススメ) and "filled with information" (情報満載). Finally, we record the positively oriented words in the link lexicon. The link lexicon contains 135 positively oriented words.

Table 1. Examples of polar sentences

Polarity	Polar sentence
positive	[original] このブログはマジでオススメである。 このブログは情報満載なのでリンクさせて頂きます。 [translation] This blog is strongly recommended. I link to this information-rich blog site.
other	[original] このブログは、オレの気の向くままに運営しております。 このブログは参加しにくい。 [translation] I administer this blog as my fancy dictates. It is difficult to participate in this blog.

4   Experiments

To investigate the effectiveness of our methods, we conducted several experiments.

4.1   Data Sets and Experimental Setting

We randomly selected 840 links in blog entries, manually classified the link polarity, and used them for our examination. Table 2 shows the number of positive links and other links. We used precision, recall and F-measure as evaluation measures.

Table 2. The number of hyperlinks for each type

Link polarity	Positive	Other	Total
The number of links	378	462	840

---

[2] http://www.gsk.or.jp/catalog/GSK2007-C/catalog.html

In this paper, we propose two methods for building the link lexicon for classification of link polarity and extraction of citing areas. Thus, we considered and evaluated two different approaches. First, to evaluate the link lexicon, we use blog data A with manual extraction of citing areas. Second, to evaluate the method for extraction of citing areas, we use blog data B with automatic extraction of citing areas by our method.

4.2   Evaluation of the Link Lexicon

To evaluate the link lexicon, we use blog data A with manual extraction of citing areas manually. There are two main methods.

[Methods based on Kale *et al.*'s words]

- Kale_LinkLex (our method): Use the link lexicon.
- Kale_SentLex[θ]: Use the sent lexicon. By using polarity value and threshold $\theta(>0)$, we decided whether a word is a polar word or not. If the polarity value of the words is greater than θ, we regard the word as a positive word. Similarly, if the polarity value of the words is less than $-\theta$, we regard the word as a negative word. We calculate link polarity using equation (1).

[Methods based on Martineau's method]

We used the TinySVM (http://chasen.org/~taku/software/TinySVM/) software as the machine-learning package. We performed a four-fold cross-validation test.

- Martineau_LinkLex (our method): Use the link lexicon as features.
- Martineau_SentLex (our method): Use the sent lexicon as features.
- Martineau_Base: Use the words as features.

Results and Discussion

The evaluation results are shown in Table 3. Kale_SentLex[θ] achieved a high precision with large θ. Kale_SentLex[θ] achieved a high recall with low θ. We found Kale_SentLex[11] achieved the highest precision and Kale_SentLex[3] achieved the highest recall, as shown in Table 3. Kale_LinkLex obtained a high precision and recall in comparison with the baseline method based on polar words. Finally, Martineau_LinkLex obtained a high precision and recall in comparison with the baseline method based on Martineau's method. The methods using the link lexicon obtained the highest precision and recall. Therefore, we have confirmed the effectiveness of the link lexicon that we built for classification of link polarity. In addition, Kale_LinkLex obtained the best performance.

Table 3. Evaluation results for link lexicon

		Precision (%)	Recall (%)	F-measure (%)
Methods based on Kale *et al.*'s method	**Kale_LinkLex (our method)**	**85.2**	**90.3**	**87.8**
	Kale_SentLex[11]	72.0	9.5	40.8
	Kale_SentLex[3]	55.9	67.7	61.8
Methods based on Martineau's method	**Martineau_LinkLex (our method)**	**81.7**	**74.8**	**78.3**
	Martineau_SentLex (our method)	**78.8**	**74.1**	**76.5**
	Martineau_Base	78.0	71.6	74.8

The performances of methods based on Kale's method were better than the performances of methods based on machine learning. When blog authors introduce a linked blog, they use particular words, such as "recommended" (オススメ) and "nice" (ステキ). There are a few variations of polar words in blogs. Therefore, we obtained higher performances with methods based on Kale *et al.*'s method than Martineau's method. We show the number of correct and incorrect classification in Table 4. There are many differences in the incorrect number between Kale_LinkLex and Martineau_LinkLex. If we combine the methods based on Kale *et al.*'s and Martineau's methods, and we can further improve the performance of link polarity classification.

Table 4. Evaluation results of Kale_LinkLex and Martineau_LinkLex

		Martineau_LinkLex		Total
		correct	incorrect	
Kale_LinkLex	correct	640	105	745
	incorrect	44	51	95
Total		684	156	840

With Kale_LinkLex, there were two typical errors in the classification of link polarity: (1) the lack of polar words in the link lexicon and (2) the limitation of the method depending on polar words. We describe these errors as follows.

(1) Lack of cues
For classification of link polarity, we manually collected polar words and built the link lexicon, as described in Subsection 3.3. To improve the coverage of polar words, a statistical approach, such as applying n-gram statistics to a larger blog corpus, will be required.

(2) Limitation of the method depending on polar words
Our method mistakenly classified the following example as positive. We proposed the method based on taking account of polar words in citing areas for classification of link polarity. In the example, "nice" (ステキ), which was collected as a polar word, appears in the citing areas, so our method classified it as positive. The blog author said that the present was nice. However, the blog author did not say the linked blog is nice. Our method cannot analyze what is nice. To improve the performance of classification of link polarity, we need to consider language structure.

[original]
先日いつも仲良くしていただいている
<a href="http://blogs.yahoo.co.jp/tenmomomini/" target=_blank>テンファミリー＋オチビ</a>の「もりりんさん」から、　ステキなプレゼントが届きました～～
[translation]
The other day, I received a nice present from Ms. Momorin who was the blog author   of   <a   href="http://blogs.yahoo.co.jp/tenmomomini/" target=_blank>Tenfamily and Ochibi</a>

Fig. 2. An example of a failure

4.3   Evaluation of Citing Areas

To evaluate our methods for extraction of citing areas, we use blog data B with automatic extraction of citing areas by our method.

● Kale_LinkLex_Auto (Our method): Use the link lexicon as features.

Results and Discussion
The evaluation results are shown in Table 5. Kale_LinkLex_Auto used blog data B with automatic extraction of citing areas by our method. Kale_LinkLex used blog data A with manual extraction of citing areas. The result of Kale_LinkLex is from Table 3. Kale_LinkLex_Auto shows the same precision and recall as Kale_LinkLex. Therefore, we have confirmed the effectiveness of our method for the extraction of citing areas. In this paper, we proposed the method for extraction of citing areas using cues. In our future work, we will consider linguistic information from the linked site and sentences surrounding the link and improve the performance of extraction of citing areas.

Table 5. Evaluation results for citing areas

	Precision (%)	Recall (%)	F-measure (%)
Kale_LinkLex_Auto (our method)	**86.5**	**87.6**	**87.1**
Kale_LinkLex (our method)	**85.2**	**90.3**	**87.8**

5   Conclusion

In this paper, we proposed methods for classification of link polarity in blogs. First, we collected polar words and built the link lexicon. Next, we extracted citing areas from blog entries and classified the link polarity in the blog as positive or other using the link lexicon. In the evaluation of the link lexicon, Kale_LinkLex obtained precision and recall scores of 85.2% and 90.3%. Martineau_LinkLex obtained precision and recall scores of 81.7% and 74.8%. Kale_LinkLex obtained the best performance. In the evaluation of our method for the extraction of citing areas, Kale_LinkLex_Auto shows the same precision and recall as Kale_LinkLex with manual extraction of citing areas. The experimental results confirmed the effectiveness of our methods.

References

1. Brin, S., Page, L.: The anatomy of a large-scale hypertextual web search engine. Computer Networks and ISDN Systems Archive 30(1-7), 107–117 (1998)
2. Nanno, T., Fujiki, T., Suzuki, Y., Okumura, M.: Automatically Collecting, Monitoring, and Mining Japanese Weblogs. In: 13th International World Wide Web Conference, Poster Session (2004)
3. Turney, P.D.: Thumbs Up or Thumbs Down? Semantic Orientation Applied to Unsupervised Classification of Reviews. In: 40th Annual Meeting of the Association for Computational Linguistics, pp. 417–424 (2002)
4. Pang, B., Lee, L., Vaithyanathan, S.: Thumbs up? Sentiment Classification using Machine Learning Techniques. In: Conference on Empirical Methods in Natural Language Processing, pp. 76–86 (2002)
5. Kaji, N., Kitsuregawa, M.: Building Lexicon for Sentiment Analysis from Massive Collection of HTML Documents. In: Conference on Empirical Methods in Natural Language Processing (EMNLP-CoNLL 2007), pp. 1075–1083 (2007)
6. Kale, A., Karandikar, A., Kolari, P., Java, A., Finin, T., Joshi, A.: Modeling Trust and Influence in the Blogosphere Using Link Polarity. In: International Conference on Weblogs and Social Media (2007)
7. Martineau, J., Hurst, M.: Blog Link Classification. In: International Conference on Weblogs and Social Media (2008)
8. Guha, R.V., Kumar, R., Raghavan, P., Tomkins, A.: Propagation of Trust and Distrust. In: 13th WWW (2004)
9. Leskovec, J., Huttenlocher, D., Kleinberg, J.: Predicting Positive and Negative Links in Online Social Networks. In: 10th WWW (2010)
10. Nanba, H., Okumura, M.: Towards Multi-paper Summarization Using Reference Information. In: 16th International Joint Conferences on Artificial Intelligence, pp. 926–931 (1999)
11. Ishino, A., Nanba, H., Takezawa, T.: Automatic Compilation of an Online Travel Portal from Automatically Extracted Travel Blog Entries. In: ENTER 2011 (2011)
12. Nanba, H., Taguma, H., Ozaki, T., Kobayashi, D., Ishino, A., Takezawa, T.: Automatic Compilation of Travel Information from Automatically Identified Travel Blogs. In: The Joint Conference of the 47th Annual Meeting of the Association for Computational Linguistics and the 4th International Joint Conference on Natural Language Processing, Short Paper, pp. 205–208 (2009)

Feasibility Study for Procedural Knowledge Extraction in Biomedical Documents

Sa-Kwang Song[1,2], Yun-Soo Choi[1], Heung-seon Oh[2], Sung-Hyon Myaeng[2], Sung-Pil Choi[1], Hong-Woo Chun[1], Chang-Hoo Jeong[1], and Won-Kyung Sung[1]

[1] Korea Institute of Science and Technology Information, Korea
[2] Korea Advanced Institute of Science and Technology, Korea
{esmallj,armian}@kisti.re.kr, {ohs,myaeng}@kaist.ac.kr,
{spchoi,hw.chun,chjeong,wksung}@kisti.re.kr

Abstract. We propose how to extract procedural knowledge rather than declarative knowledge utilizing machine learning method with deep language processing features in scientific documents, as well as how to model it. We show the representation of procedural knowledge in PubMed abstracts and provide experiments that are quite promising in that it shows 82%, 63%, 73%, and 70% performances of purpose/solutions (two components of procedural knowledge model) extraction, process's entity identification, entity association, and relation identification between processes respectively, even though we applied strict guidelines in evaluating the performance.

Keywords: Information Extraction, Text Mining, Procedural Knowledge Modeling, Procedural Knowledge Extraction.

1 Introduction

Technology Intelligence is an activity helping companies or organizations to make better decisions by gathering and providing information about the state-of-the-art technologies [1]. Recently, the systems supporting Technology Intelligence have been actively developed to assist researchers and practitioners to make strategic technology plans [2]. Usually, these systems import text mining methodologies to analyze tacit information inside company or on the Internet. However, they focused on extracting *declarative knowledge*, which describes objects and events by specifying the properties which characterize them; it does not pay attention to extract the actions needed to obtain a result, but only on its properties [3]. Therefore, we propose a methodology that enables to build *procedural knowledge* using text mining technique based on deep language processing. In general, *procedural knowledge* has been considered as *knowledge of how to do something* or *knowledge of skills* [4]. It is contrasted with *propositional knowledge* or *declarative knowledge*. Even though two kinds of knowledge have been defined differently in different domain, Sahdra and Thagard [4] summarized them as shown in Table 1.

M.V.M. Salem et al. (Eds.): AIRS 2011, LNCS 7097, pp. 519–528, 2011.

Table 1. Different terms used with respect to knowledge-how and knowledge-that

	Knowledge-that	*Knowledge-how*
Philosophy	propositional knowledge	procedural knowledge, abilities
Psychology	Explicit knowledge, declarative knowledge	Implicit knowledge, tacit abilities, skills
Artificial Intelligence	Declarative knowledge	Procedural knowledge

If we could build the *Knowledge-how* information from documents, there could be a lot of application analyzing such highly organized procedural knowledge. As an example, the procedural knowledge in the biomedical domain enables doctors or researchers find state-of-the-art technologies and their detailed procedures conveniently. So, they can improve the quality of medication services as well as technology enhancements. Moreover, it is also beneficial to the policy makers in governments or companies on building new plans preparing for the upcoming highly diversified world. We explain related work in section 2 and describe how to model and extract the procedural knowledge in PubMed[1] abstracts in section 3. Section 4 shows two major experiments; purpose/solution sentence classification and unit procedure identification. The results on how to extract procedural knowledge using text mining methodologies are followed at section 5. At last, we summarize and conclude in section 6.

2   Related Work

A lot of research on extracting information like terminology, entity, and concept using various resources such as dictionary, thesaurus, or ontology has been published continuously until now [5-7]. The research on relation or event extraction between them also has been popular these days . However, those works have been focusing on knowledge-that instead of knowledge-how. Even though Jung, et al. [8] extracted procedural knowledge and built ontology from the web documents like eHow[2] and wikiHow [3], their target documents are already structured (listed) in a bulleted sequential form. For an example in wikiHow, there is an article labeled "How to Celebrate National Egg Month". It contains 6 sequential instructions which are imperative sentences. From the article, he extracted sequential actions and built ontology for further usage. The sequential instructions are structured by the wiki-authors. In addition, parsing the sentences is straightforward since almost of them are simple sentences rather than compound or complex sentences.

---

[1] PubMed: http://www.ncbi.nlm.nih.gov/pubmed/
[2] eHow URL: http://www.ehow.com
[3] wikiHow URL: http://www.wikihow.com

3   Methodology for Procedural Knowledge Extraction

3.1   Modeling

We modeled the procedural knowledge which is structured to solve a specific purpose or goal. That is, procedural knowledge in a document consists of a set of unit procedures and each unit has a common purpose to be resolved by the procedures. So, the target document could be represented as a pair of purpose and its solution(s) which consists of a set of graphs of unit procedures. It can be depicted as following Fig. 1.

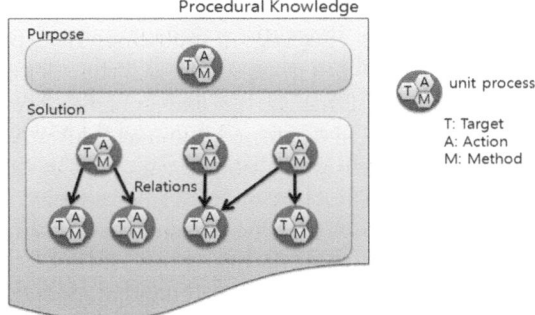

Fig. 1. Graphical representation of procedural knowledge in PubMed abstract: T, A, and M are acronyms of Target, Action, and Method respectively

As depicted in Fig. 1, we defined procedural knowledge as a combination of a purpose and a corresponding solution. And the solution consists of one or more unit procedures having relationships with each other. The unit procedure is a triple combination of *Target*, *Method*, and *Action*; *Target* is defined as diseases, symptoms, objects, organs, and so on. *Method* is treatments, operations, medications, etc. *Action* is a predicate part connecting or relating *Target* with *Method* to explain how to apply a *Method* to treat a *Target* disease or symptom. This modeling has been carried out with medical doctors who have supported us because of their professional knowledge in the medical domain and being one of the best benefit recipients from this research.

3.2   Extraction Procedures

According to the model constructed in subsection 3.1, we designed how the procedural knowledge could be extracted. The major four steps to build procedural knowledge are as follows. 1) Preprocessing target documents by extracting possible lexical, syntactic, or semantic features using various natural language processing techniques; such POS tagging, syntactic parsing, predicate-argument structure tagging, and ontology based terminology identification. 2) Classifying the purpose and solution sentences among the entire sentences belonging to a document. The features gathered in the step 1 are supplied to machine learning algorithms that actually classify the sentences into one of the three categories (purpose/solution/other). 3) Identifying the unit procedures in each purpose or solution sentences. A unit procedure consists of three basic entities, Target/Action/Method. Unit procedure must have at least two entities except that in purpose sentence. The triple from the purpose sentence is considered as not an actual procedure but a nominal procedure. So, the triple is used only for simplifying purpose sentence. 4) Assigning the relationship between two unit processes. The relationship could be sequential, parallel, casual, etc.

3.3   Target Documents

In this paper, the target document for procedural knowledge is confined to the areas of Gastric Cancer and Spinal Disease by the help of medical doctors having been working together. That's because those diseases have more probability of sentences containing appropriate procedural knowledge as well as they are popular and familiar topics people are interested in. Most of review papers or case study papers popular in the biomedical domain are not appropriate for procedural knowledge extraction since they do not include experiments or methodologies which contain procedural information. The target document is semantically divided into several blocks by authors on submitting their papers. The blocks are classified as OBJECTIVE, BACKGROUND, METHODS, RESULTS, and CONCLUSIONS in general. Sometimes, one or more block are omitted or merged. By the way, we are focusing on classifying purpose and solution parts which include two or more out of the three entities. So, it is differentiated with other researches on sentence classification [9], [10] because we only select the sentences possibly containing one or more entities (Target, Action, and Method). In general, the solution part consists of one or more methodological sentences.

3.4   Training Corpus

We developed a training corpus for extracting procedural knowledge by the help of two medical doctors. Total 1309 documents are tagged with purpose/solution labels which contain one or more unit processes (Triple: Target, Action, and Method). In addition, the relationship between two unit processes is also marked. After tagging, the two doctors carried out cross-validation of the tagged corpus. The corpus consists of two domains: Spinal Disease (949 documents) and Gastric Cancer (360 documents).

4   Experiments

The experiments are divided into two parts; purpose/solution classification, unit process identification. The former classifies sentences into one of three classes: purpose, solution, and others, but the latter identifies the triple, Target/Action/Method, in the purpose/solution sentences. For preparing the two experiments, several text mining techniques are applied to the target documents explained in the next subsection.

4.1   Preprocessing

The target documents are preprocessed with Part Of Speech (POS) Tagging, Syntactic Parsing, Predicate-Argument Structure Tagging, and Ontology Mapping. The POS tagging has been applied using Enju parser[4]. Predicate-argument structure [11] is applied, which is a representation of the meaningful relationships of words in a sentence according to the relation between predicate and its arguments. At last, the ontology mapping for terminology identification is added. The terms corresponding to ontology item in UMLS[5], UniProt[6], or GO(Gene Ontology)[7] are marked. This deep processed information is utilized on training or testing the machine learning based algorithms explained in subsection 4.2.

4.2   Purpose/Solution Sentence Classification

Extracting purpose/solution sentences from an abstract could be regarded as a classification problem selecting one category out of three categories such as purpose, solution, and other. For this task, we utilized two machine learning approaches, Support Vector Machines (SVMs)[8] and Conditional Random Fields (CRFs)[9]. The reason why we applied CRFs, frequently used in sequence labeling problem, in addition to the SVMs is that the order of the semantic blocks in abstract are sequential.

The features for this experiment consist of four kinds of items. 1) Content features: unigrams and bigrams in target sentence. Stemming and Stopwords elimination are applied. 2) Position features: sentence number of target sentence in the abstract. The purpose sentence tends to be located at the first few sentences and the solution sentences are rather later part of the abstract. 3) Neighbor features: content features of previous and next k sentences of the target sentence. 4) Ontological features: ontology terms in the UMLS, UniProt, and GO.

4.3   TAM Identification

This experiment is to extract the three entities as Target, Action, and Method (abbreviated as TAM) using CRFs algorithm with 4 kinds of features as follows. 1) Word features: word, word lemma, POS tag, whether first character is capital or not, whether all characters are capital or not. 2) Context features: words and POS tags of previous and next k words of the target word. 3) Predicate-argument structure: predicate type and its argument words and POS tags. 4) Ontological features: ontology terms in the UMLS, UniProt, and GO. This task is to find the boundary of the word or phrase that is recognized as Target, Action, or Method. Therefore, we

---

[4] Enju Parser, http://www-tsujii.is.s.u-tokyo.ac.jp/enju/
[5] UMLS, http://www.nlm.nih.gov/research/umls/
[6] UniProt, http://www.uniprot.org/
[7] Gene Ontology, http://www.geneontology.org/
[8] LIBSVM v3.0, http://www.uniprot.org/
[9] Mallet 2.0 for CRFs, http://mallet.cs.umass.edu/

used most widespread representation so-called IOB tags for chunking of each entity. The B and I tags are suffixed with the entity type, e.g. B-Target, I-Target, B-Action, I-Action, B-Method, and I-Method. Of course, it is not necessary to specify a chunk type for tokens that appear outside an entity, so these are just labeled O. An example of this scheme is shown in Fig. 2.

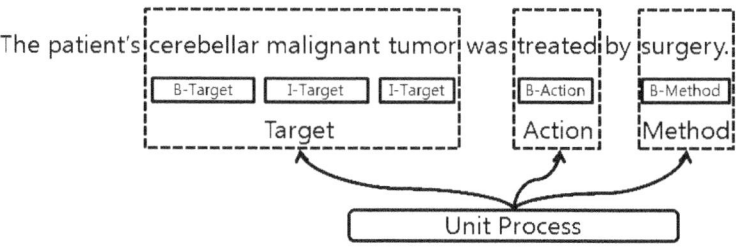

Fig. 2. Tagging TAM entities

4.4 TAM Association

In previous sub-section, we identified all entities in each sentence. However, a sentence may contain one or more TAM triples. In some cases, one or more entities between two TAMs could be shared. In an example sentence, "T1 and T2 was A1 by M1 and M2", many triples such as <T1, A1, M1>, < T1, A1, M2>, <T2, A1, M1>, etc. could be extracted. So, the entities in a sentence should be engaged with each other in a way that each action tries to connect to each Target or Method and decides whether it is appropriate to be a Target or Method of the Action, as depicted in Fig. 3. There are one Target, one Action, and two Methods in the example sentence. Based on the Action 'analyzed', two triples could be extracted, like <'simulated electrograms', 'analyzed', 'basic time-domain method'> and <'simulated electrograms', 'analyzed', 'frequency-domain method'>. This task could be regarded as binary classification at the Action entity since it is a decision problem of whether each link from Action to either Target or Method is appropriate or not.

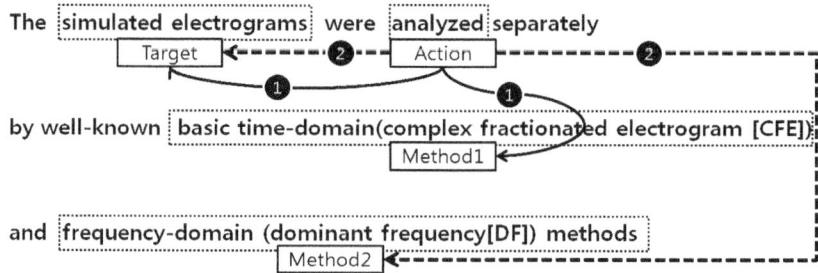

Fig. 3. Multiple TAMs in a sentence

Therefore, we applied SVMs binary classification algorithm so as to decide whether each link from Action to Target (or Method) is feasible or not, using the following three features. 1) Position features: position (previous: negative value, next: positive value) of Target (or Method) from Action. 2) Context features: words and POS tags between two entities. 3) Predicate-argument structure: predicate type and predicate word between two entities, whether the Target (or Method) is related through predicate-argument structure. The Target or Method entity could be absent while the Action should exist. So, the triple <Target, Action, > or <, Action, Method> is possible, but <Target, , Method> is not.

4.5   Relation Extraction

After associating TAMs to identify a unit process in a sentence, we need to associate each TAM to the other based on their relationships. In Fig. 3, there are two unit processes consisting of a triple entity, TAM, as described in subsection 4.4. The two processes have parallel relationship with each other because the two methods (CFE and DF) are carried out separately and in parallel, according to the example sentence. Fig. 4 shows an additional example of a sequential relation. There are three unit processes and they are engaged in sequential manner according to the clue words, *'first'*, *'then'*, and *'finally'*.

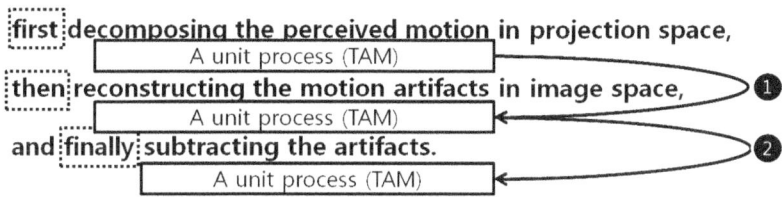

Fig. 4. An example of sequential relationship between two unit processes from a snippet text

This task could be thought of as a multi-class classification because we have to find one among several relationships between the pair of unit processes in a sentence or adjacent (previous and next) sentences, using various clue words. For this task, we defined only two relationships such as sequential and parallel relationship as a feasibility task. And we utilized SVMs for this binary classification task using the following three feature groups. 1) Position features: position (previous, next) of the target unit process, clue words (first, second, finally, parallel, one, the other, and, both, as well as, …) and their relative positions to the unit process. 2) Context features: words and POS tags between the processes, POS tag of Action word in the unit process, the composition of the unit process such as <T,A,>, <,A,M>, or <T,A,M>. 3) Predicate-argument structure: predicate type and predicate word related to the two processes.

5   Results

5.1   Results on Purpose/Solution Sentence Classification

The training and test set are divided in the ratio 8:2 (leave-two-out method) and CRFs and SVMs methods are applied to train purpose/solution sentence classification models. The F-1 score of purpose sentence classification using CRFs achieved 85% while it is relatively low (69%) in solution sentence classification. The reason why the performance is rather bad in solution sentence classification is that there are quite a few sentences that have not at least two entities out of the TAM, even though the sentence sequences of the abstract affect the performance in assigning categories of the sentences.

Table 2. Purpose/Sentence Classification Results

	Precision		Recall		F-1	
	CRFs	SVMs	CRFs	SVMs	CRFs	SVMs
Purpose	0.8326	0.8462	0.8578	0.9009	0.8450	**0.8727**
Solution	0.6923	0.8333	0.6913	0.7610	0.6918	**0.7955**
Total	0.7279	0.8369	0.7326	0.7957	0.7303	**0.8158**

However, the result using SVMs is quite promising since the F-1 scores of the two tasks are 87% and 80% respectively in Table 2. Recall that the model for this experiment is not only for a sentence classification but also for checking whether the sentence contains TAM or not. Actually, some sentences in METHODS block could not be assigned to solution category because they only have at most one component of TAM. So it is rather different from the general sentence classification [9], [10] which performs over 0.90 in their F-1 scores. The both machine learning methods show in common that performance on purpose is better than that on solution because of the consistency in writing the purpose sentences. Usually, '*to ~*', '*the aim of this study ~*', and '*the goal is ~*' are the sentence patterns frequently observed in purpose sentences, while it is hard to find the common pattern in solution sentences.

5.2   Results on TAM Identification

For this experiment, the training and the test set are also divided in the ratio 8:2 (leave-two-out method) and only the CRFs method is applied to train TAM identification model. As we mentioned previously, the performance below does not include partial matching in multi-word entities since most of the medical terms are very sensitive in the semantic perspective according to medical experts. For example, the substring such as '*cooperative ataxia rating scale*', '*ataxia rating scale*', or '*rating scale*' is not regarded as the correct one in the Method term, '*international cooperative ataxia rating scale*', shown in subsection 3.3. The result on Action entity shows high compared to the other two because the number of words in Action entity is at most 2-3 and the main word is verb or verb equivalent. On the contrary, Target and Method entities are large in their length and they contain relatively more adverbs/adjectives as well as composite nouns.

Table 3. TAM Identification using CRFs

Entity	Precision	Recall	F-1
Target	0.5212	0.5696	0.5443
Action	0.7878	0.7753	0.7815
Method	0.6014	0.5078	0.5507
Total	0.6401	0.6102	0.6248

5.3 Results on TAM Association and Relation Extraction

We also used leave-two-out method with SVMs classification method and got a result like Table 4. According to the result, the performance of Action-Target Association is superior to Action-Method. It's because the syntactic variation in sentence of the Method components is much complicated than that of the Target which is usually located in the beginning part of sentence. Additionally, we also performed relation identification experiment using SVMs classification method with leave-two-out method. We only focused on two relations (Sequential and Parallel) as a feasibility task. The parallel relation identification is better than the sequential one since its clue is more direct that the sequential clue because the parallel clues are 'both', 'and', 'as well as', 'at the same time', and so on.

Table 4. Unit process association using SVMs

Association	Precision	Recall	F-1
Action→Method	0.7455	0.6823	0.7125
Action→Target	0.7921	0.7087	0.7481
<T,A,M> match	0.6305	0.6003	0.6150

Table 5. Relation Identification using SVMs

Relation	Precision	Recall	F-1
Sequential	0.6799	0.6911	0.6855
Parallel	0.7122	0.7210	0.7166
Total	0.6961	0.7061	0.7010

6 Conclusion

We proposed a procedural knowledge modeling and extraction method for Technology Intelligence based on machine learning approaches with deep language processing analysis. The experiments showed that the proposed approach is quite promising because it shows 63%~82% in each step of the procedural knowledge extraction steps, even though we applied strict guidelines in evaluating the performance. In addition, we built a handcrafted valuable training corpus with two medical doctors, which have 1309 PubMed abstracts categorized into 8 diseases from both gastric cancer and spinal disease. For future work, we plan to apply the approach to the full-text of documents such as papers, patents, and/or reports.

Acknowledgments. This research was partially supported by the MKE(The Ministry of Knowledge Economy), Korea, under the ITRC(Information Technology Research Center) support program supervised by the NIPA(National IT Industry Promotion Agency) (NIPA-2011-C1090-1111-0008).

References

[1] Mortara, L., Kerr, I.V.C., Phaal, R., Probert, D.: Technology Intelligence Practice in UK Technology-based Companies. International Journal of Tehcnology Management 48(1), 115–135 (2009)

[2] Yoon, B.: On the development of a technology intelligence tool for identifying technology opportunity. Expert Systems with Applications 35(1-2), 124–135 (2008)

[3] Turban, E., Aronson, E.: Decision Support Systems and Intelligent Systems. Prentice Hall, Inc., Upper Saddle River (1988)

[4] Sahdra, B., Thagard, P.: Procedural knowledge in molecular biology. Philosophical Psychology 16(4), 477–498 (2003)

[5] Kazuhiro, Y., Junichi, T.: Reranking for Biomedical Named-Entity Recognition. In: Biomedical Natural Language Processing, BioNLP (2007)

[6] Yoshimasa, T., Junichi, T., Sophia, A.: FACTA: a text search engine for finding associated biomedical concepts. Bioinformatics 24(21), 2559–2560 (2008)

[7] Sophia, A., Carol, F., Junichi, T.: Introduction: named entity recognition in biomedicine. Biomedical Informatics 37, 393–395 (2004)

[8] Jung, Y., Ryu, J., Kim, K.-M., Myaeng, S.-H.: Automatic construction of a large-scale situation ontology by mining how-to instructions from the web. Web Semantics: Science, Services and Agents on the World Wide Web 8(2-3), 110–124 (2010)

[9] Hirohata, K., Okazaki, N., Ananiadou, S., Ishizuka, M., Biocentre, M.I.: Identifying sections in scientific abstracts using conditional random fields. In: Proc. of 3rd International Joint Conference on Natural Language Processing, pp. 381–388 (2008)

[10] Ruch, P., et al.: Using argumentation to extract key sentences from biomedical abstracts. International Journal of Medical Informatics 76(2-3), 195–200 (2007)

[11] Yakushiji, A., Miyao, Y., Ohta, T., Tateisi, Y., Tsujii, J.: Automatic construction of predicate-argument structure patterns for biomedical information extraction. In: Proceedings of the 2006 Conference on Empirical Methods in Natural Language Processing - EMNLP 2006, p. 284 (July 2006)

Small-Word Pronunciation Modeling for Arabic Speech Recognition: A Data-Driven Approach

Dia AbuZeina, Wasfi Al-khatib, and Moustafa Elshafei

King Fahd University of Petroleum and Minerals
{abuzeina,wasfi}@kfupm.edu.sa, shafei@mit.edu

Abstract. Incorrect recognition of adjacent small words is considered one of the obstacles in improving the performance of automatic continuous speech recognition systems. The pronunciation variation in the phonemes of adjacent words introduces ambiguity to the triphone of the acoustic model and adds more confusion to the speech recognition decoder. However, small words are more likely to be affected by this ambiguity than longer words. In this paper, we present a data-driven approach to model the small words problem. The proposed method identifies the adjacent small words in the corpus transcription to generate the compound words. The unique compound words are then added to the expanded pronunciation dictionary, as well as to the language model as a new sentence. Results show a significant improvement of 2.16% in the word error rate compared to that of the Baseline speech corpus of Modern Standard Arabic broadcast news.

Keywords: Speech recognition, pronunciation variation, small-word, phonetic dictionary, language model, Modern Standard Arabic.

1   Introduction

Automatic Speech Recognition (ASR) is a key technology for a variety of applications. Speech recognition is often used as the front-end for many natural language processing (NLP) applications. Some of these typical applications include voice dialing, call routing, data entry and dictation, command and control, and computer-aided language learning. Intuitively, improving the ASR's performance will improve the related NLP applications. In fact, speech communication with computers, PCs, and household appliances is envisioned to be the dominant human-machine interface in the near future.

The substantial goal of ASRs is to enable people to communicate more naturally and effectively. But this ultimate dream faces many obstacles such as different speaking styles, continuous speech, and small words. The small words problem appears when augmenting two adjacent small words to form one compound word that does not exist in the pronunciation dictionary list. This ambiguity leads to the Out-Of-Vocabulary (OOV) problem. OOV is a percentage of the unsatisfied requests among all queries to the dictionary. In the case of unsatisfied request, another word with a sequence of nearest match pronunciation (might be wrong) will be chosen, consequently increasing errors and reducing performance. Initiatively, to ameliorate the

M.V.M. Salem et al. (Eds.): AIRS 2011, LNCS 7097, pp. 529–537, 2011.
© Springer-Verlag Berlin Heidelberg 2011

ASRs performance, OOV should be reduced as much as possible. This reduction in OOV will alleviate the difficulties that may rise during the decoding process. OOV problem is partially solved by adding more candidate pronunciation variations (e.g. compound words) to the dictionary.

Augmenting small words is like the cross-word modeling, a well known approach to augment words according to the phonological rules. As a comparison, the cross-word method does not pay attention to the words length, unlike the small word modeling which is based exclusively on the words length, without caring about phonology. Cross-word modeling demonstrated in many publications, and for many languages. Phonological rules are used for cross-word modeling in [1], [2], [3], and [4] for US English. Research work for other languages such as Dutch, French, Malay, Mandarin, Korean, etc., also exists.

In contrast with phonological rules approach, a data-driven approach can also be used to augment small words. The Data-driven approach depends solely on the corpus transcription to generate compound words. In [5], a significant enhancement was achieved when applying the data-driven approach to model small-words for US English. To the authors' knowledge, the small-word problem has not been tackled in modern standard Arabic (MSA), which is the recently spoken Arabic in news casts and formal speeches.

This paper is organized as follows: Section 2 elaborates on the motivation. We describe the Baseline system in Section 3. In Section 4, we present the Arabic phoneme set. Next, Section 5 provides a description of the Baseline pronunciation dictionary. After that, Section 6 presents our proposed method. We, then, show the results in Sections 7 and finally conclude the work in section 8.

2   Motivation

Unlike isolated speech, continuous speech is known to be a source of augmenting words. This augmentation depends on many factors such as the phonology of the language and the lengths of the words. Our work is focused on adjacent small words being a source of this merging of words. During our previous research work in Arabic speech recognition, it became evident that adjacent small words contribute negatively to achieving less than optimal performance. Table 1 presents an example of the small-word problem.

Table 1. A small-word problem explanation

A speech sentence to be tested	
	وَمُمَثِّلِين عَن الـدُّوَل مِن عَدَد الأوروبِيَّة
Recognized as (Baseline):	وَمُمَثِّلِين عَن إنَّ الـدُّوَل الأوروبِيَّة

Table 1 shows that small words were negatively affected by the concatenations. The decoder mistakenly recognized two separated small words as one word, although it recognized longer words correctly. So, we expect that if we compound the small words as one word, a better performance will be achieved.

3   The Baseline System

The Baseline system used is the Arabic speech recognition system developed at KFUPM. The system was built using CMU Sphinx 3 speech recognition engine by Ali M. [9]. The engine uses 3-emmiting states Hidden Markov Models (HMM) for triphone-based acoustic models. The state probability distribution uses a continuous density of 8 Gaussian mixture distributions. The Baseline system is trained using audio files recorded from several TV news channels at a sampling rate of 16 k samples per seconds. Our corpus includes of 249 business/economics and sports stories (144 by male speakers, 105 by female speakers), summing up to 5.4 hours of speech. The 5.4 hours were split into 4572 files with an average file length of 4.5 seconds. The length of wave files ranges from 0.8 seconds to 15.6 seconds. An additional 0.1 second silence period is added to the beginning and end of each file. Although care was taken to exclude recordings with background music or excessive noise, some of the files still contain background noise such as low level or fainting music, environmental noise such as that of a reporter in an open area, e.g., a stadium or a stock market, and low level overlapping foreign speech, occurring when a reporter is translating foreign statements. The 4572 wav files were completely transcribed with fully diacritized text.

The transcription is meant to reflect the way the speaker has uttered the words, even if they were grammatically wrong. It is a common practice in MSA and most Arabic dialects to drop the vowels at the end of words; this situation is represented in the transcription by either using a silence mark (Sukun) or dropping the vowel, which is considered equivalent to the silence mark. The transcription file contains 39,217 words. The vocabulary list contains 14,234 words. The Baseline WER is 13.39%.

4   Arabic Phoneme Set

A phoneme is the basic unit of the speech that is used in ASR systems. Table 2 shows the listing of the Arabic phoneme set used in the training, and the corresponding phoneme symbols. The table also shows illustrative examples of vowel usage. This phoneme set is chosen based on the previous experience with Arabic text-to-Speech systems in [6],[7],[8], and the corresponding phoneme set which is successfully used in the CMU English pronunciation dictionary.

Table 2. The complete phoneme set used in the training

Phoneme	Example ◀ Letter	Phoneme	Letter
/AE/	بَ ◀ - Fatha	/DH/	ذ (Thal)
/AE:/	بَاب ـَ	/R/	ر (Raa)
/AA/	خَ ◀ - Hard Fatah	/Z/	ز (Zain)

Table 2. (*continued*)

/AH/	دَ ◄ - Soft Fatah	/S/	س (Seen)
/UH/	بُ ◄ - Damma	/SH/	ش (Sheen)
/UW/	دُون ◄ ـُو	/SS/	ص (Sad)
/UX/	غُضن ◄ ـُ	/DD/	ض (Dad)
/IH/	بنت ◄ - Kasra	/TT/	ط (Taa)
/IY/	فيل ◄ ـِي	/DH2/	ظ (Thaa)
/IX/	صِنف ◄ -	/AI/	ع (Ain)
/AW/	لوم ◄ ـُو	/GH/	غ (Ghain)
/AY/	ضيف ◄ ـَي	/F/	ف (Faa)
/E/	ء (Hamza)	/Q/	ق (Qaf)
/B/	ب (Baa)	/K/	ك (Kaf)
/T/	ت (Taa)	/L/	ل (Lam)
/TH/	ث (Thaa)	/M/	م (Meem)
/JH/	جيم فصحة (Jeem)	/N/	ن (Noon)
/HH/	ح (Haa)	/H/	ه (Haa)
/KH/	خ (Khah)	/W/	و (Waw)
/D/	د (Dal)	/Y/	ي (Yaa)

Ali M. In [9] contains more elaborate information about the Arabic phoneme set (40 phonemes) used in this work.

5   Phonetic Dictionary

Phonetic dictionaries are essential components of large vocabulary natural language speaker-independent speech recognition system. For each transcription word, the phonetic dictionary contains its pronunciation in terms of phonemes sequence. Ali M. in [9] developed an application to generate a dictionary for transcriptions. We utilized this tool to generate the Enhanced dictionary as it will be explained in section 6, the proposed method. Ali M. dictionary already takes care for some of within-word variation such as:

- The context in which the words are uttered, For example, Hamzat al-wasl (ا) at the beginning of the word and the Ta' al marbouta (ة) at the end of the word.
- Words that have multiple readings due to dialect issues.

The following example shows the within-word variants of (أُدِنبرة), in the Baseline dictionary:

```
أُدِنبرة       E AE D IH M B R AA H (default)
أُدِنبرة (2) E AE D IH M B R AA T
أُدِنبرة (3) E AE D IH N B R AA H
أُدِنبرة (4) E AE D IH N B R AA T
```

The Baseline dictionary contains 14234 words (without variants) and 23840 words (with variants)

6   The Proposed Method

Modeling the small-word problem is a data-driven approach in which a compound word is distilled from the corpus transcription. The compound word length is the total length of the two adjacent small words that form the corresponding compound word. The small word's length could be 2, 3, 4 letters, or more. During training, several experiments were made to choose the best small word's length. As an illustrative example, suppose as shown in Figure 1 that the sentence has many words, and that w2 and w3 are small words. According to our method, w2 and w3 will be merged to generate a compound word. It is worth mentioning that no phonological rules or any kind of knowledge based approaches are involved in this merging. Figure 1 also shows that the boundary appearing between word 2 and word 3disappears after merging.

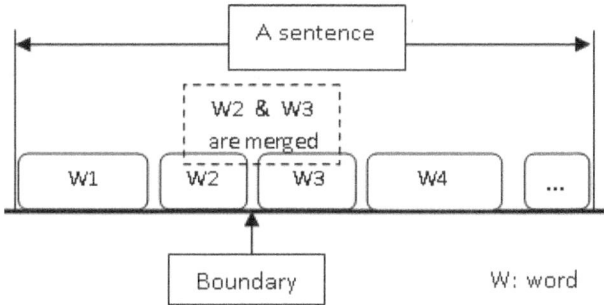

Fig. 1. The concept of modeling small-word

The generated compound words are then filtered to remove all duplicates. Finally, the unique compound words are added to the dictionary and to the language model. The process can be explained in the following example:

The first sentence is from the Baseline corpus transcription sentences, where the text in bold represent two words, one 2-letter word followed[1] by one three-letter word. The second one shows that the two small words found in the first sentence were merged to generate the new compound word. In this example the total length of the small words is 9, as the diacritics are included in computing the length. Both sentences will be appended during corpus transcription to generate the enhanced pronunciation dictionary and the enhanced language model. The expansion of the pronunciation dictionary and the language model depends on the length of small

---

[1] Arabic is read from right to left.

words chosen for merging. As it gets larger, the dictionary and the language model expand more. The proposed method can be described by the following steps:

<u>Small-word pronunciation modeling algorithm</u>

```
While there are more corpus transcription
  Read next two adjacent words
  If their length is less than or equal certain threshold
      Merge them into a new compound word
      Add the compound word to the Baseline dictionary
      Represent the compound word in the related sentence
      Add the generated sentence to corpus transcription
Build the language model
```

After finishing the recognition process, the results will be scanned for decomposing compound words to return them back to their original form. This process can be done using a lookup table. For example: فِـيـبَـنـكِ will be reverted back to فِـي بَـنـكِ.

7   Testing and Evaluation

In order to test our proposed method, we split the audio recordings into two sets: a training set and a testing set. The training set contains around 4.3 hours of audio while the testing set contains the remaining 1.1 hour. We use the CMU language toolkit to build the Baseline language model from the transcription of the fully diacritized text of 5.4 hours of audio. The enhanced language model is built after expanding the Baseline transcription. In order to analyze the effect of the length of the small words on the system performance, we compare the results of our approach when applied on compound words of lengths 5,6,7,8,9,10,11,12 and 13. Table 3 summarizes the results of executing the 9 experiments. We use the following shorthand for the keys in Table 3: TL: Total Length of the two adjacent small words.

TC: Total Compound words found in the corpus transcription.
TU: Total Unique compound words without duplicates.
TR: Total Replaced words after recognition process.
AC: Accuracy achieved.
EN: enhancement achieved. It is also the reduction in word error rate (WER).

Table 3. Results for different small word lengths

TL	TC	TU	TR	AC (%)	EN (%)
5	8	6	25	87.80	0.01
6	103	48	41	88.23	0.44
7	235	153	51	88.53	0.74

Table 3. (*continued*)

8	794	447	132	89.42	1.63
9	1618	985	216	89.74	1.95
10	3660	2153	374	89.95	2.16
11	5805	3687	462	89.69	1.90
12	8518	5776	499	89.68	1.89
13	11785	8301	510	88.92	1.13

Table 3 shows that the best reduction of 2.16% in (WER) is achieved when the length of the compound word is 10. It also shows that performance noticeably decreases when the number of characters in the compound words exceeds 10. Figure 2 shows the accuracy of the system with respect to the words length.

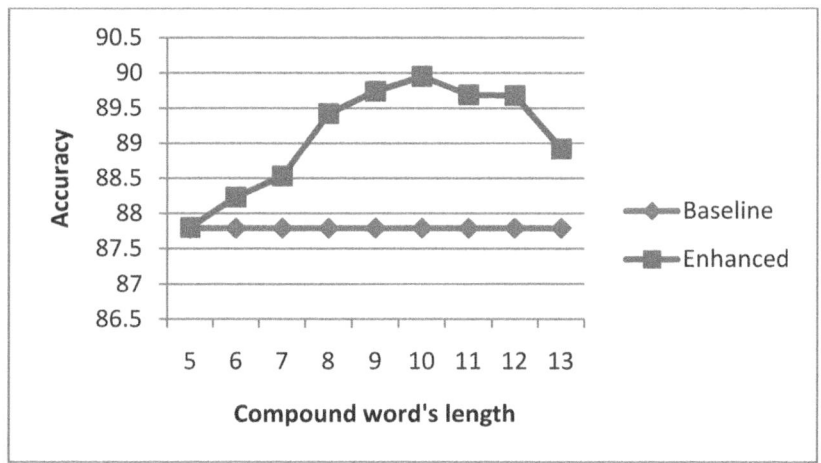

Fig. 2. Comparison of achieved accuracy for different lengths of compound words

With 87.79% accuracy of the Baseline system, Figure 2 shows that the accuracy of the Enhanced system starts increasing until a specific compound word's length (10), and then starts decreasing. The reason of this reduction in accuracy is the confusion introduced in the language model. Figure 3 shows that using a high number of compound words does not openly increase the performance. There is a maximum limit to utilize these compound words, after this limit the performance start decreasing according to the ambiguity occurred in the language model. Figure 3 shows that 510 compound words used (see table 3, TL=13) do not help to maintain the performance.

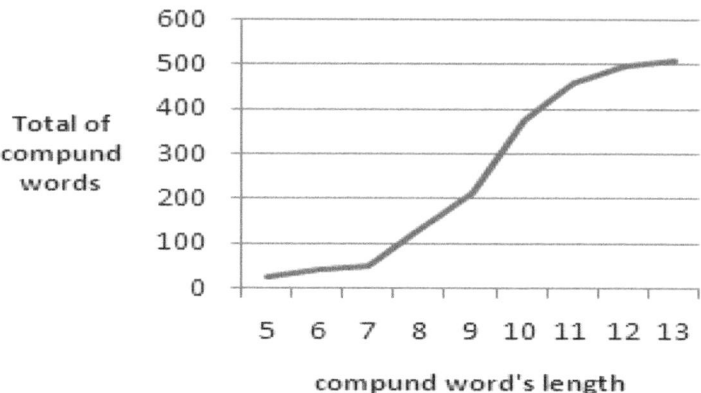

Fig. 3. Compound words usage

The standard measure for language model quality is perplexity. The perplexity for the Baseline language model is 32.88 which is based on 9288 words (testing set words) words. For the Enhanced system, the perplexity is 7.14 computed based on the same testing set words (9288 words). This means that the performance of the Enhanced system is better than the Baseline system since it has a lower perplexity value. With regard to the enhancement, we used the performance detection method suggested in [10] to investigate the significance of the achieved enhancement. We used a level of confidence of 95% to specify the confidence intervals. In addition, we used the total number of words in the Baseline corpus and the word accuracy of the Baseline system. We found the boundaries of the confidence interval to be (+/-) 0.61%. Since our achieved enhancement is greater (2.16%), we can conclude that it is a statistically significant improvement.

The great impact on the perplexity could be understood in two ways: first, the robustness occurred in the language model increases the probability of the testing set $W=w1,w2,...,w_N.$, therefore reducing the perplexity according the perplexity formula:

$$PP(W) = N\sqrt{\frac{1}{P(w1,w2,...,w_N)}}$$

Where PP is the perplexity, P is the probability, and N is the total number in the testing set. Second, the 2153 compound words (see table 3, TL=10) added to the transcription as new words have an extremely low perplexity. For example: consider the two words (من) and (بعد). These two words have an average certain perplexity. When the compound word (منبعد) represented in the language model, it will share others with its very low perplexity, so reducing the overall perplexities. For more information about the effect of compound words, Saon and Padmanabhan in [5] showed mathematically that compound word will enhance the performance. They demonstrated that compound words have the effect of incorporating a trigram in dependency in a bigram language model, as an example.

Finally, our method is implemented as a preprocessing step to extend the span of the dictionary and the language model. The training stage is not involved; i.e. the acoustic model of all training utterances has not been changed during the experiments.

8   Conclusion

We proposed and tested a data-driven approach to model the small words for modern standard Arabic MSA. Our method depends on augmenting adjacent small words and have them represented in the pronunciation dictionary and the language model as well. The results show that our method significantly improves the performance, particularly for compound words of lengths 10. The perplexity of the language model was also enhanced.

Acknowledgements. This work has been supported by Saudi Arabia National Science and Technology Program grant # NSTP 08-INF100-04. The authors would like also to acknowledge the support of King Fahd University of Petroleum and Minerals for its support of this research work.

References

1. Tajchman, G., Fosler, E., Jurafsky, D.: Building multiple pronunciation models for novel words using exploratory computational phonology. In: EUROSPEECH 1995, Madrid, Spain, pp. 2247–2250 (1995)
2. Finke, M., Waibel, A.: Speaking mode dependent pronunciation modeling in large vocabulary conversational speech recognition. In: Proc. of EuroSpeech 1997, Rhodes, pp. 2379–2382 (1997)
3. Beulen, K., Ortmanns, S., Eiden, A., Martin, S., Welling, L., Overmann, J., Ney, H.: Pronunciation modelling in the RWTH large vocabulary speech recognizer. In: Proc. of the ESCA Workshop Modeling Pronunciation Variation for Automatic Speech Recognition, pp. 13–16 (1998)
4. Fosler-Lussier, E., Greenberg, S., Morgan, N.: Incorporating Contextual Phonetics into Automatic Speech Recognition. In: International Congress of Phonetic Sciences (ICPhS 1999), San Francisco, California, pp. 611–614 (1999)
5. Saon, G., Padmanabhan, M.: Data-driven approach to designing compound words for continuous speech recognition. IEEE Transactions on Speech and Audio Processing 9(4), 327–332 (2001), doi:10.1109/89.917678
6. Elshafei Ahmed, M.: Toward an Arabic Text-to-Speech System. The Arabian Journal of Science and Engineering 16(4B), 565–583 (1991)
7. Alghamdi, M., Almuhtasib, H., Elshafei, M.: Arabic Phonological Rules. King Saud University Journal: Computer Sciences and Information 16, 1–25 (2004)
8. Elshafei, M., Almuhtasib, H., Alghamdi, M.: Techniques for High Quality Text-to-speech. Information Science 140(3-4), 255–267 (2002)
9. Ali, M., Elshafei, M., Al-Ghamdi, M., Al-Muhtaseb, H., Al Najjar, A.: Generation of Arabic phonetic dictionaries for speech recognition. In: 5th International Conference on Innovations in Information Technology, United Arab Emirates (December 2008)
10. Plötz, T.: Advanced stochastic protein sequence analysis, Ph.D. Thesis, Bielefeld University (June 2005)

The SALAH Project:
Segmentation and Linguistic Analysis
of ḥadīṯ Arabic Texts

Marco Boella[1], Francesca Romana Romani[2], Anjela Al-Raies[2],
Cristina Solimando[2], and Giuliano Lancioni[2,*]

[1] University "La Sapienza", Rome, Italy
marco.boella@alice.it
[2] Roma Tre University, Rome, Italy
{francescaromana.romani,anjela.alraies}@gmail.com,
{csolimando,lancioni}@uniroma3.it

Abstract. A model for the unsupervised segmentation and linguistic analysis of Arabic texts of Prophetic tradition (*ḥadīṯ*s), SALAH, is proposed. The model automatically segments each text unit in a transmitter chain (*isnād*) and a text content (*matn*) and further analyses each segment according to two distinct pipelines: a set of regular expressions chunks transmitter chains in a graph labeled with the relation between transmitters, while a tailored, augmented version of the AraMorph morphological analyzer (RAM) analyzes and annotates lexically and morphologically the text content. A graph with relations among transmitters and a lemmatized text corpus, both in XML format, are the final output of the system, which can further feed the automatic generation of concordances of the texts with variable-sized windows. The model results can be useful for a variety of purposes, including retrieving information from *ḥadīṯ* texts, verify the relations between transmitters, finding variant readings, supplying lexical information to specialized dictionaries.

Keywords: segmentation, Arabic text treatment, information retrieval, morphological analyzer, hadith, regular expressions, graph.

1 Introduction

Information retrieval in Arabic has often pivoted on contemporary texts, for obvious reasons: electronic availability, usefulness of information, analogy with work done in other linguistic domains. However, Classical texts are much more important in contemporary Arabic culture than in most Western countries, as witnessed by the large diffusion of websites which make Middle Ages books available not only to scholars, but also - and most important - to laymen interested in such texts.

---

* All authors have contributed equally to this work , but since it refers to a modular project, Boella should be mainly credited for sec. 3, Romani for sec. 2, Al-Rajes for sec. 5, Solimando for sec. 4.1, Lancioni for secs. 1, 4.2 and 6.

M.V.M. Salem et al. (Eds.): AIRS 2011, LNCS 7097, pp. 538–549, 2011.
© Springer-Verlag Berlin Heidelberg 2011

A special role in this context is played by *ḥadīṯ* texts, the set of narratives on the life and deeds of the Prophet that altogether constitute the *sunna*, or Islamic Tradition (see Section 2). These texts do not have only a historical importance, they are the cornerstone of Muslim law and a favored reading of most Muslims around the world, and their presence in contemporary written Arabic is widespread.

Notwithstanding their importance, Classical texts have not been - at least to the best of our knowledge - the subject of any scholar research project as far as information retrieval is regarded: to search the texts, most scholars still refer to older, paper resources such as Wensinck's concordances [1].

On the contrary, *ḥadīṯ* texts are a privileged fields for information retrieval texts. Their structure, which couples a text with a preceding chain of transmitters that assures the validity of the tradition, or *isnād*, is already (if informally) organized in such a way that readers are able to detect information with a relatively small amount of ambiguity. Yet, notwithstanding the importance of relations among transmitters in order to ascertain the legal validity of a tradition, such data are still managed in a rather haphazard way, by recurring to traditional resources such as prosopographical repertories and by evaluating transmission relations in a mostly impressionistic way.

The same is true of the lexical and grammatical content of traditions: in most cases, interpreters analyze each *ḥadīṯ* on its own merit, making few, if any, recur to cross-textual regularities and collocations.

Our research project aims to devise methods and algorithm to extract as much information as possible from such texts in an automatic way. The subject matters on which we started working are the automatic segmentation of *isnād* and narrative text (or *matn*: see Section 3), the reconstruction of chains of transmitters through graphs, the creation of (semi-)automatic lexical concordances and the prospective development of a grammar suitable to (semi-)automatically interpret texts and to build semantic representation which can further be employed in inference (by modeling a classical method used by Islamic law scholars). Preliminary results of a morphological analyzer and lemmatizer (see Section 4) are discussed (see Section 5).

2   Contents and Structure of the Corpus: The *Ḥadīṯ*s

Ḥadīṯ, lit. 'narrative, talk', is the term used to indicate each member of the set of shorter or longer narratives on the life and deeds of the Prophet Muḥammad (571-632) that report what he said or did, or of his tacit approval of something said or done and by itself define what is considered good, by providing details to regulate all aspects of life in this world and to prepare people for the beyond, clarifying the Koranic shades; *ḥadīṯ* texts constitute the *sunna*, lit. 'way of life', or Islamic Tradition, that in Muslim culture is considered second in authority only to the Koran: other sources of the Islamic Law (*uṣūl al-fiqh*), *ijmā'* 'consensus' and *qiyās* 'analogical reasoning' have generally a lower rank.[1]

---

[1] Since *ḥadīṯ*s become sources of rules of conduct as authoritative example of the Prophet's behavior, then very badly wanted: in fact the Companions of the Prophet, or simply those who had known him, should have much to tell about him and the new converts wished to learn what he said or did to imitate him, to conform to his traditional standards of behavior, as a rule, in name of the *taqlīd* (the so called *imitatio Muḥammadica* or 'imitation of Muḥammad').

Ḥadīṯs structure is a sequence of binary elements: a text of the narrative, *matn*, with a preceding chain of transmitters (*isnād*, literally 'support'), that have transmitted the narrative, and that assures the validity of the tradition, following one another until the first one who saw or heard Muḥammad.[2]

For the selection of input data computational and linguistics criteria were privileged, rather than philological ones. Among the canonical collection of *ḥadīṯs*, it has been chosen an on-line edition of the collection known as *Ṣaḥīḥ Al-Buḫārī*, compiled by Muḥammad ibn Isma'īl al-Buḫārī [3]. Its features, namely full digitalization and vocalization, allow a wide range of investigations without any needing of manual intervention or preparation. The text has been processed as is, and a systematic control of orthographical and philological coherence has been postponed as not relevant at this stage of project's architecture's implementation.

3   Extracting Surface's Information: From Segmentation to Representation

The automatic segmentation is a process that assigns segment boundaries to get discrete objects from a non-discrete continuum [4]. This approach aims to avoid or at least to limit drastically the supervised intervention, which is rather resource-consuming in time and human involvement, especially considering large amount of data [4-6].

3.1   *Ḥadīṯ*'s Segmentation: Pairing Explicit and Implicit Information

The task of segmenting *ḥadīṯ* texts is in many respects analogous to other cases of semi-structured texts in (pseudo-)natural languages, such as semi-formal texts in descriptions of mathematics. Wolska and Kruijff-Korbayová [7] approached the analysis and formalization of symbolisms and formulas used in mathematics manuals, and drafted a model that: (i) finds regularities in a text; (ii) employs them as patterns to extract textual and meta-textual information; (iii) conceives a set of rules based on these patterns in order to automatically translate extensive verbal expression in math's formulas and vice-versa. This study and others [8], point out that segmentation could be used in textual analysis not only to identify discrete strings, but to try to assign, through an analysis of regularities and recurrences, a global structure to the text itself as well. This structure could be seen as governed by a sort of contextual "grammar of rules", which also controls connections between content's information and its textual organization.

As written above, the *ḥadīṯ* collections fit undoubtedly well the definition of "structured texts". The organization of a *ḥadīṯ* is, in fact, almost rigorous and dependant on a

---

2   *Isnād* is a guarantee of truth for *ḥadīṯ*, through the reputation and the good faith of the transmitters who handed down the narratives orally [2]: Islamic civilization was built upon the supremacy of the spoken word and hearing, the written fixation has only a support role, as prescriptive and restrictive measure, against the aptitude to establish false chains of traditions, considered valid.

set of recurrent "functional expressions" (based on verbs and prepositions in particular) that bound, define and sometimes nest different kind of content [9-10]. The text's continuum could therefore read as formed by two levels, the first one containing information and the other which assumes, beside its textual value, a meta-textual function which organize and define the first level. This seem to show, although in a linear way, somehow a similar structure of that employed in databases, in which records contain information defined by fields. The parallelism with mathematics and information science is far from perfect, however. In fact, a look at literature about the structure of ḥadīṯ collections shows that there is not a general agreement about the original value, meaning and translation of these "functional expressions" [11-12-13] and a complete set of them has not been jet fully defined. However, they could be undoubtedly considered as provided with some extra meanings beside the merely linguistic ones: (i) they separate one transmitter from another; (ii) they specify the authority and typology of transmission; (iii) sometimes they show the "direction" of the transmission. An automatic recognition of these elements in text and their role seems able to draft a rather solid structure of relationships and meanings.

3.2   Extraction and Organization: The HadExtractor Program

In order to test the above-mentioned models of segmenting and structuring texts, a specific program named HadExtractor (HE) has been designed to deal with ḥadīṯ corpora, aiming to: (i) read the full collection and identify single ḥadīṯs; (ii) segment for each ḥadīṯ isnād from matn; (iii) extract from each isnād all transmitters' names together with relative supplementary information (position, typology and direction of transmission). HE was written and implemented in Python [14]. At the present HE has designed as rather close system, as specifically requires as input ḥadīṯ texts only and not jet other Classical Arabic textual structures.

Direct processing of Arabic script in programming languages is possible in theory but hard to manage, especially for switching direction (right to left – left to right) among strings in different characters' systems. The original Arabic script has been therefore converted by using a set of characters based on Buckwalter transliteration system [15] and modified by us in order to fit to Python and regular expressions' constraints on special reserved characters. This transliteration uses ASCII characters only and substitutes usual diacritics employed academically in Latin characters with capital letters and, where needed, supplying with non-alphabetic characters. We implemented the conversion by employing a specific program that allows back transliteration to Arabic at every stage and processes together either vocalized and not vocalized strings [16].

The core of HE is based on Regular Expressions Syntax, conceived in the 1950s by Kleene as tool of automata theory to describe formal languages and developed afterwards by Thompson to be used in programming languages [17]. A regular expression (regex) consists of a formally structured text string for describing complex search patterns, which could be applied to other strings in order to find for matches. The complexity of constants and operators employed allows regexs to be powerful and deeply expressive instruments to retrieve textual segments [18]. HE has been built mainly on three regexs: the first one identifies single *ḥadīṯ*, the second one separates *isnād* from *matn*, the last one catches the transmitters' names.

After the identification of single *ḥadīṯ*s, HE looks for the textual separation point between *isnād* and *matn*, through a regex that models a pattern containing as variable the above-mentioned "functional expressions" and some look-back and look-forwards operators that verify the context to detect effectively the "functional" value of the expression as, obviously, the same word could recur in other contexts, for example inside *matn* without any particular meta-textual role. Once that all *isnād*s are obtained, another regex, working in similar way but referring to a larger list of "functional expressions", extract all transmitters' names pairing them with the corres pondent "functional expression". Concerning the extracted *matn*s, they were paired with a digitalized English translation [19-20], which was processed by a tailored version of HE.

Once HE has implemented all regex routines and related tasks, it produces as output an XML file, in which all extracted information is organized, automatically tagged and nested. The following lines show an excerpt from output XML file, related to a single *ḥadīṯ* (Arabic script is in Buckwalter modified transliteration):

```
<hadith id_ar="7296" id_cor="">
  <source_info> <vol>9</vol>
                <num>7554</num></source_info>
  <isn>   <trasm type="Had+aCaniy">muHam+adu b_nu Eabiy GaAlibI</trasm>
     <trasm type="Had+aCanaA">muHam+adu b_nu eis_maAciyla</trasm>
     <trasm type="Had+aCanaA">muc_tamirU</trasm>
     <trasm type="samic_tu">Eabiy</trasm> […]
     <trasm type="Ean+a">Eabiy raAficI</trasm>
     <trasm type="Had+aCahu Ean+ahu samica">Eabuw huray_raoa</trasm>
  </isn>
  <mat>yaquwlu […] camalNA</mat></hadith>
```

3.3   Representation: Transmitters' Chains and Graphs

Once HE has been applied to the *ḥadīṯ* corpus, a large amount of automatically extracted information was available for further investigations, most of them dealing with extracted *matn* (see Section 4). Focusing instead on the *isnād*, a smart example of representation of information about transmitters is given here below, with the aim to focus on objects' relationships rather then objects themselves.

We structured accordingly the features of extracted *isnād*'s information in the following categories: (i) name of transmitter; (ii) its position in the single *ḥadīṯ* transmission chain, which starts usually from the collector and arrives to the Prophet Muḥammad); (iii) the typology of transmission (see Section 3.1). These categories were read as pertaining to a simple model in which "objects" have different kinds of "relationships" among them. This model is undoubtedly similar to the ones coming from the graph theory, in which a graph is defined as a structure of nodes and edges to model relations among objects from a given collection [21]. These nodes and edges are drawn on a bi-dimensional grid through specific algorithms, in order to graphically visualize the above-mentioned relationships [22]. It was therefore clear that the graph theory could be usefully applied to our data in order to try a graphical representations of transmitters' chains. This kind of representation aims to offer a sophisticate and quantitative-based instrument in a field of research traditionally characterized by analogical and human-based approaches [23] [11].

On the basis of fundamental literature on graph drawing [24-25] we have conceived and implemented in Python another specific program, named ChainViewer. This application, by using existing Python libraries for graph drawing: (i) gets all information about transmitters stored in the XML file containing previously extracted data through HE); (ii) for each *ḥadīt* assigns the transmitters' names to nodes , the relationships among each transmitter and the previous/next one to edges(i.e. the chains), the typology of transmission to edges' types; (iii) through an algorithm is able to generate graphs for single chains or joins together multiple chains in the same graph (in this case if a transmitter's name appears twice or more is shown once but with multiple edges).

At the current stage of development, ChainViewer works well with limited number of chains only (see fig. 1), but could virtually be applied to all chains at the same time, in order to automatically gather in a unique graph all the transmitters of an *ḥadīt* collection together with the complete set of their transmission's relationships. This task obviously presents new problems to deal with, namely: (i) the needing of a semi-automatic instrument able to disambiguate homonyms, unify various inflected forms of the same name, identify nicknames and aliases; (ii) a specifically designed drawing's algorithm that could deal with thousands of nodes and edges, and dynamically represent them with expansion/compression tools.

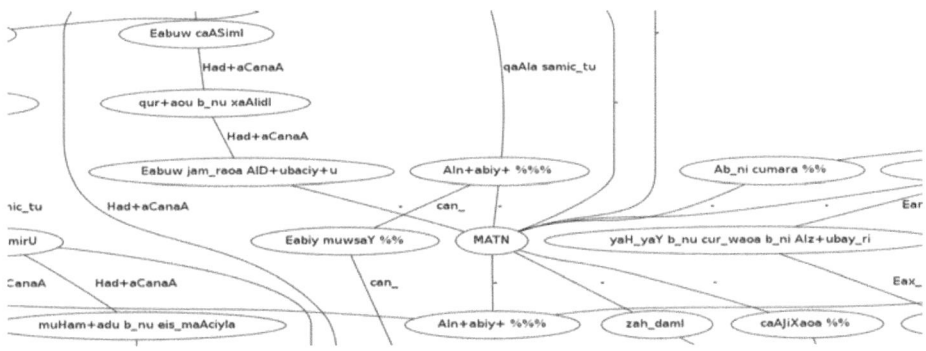

Fig. 1. A portion of a not-directional graph of transmission chains obtained by processing 11 *ḥadīt*s together (Arabic script is in Buckwalter modified transliteration)

4   Analyzing the Corpus: The Revised AraMorph Analyzer

4.1   The Original AraMorph Implementation

As a starting point for the implementation of the text analysis module of the SALAH project, the morphological analyzer and lemmatizer AraMorph (AM) by Tim Buckwalter

[26] was chosen. The main reasons for this choice are the simplicity of design of the model, its high performance level even in unsupervised environment and the easiness of its maintaining and extending.[3]

Opposite to a long-standing linguistic and computer science tradition which emphasized the need for complex, multi-layered, "deep" morphological components in order to analyze properly Arabic texts and to account for the apparent lack of linearity of many Arabic morphemes —see examples and discussion in [26][28-30], — AM chooses to treat Arabic words (in the rather naive, but computationally efficient sense of "any sequence of characters separated by spaces") as elements linearly decomposable in three sub-elements: a prefix, a stem, and a suffix, the stem being the only necessary sub-element (in fact, zero prefixes and suffixes are postulated, sometimes adding some grammatical information to the stem according to the time-honoured morphological concept of "zero morpheme").

This simple account is straightforwardly implemented in the (possibly) simplest way, by feeding the system with three lookup lists of, respectively, (a) prefixes, (b) stems and (c) suffixes, together with three compatibility tables between, respectively, (d) prefixed and stems, (e) prefixes and suffixes and (f) stems and suffixes. Entries in the lookup lists are made up of four fields: (i) unvocalized and (ii) vocalized forms of the morpheme, (iii) grammatical category and (iv) English gloss; compatibility tables just list couples of compatible morphemes, all other combinations being incompatibles. Supplementary pieces of information, not employed in the analysis proper but potentially useful for glossing the texts (root and lemma for a group of morphemes), are provided in the stem lookup list in the form of pseudo-comments.

The analysis, both in the original Buckwalter model (implemented in Perl) and in the Java implementation by the AM project, is performed through a brute-force search of every possible decomposition of words into prefixes, stems and suffixes, by looking up for prefixes from 0 to 4 letters long, stems from 1 letter upwards, and suffixes from 0 to 6 letters long. Only the unvocalized form of words is taken into account (short vowels and other diacritics are stripped before looking up): candidate prefix-stem-suffix decompositions are first matched against the first fields of the respective lookup lists and discarded if any of the elements is missing, then the grammatical categories of the surviving combinations are matched against the compatibility tables and discarded if any of the combination is not present. As a result, each word of the input text can be labeled as (i) unrecognized, if no possible analysis passes the text, (ii) unambiguous or (iii) ambiguous if, respectively, one ore more analyses arc licensed.

This model, whose beauty lies in its very simplicity, is a good starting point for a successful morphological analysis, but does not fits our needs for a plurality of reasons. First, while the emphasis on the unvocalized form of the word is relatively justified for the ideal text genre targeted by Buckwalter —newspaper texts and other Modern Standard Arabic (MSA) non-literary texts that largely comprise the LDC

---

[3]   Buckwalter's system, has been used in many different projects, mostly in its Java implementation; it is, for instance, included as a morphological analysis tool in the Arabic WordNet Project [27].

Arabic corpus— it is far from ideal for other types of texts, first of all fully vocalized ones like ḥadīṯ corpora, but also sparsely vocalized texts: each diacritic added to the consonantal skeleton of a text reduces ambiguity, and thus a system that, like the original AM model, deliberately chooses to ignore this information accepts to live with a higher degree of (morphological) ambiguity and automatically passes a number of wrong analyses that would instead be ruled out by taking into account diacritics present in the text.

The second weak point in the original model lies in the fact that the lookup lists and the relative compatibility tables were built from a sample of the text corpora Buckwalter worked on: again, only morphemes attested in a subset of MSA texts and their combinations are included in the lookup lists and the combination tables, which unavoidably brings to reject or analyze wrongly many words attested in other textual types.

A third weakness in the original AM implementation, which is linked to the previous one, lies in the lack of any stylistic or chronological information in the lookup lists. This way, many morphemes that are virtually exclusive of MSA texts —for instance, a not negligible number of transliterated foreign named entities which cannot be found in Classical texts and which are relatively rare in modern literary texts as well— are included in the lists (and more ore less properly vocalized —foreign proper names are never vocalized in real-world Arabic texts— in order to respect the field structure of the lists themselves) and are likely to give rise to a number of false positives in the analysis of some textual genres.

4.2   Modifications to the Algorithm

In order to overcome the weaknesses listed in the previous section, a number of modifications to the original AM algorithm were devised that tackle the single problems detected above; the new algorithm has been dubbed "Revised AraMorph" (RAM). The first modification is about the token identification mechanism: instead than discarding vocalization, our revised lemmatizer uses it to reduce the number of false positives by taking into account all the vowels present in the text. The comparison phase is less trivial than it might seem, since it must proceed on a three-stage level: (i) the token is segmented in consonants and diacritics (where everything between two characters marked as a consonant is a diacritic); (ii) consonants must match exactly — in fact some qualifications are orders which take into account current practice, e.g. an *alif* with *hamza* above or below matches a simple *alif* (which the original AM accounts for pragmatically, but rather inefficiently, by multiplying entries), and some more are required to reflect idiosyncrasies in the ḥadīṯ orthography;— (iii) diacritics present in the token must not contradict the full vocalized form in the lexicon (that is, e.g., missing vowels are ok, but a vowel cannot match a different one).[4]

---

[4] Some parameters were introduced in the experiments to test the impact of full vocalization in ḥadīṯ texts: since the texts were full vocalized, it is meaningful not to allow additional vowels nor a *tašdīd* (reduplication) symbol if not present in the text. Both the original algorithm and our implementation were rewritten in Python 2.6 in order to profit from other existing tools and to have the possibility to treat directly Arabic texts in Unicode format (even if this is not expedient in some cases, see also Section 3.2).

To tackle the second weak point, namely the partial and unbalanced coverage of Arabic lexicon in the original AM implementation, a file with additional stems automatically extracted from Anthony Salmoné Arabic-English dictionary [31] (a work from the end of the 19th century encoded in TEI-compliant XML format within the Perseus project) was added to the system. Moreover, an analysis of most frequent types of unrecognized tokens allowed to add a limited, but important, number of additional lists of prefixes and suffixes together with the relative combination tables. The single more important addition was the set of prefixes, suffixes and combinatory rules for verb imperatives, a category entirely missing from the original AM implementation —perhaps on purpose, since Arabic imperatives are morphologically complex and quite rare in newspaper texts— and relatively frequent in *ḥadīṯ* texts, given the abundance of prescriptions and performative contexts in the latter.

To reduce the third problem detected, namely the genre and style indistinctness in the AM lexicon, we experimented with automatically remove items that are likely to correspond to contemporary foreign named entities, especially proper names and place-names.[5] In order to do so, we first extracted a list of potential named entities by exploiting a suggestion by Tim Buckwalter himself that in most cases a gloss starting with an uppercase letter is a named entities in 99% of cases; we after perform a full-text search for each word in Salmoné's dictionary and retain only words found there. This way, we are likely to exclude most foreign contemporary named entities by retaining Arabic proper names and place-names which can be found in Classical texts and which are often (but unfortunately not always) included by Salmoné.

5   Results and Evaluation

Results of both HE and RAM have been submitted to standard practice of evaluation [32] through division of the corpus in a training (95%) and testing (5%) section; the testing section has been held relatively small in consideration of the homogeneity of the corpus and the necessity to manually annotate the test sentences. At the present stage of development, the results obtained through both modules are brilliant but quite faceted. The total number of processed *ḥadīṯ* was 7305, and the segmentation produced outputs for 7135 of them, showing an effectiveness' rate of 97.7%. A manual screening of the testing *ḥadīṯ* sample showed a rate of 7.7% incorrect, of which 6.8% are false negatives and 0.9% false positives.

---

[5]   The original AraMorph implementation, true to its newspaper-based spirit (in this case, the source was the AFP corpus), included pretty contemporary items such as the Arabized version of the names of the Belgian tennis player Sabine Appelmans or the Czech soccer team Sigma Olomouc. In some cases, confusion with Arabic words is in fact possible, especially if we take into account the fact that nothing like capitalization is available in the Arabic script: for instance, the transcription of the English first name 'John' (*juwn*) is indistinguishable from *jūn* 'inlet, bay'.

Table 1. Summary of HE results

segmentation		testing of segmented data	
effective	97.7%	error rate	7.7%
wrong	2.3%	precision	99.2%
		recall	93.1%
		F measure	96.1%

The accuracy is going to be improved mainly by refining the above-mentioned operators, and secondly by raising human control on outputs whereas automatic recognition is still impeded.

As to RAM, the system was applied only on the effectively segmented *matn* text output by HE. We obtained a corpus that gathers only *matn*'s section and consists of 382,700 words. Then we applied the original AM analyzer to get a preliminary system output; the results were then compared to the output of the RAM analyzer with different parameter settings.

Table 2. Summary of RAM results

recognition				
	original AM	RAM with vocalization	RAM with added entries	RAM with contemporary NEs removed
unanalyzed	10.36%	12.55%	7.23%	8.12%
univocal	29.45%	58.98%	62.52%	67.79%
ambiguous	60.19%	28.47%	30.25%	24.09%
testing of segmented data				
error rate	60.54%	32.77%	27.65%	24.58%
precision	64.90%	74.57%	81.47%	83.37%
recall	74.56%	92.66%	90.88%	92.05%
F measure	69.40%	82.64%	85.92%	87.50%

The RAM system with vocalization fares far better than the original AM in univocal token recognition, even if the rate of unanalyzed token is slight higher. In fact, the result is equivocal, since AM gets a better result at the price of a higher number of false positives (which RAM discriminates through vocalization).

6   Further Research

Both HE and RAM can be seen as starting points for future research. HE can be extended and generalized to other domains within Classical Arabic culture where texts are arranged according to semi-formal criteria: genealogical repertories, specialized dictionaries, definition lexica (as opposed to lexical encyclopedia). RAM can be further extend to cope with a larger domain of textual genres, especially if coupled with some reasonably well performing system of Arabic Named Entities recognition. As showed by the flowchart in fig. 2 and alluded in the Introduction, this might well feed other, higher-level systems of text analysis and information retrieval.

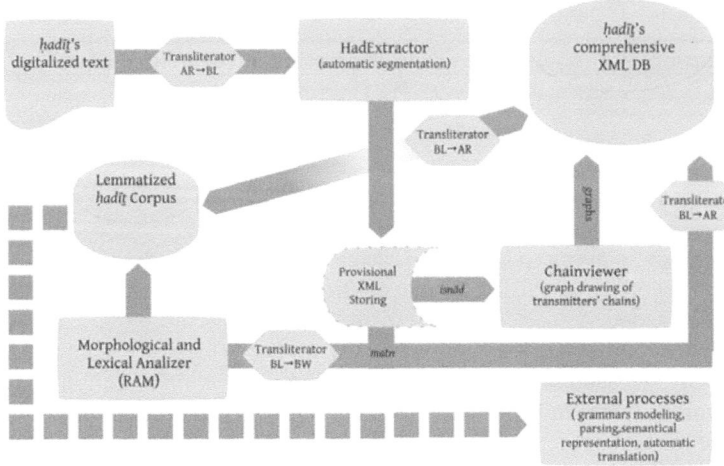

Fig. 2. The SALAH process flowchart

References

1. Wensinck, A.J., et al.: Concordance et indices de la tradition musulmane: Le six Livres, le Musnad d'al-Dārimī, le Muwaṭṭa de Malik, le Musnad de Aḥmad ibn Ḥanbal. Brill, Leiden (1933)
2. Brown, J.: The Canonization of al-Bukhārī and Muslim: the Formation and Function of the Sunnī Ḥadīṯ Canon, Brill, Leiden (2007)
3. al-Bukhārī, M.b.I.: Ṣaḥīḥ al-Bukhārī. Dār Ṭūq al-Najāḥ, Riyaḍ (1990)
4. Gibbon, D., Moore, R., Winski, R. (eds.): Handbook of Standards and Resources for Spoken Language Systems. Mouton de Gruyter, Berlin (1997)
5. Jackson, P., Moulinier, I.: Natural Language Processing for Online Applications: Text Retrieval, Extraction & Categorization. John Benjamins, Amsterdam (2002)
6. Abu El-Khair, I.: Arabic Information Retrieval. Annual Review of Information Science and Technology 41(1), 505–533 (2007)
7. Wolska, M., Kruijff-Korbayová, I.: Analysis of mixed natural and symbolic language input in mathematical dialogs. In: Proceedings of the 42nd Annual Meeting on Association for Computational Linguistics. Association for Computational Linguistics, Stroudsburg (2004)
8. Bird, S., Klein, E.: Regular Expressions for Natural Language Processing. University of Pennsylvania (2006)
9. Robson, J.: Standard applied by Muslim traditionists. Bulletin of the John Rylands Library 63 (1961)
10. Juynboll, G.H.A.: Encyclopedia of Canonical Hadith. Brill, Leiden (2007)
11. Sezgin, F.: Geschichte des Arabischen Schrifttums, vol. 1. Brill, Leiden (1967)
12. Ḥadīth, R.J.: Enciclopaedia of Islam, vol. 3, pp. 23–28. Brill, Leiden (1978)
13. Günther, S.: Assessing the Sources of Classical Arabic Compilations: The Issue of Categories and Methodologies. British Journal of Middle Eastern Studies 32(1), 75–98 (2005)
14. van Rossum, G.: An Introduction to Python for UNIX/C Programmers. In: Proceedings of the NLUUG Najaarsconferentie (1993)

15. Buckwalter, T.: Buckwalter Arabic transliteration (undated),
    `http://qamus.org/transliteration.htm`
16. Lancioni, G.: An Adaptation of Buckwalter Transcription Model to XML and Regular Expression Syntax. Technical report, Roma Tre University, r3a (2011)
17. Aho, A.V.: Algorithms for finding patterns in strings. In: van Leeuwen, J. (ed.) Handbook of Theoretical Computer Science. Algorithms and Complexity, vol. A, pp. 255–300. The MIT Press (1990)
18. Goyvaens, J., Levitan, S.: Regular Expressions Cookbook. O'Reilly, Sebastopol (2009)
19. Khan, M.M.: The English Translation of Sahih Al Bukhari. Al-Saadawi Publications, Alexandria (1984)
20. Al-ʿAsqalānī, A.: Fatḥ al-bārī bi-sharḥ Ṣaḥīḥ al-Bukhārī. Bayrūt, Dār al-Maʿrifah (1959)
21. Berge, C.: Théorie des graphes et ses applications, Dunod, Paris. Collection Universitaire de Mathématiques, vol. II (1958)
22. Bondy, J.A., Murty, U.S.R.: Graph Theory. Springer, Heidelberg (2008)
23. Fück, J.: Beiträge zur Überlieferungsgeschichte von Buḫāris Traditionssammlung. Zeitschrift der Deutschen Morgenländischen Gesellschaft, 60–87 (1938)
24. Di Battista, G., Eades, P., Tamassia, R., Tollis, I.G.: Graph Drawing: Algorithms for the visualization of graphs. Prentice Hall, Upper Saddle River (1999)
25. Kaufmann, M., Wagner, D.: Drawing Graphs: Methods and Models. Springer, Heidelberg (2001)
26. Buckwalter, T.: Buckwalter Arabic Morphological Analyzer Version 1.0. Linguistic Data Consortium, Philadelphia (2002)
27. Black, W., Elkateb, S., Rodriguez, H., Alkhalifa, M., Vossen, P., Pease, A., Fellbaum, C.: Introducing the Arabic WordNet Project. In: Sojka, Choi, Fellbaum, Vossen (eds.) Proceedings of the Third International WordNet Conference (2006)
28. Al-Sughaiyer, I.A., Al-Kharashi, I.A.: Arabic morphological analysis techniques: A comprehensive survey. Journal of the American Society for Information Science and Technology 55(3), 189–213 (2004)
29. Bebah, M., Belahbib, R., Boudlal, A., Lakhouaja, A., Mazroui, A., Meziane, A.: A Markovian Approach for Arabic Root Extraction. The International Arab Journal of Information Technology 8(1) (2011)
30. Habash, N., Rambow, O.: Arabic Tokenization, Part-of-Speech Tagging and Morphological Disambiguation in One Fell woop. In: Proceedings of the 43rd Annual Meeting of the ACL, Ann Arbor, pp. 573–580 (2005)
31. Salmoné, H.A.: An Advanced Learner's Arabic-English Dictionary. Librairie du Liban, Beirut (1889)
32. van Rijsbergen, C.J.: Information Retrieval, Butterworths, London (1979)

Exploring Clustering for Multi-document Arabic Summarisation

Mahmoud El-Haj, Udo Kruschwitz, and Chris Fox

Computer Science and Electronic Engineering
University of Essex, United Kingdom
{melhaj,udo,foxcj}@essex.ac.uk

Abstract. In this paper we explore clustering for multi-document Arabic summarisation. For our evaluation we use an Arabic version of the DUC-2002 dataset that we previously generated using Google Translate. We explore how clustering (at the sentence level) can be applied to multi-document summarisation as well as for redundancy elimination within this process. We use different parameter settings including the cluster size and the selection model applied in the extractive summarisation process. The automatically generated summaries are evaluated using the ROUGE metric, as well as precision and recall. The results we achieve are compared with the top five systems in the DUC-2002 multi-document summarisation task.

1   Introduction

Multi-document summarisation is the process of producing a single summary of a collection of related documents. Much of the current work on multi-document text summarisation is concerned with the English language; relevant resources are numerous and readily available. Arabic multi-document summarisation is still in its infancy. One of the obstacles to progress is the limited availability of Arabic resources to support this research. We are not aware of any publicly available Arabic multi-document gold-standard summaries apart from what we have created ourselves in earlier work, translating the Document Understanding Conference (DUC) 2002 English dataset [5] into Arabic [7].[1] The original dataset provided gold-standard extractive and abstractive summaries in English, both human and machine generated.

Our overall aim is to improve the state of the art in Arabic multi-document summarisation. This requires advances in at least two areas. First, appropriate Arabic test collections are needed. Second, experiments need to be conducted with different approaches to the summarisation process to find techniques that produce good quality Arabic summaries.

---

[1] Although we anticipate this will change soon, at the time of writing the only competition we know of which included Arabic was the DUC-2004 task 3, where noisy Arabic-to-English machine translated articles were provided. The task was to produce very short cross-lingual single-document summaries. The summaries were required to be generated in English.

M.V.M. Salem et al. (Eds.): AIRS 2011, LNCS 7097, pp. 550–561, 2011.

We have addressed the first issue in earlier work by creating a parallel English-Arabic version of the commonly used DUC-2002 dataset [7]. The dateset allows Arabic summarisers to be compared with English summarisers. The current paper is primarily concerned with the summarisation process itself, in particular the use of clustering. We explore different ways of using clustering, and compare our results to other systems.

The paper is structured as follows. We will start with a discussion of related work in Section 2. We will then reprise how we created an Arabic summarisation test collection in Section 3. Section 4 will describe the clustering approaches we are working with together with a description of our summarisers and the experimental setup. Results are discussed in Section 5, and we conclude in Section 6.

2   Related Work

First we will describe multi-document summarisation, the application of clustering to summarisation, our test collection, and the evaluation methodology.

2.1   Multi-document Summarisation

The analysis for multi-document summarisation is usually performed at the level of sentences or documents. Multi-document summarisation systems follow two approaches: *extractive* or *abstractive*. There has been only limited work on abstractive summarisation; it requires natural language analysis and generation techniques that do not yet appear to be sufficiently robust for unconstrained data. For this reason, most of the current summarisation systems rely on the extractive approach.

One early multi-document summariser used information extraction (IE) to identify similarities and differences between documents [16]. Later systems combined IE with a process that regenerates the extracted units in order to improve the quality of the summarisation [1]. Zhao et. al [29] describe a method for query-focused multi-document summarisation.

In our own work we will not consider query-based summarisation as we are focusing on query-independent summarisation.

Summarisation of Arabic documents has not advanced as fast as work in other languages such as English. The summariser "Lakhas" [4] was developed using extraction techniques to produce ten-words summaries of news articles. Turchi et. al [26] presented a method for evaluating multilingual, multi-document, extractive summarisation, using a parallel corpus of seven languages. In their approach, the most important sentences in a document collection were manually selected in one language. This gold-standard summary was then projected into the other languages in the parallel corpus.

Our work aims at making progress in Arabic multi-document summarisation which has started to attract more attention in the research community. An example for that trend is the inclusion of Arabic as one of the languages in the new TAC MultiLing pilot track[2] this year.

2.2 Clustering for Summarisation

Data clustering is the assignment of a set of observations into subsets, so called clusters. As a method of unsupervised learning, clustering has received a lot of attention in past years to improve information retrieval (IR) or to enhance the quality of multi-document summaries [6]. Clustering has been applied to many document levels starting from the document itself down to sentences and words. Clustering can broadly be grouped into hierarchical clustering and partitional clustering.

In our work we focus on the partitional clustering technique, in particular "centroid-based" clustering. In centroid-based multi-document summarisation, similarity to the cluster centroid is used as a measure to rank sentences. The centroid is defined as a pseudo-document consisting of words with TF*IDF scores greater than a predefined threshold [19,20]. It is this centroid-based clustering that we will also focus on here.

Liu et al. [15] proposed a Chinese multi-document summariser which is based on clustering paragraphs of the input articles, rather than sentences. The number of clusters used changes automatically depending on the number of paragraphs.

Wan et al. [27], proposed a multi-document summarisation technique using cluster-based link analysis. In their work they used three clustering detection algorithms including k-means [12], agglomerative and divisive clustering. Their system seeks to cluster the sentences into different themes (subtopics), the number of clusters is defined by taking the absolute square root of the number of all sentences in the document set. In contrast, we use a fixed number of clusters for each run, although we do explore the impact of using different numbers of clusters.

Sarkar [23] presented a multi-document summariser which used sentence-based clustering. The system adopted the incremental clustering method that has been used for web clustering in [11]. In Sarkar's work the clusters are re-ordered according to the number of sentences they contain. This is based on the assumption that size of a cluster correlates with its importance.

To the best of our knowledge, little work has been reported on applying clustering for Arabic multi-document summarisation. Schlesinger et al. [24] present CLASSY, an Arabic/English query-based multi-document summariser system. They use an unsupervised modified k-means method to iteratively cluster multiple documents into different topics (stories). They rely on the automatic translation of an Arabic corpus into English. At the time of their experiments, the quality of machine translation was not high. This led to difficulties in reading and understanding the translated dataset. The translation resulted in inconsistent sentences; core keywords may have been dropped when translating. Errors in tokenisation and sentence-splitting were among the main challenges.

---

[2] http://www.nist.gov/tac/2011/Summarization/

2.3   Test Collections

In our work we also rely on machine translation. This is just to overcome the lack of Arabic multi-document datasets for multi-document summarisation. The DUC–2002 English dataset [5] provides articles and multi-document gold standards for extractive summaries. We performed sentence-by-sentence translation of this dataset into Arabic [7]. The translation is not part of the summarisation process; it is merely intended to allow us to evaluate Arabic summaries against English gold standards. The technique involved translating the DUC-2002 dataset into Arabic using Google Translate[3].

One reason we created our own parallel translated dataset is because the well-known freely available parallel corpora such as Europarl [13] and JRC-Acquis [25] do not include Arabic. Although Salhi [21] provides an Arabic/English parallel corpus (The English-Arabic Parallel Corpus Of United Nations Texts — EAPCOUNT), we decided not to use it for a number of reasons. The corpus contains 341 texts aligned by paragraphs rather than sentences. We also required a corpus that was divided into groups of related articles, which is not the case with EAPCOUNT.

We also wished to be able to compare the performance of our Arabic multi-document summariser against the results of top performing (English) summarisers as published in conferences such as DUC and TAC. Having an Arabic-only corpus would make it extremely difficult to compare our summariser against the best performing English summarisers. Our parallel English–Arabic dataset provides a solution to this problem, and allows us to compare Arabic summaries with gold-standard English summaries.

2.4   Evaluation

Evaluating the quality and consistency of a generated summary has proven to be a difficult problem [9]. This is mainly because there is no obvious ideal, objective summary. Two classes of metrics have been developed: form metrics and content metrics. Form metrics focus on grammaticality, overall text coherence, and organisation. They are usually measured on a point scale [3]. Content metrics are more difficult to measure. Typically, system output is compared sentence by sentence or unit by unit to one or more human-generated ideal summaries. As with information retrieval, the percentage of information presented in the system's summary (precision) and the percentage of important information omitted from the summary (recall) can be assessed.

There are various models for system evaluation that may help in solving this problem. Automatic evaluation metrics such as ROUGE [14] and BLEU [18] have been shown to correlate well with human evaluations for content match in text summarisation and machine translation. Other commonly used evaluations include assessing readers' understanding of automatically generated summaries. Human-performed evaluation may be preferable to automatic methods, but the cost is high.

---

[3] http://translate.google.com

3   Arabic Translation of the DUC–2002 Dataset

As indicated earlier, we chose to use the DUC-2002 dataset [5] to create an Arabic test collection.[4] This dataset contains 59 collections of related newswire and newspaper articles. On average, each collection contains ten related articles.

The dataset includes fully-automatic, multi-document gold-standard summaries, each on a single subject. The summaries were of approximately either 200 or 400 words (white-space delimited tokens), or less.

All the summaries were extractive, consisting of some subset of the sentences predefined by NIST in the sentence-separated document set. Each predefined sentence was used in its entirety or not at all in constructing an extract[5].

The main objective of experimenting with DUC-2002 is to be able to compare our Arabic multi-document summariser with the currently available English summarisers. In order to do this, we first had to translate the sentences in each article into Arabic. This was done using the Java version of the Google Translate API. A total of 17,340 sentences were translated. The public, online version of Google Translate imposes limits on the size of text that can be translated, and on the number of translations that can be requested within a given period. Our software had to work within these limits, making just one translation request every half second. The above process resulted in a parallel sentence-by-sentence Arabic/English version of the DUC-2002 dataset. The summaries for the corpus were created using the applicable English or Arabic version of our multi-document summariser system. For our purposes, it was not necessary to translate the gold-standard summaries themselves into Arabic. Given that we had a sentence-by-sentence translation, we could compare the results of our Arabic summariser (under different settings such as cluster sizes and summarisation approaches) with the top performing systems in the DUC-2002 multi-document English summarisation competition as well as against each other and sensible baselines simply by using the sentence identifiers.

4   Cluster-Based Summarisation

For all our experiments we are using our generic multi-document summarisers that have been implemented for both Arabic and English (using identical processing pipelines for both languages). These summarisers take a set of related articles and generate a summary by selecting sentences from these articles. Having summarisers for both English and Arabic allows us to confirm that the summarisation technique is working and that the impact of using Google Translate to create the parallel dataset is minimal. In the future, it will also allows us to contrast the impact of related techniques, such as stemming, when applied to different languages.

We will now describe the clustering methods employed in our experiments, the actual summarisation process and the experimental setup. We explored clustering in two different ways, first of all by clustering all sentences prior to selecting

---

[4] The new resource is available from the authors.

[5] http://www-nlpir.nist.gov/projects/duc/guidelines/2002.html

a summary. In the second experiment we applied clustering as part of the redundancy elimination step following an initial selection of sentences.

4.1 K-means Clustering

K-means clustering [12] is a partitional centroid-based clustering algorithm. The algorithm randomly selects a number of sentences as the initial centroids, the number of sentences is dependent on the cluster size assigned. The algorithm then iteratively assigns all sentences to the closest cluster, and recalculates the centroid of each cluster, until the centroids no longer change. For our experiments, the similarity between a sentence and a cluster centroid is calculated using the standard cosine measure applied to tokens within the sentence.

4.2 Experiment 1: Clustering All Sentences

In this experiment we treat all documents to be summarised as a single bag of sentences. The sentences are clustered using different cluster sizes and we then select a summary using two approaches:

1. Select sentences from the biggest cluster only
2. Select sentences from all clusters.

The intuition for the first approach is the assumption that a single cluster will give a coherent summary all centered around a single theme, whereas the second approach is expected to result in summaries that contain more aspects of the topics discussed in the documents and therefore a summary that gives a broader picture.

4.3 Experiment 2: Clustering for Redundancy Elimination

Redundancy elimination is an important part of automatic summarisation. A summary that contains very similar sentences drawn from different documents is not ideal. Previously we used redundancy elimination algorithms applied to an initial set of selected sentences. Experiments demonstrated that this has worked effectively [7]. We used the vector space model (VSM), latent semantic analysis (LSA) and Dice's coefficient. We expect that clustering of the pre-selected sentences has the potential of improving the summarisation quality further.

In contrast to query-based summarisation [2,10], in generic summarisation there is no query that can be used to identify relevant sentences. Instead, we use the first sentence of each article to provide the relevant selection criteria. This is based on the assumption, born out in practice, that the first sentence of a well-written article provides a good starting point for generic summarisation [8,28].

Together with the first sentence from each document we select from each document a second sentence, the one most similar to the first one using the VSM [22]. We are effectively performing an initial filtering step, as a result of which we have a similar (but smaller) bag of sentences compare to those of Experiment 1. All the subsequent steps are then the same as in that experiment.

4.4   Experimental Setup

We run k-means clustering using different numbers of clusters, the number of clusters used is 1, 2, 5, 10, 15 and 20. Clustering using a single cluster essentially results in a list of sentences which can be ranked according to the centroid of all sentences.

Two selection methods are used to pick sentences for inclusion in the summary:

1. Select the *first* sentence of each cluster;
2. Select all the sentences in the *biggest* cluster.

The ranking of sentences is done according to similarity to the centroid. The biggest cluster is defined as the one comprising the largest number of sentences. In the resulting summary we keep the order of sentences as they appear in the clusters (i.e. a sentence very similar to the centroid appears earlier on in the summary than one that is less similar).

There were two parts to each of the summarisation experiments. First, we created summaries for the DUC-2002 dataset using the English version of our multi-document summariser. The results were evaluated against the English gold-standard summaries, using ROUGE. Second, we summarised the Arabic parallel translation version of using our Arabic multi-document summariser. These Arabic summaries were evaluated by re-constructing the corresponding English translations using the sentence identifiers. We were then able to evaluate the "translated summaries" using ROUGE. Running both English and Arabic experiments acts as a check on the automatic translation process, and provides some assurance that the underlying summarisation algorithm itself is sensible.

Note that in our experiments we do not trim the resulting summary to a particular length. ROUGE will do this automatically for summaries that are too long. Alternatively, we could produce a summary that does not exceed a fixed maximum length.

ROUGE-1 was used as the evaluation metric. This is for two reasons. First, it was the metric used in evaluating the top performing summarisers in DUC-2002, with which we wish to compare our approach. Second, it has been shown to work well in multi-document summarisation tasks [14]. The ROUGE evaluation was performed having N-gram 1:1 and confidence interval of 95%, which was believed to give results close to those of human evaluation.

Where appropriate, we determined significant differences by performing pairwise t-tests ($p < 0.05$) using the R statistics package.[6]

5   Results and Discussion

Tables 1 and 2 illustrate the results of our Arabic summarisers applying k-means clustering to *all* sentences using the two different selection methods. The highest ROUGE scores are displayed in bold font. Similarly, Tables 3 and 4 represent the results of clustering for redundancy elimination.

---

[6] *R Project for Statistical Computing*, http://www.r-project.org/

Table 1. Clustering all Sentences (Biggest Cluster)

Number of Clusters	Rouge-1	Precision	Recall
1	**0.3899**	0.3849	0.3952
2	0.3701	0.3660	0.3744
5	0.3736	0.3697	0.3778
10	0.3848	0.3800	0.3899
15	0.3882	0.3839	0.3930
20	0.3898	0.3853	0.3946

Table 2. Clustering all Sentences (First Sentence from each Cluster)

Number of Clusters	Rouge-1	Precision	Recall
1	0.1725	0.6733	0.0997
2	0.2322	0.5495	0.1496
5	0.3152	0.4516	0.2469
10	0.3822	0.3954	0.3729
15	**0.3878**	0.3868	0.3898
20	0.3848	0.3819	0.3879

We observe the following results.

1. The best overall ROUGE score is obtained by clustering *all* sentences and creating a single "cluster", i.e. by selecting sentences to form the summary that are most similar to the centroid of all sentences. In fact, when selecting the biggest cluster of *all* sentences we observe that the number of clusters in our experiments does not have a significant effect on the ROUGE scores. All results are fairly consistent with no significant difference measurable between any pair of results.
2. An alternative way to obtain top ROUGE scores is to use clustering for redundancy elimination (as illustrated in Tables 3 and 4). However, in this case the experimental settings do have a significant impact. If the cluster size is small (1 or 2) when clustering a pre-selected set of sentences and then using the biggest cluster to form the summary, we get a ROUGE score that is significantly better than using clusters of size 10 or above. Inversely, we get a ROUGE value marginally smaller than the top overall score by selecting the centroids from a large set of clusters. Given our approach to redundancy elimination we would in fact expect the ROUGE scores (as well as precision and recall) to vary across different cluster sizes and selection methods.
3. We also observe that for certain experimental settings (as just described) the cluster-based redundancy elimination approach can (marginally) improve our previously applied redundancy elimination steps which are using no clustering. Table 5 reproduces the results of our earlier work [7].

Table 3. Redundancy Elimination (Biggest Cluster)

Number of Clusters	Rouge-1	Precision	Recall
1	0.3806	0.3761	0.3852
2	**0.3817**	0.3788	0.3853
5	0.3591	0.3794	0.3504
10	0.3346	0.3991	0.3134
15	0.2718	0.4270	0.2290
20	0.1704	0.4614	0.1149

Table 4. Redundancy Elimination (First Sentence from each Cluster)

Number of Clusters	Rouge-1	Precision	Recall
1	0.1421	0.4680	0.0848
2	0.2279	0.4267	0.1570
5	0.3638	0.3848	0.3470
10	0.3645	0.3616	0.3676
15	0.3676	0.3638	0.3717
20	**0.3820**	0.3777	0.3865

4. Throughout the experiments we observe that high precision goes hand in hand with relatively low ROUGE score, i.e. very low recall values.
5. We also found that *none* of the pairwise comparisons of ROUGE scores when comparing the Arabic summariser with the corresponding English summariser are significant. This is an indication that the automatic translation process did not affect the summarisation quality.

The ROUGE-1 results of the five best performing summarisers in DUC-2002 are given in [17], where the authors of that paper compared their summariser to those systems. The results are reproduced in Table 6. These results were produced using an earlier version of ROUGE. For this reason, we sought to replicate their baseline experiment to ensure comparability of the results. The baseline summariser simply selects the first sentence of each article of the set of related articles in the dataset and combines them. The baseline results we reproduced for the DUC-2002 dataset were comparable to those published in [17], with just a slight increase in the ROUGE-1 evaluation.

Comparing the ROUGE-1 results of our systems with those of the top five summarisers in DUC-2002 competition, we observe that our multi-document summarisers achieve slightly higher scores than the top systems reported at DUC-2002. This is true for the English system (top ROUGE score 0.3856), but more importantly for this paper, also for our Arabic multi-document summariser (working on the Arabic translation of the dataset).

Table 5. Summarisation with Redundancy Elimination (no Clustering)

System	Rouge-1	Precision	Recall
Dice–Sum–Arb	0.37085	0.37366	0.36915
Dice–Sum–Eng	0.36945	0.36950	0.36958
VSM–Sum–Arb	0.36411	0.39115	0.34784
VSM–Sum–Eng	0.35666	0.37998	0.34220
Our Baseline	0.29457	0.29505	0.29421
LSA–Sum–Eng	0.28986	0.39071	0.23810
LSA–Sum–Arb	0.28734	0.41296	0.22986

Table 6. Top 5 Systems in DUC-2002

System	Rouge-1
System26	0.3578
System19	0.3450
System28	0.3435
System29	0.3264
System25	0.3056

The main finding of our experiments appears to be the fact that a simple centroid-based similarity clustering with a single "cluster" when performing summarisation could be considered an alternative to using several cluster. The work by Sarkar and by Radev et al. [19,20,23] had a variable number of clusters when performing clustering, depending on the number of sentences, whereas our experiments demonstrate that for the given test collection the closeness to the centroid to identify the important sentences can produce summaries with similar quality.

6   Conclusion

In this paper we explored clustering for multi-document Arabic summarisation. We investigated how clustering can be applied to multi-document summarisation as well as for redundancy elimination within this process. We used different parameter settings including the cluster size and the selection model applied in the extractive summarisation process. Using ROUGE, precision and recall we could measure the different effects objectively. One of our main findings is that selecting sentences similar to the centroid of all sentences in the collection of related documents gives the highest ROUGE scores. We also showed that the Arabic (as well as English) summarisation system we developed has comparable performance with the top performing (English summarisation) systems at DUC-2002.

Running our summarisation experiments in parallel on Arabic as well as an English version of our summariser has allowed us to confirm that the automatic translation process does not affect the quality of the result. In future it will allow us to compare the impact of similar techniques on the two languages.

We intend to explore the application of more fine-tuned clustering to improve results further. We also aim to experiment with more language-specific features (such as light stemming) which we suspect will be more effective for Arabic than for English.

References

1. Funk, A., Maynard, D., Saggion, H., Bontcheva, K.: Ontological integration of information extracted from multiple sources. In: In the Multi-source Multilingual Information Extraction and Summarization (MMIES) Workshop at Recent Advances in Natural Language Processing (RANLP 2007), Borovets, Bulgaria (2007)
2. Berger, A., Mittal, V.O.: Query-relevant summarization using FAQs. In: Proceedings of the 38th Annual Meeting of the Association for Computational Linguistics, ACL 2000, pp. 294–301. Association for Computational Linguistics, Stroudsburg (2000)
3. Brandow, R., Mitze, K., Rau, L.F.: Automatic condensation of electronic publications by sentence selection. Inf. Process. Manage. 31, 675–685 (1995)
4. Douzidia, F.S., Lapalme, G.: Lakhas, an Arabic summarising system. In: In the Proceedings of the Document Understanding Conferences (DUC) Workshop, pp. 128–135. DUC (2004)
5. Document Understanding Conference (DUC) dataset (2002),
   http://duc.nist.gov/
6. Dunlavy, D.M., O'Leary, D.P., Conroy, J.M., Schlesinger, J.D.: Qcs: A system for querying, clustering and summarizing documents. Inf. Process. Manage. 43, 1588–1605 (2007)
7. El-Haj, M., Kruschwitz, U., Fox, C.: Multi-document Arabic text summarisation. In: Proceedings of the third Computer science and Electronic Engineering Conference. IEEE, Colchester (2011)
8. Fattah, M.A., Ren, F.: Automatic text summarization. Proceedings of World Academy of Science 27, 192–195 (2008)
9. Fiszman, M., Demner-Fushman, D., Kilicoglu, H., Rindflesch, T.C.: Automatic summarization of MEDLINE citations for evidence-based medical treatment: A topic-oriented evaluation. Jouranl of Biomedical Informatics 42(5), 801–813 (2009)
10. Gong, Y., Liu, X.: Generic text summarization using relevance measure and latent semantic analysis. In: Proceedings of the 24th Annual International ACM SIGIR Conference on Research and Development in +Information Retrieval, SIGIR 2001, pp. 19–25. ACM, New York (2001)
11. Hammouda, K.M., Kamel, M.S.: Efficient phrase-based document indexing for web document clustering. IEEE Trans. on Knowl. and Data Eng. 16, 1279–1296 (2004)
12. Hartigan, J.A., Wong, M.A.: Algorithm as 136: A k-means clustering algorithm. Journal of the Royal Statistical Society. Series C (Applied Statistics) 28(1), 100–108 (1979)
13. Koehn, P.: Europarl: A parallel corpus for statistical machine translation. In: X Machine Translation Summit, pp. 79–86. Phuket, Thailand (2005)
14. Lin, C.: Rouge: A package for automatic evaluation of summaries. In: Proceedings of the Workshop on Text Summarization Branches Out (WAS 2004), pp. 25–26 (2004)

15. Liu, S., Lindroos, J.: Towards fast digestion of IMF staff reports with automated text summarization systems. In: Proceedings of the 2006 IEEE/WIC/ACM International Conference on Web Intelligence, pp. 978–982. IEEE Computer Society (2006)
16. McKeown, K., Radev, D.R.: Generating summaries of multiple news articles. In: Proceedings of the 18th Annual International ACM SIGIR Conference on Research and Development in Information Retrieval, SIGIR 1995, pp. 74–82. ACM, New York (1995)
17. Mihalcea, R., Tarau, P.: Multi-Document Summarization with Iterative Graph-based Algorithms. In: The First International Conference on Intelligent Analysis Methods and Tools (IA 2005), McLean, VA (2005)
18. Papineni, K., Roukos, S., Ward, T., Zhu, W.J.: Bleu: a method for automatic evaluation of machine translation. In: Proceedings of the 40th Annual Meeting of the Association for Computational Linguistics, ACL 2002, pp. 311–318. Association for Computational Linguistics, Stroudsburg (2002)
19. Radev, D.R., Jing, H., Budzikowska, M.: Centroid-based summarization of multiple documents: sentence extraction, utility-based evaluation, and user studies. In: Proceedings of the 2000 NAACL-ANLP Workshop on Automatic Summarization, NAACL-ANLP-AutoSum 2000, vol. 4, pp. 21–30. Association for Computational Linguistics, Stroudsburg (2000)
20. Radev, D.R., Jing, H., Sty, M., Tam, D.: Centroid-based summarization of multiple documents. Information Processing and Management 40, 919–938 (2004)
21. Salhi, H.: Small parallel corpora in an English-Arabic translation classroom: No need to reinvent the wheel in the era of globalization. In: Globalisation and Aspects of Translation, pp. 53–67. Cambridge Scholars Publishing, Newcastle (2010)
22. Salton, G., Wong, A., Yang, S.: A vector space model for automatic indexing. Proceedings of the Communications of the ACM 18(11), 613–620 (1975)
23. Sarkar, K.: Centroid-based summarization of multiple documents. TECHNIA — International Journal of Computing Science and Communication Technologies 2 (2009)
24. Schlesinger, J.D., O'Leary, D.P., Conroy, J.M.: Arabic/English Multi-document Summarization with CLASSY—The Past and the Future. In: Gelbukh, A. (ed.) CICLing 2008. LNCS, vol. 4919, pp. 568–581. Springer, Heidelberg (2008)
25. Steinberger, R., Pouliquen, B., Widiger, A., Ignat, C., Erjavec, T., Tufis, D., Varga, D.: The jrc-acquis: A multilingual aligned parallel corpus with 20+ languages. In: Proceedings of LREC, Genova, Italy, pp. 24–26 (2006)
26. Turchi, M., Steinberger, J., Kabadjov, M., Steinberger, R.: Using Parallel Corpora for Multilingual (Multi-document) Summarisation Evaluation. In: Agosti, M., Ferro, N., Peters, C., de Rijke, M., Smeaton, A. (eds.) CLEF 2010. LNCS, vol. 6360, pp. 52–63. Springer, Heidelberg (2010)
27. Wan, X., Yang, J.: Multi-document summarization using cluster-based link analysis. In: Proceedings of the 31st Annual International ACM SIGIR Conference on Research and Development in Information Retrieval, SIGIR 2008, pp. 299–306. ACM, New York (2008)
28. Yeh, J.Y., Ke, H.R., Yang, W.P.: iSpreadRank: Ranking sentences for extraction-based summarization using feature weight propagation in the sentence similarity network. Expert Systems with Applications 35(3), 1451–1462 (2008)
29. Zhao, L., Wu, L., Huang, X.: Using query expansion in graph-based approach for query-focused multi-document summarization. Inf. Process. Manage. 45, 35–41 (2009)

ZamAn and Raqm: Extracting Temporal and Numerical Expressions in Arabic

Iman Saleh[1], Lamia Tounsi[2], and Josef van Genabith[2]

[1] Faculty of Computers & Information, Cairo University
iman.saleh@fci-cu.edu.eg
[2] NCLT, School of Computing, Dublin City University
{ltounsi,josef}@computing.dcu.ie

Abstract. In this paper we investigate automatic identification of Arabic temporal and numerical expressions. The objectives of this paper are 1) to describe *ZamAn*, a machine learning method we have developed to label Arabic temporals, processing the functional dashtag -TMP used in the Arabic treebank to mark a temporal modifier which represents a reference to a point in time or a span of time, and 2) to present *Raqm*, a machine learning method applied to identify different forms of numerical expressions in order to normalise them into digits.

We present a series of experiments evaluating how well *ZamAn* (resp. *Raqm*) copes with the enriched Arabic data achieving state-of-the-art results of F1-measure of 88.5% (resp. 96%) for bracketing and 73.1% (resp. 94.4%) for detection.

Keywords: Temporal Expression, Numerical Expression, Temporal Phrase, Machine Learning, Recognition, Extraction, Labeller.

1 Introduction

The first task addressed in this paper concentrates on extracting temporal expressions (TMPs). Extracting TMPs has a considerable importance in a variety of applications such as event detection, question answering and text summarization. For event detection, identifying TMPs can be useful since an event is defined as something that happens at a particular time and place. In question answering, it is useful to resolve temporal references such as the date in the sentence "Prince William and Kate Middleton are to marry on April 29, 2011" to answer the question "When is the royal wedding?". In summarization, TMPs are used to establish a time line for all events in multiple documents to create a coherent summary. Certain characteristics of Arabic such as the large number of distinct word-forms, the relative freedom with respect to word order, and the syntactic information expressed at the level of words increase the complexity of detecting temporal phrases and make this task very challenging. For instance, the list of names representing months in Arabic includes names from the solar calendar (e.g. January), the lunar calendar (e.g. Muharram), and also from the Syriac calendar (e.g. Kaanoon Althaanee). In addition, Arabic speakers tend to

M.V.M. Salem et al. (Eds.): AIRS 2011, LNCS 7097, pp. 562–573, 2011.

use the five daily Muslim prayers as a reference of time (e.g. meeting before the sunset prayer). However, prayer times changes every day as it depends on the sun hours and the geography.

Our focus in this paper, *in contrast to previous work*, is on detecting *temporal phrases* rather than *temporal expressions* (e.g. "in 1942" vs. "1942") present in the Arabic Penn Treebank (ATB)[22]. One of the aims of this work is to investigate how labelling temporal phrases improves statistical parsing results. More future work (not discussed in this paper) is the transformation/normalisation of temporals into a formal semantic knowledge representation (e.g. on next Saturday before lunch time → 23/07/2011 [10am-12pm] GMT) in order to provide an input to powerful tools developed for real world applications [28].

The ATB is a treebank containing syntactic and semantic annotations including temporal annotation, namely those phrases that hold a -TMP dashtag. According to the ATB annotation guidelines[1], dashtags are used only if they are relevant, for instance -TMP cannot be associated with -SBJ or -OBJ dashtags. As our research focuses on labelling -TMP, we have manually annotated the rest of -TMP not appended in the development and test sets. To the best of our knowledge, a technique for identifying temporal phrases has not been presented in the literature on Arabic Natural Language Processing. The method used for labelling temporals, named *ZamAn*, is based on a Support Vector Machine (SVM) classifier trained on existing -TMP phrases in the ATB using various lexical, syntactic, semantic and external features. Syntactic features include head word features for which we developed our own head-finding rules inspired by the Bikel parser [6].

The second task addressed in the paper is the identification and normalisation of numerical expressions (NUMs). This task is very useful in machine translation, speech recognition and text-to-speech applications. In machine translation, NUMs in the source language are normalised into a digital form and then used to generate them appropriately in the target language. Text to speech and speech recognition systems need to identify NUMs in order to convert the digits into words and generate a correct output for numerical expressions. This task is challenging because numbers can be represented in different forms: digits, multi-word sequences or a mix of both [13]. Moreover, the strong inflective nature of Arabic generates various forms for numeric words in temporal expressions such as "العشرين","twenty" which can be written as "العشرون","twenty" depending on its position in the sentence.

The identification of NUMs is often merged with the identification of Named Entities [27]. However, to the best of our knowledge, the only method dedicated to identifying and normalising Arabic NUMs is a rule-based method [13]. Our approach to detecting NUMs, named *Raqm*, is based on a SVM classifier trained on NUMs extracted from the ATB. Unlike temporal expressions, numerical expressions are not annotated in the ATB. Therefore, we semi-automatically annotate NUMs present in the ATB based on their Part Of Speech (POS) tags and we exploit this annotated data to train a classifier with lexical features as learning

---

[1] `http://projects.ldc.upenn.edu/ArabicTreebank/`

features. Normalising NUMs is done using an improved/extended version of the
normalisation method proposed by [13].

2 General Background

2.1 Temporal Expressions (TMPs)

Guidelines for the identification of TMPs in multilingual corpora were developed
for this specific task. There are two main guidelines that define temporal expressions: TimeML and TIMEX2. TimeML [26] is a specification language designed
to address events and TMPs in natural language. TIMEX2 focuses more on
TMPs and covers both their detection and normalization [9]. Several TimeML
corpora were created for English and other languages including TimeBank [25],
TempEval and AQUAINT TimeML[2]. The Automatic Content Extraction corpus
(ACE)[3], in which TMPs are annotated based on TIMEX2 specifications, is the
only corpus that contains annotated Arabic temporal expressions. However, the
total number of temporal expressions is only 865. This number is not sufficient
to train a machine-learning-based system.

The ATB[4] contains 5,496 non-overlapping temporal phrases and hence it is
more appropriate to train a machine-learning-based system. In addition, the
ATB provides syntactic annotation for constituency parsing. The following are
some examples of temporal phrases extracted from the ATB: "منذ ٨٣ سنة =
since 38 years", "في الشهر ذاته = in the same month", "في مناسبة الليلة الأولى من
ليالي شهر محرم = on the occasion of the first night of the nights of the month
Muharram".

The dashtag -TMP is appended as a temporal modifier which refers to a
point in time or a span of time. This dashtag marks phrases that answer the
questions *when*, *how often*, or *how long*. Possible temporal modifiers are NP-
TMP, PP-TMP, S-NOM-TMP, SBAR-NOM-TMP, ADVP-TMP, S-TMP, and
SBAR-TMP. The dashtag -TMP is not specified for other scenarios (e.g. if a
subject is also a temporal, the dashtag -SBJ has priority and -TMP will not be
added). Table 1 describes the presence of -TMP tags in the ATB. For instance,
in the training set, there are 2,771 temporal phrases marked -TMP of length
between 2-5 tokens.

2.2 Numerical Expressions (NUMs)

The Message Understanding Conference introduced the NUMEX guidelines
to annotate NUMs only for currency and percentages [10]. The ACE data
is annotated for NUMs. Although different types of values exist including
phone numbers, currency and percentages, the ACE contains only 13 Arabic NUMs. Therefore, we used the ATB as our data set for extracting NUMs

---

[2] http://www.timeml.org/site/timebank/timebank.html
[3] LDC Catalog No. LDC2006T06.
[4] Unvocalised version of the Penn Arabic Treebank (ATB), Part3v3.2 [22].

(see Table 1, ATB contains 6,890 NUMs). Unlike temporals, numerical expressions are not annotated in the ATB. Instead, numerical words are only given a POS tag indicating whether the word is a cardinal number or an ordinal number. Therefore, we semi-automatically annotated NUMs using their POS tags in addition to some rules. The following are examples of NUMs extracted from the ATB: "مليون ١,٢ = 1,2 million", "في المائة ٨.٤" = 4.8 percent", "الثاني" = the second", "واحد وعشرين" = twenty one".

Table 1. Distribution of the functional dashtag -TMP and NUMs in the ATB

ATB	Train	Dev	Test
Size (sentences)	10,540	1,326	1,335
Number of NUMs	6,855	860	948
Number of -TMP	5,496	686	708
|TMP| = 1 token	1,115	148	139
1 token < |TMP| ≤ 5 tokens	2,771	360	367
5 tokens < |TMP| ≤ 10 tokens	801	94	102
|TMP| > 10 tokens	809	84	100

2.3 Evaluation of Temporal and Numerical Expressions

In the literature, two evaluation metrics are used to evaluate TMPs extraction [29] : detection performance (DP) and bracketing performance (BP). DP is generous in the sense that it considers any system output correct as long as at least one word overlaps with the reference expression. BP measures the ability of the system to correctly determine the extent of the temporal expression where the output of the system and the reference expression must match exactly.

[13] introduces three evaluation measures to evaluate the identification of NUMs: Full-Match, CoreMatch, and SpanID. Full-Match requires the system output and reference value to match exactly in value, type (plural, ordinal or cardinal) and span. Core match requires the numerical value and the span to match only. SpanID evaluates the identification of the span of a numerical expression. Since we do not need to identify the type of a numeral and we are concerned only about detecting and normalizing NUMs, we found that the use of the same evaluation metrics as for temporals is more appropriate for our work.

2.4 Related Work

TMPs Detection: There are two main approaches to detect temporal expressions: rule-based approaches and machine-learning-based approaches.

For English, [21] uses a rule-based system achieving an F-score of 96.2%, [23] handles conflicts between possibly overlapped tags achieving 92.6% DP and 87.2% BP and [7] uses a rule-based system based on a cascade of finite-state grammars achieving 89.6% DP and 81.7% BP. Machine learning approaches used to detect temporal expressions are either based on tokens or phrases.

Token level approaches use token features such as POS tag, word case, and whether a word belongs to a certain class of tokens. [14] considers the problem of detecting time expressions as a tagging problem at the token level. They use lexical, syntactic, semantic as well as external features from a rule based time expression tagger achieving 93.5% DP and 87.8% BP for English (resp. 90.5% DP and 78.6% BP for Chinese). [1] uses a Conditional Random Fields classifier and [24] uses SVM and an Inductive Logic Programming system. Phrase based approaches use features of phrases such as the syntactic annotation of the phrase and head rules. The latter approaches usually use a parser to extract phrases from text. These approaches are very similar to chunking approaches and named-entity recognition systems [5]. [2] treats TMP recognition as a binary phrase classification. Using the Charniak parser, 90.2% of the TMPs were exactly aligned with a parse. They use SVM and various features including parse-based features and report 84.4% DP and 78.7% BP. [16] shows that binary constituent based classification of chunk phrases outperforms a token-by-token classification system with 85.2% DP and 82.8% BP.

To the best of our knowledge, there is not much research specifically addressing the problem of detecting temporal expressions in Arabic. [15] developed TREX, a rule-based temporal reference extractor for Arabic texts based on stemmed temporal Arabic words achieving an f-score of 92%. However, the corpus used was small and lacking in temporals. Another linguistically motivated approach was proposed by [4] on spoken Jordanian time expressions. In this study, time expressions are classified into three types : precise, semi-precise and imprecise. Precise time expressions include a specific clock time or calendar day. Semi-precise time expressions are the ones similar to the word "لحظة" = "moment" in "لحظة من فضلك" = "one moment please" and imprecise time expressions are expressions whose temporal specifications are vague such as "وستظهر الحقيقة في النهاية" = "the truth will come out *eventually*". This research however focuses only on the linguistic features of TMPs rather than on their detection.

NUMs Detection: [3] presents two grammars for converting Arabic numbers from digits to word forms and vice versa. [8] developed an implementation of the Arabic numeral system in the Grammatical Framework programming language. [13] defines the task of identifying numbers in a natural context and develops a corpus to evaluate their efforts in identifying Arabic numbers achieving 77.9% DP and 73.3% BP. However, all previously mentioned work is based on rule based methods. Our numerical expressions identification system, *Raqm*, is a machine learning system trained on the ATB corpus.

3   ZamAn: Temporal Expressions Labeller

We use Yamcha (Yet Another Multipurpose Chunk Annotator) to detect temporal expressions [19,17,18]. Yamcha uses an SVM learning algorithm. We search for non-overlapping temporal structures because the number of overlapping entities present in the training data is much lower than the non-overlapping structures; and secondly, the search for non-overlapping representations has the potential to correctly label temporal phrases of any length, unlike overlapping structures, for which an upper bound on the length must be fixed for tractability. In addition, apart from [14], all machine-learning-based TIMEX systems reduce the identification task to identifying non-overlapping structures only.

We also develop a post processing rule-based system to identify instances of -TMP common in Arabic but not detected by *ZamAn*.

3.1   Preprocessing the ATB

Preprocessing the raw ATB trees is an important step in order to smooth and prepare the data for temporal labelling. The preprocessing consists of:

1. **Spelling variation:** We collapse all possible variants of the character Alif to normalise the data. The variation can be seen as a level of vocalisation and also because in Arabic raw text the different forms of the Alif are very often misspelled.
2. **Reducing the tagset:** The ATB tagset consists of 492 fine-grained POS tags encoding morphological features. For example, Al+dawol+atayoni (the states) receives the following tag DET+NOUN+NSUFF_FEM_DU_GEN, annotating a definite, feminine, dual noun in the genitive case. Due to the large size of this tagset, we use the Bikel-Bies POS mapping provided with the ATB, which maps the original ATB tagset onto a small PTB-style tagset (28 tags), discarding almost all morphological features that are not also present in the English PTB (except for definiteness information).
3. **Splitting the treebank:** We apply the usual treebank split (80% training, 10% development, 10% test [20]).
4. **Manual annotation for -TMP:** as our research focuses on labelling -TMP, we decided to annotate manually the remaining -TMP not appended in the development and test set in order to reduce the number of false positive.

3.2   Classification Features

Our method is comparable to the work of [14] in terms of the detection method and the features used in detection. However, the results are not comparable because we are not searching for the same entities. In fact, this paper describes temporal phrase labelling and [14] describes temporal expression labelling (e.g. "in 1942" vs. "1942"). The dataset is also different as well as the languages. We treat the problem of detecting temporal phrases as a chunking problem at the level of tokens.

- **Tokens:** Unvocalised and gold tokenized tokens are used as a classification feature where each token in the data set is considered to be a feature. In fact, tokenisation is necessary in Arabic as a single word can comprise up to four independent tokens.
- **POS tags:** We tag the data (development and test sets) using MADA [12], a state-of-the-art tool for morphological processing[5]. We used two different

---

[5] Tested on the ATB, MADA is reported to achieve high accuracy for POS tagging, achieving more than 93%. MADA performs tokenisation, lemmatisation, diacritization, POS tagging, and disambiguation of Arabic texts.

tagsets in our experiments. The first is the full tagset of the ATB which includes morphological information (492 tags), and the second is the Bikel-Bies reduced tagset (28 tags). We include POS tags as a feature in our system to help the disambiguation of tokens as Arabic is a morphologically complex language. In the ATB, 492 morphological tags are used to cover the data but theoretically 333,000 different complex tags are possible. In contrast, English morphological tagsets use 50 tags to cover all morphological variation [11].

- **Token frequency:** We use two different configurations with counts created from the training data: the first configuration counts the frequency of the tokens in the training data and the second counts the frequency of both the token and the POS tag. The second configuration aims at reducing the ambiguity at the lexical level (e.g. "ملك" is a noun (king), verb (own), adj).

- **Lemmas:** Arabic is known for its morphological richness leading to the high number of distinct word-forms, and their inflections. To reduce this data sparsity in labelling TMPs, we map tokens to their associated lemmas. (The lemma of both "الطبيب" = "the physician" and "أطباء" = "physicians" is "physician"). We report the result of using lemmas (obtained using MADA) as features instead of tokens.

- **Token patterns:** We identify five specific patterns for temporal expressions as follows: 1) is the token a digit? 2) does the token belong to a day of the week? 3) is the token a name of a month? 4) is the token an atomic temporal expression (e.g. "أمس" = "yesterday") ? 5) is the token a numerical word (e.g. "ثلاثة" = "three", "عشرة" = "ten")? We construct a list of all possible days of the week and another list for all months (single word months as well as multiple words months such as "تشرين الثاني" = "November"). For patterns 4 and 5 we manually build lexicons that contain the lemmas of temporal words and numerical words that consist of a single token.

- **Constituency relations:** We parse our data to identify the label of the immediate syntactic mother node of each token, namely the types of phrases or clause: NP, VP, PP, S, SBAR, SQ, etc. We decided to apply constituency parsing to include the head-word feature as well but base phrase chunking can be used instead.

- **Head words:** We specify whether each token is the head word of its mother node, it appears to the left of the head word or to the right of the head word. For this task, we build our own Arabic head-rules starting from the list present in [6].

- **Dynamic features:** dynamic features refer to the use of the output tags, predicted by *ZamAn*, of a window of n ($3 \le n \le 5$) tokens preceding the current token as features to predict the label of the current token.

- **Contextual features:** a window of size m ($3 \le m \le 5$) is specified such that the m tokens preceding the current token and the m tokens succeeding the current token and their associated features are used as features to identify the tag of the current token.

3.3 ZamAn: Experimental Results

Table 3 describes the classification results we obtained using various feature combinations. We notice from Table 1 that 85% of -TMP present in the training set are composed of 10 tokens or less, thus, we experiment with the identification of -TMP of <5 (5 tokens or less), of <10 (10 tokens or less), and of all lengths (length is not specified).

Run 1 represents our baseline (83.6% DP and 63.6% BP), where only tokens are used as a feature to train *ZamAn*. Run 11 provides the best results (90.7% DP and 76% BP), when all features are used together with the reduced Bies-Bikel tagset and a post processing rule-based system we designed to identify instances of -TMP not detected by *ZamAn* (see Table 3).

We experiment with the use of lemmas instead of the tokens themselves (runs 12-15) but the results are generally lower, especially when detecting -TMP of any length. The use of lemmas hurts the bracketing results considerably because it increases the ambiguity at the lexical level. We also experiment with a version of the labeller which uses the full tagset (492 tags) to make use of the morphological information present in the tags (runs 16-18). However, the results are lower as the tagset is too fine-grained.

The error analysis shows that the number of undetected -TMPs can be decreased using a rule based labeller because missing phrases are often common Arabic expressions that are very short but not frequent in the training set such as: سلفاً، تالياً، في الماضي، في الشهر ذاته. Using the literature on Arabic temporal expressions [15], we built our own rule-based system to detect common Arabic temporal phrases and add the detected temporal phrases to *ZamAn's* output and re-evaluate the new output (see Table 2, best results on dev and test sets). The remaining missing -TMP are mainly due to inconsistencies in the original ATB annotation. For instance, the phrase 21/08/2001 is tagged in the ATB incorrectly as 21/(08/2001). Since *ZamAn* detects the whole structure as one entity, the evaluation classifies the labelling as incorrect compared to the gold. In addition the rule based labeller does not search for embedded structures. As expected from the work of [14], using all features gives the best performance. Arabic results (90.7% DP and 76% BP) for temporal phrases recognition are close to those obtained for the Chinese system for temporal expression recognition (90.5% DP and 78.6% BP) even though the results are not directly comparable.

As the ACE Corpus contains annotations for Arabic temporal expressions, we decided to evaluate the performance of *ZamAn* using this data. We preprocessed the temporal expressions already annotated in this corpus in order to convert them to temporal phrases similar to the ones in ATB. This preprocessing mainly consists of adding prepositions such as "في=in/at/on/etc" and "منذ=since" to the extent of an expression. We used the model trained on ATB (run 1, -TMP of length 10 as this was the average length of TMPs in the ACE) for this evaluation. Although, we detect almost all temporals present in the ACE, the results we obtained were low: 43.3% DP and 24% BP due to the large number of unannotated temporal phrases (1,136 -TMP) in the ACE detected by *ZamAn* (e.g. "في سبتمبر أيلول" = "in September"), and also, due to incomplete annotation in the ACE such as "Friday evening" where only "Friday" in annotated (*ZamAn* detects the whole expression).

4   Raqm: Numerical Expressions Labeller

We annotate NUMs in the ATB semi-automatically to create training data for *Raqm* using POS tags and lexical information. Special treatment is required 1) for multi-word expressions containing the conjunction "و= and" or a comma "," in order to decide when "و" is used between two individual expressions "بين ١٥١٤,٠٠ و ١٥١٣,٥٠ = between 1513,50 and 1514,00 " or is a part of one multi-word expression "واحد وعشرين = twenty one" and 2) for splitting two independent consecutive digits "بلغ عام ١٩٨٠ ١١,٩٨ في المائة = it reaches in 1980 11.98 per cent".

Due to a high level of ambiguity, we exclude the words "الأول= first" and "الثاني= second" if they are part of a month name as in "كانون الأول=December". We also exclude the word "أحد=one/anybody" because it is always labelled as a number by MADA even if it is not a number as in "لن يبقى لأحد من الناس" "أحد'means in this case "anybody".

To prepare the training data, we convert all the digits into numeric words using a lexicon of digits and their corresponding words to run the Yamcha SVM classifier using tokens, dynamic and contextual features (see Section 3.2).

4.1   Normalisation of NUMs

The next step is the normalization of NUMs. We use the same algorithm proposed by [13] with two extensions.

- Extension 1: [13] do not mention the processing of fractions. For instance, the ATB contains NUMs indicating a fraction such as "نصف" = "half" and "ربع" = "one quarter". If a numerical expression contains one of those words followed by another numeric word, we multiply the fraction by the next number e.g. "مليونا ونصف مليون= one million and half million" is converted to "1000,000 w 500,000". Then we convert the expression to "1,500,000".
- Extension 2: [13] fails to correctly normalize long NUMs containing several scales[6]. For instance, "six hundred and twelve million and five hundred thousand" is converted to 612,000,500,000 instead of 612,500,000. To solve this issue, we modify the algorithm to rank the scales and use an additional stack to save the number obtained when a value is multiplied by its scale.

4.2   Raqm: Experimental Results

The results of Raqm on the dev and test sets are shown in Table 2. We achieved state-of-the-art performance in identifying NUMs (96% DP and 94.4% BP). The remaining non labelled NUMs were due to the ambiguity in using the conjunction

---

[6] A scale can have the value of thousand, million, billion, etc.

"و= and" in a multi-word expression vs. using it to split two different NUMs. The performance of our systems *ZamAn* and *Raqm* has been also evaluated and validated using 3-fold cross validation.

Table 2. Best Performance of *Raqm* and *ZamAn* systems on dev and test sets using full length expressions for temporals

	Raqm		ZamAn	
	DP	BP	DP	BP
Dev set	94.9	92.6	90.8	77.5
Test set	96	94.4	88.5	73.1
ACE	n/a	n/a	43.3	24

Table 3. Performance of *ZamAn* System at different feature combinations: tok: tokens, lem: lemma, tf: term frequency, tpf: frequency of token and its POS tag, rpos: reduced POS tags, fpos: ATB POS tag set, ddm: whether the token is a digit, day of week or month, ddmt: ddm + whether the token is an atomic temporal token or not, ddmtn: ddmt + whether the token is a numeric word or not, const: constituency relations, head: head words information

Run	tok	lem	tf	tpf	rpos	fpos	ddm	ddmt	ddmtn	const	head	|TMP| <= 5		|TMP| <= 10		TMP All	
												DP	BP	DP	BP	DP	BP
1	+											69.5	62.3	75.9	64.1	83.6	63.6
2	+		+									68	61.1	73	60.8	81.6	60.8
3	+		+		+							68.6	64	72.4	62.8	81	64
4	+			+	+							67.4	63.5	72.2	62	80	61.1
5	+			+	+	+						70.6	66.7	74.9	63.5	82.8	64.6
6	+			+	+	+	+					70.1	66.3	75.5	64.6	83.4	64.5
7	+			+	+	+	+	+				67.4	63.5	72.2	62	80	61.1
8	+	+		+						+		67.8	64.5	71.7	61.6	79.4	61.4
9	+			+	+					+	+	67.1	63.6	72.3	62.1	80.2	63.4
10	+			+	+	+	+	+	+	+	+	69.8	66.2	74.5	64.3	83.4	66.6
11	+	+		+		+	+	+	+	+	+	**70.8**	**66.6**	**74.5**	**65.4**	**84.1**	**67.4**
12		+										63.7	57.5	71.6	58.7	80.9	58.8
13		+	+									63.9	58.3	70.7	58.3	79.5	57.6
14		+	+		+							68.5	63.7	73.4	62.6	80.6	61.9
15		+		+	+							67.4	62.4	74	63.2	80.5	61.4
16	+		+			+						66.3	61	71.5	61.6	80.7	62.6
17	+			+		+						61.5	55.5	68.2	56.7	76.8	57.8
18	+		+			+	+	+	+	+	+	68.8	64.8	73.5	64.4	83.8	66.8
Run 11 + post processing common -TMP not detected by *ZamAn*																**90.8**	**77.5**

5   Conclusion

We develop the *ZamAn* system to label temporal structures and the *Raqm* system to detect numerical expressions achieving state-of-the-art results with an

F-score of 88.5% (resp. 96%) for bracketing and 73.1% (resp. 94.4%) for detection. As far as we are aware, *ZamAn* is the first robust and accurate system for recognition and extraction of temporal phrases for Arabic. In the future, we aim to use this system for normalisation of temporals in Arabic texts in order to improve Arabic statistical parsing.

Acknowledgements. This research is funded by the Faculty of Engineering and Computing at Dublin City University. We thank Jennifer Foster and our reviewers for their very helpful suggestions.

References

1. Ahn, D., Adafre, S.F., de Rijke, M.: Extracting temporal information from open domain text: A comparative exploration. In: JDIM, pp. 14–20 (2005)
2. Ahn, D., van Rantwijk, J., de Rijke, M.: A cascaded machine learning approach to interpreting temporal expressions. In: HLT-NAACL, pp. 420–427. The Association for Computational Linguistics (2007)
3. Al-Anzi, F.S.: Sentential count rules for arabic language. Computers and the Humanities 35(2), 153–166 (2001)
4. Al-Rashdan, B.: Extracting temporal information from open domain text: A comparative exploration. Jordan Journal of Modern Languages and Literature 1(1), 61–80 (2008)
5. Benajiba, Y., Diab, M., Rosso, P.: Arabic named entity recognition using optimized feature sets. In: Proceedings of the Conference on Empirical Methods in Natural Language Processing, EMNLP 2008, pp. 284–293. Association for Computational Linguistics, Stroudsburg (2008)
6. Bikel, D.: On the Parameter Space of Generative Lexicalized Parsing Models. PhD thesis, University of Pennslyvania (2004)
7. Boguraev, B., Ando, R.K.: Timebank-driven timeml analysis. In: Dagstuhl Seminars (2005)
8. Dada, A.: Implementation of the arabic numerals and their syntax in gf. In: Proceedings of the 2007 Workshop on Computational Approaches to Semitic Languages: Common Issues and Resources, Semitic 2007, pp. 9–16. Association for Computational Linguistics, Stroudsburg (2007)
9. Ferro, L.: Instruction manual for the annotation of temporal expressions. Technical Report MTR 01W0000046, The MITRE Corporation (2001)
10. Grishman, R., Sundheim, B.: Design of the muc-6 evaluation. In: Proceedings of the 6th Conference on Message Understanding, MUC6 1995, pp. 1–11. Association for Computational Linguistics, Stroudsburg (1995)
11. Habash, N., Rambow, O.: Arabic tokenization, part-of-speech tagging and morphological disambiguation in one fell swoop. In: Proceedings of the 43rd Annual Meeting on Association for Computational Linguistics, ACL 2005, pp. 573–580. Association for Computational Linguistics, Stroudsburg (2005)
12. Habash, N., Rambow, O., Roth, R.: MADA+TOKAN: A toolkit for Arabic tokenization, diacritization, morphological disambiguation, pos tagging, stemming and lemmatization. In: Proceedings of the 2nd International Conference on Arabic Language Resources and Tools, MEDAR (2009)
13. Habash, N., Roth, R.: Identification of naturally occurring numerical expressions in arabic. In: Proceedings of the Sixth International Language Resources and Evaluation (LREC 2008), Marrakech, Morocco (May 2008)

14. Hacioglu, K., Chen, Y., Douglas, B.: Automatic Time Expression Iabeling for English and Chinese Text. In: Gelbukh, A. (ed.) CICLing 2005. LNCS, vol. 3406, pp. 548–559. Springer, Heidelberg (2005)
15. Haraty, R.A., Khatib, S.A.: Trex: a temporal reference extractor for arabic texts. In: Proceedings of the ACS/IEEE 2005 International Conference on Computer Systems and Applications. IEEE Computer Society, Washington, USA (2005)
16. Kolomiyets, O., Moens, M.-F.: Comparing Two Approaches for the Recognition of Temporal Expressions. In: Mertsching, B., Hund, M., Aziz, Z. (eds.) KI 2009. LNCS, vol. 5803, pp. 225–232. Springer, Heidelberg (2009)
17. Kudo, T., Matsumoto, Y.: Chunking with support vector machines. In: Proceedings of NAACL (2001)
18. Kudo, T., Matsumoto, Y.: Use of support vector learning for chunk identification. In: Proceedings of the 2nd Workshop on Learning Language in Logic and the 4th Conference on Computational Natural Language Learning, ConLL 2000, Stroudsburg, USA, vol. 7, pp. 142–144 (2000)
19. Kudo, T., Matsumoto, Y.: Fast methods for kernel based text analysis. In: Proceedings of ACL, Sapporo, Japan (July 2003)
20. Kulick, S., Gabbard, R., Marcus, M.: Parsing the Arabic Treebank: Analysis and improvements. In: Proceedings of the Treebanks and Linguistic Theories Conference (TLT 2006), Prague, Czech Republic, pp. 31–42 (2006)
21. Mani, I., Wilson, G.: Robust temporal processing of news. In: Proceedings of the 38th Annual Meeting on Association for Computational Linguistics, ACL 2000, Stroudsburg, PA, USA, pp. 69–76 (2000)
22. Mohamed, M., Bies, A., Kulic, S.: Creating a methodology for large-scale correction of treebank annotation: The case of the arabic treebank. In: MEDAR Second International Conference on Arabic Language Resources and Tools, Egypt (2009)
23. Negri, M., Marseglia, L.: Recognition and normalization of time expressions: Itc-irst at tern 2004. Technical report, ITC-irst (2004)
24. Poveda, J., Surdeanu, M., Turmo, J.: A comparison of statistical and rule-induction learners for automatic tagging of time expressions in english. In: Proc. of the 14th International Symposium on Temporal Representation and Reasoning (TIME 2007), pp. 141–149. IEEE (2007)
25. Pustejovsky, J., Hanks, P., Sauri, R., See, A., Gaizauskas, R., Setzer, A., Radev, D., Sundheim, B., Day, D., Ferro, L., Lazo, M.: The timebank corpus. In: Proceedings of Corpus Linguistics (March 2003)
26. Pustejovsky, J., Castao, J.M., Ingria, R., Sauri, R., Gaizauskas, R.J., Setzer, A., Katz, G., Radev, D.R.: Timeml: Robust specification of event and temporal expressions in text. In: New Directions in Question Answering, pp. 28–34 (2003)
27. Shaalan, K., Raza, H.: Nera: Named entity recognition for arabic. J. Am. Soc. Inf. Sci. Technol. 60, 1652–1663 (2009)
28. Tao, C., Wei, W.Q., Solbrig, H.R., Savova, G., Chute, C.G.: Cntro: A semantic web ontology for temporal relation inferencing in clinical narratives. In: AMIA 2010 Symposium, pp. 787–791 (2010)
29. TERN. The tern 2004 evaluation plan (draft): Time expression recognition and normalization. Technical report, The MITRE Corporation (2004)

Extracting Parallel Paragraphs and Sentences from English-Persian Translated Documents*

Mohammad Sadegh Rasooli, Omid Kashefi, and Behrouz Minaei-Bidgoli

Department of Computer Engineering, Iran University of Science and Technology
rasooli@comp.iust.ac.ir, kashefi@{ieee.org,iust.ac.ir},
b_minaei@iust.ac.ir

Abstract. The task of sentence and paragraph alignment is essential for preparing parallel texts that are needed in applications such as machine translation. The lack of sufficient linguistic data for under-resourced languages like Persian is a challenging issue. In this paper, we proposed a hybrid sentence and paragraph alignment model on Persian-English parallel documents based on simple linguistic features as well as length similarity between sentences and paragraphs of source and target languages. We apply a small bilingual dictionary of Persian-English nouns, punctuation marks, and length similarity as alignment metrics. We combine these features in a linear model and use genetic algorithm to learn the linear equation weights. Evaluation results show that the extracted features improve the baseline model which is only a length-based one.

Keywords: Sentence Alignment, Paragraph Alignment, Parallel Corpus, Bilingual Corpus, Persian, English, Machine Translation.

1 Introduction

A parallel corpus is made up of sentences of two languages that each sentence from one language is a translation of the sentence in the other one [1]. Sentence aligned corpora are necessary for the task of statistical machine translation. Furthermore, there are many other applications of parallel corpora such as lexicography and language analysis [2]. The importance of parallel corpus increased since the IBM translation models [3] has been introduced. One of the main steps of building a parallel corpus is sentence alignment [4], so in order to obtain a sentence aligned corpus, it is necessary to perform sentence alignment on bitext documents. Sentence alignment is the task of mapping each sentence in the source language to its corresponding sentence (or sentences) in the target language [5]. In addition, if the parallel data are not aligned in the paragraph level, it is essential to align paragraphs before the sentence alignment task, because knowing paragraph boundaries help to reduce candidate space for sentence alignment. There are some ways to collect bilingual texts. Using widely translated books such as holy religious books [6], multi-lingual catalogues [7], or getting parallel texts from web [8] are some of the ways for

---

* This paper is funded by Computer Research Center of Islamic Sciences (CRCIS).

M.V.M. Salem et al. (Eds.): AIRS 2011, LNCS 7097, pp. 574–583, 2011.

collecting bitext data. Therefore, building appropriate parallel text is a hardship for under-resourced languages similar to Persian language.

Persian is a variation of Arabic-script language that is mostly spoken in Iran, Afghanistan, Tajikistan and some parts of India and Pakistan. One of the most challenging issues of Persian language processing and information retrieval (IR) is the lack of feasible corpus. For example, to our knowledge, there is no syntactically and/or semantically tagged corpus in Persian. There are some bilingual Persian-English corpora. One of them is introduced in [9]. Another Persian-English parallel corpus is introduced in [10] that is made of Persian-English aligned movie subtitles wherein the Persian parts are in colloquial shape instead of official Persian language that is used in academic and governmental organizations and medias. Therefore, it is required to have a corpus of official Persian language. The process of manually aligning corpora is a time consuming task, where intelligent computer programs may help reduce the time, so automatically building parallel Persian-English corpora made sentence is a hot topic.

In this paper, considering the need for a feasible parallel Persian-English corpus, we propose a hybrid approach to extract aligned paragraphs and sentences from translated documents. We used some clues such as paragraph or sentence length, punctuation marks and a small bilingual lexicon of simple one-word nouns. In the following sections, after reviewing some related works on sentence alignment in Section 2, we describe our proposed method in Section 3. We have done two experiments on Persian-English data, one for paragraph alignment and one for sentence alignment that are described in Section 4. Finally, Section 5 concludes the paper.

2 Related Works

There have been many sentence alignment models in recent years that can be categorized into three model types: 1) length-based approaches, 2) Lexical matching approaches, and 3) Hybrid approaches. Furthermore, there are some other approaches such as measuring cognate similarity between sentences. Most of the works on sentence alignment models are based on the assumption that paragraph anchors are aligned [11-15]. Even though, in some works, candidate sentences are chosen based on a window size of adjacent sentences in the text without any information about paragraph anchors [16].

The first attempts on sentence alignment were done based on length-based models. In the length based approach, it is assumed that the sentence pairs of source-target languages are similar in their length. For example, the sentences in German-English parallel corpus have a correlation of 0.91 [17]. The first attempts on length-based approaches were in [11, 12, 17]. Not only this model is very simple and language independent, but also it can gain global optimum [18]. On the other hand, small deletions and insertions decrease the accuracy drastically [14] and error propagation may happen [18]. The second approach is based on lexical matching. In this approach, bilingual lexicons are used as guides to alignment. The early works on this approach were done by [14, 15, 19, 20].

In the third approach, the combination of statistical and linguistic features (such as bilingual lexical matching and simple linguistic clues) is used as a guide to the alignment task. The main reason to use this approach is that for many languages, simple statistical approaches do not gain enough accuracy and there is a need for new approaches to overcome this problem [21]. Most of the recent works on sentence alignment is based on hybrid models. In [13], three phases were used in order to extract aligned sentences. In the first phase, some aligned sentences were extracted via length based models. In this phase, a threshold was considered on length similarity to select only reliable parallel sentences in the corpora. In the second phase, IBM model 1 was used on the extracted aligned corpora and a bilingual lexicon was built. In the last phase, the program used both the gained lexical information and length similarity to find aligned sentences. This work became state of the art in its time and many other researchers improved this model. In [22], a combination of dynamic programming (DP) and divisive clustering was used to improve Moore's model[13].In this work, DP allows many-to-many alignments and divisive clustering refines those alignments with iterative binary splitting. In [23], a two step clustering approach was used to improve both accuracy and efficiency of Moore's model. In the first step the program finds a model-optimal alignment made up of possible 0/1 to 0/1 alignments and in the second step, it merges those alignments into larger ones. That method was 550 times faster than the work in [22].

In [24] cognates similarity (similarity based on transliteration) was used as a measure of similarity. In [25], the order of punctuation marks in bitext and lexical information were combined to achieve aligned sentences. In [26], the combination of punctuation marks, cognates and length similarity was used to find better alignments. In addition, probabilistic neural network (P-NNT) and Gaussian mixture models (GMM) were used to combine those features. In [5], a modification of the Champollion in [27] was proposed. This approach was based on a hybrid model that optimized the process of splitting the bilingual texts into small parts for alignment. In [28], an iterative model was used to improve alignment accuracy. That work was an extension of [29] that was used for aligning OCR generated texts. In that work, a length based approach was used in the initialization phase. In the next iterations, a statistical machine translation (SMT) model was built and based on that model; bleu measure was used in the next iterations to compare the translated text by the translation model to the candidate sentences. In the last iteration, the final SMT model was built from the sentence aligned corpora. In [30], a bootstrapping algorithm was done on bitext based on cosine similarity measure to measure similarity of the documents based on TF-IDF. In [31], Wikipedia was considered as a good source of multilingual data with many noises such as sentences without translations. Some features such as date matching, same pictures in the pages and a little manually aligned train data for building bilingual lexicon was used as guides to alignment. In [32], based on the assumption that parallel web pages have similar page structures and translators respect the original structure of the document, an HTML tree alignment model was proposed using dynamic programming. In [33], a language independent context model based on Zipfian word vector was proposed to improve sentence alignment problem. The Zipfian word vector is a vector of values in the sentences based on the logarithmic division of the word frequency in the sentence context and a threshold. In that work, dynamic programming was used to align sentences. There are also some alignment works on movie subtitles based on simple clues such as sentence length and time overlap that some of these works are proposed in [34, 35].

3   Extracting Parallel Paragraphs and Sentences

One of the main problems we faced in Persian-English parallel corpus extraction is the lack of paragraph aligned corpora. Indeed, in this paper we focus on introducing a new model that can be used both for sentence alignment and paragraph alignment. In the study we have done on Persian-English bitext, we found that the most of the paragraph lengths are in the similar length order and using pure length-based models results in unreliable alignment. On the other hand, the methods used in [13] is employed IBM model 1 [3] which is not efficient for long paragraphs. The length of paragraphs in Persian is about 100 words (based on our test bed). The IBM model performance is proportional to the sentence length, where for long sentences the space of candidates and processing time increases. We chose three most efficient ones as: 1) length similarity, 2) punctuation mark similarity, and 3) semantic similarity between words of source and target paragraphs or sentences that is calculated exploiting a bilingual dictionary of nouns.

3.1   Feature Similarities

For the length based similarity we used Poisson distribution that is employed in [13, 26, 31]. This distribution has only one parameter and is simpler than Gaussian distribution used in [11]. The Poisson distribution only needs the length rate between the source and target sentences as shown in (1) where l_t and l_s are sentence length of the target and source languages and r is the sentence length rate.

$$p_{length(S, T)} = \frac{e^{-l_s r}.(l_s.r^{l_t})}{l_t!} \tag{1}$$

For the punctuation similarity, we chose 11 different punctuation types that are comparable in Persian and English[1]. Equation (2) calculates the punctuation similarity score of each punctuation mark (punc$_i$), where min(s,t) and max(s,t) are the minimum and maximum number of occurrence of punctuation punc$_i$ in the source and target sentence or paragraph respectively.

---

[1] We replace brackets with their open form; as an example, close parenthesis mark is replaced with open parenthesis mark. Therefore, punctuation marks in the text are mapped into the (, ; ? ! . - { [" : set. Some Persian punctuation marks are different but corresponding with English ones such as "،" in Persian that corresponds to "," in English, or "؟" in Persian corresponds "?" in English.

$$p(punc_i) = \frac{\min(s,t)}{\max(s,t)}, \max(s,t) > 0 \tag{2}$$

So, the overall punctuation probability will be as equation (3), where np is the number of distinct punctuations.

$$p_{punc}(S,T) = \frac{\sum_{i}^{np} p(punc_i)}{np} \tag{3}$$

We used a small dictionary of Persian one-word nouns and their English translations. In order to calculate the semantic similarity of sentences or paragraphs, we count the co-occurrence of translated nouns (either in word form or stem form) in source and target sentences or paragraphs as shown in (4) where $dic(s_i)$ and $dic(t_i)$ are the number of translated word occurrences. In addition, *count(S)* and *count(T)* are the number of distinct words in the source and target languages respectively. In other words, for each noun in the source sentence or paragraph, if the corresponding translated word exists in the target sentence or paragraph, we count a co-occurrence score for that word.

$$p_{dic}(S,T) = \frac{\sum_{i,j} \frac{\min(dic(s_i),dic(t_i))}{\max(dic(s_i),dic(t_i))}}{\max(count(S),count(T))}, \max(dic(s_i),dic(t_i)) > 0 \tag{4}$$

3.2   Combining Similarities

In order to combine similarity scores, we used a linear model similar to the mathematical union, in which each part of the equation is weighted by a coefficient as in (5) where w_1 and w_2 are in [0, 1] and the other coefficients are in [-1, 1]. The reason that we used this form is that it is simple to use and with this type of equation all types of linear combinations of variables is considered, so the learner adapts appropriate weight to each coefficient.

$$\begin{aligned} p_{align}(S,T) = & \, w_1 \, p_{punc}(S,T) + w_2 \, p_{length}(S,T) + w_3 \, p_{dic}(S,T) \\ & + w_4 \, p_{punc}(S,T) \cdot p_{length}(S,T) + w_5 \, p_{punc}(S,T) \cdot p_{dic}(S,T) \\ & + w_6 \, p_{dic}(S,T) \cdot p_{length}(S,T) + w_7 \, p_{punc}(S,T) \cdot p_{dic}(S,T) \cdot p_{length}(S,T) \end{aligned} \tag{5}$$

With this assumption, it is needed to find 18 unknown weight coefficients in (5).

3.3   Using Genetic Algorithm for Weight Learning

We used genetic algorithm to find the unknown weight coefficients. It is worth to say that in[36], genetic algorithm was also used for sentence alignment. Elitism is used in the genetic algorithm in order to keep good chromosome in each generation. The fitness function is shown in (6). The precision of the weights of the chromosome in the training data is used in the equation to find the fitness.

$$\text{Fitness(cromosome)} = \frac{e^{\frac{\text{precision(cromosome)}}{10}}}{\sum e^{\frac{\text{precision(cromosome)}}{10}}} \tag{6}$$

4   Experiments and Results

We evaluate the effectiveness of our proposed alignment method for both sentence and paragraph alignment. In the *experiment 1*, we evaluated our method through paragraph alignment. The length rate for Poisson distribution has been chosen as 1. Uniform distribution was used to choose each chromosome gene for crossover in learning part using genetic algorithm. The uniform probability is chosen as 0.5 and the crossover rate as 0.9. The mutation rate is chosen as 0.01. In the *experiment 2*, we used our method for sentence alignment. To stem English words, we used an open-source code of Porter stemmer [37] and for Persian words, we used lemmatization code proposed in [38][2].

4.1   Experiment 1: Paragraph Alignment

In order to evaluate the effectiveness of semantic similarity and exploiting a bilingual dictionary, we relaxed (5) to (7) and compare the results.

$$p_{\text{align}} (S, T) = w_1\, p_{\text{punc}} (S, T) + w_2\, p_{\text{length}} (S, T)$$
$$+ w_3\, p_{\text{punc}} (S, T) . p_{\text{length}} (S, T) \tag{7}$$

In this experiment, we extracted about 400 lecture paragraphs from the Iran supreme leader official website[3] that provides both English and Persian edition of lectures. The translation mode in this website is free and there are many free deletions and insertions in the translations. There are also some Arabic sentences (like holy Quran verses) during lectures. In order to find parallel documents, we matched the dates of the lecture pages and aligned the paragraphs manually. Finally, we prepared a paragraph level aligned corpus with about 300 Persian paragraphs and about 350 corresponding English

---

[2] The Persian lemmatizer code is available in the Virastyar software pack in:
   http://sourceforge.net/projects/virastyar/.
[3] http://www.khamenei.ir

Paragraphs. The reason of difference between the numbers of Persian and English paragraphs in the manually aligned corpus is a source paragraph may be translated into two or more paragraphs in target language as we have one-to-one, one-to-two, one-to-three, and one-to-four alignment in our corpus. To overcome the small training data issue, we used 6-fold cross-validation. The results of paragraph alignment using our proposed method are shown in Table 1. In order to compare the results with the baseline length-based model, we tested the baseline on this data and did not get alignment precision more than 35% considering only the length of paragraphs, which means that the pure length-based model is not appropriate enough for sentence alignment.

Table 1. Results of paragraph alignment experiment

Fold #	Precision with Semantic Similarity	Precision without Semantic Similarity
1	100	71.93
2	94.12	79.41
3	100	44.12
4	90	70
5	89.01	47.26
6	81.25	59.37
Average	92.40	62.01

4.2   Experiment 2: Sentence Alignment

After experiment 1, we developed a visual user-friendly software for manually aligning sentences. We used translated novels as our initial data. In order to build a binary classifier, all bilingual sentences pairs are labeled by our method as "are-aligned" or "not-aligned". In this way, we are able to calculate both precision and recall. We also evaluate the effectiveness of punctuation similarity relaxing (5) to (8).

$$
\begin{aligned}
p_{align}(S,T) &= w_1 \, p_{dic}(S,T) + w_2 \, p_{length}(S,T) \\
&+ w_3 \, p_{dic}(S,T).p_{length}(S,T)
\end{aligned}
\tag{8}
$$

The dataset of our test is composed of 26,108 aligned sentences (13,054 sentences in each language). The results are shown in Table 2, the punctuation similarity improved the model performance, but the improvement is not significant. The reason is that for sentences, numbers of punctuation marks are very small and it does not guide the model very well. On the other hand, in the paragraph level, there are much more punctuation marks which help the model predict better alignments.

Table 2. Results of sentence alignment experiment

Method	Precision	Recall	F_1-Measure
Precision with punctuation similarity	96.63	79.42	86.48
Precision without punctuation similarity	92.25	79.45	86.03

5 Conclusion

We faced some problems in using parallel data; as an instance, we tried to use texts extracted by OCR but due to poor performance of Persian OCR software, rate of erroneous recognized words was very high. The problem of copyright in translations, un-uniform Persian characters, the lack of bilingual electronic texts and colloquial language typography in many Persian texts made the task in Persian harder than languages like English.

As seen in the experiments, the performance of the model depends on the task. If it used to paragraph alignment, the punctuation mark and bilingual dictionary significantly improve the accuracy of the pure length-based model result about 35%. But in the sentence alignment, the length and number of words are small enough to suppress other features. Furthermore, the cognate similarity is not applicable to Persian-English texts, because the alphabet types of these two languages are different.

Acknowledgement. This paper is funded by Computer Research Center of Islamic Sciences (CRCIS). I would also like to thank Dr. Shahram Khadivi, Dr. Heshaam Faili, Maryam Aminian, Sina Iravaninan, Dr. Morteza Analoui, Mehrdad Senobari and other people who help me on this work.

References

1. Frankenberg-Garcia, A.: Compiling and using a Parallel Corpus for Research in Translation. International Journal of Translation (2009)
2. Tiedemann, J.: Recycling Translations: Extraction of Lexical Data from Parallel Corpora and their Application in Natural Language Processing. In: Faculty of Languages, Department of Linguistics. Uppsala University, Uppsala (2003)
3. Brown, P., et al.: The Mathematics of Statistical Machine Translation: Parameter Estimation. Computational Linguistics 19(2), 263–311 (1993)
4. Simões, A., Almeida, J.J.: Parallel Corpora based Translation Resources Extraction. Procesamiento del Lenguaje Natural, 265–272 (2007)
5. Li, P., Sun, M., Xue, P.: Fast-Champollion: a fast and robust sentence alignment algorithm. In: 23rd International Conference on Computational Linguistics: Posters, pp. 710–718. Association for Computational Linguistics, Beijing (2010)
6. Resnik, P., Olsen, M.B., Diab, M.: The Bible as a parallel corpus: Annotating the "Book of 2000 Tongues". Computers and the Humanities 33, 129–153 (1999)
7. Menezes, A., Richardson, S.D.: A best-first alignment algorithm for automatic extraction of transfer mappings from bilingual corpora. In: Proceedings of the ACL 2001 Workshop on Data-Driven Methods in Machine Translation, Toulouse, France, pp. 39–46 (2001)
8. Resnik, P., Smith, N.A.: The Web as a Parallel Corpus. Computational Linguistics 29, 349–380 (2003)
9. Miangah, T.M.: Constructing a large-scale english-persian parallel corpus. Meta: Translators' Journal 54(1), 181–188 (2009)
10. Pilevar, M.T., Faili, H., Pilevar, A.H.: TEP: Tehran English-Persian Parallel Corpus. In: Gelbukh, A. (ed.) CICLing 2011, Part II. LNCS, vol. 6609, pp. 68–79. Springer, Heidelberg (2011)

11. Brown, P.F., Lai, J.C., Mercer, R.L.: Aligning Sentences in Parallel Corpora. In: Proceedings of the 29th Annual Meeting of the Association for Computational Linguistics, Berkeley, California, pp. 169–176 (1991)
12. Gale, W.A., Church, K.W.: A program for Aligning Sentences in Bilingual Corpora. In: Proceedings of the 29th Annual Meeting of the Association for Computational Linguistics, Berkeley, California, pp. 177–184 (1991)
13. Moore, R.C.: Fast and Accurate Sentence Alignment of Bilingual Corpora. In: Richardson, S.D. (ed.) AMTA 2002. LNCS (LNAI), vol. 2499, pp. 135–144. Springer, Heidelberg (2002)
14. Chen, S.F.: Aligning Sentences in Bilingual Corpora Using Lexical Information. In: Proceedings of the 31st Annual Meeting of the Association for Computational Linguistics, Columbus, Ohio, pp. 9–16 (1993)
15. Wu, D.: Aligning a Parallel English-Chinese Corpus Statistically with Lexical Criteria. In: Proceedings of the 32nd Annual Meeting of the Association for Computational Linguistics, Las Cruces, New Mexico, pp. 80–87 (1994)
16. Meyers, A., Kosaka, M., Grishman, R.: A Multilingual Procedure for Dictionary-Based Sentence Alignment. In: Farwell, D., Gerber, L., Hovy, E. (eds.) AMTA 1998. LNCS (LNAI), vol. 1529, pp. 187–198. Springer, Heidelberg (1998)
17. Gale, W.A., Church, K.W.: A Program for Aligning Sentences in Bilingual Corpora. Computational Linguistics 19(1), 75–102 (1993)
18. Tiedemann, J.: Bitext Alignment. Synthesis Lectures on Human Language Technologies 4, 1–165 (2011)
19. Kay, M., Röscheisen, M.: Text-translation alignment. Computational Linguistics 19(1), 121–142 (1993)
20. Melamed, I.D.: A geometric approach to mapping bitext correspondence. In: Conference on Empirical Methods in Natural Language Processing, EMNLP (1996)
21. Haruno, M., Yamazaki, T.: High-Performance Bilingual Text Alignment Using Statistical and Dictionary Information. Natural Language Engineering 3(1), 131–138 (1997)
22. Deng, Y., Kumar, S., Byrne, W.: Segmentation and alignment of parallel text for statistical machine translation. Natural Language Engineering 13(03), 235–260 (2007)
23. Braune, F., Fraser, A.: Improved unsupervised sentence alignment for symmetrical and asymmetrical parallel corpora. In: Proceedings of the 23rd International Conference on Computational Linguistics: Posters, pp. 81–89. Association for Computational Linguistics, Beijing (2010)
24. Simard, M., Foster, G.F., Isabelle, P.: Using cognates to align sentences in bilingual corpora. In: Fourth International Conference on Theoretical and Methodological Issues in Machine Translation (TMI 1992), Montreal, Canada, pp. 67–81 (1992)
25. Chuang, T.C., Wu, J.-C., Lin, T., Shei, W.-C., Chang, J.S.: Bilingual Sentence Alignment Based on Punctuation Statistics and Lexicon. In: Su, K.-Y., Tsujii, J., Lee, J.-H., Kwong, O.Y. (eds.) IJCNLP 2004. LNCS (LNAI), vol. 3248, pp. 224–232. Springer, Heidelberg (2005)
26. Fattah, M.A., et al.: Sentence alignment using P-NNT and GMM. Computer Speech & Language 21(4), 594–608 (2007)
27. Ma, X.: Champollion: A robust parallel text sentence aligner. In: LREC 2006: Fifth International Conference on Language Resources and Evaluation, pp. 489–492 (2006)
28. Sennrich, R., Volk, M.: Iterative, MT-based Sentence Alignment of Parallel Texts. In: 18th Nordic Conference of Computational Linguistics, NODALIDA 2011 (2011)

29. Sennrich, R., Volk, M.: MT-based Sentence Alignment for OCR-generated Parallel Texts. In: The Ninth Conference of the Association for Machine Translation in the Americas (AMTA 2010), Denver, Colorado (2010)
30. Sarikaya, R., et al.: Iterative Sentence-Pair Extraction from Quasi-Parallel Corpora for Machine Translation. In: 10th Annual Conference of the International Speech Communication Association (INTERSPEECH 2009), Brighton, United Kingdom (2009)
31. Smith, J.R., Quirk, C., Toutanova, K.: Extracting parallel sentences from comparable corpora using document level alignment. In: Human Language Technologies: The 2010 Annual Conference of the North American Chapter of the Association for Computational Linguistics, pp. 403–411. Association for Computational Linguistics, Los Angeles (2010)
32. Shi, L., Zhou, M.: Improved sentence alignment on parallel web pages using a stochastic tree alignment model. In: Proceedings of the Conference on Empirical Methods in Natural Language Processing, pp. 505–513. Association for Computational Linguistics, Honolulu (2008)
33. Biçici, E.: Context-Based Sentence Alignment in Parallel Corpora. In: Gelbukh, A. (ed.) CICLing 2008. LNCS, vol. 4919, pp. 434–444. Springer, Heidelberg (2008)
34. Tiedemann, J.: Improved sentence alignment for movie subtitles. In: Conference on Recent Advances in Natural Language Processing (RANLP 2007), Borovets, Bulgaria, pp. 582–588 (2007)
35. Tiedemann, J.: Synchronizing translated movie subtitles. In: 6th International Conference on Language Resources and Evaluation, LREC 2008 (2008)
36. Gautam, M., Sinha, R.M.K.: A Hybrid Approach to Sentence Alignment Using Genetic Algorithm. In: International Conference on Computing: Theory and Applications (ICCTA 2007), pp. 480–484. IEEE Computer Society (2007)
37. Porter, M.F.: An algorithm for suffix stripping. Program 14, 130–137 (1980)
38. Kashefi, O., Nasri, M., Kanani, K.: Automatic Spell Checking in Persian Language. Supreme Council of Information and Communication Technology (SCICT), Tehran (2010)

Effect of ISRI Stemming on Similarity Measure for Arabic Document Clustering

Qusay Walid Bsoul and Masnizah Mohd

Knowledge Technology Research Group, Faculty of Information Science and Technology,
University Kebangsaan Malaysia, Selangor, 43600 Bangi, Malaysia
qusaya068@gmail.com, mas@ftsm.ukm.my

Abstract. Arabic Document Clustering has increasingly become an important task for obtaining good results with the unsupervised learning task. This paper aims to evaluate the impact of the five measures (Cosine similarity, Jaccard coefficient, Pearson correlation, Euclidean distance and Averaged Kullback-Leibler Divergence) for Document Clustering with two types of pre-processing morphology-based The Information Science Research Institute (ISRI) is equivalent to the root-based stemmer and light stemmer; and without stemming without morphology) for an Arabic dataset. Stemming is known as a computational process used to reduce words to their stems. For classification, it is categorised as a recall-enhancing or precision-enhancing component. It is concluded that the method of ISRI for words is proved to be better than without stemming methods which use a five similarities/distance measures for Document Clustering.

Keywords: Similarity measures, partitional clustering, document clustering, Stemming.

1 Introduction

In the recent past, the world has been witnessing a growing increase in the area of information search and retrieval due to the spread of the Internet. However, it is unfortunate that, this increasing growth of information is still restricted to the English language as this is the most dominant language used. Attempts and efforts devoted to improve the area of information search and retrieval in Arabic language are still limited and simple compared to the efforts done in the same area in other languages.

As indicated by Khoja [1] the Arabic language is different from other indo-European languages in terms of its syntax, morphology and semantics. Since the morphological nature of the Arabic language is characterised by being complex, a wide of body of research has been conducted in this area, and more specifically, focusing on the impact of Arabic morphology on Arabic Document Clustering. It is indicated that the Arabic morphology in Document Clustering aims to conflate words of similar or related meanings. Moreover, it is suggested that the retrieval effectiveness over the use of words or stems is increasingly and significantly enhanced by indexing Arabic text using roots [2].

M.V.M. Salem et al. (Eds.): AIRS 2011, LNCS 7097, pp. 584–593, 2011.
© Springer-Verlag Berlin Heidelberg 2011

In defining Document Clustering, it is categorised as an unsupervised learning task, which does not demand pre-defined categories and labelled documents. It aims to group or put text documents with high intra-group similarities and low inter-group similarities in a group [3].

The first section of this study is to review related work. The second section is methodology, which is divided into four parts; the first part indicates pre-process tokenisation and normalisation, the second part which has method for stemming the data named ISRI, the third part provides a discussion of the representation process of the Arabic dataset. The forth part is concerned with the similarity/distance measure. The third section of the study is evaluation which is divided into two parts, which are; explaining and discussing cluster data by using K-mean algorithm, describing of Arabic data set and viewing of performance measure. The forth section is experiments and result. Finally, a conclusion and future work is provided.

2   Related Work

The two most effective Arabic stemmers are Khoja [4] based on root-extraction stemmer and Larkey's light stemmer [5], [6]. Moreover, Duwairi [7], El-Kourd et al [8] and Mustafa et al. [9] discovered that N-gram stemming technique is not efficient for Arabic Text processing.

In a study by Larkey's [5] the researcher proposed several light stemmers based on heuristics and a statistical stemmer based on co-occurrence for Arabic retrieval. It was determined that the effectiveness of the best light stemmer was better for cross-language retrieval than that of a morphological stemmer which attempted to find the root for each word. Darwish [10] examined the effect of improved morphological analysis, in particular, the focus was on the effect of the context sensitive morphology on monolingual Arabic Information Retrieval (IR). Based on the results of a comparison of the effect of context sensitive morphology with that of non-context sensitive morphology, it was revealed that better coverage and improved correctness dramatically impacted the IR effectiveness and that context sensitive morphology further enhanced the retrieval effectiveness.

Taghva et al [11] indicated that by using a root-extraction stemmer for Arabic which is similar to the Khoja stemmer but without a root dictionary, in comparison with Khoja stemmer, the performance of the stemmer was equivalent to the Khoja stemmer. It was also found that it is equivalent to the so-called light stemmers in monolingual document retrieval tasks. Froud et al. [12] assessed the effect of stemming on Arabic Text Document Clustering by using similarity/distance measures. It was found that the Euclidean Distance, the Cosine Similarity and the Jaccard measures obtained comparable effectiveness for the partitional Arabic Document Clustering of tasks for finding more coherent clusters in cases when used without stemming and without stemming better than Khoja and light stemmers. On the other hand, the Pearson Correlation and averaged KL Divergence obtained quite similar results when using Khoja and light stemmers.

In this paper we evaluated same five similarities/distances measure for Document Clustering, with morphology-based ISRI stemming and without stemming, where ISRI stemmer does not need a root dictionary in order to make clustering faster.

3   Methodology

This section comprises four parts: Arabic text pre-processing, stemming proposed, term representation and similarity/distance measures.

3.1   Arabic Text Pre-processing

In the first phase, processing of the Arabic document is conducted according to the most common Arabic tokenising and normalising methods. This is a part of the text pre-processing methods which are applied as the followings:

Step 1: Conversion of the text to Unicode.
Step 2: Removal of digits, diacritics and punctuation marks for each text in the Arabic dataset as example Remove (', ',': ','? ',' \ ' …).
Step 3: Filtering the non Arabic words.
Step 4: Splitting the text into tokens which usually consist only of letters.
Step 5: Normalize آ ,أ ,إ to (ا). Removing the hamza in this case does not affect the root.

3.2   Arabic Stemming Algorithm

According to the report of The Information Science Research Institute's (ISRI) generated by Taghva et al [11] the Arabic stemmer is proved to have common features with the Khoja stemmer. However, the main difference between the two stemmers is that no root dictionary is used in the Arabic stemmer. As we indicate in the following table:

Table 1. Summary of ISRI Algorithm

ISRI
Input: Arabic documents.
Phase one: remove stop word.
Phase two: Remove length three and length two prefixes in that order.
Phase three: Remove connector وif it precedes a word beginning with و.
Phase four: Return stem if less than or equal to three. Attempting to shorten stems further results in ambiguous stems.
Phase five: Length = 4: If the word matches one of the patterns 4, extract the relevant stem and return. Otherwise, attempt to remove length-one suffixes and prefixes in that order provided the word is not less than length three.
Phase six: Length = 5: Extract stems with three characters for words that match patterns 5. If none are matched, attempt to remove suffixes and prefixes, otherwise the relevant length-three stem is returned. If the word is still five characters in length, the word is matched patterns 5 to determine if it contains any stems of length 4. The relevant stem is returned if found.

Table 1. (*continued*)

Phase seven: Length = 6. Extract stems of length three if the word matches a pattern 6. Otherwise, attempt to remove suffixes. If a suffix is removed and a resulting term of length five results, send the word back through Phase six. Otherwise, attempt to remove one character prefixes, and if successful, send the resulting length-five term to Phase six.

Phase eight: Length = 7. Attempt to remove one-character suffixes and prefixes. If successful, send the resulting length-six term Phase seven.

Output: stemmed documents

3.3   Term Representation

We used the bag-of-words (BOW) methodology which is commonly used in the previous studies of Document Clustering. This has been found to be efficient as a language-independent method since they are independent of the meaning of the language and perform well in case of noisy text [13]. In the current study the term weighting method based on Boolean weighting was used, this was considered as the simplest and easiest method for term weighting. In applying this approach, 1 is the matrix to which the weight of a term is assigned in the case when the term appears in the document and 0 is the matrix to which the term the weight of a term is assigned in case when the term does not appear in the document. Weighting each term will be conducted by using the Term Frequency × Inverse Document Frequency (TF×IDF) weighting. In this approach, proportional assigning of the weight of term i in document d is to the number of times of occurrence of the term in the document is done, and in inverse proportion, it is assigned to the number of documents in the corpus in which the term occurs.

$$w_i = tf_i \cdot \log\left(\frac{N}{df}\right). \tag{1}$$

3.4   Similarity Measures

For the current study, the similarity/distance measures are Cosine Similarity, Jaccard Coefficient, Pearson Correlation Coefficient, Euclidean Distance and Averaged Kullback-Leibler Divergence, where not every distance measure is a metric. To qualify as a metric, a measure d must satisfy the following four conditions.

Let x and y be any two objects in a set and d(x, y) be the distance between x and y.

1. The distance between any two points must be nonnegative, that is, $d(x, y) \geq 0$.
2. The distance between two objects must be zero if and only if the two objects are identical, that is, $d(x, y) = 0$ if and only if x = y.
3. Distance must be symmetric, that is, distance from x to y is the same as the distance from y to x, i.e. $d(x, y) = d(y, x)$.
4. The measure must satisfy the triangle inequality, which is $d(x, z) \leq d(x, y) + d(y, z)$.

A. Cosine Similarity

Cosine similarity is one of the most well-known similarity measures which is applied to text documents in numerous information retrieval [14] and clustering applications [15]. In measuring the given two documents $\vec{t_a}$ and $\vec{t_b}$ their cosine similarity is:

$$SIM_C(\vec{t_a}, \vec{t_b})\ \frac{\vec{t_a}.\vec{t_b}}{|\vec{t_a}|\times|\vec{t_b}|}. \tag{2}$$

Where $\vec{t_a}$ and $\vec{t_b}$ are seen as m-dimensional vectors over the term set $T\{t_1 \ldots t_m\}$. Each term with its weight in the document is represented by a particular dimension, which is non-negative. Therefore, the cosine similarity is non-negative and bounded between [0, 1]. The cosine similarity is independent of document length, which is an important property of the cosine similarity that makes it distinguished. For instance, in measuring the cosine similarity between two identical copies of a document d which are combined to get a new pseudo document d_0, the cosine similarity between d and d_0 is 1, thus, implying that these two documents are regarded to be identical.

B. Jaccard Coefficient

The Jaccard coefficient, which is another similarity measure, is used to measure the similarity in the intersection divided by the union of the objects. For text documents, the use of Jaccard coefficient is to make a comparison of the sum weight of shared terms and the sum weight of terms presented in either of the two documents but on condition that they are not the shared terms. The formal definition is as follows:

$$SIM_J\left(\vec{t_a}, \vec{t_b}\right) = \frac{\vec{t_a}.\vec{t_b}}{|\vec{t_a}|^2\times|\vec{t_b}|^2-\vec{t_a}.\vec{t_b}}. \tag{3}$$

According to the Jaccard coefficient, the similarity measure ranges between 0 and 1. It is 1 when the $\vec{t_a}=\vec{t_b}$ and 0 when $\vec{t_a}$ and $\vec{t_b}$ are disjointed. The corresponding distance measure is $D_J=1-SIM_J$ and D_J is used instead in subsequent experiments.

C. Pearson Correlation Coefficient

The Pearson's correlation coefficient is another similarity measure. It functions to measure the degree of relationship between two vectors. It is formulated in different forms. Given the term set $T=\{t_1 \ldots t_m\}$ a popularly used form is:

$$SIM_P\left(\vec{t_a}, \vec{t_b}\right) = \frac{m\sum_{t=1}^m w_{t,a}- w_{t,b}-TF_a\times TF_b}{\sqrt{[m\sum_{t=1}^m w_{t,a}^2-TF_a^2][m\sum_{t=1}^m w_{t,b}^2-TF_b^2]}} \tag{4}$$

Where $TF_a=\sum_{t=1}^m w_{t,a}$ and $TF_b=\sum_{t=1}^m w_{t,b}$, Like other measures, the Pearson's correlation coefficient is also a similarity measure. However, it differs from other measures in that it ranges from -1 to +1 and it is 1 when $\vec{t_a}=\vec{t_b}$ In subsequent experiments, the corresponding distance measure, which is $D_p=1-SIM_p$ is used when $SIM_p \geq 0$ and $D_p=|SIM_p/$ when $SIM_p < 0$.

D. Euclidean Distance

The Euclidean distance is a distance measure which is used in clustering problems, including Document Clustering. It is also identified as a measure for determining the default distance used with the K-means algorithm. In measuring the distance between text documents, given two documents d_a and d_b are represented by their term vectors $\vec{t_a}$ and $\vec{t_b}$ respectively. The Euclidean distance of the two documents is defined as follows:

$$D_E(\vec{t_a}, \vec{t_b}) = \left(\sum_{i=1}^{m} |w_{t,a} - w_{t,b}|^2\right)^{1/2}. \tag{5}$$

According to this, the term set is $T\{t_1 \ldots t_m\}$ and as previously mentioned the tfidf value is used as term weights, that is $w_{t,a} = \text{tfidf}(d_a, t)$.

E. Averaged Kullback-Leibler Divergence

In Document Clustering which is based on information theory, a document is recognised or identified as a probability distribution of terms. The distance between the two corresponding probability distributions is known as the similarity of two documents which has to be measured. The Kullback-Leibler divergence (KL divergence). Given two distributions P and Q, the KL divergence from distribution P to distribution Q is defined as:

$$D_{KL}(P||Q) = P \log\left(\frac{P}{Q}\right). \tag{6}$$

In such document scenario, the divergence between two distributions of words is:

$$D_{KL}(\vec{t_a}||\vec{t_b}) = \sum_{t=1}^{m} w_{t,a} \times \log\left(\frac{w_{t,a}}{w_{t,b}}\right). \tag{7}$$

In contrast to the previously discussed measures, the KL divergence is not symmetric, i.e. D_{KL} (P||Q) $\neq D_{KL}$ (Q||P) Therefore, it is not a true metric. Thus, the averaged KL divergence is used instead in the current study, and which is defined as:

$$D_{AvgKL}(P||Q) = \pi_1 D_{KL}(P||M) + \pi_2 D_{KL}(Q||W). \tag{8}$$

Where $\pi_1 = \frac{P}{P+Q}$, $\pi_2 = \frac{Q}{Q+P}$ and $M = \pi_1 P + \pi_2 Q$ for documents, the following formula provides or shows computation of averaged KL divergence:

$$D_{AvgKL}(\vec{t_a}||\vec{t_b}) = \sum_{t=1}^{m}(\pi_1 \times D_{KL}(w_{t,a}||w_t) + \pi_2 \times D_{KL}(w_{t,b}||w_t)). \tag{9}$$

Where $\pi_1 = \frac{w_{t,a}}{w_{t,a}+w_{t,b}}, \pi_2 = \frac{w_{t,b}}{w_{t,b}+w_{t,a}}$ and $w_t = \pi_1 \times w_{t,a} + \pi_2 \times w_{t,b}$. It is also stated that there is a symmetry ensured by the average weighting between two vectors. In other words, the divergence from document i to document j is symmetrical to the divergence from document j to document i.

4   Evaluation

For the testing dataset, we experimented with three similarities and two distance measures using two methods: ISRI using the morphological analyser from Taghva et al [11] and without stemming.

4.1   Clustering

Cluster analysis [16] refers to the process of classifying objects into groups of similar objects based on a similarity/distance measure. It has been applied in a wide number of different fields including text mining, information retrieval and machine learning. K-means is a method which has been widely used for partitional clustering with a linear time complexity [17]. It has proved to be useful and workable when it is used with distance measures which principally aim at minimising the within-cluster distances.

The Euclidean distance and the averaged KL divergence are defined as measures used for distance whereas the cosine similarity, Jaccard coefficient and Pearson coefficient are measures used for similarity. In the current study, a simple transformation was used for the purpose of converting the similarity measure to distance values. Since both cosine similarity and Jaccard Coefficient are bounded in $(0, 1)$ and monotonic, $D = 1$-SIM is taken as value of the corresponding distance. As far as Pearson coefficient, which ranges from -1 to $+1$ is concerned, $D=1$-SIM is taken when SIM ≥ 0 and $D = |$SIM$|$ when SIM < 0.

4.2   Data Description and Performance Measure

The testing dataset consisted of 4 categories namely art, economics, politics and sport articles, and each contains documents taken from Al-salemi and Aziz [18] 1680 documents where used in testing the dataset.

For measuring the external quality in the current study, the overall purity and other measures called overall F-measure, which is the most well-known and used measure in Document Clustering, was used for this purpose [17] where the higher overall purity and F-measure means the best cluster.

5   Experiments and Result

Three different sets of experiments were created. Moreover, each experiment was run 5 times and the results are the averaged value over 5 runs. Each run had different initial seed sets.

The first experiment aimed at Clustering Documents which are clustered under two different groups, based on Table 2 the distance measure better than similarity measure as a general with or without stemming, In contrast the similarity with ISRI stemming is better than without, where the best value was (0.75%) with $D_{Avg\ KL}$ as a distance measure and (0.7%) with cosine similarity measure.

The second experiment was conducted to Cluster Documents which belonged to three groups, based on Table 3 the results are equivalent between cosine, Jaccard and $D_{Avg\ KL}$ In addition, the ISRI method was better than without stemming with similarity/distance measure.

The last experiment was carried out to examine the effect on clustering of the four different groups. The aim of carrying out the later experiments was to examine the effect on highly relevant documents, based on Table 4 the cosine and Jaccard was better than Pearson as similarity measure and Euclidean better than $D_{Avg\ KL}$ measure, on the another hand ISRI stemming was better than without stemming.

Table 2. Overall purity and F-measure evaluation on 2 categories (art, economics)

Stemming	evaluations	Cosine	Jaccard	Pearson	Euclidean	DAvg KL
ISRI	Purity	**0.7**	0.65	0.6	0.6	**0.7**
	F-measure	0.67	0.6	0.58	0.52	**0.69**
Without	Purity	0.7	0.6	0.45	0.6	**0.75**
	F-measure	0.67	0.52	0.37	0.52	**0.74**

Table 3. Overall purity and F-measure evaluation on 3 categories (art, economics, politics)

Stemming	evaluations	Cosine	Jaccard	Pearson	Euclidean	DAvg KL
ISRI	Purity	**0.6**	**0.6**	0.53	0.47	**0.6**
	F-measure	0.57	0.58	0.52	0.39	**0.62**
Without	Purity	**0.6**	**0.6**	0.4	0.43	0.57
	F-measure	0.55	0.51	0.38	0.37	**0.59**

Table 4. Overall purity and F-measure evaluation on 4 categories (art, economics, politics, sport)

Stemming	evaluations	Cosine	Jaccard	Pearson	Euclidean	DAvg KL
ISRI	Purity	**0.57**	**0.57**	0.2	0.42	0.2
	F-measure	0.52	**0.54**	0.29	0.38	0.28
Without	Purity	0.5	**0.55**	0.2	0.42	0.2
	F-measure	**0.49**	**0.49**	0.21	0.35	0.28

6   Conclusion and Future Work

We concluded based on our experiments which are detailed above, that the similarity/distance measure is more effective on ISRI stemming as morphological words than without stemming; especially when they group documents on similarity measures more than distance measures.

The main reason for determining the effectiveness of the three similarities and two distances with the ISRI method, is that without a stemming method have under-stemming error in which some terms that should be stemmed to one root are not, which leads to create similarities among the unrelated documents containing the same roots for different words, as shown in Table 5. The ISRI method have over-stemming error that means there are two words with different stems which are stemmed to the same root as displayed in Table 5. These errors of stemming decrease the effectiveness on the similarities and distances measure. The similarities and distances calculate the term frequency for each word, and within the previous errors, the frequency of the terms split or merge. So that, the ambiguity with the similarities and distances measure can be done as shown in our results, and also a lot of terms have under-stemming more than over-stemming so that the ISRI method has on the result better than without stemming. The cosine as specific and jaccard as general similarity have proved to be better than the Pearson in our experiments.

Table 5. Example of under-stemming and over-stemming errors

Words		methods	
	root	**Without stemming**	**ISRI**
يشرب, تشرب,أشرب,يشربان	شرب	يشرب, تشرب, أشرب,يشربان	شرب
الفيتامين	فيتامين	الفيتامين	يتم يتام
السعودية	سعودية	السعودية	سعد
سعيدة	سعد	سعيدة	سعد

There are some suggestions are mentioned for future research. Firstly, a combination of different models is planned as to make a representation of the documents. So that, we propose another similarities/distances measure such as Probabilistic Models as BM25 and Language models. Secondly, using another stemming method or generating new one which is based on the morphological and syntactic structure same as generated by [19]. These are due to the idea of stemming and stop word to decrease the words which are not important to the documents.

References

1. Khoja, S.: APT: Arabic part-of-speech tagger. In: Proceedings of the Student Workshop at NAACL, pp. 20–25 (2001)
2. Abu-Salem, H., Al-Omari, M., Evens, M.W.: Stemming methodologies over individual query words for an Arabic information retrieval system. J. Am. Soc. Inf. Sci. 50(6), 524–529 (1999)
3. Ozgür, A.: Supervised and unsupervised machine learning techniques for text document categorization (2004)
4. Khoja, S., Garside, R.: Stemming Arabic Text. Computing Department, Lancaster University, Lancaster (1999)
   http://www.comp.lancs.ac.uk/computing/users/khoja/stemmer.Ps
5. Larkey, L., Larkey, L.S., Connell, M.E.: Arabic information retrieval at UMass in TREC-10. In: Proceedings TREC, pp. 562–570 (2001)
6. Larkey, L.S., Ballesteros, L., Connell, M.E.: Improving stemming for Arabic information retrieval: light stemming and co-occurrence analysis. In: Proceedings of the 25th Annual

International ACM SIGIR Conference on Research and Development in Information Retrieval, Finland (2002)

7. Duwairi, R.: A Distance-based Classifier for Arabic Text Categorization. In: Proceedings of DMIN, pp. 187–192 (2005)
8. El Kourdi, M., Bensaid, A., Rachidi, T.: Automatic Arabic document categorization based on the Naïve Bayes algorithm. In: Proceeding Semitic 2004 Proceedings of the Workshop on Computational Approaches to Arabic Script-based Languages, pp. 51–58 (2004)
9. Mustafa, S.H., Al-Radaideh, Q.A.: Using N-grams for Arabic text searching. J. Am. Soc. Inf. Sci. Technol. 55(11), 1002–1007 (2004)
10. Darwish, K., Hassan, H., Emam, O.: Examining the effect of improved context sensitive morphology on Arabic information retrieval. In: Proceedings of the ACL Workshop on Computational Approaches to Semitic Languages, pp. 25–30 (2005)
11. Taghva, K., Elkhoury, R., Coombs, J.: Arabic stemming without a root dictionary. In: ITCC International Conference on Information Technology: Coding and Computing, vol. 1, pp. 152–157 (2005)
12. Froud, H., Benslimane, R., Lachkar, A., Ouatik, S.A.: Stemming and similarity measures for Arabic Documents Clustering. In: 5th International Symposium on I/V Communications and Mobile Network (ISVC), pp. 1–4 (2010)
13. Khreisat, L.: Arabic text classification using N-gram frequency statistics a comparative study. In: DMIN, pp. 78–82 (2006)
14. Baeza-Yates, R., Ribeiro-Neto, B.: Modern information retrieval. ACM press, New York, Key: citeulike:532542 (1999)
15. Larsen, B., Aone, C.: Fast and effective text mining using linear-time document clustering, pp. 16–22 (1999)
16. Jain, A.K., Dubes, R.C.: Algorithms for clustering data (1988)
17. Steinbach, M., Karypis, G., Kumar, V.: A comparison of document clustering techniques. In: KDD Workshop on Text Mining, pp. 525–526 (2000)
18. Al-Salemi, B., Ab Aziz, M.J.: Statistical Bayesian Learning for Automatic Arabic Text Categorization. J. Comput. Sci., 39–45 (2011)
19. Al-Shammari, E., Lin, J.: Towards an error-free Arabic stemming. In: Proceeding of the 2nd ACM Workshop on Improving non English Web Searching, pp. 9–16. ACM, Napa Valley (2008)

A Semi-supervised Approach for Key-Synset Extraction to Be Used in Word Sense Disambiguation

Maryam Haghollahi and Mehrnoush Shamsfard

NLP Research Lab, Faculty of ECE, Shahid Beheshti University, Tehran, Iran
`maryam.haghollahi@yahoo.com`, `m-shams@sbu.ac.ir`

Abstract. Nowadays, although many researches is being done in the field of word sense disambiguation in some languages like English, still some other languages like Persian have many things to be done. Some difficulties are in this way which might have made it less interactive for researchers. For example, Persian WordNet or FarsNet is newly developed and there is no sense tagged corpus developed based on it yet. So we propose a semi-supervised approach for extending FarsNet with some new relations and then use it for WSD. Also a method to extract semantic keywords or key-concepts from textual documents is used. As the key-concepts are extracted exploiting FarsNet, we call them Key-synsets. In fact Key-synsets of a document are those synsets which are semantically related to the main subjects of that document. This method is exploited to improve the precision of the proposed WSD. Although our approach is tested on Persian it can be easily adopted for other languages such as English.

1   Introduction

Word sense disambiguation is a critical task in many applications like translation and semantic search. Many approaches are implemented in order to facilitate it; some of them are supervised and some others are unsupervised. In English some sense tagged corpora which are tagged with WordNet can be found, like SemCor [1]. Thus supervised approaches have the feasibility to be implemented. But in some languages like Persian, there isn't any sense tagged corpus to be used for the learning phase.
In this paper, we introduce a software system which extracts the candidate senses (and so synsets) of the words within a context using FarsNet relations; and then with respect to the relations between these candidate synsets, it finds the Key-synsets of the context to make the approach more precise. In this way, some new relations are extracted semi-automatically and added to FarsNet. The results show a significant improvement in precision with these new relations.

2   Related Work

This paper discusses a key concept extraction method to be used in word sense disambiguation. Thus in this section we briefly point to the related work on WSD and keyword extraction.

M.V.M. Salem et al. (Eds.): AIRS 2011, LNCS 7097, pp. 594–603, 2011.
© Springer-Verlag Berlin Heidelberg 2011

According to [2] word sense disambiguation methods are categorized into three categories; supervised, unsupervised and knowledge based. Some other researchers have different categorizations. Tsatsaronis and colleagues [3] categorize WSD methods into supervised and unsupervised and say that unsupervised WSD methods comprise corpus-based [4], knowledge-based such as Lesk-like [5] and graph-based [6] methods, as well as ensembles [7] that combine several methods.

Supervised approaches (such as [8]) use corpora which are tagged with the concepts of ontologies, for their training phase. While semi-supervised approaches restrict the need to such a resource. For instance Tang and colleagues [9] use the examples of an ontology and some raw text resources and extract their subject-predicates and predicate-objects as collocation words. Then the training data are obtained with two approaches: SRP which means that within all possible senses sets of a word collocation, the one which has the most redundant information between senses is the best; and PRP which is calculated by exchanging the words with their synonyms, etc. and finding the most co-occurring ones.

Unsupervised approaches often are used for languages which have no or less concept tagged corpus for training phase. Results show that supervised approaches are often more precise but limited to those words that have sense tagged data [10 quoting [11]. For example, Tsatsaronis and colleagues [12] utilize WordNet, and use neural network and a spread activation method to disambiguate the words of sentences. Tran and colleagues [13] construct a wide tree of word relations with their weights using many internet web pages. Then for disambiguation of a word in a context, the glosses of all of its senses and the context of the to-be-disambiguated word are parsed and using the constructed tree each gloss obtains a score. The sense whose gloss obtains the most score will be selected for that word.

Knowledge based approaches use some knowledge resources like dictionaries or thesauri for WSD task. As Navigli [2] mentioned, their precision is less than supervised methods but their coverage is more expanded because of the expanded resources they have. Lesk [5] and extended Lesk [14] are two of these methods. They use the glosses of the senses in WordNet to disambiguate the words.

Recent research results [2] show that "the accuracy of the state of the art supervised WSD methods is above 60% with an upper bound reaching 70% for all words, fine-grained WSD for English, while the accuracy of unsupervised methods is usually between 45 – 60%". There are some known Baselines which can be used for evaluation phase of WSD works. The best known of them is the First-Sense approach. Also Lesk method can be used as a baseline, as Navigli [2] mentioned.

In Persian, as there is no corpus tagged by word senses, there is no supervised work on WSD. Saedi and Shamsfard [15] propose a knowledge based WSD method to be used in a Persian to English machine translation system. Faili [16] introduces an English to Persian translation method which has a WSD approach on English texts that uses a parallel corpus for its training phase. There is no work which assigns senses according to a Persian WordNet so far.

Keyword extraction is another field related to the subject of this paper. Many features are used for keyword extraction process. For example Xu and colleagues [17] use Wikipedia to derive a set of novel word features which reflect the document's background knowledge. These features are the inlink, outlink, category and infobox information of the document's related articles in Wikipedia. Ercan and colleagues

[18] concern the relations between the words of the document to extract the keywords. In fact, using a supervised method, a lexical chain is developed from each document by using WordNet to be used for keyword extraction.

Hulth [19] explains some methods for extracting the keywords. Some methods use the syntactic information of words (such as [20]), some have supervised learning phase (such as [21]) and some others are statistical (such as [22]).

3   Word Sense Disambiguation Approach

The proposed method is composed of three essential processes: Stemming and tokenizing, Word sense disambiguation and Key-synset extraction. Stemming and tokenization of Persian documents are done by STeP-1 software [23]. In the rest of this section, we will discuss WSD and Key-synset extraction approaches in more details.

3.1   Persian Word Sense Disambiguation

FarsNet [24] is a Persian WordNet, recently developed in NLP lab of Shahid Beheshti University. In this lexical ontology, various kinds of relations are defined between synsets including: Hypernym, Hyponym, Meronym, Holonym, Antonym and Cause. Many researchers have used these kinds of relations to disambiguate the words senses. For example Fragos et al. [25] proposed a method to find the words senses using WordNet relations. Here, we have used synsets' relations of FarsNet to find the senses of words.

Our experiments showed that the above relations are not enough to find the word senses, because some combinations of related words haven't got any of these kinds of relations. For instance, human's mind comprehends a semantic relation between "شیر" (shir, means: lion) and "جنگل" (jangal, means: jungle). But this relation is not among the above kinds. We call these relations just as "Is related to" without putting any specific name or label on them. Neither FarsNet nor most of the other WordNets include this kind of relation. We have extended FarsNet relations with a few new "Is related to" relations for some concepts semi-automatically. Results of using these new relations for disambiguation showed that they really increase the precision.

Semi-automatic extraction of semantic relations. As it was described before, extracting semantic relations between synsets can improve the precision of search. Here, we have used a semi-automatic approach to do it. To find the words which are related to a target word, first we search it via Hamshahri-1 corpus [26] with tf-idf method and retrieve some highest ranked documents. In each document, we extract the words within a 5-words sized window around the target word which are not stop words. Then, all possible synsets of these adjacent words and their hypernyms up to two levels are extracted from FarsNet and added to a list. The frequency of occurrence of each of these synsets in the obtained list is considered as its rank. "n" best synsets with respect to their ranks are semantically related synsets with the target word. It should be considered that the synset of the target word is assigned manually, but the synsets of its co-occurring words are obtained automatically. Thus the approach is semi-automatic.

Some of the co-occurring words of our target words are not presented in FarsNet and though with this approach we will lose them. Thus, we introduce a new relation type which is between a synset and an unkown word (word which is not in FarsNet). It means that although we haven't got that word in FarsNet and don't know which synset it can occurs in, but we know that this word is co-occurring with some specific sense of the target word. These relations will be considered as direct relations in FarsNet in the process of disambiguation. Our results show some improvements in precision by adding these new relations.

Using FarsNet for ranking the synsets' semantic relations. Having FarsNet relations, we can find the weight of the relation between any two FarsNet synsets. Equation (1) shows how to calculate this weight. In this equation distance(S_i,S_j) is the number of relations that should be passed from S_i to arrive to S_j.

$$\text{Weight}(S_i , S_j) = \log_{10}\left(\frac{1}{\text{distance}(S_i,S_j)}\right) + c. \qquad (1)$$

In the next part we will explain the use of this weighting process for our disambiguation algorithm.

Finding the set of words' synsets of each block of content. To disambiguate the words of each document, we need to split it to smaller blocks. Then, in each block we can find the relatedness weight of any two synsets of any two words to find the best synsets of the block's words. In this work, we examined some different number of words within blocks to know which one is better. The results and comparisons are presented in our results section. Also as it is mentioned in our results section, the jump number of blocks has an important effect on the efficiency of the approach. Jump number is the number of words that we will pass after disambiguation of a block, in order to obtain the next block. For example, if we set the block's number of words to 4 words and set the jump number to 2, the sentence "What is the past tense of split?" will be split to these three blocks (here without omission of stop words):

"What is the past", "the past tense of" and "tense of split - ".

Having each block of "n" words, each possible sense (and so synset) of each word is found from FarsNet, and then the relatedness weight of any two synsets of them will be calculated as described before. Here, we only consider direct relations of synsets and ignore indirect ones to reduce the computation time. Now we have a set of synset pairs with their similarity weights. Equation (2) shows how to calculate the total score of each pair. In this equation, pos(W_j) and pos(W_i) are respectively the position of second and first words in the current block.

$$\text{TotalScore}(i,j) = \text{coef}(i,j) * \text{Weight}(i,j). \qquad (2)$$

$$\text{coef}(i,j) = \log_{10}\left(\frac{1}{\text{pos}(W_j) - \text{pos}(W_i)}\right) + c.$$

After acquiring the set of pair synsets with their total scores, each two pairs which are not mutex will be merged together to build a bigger set. Two pairs are mutex if they contain at least a similar word with different senses. After merging the pairs, the score of the new collection is the addition of the scores of its components. This process will be continued until no new collection can be added. After that, the collections will be sorted by their scores, and better collections are extracted for the next phase.

3.2   Key-Synset Extraction Approach

The main idea of this method of key synset extraction is taken from a general principle about information density. For example, in clustering algorithms, the points which are inside a dense part have more probability to be a cluster, and points which have less density may be noises.

Inspired by this idea, we claim that those senses of a document which have more valuable relations with other senses are more probable to be key senses. In continuation of this section, this method will be described in more details.

Calculating the pseudo-frequency. As we described in previous sections, better combinations of synsets of each block of "n" words will be used to find the key synsets. Actually, each synset which occurs in any of the best combinations can be a candidate to be a key synset. So, we will calculate a rank for each of them to find better ones. To calculate this rank, first using equation (3) we will compute a pseudo-frequency for each synset, which *somehow* shows the amount of its occurrences in the document. In this equation, the pseudo-frequency of ith synset is calculated. $ColSet_i$ is the set of collections that i^{th} synset occurs in them and $Score(Col_k)$ is the score of kth collection.

$$SF_i = \sum_{k \in ColSet_i} Score(Col_k). \tag{3}$$

Calculating the total ranks. The last thing we need is to compute the total rank of each synset in each document, which is the main criterion to find the Key-synsets. What we need, is to calculate the relation score between each two candidate synsets of the document as described before. Here, we used indirect relations with the threshold of "at most ten fathers" in addition to direct ones for the process of relation scoring. We used this parameter (relation score) before for some other reasons. Here we are using it to calculate the importance of each synset in the document, but before, we used it to disambiguate the senses of the words of "n" words blocks.

Finally we have everything we need! Using Equation (4) we can calculate the total rank of each synset to find Key-synsets of each document's words. Here, "k" is the number of candidate synsets.

$$TR_i = SF_i * \sum_{j=1:k} SF_j * Weight(S_i, S_j). \tag{4}$$

4   Experimental Results

As there is no Persian sense tagged corpus being tagged with FarsNet, we had to make our own test corpus. We have used Hamshahri-1 corpus for this task. First of all, we divided the corpus into %70 training and %30 testing corpora randomly. Using these corpora, we can find the new relations and evaluate our approach.

4.1   Training Phase

First we chose three ambiguous words that more than one of their senses occur in the corpus. These words are "شیر" ("shir" means: lion, faucet, breastfeeding, milk), "سیر" ("sir" means: garlic, full. or "seir" means: process, travel) and "گل" ("gol" means: flower, goal). Then these three words have been searched within training corpus with tf-idf measure. For our search, we used Lucene[1] search engine that uses tf-idf measure in order to rank the results of search. Within the highest ranked documents, we extracted the co-occurring words. Then with the explained method, new relations between synsets and between unknown words and synsets were extracted.

4.2   Building the Test Corpus

In order to build the test corpus, those three words were searched within test corpus and about 600 "nearly 200 character phrases" around our specified words were extracted from the retrieved documents to build a test corpus. Our three words were tagged manually within the test set and the content of each phrase was stemmed with STep-1. Then the proposed method was evaluated via this built corpus over the tagged words.

4.3   Evaluation of the Method

The proposed method was tested over the built test corpus with different states. Each state will be described below and its results will be presented. The programs are written in java and the tests are being done on an ordinary PC with 2GB RAM and 2.66GHz CPU.

Precision of baselines. Lesk and extended Lesk approaches were implemented as our baselines, because there was no Persian WSD method working with FarsNet to be compared with our method. Also as there were no sense tagged corpora, the precision of MFS couldn't be calculated, but here, we calculated the best case of MFS within the test corpus and used it as another baseline. Table 1 shows the precision of baselines.

---

[1] http://lucene.apache.org/

Table 1. Precision of baselines

Method	Precision
First Sense (In best case)	64%
Lesk	10%
Extended Lesk	16%

Precision of our approach with different parameters and features. The proposed approach has some kinds of parameters and features. Results show that changing them have a significant effect over the precision. Here, first we will show the effect of these parameters, then we will compare our approach with baselines and eventually we will show the effect of features.

Word number and jump number effect. We have calculated the precision of our approach over the test corpus with different words number of blocks and jump number. Fig. 1 and Fig. 2 show the experimental results.

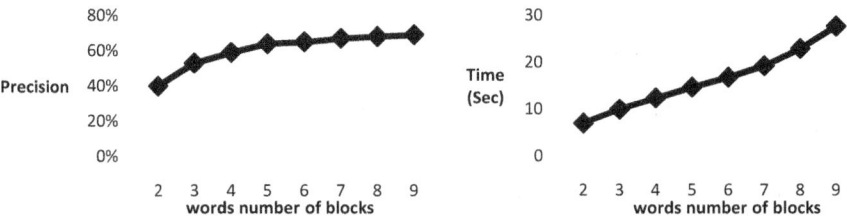

Fig. 1. The left curve shows the Precision of proposed approach with different words number of blocks and jump number of one. The right curve shows the computation time of each complete phrase (each "nearly 200 character phrase") in test set with mentioned conditions.

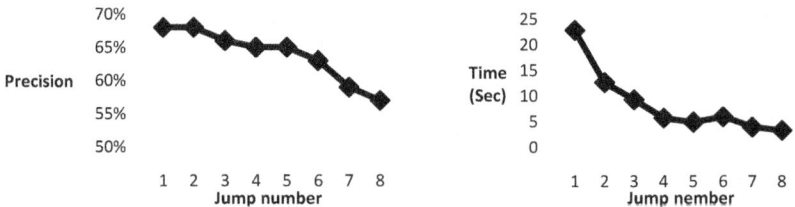

Fig. 2. Left curve shows the precision of proposed approach with 8 words in each block and different jump numbers. Right curve shows the computation time for each phrase with mentioned conditions.

Fig. 3 shows the comparison of the precision of our approach with the base lines. Our approach's precision is calculated with different words number of blocks and jump number of one. Results show that our approach has out-performed the baselines.

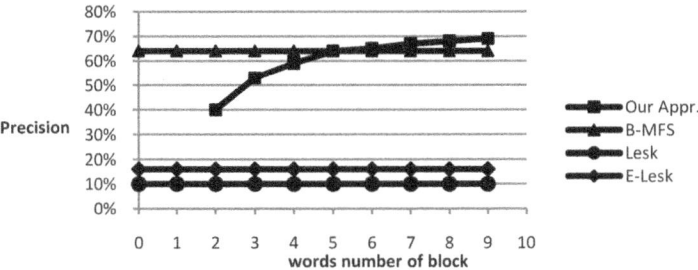

Fig. 3. comparison of our approach with baselines. B-MFS stands for Best case of MSF and E-Lesk stands for Extended Lesk.

Different features effect. We have calculated the precision of our approach with and without each of the proposed features. Table 2 shows the results of the test.

Table 2. Precision of our approach with different features. Here "+ something" means that the state has that thing and "- something" means that it doesn't have it. Also syn-syn relation means the added relations which are between two synsets and unk-syn relations are the added relations that are between an unknown word and a synset. Our approach's precision is calculated with concerning 4 words inside each block and jump number of 2 as an example. As it can be understood from this table, without the extraction of new relations for FarsNet, this approach is so inefficient, but with adding new relations, the precision grows much. Also adding the relations between unknown words and synsets will improve the method by about 8 percent. In addition, without the post-process of Key-synset extraction, the results show a worse precision.

Feature			Precision
- syn-syn rels.	- unk-syn rels.		7%
+ syn-syn rels.	- unk-syn rels.		48%
- syn-syn rels.	+ unk-syn rels.		23%
+ all other features			55%
+ syn-syn rels.	+ unk-syn rels.	- Key-synset	44%

Complexity of the method. If we assume that the words number of block is "n", the average senses of each word is "m", the jump number is "j" and the average number of words in each context is "c", then the complexity of finding Key-synsets of each document in its worst case is calculated in equation (5). In this equation $\frac{c}{j}$ is the number of blocks in the document and $\binom{m.n}{2}*\binom{m.n}{2}*\binom{m.n}{2}$ is the complexity of disambiguating each block.

$$O\left(\frac{c}{j}*\binom{m.n}{2}*\binom{m.n}{2}*\binom{m.n}{2}\right)= O\left(\frac{c*m^6*n^6}{j}\right).$$ (5)

5   Conclusion and Future Works

We have presented an approach for WSD with FarsNet on Persian texts. Our approach uses both FarsNet relations and some other relations that are extracted by a semi-supervised approach and added to FarsNet. In addition, we made an automatic overall revise over the detected synsets of words to make them more precise, or in other words, we found the Key-synsets of the distinct words within the context. The results show improvement in the precision with adding these new features. Also our approach out-performs First sense, Lesk and extended Lesk with respect to the experimental results. The most important disadvantage of our approach is its computation time which can be better with making some optimizations in approach. Of course it should be considered that the computation times are calculated within our runs over an ordinary PC and they would be much better over a stronger server.

For our future work we might optimize our approach to make it faster. Then we will develop a semantic search engine to use this approach for indexing the documents. We think that common search engines don't work well in Persian and have no sense to the semantic of queries and documents. Thus, we are going to apply these semantic methods over search engines to make the results more admissible.

References

1. Miller, G.A., Leacock, C., Tengi, R., Bunker, R.T.: A semantic concordance. In: Proc. of the ARPA Workshop on Human Language Technology, pp. 303–308 (1993)
2. Navigli, R.: Word Sense Disambiguation: A Survey. ACM Computing Surveys 41(2), Article 10 (February 2009)
3. Tsatsaronis, G., Varlamis, I., Nørvåg, K.: An Experimental Study on Unsupervised Graph-based Word Sense Disambiguation. In: Gelbukh, A. (ed.) CICLing 2010. LNCS, vol. 6008, pp. 184–198. Springer, Heidelberg (2010)
4. Yarowsky, D.: Word-sense disambiguation using statistical models of roget's categories trained on large corpora. In: Proc. of COLING, pp. 454–460 (1992)
5. Lesk, M.: Automatic sense disambiguation using machine readable dictionaries: How to tell a pine cone from an ice cream cone. In: Proc. of the 5th SIGDOC, pp. 24–26 (1986)
6. Agirre, E., Soroa, A.: Personalizing pagerank for word sense disambiguation. In: Proc. of EACL, pp. 33–41 (2009)
7. Brody, S., Navigli, R., Lapata, M.: Ensemble methods for unsupervised wsd. In: Proc. of COLING/ACL, pp. 97–104 (2006)
8. Agirre, E., Martínez, D.: Exploring automatic word sense disambiguation with decision lists and the Web CoRR cs.CL/0010024 (2000)
9. Tang, X., Chen, X., Qu, W., Yu, S.: Semi-Supervised WSD in Selectional Preferences with Semantic Redundancy. In: COLING (Posters), pp. 1238–1246 (2010)
10. Brody, S.: Closing the Gap in WSD: Supervised Results with Unsupervised Methods. Doctor of Philosophy thesis, Institute for Communicating and Collaborative Systems. School of Informatics. University of Edinburgh (2009)
11. Yarowsky, D., Radu, F.: Evaluating sense disambiguation across diverse parameter spaces. Natural Language Engineering 9(4), 293–310 (2002)
12. Tsatsaronis, G., Vazirgiannis, M., Androutsopoulos, I.: Word sense disambiguation with spreading activation networks generated from thesauri. In: Proc. of IJCAI, pp. 1725–1730 (2007)
13. Tran, A., Bowes, C., Brown, D., Chen, P., Choly, M., Ding, W.: TreeMatch: A Fully Unsupervised WSD System Using Dependency Knowledge on a Specific Domain. In: Proc. of the 5th International Workshop on Semantic Evaluation, ACL 2010, Uppsala, Sweden, July 15-16, pp. 396–401 (2010)

14. Banerjee, S., Pedersen, T.: Extended gloss overlaps as a measure of semantic relatedness. In: Proc. of the 18th International Joint Conference on Artificial Intelligence, IJCAI, Acapulco, Mexico, pp. 805–810 (2003)
15. Saedi, C., Shamsfard, M.: Translating Persian documents into English using knowledge based WSD. In: ICDIM 2009, pp. 229–234 (2009)
16. Faili, H.: An Experiment of Word Sense Disambiguation in a Machine Translation System. In: Proc. of 2008 IEEE International Conference on Natural Language Processing and Knowledge Engineering (NLP-KE 2008), pp. 28–35 (2008)
17. Xu, S., Yang, S., Lau, F.: Keyword Extraction and Headline Generation Using Novel Word Features. In: Proc. of the Twenty-Fourth AAAI Conference on Artificial Intelligence (2010)
18. Ercan, G., Cicekli, I.: Using lexical chains for keyword extraction. Information Processing and Management 43(6), 1705–1714 (2007)
19. Hulth, A.: Automatic Keyword Extraction. VDM Verlag Dr. Mueller, E.K. Binding: Paperback (2008) ISBN: 363903855X, ISBN-13: 9783639038552
20. Barker, K., Cornacchia, N.: Using Noun Phrase Heads to Extract Document Keyphrases. In: Canadian Conference on AI 2000, pp. 40–52 (2000)
21. Turney, P.D.: Learning algorithms for keyphrase extraction. Information Retrieval 2(4), 303–336 (2000) (NRC #44105)
22. Wartena, C., Brussee, R., Slakhorst, W.: Keyword Extraction Using Word Co-occurrence. In: Workshops on Database and Expert Systems Applications, Bilbao, Spain, August 30-September 03 (2010) ISBN: 978-0-7695-4174-7
23. Shamsfard, M., Jafari, H., Ilbeygi, M.: STeP-1: A Set of Fundamental Tools for Persian Text Processing. In: LREC (2010)
24. Shamsfard, M., Hesabi, A., Fadaei, H., Mansoory, N., Famian, A., Bagherbeigi, S., Fekri, E., Monshizadeh, M., Assi: Semi Automatic Development of FarsNet; The Persian WordNet. In: Assi: Semi Automatic Development of FarsNet; The Persian WordNet. 5th Global WordNet Conference (GWA 2010), Mumbai, India (2010)
25. Fragos, K., Maistros, I., Skourlas, C.: Word Sense Disambiguation using WordNet relations. In: Proc. of 1st Balkan Conference in Informatics, October 20-22 (2003)
26. AleAhmad, A., Amiri, H., Darrudi, E., Rahgozar, M., Oroumchian, F.: Hamshahri: A standard Persian text collection. Journal of Knowledge-Based Systems 22(5), 382–387 (2009)

Mapping FarsNet to Suggested Upper Merged Ontology

Aynaz Taheri and Mehrnoush Shamsfard

NLP Research Lab, Computer Engineering Dept.,
Shahid Beheshti University, Tehran, Iran
ay.taheri@mail.sbu.ac.ir, m-shams@sbu.ac.ir

Abstract. FarsNet is a lexical ontology for Persian language. SUMO is an important upper level ontology that contains global knowledge. Mapping of lexical to general knowledge of these two ontologies will be beneficial for Persian language. Producing a mapping of FarsNet to SUMO began after development of the first phase of FarsNet. Since we had mapping to WordNet for some FarsNet synsets, the mapping of FarsNet to SUMO was started with contribution of mapping FarsNet to WordNet. Obviously, there are some gaps between two languages of Persian and English such as lexical gap. So, there is no compulsion to obtain mapped SUMO concepts only through the WordNet mapping directly. Therefore, for covering the gaps, we take advantage of hierarchy relations in FarsNet. Hence, the bias of our mapping to English WordNet will be reduced. In this paper we propose the methodology of our semi automatic mapping method.

Keywords: Mapping, Ontology, Princeton WordNet, Suggested Upper Merged Ontology, FarsNet.

1 Introduction

Nowadays semantic processing of natural languages has became more important and we can see significant progresses in this domain. Semantic lexicons and lexical ontologies are important resources in semantic processing of natural languages. They are used in various tasks and applications especially where semantic processing is evolved such as question answering, machine translation, text understanding, information retrieval, content management, text summarization and search engines [1]. Because of these reasons an effort for development of a Persian WordNet was begun [1] and the result was called FarsNet [2].

Our motivation from mapping FarsNet to SUMO was to extend the ability of Persian language semantic processing by integrating the lexical knowledge of FarsNet and common knowledge of SUMO. SUMO is an influential upper level ontology and has excellent power in description of general concepts in the world. Merging SUMO and FarsNet provide an appropriate source for those who need global knowledge in conjunction with Persian language processing.

There are two approaches for mapping FarsNet to SUMO but there are difficulties with them. The first approach is mapping FarsNet to SUMO in a straightforward

M.V.M. Salem et al. (Eds.): AIRS 2011, LNCS 7097, pp. 604–613, 2011.

manner and without utilizing any external source. In this approach it is better to make an automatic mapping but it is so complicated and perhaps impossible. Therefore, the only solution in this approach is manual mapping. This kind of mapping is obviously a time consuming process. Accordingly, the first approach is not a good choice.

The Second approach is utilizing a third party. Applying lexical resources such as WordNet seems suitable. There are some problems in this method, too. First of all, because of some reasons like lexical gaps between Persian and English languages, some FarsNet synsets were not mapped to WordNet. In the second place, there is a certain amount of risk in biasing the result of mapping to English language. In spite of these complications we selected the second approach for our mapping. But we attempted to cover the gaps and avoid from biasing to English language. Hierarchical relations in FarsNet helped us to keep the mapping away from troubles.

In this paper we propose our methodology for mapping FarsNet to SUMO via WordNet. We have mappings of some FarsNet synsets to WordNet synsets and on the other hand, we use mapping of WordNet 3.0 to SUMO. In other words, we use WordNet as an intermediator. The mapping of FarsNet to WordNet has been created before [2,3]. In our project we now benefit from their work and also we reviewed it and made a few slight changes in some details. In section 2 we introduce our resources that are necessary for producing the mapping such as WordNet, FarsNet and SUMO. In the third section the methodology of our approach is explained and in section 4 we analyze our results.

2   Resources

We have three important resources that will be introduced in this section.

2.1   Princeton WordNet

Princeton WordNet (PWN) [4] is an electronic lexical database that is designed in Princeton University for English language. It has been first developed by Miller in a hand-crafted way. PWN uses synonymous sets, called synset. The latest version of WodrNet contains 155,287 words organized in 117,659 synsets [5]. PWN includes nouns, adjectives, verbs and adverbs. Synsets in PWN are connected to each other with semantic relation such as: synonymy, antonymy, hyponymy, hypernymy, meronymy, troponymy, etc.

2.2   FarsNet

FarsNet [2] is a lexical ontology for Persian language. FarsNet 1.0 includes the lexical, syntactic and semantic knowledge about more than 17000 persian words and phrases organized in about 10000 synsets of nouns, adjectives and verbs [2]. FarsNet has different kinds of relations such as: synonymy, hypernymy and hyponymy, meronymy, antonymy and causes. FarsNet has three parts: nouns, adjectives and verbs and about %70 of it is mapped to WordNet 3.0.

2.3 SUMO

SUMO (Suggested Upper Merged Ontology) [6] is an important upper level ontology. SUMO and its domain ontologies form the largest free, formal ontology set [7]. SUMO has a MID-Level Ontology (MILO) and some domain ontologies. Altogether they have about 20000 terms and 70000 axioms. SUMO was created at Teknowledge Corporation and owned by IEEE[6]. The Knowledge representation language is a version of KIF. SUMO is mapped to PWN and thus two significant resources of lexical knowledge and global knowledge can benefit from each other.

3   Our Mapping Methodology

According to our study on mapping FarsNet to PWN and mapping PWN to SUMO, FarsNet synsets can be categorized into different groups of nouns, adjectives and verbs. In this section we first discuss these groups and their features and then mine some general rules for mapping FarsNet to SUMO according to these categories.

3.1   Adjectives

In this part we confront with three different categories of adjectives which is necessary to adopt new techniques for processing each of them:
1. The first group has no difficulty and complication in mapping to SUMO and is mapped to SUMO straightforwardly via PWN. There is an illustration of this group in table 1.

Table 1. An example of the first group of adjectives

FarsNet	WordNet	SUMO
'afsordeh' :افسرده 'ghamgin' :غمگین، 'mahzoun' :محزون،	01361863: sad | experiencing or showing sorrow or unhappiness	Unhappiness

Relation: Equivalence   Relation: Hypernymy

Relation: Hypernymy

2. There is a category of FarsNet adjectives that are (or should be) mapped into PWN nouns instead of adjectives. This is one of the interesting differences of the Persian and English languages. These words are used in forms of adjective in Persian while, they are nouns in PWN. Accordingly, we use mapping of PWN nouns to SUMO for this category. An example of this group is shown in table 2.

Table 2. An Example of the second group of adjectives

FarsNet	WordNet	SUMO
نسیه: 'nesiyeh'	05803747: credit_rating , credit|an estimate, based on previous dealings, of a person's or an organization's ability to fulfill their financial commitments	Subjective Assessment Attribute

Relation: near-equivalent   Relation: Hypernymy

Relation: Hypernymy

2.1. There is a subgroup in the second category. This subgroup makes a large part of FarsNet adjectives that has no equivalent in PWN adjectives. There is a gap between adjectives in Persian and English, but this gap is not a lexical gap. There is not such an adjective in English. These kinds of adjectives are made by combining a noun and a latter: 'ی'. The meanings of these adjectives present a pertaining to a noun. We define a kind of new relation between these adjectives and their corresponding nouns in PWN and named it 'Characteristic of'. In the following you can see two illustration of these adjectives:

Noun + character = دبیرستان: dabirestan 'high school'+ 'ی' = دبیرستانی: dabirestani , pertaining to high school

Noun + character = بانک: bank 'bank'+ 'ی' = بانکی: banki, pertaining to bank

An example of the relations in this group is showed in table 3.

Table 3. An Example of the subgroup in the second group of adjectives

FarsNet	WordNet	SUMO
دبیرستانی 'dabirestani'	08409617: senior_high_school , senior_high , high , highschool 0 high_school | a public secondary school usually including grades 9 through 12	Secondary School

Relation: Characteristic of   Relation: Hypernymy

Relation: Partial Characteristic of

3. The next group of adjectives has no equivalence in PWN. In other words, these adjectives haven't been lexicalized in English and there is a lexical gap between two languages. So there is no compulsion for mapping this group to PWN.

Therefore, they will be mapped to SUMO directly and without any intermediator. Some examples of this category is illustrated in Table 4.

Table 4. An Example of the third group of adjectives

FarsNet	Meaning	SUMO
چندرغاز: 'chender-ghaz'	The property of little money	Subjective Assessment Attribute
پدر آمرزیده : 'pedar-amorzide'	(of) someone whose father has had salvation, son of a blessed man	Subjective Assessment Attribute
خوش خبر: 'khosh-khabar'	Bringing goodness , harbinger of good things	Subjective Assessment Attribute

Relation: Hypernymy

3.2 Nouns

In this section, the method of mapping nouns is explained. We divide them into three categories and handle them differently. There are four kinds of relations between noun synsets of FarsNet: hypernymy, hyponymy, holonymy and antonymy. In this step in order to make mapping, we sometimes use these relations. There are three different groups of nouns:

1. Similar to the first group of adjectives, the first group of nouns became mapped to SUMO via PWN automatically without any problem.(table 5).

Table 5. An example of the first group of nouns

FarsNet	WordNet	SUMO
تهران : 'Tehran'	08911421:Teheran, Tehran, capital_of_Iran, Iranian_capital	City

Relation: Equivalence   Relation: Instantiation

Relation: Instantiation

2. The next group of nouns is not lexicalized in English. But the difference of this group with the third group of adjectives is that, we have relations for them in FarsNet and it is possible to find their mapping via these relations automatically. In the following example we show how relations facilitate our method of mapping for some synsets that don't have any mapping to PWN.

   The word 'جاری : Jari' in Persian means: womans whose husbands are brothers. There is a synset for this word in FarsNet. This word is not lexicalized in English and therefore there is no equivalent for it in PWN. This synset in FarsNet has hypernym relation with the synset 'اقوام، خویشاوند:khishavand, aghvam', which means: a person having kinship with another or others. The new synset that was obtained from hypernymy relation is mapped to a synset of PWN with an equivalence relation and the synset of PWN was mapped to 'familyRelation' concept of SUMO with a synonym relation. So a hypernym relation is resulted from 'جاری: jari' to 'familyRelation' (Figure 1).

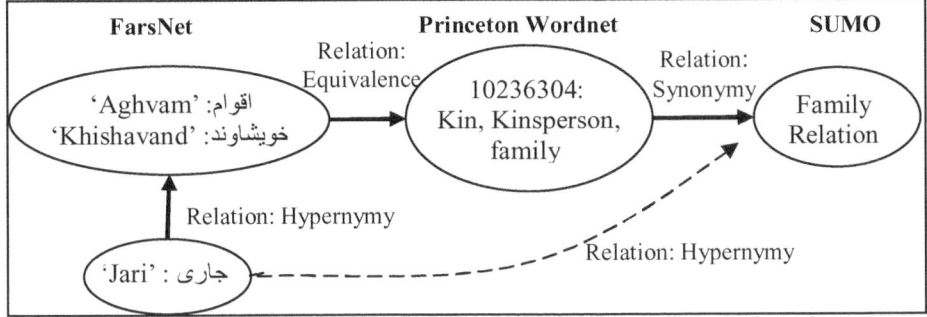

Fig. 1. An example of the first group of nouns

3. The members of the third group of FarsNet nouns don't have any equivalent in PWN. The reason of that is not because of lexicalization issues, however it is because of cultural or religious issues. We separate these synsets of FarsNet. For some of these synsets that there are hierarchical relations for them in FarsNet, we made mapping in the same way as the previous group. The noun synset 'فطریه: fetriyeh' in Persian language that means : alms given at the festival of the end of fasting month, has hypernymy relation with a synset 'فریضه، واجب: vajeb, farizeh' that means: duties that attention to them is mandatory. 'فطریه: fetriyeh' doesn't have mapping to PWN but 'فریضه، واجب: vajeb, farizeh' has an Equivalence relation to PWN and this synset of PWN has a hypernymy relation to 'ReligiousProcess' concept of SUMO. So it is possible to build a hypernymy relation from 'فطریه : fetriyeh' to 'ReligiousProcess' concept of SUMO via the father of synset 'فطریه: fetriyeh' at FarsNet (Figure 2).

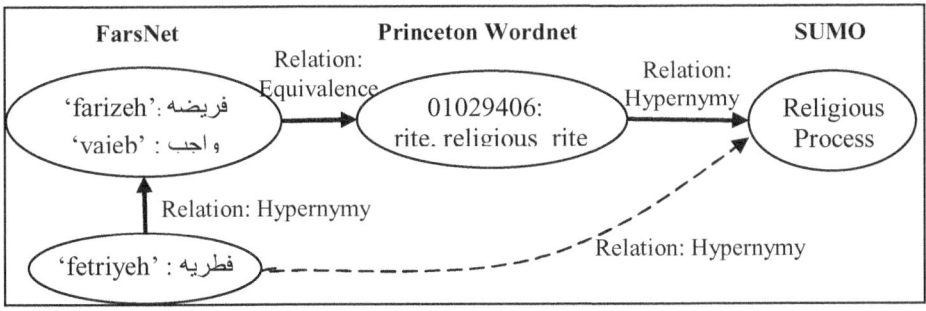

Fig. 2. An example of the second group of nouns

3.3 Verbs

Fortunately, many of the verbs in FarsNet have been mapped to PWN and so, automatic mapping of FarsNet to SUMO can be done directly via PWN for many verbs of FarsNet. There are different relations for the verbs in FarsNet such as antonymy, causes, hypernymy. We use these relations for those verbs that don't have mapping to PWN, in the same way as what we do for the nouns that don't have mapping to PWN. Figure 3 shows an example for mapping of a verb that doesn't have equivalence in PWN and has mapped to SUMO via his father in FarsNet. The meanings of the verbs in figure 3 are explained in the following:

'آهن ربا کردن : ahan roba kardan' = magnetize , make magnetic

'آهن ربا شدن : ahan roba shodan'= to became magnetized

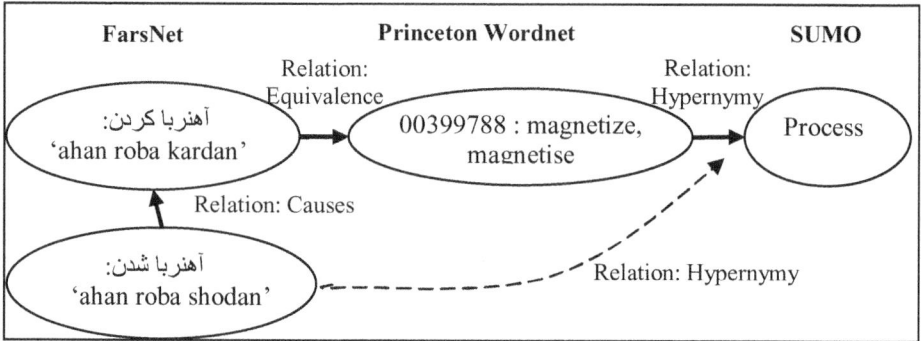

Fig. 3. An example of mapping verbs synsets of FarsNet to SUMO

4   Discussion

In this section the result of mapping is explained and mapping rules for our methodology are described.

4.1 Mapping Output

The outcome of the mapping is available in form of some textual records in three separate parts for nouns, adjectives and verbs. This format is similar to the result of mapping PWN 3.0 to SUMO [8]. The following example is a record of mapping for a FarsNet synset:

2751 *3453,جمله,* WordNet:06285090, Sentence=

The first number shows the ID of FarsNet synset. The expression between two '*' declares senses of this synset. The next term is synset of PWN that FarsNet synset is mapped to and the last one is SUMO concept that the PWN synset is mapped to it with a relation. The '=' sign, makes clear the kind of relation between FarsNet synset and

SUMO concept. We used three signs same as [8], for representing the kinds of relations: '=' for synonymy, '+' for hypernymy and '@' for instantiation. Moreover we used '-' for characteristic of, '&' for causes, '#-' for partial characteristic of and '#&' for partial causes.

But this is not enough for us. we know that there are some FarsNet synsets that are not mapped to SUMO via PWN. Moreover there is nothing about explanation of relations between FarsNet and SUMO. So we showed this record in the form of something like the following XML format in figure 4 that belongs to synset "توقف" (stop) in FarsNet.

```
<SYNSET>
 <ID>3129</ID>
 <UID>5100000877</UID>
 <POS>Noun</POS>
  <SemanticCategory>Nothing</SemanticCategory>
 <EXAMPLE>توقف جریان آب باعث ایجاد سیل گردید</EXAMPLE>
 <GLOSS>عمل باز داشتن چیزی</GLOSS>
 <STATUS>DEFAULT</STATUS>
 <REVISE>UnRevised</REVISE>
 <COMMENT/>
 <MappedWNsynId>
  <OFFSET>1076046</OFFSET>
  <Type>Equivalence</Type>
 </MappedWNsynId>
 <MappedSUMO>
  <CONCEPT>Prevents</CONCEPT>
  <Type>Hypernym</Type>
 </MappedSUMO>
 <SENSES>
  <SENSE>
   <WID>3220</WID>
   <SID>1</SID>
   <STATUS>DEFAULT</STATUS>
  </SENSE>
 </SENSES>
</SYNSET>
```

Fig. 4. XML format of mapping result

4.2 Mapping Rules

There are several rules for making relations from FarsNet synsets to SUMO concepts. We introduce them corresponding to the relations that are from FarsNet to PWN and PWN to SUMO. These rules are described in table 6. The second column in the table displays all kinds of relations that there are in FarsNet to WordNet 3.0 mapping. The third column displays all kinds of relations that there are in WordNet 3.0 to SUMO mapping. The last columns express all kinds of relations that may be produced by combining two prior columns.

The rule numbers 1, 2, 3 have used in noun, adjective and verb synsets of FarsNet. But 4,5,6 have applied in only adjectives of FarsNet. The relations from second column and in the seventh till twelfth rows of table have composed from a hypernymy or causes relation in addition to an equivalence relation. We display the sign '+' for those FarsNet synsets that don't have direct relations to PWN. The first operand of '+' explains a relation between two FarsNet synsets and the second operand explains a relation between FarsNet synset to PWN synset. These kinds of relations belong to noun or verb synsets which don't have mapping to PWN.

Table 6. Rules for mapping FarsNet to SUMO

Rule Number	Relations from FarsNet to WordNet	Relations from WordNet to SUMO	Relations from FarsNet to SUMO
1	Equivalence	Hypernymy	Hypernymy
2	Equivalence	Instantiation	Instantiation
3	Equivalence	Synonymy	Synonymy
4	Characteristic of	Hypernymy	(partial) Characteristic of
5	Characteristic of	Instantiation	(partial) Characteristic of
6	Characteristic of	Synonymy	Characteristic of
7	Hypernymy + Equivalence	Hypernymy	Hypernymy
8	Hypernymy + Equivalence	Instantiation	Instantiation
9	Hypernymy + Equivalence	Synonymy	Hypernym
10	Causes +Equivalence	Hypernymy	(partial)Causes
11	Causes +Equivalence	Instantiation	(partial)Causes
12	Causes +Equivalence	Synonymy	Causes

The table can be summarized in the following rules:
Considering a synset F in FarsNet;

- A: If F is mapped to synset P in PWN with relation R1 and P is mapped to concept S in SUMO with relation R2 then there is a relation R3 between F and S in such a way that:
    1. Rule 1: if R1 (or R2) is equivalence or near-equivalence or synonymy then R3=R2 (or R1)
    2. Rule 2: else if R1=R2 and R1 is transitive then R3=R1.
    3. Rule 3: else if R2 is hypernymy then R3= (partial)R1
- B: otherwise if F is not mapped to PWN but is related to F' by R' and F' is mapped to synset P in PWN by R1 and P is mapped to concept S in SUMO by R2, then there is a relation R3 mapping F to S in such a way that:
    4. Rule 4: if one of R', R1 or R2 is equivalence or near-equivalence or synonymy and the other two are equal and transitive then R3 is equal to the other two.
    5. Rule 5: else if two of R', R1 or R2 are equivalence or near-equivalence or synonymy then R3= the third one.
    6. else if R' is hypernymy and R1 is equivalence or near-equivalence then R3= (partial)R2
    and we can even summarize rules 4 and 5 into
    7. Rule7: if any of R1 or R2 are equivalence or near-equivalence or synonymy then we can ignore this relation and treat the others as we are in case A.

4.3 Mapping Rates

In the table 7, ten concepts of SUMO that have greatest number of mappings from noun, adjective and verb synsets of FarsNet are declared respectively. In the rate columns of table, we represent the rate of mapping FarsNet synsets to the SUMO concepts that are in the concept columns. There is an interesting observation from the result of this table. 'SubjectiveAssessmentAttribute' is the only concept of SUMO that is mapped by three groups of FarsNet synsets: noun, adjective and verb.

Table 7. Mapping rate for noun, verb and adjective synsets

Noun		Verb		Adjective	
Concept	Rate	Concept	Rate	Concept	Rate
Subjective Assessment Attribute	6.43%	Intentional Psychological Process	5.1%	Subjective Assessment Attribute	29.84%
Device	2.12%	IntentionalProcess	4.78%	Nation	4.29 %
Artifact	1.9%	Process	4.19%	Capability	2.89%
Position	1.65%	Communication	3.77%	NormativeAttribute	2.59%
SocialRole	1.42%	SubjectiveAssessment Attribute	2.92%	ShapeAttribute	1.99%
Region	1.32%	Removing	2.76%	TraitAttribute	1.39%
Text	1.22%	Putting	2.07%	FieldOfStudy	1.29%
Process	1.17%	SocialInteraction	2.02%	ColorAttribute	0.99%
Human	1.12%	Motion	1.59%	EmotionalState	0.99%
BodyPart	1.07%	BodyMotion	1.54%	Human	0.89%

References

1. Shamsfard, M.: Developing FarsNet: A Lexical Ontology for Persian. In: 4th Global WordNet Conference, Szeged, Hungary (2008)
2. Shamsfard, M., Hesabi, A., Fadaie, H., Famian, A., Bagherbeigi, S., Mansoory, N., Fekri, E., Monshizadeh, M., Assi, S.M.: Semi Automatic Development of FarsNet: The Persian WordNet. In: 5th Global WordNet Conference, Mumbai, India (2010)
3. Dehkharghani, R., Shamsfard, M.: Mapping Persian Words to WordNet Synsets. International Journal of Artificial Intelligence and Interactive Multimedia 1(1) (2010)
4. Fellbaum, C.: WordNet: An Electronic Lexical Database. MIT Press (1998)
5. WordNet 3.0 database statistics,
   http://wordnet.princeton.edu/wordnet/man/
   wnstats.7wn.html
6. Niles, I., Pease, A.: Toward a standard Upper Ontology. In: 2nd International Conference on Formal Ontology in Information Systems, Ogunquit, Maine (2001)
7. Suggested Upper Merged Ontology (SUMO),
   http://www.ontologyportal.org
8. Niles, I., Pease, A.: Linking Lexicons and Ontologies: Mapping WordNet to the Suggested Upper Merged Ontology. In: International Conference on Information and Knowledge Engineering, Las Vegas (2003)

Topic Detection and Multi-word Terms Extraction for Arabic Unvowelized Documents

Rim Koulali and Abdelouafi Meziane

LARI Laboratory, Mohammed 1 University, Sciences College,
Hay al quods, Oujda, Morocco
rim.koulali@gmail.com, abdelouafi_meziane@yahoo.fr

Abstract. This paper focuses on Topic Detection (TD) for Arabic Unvowelized documents. Our topic detection system was implemented using two different metrics: adapted TF-IDF and Jaccard indicator. The experiments were conducted while studying the impact of working with stems or roots of words, all the words or nouns only. To enhance the TD system we developed The MWTs extraction prototype to generate MWTs vocabularies. To the best of our knowledge MWTs vocabulary has never been used in arabic documents topic's detection. In this paper we investigate the impact of such use on the quality of topic detection. We used the standard measures: Recall, Precision and F-measure to evaluate the performance of the realized systems on Wattan; an Arabic newspaper corpus.

Keywords: Topic Detection, Topic Oriented Vocabulary, Mutual Information, Jaccard Indicator, TF-IDF, Multi-Word Terms Extraction, C-value, LLR.

1 Introduction

The continuous growth of Arabic electronic documents leads to increasing attention in researches and improvement covering the Arabic Natural Language Processing (ANLP) tasks such as : topic detection and Multi-Word Terms Extraction, taking in consideration, particularities and complex morphological composition of the Arabic language.

Topic detection is a very important task in NLP. Many researches [1–3] have been conducted to improve it and use the obtained results in other NLP tasks such as: categorization, classification, QA system, summarizing; speech recognition... Topic detection is based on supervised machine learning using a training corpus to represent each topic with a specific content obtained by methods including: statistical methods(TF-IDF, Mutual Information, Information gain ...), data mining methods(decision tree, neural networks, SVM) and others NLP tasks (summarizing , MWTs extraction).

MWTs Extraction is an important task of Automatic term recognition and is employed in numerous NLP fields such as: text mining[4], syntactic parsing [5, 6], Machine Translation[7] and text classification[8]. The MWTs extraction task

M.V.M. Salem et al. (Eds.): AIRS 2011, LNCS 7097, pp. 614–623, 2011.

covers detection and extraction of a consecutive set of semantically related words. The technics used in MWTs extraction can be classified into four categories:

- Statistical approaches based on frequency and co-occurrence measures[9].
- Symbolic approaches using parsers, morphological analysis and patterns[10].
- Hybrid approaches combining statistical and morphological methods[11, 12].
- Word alignment approaches[13].

The hybrid approaches are wildly used since they combine the benefits of statistical and symbolic methods.

Our work is part of the semantic processing of unvowelized Arabic documents and aims to develop a topic detection system for Arabic texts based on machine learning. We constructed Topic Oriented Vocabularies (TOV) based on Mutual Information and classified documents on the basis of Jaccard indicator and an adaptation of the TF-IDF classifier to topic detection. We also implemented a MWTs Extraction prototype based on the hybrid approach, and used MWTs vocabulary in the topic identification.

This paper is organized as follows: In section 2 we present related works. In section 3; our topic detection system is explained. We present the MWTs extraction tool in Section 4. Section 5 details the conducted experiments and obtained results. The last section concludes the paper and announces future works.

2   Related Works

Works on topic detection for Arabic documents take into consideration the complex morphological specificities of the Arabic language. [14] realized a classification and clustering document system based on the extraction of key phrases using statistical methods such as Mutual Information and maximum entropy modelling. [15] used Naïve bayes algorithm to classify unvowelized Arabic web documents and achieved 68.78% in cross validation and 62% in evaluation set experiments. A comparative study between two methods of classification: TF-IDF and SVM by [16] shows that both methods give good results for Arabic documents: TF-IDF achieves results of 90.95% in terms of F-measure. SVM classifier outperforms the results obtained by TF-IDF by more than 7.5%. The experiments of [17] included the study of three statistical methods: the TF-IDF classifier, the SVM method and the Topic Unigram Language Model, and showed the superiority of the SVM classifier and its high capability to distinguish topics. [18] proposed a study of topic identification for Arabic language by using two methods: the k Nearest Neighbors which is used as a baseline and the TR-Classifier which gives the best performances using reduced size of topic vocabularies with 90% in terms of recall.

Numerous MWTs extraction systems and prototypes have been developed for various languages such as: English, French, Chinese, Turkish, Dutch, urdu... Few researches have been dedicated to the Arabic language. The authors of [19] explore three approaches: the first one based on crossing correspondence asymmetries between Arabic Wikipedia titles and titles in 21 different languages, the second approach uses translated English MWTs to Arabic language and proceeds to validation. The last one benefits from large corpora and lexical association measures. These approaches prove to be very efficient for large-scale extraction of Arabic MWTs. [11] created a MWTs extraction tool by adopting the standard approach that combines grammatical patterns and several statistical scores: T-Score, FLR, Mutual Information and LLR which gave the best result: 85% in terms of precision. A similar work was presented in[20]: a larger set of patterns was used by introducing the noun definition notion, and using the c-value and LLR metrics. The experiments shows that using a combination of the two previous metrics in ranking MWTs, gives better results than using only one of them.

3   Topic Detection System

Our developed topic detection system relies on topic oriented vocabularies to classify Arabic documents. Each topic is described by a vector of words (vocabulary) that represents specifically and accurately the topic [21]. Various methods for generation of vocabulary features were proposed in the literature. Although the word frequency is the most intuitive approach, other methods based on probabilistic measurements are more efficient such as: mutual information, gain information, unigram model... The generated topic oriented vocabularies describe a semantic relationship between words in documents of the same topic. The topic detection System matches documents against the generated TOVs to determine the general topic of each document. The implementation of our system is based on several phases to identify and classify the texts according to their general topic.

3.1   Documents Pre-treatment

This phase is crucial in order to extract relevant information. It consists of the following steps:

- **Documents encoding unification:** The unification of the encoding objective is to represent all the documents in the same encoding system. We adopted the UTF-8 that supports the Arabic language.
- **Documents normalization:** Suppression of: Latin words, symbols, numbers, markers, special characters...
- **Stop words elimination:** Suppression of noisy words by comparing each word with the elements of a handmade list of over 600 noisy words including: prepositions, demonstrative pronouns, identifiers, logical connectors...
- **Roots and stems extraction:** Although several articles on classification estimate that working with words roots favors the obtention of efficient results due to the reduction of the noise and best qualification of words, we conduct a comparative study between roots and stems to evaluate which is the most effective for the Arabic language. To achieve that, we used the morphological analyzer: Alkhalil[22]. We adapted Alkhalil to recover for each

document two lists: one for stems and the other for roots. Alkhalil realizes morphological analysis for each word in the corpus and returns among other morphological information all possibly related stems and roots to the considered word. So, we implemented a Viterbi algorithm to keep only the stems and roots that are relevant to the context.

3.2   Vocabulary Oriented Topic Generation

Our approach is based on the generation of a specific vocabulary for each topic. The vocabulary is composed by features which define specifically the topic. Various methods are used to create these vocabularies, we used the mutual information method which measures the influence of a word w on each topic t and attributes the word to the appropriate topic (the one that has the greatest IM value) based on the formula:

$$IM(w,t) = log(P(w|t)) - log(P(w)) \qquad (1)$$

We calculate the mutual information for each word in the test corpus with each topic. The word will be affected to the topic which gives the maximal value [23]. Finally, we generate six vocabularies specific to each topic of the training corpus by considering for each one the affected words ordered by decreasing mutual information. We considered vocabularies of various lengths: 50, 100, 150, 200, 250, 1000 and 4000.

3.3   Topic Detection

During this phase, we tested our system on the test corpus with 2035 articles belonging to six topics. Each document is represented with a vector of words composing it and each topic is represented by its vocabulary vector. The topic detection of a new document consists in calculating the similarity between the vector representing the document and those representing the topics. We based our similarity calculus on two approaches: Jaccard indicator and TF-IDF.

Jaccard indicator: The Jaccard indicator[24] measures the degree of similarity between two documents. The indicator is expressed as:

$$s_j = \frac{m_c}{m_d + m_{d'} - m_c} \qquad (2)$$

Where m_c is the total number of words of the first document, m_d represents the total number of words of the second document and $m_{d'}$ is the number of common words between the two documents.

The choice of Jaccard indicator is motivated by the fact that it employs words in their brut state and thus semantic information is accounted for.

TF-IDF: We adapted The classic TF-IDF [25, 26] to construct topics vectors. Each Topic is represented by a vector that contains the weights of the topic vocabulary words. the weight of the kth vocabulary word of topic j is

$t_{jk} = nf_k^j * idf_k$, where nf_k^j is the frequency of the word k in documents of the training corpus relative to topic j . Let d_{fk} be the number of documents not relative to topic j in which the word k appears at least once and N the total number of corpus documents. idf_k, the inverse document frequency, is given by: $idf_k = log(\frac{N}{df_k})$. Test documents are also represented by vectors containing weights of their words. To judge the similarity between a topic t and a document d, we used the cosine similarity:

$$cos(\theta) = \frac{\sum_{k=1}^{n} d_{ik} t_{jk}}{\sum_{k=1}^{n} d_{ik}^2 \sum_{k=1}^{n} t_{jk}^2} \tag{3}$$

The smaller the θ is the bigger is the similarity between a test document d and topic t. The topic of highest similarity will be assigned to the test document.

4   Multi-word Terms Extraction System

Our system is based on the hybrid approach and performs in two major steps:

4.1   Linguistic Filter

The linguistic filter has a major importance due to its contribution in the very early selection of MWTs candidate terms. To establish the linguistic filter, we follow these steps:

Documents pre-treatment: This task covers the unification of documents encoding to avoid any ambiguity, elimination of Latin words, symbols, numbers, Roman numeral, special characters ... In this step, no morphological treatment is applied on words.

Stop-word elimination: We eliminate the stop-words using the same stop-word list used previously in the topic detection system. As we are using words with no stemming , we implemented a program that assigns connectors like: 'ف','ل','ب','و','ك' to stop-words in order to create all the variations of words in the list. For example: 'هذه'→'هذه','لهذه','وهذه','كهذه','فهذه '. Thus, we can delimit approximately stop-words appearing in documents with variation.

Sentence boundary determination: In order to extract MWTs from documents, We implemented a program that breaks up the corpus documents to sentences. The full stop is considered to be the sentence delimiter.

Document POS-tagging: We assign morphological tickets to the corpus documents sentences using The Stanford Arabic POS Tagger. This step will help us to detect possible MWTs following the patterns bellow:

- $[Noun]^+$.
- $Noun, [Adjective]^+$.
- $Noun, Preposition, Noun$.

In order to extract MWTs vocabulary for each topic, we compared each two consecutive sentences to detect the set of words repeated in both of them. If the document was composed with one sentence only, we extract the set of words repeated more than once in this sentence. The linguistic filter permit to extract MWTs candidates with various sizes; Bigrams, Trigrams and Four-grams.

4.2   Statistical Filter

In order to reduce linguistic ambiguities and increase the ratio of correct extracted MWTs, we adopted a combination of two well known methods for their high effectiveness in MWTs extraction:

- **C-value metric[27]:** a termhood statistical method based on the frequency of occurrence that gives best results for nested MWTs ranking.The C-value measure comes together with a computationally efficient algorithm, which scores candidate multi-token terms according to the measure.
- **LLR [28]:** a unithood method used to qualify the association between two words in Bigrams by calculating the ratio between two likelihoods : the probability of observing one component of a collocation given the other is present and the probability of observing the same component of a collocation in the absence of other.

We used The C-value metric for the nested words and their variations, the LLR metric was used for the remaining MWTs Bigrams.

5   Experimentation and Results

For the realization of our system, we used a corpus of over 20.291 articles, collected from the Arabic newspaper Wattan of the year 2004[18]. The corpus contains about 20000 articles covering the six following topics: culture, economics, international, local, religion and sport. We used 90% as a training corpus and 10% to test the prototype. We conducted several experimentations on the test corpus to study and compare the influence of factors such as: the variation of the vocabulary size, the morphological nature of words i.e: stem or root and the use of nouns only.

We used Precision, Recall and F-measure metrics to evaluate the performances of our realized systems:

$$Recall = \frac{\text{Number of correctly labelled documents}}{\text{number of topic's documents}}. \tag{4}$$

$$Precision = \frac{\text{Number of correctly labelled documents}}{\text{number of labelled documents}}. \tag{5}$$

$$F - measure = \frac{2 \times \text{Precision} \times \text{Recall}}{\text{Precision} + \text{Recall}}. \tag{6}$$

620   R. Koulali and A. Meziane

5.1   Topic Detection

Fig.1 depicts the F-measure metric realized by our proposed topic identification system using Jaccard Indicator over two morphologically distinct corpora. The former was obtained by stemming Wattan documents, while the latter's documents contain only the roots of words. It shows that the developed system realizes higher performances when using the stemmed corpus. The best performance is obtained for the Local topic with 14.87% of enhancement. To study the impact of removing non-noun words from the corpus on the system performances we used Stanford POS-tagger for Arabic language to identify nouns contained in our corpus and stripped documents from other words.

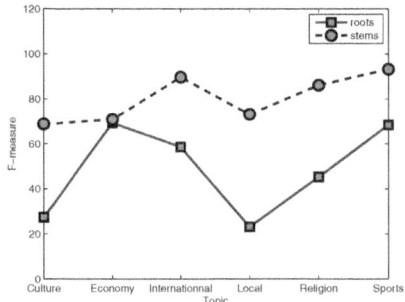 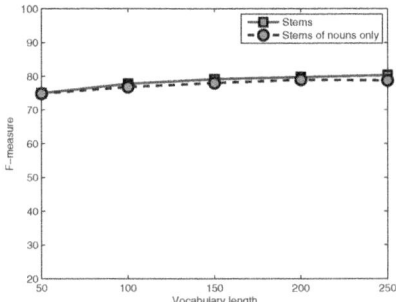

Fig. 1. F-measure, Stems vs Roots vocabulary of 250 words

Fig. 2. F-measure, Stems vs stems of nouns only

Fig.2 indicates that the system performances are not affected by this operation. The performance decrease is around 1%. Thus, we conclude that nouns in the Arabic language are more useful to construct VOT than verbs, adjectives, adverbs... as they hold the essential of semantic information. Statistical classifiers such as TF-IDF are widely used in topic detection literature. We compare the performance of the developed system for TF-IDF and Jaccard indicator.

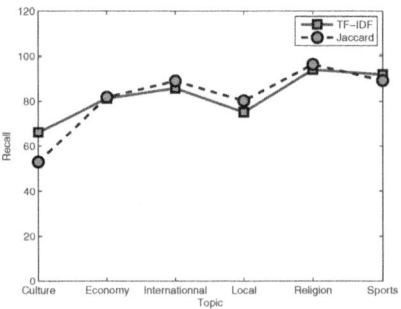  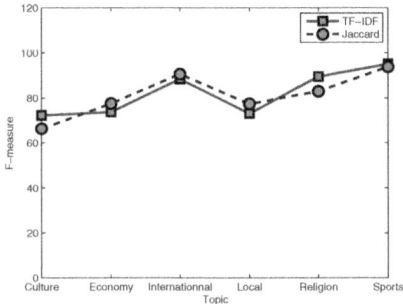

Fig. 3. Recall, TF-IDF vs Jaccard for a vocabulary of 1000 stemmed words

Fig. 4. F-measure, TF-IDF vs Jaccard for a vocabulary of 1000 stemmed words

Fig.3 indicates that the recall of our system using Jaccard indicator on stemmed documents outperforms the results obtained by TF-IDF except for the culture topic. This can be explained by the fact that the culture topic contains numerous words that are shared with other topics and this fact results in classification errors. The F-measure results depicted in Fig.4 correlate with the Recall ones.

5.2 MWTs Extraction

To test the performance of our MWTs extraction prototype, we used the manual validation of extracted terms.

The results are showed by Fig.5. Which gives the precision of the developed system according to several size of terms of the 200 first MWTs extracted with the highest score raking by the combination between C-value and LLR. The average precision of our MWTs extraction system is 90.25%.

To the best of our knowledge, it's the first time an Arabic detection topic system employs MWTs vocabulary. As shown in Fig.6, the system achieves higher performances for: religion and sports topics. This can be explained by the specificity of the MWTs extracted for these topics and the literature nature of the others topics which produces some ambiguity. The average Precision of our topic detection system with MWTs vocabularies is 84.10%.

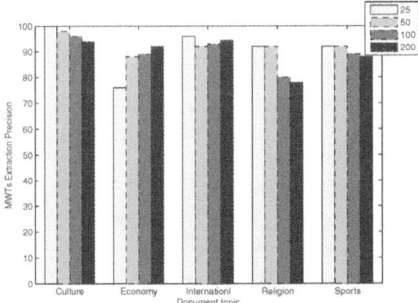

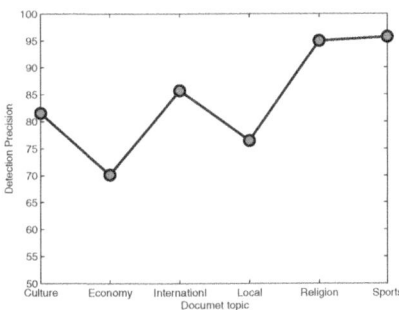

Fig. 5. Precision of the MWTs Extraction tool

Fig. 6. Precision for topic detection system using MWTs vocabulary

6 Conclusion

We developed a topic detection system based on topic oriented vocabulary for the Arabic Language using Jaccard indicator and an adapted TF-IDF classifier. Our system manages to achieve 84% of correctly classified documents over Wattan corpus. We conducted several experiments to improve our system performances.

We showed that working with the stemmed corpus is more efficient than using roots. Also, we found that stripping Wattan corpus from non-noun words does not affect our system performances, the resulting performances decrease is only of 1%. The obtained results prove that the use of Jaccard indicator is judicious in Arabic Language topic detection.

To study the impact of MWTs on topic detection, we built a hybrid prototype that reaches over 90.25% in term of MWTs extraction precision and cerated a MWTs vocabulary for Topic detection. We were able to extract words with bigrams, trigrams and four-grams. Our Topic detection system achieved a performance of 84.10% in term of F-measure when using MWTs vocabulary.

As future works we intend to explore other features such as: Named Entities (NE) to enhance our Topic detection system performances.

References

1. Martin, S.C., Liermann, J., Ney, H.: Adaptive topic-dependent language modelling using word-based varigrams. In: Proceedings of Fifth European Conference on Speech Communication and Technology, vol. 3, pp. 1447–1450 (1997)
2. Allan, J., Carbonell, J., Doddington, G., Yamron, J., Yang, Y., et al.: Topic detection and tracking pilot study: Final report. In: Proceedings of the DARPA Broadcast News Transcription and Understanding Workshop, pp. 194–218 (1998)
3. Yang, Y., Zhang, J., Carbonell, J., Jin, C.: Topic-conditioned novelty detection. In: Proceedings of the Eighth ACM SIGKDD International Conference on Knowledge Discovery and Data Mining, pp. 688–693 (2002)
4. SanJuan, E., Ibekwe-SanJuan, F.: Text mining without document context. Information Processing & Management 42(6), 1532–1552 (2006)
5. Nivre, J., Nilsson, J.: Multiword units in syntactic parsing. In: Workshop on Methodologies and Evaluation of Multiword Units in Real-world Applications (LREC 2004), pp. 37–46 (2004)
6. Attia, M.A.: Accommodating Multiword Expressions in an Arabic LFG Grammar. In: Salakoski, T., Ginter, F., Pyysalo, S., Pahikkala, T. (eds.) FinTAL 2006. LNCS (LNAI), vol. 4139, pp. 87–98. Springer, Heidelberg (2006)
7. Deksne, D., Skadiņš, R., Skadiņa, I.: Dictionary of multiword expressions for translation into highly inflected languages. In: Proceedings LREC Marrakech, pp. 1401–1405 (2008)
8. Zhang, W., Yoshida, T., Tang, X.: Text classification based on multi-word with support vector machine. Knowledge-Based Systems 21(8), 879–886 (2008)
9. Van de Cruys, T., Moirón, B.V.: Lexico-semantic multiword expression extraction. In: Proceedings of the 17th Meeting of Computational Linguistics in the Netherlands (CLIN), pp. 175–190 (2007)
10. Vintar, Š., Fišer, D.: Harvesting multiword expressions from parallel corpora. In: Proceedings of the 6th International Conference on Language Resources and Evaluation, LREC 2008 (2008)
11. Boulaknadel, S., Daille, B., Aboutajdine, D.: A multi-word term extraction program for arabic language. In: Proceedings of the 6th International Conference on Language Resources and Evaluation, LREC, pp. 1485–1488 (2008)

12. Duan, J., Zhang, M., Tong, L., Guo, F.: A Hybrid Approach to Improve Bilingual Multiword Expression Extraction. In: Theeramunkong, T., Kijsirikul, B., Cercone, N., Ho, T.-B. (eds.) PAKDD 2009. LNCS, vol. 5476, pp. 541–547. Springer, Heidelberg (2009)

13. Moirón, B.V., Tiedemann, J.: Identifying idiomatic expressions using automatic word-alignmen. In: Workshop on Multi-wordexpressions in a Multilingual Context (EACL 2006), pp. 33–40 (2006)

14. Sawaf, H., Zaplo, J., Ney, H.: Statistical classification methods for arabic news articles. In: Natural Language Processing in ACL 2001, Toulouse, France (2001)

15. El Kourdi, M., Bensaid, A., Rachidi, T.: Automatic arabic document categorization based on the naive bayes algorithm. In: Workshop on Computational Approaches to Arabic Script-Based Languages, pp. 51–58 (2004)

16. Abbas, M., Smaili, K.: Comparison of topic identification methods for arabic language. In: Proceedings of International Conference on Recent Advances in Natural Language Processing, RANLP, pp. 14–17 (2005)

17. Abbas, M., Berkani, D.: Topic identification by statistical methods for arabic language. WSEAS Transactions on Computers 5(9), 1908–1913 (2006)

18. Abbas, M., Smaili, K., Berkani, D.: Tr-classifier and knn evaluation for topic identification tasks. The International Journal on Information and Communication Technologies (IJICT) 3(3), 65–74 (2010)

19. Attia, M., Tounsi, L., Pecina, P., van Genabith, J., Toral, A.: Automatic extraction of arabic multiword expressions. In: Workshop on Multiword Expressions:from Theory to Applications (MWE 2010), pp. 19–27 (2010)

20. Al Khatib, K., Badarneh, A.: Automatic extraction of arabic multi-word terms. In: Proceedings of the International Multiconference on Computer Science and Information Technology, pp. 411–418 (2010)

21. Brun, A., Smaili, K., Haton, J.P.: Nouvelle approche de la sélection de vocabulaire pour la détection de thème. In: Traitement Automatique des Langues Naturelles (TALN 2003), pp. 45–54 (2003)

22. Mazroui, A., Meziane, A., Ould Abdallahi Ould Bebah, M., Boudlal, A., Lakhouaja, A., Shoul, M.: Alkhalil morphosys: Morphosyntactic analysis system for non vocalized arabic. In: Seventh International Computing Conference in Arabic (2011)

23. Yang, Y., Pedersen, J.O.: A comparative study on feature selection in text categorization. In: Proceedings 14th International Conference on Machine Learning (ICML 1997), pp. 412–420 (1997)

24. Real, R., Vargas, J.M.: The probabilistic basis of jaccard's index of similarity. Systematic Biology 45(3), 380 (1996)

25. Salton, G.: Developments in automatic text retrieval. Science 253(5023), 974–979 (1991)

26. Seymore, K., Rosenfeld, R.: Using story topics for language model adaptation. In: Proceedings of European Conference on Speech Communication and Technology, vol. 97, pp. 1987–1990 (1997)

27. Frantzi, K.T., Ananiadou, S.: Extracting nested collocations. In: Proceedings of the 16th Conference on Computational Linguistics, vol. 1, pp. 41–46 (1996)

28. Dunning, T.: Accurate methods for the statistics of surprise and coincidence. Computational Linguistics 19(1), 61–74 (1993)

Author Index

GPSR Compliance

*The European Union's (EU) General Product Safety Regulation (GPSR)
is a set of rules that requires consumer products to be safe and our
obligations to ensure this.*

*If you have any concerns about our products, you can contact us on
ProductSafety@springernature.com*

In case Publisher is established outside the EU, the EU authorized
representative is:

Springer Nature Customer Service Center GmbH
Europaplatz 3
69115 Heidelberg, Germany

Batch number: 09473851

Printed by Printforce, the Netherlands